THEORY OF LINEAR AND
INTEGER PROGRAMMING

WILEY-INTERSCIENCE
SERIES IN DISCRETE MATHEMATICS AND OPTIMIZATION

ADVISORY EDITORS

RONALD L. GRAHAM
AT & T Bell Laboratories, Murray Hill, New Jersey, U.S.A.

JAN KAREL LENSTRA
Department of Mathematics and Computer Science,
Eindhoven University of Technology, Eindhoven, The Netherlands

ROBERT E. TARJAN
Princeton University, New Jersey, and
NEC Research Institute, Princeton, New Jersey, U.S.A.

THEORY OF LINEAR AND INTEGER PROGRAMMING

ALEXANDER SCHRIJVER

**Centrum voor Wiskunde en Informatica,
Amsterdam**

A Wiley-Interscience Publication

JOHN WILEY & SONS

Chichester • New York • Weinheim • Brisbane • Singapore • Toronto

A. Schrijver,
CWI, Kruislaan 413,
1098 SJ Amsterdam,
The Netherlands

First published as paperback by John Wiley, 1998

Reprinted January 1999, April 2000

Other Wiley Editorial Offices

John Wiley & Sons, Inc., 605 Third Avenue,
New York, NY 10158-0012, USA

Wiley-VCH Verlag GmbH, Pappelallee 3,
D-69469 Weinheim, Germany

Jacaranda Wiley Ltd, 33 Park Road, Milton,
Queensland 4064, Australia

John Wiley & Sons (Asia) Pte Ltd, 2 Clementi Loop #02-01,
Jin Xing Distripark, Singapore 0512

John Wiley & Sons (Canada) Ltd, 22 Worcester Road,
Rexdale, Ontario M9W 1LI, Canada

British Library Cataloguing in Publication Data

A catalogue record for this book is available from the British Library

ISBN 0 471 98232 6

Preface

There exist several excellent books on linear and integer programming. Yet, I did not feel it superfluous to write the present book. Most of the existing books focus on the, very important, algorithmic side of linear and integer programming. The emphasis of this book is on the more theoretical aspects, and it aims at complementing the more practically oriented books.

Another reason for writing this book is that during the last few years several interesting new results have been obtained, which are not yet all covered by books: Lovász's basis reduction method, Khachiyan's ellipsoid method and Karmarkar's method for linear programming, Borgwardt's analysis of the average speed of the simplex method, Tardos' and Megiddo's algorithms for linear programming, Lenstra's algorithm for integer linear programming, Seymour's decomposition theorem for totally unimodular matrices, and the theory of total dual integrality.

Although the emphasis is on theory, this book does not exclude algorithms. This is motivated not only by the practical importance of algorithms, but also by the fact that the complexity analysis of problems and algorithms has become more and more a theoretical topic as well. In particular, the study of polynomial-time solvability has led to interesting theory. Often the polynomial-time solvability of a certain problem can be shown theoretically (e.g. with the ellipsoid method); such a proof next serves as a motivation for the design of a method which is efficient in practice. Therefore we have included a survey of methods known for linear and integer programming, together with a brief analysis of their running time. Our descriptions are meant for a quick understanding of the method, and might be, in many cases, less appropriate for a direct implementation.

The book also arose as a prerequisite for the forthcoming book *Polyhedral Combinatorics*, dealing with polyhedral (i.e. linear programming) methods in combinatorial optimization. Dantzig, Edmonds, Ford, Fulkerson, and Hoffman have pioneered the application of polyhedral methods to combinatorial optimization, and now combinatorial optimization is dissolubly connected to (integer) linear programming. The book *Polyhedral Combinatorics* describes these connections, which heavily lean on results and methods discussed in the present book. For a better understanding, and to make this book self-contained, we have illustrated some of the results by combinatorial applications.

v

Several friends and colleagues have helped and inspired me in preparing this book. It was Cor Baayen who stimulated me to study discrete mathematics, especially combinatorial optimization, and who advanced the idea of compiling a monograph on polyhedral methods in combinatorial optimization. During leaves of absence spent in Oxford and Szeged (Hungary) I enjoyed the hospitality of Paul Seymour and Laci Lovász. Their explanations and insights have helped me considerably in understanding polyhedral combinatorics and integer linear programming. Concurrently with the present book, I was involved with Martin Grötschel and Laci Lovász in writing the book *The Ellipsoid Method and Combinatorial Optimization* (Springer-Verlag, Heidelberg). Although the plans of the two books are distinct, there is some overlap, which has led to a certain cross-fertilization. I owe much to the pleasant cooperation with my two co-authors. Also Bob Bixby, Bill Cook, Bill Cunningham, Jack Edmonds, Werner Fenchel, Bert Gerards, Alan Hoffman, Antoon Kolen, Jaap Ponstein, András Sebö, Éva Tardos, Klaus Trümper and Laurence Wolsey have helped me by pointing out to me information and ideas relevant to the book, or by reading and criticizing parts of the manuscript. The assistance of the staff of the library of the Mathematical Centre, in particular of Carin Klompen, was important in collecting many ancient articles indispensable for composing the historical surveys.

Thanks are due to all of them. I also acknowledge hospitality and/or financial support given by the following institutions and organizations: the Mathematical Centre/Centrum voor Wiskunde en Informatica, the Netherlands organization for the advancement of pure research Z.W.O., the University of Technology Eindhoven, the Bolyai Institute of the Attila József University in Szeged, the University of Amsterdam, Tilburg University, and the Institut für Ökonometrie und Operations Research of the University of Bonn.

Finally, I am indebted to all involved in the production of this book. It has been a pleasure to work with Ian McIntosh and his colleagues of John Wiley & Sons Limited. In checking the galley proofs, Theo Beekman, Jeroen van den Berg, Bert Gerards, Stan van Hoesel, Cor Hurkens, Hans Kremers, Fred Nieuwland, Henk Oosterhout, Joke Sterringa, Marno Verbeek, Hein van den Wildenberg, and Chris Wildhagen were of great assistance, and they certainly cannot be blamed for any surviving errors.

ALEXANDER SCHRIJVER

Contents

PART IV: INTEGER LINEAR PROGRAMMING   227

1

Introduction and preliminaries

After the introduction in Section 1.1, we discuss general preliminaries (Section 1.2), preliminaries on linear algebra, matrix theory and Euclidean geometry (Section 1.3), and on graph theory (Section 1.4).

1.1. INTRODUCTION

The structure of the theory discussed in this book, and of the book itself, may be explained by the following diagram.

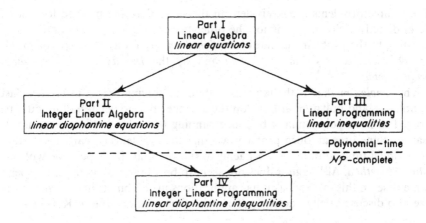

In Part I, 'Linear Algebra', we discuss the theory of linear spaces and of systems of linear equations, and the complexity of solving these systems. The theory and methods here are to a large extent standard, and therefore we do not give an extensive treatment. We focus on some less standard results, such as sizes of solutions and the running time of the Gaussian elimination method. It is shown that this method is a *polynomial-time* method, i.e. its running time is bounded by a polynomial in the size of the input data.

In Part II, 'Lattices and Linear Diophantine Equations', our main problem is to solve systems of *linear diophantine equations*, i.e. to solve systems of linear equations in integer variables. The corresponding geometric notion is that of a *lattice*. The existence of solutions here is characterized with the help of the *Hermite normal form*. One linear diophantine equation can be solved in polynomial time with the classical *Euclidean algorithm*. More generally, also *systems* of linear diophantine equations can be solved in polynomial time, with methods due to Frumkin and Votyakov, von zur Gathen and Sieveking, and Kannan and Bachem.

Also in Part II we discuss the problem of *diophantine approximation*. The *continued fraction method* approximates a real number by a rational number with low denominator, and is related to the Euclidean algorithm. Its extension to more dimensions, i.e. approximating a real vector by a rational vector whose entries have one common low denominator, can be done with Lovász's *basis reduction method* for lattices. These techniques are also useful in linear and integer programming, as we shall see in Parts III and IV.

In Part III, 'Polyhedra, Linear Inequalities, and Linear Programming', our main problems are the following:

(1)   – solving systems of linear inequalities;
        – solving systems of linear equations in nonnegative variables;
        – solving *linear programming* problems.

These three problems are equivalent in the sense that any method for one of them directly yields methods for the other two. The geometric notion corresponding to the problems is that of a *polyhedron*. Solutions of the problems (1) are characterized by *Farkas' lemma* and by the *Duality theorem of linear programming*.

The *simplex method* is the famous method for solving problems (1); it is fast in practice, and polynomial-time 'on the average', but no version of it could be proved to have polynomially bounded running time also in the worst case. It was for some time an open problem whether the problems (1) can be solved in polynomial time, until in 1979 Khachiyan showed that this is possible with the *ellipsoid method*. Although it does not seem to be a practical method, we spend some time on this method, as it has applications in combinatorial optimization. We also discuss briefly another polynomial-time method, due to Karmarkar.

The problems discussed in Parts I–III being solvable in polynomial time, in Part IV 'Integer Linear Programming' we come to a field where the problems in general are less tractable, and are $\mathcal{NP}$-*complete*. It is a general belief that these problems are not solvable in polynomial time. The problems in question are:

(2)   – solving systems of linear diophantine inequalities, i.e. solving linear
          inequalities in integers;

– solving systems of linear equations in nonnegative integer variables;
– solving *integer linear programming* problems.

Again, these three problems are equivalent in the sense that any method for one of them yields also methods for the other two. Geometrically, the problems correspond to the intersection of a lattice and a polyhedron. So the problems discussed in Parts II and III meet here.

The theory we shall discuss includes that of characterizing the convex hull P_I of the integral vectors in a polyhedron P. The case $P = P_I$ generally gives rise to better-to-handle integer linear programming problems. This occurs when P is defined by a *totally unimodular* matrix, or, more generally, by a *totally dual integral* system of inequalities. Inter alia, we shall discuss (but not prove) a deep theorem of Seymour characterizing total unimodularity.

If P is not-necessarily equal to P_I, we can characterize P_I with the *cutting plane method*, founded by Gomory. This method is not a polynomial-time method, but it yields some insight into integer linear programming. We also discuss the result of Lenstra that for each fixed number of variables, the problems (2) are solvable in polynomial time.

The theory discussed in Part IV is especially interesting for combinatorial optimization.

Before Parts I–IV, we discuss in the present chapter some preliminaries, while in Chapter 2 we briefly review the complexity theory of problems and algorithms. In particular, we consider polynomiality as a complexity criterion.

1.2. GENERAL PRELIMINARIES

Some general notation and terminology is as follows. If α is a real number, then

(3) $\lfloor \alpha \rfloor$ and $\lceil \alpha \rceil$

denote the lower integer part and the upper integer part, respectively, of α.

The symbols $\mathbb{Z}, \mathbb{Q}$, and $\mathbb{R}$ denote the sets of integers, rationals, and real numbers, respectively. $\mathbb{Z}_+, \mathbb{Q}_+$ and $\mathbb{R}_+$ are the restrictions of these sets to the nonnegatives. We denote, for real numbers α and β,

(4) $\alpha | \beta$ if and only if α divides β, i.e. if and only if $\beta = \lambda\alpha$ for some integer λ.

Moreover, $\alpha \equiv \beta \pmod{\gamma}$ means $\gamma | (\alpha - \beta)$.

If $\alpha_1, \ldots, \alpha_n$ are rational numbers, not all equal to 0, then the largest rational number γ dividing each of $\alpha_1, \ldots, \alpha_n$ exists, and is called the *greatest common divisor* or *g.c.d.* of $\alpha_1, \ldots, \alpha_n$, denoted by

(5)   g.c.d.$\{\alpha_1, \ldots, \alpha_n\}$

(so the g.c.d. is always positive). The numbers $\alpha_1, \ldots, \alpha_n$ are *relatively prime* if g.c.d. $\{\alpha_1, \ldots, \alpha_n\} = 1$.

We write $f(x) = O(g(x))$ for real-valued functions f and g, if there exists a constant C such that $|f(x)| \leqslant Cg(x)$ for all x in the domain.

If we consider an optimization problem like

(6) $\max\{\varphi(x)|x\in A\}$

where A is a set and $\varphi: A \to \mathbb{R}$, then any element x of A is called a *feasible solution* for the maximization problem. If A is nonempty, problem (6) is called *feasible*, otherwise *infeasible*. Similarly, a set of conditions is *feasible* (or *solvable*) if the conditions can be fulfilled all at the same time. Otherwise, they are called *infeasible* (or *unsolvable*). Any instance satisfying the conditions is called a *feasible solution*.

If the maximum (6) is attained, we say that the maximum *exists, is finite*, or *is bounded*. A feasible solution attaining the maximum is called an *optimum* (or *optimal*) *solution*. The maximum value then is the *optimum value*.

Similar terminology is used for minima.

A constraint is *valid* for a set S if each element in S satisfies this constraint.

'Left-hand side' and 'right-hand side' are sometimes abbreviated to *LHS* and *RHS*.

1.3. PRELIMINARIES FROM LINEAR ALGEBRA, MATRIX THEORY, AND EUCLIDEAN GEOMETRY

We assume familiarity of the reader with the elements of linear algebra, such as linear (sub)space, linear (in)dependence, rank, determinant, matrix, non-singular matrix, inverse, etc. As background references we mention Birkhoff and Mac Lane [1977], Gantmacher [1959], Lancaster and Tismenetsky [1985], Lang [1966a], Nering [1963], Strang [1980].

If $a = (\alpha_1, \ldots, \alpha_n)$ and $b = (\beta_1, \ldots, \beta_n)$ are row vectors, we write $a \leqslant b$ if $\alpha_i \leqslant \beta_i$ for $i = 1, \ldots, n$. Similarly for column vectors. If A is a matrix, and x, b, y, and c are vectors, then when using notation like

(7) $Ax = b, \quad Ax \leqslant b, \quad yA = c$

we implicitly assume compatibility of sizes of A, x, b, y, and c. So as for (7), if A is an $m \times n$-matrix, then x is a column vector of dimension n, b is a column vector of dimension m, y is a row vector of dimension m, and c is a row vector of dimension n.

Similarly, if c and x are vectors, and if we use

(8) cx

then c is a row vector and x is a column vector, with the same number of components. So (8) can be considered as the inner product of c and x.

An *n-vector* is an n-dimensional vector.

If a is a row vector and β is a real number, then $ax = \beta$ and $ax \leqslant \beta$ are called a *linear equation* and a *linear inequality*, respectively. If vector x_0 satisfies a linear inequality $ax \leqslant \beta$, then the inequality is called *tight* (*for x_0*) if $ax_0 = \beta$.

If A is a matrix, and b is a column vector, we shall call $Ax = b$ a *system of linear equations*, and $Ax \leqslant b$ a *system of linear inequalities*. The matrix A is called the *constraint matrix* of the system.

A system of linear inequalities can have several alternative forms, like

(9) $Ax \geqslant b$ (for $(-A)x \leqslant -b$)

$Ax \leqslant b, Cx \leqslant d$ $\left(\text{for } \begin{bmatrix} A \\ C \end{bmatrix} x \leqslant \begin{pmatrix} b \\ d \end{pmatrix} \right)$

$Ax = b$ (for $Ax \leqslant b, -Ax \leqslant -b$)

and so on.

If $A'x \leqslant b'$ arises from $Ax \leqslant b$ by deleting some (or none) of the inequalities in $Ax \leqslant b$, then $A'x \leqslant b'$ is called a *subsystem* of $Ax \leqslant b$. Similarly for systems of linear equations.

The identity matrix is denoted by I, where the order usually is clear from the context. If δ is a real number, then an *all-δ vector* (*all-δ matrix*) is a vector (matrix) with all entries equal to δ. So an *all-zero* and an *all-one vector* have all their entries equal to 0 and 1, respectively. **0** and **0** stand for all-zero vectors or matrices, and **1** stands for an all-one vector, all of appropriate dimension.

The transpose of a matrix A is denoted by A^{T}. We use $\|\cdot\|$ or $\|\cdot\|_2$ for the *Euclidean norm*, i.e.

(10) $\|x\| := \|x\|_2 := \sqrt{x^{\mathsf{T}}x}.$

$d(x, y)$ denotes the *Euclidean distance* of vectors x and y (i.e. $d(x, y) := \|x - y\|_2$), and $d(x, P)$ the *Euclidean distance* between x and a set P (i.e. $d(x, P) := \inf\{d(x, y) | y \in P\}$).

The *ball* with *centre* x and *radius* ρ is the set

(11) $B(x, \rho) := \{y | d(x, y) \leqslant \rho\}.$

A point $x \in \mathbb{R}^n$ is an *internal point* of $S \subseteq \mathbb{R}^n$ if there exists an $\varepsilon > 0$ such that

(12) $B(x, \varepsilon) \subseteq S.$

Other norms occurring in this text are the l_1- and the l_∞-norms:

(13) $\|x\|_1 := |\xi_1| + \cdots + |\xi_n|$
        $\|x\|_\infty := \max\{|\xi_1|, \ldots, |\xi_n|\}$

for $x = (\xi_1, \ldots, \xi_n)$ or $x = (\xi_1, \ldots, \xi_n)^{\mathsf{T}}$.

An $m \times n$-matrix A is said to have *full row rank* (*full column rank*, respectively) if rank $A = m$ (rank $A = n$, respectively).

A *row submatrix* of a matrix A is a submatrix consisting of some rows of A. Similarly, a *column submatrix* of A consists of some columns of A.

A matrix $A = (\alpha_{ij})$ is called *upper triangular* if $\alpha_{ij} = 0$ whenever $i > j$. It is *lower triangular* if $\alpha_{ij} = 0$ whenever $i < j$. It is *strictly upper triangular* if $\alpha_{ij} = 0$ whenever $i \geqslant j$. It is *strictly lower triangular* if $\alpha_{ij} = 0$ whenever $i \leqslant j$. It is a *diagonal matrix* if $\alpha_{ij} = 0$ whenever $i \neq j$. The square diagonal matrix of order n, with the numbers $\delta_1, \ldots, \delta_n$ on its main diagonal, is denoted by

(14) $\text{diag}(\delta_1, \ldots, \delta_n).$

For any subset T of $\mathbb{R}$, a vector (matrix) is called a *T-vector* (*T-matrix*) if its entries all belong to T. A vector or matrix is called *rational* (*integral*, respectively) if its entries all are rationals (integers, respectively).

A linear equation $ax = \beta$ or a linear inequality $ax \leqslant \beta$ is *rational* (*integral*) if a and β are rational (integral). A system of linear equations $Ax = b$ or inequalities $Ax \leqslant b$ is *rational* (*integral*) if A and b are rational (integral). A *rational polyhedron* is a polyhedron determined by rational linear inequalities, i.e. it is $\{x \in \mathbb{R}^n \mid Ax \leqslant b\}$ for some rational system $Ax \leqslant b$ of linear inequalities.

Lattice point is sometimes used as a synonym for integral vector. A vector or matrix is $1/k$-*integral* if its entries all belong to $(1/k)\mathbb{Z}$, i.e. if all entries are integral multiples of $1/k$.

Scaling a vector means multiplying the vector by a nonzero real number.

For any finite set S, we identify the function $x : S \to \mathbb{R}$ with the corresponding vector in $\mathbb{R}^S$. If $T \subseteq S$, the *incidence vector* or *characteristic vector* of T is the $\{0, 1\}$-vector in $\mathbb{R}^S$, denoted by χ_T, satisfying

(15)
$$\chi_T(s) = 1 \quad \text{if } s \in T$$
$$\chi_T(s) = 0 \quad \text{if } s \in S \setminus T.$$

If S and T are finite sets, an $S \times T$-*matrix* is a matrix with rows and columns indexed by S and T, respectively. If A is an $S \times T$-matrix and $b \in \mathbb{R}^T$, the product $Ab \in \mathbb{R}^S$ is defined by:

(16)
$$(Ab)_s := \sum_{t \in T} \alpha_{s,t} \beta_t$$

for $s \in S$ (denoting $A = (\alpha_{s,t})$ and $b = (\beta_t)$).

If $\mathscr{C}$ is a collection of subsets of a set S, the *incidence matrix* of $\mathscr{C}$ is the $\mathscr{C} \times S$-matrix M whose rows are the incidence vectors of the sets in $\mathscr{C}$. So

(17)
$$M_{T,s} = 1 \quad \text{if } s \in T$$
$$M_{T,s} = 0 \quad \text{if } s \notin T$$

for $T \in \mathscr{C}$, $s \in S$.

The *support* of a vector is the set of coordinates at which the vector is nonzero.

The *linear hull* and the *affine hull* of a set X of vectors, denoted by lin.hull X and aff.hull X, are given by

(18)   lin.hull $X = \{\lambda_1 x_1 + \cdots + \lambda_t x_t \mid t \geqslant 0; x_1, \ldots, x_t \in X; \lambda_1, \ldots, \lambda_t \in \mathbb{R}\}$

aff.hull $X = \{\lambda_1 x_1 + \cdots + \lambda_t x_t \mid t \geqslant 1; x_1, \ldots, x_t \in X; \lambda_1, \ldots, \lambda_t \in \mathbb{R};$
$$\lambda_1 + \cdots + \lambda_t = 1\}.$$

A set C of vectors is *convex* if it satisfies:

(19)   if $x, y \in C$ and $0 \leqslant \lambda \leqslant 1$, then $\lambda x + (1 - \lambda)y \in C$.

The *convex hull* of a set X of vectors is the smallest convex set containing X, and is denoted by conv.hull X; so

(20)   conv.hull $X = \{\lambda_1 x_1 + \cdots + \lambda_t x_t \mid t \geqslant 1; x_1, \ldots, x_t \in X;$
$$\lambda_1, \ldots, \lambda_t \geqslant 0; \lambda_1 + \cdots + \lambda_t = 1\}.$$

A (*convex*) *cone* is a nonempty set of vectors C satisfying

(21)   if $x, y \in C$ and $\lambda, \mu \geqslant 0$, then $\lambda x + \mu y \in C$.

The *cone generated by* a set X of vectors is the smallest convex cone containing X, and is denoted by cone X; so

(22)   cone $X = \{\lambda_1 x_1 + \cdots + \lambda_t x_t | t \geqslant 0; x_1, \ldots, x_t \in X; \lambda_1, \ldots, \lambda_t \geqslant 0\}$.

If $S \subseteq \mathbb{R}^n$, then a function $f : S \to \mathbb{R}$ is *convex* if S is convex and $f(\lambda x + (1 - \lambda)y) \leqslant \lambda f(x) + (1 - \lambda)f(y)$ whenever $x, y \in S$ and $0 \leqslant \lambda \leqslant 1$. f is *concave* if $-f$ is convex.

Pivoting

If A is a matrix, say

(23) $A = \begin{bmatrix} \alpha & b \\ c & D \end{bmatrix}$

where α is a nonzero number, b is a row vector, c is a column vector, and D is a matrix, then *pivoting* over the *pivot element* $(1, 1)$ means replacing A by the matrix

(24) $\begin{bmatrix} -\alpha^{-1} & \alpha^{-1}b \\ \alpha^{-1}c & D - \alpha^{-1}cb \end{bmatrix}$.

Pivoting over any other element of A is defined similarly.

Some inequalities

We recall the following well-known (in)equalities (cf. Beckenbach and Bellman [1983]). First the *Cauchy–Schwarz inequality*: if $c, d \in \mathbb{R}^n$ then

(25) $c^{\mathsf{T}} d \leqslant \|c\| \cdot \|d\|$.

If $b_1, \ldots, b_m$ are column vectors in $\mathbb{R}^n$, and B is the $n \times m$-matrix with columns $b_1, \ldots, b_m$, then

(26) $\sqrt{\det B^{\mathsf{T}} B}$ = the area of the parallelepiped spanned by $b_1, \ldots, b_m$.

This implies the *Hadamard inequality*:

(27) $\sqrt{\det B^{\mathsf{T}} B} \leqslant \|b_1\| \cdots \cdot \|b_m\|$.

In particular, if B is a square matrix, then

(28) $|\det B| \leqslant \|b_1\| \cdots \cdot \|b_m\|$.

(26) also implies that if A denotes the matrix with columns $b_1, \ldots, b_{m-1}$, and c is a vector orthogonal to $b_1, \ldots, b_{m-1}$, where c is in the space spanned by $b_1, \ldots, b_m$, then

(29) $\sqrt{\det B^{\mathsf{T}} B} = \dfrac{|c^{\mathsf{T}} b_m|}{\|c\|} \sqrt{\det A^{\mathsf{T}} A}$.

Positive definite matrices

A real-valued matrix D is called *positive definite* if D is symmetric and its eigenvalues all are positive. The following are equivalent for a real-valued symmetric matrix D:

(30)   (i) D is positive definite;
             (ii) $D = C^{\mathsf{T}}C$ for some nonsingular matrix C;
             (iii) $x^{\mathsf{T}}Dx > 0$ for each nonzero vector x.

Let D be a positive definite matrix, and let $D = C^{\mathsf{T}}C$, for nonsingular C. Two vectors c and d are *orthogonal* (*relative to the inner product defined by* D) if $c^{\mathsf{T}}Dd = 0$. The *norm* $\|\cdot\|$ *defined by* D is given by

(31) $\|c\| := \sqrt{c^{\mathsf{T}}Dc}.$

By replacing $b_1, \ldots, b_m$ by $Cb_1, \ldots, Cb_m$, c by $(C^{\mathsf{T}})^{-1}c$, and d by Cd, the (in)equalities (25), (27), and (29) become:

(32) $|c^{\mathsf{T}}d| \leqslant \sqrt{c^{\mathsf{T}}D^{-1}c}\,\sqrt{d^{\mathsf{T}}Dd}$

(33) $\sqrt{\det B^{\mathsf{T}}DB} \leqslant \sqrt{b_1^{\mathsf{T}}Db_1} \cdots \sqrt{b_n^{\mathsf{T}}Db_n}$

(34) $\sqrt{\det B^{\mathsf{T}}DB} = \dfrac{|c^{\mathsf{T}}b_m|}{\sqrt{c^{\mathsf{T}}D^{-1}c}}\sqrt{\det A^{\mathsf{T}}DA}.$

Balls and ellipsoids

For $S \subseteq \mathbb{R}^n$ and $\varepsilon \geqslant 0$, we denote

(35) $B(S, \varepsilon) := \{y \in \mathbb{R}^n \mid \|x - y\| \leqslant \varepsilon \text{ for some } x \text{ in } S\}.$

For $x \in \mathbb{R}^n$, $B(x, \varepsilon) := B(\{x\}, \varepsilon)$ is the *ball* of *radius* ε with *centre* x.

A set E of vectors in $\mathbb{R}^n$ is called an *ellipsoid* if there exists a vector $z \in \mathbb{R}^n$ and a positive definite matrix D of order n such that

(36) $E = \mathrm{ell}(z, D) := \{x \mid (x - z)^{\mathsf{T}}D^{-1}(x - z) \leqslant 1\}.$

Here the parameters z and D are uniquely determined by E. The vector z is called the *centre* of E. It follows from (30) that a set E is an ellipsoid if and only if E is an affine transformation of the unit ball $B(0, 1)$.

1.4. SOME GRAPH THEORY

In this book we will describe occasionally some combinatorial applications, as illustrations of the theory. These applications are in terms of graphs. Moreover, graphs will be used sometimes as a tool in some of the proofs and algorithms, especially in Chapter 20 on totally unimodular matrices. Therefore, in this section we give a brief review of some elementary concepts, results, and problems in graph theory. The reader not familiar with graphs could read, for example, the first chapters of Wilson [1972] or Bondy and Murty [1976].

Undirected graphs

An (*undirected*) *graph* is a pair $G = (V, E)$, where V is a finite set, and E is a family of unordered pairs of elements of V. The elements of V are called the *vertices* or *points* of G, and the elements of E are called the *edges* or *lines* of G.

The term 'family' in the definition of graph means that a pair of vertices may occur several times in E. A pair occurring more than once in E is called a *multiple edge*. So distinct edges may be represented in E by the same pair. Nevertheless, we shall often speak of 'an edge $\{v, w\}$' or even of 'the edge $\{v, w\}$', where 'an edge of type $\{v, w\}$' would be more correct. Graphs without multiple edges are called *simple*. Sometimes also *loops* are allowed, i.e. edges of the form $\{v, v\}$.

We shall say that an edge $\{v, w\}$ *connects* the vertices v and w. The vertices v and w are *adjacent* if there is an edge connecting v and w. The edge $\{v, w\}$ is said to be *incident with* the vertex v and with the vertex w, and conversely. The vertices v and w are called the *ends* of the edge $\{v, w\}$.

The number of edges incident with a vertex v is called the *valency* or *degree* of v, usually denoted by $d_G(v)$. The maximum and minimum degree of the vertices of G are denoted by $\Delta(G)$ and $\delta(G)$.

The *complementary graph* of G, denoted by $\bar{G}$, is the simple graph with the same vertex set as G, and with edges all pairs $\{v, w\}$ of vertices which are not in E. A simple graph is *complete* if E is the set of all pairs of vertices. The complete graph with n vertices is denoted by K_n. The *line graph* of G, denoted by $L(G)$, is the simple graph with vertex set E, in which two elements of E are adjacent if and only if they intersect.

A graph $G' = (V', E')$ is a *subgraph* of $G = (V, E)$ if $V' \subseteq V$ and $E' \subseteq E$. If E' is the family of all edges of G which have both ends in V', then G' is said to be *induced by* V', and we denote G' by $\langle V' \rangle$.

Usually, the vertices of a graph are represented by dots, or small circles, in the plane. Each edge, $\{v, w\}$ say, then is represented by a (curved) line segment connecting the two dots representing v and w, and not meeting any other dot. Two of such line segments may meet each other in or outside dots. If a graph can be represented in such a way that no two line segments meet each other outside dots, the graph is called *planar*. Two graphs which are nonplanar are the graphs denoted by K_5 and $K_{3,3}$, represented by:

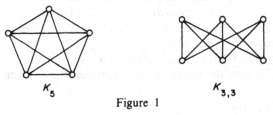

K_5 $K_{3,3}$

Figure 1

There is the famous characterization of Kuratowski [1930] that a graph is planar if and only if it has no subgraph which comes from K_5 or from $K_{3,3}$ by adding zero or more extra dots on each of the line segments.

A *path* in the graph $G = (V, E)$ *from* v_0 *to* v_t is a sequence of the form

(37) $(v_0, e_1, v_1, e_2, \ldots, v_{t-1}, e_t, v_t)$

where $v_0, \ldots, v_t$ are vertices and $e_1, \ldots, e_t$ are edges, such that $e_i = \{v_{i-1}, v_i\}$ for $i = 1, \ldots, t$. The vertices v_0 and v_t are the *starting point* and the *end point* of the path, respectively, or just the *end points*. We call path (37) a v_0–v_t-*path*, and it is said to *connect* v_0 and v_t. Path (37) is *simple* if all vertices and edges in (37) are different.

The *length* of path (37) is t. The *distance* between two vertices r and s in a graph is the minimum length of a path connecting r and s. Note that there is an easy algorithm for finding the distance between r and s. It consists of recursively determining the sets $V_0, V_1, V_2, \ldots$, where V_i is the set of vertices v of distance i to r:

(38)   (i) let $V_0 := \{r\}$;
            (ii) let $V_1 := \{v \in V \setminus \{r\} \mid v$ is adjacent to $r\}$;
            (iii) if $V_0, V_1, \ldots, V_i$ have been determined, let $V_{i+1} :=$
                  $\{v \in V \setminus (V_0 \cup \cdots \cup V_i) \mid v$ is adjacent to at least one vertex in $V_i\}$.

Step (iii) is repeated until we have found a set V_d containing s. Then d is the distance between r and s, and an r–s-path of shortest length is easily derived from $V_0, \ldots, V_d$. In the case where there is no path from r to s at all, this is detected by arriving with (iii) at $V_{i+1} = \varnothing$ without having s in $V_0 \cup \cdots \cup V_i$. Algorithm (38) is a 'polynomial-time' algorithm—see Section 2.4.

If $v_0 = v_t$, path (37) is called *closed*. A closed path of length at least one and without repeated edges or vertices (except for the end points) is called a *circuit*. An edge connecting two vertices of a circuit which are not connected by an edge of the circuit is called a *chord* of the circuit.

A *Hamilton circuit* is a circuit containing each vertex of the graph exactly once (except for the end points). Famous is the following *traveling salesman problem*: given a graph $G = (V, E)$ and a 'length' function $l : E \to \mathbb{Q}_+$, find a Hamilton circuit of minimum length. (The *length* of a circuit is the sum of the lengths of its edges.) The *traveling salesman polytope* is the convex hull of the incidence vectors of the Hamilton circuits (so it is a subset of $\mathbb{R}^E$).

A graph is *connected* if each two vertices of the graph are connected by a path. Connectedness by paths induces an equivalence relation on the vertices. Its classes are called the (*connected*) *components* of the graph. By extending algorithms (38) it is easy to determine the components of a given graph $G = (V, E)$ 'in polynomial time'.

A *forest* is a graph having no circuits, and a *tree* is a connected forest. It is not difficult to see that the following are equivalent for a given simple graph $G = (V, E)$:

(39)   (i) G is a tree;
            (ii) G contains no circuits and $|E| = |V| - 1$;
            (iii) G is connected and $|E| = |V| - 1$;
            (vi) any two vertices of G are connected by exactly one simple path.

If we add one new edge connecting two vertices of the tree, we obtain a graph with a unique circuit. Each tree with at least two vertices has a vertex of degree one.

A subgraph $G' = (V', E')$ of $G = (V, E)$ is a *spanning (sub)tree* of G if $V' = V$ and G' is a tree. Then G has a spanning subtree if and only if G is connected. A *maximal* forest in $G = (V, E)$ is a subgraph (V, E') which is a forest, where E' is as large as possible. This implies that (V, E') has the same components as (V, E).

A *clique* (*coclique*, respectively) in the graph $G = (V, E)$ is a set of pairwise adjacent (nonadjacent, respectively) vertices of G. The (*maximum*) *clique problem* is the problem: given a graph $G = (V, E)$, find a clique of maximum size. Similarly for the (*maximum*) *coclique problem*. The *clique polytope* of G is the convex hull of the incidence vectors of the cliques in G (so it is a subset of $\mathbb{R}^V$). Similarly, the *coclique polytope* of G is the convex hull of the incidence vectors of the cocliques in G.

A *matching* is a set of pairwise disjoint edges. A matching covering all vertices is called a *perfect matching* or a *1-factor*. The *matching polytope* is the convex hull of the incidence vectors of matchings (so it is a subset of $\mathbb{R}^E$).

A graph $G = (V, E)$ is called *bipartite* if V can be partitioned into two classes V_1 and V_2 such that each edge of G contains a vertex in V_1 and a vertex in V_2. The sets V_1 and V_2 are called *colour classes*. It is easy to see that a graph G is bipartite if and only if G contains no circuits of odd length. Again by an extension of algorithm (38) one can test in polynomial time whether a given graph $G = (V, E)$ is bipartite. If G is connected, choose an arbitrary vertex r, and determine the classes $V_0, V_1, V_2, \ldots$ as in (38), as long as they are nonempty. It is not difficult to see that G is bipartite if and only if each of the classes $V_0, V_1, V_2, \ldots$ is a coclique. In that case, the sets $V_0 \cup V_2 \cup V_4 \cup \cdots$ and $V_1 \cup V_3 \cup V_5 \cup \cdots$ form the colour classes for G. If G is not connected, we can apply this algorithm to each connected component of G.

A *complete bipartite* graph is a simple bipartite graph, with colour classes say V_1 and V_2, so that each vertex in V_1 is adjacent to each vertex in V_2. A complete bipartite graph with colour classes of sizes m and n is denoted by $K_{m,n}$ (cf. Figure 1).

Let $G = (V, E)$ be a graph, and let $e = \{v, w\}$ be an edge of G. *Deletion* of the edge e means replacing G by $G' := (V, E \setminus \{e\})$. *Contraction* of the edge e means replacing G by $G' := (V \setminus \{w\}, E')$, where E' consists of all edges of G contained in $V \setminus \{w\}$, together with all edges $\{x, v\}$, where $\{x, w\}$ is an edge of G different from e.

The *adjacency matrix* of an undirected graph $G = (V, E)$ is the matrix with both rows and columns indexed by V, where the entry in position (v, w) is the number of edges connecting v and w. So G is simple if and only if its adjacency matrix is a $\{0, 1\}$-matrix. The *incidence matrix* of G is the $\{0, 1\}$-matrix with rows and columns indexed by V and E, respectively, where the entry in position (v, e) is 1 if and only if vertex v is incident with edge e.

Directed graphs

A *directed graph* or a *digraph* is a pair $D = (V, A)$, where V is a finite set, and A is a finite family of ordered pairs of elements of V. The elements of V are called the *vertices* or *points*, and the elements of E are called the *arcs* of D. The vertices v and w are called the *tail* and the *head* of the arc (v, w), respectively.

So the difference with undirected graphs is that orientations are given to the pairs. Each directed graph gives rise to an *underlying* undirected graph, in which we forget the orientation of the arcs. Sometimes, when there is no fear of mis-understanding, we use 'undirected' terminology for directed graphs. The concepts of *multiple*, *simple*, and *loop* for directed graphs are analogous to those for undirected graphs.

We say that the arc (v, w) *enters* w and *leaves* v. If W is a set of vertices such that $v \notin W$ and $w \in W$, then (v, w) is said to *enter* W and to *leave* $V \setminus W$. If $W \subseteq V$, then $\delta_A^-(W)$ or $\delta^-(W)$ denotes the set of arcs in A entering W, and $\delta_A^+(W)$ or $\delta^+(W)$ denotes the set of arcs in A leaving W. $\delta^-(v)$ and $\delta^+(v)$ stand for $\delta^-(\{v\})$ and $\delta^+(\{v\})$.

A *(directed) path, from v_0 to v_t*, or a *v_0–v_t-path*, in a digraph $D = (V, A)$ is a sequence of the form

(40) $(v_0, a_1, v_1, \ldots, v_{t-1}, a_t, v_t)$

where $v_0, \ldots, v_t$ are vertices and $a_1, \ldots, a_t$ are arcs, such that $a_i = (v_{i-1}, v_i)$ for $i = 1, \ldots, t$. Path (40) is said to *start* in v_0 and to *end* in v_t. v_0 is the *starting point* and v_t is the *end point* of the path. The number t is the *length* of the path (40). A simple modification of algorithm (38) will give a polynomial-time algorithm for finding a shortest r–s-path in a directed graph.

If $l: A \to \mathbb{R}_+$, one easily extends this method to one finding a v_0–v_t-path with length $l(a_1) + \cdots + l(a_t)$ as small as possible (Dijkstra [1959]): Start with $w_0 := v_0$ and $d_0 := 0$. If vertices $w_0, w_1, \ldots, w_k$ and numbers $d_0, d_1, \ldots, d_k$ have been found, choose arc (w_i, w) attaining

(41) $\min \{d_i + l(w_i, w) | i = 0, \ldots, k; w \in V \setminus \{w_0, \ldots, w_k\}; (w_i, w) \in A\}$.

Let $w_{k+1} := w$, and let d_{k+1} be the value of (41). By induction on k one shows that d_k is the length of a shortest v_0–w_k-path.

A v_0–v_t-path is *closed* if $v_0 = v_t$. A *directed circuit* or a *cycle* is a closed path of length at least one, without repeated vertices or arcs (except for its starting and end point). A *Hamilton cycle* is a cycle containing all vertices of the digraph.

A set of arcs intersecting all r–s-paths for given vertices r and s is called an r–s-*cut*. So if $r \in W$, $s \notin W$ then $\delta_A^+(W)$ is an r–s-cut.

Given a digraph $D = (V, A)$ and vertices $r, s \in V$, a *flow from r to s*, or an r–s-*flow*, is a function $x: A \to \mathbb{R}$ satisfying:

(42) $x(a) \geq 0$ $(a \in A)$

$$\sum_{a \in \delta^-(v)} x(a) = \sum_{a \in \delta^+(v)} x(a) \qquad (v \in V; v \neq r, s).$$

The second set of constraints here means that at any vertex $v \neq r, s$, the total 'amount' of flow entering v is equal to the total 'amount' of flow leaving v—the *flow conservation law*. The *value* of the flow is the net amount of flow leaving r, i.e.

$$(43) \qquad \sum_{a \in \delta^+(r)} x(a) - \sum_{a \in \delta^-(r)} x(a).$$

It is not difficult to see that this is equal to the net amount of flow entering s. Flow x is said to be *subject* to a given 'capacity' function $c : A \to \mathbb{R}_+$ if $x(a) \leqslant c(a)$ for each arc a.

A *circulation* is a flow for which the flow conservation law holds at each vertex (i.e. it is an r–s-flow of value 0, or an r–r-flow, for arbitrary r and s).

An *undirected path* (*undirected circuit*, respectively) is a path (circuit, respectively) in the underlying undirected graph. In a natural way, an undirected path or circuit in a directed graph has *forward* arcs and *backward* arcs.

A digraph $D = (V, A)$ is called *strongly connected* if for each two vertices r and s of D there is a directed r–s-path. D is (*weakly*) *connected* if its underlying undirected graph is connected. A weakly connected digraph without undirected circuits is called a *directed tree*.

The *incidence matrix* of a digraph $D = (V, A)$ is the matrix with rows and columns indexed by V and A, respectively, where the entry in position (v, a) is -1, $+1$, or 0, if vertex v is the head of a, the tail of a, or neither, respectively.

2

Problems, algorithms, and complexity

The complexity of problems and algorithms in linear and integer programming will be one of the focuses of this book. In particular, we are interested in the solvability of problems in time bounded by a polynomial in the problem size.

Many of the problems in linear and integer programming, and in combinatorial optimization, can be easily seen to be solvable in finite time, e.g. by enumerating solutions. This generally yields an exponential-time algorithm. Solvability in polynomial time as a complexity criterion for problems was mentioned implicitly by von Neumann [1953] and explicitly by Cobham [1965] and Edmonds [1965a]. Edmonds introduced the term *good* for polynomial-time algorithms.

To indicate the significance of polynomial-time solvability, if we have an algorithm with running time 2^n (where n is the problem size), then a quadrupling of the computer speed will *add* 2 to the size of the largest problem that can be solved in one hour, whereas if we have an algorithm with running time n^2, this size will be *multiplied* by 2.

Often, if a problem was proved to be solvable in polynomial time in theory, it could be solved quickly also in practice. Moreover, from a theoretical point of view it is an interesting phenomenon that most of the problems which can be solved in polynomial time also allow a certain representation in terms of polyhedra, and conversely. In this text some openings for an explanation of this phenomenon are described (e.g. the ellipsoid method, the primal–dual method, Theorem 18.3).

After the pioneering work by Cook [1971] and Karp [1972], one generally sets the polynomially solvable problems against the so-called $\mathscr{NP}$-*complete* problems. These problems can be proved to be the hardest among the problems in a certain natural class of problems, called $\mathscr{NP}$ — hardest, with respect to a certain natural complexity ordering. Although there is no proof, as yet, that $\mathscr{NP}$-complete problems are really hard, no polynomial-time algorithms could be found to solve them, and it is a general belief that no such algorithms exist.

To study problem complexity, we should describe more or less precisely what is meant by concepts like 'problem', 'size', 'algorithm', 'running time', etc. We shall however not venture upon defining each of these notions mathematically exactly, but will appeal at several points to the reader's intuition, which will suffice to understand the greater part of this book. Here we confine ourselves to giving a brief introduction to complexity theory. For a more extensive and precise treatment, see Aho, Hopcroft, and Ullman [1974], Garey and Johnson [1979], and Savage [1976]. For an introduction, see Karp [1975].

2.1. LETTERS, WORDS, AND SIZES

Ground objects when formalizing problem complexity are symbols and strings of symbols. Let Σ be a finite set (often $\Sigma = \{0, 1\}$). Σ is called the *alphabet* and its elements are called *symbols* or *letters*. An ordered finite sequence of symbols from Σ is called a *string* (*of symbols*) or a *word*. Σ^* stands for the collection of all strings of symbols from Σ. The *size* of a string is the number of its components. The string of size 0 is the *empty string*, denoted by $\varnothing$.

Strings can have the form of (finite sequences of) rational numbers, vectors, matrices, graphs, (systems of) linear equations or inequalities, and so on. There are some standard ways of transformation in order to encode these objects uniformly as proper strings of symbols from some fixed alphabet like $\{0, 1\}$. Depending on the chosen transformation, this induces a concept of size for these objects. To fix one (which will be used in this text), the *sizes* of a rational number $\alpha = p/q$ (where p and q are relatively prime integers), of a rational vector $c = (\gamma_1, \ldots, \gamma_n)$ and of a rational matrix $A = (\alpha_{ij})_{i=1}^m \,_{,j=1}^n$ are:

(1) $\text{size}(\alpha) := 1 + \lceil \log_2(|p| + 1) \rceil + \lceil \log_2(|q| + 1) \rceil$

   $\text{size}(c) := n + \text{size}(\gamma_1) + \cdots + \text{size}(\gamma_n)$

   $\text{size}(A) := mn + \sum_{i,j} \text{size}(\alpha_{ij}).$

The *size* of a linear inequality $ax \leqslant \beta$ or equation $ax = \beta$ is equal to $1 + \text{size}$ $(a) + \text{size}(\beta)$. The size of a system $Ax \leqslant b$ ($Ax = b$) of linear inequalities (equations) is $1 + \text{size}(A) + \text{size}(b)$. The *size* of a (directed or undirected) graph is equal to the size of its incidence matrix.

(As said, our definition of size is one choice, convenient for our treatment—other authors use other definitions. However, generally most of the size functions are 'linearly equivalent', in the sense that if size_1 and size_2 denote two size functions, then $\text{size}_1 = O(\text{size}_2)$ and $\text{size}_2 = O(\text{size}_1)$.)

2.2. PROBLEMS

Informally, a problem can have the form of a question or a task. Mathematically, a (*search*) *problem* is a subset Π of $\Sigma^* \times \Sigma^*$, where Σ is some alphabet. The corresponding metamathematical problem then is:

(2)   given string $z \in \Sigma^*$, find a string y such that $(z, y) \in \Pi$, or decide that no such string y exists.

Here the string z is called an *instance* or the *input* of the problem, and y is a *solution* or the *output*.

Problem Π is called a *decision problem* or a *yes/no problem* if, for each (z, y) in Π, y is the empty string $\varnothing$. In that case, the problem is often identified with the set $\mathscr{L}$ of strings z in Σ^* for which $(z, \varnothing)$ belongs to Π. The problem, in metamathematical language, can have forms like:

(3)   'given string $z \in \Sigma^*$, decide whether z is in $\mathscr{L}$', or:
   'does a given $z \in \Sigma^*$ belong to $\mathscr{L}$?'

Examples of search problems are:

(4)   (i) $\{((A,b),\varnothing)|A$ is a matrix, b is a column vector, such that $Ax \leqslant b$ for at least one column vector $x\}$;
          (ii) $\{((A,b),x)|A$ is a matrix, b and x are column vectors, and $Ax \leqslant b\}$;
         (iii) $\{((A,b,c),k)|A$ is a matrix, b is a column vector, c is a row vector, and $k = \max\{cx|Ax \leqslant b\}\}$;
          (iv) $\{((A,b),\varnothing)|Ax \leqslant b$ is a system of linear inequalities with at least one integral solution$\}$;
          (v) $\{(A,\varnothing)|A$ is the adjacency matrix of an undirected graph with at least one perfect matching$\}$.

Corresponding informal forms of these problems are:

(5)   (i) 'Given a system $Ax \leqslant b$ of linear inequalities, does it have a solution?';
          (ii) 'Given a system $Ax \leqslant b$ of linear inequalities, find a solution, if there is one, or decide that no solution exists';
         (iii) 'Given matrix A, column vector b, and row vector c, determine $\max\{cx|Ax \leqslant b\}$';
          (iv) 'Given a system $Ax \leqslant b$ of linear inequalities, does it have an integral solution?';
          (v) 'Given an undirected graph, does it have a perfect matching?'

2.3. ALGORITHMS AND RUNNING TIME

We assume some intuitive feeling on the part of the reader of what algorithms are, and how to operate with them (cf. Knuth [1968: Ch. 1]). An algorithm is a list of instructions to solve a problem. Following Turing [1936–7], an algorithm can be formalized in terms of a *Turing machine*—see Aho, Hopcroft, and Ullman [1974] or Garey and Johnson [1979]. In Section 18.1 we shall discuss the concept of algorithm more formally.

For a given input $z \in \Sigma^*$, an algorithm for problem $\Pi \subseteq \Sigma^* \times \Sigma^*$ determines an output y such that (z,y) is in Π, or stops without delivering an output if there exists no such y. An algorithm can have the shape of a computer program, which is formally just a finite string of symbols from a finite alphabet. Mathematically, an algorithm can be defined as a finite string A of 0's and 1's. One says that A *solves* problem Π or A *is an algorithm for* Π, if for any instance z of Π, when giving the string (A,z) to a 'universal Turing machine', the machine stops after a finite number of steps, while delivering a string y with $(z,y) \in \Pi$, or delivering no string in the case where such a string y does not exist.

There are several ways of defining a concept of running time, to indicate the number of 'elementary bit operations' in the execution of an algorithm by a computer or computer model. In practice, this time will depend on the eventual implementation of the algorithm. Mathematically, the *running time* of an algorithm A for a certain problem instance z can be defined as the number of moves the 'head' of a universal Turing machine makes before stopping, when it is given the input (A, z). We define the *running time (function)* of an algorithm A as the function $f: \mathbb{Z}_+ \to \mathbb{Z}_+$ with

(6) $f(\sigma) := \max_{z, \text{size}(z) \leqslant \sigma}$ (running time of A for input z)

for $\sigma \in \mathbb{Z}_+$.

2.4. POLYNOMIAL ALGORITHMS

If $f, g_1, \ldots, g_m$ are real-valued functions (possibly multi-variable or 'zero-variable', i.e. constant), then f is said to be *polynomially bounded by* $g_1, \ldots, g_m$ if there is a function ϕ such that $\phi \geqslant f$ and such that ϕ arises by a sequence of compositions from the functions $g_1, \ldots, g_m$ and from some polynomials.

In the special case that $g_1, \ldots, g_m$ are polynomials, it follows that if f is polynomially bounded by $g_1, \ldots, g_m$, then f is bounded above by a polynomial. In that case, f is called a *polynomially bounded* function.

Generally, the functions to which we will apply the concept of polynomially bounded are monotone nondecreasing (as is the running time function of an algorithm). Then we know that if f is polynomially bounded by $g_1, \ldots, g_m$, and each g_i is polynomially bounded by $h_1, \ldots, h_n$, then f is polynomially bounded by $h_1, \ldots, h_n$.

An algorithm is called *polynomial-time* or *polynomial* if its running time function is polynomially bounded. A problem is said to be *solvable in polynomial time* or *polynomially solvable* if the problem can be solved by a polynomial-time algorithm. We are interested mostly in the asymptotic behaviour of the running time of an algorithm. Therefore, one often says that the running time is $O(g(\sigma))$, for some function $g(\sigma)$, meaning that there is a constant C such that the running time is upper bounded by $Cg(\sigma)$.

The *elementary arithmetic operations* are: adding, subtracting, multiplying, dividing, and comparing numbers. It is easy to see that, in rational arithmetic, they can be executed by polynomial-time algorithms. Therefore, for deriving the polynomiality of an algorithm which performs a sequence of elementary arithmetic operations, it suffices to show that the total number of these operations is polynomially bounded by the size of the input, and that the sizes of the intermediate numbers to which these elementary arithmetic operations are applied are polynomially bounded by the size of the input of the algorithm.

Note: We assume rationals are given as p/q, with p and q relatively prime integers, $q \geqslant 1$. The elementary arithmetic operations thus require that a quotient x/y can be reduced to p/q, with p and q relatively prime integers. This can be done in polynomial time with the Euclidean algorithm—see Section 5.1.

Usually in this text, at any occurrence of the term 'polynomially bounded', we could specify the above function ϕ explicitly by a form containing no unspecified terms other than $g_1, \ldots, g_m$. We could have given these explicit expressions if we were diligent enough in making the necessary calculations, but we think this would not contribute to the readability of the text. Moreover, in some cases the polynomial boundedness of functions is, as yet, of theoretical value only, especially in Chapters 13 and 14 on the ellipsoid method, where the coefficients and degrees of the polynomials are too high for practical application.

2.5. THE CLASSES $\mathscr{P}, \mathscr{NP}$, AND co-$\mathscr{NP}$

The class of decision problems solvable in polynomial time is denoted by $\mathscr{P}$. Another, possibly larger, 'complexity class' is the class $\mathscr{NP}$. [These letters stand for 'solvable by a Non-deterministic Turing machine in Polynomial time', and *not* for nonpolynomial.] The class $\mathscr{NP}$ was first studied by Edmonds [1965a, b], Cook [1971], and Karp [1972] (cf. Garey and Johnson [1979]). Most of the problems occurring in this text, and many combinatorial optimization problems, turn out to belong to $\mathscr{NP}$.

Informally, the class $\mathscr{NP}$ can be described as the class of those decision problems $\mathscr{L}$ satisfying:

(7)   for any $z \in \mathscr{L}$, the fact that z is in $\mathscr{L}$ has a proof of length polynomially bounded by the size of z.

More formally, a decision problem $\mathscr{L} \subseteq \Sigma^*$ belongs to $\mathscr{NP}$ if there exists a polynomially solvable decision problem $\mathscr{L}' \subseteq \Sigma^* \times \Sigma^*$ and a polynomial ϕ such that for each z in Σ^*:

(8) $z \in \mathscr{L} \Leftrightarrow \exists y \in \Sigma^* : (z, y) \in \mathscr{L}'$ and size $(y) \leqslant \phi(\text{size}(z))$.

As an interpretation, y here fulfils the role of a 'polynomial-length proof' of the fact that z is in $\mathscr{L}$. This 'proof' can be checked in polynomial time, as $\mathscr{L}'$ is polynomially solvable. The crucial point is that it is not required that y in (8) is to be found in polynomial time. The string y can be seen as a 'certificate' for z, which can be given to the 'boss' in order to convince him that z is in $\mathscr{L}$. To test whether z belongs to $\mathscr{L}$, we could 'guess' a string y such that (z, y) is in $\mathscr{L}'$—this guess corresponds to the $\mathscr{N}$ of nondeterministicity in the name $\mathscr{NP}$. Since there are at most $(|\Sigma| + 1)^{\phi(\text{size}(z))}$ such guesses, it follows that if $\mathscr{L}$ is in $\mathscr{NP}$, there is a polynomial $\psi(\sigma)$ and an algorithm which solves the problem $\mathscr{L}$ in time $O(2^{\psi(\sigma)})$, where σ is the input size.

Each of the problems (4) can be shown to belong to $\mathscr{NP}$. As for (i), in Chapter 10 we shall see that if a system $Ax \leqslant b$ of linear inequalities has a solution, it has one of size polynomially bounded by the sizes of A and b. So we can take as the above decision problem $\mathscr{L}'$:

(9) $\mathscr{L}' := \{((A, b), x) | A \text{ is a matrix, } b \text{ and } x \text{ are column vectors, } Ax \leqslant b\}$.

It is easy to see that this defines a polynomially solvable problem, satisfying (8). So any solution x of $Ax \leqslant b$ is a 'certificate' for the fact that $Ax \leqslant b$ has a solution. For problem (ii) of (4) we can take

(10) $\mathscr{L}' := \{((A, b, x), \varnothing) \mid A$ is a matrix, b and x are column vectors, $Ax \leqslant b\}$.

It is clear that $\mathscr{P}$ is a subset of $\mathscr{NP}$: if problem $\mathscr{L}$ belongs to $\mathscr{P}$, we can take as decision problem $\mathscr{L}'$:

(11) $\mathscr{L}' := \{(z, \varnothing) \mid z \in \mathscr{L}\}$.

The *complement* of a decision problem $\mathscr{L} \subseteq \Sigma^*$ is the decision problem $\Sigma^* \backslash \mathscr{L}$. The class of decision problems $\mathscr{L}$ whose complement is in $\mathscr{NP}$ is denoted by co-$\mathscr{NP}$. So co-$\mathscr{NP}$ consists of those decision problems $\mathscr{L}$ for which the fact that a certain string z is *not* in $\mathscr{L}$ has a proof of length polynomially bounded by size(z). Since the complement of every polynomially solvable decision problem is trivially polynomially solvable again, we know that $\mathscr{P} \subseteq$ co-$\mathscr{NP}$, and hence $\mathscr{P} \subseteq \mathscr{NP} \cap$ co-$\mathscr{NP}$.

The class $\mathscr{NP} \cap$ co-$\mathscr{NP}$ consists of those decision problems for which both a positive answer and a negative answer have a proof of polynomial length. That is, it consists of all problems $\mathscr{L} \subseteq \Sigma^*$ for which there exist polynomially solvable decision problem $\mathscr{L}'$, $\mathscr{L}''$ and a polynomial ϕ, such that for each string $z \in \Sigma^*$:

(12) $z \in \mathscr{L} \Leftrightarrow (z, x) \in \mathscr{L}'$ for some string x with size $(x) \leqslant \phi(\text{size}(z))$
          $z \notin \mathscr{L} \Leftrightarrow (z, y) \in \mathscr{L}''$ for some string y with size $(y) \leqslant \phi(\text{size}(z))$.

A problem is called *well-characterized* if it belongs to $\mathscr{NP} \cap$ co-$\mathscr{NP}$. To any well-characterized problem there corresponds a *good characterization*, which is the theorem asserting that, in the notation as above:

(13) $\exists x : (z, x) \in \mathscr{L}'$ if and only if $\forall y : (z, y) \notin \mathscr{L}''$,

where $\mathscr{L}'$ and $\mathscr{L}''$ satisfy, for a certain polynomial ϕ:

(14)   if $(z, x) \in \mathscr{L}'$ then $(z, x') \in \mathscr{L}'$ for some string x' with size$(x') \leqslant \phi(\text{size}(z))$
          if $(z, y) \in \mathscr{L}''$ then $(z, y') \in \mathscr{L}''$ for some string y' with size$(y') \leqslant \phi(\text{size}(z))$.

As an example, it is well-known from linear algebra that

(15)   a system $Ax = b$ of linear equations has a solution if and only if $yA = 0$ implies $yb = 0$

(cf. Section 3.1). Now let

(16) $\mathscr{L} := \{(A, b) \mid A$ is a matrix, b is a column vector, such that $Ax = b$ has a solution$\}$;

          $\mathscr{L}' := \{((A, b), x) \mid A$ is a matrix, b and x are column vectors, $Ax = b\}$;

          $\mathscr{L}'' := \{((A, b), y) \mid A$ is a matrix, b is a column vector, y is a row vector, such that $yA = 0$ and $yb \neq 0\}$.

So for this choice of $\mathscr{L}'$ and $\mathscr{L}''$, (13) is equivalent to (15). Moreover, in Section 3.2 we shall see that $\mathscr{L}'$ and $\mathscr{L}''$ satisfy (14). Therefore, (12) holds, and (15) gives a good characterization.

It is as yet a big open problem in complexity theory whether the class $\mathscr{P}$ is equal to the class $\mathscr{NP}$. It is generally believed that $\mathscr{P} \neq \mathscr{NP}$.

The traveling salesman problem is one of the many problems in $\mathscr{NP}$ for which no polynomial-time algorithm has been found as yet. Most of these problems turned out to belong to the hardest in $\mathscr{NP}$, the so-called $\mathscr{NP}$-complete problems—see Section 2.6.

Similarly, it is unknown whether $\mathscr{NP} = $ co-$\mathscr{NP}$ or $\mathscr{P} = \mathscr{NP} \cap$ co-$\mathscr{NP}$. Thus we have the following Venn diagram:

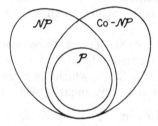

Figure 2

where it is unknown which of the regions is empty. Clearly, as soon as $\mathscr{NP} \subseteq$ co-$\mathscr{NP}$ or co-$\mathscr{NP} \subseteq \mathscr{NP}$, then $\mathscr{NP} = $ co-$\mathscr{NP} = \mathscr{NP} \cap$ co-$\mathscr{NP}$. If $\mathscr{P} = \mathscr{NP}$, then all sets drawn coincide.

2.6. $\mathscr{NP}$-COMPLETE PROBLEMS

It can be shown that certain problems in the class $\mathscr{NP}$ are the hardest among all problems in $\mathscr{NP}$, under a certain ordering of problems by difficulty. Several well-known canonical problems are among these hardest problems, like the integer linear programming problem, the traveling salesman problem, and the maximum clique problem.

A decision problem $\mathscr{L} \subseteq \Sigma^*$ is called *Karp-reducible* to decision problem $\mathscr{L}' \subseteq \Sigma^*$ if there is a polynomial-time algorithm A such that for any input string z in Σ^*, A delivers a string x as output, such that

(17) $z \in \mathscr{L}$ if and only if $x \in \mathscr{L}'$.

Trivially, if $\mathscr{L}'$ is polynomially solvable, and $\mathscr{L}$ is Karp-reducible to $\mathscr{L}'$, then also $\mathscr{L}$ is polynomially solvable. Similarly, if $\mathscr{L}'$ belongs to $\mathscr{NP}$, and $\mathscr{L}$ is Karp-reducible to $\mathscr{L}'$, then also $\mathscr{L}$ belongs to $\mathscr{NP}$. The same applies to co-$\mathscr{NP}$.

A problem $\mathscr{L}$ is called $\mathscr{NP}$-*complete* if it is in $\mathscr{NP}$ and each problem in $\mathscr{NP}$ is Karp-reducible to $\mathscr{L}$. So if any $\mathscr{NP}$-complete problem is polynomially solvable, then all problems in $\mathscr{NP}$ are polynomially solvable, and hence $\mathscr{P} = \mathscr{NP}$. Similarly, if any $\mathscr{NP}$-complete problem has a good characterization (i.e. is in $\mathscr{NP} \cap \text{co-}\mathscr{NP}$), then $\mathscr{NP} = \text{co-}\mathscr{NP} = \mathscr{NP} \cap \text{co-}\mathscr{NP}$.

Note that if $\mathscr{L}$ is Karp-reducible to problem $\mathscr{L}' \in \mathscr{NP}$, and $\mathscr{L}$ is $\mathscr{NP}$-complete, then also $\mathscr{L}'$ is $\mathscr{NP}$-complete.

It is not difficult to see that $\mathscr{NP}$-complete problems indeed exist. More surprising, and more difficult to show, is that several well-known, simply formulated problems are $\mathscr{NP}$-complete, like the integer linear programming problem, the traveling salesman problem, the maximum clique problem, and many other combinatorial optimization problems. This was shown by the pioneering work of Cook [1971] and Karp [1972]. For an encyclopedia of known $\mathscr{NP}$-complete problems, see Garey and Johnson [1979] (supplements are given by D. S. Johnson in his periodical '$\mathscr{NP}$-completeness column' in the *Journal of Algorithms*). In fact, most of the combinatorial optimization problems could be proved either to be polynomially solvable or $\mathscr{NP}$-complete, which might ground the assumption $\mathscr{P} \neq \mathscr{NP}$.

A more general form of reducibility is as follows. Problem $\Pi \subseteq \Sigma^* \times \Sigma^*$ is *polynomially reducible* or *Turing-reducible* to problem $\Pi' \subseteq \Sigma^* \times \Sigma^*$, if there exists an algorithm A such that: if A is given the input (z, A'), where $z \in \Sigma^*$ and A' is an algorithm for Π', then A solves problem Π for the input z, in time polynomially bounded by the size of z and the running time function of A'. So algorithm A' for Π' can be used as a subroutine in algorithm A for Π.

Again, if Π is polynomially reducible to Π', and Π' is polynomially solvable, then Π is polynomially solvable. Moreover, if $\mathscr{L}$ is Karp-reducible to $\mathscr{L}'$, then $\mathscr{L}$ is also polynomially reducible to $\mathscr{L}'$. Two problems are called *polynomially equivalent* if they are polynomially reducible to each other.

SOME HISTORICAL NOTES

We here quote some historical passages on the criterion of polynomiality.

Lamé [1844], in a paper showing that the classical Euclidean g.c.d. algorithm is a polynomial-time method:

> Dans les traités d'Arithmétique, on se contente de dire que le nombre des divisions à effectuer, dans la recherche du plus grand commun diviseur entre deux entiers, *ne pourra pas surpasser la moitié du plus petit*. Cette limite, qui peut être dépassée si les nombres sont petits, s'éloigne outre mesure quand ils ont plusieurs chiffres. L'exagération est alors semblable à celle qui assignerait la moitié d'un nombre comme la limite de son logarithme, l'analogie devient évidente quand on connait le théorème suivant:
>
> THÉORÈME. *Le nombre des divisions à effectuer, pour trouver le plus grand commun diviseur entre deux entiers A, et B < A, est toujours moindre que cinq fois le nombre des chiffres de B.*

Von Neumann [1953], studying the optimal assignment problem:

The game matrix for this game will be $2n \times n^2$. From this it is not difficult to infer how many steps are needed to get significant approximate solutions with the method of G. W. Brown and J. von Neumann.... It turns out that this number is a moderate power of n, i.e., considerably smaller than the 'obvious' estimate $n!$ mentioned earlier.

Cobham [1965], in a paper on Turing machines and computability (denoting the size of n by $l(n)$):

> To obtain some idea as to how we might go about the further classification of relatively simple functions, we might take a look at how we ordinarily set about computing some of the more common of them. Suppose, for example, that m and n are two numbers given in decimal notation with one written above the other and their right ends aligned. Then to add m and n we start at the right and proceed digit-by-digit to the left writing down the sum. No matter how large m and n, this process terminates with the answer after a number of steps equal at most to one greater than the larger of $l(m)$ and $l(n)$. Thus the process of adding m and n can be carried out in a number of steps which is bounded by a linear polynomial in $l(m)$ and $l(n)$. Similarly, we can multiply m and n in a number of steps bounded by a quadratic polynomial in $l(m)$ and $l(n)$. So, too, the number of steps involved in the extraction of square roots, calculation of quotients, etc., can be bounded by polynomials in the lengths of the numbers involved, and this seems to be a property of simply functions in general. This suggests that we consider the class, which I will call $\mathscr{L}$, of all functions having this property.

Edmonds [1965a], showing the polynomial solvability of the matching problem:

> For practical purposes computational details are vital. However, my purpose is only to show as attractively as I can that there is an efficient algorithm. According to the dictionary, "efficient" means "adequate in operation or performance." This is roughly the meaning I want—in the sense that it is conceivable for maximum matching to have no efficient algorithm. Perhaps a better word is "good".
>
> I am claiming, as a mathematical result, the existence of a *good* algorithm for finding a maximum cardinality matching in a graph.
>
> There is an obvious finite algorithm, but that algorithm increases in difficulty exponentially with the size of the graph. It is by no means obvious whether *or not* there exists an algorithm whose difficulty increases only algebraically with the size of the graph.
>
> The mathematical significance of this paper rests largely on the assumption that the two preceding sentences have mathematical meaning. I am not prepared to set up the machinery necessary to give them formal meaning, nor is the present context appropriate for doing this, but I should like to explain the idea a little further informally. It may be that since one is customarily concerned with existence, convergence, finiteness, and so forth, one is not inclined to take seriously the question of the existence of a *better-than-finite* algorithm....
>
> One can find many classes of problems, besides maximum matching and its generalizations, which have algorithms of exponential order but seemingly none better. An example known to organic chemists is that of deciding whether two given graphs are isomorphic. For practical purposes the difference between algebraic and exponential order is often more crucial than the difference between finite and non-finite.

Edmonds [1965b], introducing the term 'good characterization', and giving a good characterization for the minimum number of linearly independent sets into which the columns of a matrix can be partitioned:

We seek a good characterization of the minimum number of independent sets into which the columns of a matrix of M_F can be partitioned. As the criterion of "good" for the characterization we apply the "principle of the absolute supervisor." The good characterization will describe certain information about the matrix which the supervisor can require his assistant to search out along with a minimum partition and which the supervisor can then use with ease to verify with mathematical certainty that the partition is indeed minimum. Having a good characterization does not mean necessarily that there is a good algorithm. The assistant might have to kill himself with work to find the information and the partition.

Part I

Linear algebra

This part consists of only one chapter, Chapter 3 on linear algebra and complexity. It is meant mainly (i) to discuss some less well-known facts on sizes and complexity in linear algebra (like: each feasible system of linear equations has a solution of polynomially bounded size; the running time of the Gaussian elimination method is polynomially bounded), and (ii) to emphasize the analogy of linear algebra with other parts of this book, viz. Part II on lattices and linear diophantine equations, and Part III on polyhedra, linear inequalities, and linear programming (like: finite basis theorems; good characterizations for feasibility).

The results on sizes and complexity will be fundamental also for other parts of this book.

3

Linear algebra and complexity

We study some complexity issues in linear algebra. First in Section 3.1 we review some well-known facts from linear algebra. In Section 3.2 we give polynomial estimates on the sizes of the determinant, of the inverse, of solutions of systems of linear equations, and the like, and we derive that the characterizations described in Section 3.1 yield good characterizations. In Section 3.3 we describe the Gaussian elimination method, and we show that it is a polynomial-time method, implying the polynomial solvability of several problems from linear algebra. Finally, in Section 3.4 we survey some other, iterative methods.

3.1. SOME THEORY

We do not aim at developing here the basics of linear algebra, but assume that elementary ideas like linear subspace, linear independence, rank, determinant, nonsingular, inverse matrix, Gaussian elimination, etc. are known. However, to put in the forefront the analogy with the theories of linear diophantine equations and of linear inequalities discussed later, we derive several (well-known) results as consequences from one fundamental theorem. Each of the results in this section holds for real data as well as if we restrict all data (matrices, vectors, variables) to rationals.

A subset L of $\mathbb{R}^n$ (respectively $\mathbb{Q}^n$) is a *linear hyperplane* if $L = \{x \mid ax = 0\}$ for some nonzero row vector a in $\mathbb{R}^n$ (respectively $\mathbb{Q}^n$).

Theorem 3.1. *Each linear subspace of $\mathbb{R}^n$ (or $\mathbb{Q}^n$) is generated by finitely many vectors, and is also the intersection of finitely many linear hyperplanes.*

Proof. The proof is well-known from linear algebra. If L is a linear subspace of $\mathbb{R}^n$, then any maximal set of linearly independent vectors in L is finite and generates L. Similarly, L is the intersection of finitely many linear hyperplanes $\{x \mid a_i x = 0\}$, where the a_i form a maximal set of linearly independent vectors generating $L^* := \{z \mid zx = 0 \text{ for all } x \text{ in } L\}$. $\qquad\square$

So for each linear subspace L of $\mathbb{R}^n$ there are matrices A and C such that $L = \{x \mid Ax = 0\} = \{x \mid x = Cy \text{ for some vector } y\}$.

Corollary 3.1a. *For each matrix A there exist vectors $x_1, \ldots, x_t$ such that: $Ax = 0$ iff $x = \lambda_1 x_1 + \cdots + \lambda_t x_t$ for certain rationals $\lambda_1, \ldots, \lambda_t$.*

Proof. Apply Theorem 3.1 to the linear subspace $\{x | Ax = 0\}$. $\square$

If the vectors $x_1, \ldots, x_t$ in Corollary 3.1a are linearly independent, they are said to form a *fundamental system of solutions* of $Ax = 0$.

The following corollary was observed by Gauss [1809:Art.180]. It is sometimes called the *Fundamental theorem of linear algebra*, the *first Fredholm theorem*, or *Fredholm's theorem of the alternatives* (after Fredholm [1903], who showed it for more general linear spaces).

Corollary 3.1b. *The system $Ax = b$ has a solution if and only if $yb = 0$ for each vector y with $yA = 0$.*

Proof. Necessity is trivial, as $yb = yAx = 0$ whenever $Ax = b$ and $yA = 0$. To see sufficiency, let $L := \{z | Ax = z$ for some $x\}$. By Theorem 3.1 there exists a matrix C such that $L = \{z | Cz = 0\}$. Hence CA is an all-zero matrix. Therefore: $(yA = 0 \Rightarrow yb = 0) \Rightarrow Cb = 0 \Rightarrow b \in L \Rightarrow \exists x : Ax = b$. $\square$

Corollary 3.1b gives a first example of a characterization in terms of necessary and sufficient conditions, where necessity is trivial, and sufficiency is the content of the characterization. Later we shall see that Corollary 3.1b yields a good characterization (cf. Section 2.5).

The following extension of Corollary 3.1a also follows as corollary. This extension is termed algebraically as *inhomogeneous* and geometrically as *affine*.

Corollary 3.1c. *If x_0 is a solution of the system $Ax = b$, then there are vectors $x_1, \ldots, x_t$ such that: $Ax = b$ if and only if $x = x_0 + \lambda_1 x_1 + \cdots + \lambda_t x_t$ for certain rationals $\lambda_1, \ldots, \lambda_t$.*

Proof. Note that $Ax = b$ iff $A(x - x_0) = 0$, and use Corollary 3.1a. $\square$

If $A = [A_1 \ A_2]$ and A_1 is nonsingular, then we can take $x_0, x_1, \ldots, x_t$ as in Corollary 3.1c so that:

$$(1) \qquad [x_0, x_1, \ldots, x_t] = \begin{bmatrix} A_1^{-1}b & A_1^{-1}A_2 \\ 0 & -I \end{bmatrix}.$$

As for any nonsingular matrix C, the (j, i)th entry of C^{-1} is given by:

$$(2) \qquad \frac{(-1)^{i+j} \det(C_{ij})}{\det C}$$

where C_{ij} arises from C by deleting the ith row and jth column, it follows that for any column vector d, the jth component of $C^{-1}d$ is given by

$$(3) \qquad \frac{\det \tilde{C}}{\det C}$$

where $\tilde{C}$ arises from C by replacing the jth column of C by d. This is the essence of *Cramer's rule* (Cramer [1750]). As a consequence, each nonzero component of each of the vectors $x_0, x_1, \ldots, x_t$ in (1) is a quotient of subdeterminants of the matrix $[A_1 \; A_2 \; b]$.

Corollary 3.1d. *Let A be a matrix, b be a column vector, c be a row vector, and δ be a number. Assume that the system $Ax = b$ has a solution. Then $Ax = b$ implies $cx = \delta$, if and only if there is a row vector y such that $yA = c$ and $yb = \delta$.*

Proof. The sufficiency of the condition is trivial. To see necessity, let x_0 be a solution of $Ax = b$, and suppose that $Ax = b$ implies $cx = \delta$. Then in particular $cx_0 = \delta$. Then: $Ax = 0 \Rightarrow A(x + x_0) = b \Rightarrow c(x + x_0) = \delta \Rightarrow cx = 0$. Therefore by Corollary 3.1b, $yA = c$ for some row vector y. This y also satisfies $yb = yAx_0 = cx_0 = \delta$.   □

Corollary 3.1d says that any implication between linear equations can be 'proved' in finite time—in fact in polynomial time, as we will see in the next section.

3.2. SIZES AND GOOD CHARACTERIZATIONS

In order to show that the results of Section 3.1 yield good characterizations, we investigate the sizes of the numbers occurring in linear algebra.

As in Section 2.1 we use the following as the sizes of a rational number $r = p/q$ ($p \in \mathbb{Z}, q \in \mathbb{N}$, p and q relatively prime), of a rational vector $c = (\gamma_1, \ldots, \gamma_n)$ and of a rational matrix $A = (\alpha_{ij})_{i=1, j=1}^{m}{}^{n}$:

(4) $\text{size}(r) := 1 + \lceil \log_2(|p| + 1) \rceil + \lceil \log_2(q + 1) \rceil$

        $\text{size}(c) := n + \text{size}(\gamma_1) + \cdots + \text{size}(\gamma_n)$

        $\text{size}(A) := mn + \sum_{i,j} \text{size}(\alpha_{ij}).$

Theorem 3.2. *Let A be a square rational matrix of size σ. Then the size of $\det A$ is less than 2σ.*

Proof. Let $A = (p_{ij}/q_{ij})_{i,j=1}^{n}$, where for each i, j, p_{ij} and q_{ij} are relatively prime integers, with $q_{ij} > 0$. Moreover, let $\det A = p/q$, where p and q are relatively prime integers, with $q > 0$. Clearly,

(5) $q \leqslant \prod_{i,j=1}^{n} q_{ij} < 2^{\sigma - 1}.$

It is immediate from the definition of a determinant that

(6) $|\det A| \leqslant \prod_{i,j=1}^{n} (|p_{ij}| + 1).$

Combining (5) and (6) gives

(7) $|p| = |\det A| \cdot q \leqslant \prod_{i,j=1}^{n} (|p_{ij}| + 1)q_{ij} < 2^{\sigma - 1}.$

Therefore, by (5) and (7),

(8) $\text{size}(\det A) = 1 + \lceil \log_2(|p| + 1) \rceil + \lceil \log_2(q + 1) \rceil < 2\sigma.$ □

Consequences are as follows.

Corollary 3.2a. *The inverse* A^{-1} *of a nonsingular rational matrix* A *has size polynomially bounded by the size of* A.

Proof. Use Theorem 3.2 and the fact that the entries in A^{-1} are quotients of (sub)determinants of A (cf. (3)).   □

Corollary 3.2b. *If the system* $Ax = b$ *of rational linear equations has a solution, it has one of size polynomially bounded by the sizes of* A *and* b.

Proof. We may assume that A has linearly independent rows, and that $A = [A_1 \ A_2]$ with A_1 nonsingular. Then by Corollary 3.2a,

(9) $x_0 := \begin{pmatrix} A_1^{-1}b \\ 0 \end{pmatrix}$

is a solution of $Ax = b$ of size polynomially bounded by the sizes of A and b.   □

Corollary 3.2b implies that Corollary 3.1b provides a good characterization.

Corollary 3.2c. *There is a good characterization for the problem: given a rational matrix* A *and a rational vector* b, *does* $Ax = b$ *have a solution?*

Proof. If the answer to the problem is positive, this has a proof of polynomial length, as by Corollary 3.2b we can specify a solution of polynomial size. If the answer is negative, by Corollary 3.1b this can be shown by specifying a vector y with $yA = 0$ and $yb = 1$. By Corollary 3.2b, again such a vector y exists with size polynomially bounded by the sizes of A and b.   □

In the next section we show that the problem described in Corollary 3.2c actually is polynomially solvable, and that an explicit solution can be *found* in polynomial time (by Gaussian elimination).

Similarly, for homogeneous equations there exists a fundamental system of solutions of polynomial size: if A is a rational matrix, there are vectors $x_1, \ldots, x_t$ such that $Ax = 0$ if and only if $x = \lambda_1 x_1 + \cdots + \lambda_t x_t$ for certain rationals $\lambda_1, \ldots, \lambda_t$, and such that the size of $[x_1, \ldots x_t]$ is polynomially bounded by the size of A. Indeed, again we may assume that $A = [A_1 \ A_2]$, with A_1 nonsingular. Then take for $x_1, \ldots, x_t$ the columns of the matrix

(10) $\begin{bmatrix} A_1^{-1}A_2 \\ -I \end{bmatrix}.$

Combined with Corollary 3.2b this gives that any feasible rational system $Ax = b$ has a fundamental system of solutions of size polynomially bounded by the sizes of A and b.

For later reference we make a more precise estimate.

Corollary 3.2d. *Let A be a rational $m \times n$-matrix, and let b be a rational column vector such that each row of the matrix $[A\ b]$ has size at most φ. If $Ax = b$ has a solution, then*

(11) $\qquad \{x\,|\,Ax = b\} = \{x_0 + \lambda_1 x_1 + \cdots + \lambda_t x_t\,|\,\lambda_1,\ldots,\lambda_t \in \mathbb{R}\}$

for certain rational vectors $x_0,\ldots,x_t$ of size at most $4n^2\varphi$.

Proof. By Cramer's rule, there are $x_0,\ldots,x_t$ satisfying (11) with all nonzero components being quotients of subdeterminants of the matrix $[A\ b]$ of order at most n. By Theorem 3.2, these determinants have size less than $2n\varphi$. Hence each component of x_i has size less than $4n\varphi$. Therefore, each x_i has size at most $4n^2\varphi$. $\qquad\square$

3.3. THE GAUSSIAN ELIMINATION METHOD

Probably the best-known algorithm in linear algebra is the so-called Gaussian elimination method, which is a polynomial-time method for solving a system of linear equations.

The idea can be described recursively. Suppose we want to solve the system

(12) $\qquad \begin{aligned} \alpha_{11}\xi_1 + \alpha_{12}\xi_2 + \cdots + \alpha_{1n}\xi_n &= \beta_1 \\ \alpha_{21}\xi_1 + \alpha_{22}\xi_2 + \cdots + \alpha_{2n}\xi_n &= \beta_2 \\ \vdots \qquad\quad \vdots \qquad\qquad \vdots \qquad\ \ \vdots \\ \alpha_{m1}\xi_1 + \alpha_{m2}\xi_2 + \cdots + \alpha_{mn}\xi_n &= \beta_m. \end{aligned}$

We may suppose $\alpha_{11} \neq 0$. Then we can subtract appropriate multiples of the first equation from the other equations so as to get:

(13) $\qquad \begin{aligned} \alpha_{11}\xi_1 + \alpha_{12}\xi_2 + \cdots + \alpha_{1n}\xi_n &= \beta_1 \\ \alpha'_{22}\xi_2 + \cdots + \alpha'_{2n}\xi_n &= \beta'_2 \\ \vdots \qquad\qquad \vdots \qquad\ \ \vdots \\ \alpha'_{m2}\xi_2 + \cdots + \alpha'_{mn}\xi_n &= \beta'_m. \end{aligned}$

Now, recursively, we solve the system of the last $m - 1$ equations in (13). Substituting the solution found into the first equation yields a value for ξ_1.

In matrix form, the Gaussian elimination method transforms a given matrix A into the form

(14) $\qquad \begin{bmatrix} B & C \\ 0 & 0 \end{bmatrix}$

where B is a nonsingular, upper-triangular matrix, by the following operations:

(15)   (i) adding a multiple of one row to another row;
           (ii) permuting rows or columns.

The method proceeds as follows. Let an $m \times n$-matrix A be given. We find matrices A_0, A_1, $A_2, \ldots$ in the following way. Let $A_0 := A$. Suppose we have found A_k, of form

(16) $A_k = \begin{bmatrix} B & C \\ 0 & D \end{bmatrix}$

where B is a nonsingular upper triangular matrix of order k. Then A_{k+1} is determined by the following procedure.

Gaussian elimination (forward) step. Given matrix A_k of form (16), choose a nonzero component of D, which will be called the *pivot element*, and permute the rows and columns of A_k so that this pivot element is in position $(1, 1)$ of D. Now add rational multiples of the first row of D to the other rows of D, in such a way that δ_{11} becomes the only nonzero element in the first column of D. This makes the matrix A_{k+1}.

This step is done for $k = 0, 1, 2, \ldots$, until in, say, A_r, the matrix D is an all-zero matrix, so that A_r has form (14).

If we start with the matrix $[A\ b]$, this procedure yields a solution for $Ax = b$, as sketched at the beginning of this section.

Gaussian elimination backward steps. Sometimes, after reaching form (14), the matrix is transformed further to form

(17) $\begin{bmatrix} \Delta & C \\ 0 & 0 \end{bmatrix}$

where Δ is a nonsingular diagonal matrix. Again this is accomplished by:

(18)   adding multiples of the kth row to the rows $1, \ldots, k - 1$

for $k = r, r - 1, r - 2, \ldots, 1$ (with $r = \text{order}(\Delta)$). These are the so-called Gaussian elimination backward steps.

Edmonds [1967b] showed that the running time of the Gaussian elimination method is polynomially bounded.

Theorem 3.3. *For rational data, the Gaussian elimination method is a polynomial-time method.*

Proof. Without loss of generality we may assume that we do not need to permute rows or columns at executing the method. It is moreover easy to see that the number of elementary arithmetic operations, like addition, multiplication, and division, in the method is polynomially bounded by the number of rows and columns of the initial matrix A, and hence by size(A). So to prove that the method is polynomial, it is only left to show that also the sizes of the numbers occurring throughout the method are polynomially bounded by size(A).

To prove this, note that if A_k has form (16), and δ_{ij} is the (i,j)th entry of D, then trivially,

(19) $\qquad \delta_{ij} = \det((A_k)_{1,\ldots,k,k+j}^{1,\ldots,k,k+i})/\det((A_k)_{1,\ldots,k}^{1,\ldots,k})$.

Here $G_{j_1,\ldots,j_t}^{i_1,\ldots,i_t}$ denotes the submatrix of matrix G induced by rows $i_1,\ldots,i_t$ and columns $j_1,\ldots,j_t$. As A_k arises from A by adding multiples of the first k rows to other rows, (19) also holds if we replace A_k by A, i.e.

(20) $\qquad \delta_{ij} = \det(A_{1,\ldots,k,k+j}^{1,\ldots,k,k+i})/\det(A_{1,\ldots,k}^{1,\ldots,k})$.

Therefore, by Theorem 3.2, the size of δ_{ij} is at most 4size(A). Since each entry of the matrices B and C in (16) has been, in a previous stage, an entry of a matrix D, it follows that throughout the forward steps each number has size at most 4size(A).

To see that also during the backward steps the entries do not grow too big in size, let E be the matrix formed by applying (18) for $k = r, r-1, \ldots, t+1$. So

(21) $\qquad E = \begin{bmatrix} B & 0 & C \\ 0 & \Delta & D \\ 0 & 0 & 0 \end{bmatrix}$

where B is a nonsingular upper-triangular $t \times t$-matrix and Δ is a diagonal $(r-t) \times (r-t)$-matrix. Split A_r accordingly, i.e.

(22) $\qquad A_r = \begin{bmatrix} B_1 & B_2 & C_1 \\ 0 & B_3 & C_2 \\ 0 & 0 & 0 \end{bmatrix}$

with B_1 and B_3 upper-triangular matrices of order t and $r-t$, respectively. Then one easily checks that $B = B_1$, $C = C_1 - B_2 B_3^{-1} C_2$, $D = \Delta B_3^{-1} C_2$, and that Δ consists of the diagonal elements of B_3. Therefore, the size of E is polynomially bounded by size(A_r), and hence by size(A). $\square$

The polynomial running time of the Gaussian elimination method implies the polynomial solvability of several basic linear algebraic problems.

Corollary 3.3a. *The following problems are polynomially solvable:*

(i) *determining the determinant of a rational matrix;*
(ii) *determining the rank of a rational matrix;*
(iii) *determining the inverse of a nonsingular rational matrix;*
(iv) *testing rational vectors for linear independence;*
(v) *solving a system of rational linear equations.*

Proof. The number of nonzero diagonal entries in the final matrix in the Gaussian elimination method is equal to the rank of the initial matrix A. If A is a square matrix, the product of the diagonal entries in the final matrix gives the determinant of A (assuming we did not permute rows or columns during the procedure—otherwise we have to multiply by the signs of the permutations).

For nonsingular A, if we start the procedure with the matrix $[A \ I]$ and we choose our pivot element always in the first column of D, then we end up with a matrix of form $[\Delta \ C]$, with Δ a diagonal matrix. Then $A^{-1} = \Delta^{-1}C$ (since

$$(23) \qquad [A \ I]\begin{bmatrix} A^{-1} \\ -I \end{bmatrix} = 0 \text{ implies } [\Delta \ C]\begin{bmatrix} A^{-1} \\ -I \end{bmatrix} = 0,$$

i.e. $A^{-1} = \Delta^{-1}C$).

Let A be a rational $m \times n$-matrix, and let $b \in \mathbb{Q}^m$. Suppose we wish to solve $Ax = b$. Let $\begin{bmatrix} \Delta & C \\ 0 & 0 \end{bmatrix}$ be the final matrix E obtained by Gaussian elimination from A (including the backward steps). Assume, without loss of generality, that we did not permute rows or columns during the method, and split A and b according to the splitting of E:

$$(24) \qquad A = \begin{bmatrix} A_1 & A_2 \\ A_3 & A_4 \end{bmatrix}, \quad b = \begin{pmatrix} b_1 \\ b_2 \end{pmatrix}.$$

Then clearly,

$$(25) \qquad Ax = b \text{ if and only if } \begin{bmatrix} \Delta & C \\ 0 & 0 \end{bmatrix}x = \begin{pmatrix} \Delta A_1^{-1}b_1 \\ b_2 - A_3 A_1^{-1}b_1 \end{pmatrix}.$$

The latter system of equations has a solution if and only if $b_2 - A_3 A_1^{-1}b_1 = 0$, in which case

$$(26) \qquad x = \begin{pmatrix} A_1^{-1}b_1 \\ 0 \end{pmatrix}$$

is a solution. Therefore, we can find a solution of a system of linear equations, or conclude its infeasibility, in polynomial time.   □

Edmonds [1967b] described a variant of the method above, so that if we start with an integral matrix A, all matrices A_k throughout the forward steps are integral. This is based on observations (19) and (20). Suppose the starting matrix A is integral. By (20) we can write, after the kth step, δ_{ij} as a quotient p/q of two integers, with $q = \det(A_{1,...,k}^{1,...,k}) = \beta_{11} \cdot ... \cdot \beta_{kk}$ (where β_{ij} and δ_{ij} are the (i,j)th entries of B and D, respectively, in the notation (16)).

Hence p can be found by elementary arithmetic operations so that we only make divisions if the result is an integer. So we have a polynomial-time algorithm in which we do not need to reduce quotients r/s so that r and s become relatively prime. Thus we avoid application of the Euclidean algorithm (Section 5.1).

Elimination by cross-multiplication (i.e. replacing in the kth step δ_{ij} by $\delta_{ij}\delta_{11}$ $-\delta_{i1}\delta_{1j}$), in order to keep all entries integral, can lead to exponential growth of the sizes of the entries. This is shown by the matrices with 2's on the main diagonal, 1's below the main diagonal, and 0's above the main diagonal (this example was given by F. Voelkle). See also Rosser [1952] and Bareiss [1968].

The Gaussian elimination method is a fast, exact method. However, when working with restricted precision calculators, round-off errors can cause problems (cf. the books on numerical methods mentioned at the end of this Part I). This may depend on the pivot choice strategy. A way of promoting the numerical stability is to choose as pivot element always that component of D which is largest in absolute value. More subtle is the currently accepted strategy of choosing at the kth step that component δ_{11} in the first column of D, for which

$$(27) \qquad \frac{|\delta_{i1}|}{\max_j|\delta_{ij}|}$$

is as large as possible.

In order to avoid the numerical instability of the Gaussion eliminatian method, other methods for approximating solutions to systems of linear equations have been proposed. Among them are the *iterative methods*, to be discussed in the next section.

Gauss–Jordan elimination method. Related to the Gaussian elimination method is the *Gauss–Jordan elimination method*, where forward and backward steps are combined. Now, given a matrix A_k of form

$$(28) \qquad A_k = \begin{bmatrix} I & C \\ 0 & D \end{bmatrix}$$

where I is the identity matrix of order k, choose a nonzero 'pivot element' in D—without loss of generality let this be δ_{11} (again we may permute rows and columns). Then divide the $(k+1)$th row of A_k by δ_{11}, and add rational multiples of the $(k+1)$th row to the other rows, in such a way that the 1 in position $(k+1, k+1)$ will be the only nonzero entry in the $(k+1)$th column. This makes A_{k+1}, and we repeat the procedure.

If we start with a matrix $A = A_0$, we end up with a matrix $\begin{bmatrix} I & C \\ 0 & 0 \end{bmatrix}$, where I is an identity matrix of order r $(:=\text{rank}(A))$. If A_k has form (28) and A is split accordingly (assuming we did not permute rows or columns):

$$(29) \qquad A = \begin{bmatrix} A_1 & A_2 \\ A_3 & A_4 \end{bmatrix}$$

then

$$(30) \qquad A_k = \begin{bmatrix} I & A_1^{-1}A_2 \\ 0 & A_4 - A_3A_1^{-1}A_2 \end{bmatrix}.$$

Hence the size of A_k is polynomially bounded by the size of A. Therefore also the Gauss–Jordan elimination method is polynomial-time, which again can be seen to imply the polynomial solvability of the linear algebraic problems mentioned above. In practice the efficiency of the Gauss–Jordan elimination method is inferior to that of the Gaussian elimination method (as, for example, for solving a system of linear equations it is enough to carry out only the forward steps in the Gaussian elimination).

Method of least squares. This method was designed by Legendre [1805] and Gauss [1809]. For a given matrix A and vector b it finds a vector x minimizing

(31) $\|Ax - b\|$.

Here $\|\cdot\|$ denotes the Euclidean norm. Basic is the observation that vector x minimizes $\|Ax - b\|$ if and only if $A^{T}Ax = A^{T}b$. So the problem is reduced to solving a system of linear equations.

Note. Aspvall and Shiloach [1980b] describe a fast algorithm for solving systems of linear equations with two variables per equation.

3.4. ITERATIVE METHODS

Iterative methods for solving a system $Ax = b$ consist of determining a sequence of vectors $x_0, x_1, x_2, \ldots$ which are hoped to converge to a solution. Such methods date back to Gauss [1823]. We here give a brief review. The methods we describe are especially suitable if the matrix A is positive semi-definite, with dominant diagonal (as in least square problems).

As a first example we describe the *Jacobi iteration method* (Jacobi [1845]). Let $Ax = b$ be a system of linear equations, with A nonsingular. Without loss of generality, all diagonal entries of A are 1. We wish to solve $Ax = b$, or equivalently, $x = (I - A)x + b$. Let $x_0 := b$, and let

(32) $x_{k+1} := (I - A)x_k + b$.

One easily checks that if the sequence converges, it converges to a solution of $Ax = b$. If all eigenvalues of $I - A$ are less than 1 in absolute value, then the process indeed converges.

A variant is the *Gauss–Seidel iteration method* (Gerling [1843], Seidel [1874]), which computes the $(i + 1)$th component of x_{k+1} with

(33) $(x_{k+1})_{i+1} := ((I - A)\tilde{x}_k + b)_{i+1}$

where

(34) $\begin{aligned} (\tilde{x}_k)_j &= (x_{k+1})_j \quad \text{if } j \leqslant i \quad \text{(which is already computed)} \\ (\tilde{x}_k)_j &= (x_k)_j \quad \text{if } j > i. \end{aligned}$

(Here $(..)_j$ means the jth component of $\ldots$). So it computes x_{k+1} by resetting x_k component by component. It means that if we write $A = L + U$, with L a lower triangular matrix and U a strictly upper triangular matrix, then

(35) $Lx_{k+1} = -Ux_k + b$.

It can be shown that the sequence $x_0, x_1, x_2, \ldots$ converges to a solution if for each row i of A one has

(36) $|\alpha_{ii}| > \sum_{j \neq i} |\alpha_{ij}|$

(α_{ij} being the (i, j)-entry of A). Compared with the Jacobi method, the Gauss–Seidel method turns out to be faster by a factor of about 2.

A further acceleration gives the method of *successive overrelaxation* (*SOR*), where (35) is replaced by

(37) $(L - \varepsilon\Delta)x_{k+1} = (-\varepsilon\Delta - U)x_k + b$

for some appropriately chosen small $\varepsilon > 0$ (Δ denotes the diagonal of A).

A related class of iterative *relaxation* methods for approximating a system $Ax = b$ is as follows.

x_0 is arbitrarily chosen (e.g. on heuristic grounds near to a solution of $Ax = b$). If x_k has been found, check if $Ax_k = b$. If so, stop: we have found a solution. If not, choose a violated equation, say $a_i x = \beta_i$ (e.g. we could choose the i with $|\beta_i - a_i x_k|$ as large as possible). Next let

$$(38) \qquad x_{k+1} := x_k + \lambda \left(\frac{\beta_i - a_i x_k}{a_i a_i^\top} \right) a_i^\top$$

for a certain λ with $0 < \lambda \leqslant 2$. (If $\lambda = 1$, x_{k+1} is the projection of x_k onto the hyperplane $a_i x = \beta_i$. If $\lambda = 2$, x_{k+1} is the reflection of x_k into $a_i x = \beta_i$.) It can be shown that if we take $0 < \lambda < 2$ and $Ax = b$ is feasible, then $x_0, x_1, x_2, \ldots$ converges to a solution (see the proof for linear inequalities in Section 12.3). A stopping rule can be incorporated, e.g. when $\sum_i |\beta_i - a_i x_k|$, or $\max_i |\beta_i - a_i x_k|$ is small enough.

There are several other methods for solving systems of linear equations—for surveys, see the literature given at the end of Part I.

Ursic and Patarra [1983] study the problem of finding an exact solution of a system of linear equations from an approximate solution, using the continued fraction method. See also Section 6.1.

Notes on linear algebra

HISTORICAL NOTES

By the second millennium B. C. special examples of linear equations were being studied by the Egyptians and Babylonians. The Babylonians also considered two simultaneous linear equations in two unknowns. The Babylonians, Greeks, and Chinese knew the idea of elimination of unknowns when solving systems of equations (linear or quadratic).

The, what is called now, Gaussian elimination method was described explicitly in the remarkable Chinese *Chiu-Chang Suan-Shu* ('Nine Books of Arithmetic'). This book dates from the early Han period (202 B.C.–A.D. 9), but describes methods which are developed probably much earlier. Book VIII contains the elimination method for problems of up to 5 linear equations in 5 unknowns, in terms of certain rectilinear arrays (now called matrices), which by column operations are transformed to triangular form (cf. Midonick [1968: pp. 182–188] and Vogel [1968] for translations).

Linear equations and elimination were also studied by Diophantos of Alexandria ($\pm$ third century A.D.), by the Hindu mathematician Āryabhata (born A.D. 476), by Al-Khowarizmi in his famous book *Al-jebr w'almuqabala* (written $\pm$ A.D. 825) and by the 13th century Chinese Yang Hui.

The Japanese mathematician Seki Kōwa (1642–1708) seems to be the first to use determinants, describing them in his manuscript *Yendan kai Fukudai No Hō* of 1683. Seki's interest came from eliminating terms in systems of algebraic equations. Further theory of determinants developed in Japan was described in 1690 in the book *Sampō Hakki* of Izeki Chishin. See Mikami [1913, 1914] for a survey of the early Japanese work on determinants.

In a letter of 28th April 1693 to l'Hospital, Leibniz [1693] described the 3×3-determinant in order to simplify the solution of $Ax = b$ where A is a 3×2-matrix. He shows by elimination of variables that if $Ax = b$ is solvable then $\det [A\ b] = 0$, and he claims that a solution can be described with determinants. The 3×3-determinant being 0 is in Leibniz's notation:

$$(1) \qquad \begin{matrix} 1_0 \cdot 2_1 \cdot 3_2 & 1_0 \cdot 2_2 \cdot 3_1 \\ 1_1 \cdot 2_2 \cdot 3_0 = 1_1 \cdot 2_0 \cdot 3_2 \\ 1_2 \cdot 2_0 \cdot 3_1 & 1_2 \cdot 2_1 \cdot 3_0 \end{matrix}$$

'qui porte sa preuve avec soy par les harmonies qui se remarquent se par tout', these harmonies being caused by Leibniz's way of using indices.

Leibniz's interest in linear equations came from analysis, where tangent and normal provide linear equations. His work on determinants became known generally only in the 19th century, after the publication of Leibniz's letters. According to Gerhardt [1891], a manuscript of Leibniz which contains already the results described above may date back to 1678 (cf. Leibniz [1678?]). Knobloch [1980] made a careful study of Leibniz's work on determinants, linear equations and elimination.

Interest in linear equations of subsequent workers originated from studying algebraic curves (e.g. to deal with questions like: how many points of a curve of given degree should one know in order to determine the coefficients of the curve?). In his *Treatise of Algebra*, MacLaurin [1748] described the method of solving linear equations in up to 4 unknowns with the help of determinants. This method was generalized to n variables by Cramer [1750] in the appendix of his book *Introduction à l'analyse des lignes courbes algébriques*. It gives the well-known 'Cramer's rule'.

Bézout [1767] showed that, for a square matrix, the system $Ax = 0$ has a nontrivial solution if and only if det $A = 0$. Vandermonde [1776], Laplace [1776], and Bézout [1779] extended and systematized the theory of determinants further. For example, Laplace described calculating a determinant by expanding a row or column.

Apparently independently, Lagrange [1775b, c] developed a theory for 3×3-determinants, in order to describe the volume of pyramids in 3-dimensional space. So here the entries of the matrix mean coordinates rather than coefficients of linear equations. One of Lagrange's results [1775b] is det $A^2 = (\det A)^2$ for 3×3-matrices.

The idea of solving linear equations by elimination of variables was a standard technique in each of the papers on determinants. Elimination also occurs in a manuscript of Lagrange [1778], where in Section 12 it is described how to find a nontrivial solution of a system of homogeneous linear equations.

The name Gaussian elimination derives from some papers of Gauss where elimination is applied. Gauss studied linear equations in order to estimate the orbits of celestial bodies, whose positions have been observed subject to errors. So again the goal is to determine or approximate the coefficients of certain algebraic curves, viz. conical sections. To this end Gauss [1809] designed the 'method of least squares', developed independently by Legendre [1805], which method gives rise to a system of linear equations to be solved. (Gauss [1809: Art. 186] claims to have used the method since 1795, before Legendre.) In Art. 180 of [1809], Gauss says that these equations can be solved by 'usual elimination' ('per eliminationem vulgarem'). Moreover, he observes that if a system $Ax = b$ of n linear equations in n unknowns is unsolvable or indeterminate (i.e. has no solution or several solutions), then $yA = 0$ for some $y \neq 0$, which is a 'duality' statement. (Fredholm [1903] extended it to more general linear spaces; therefore it is sometimes called Fredholm's theorem.)

In Art. 182, Gauss describes the, what is now called, Gram–Schmidt orthogonalization process, which decomposes a matrix A as $A = QU$, where Q is orthogonal (i.e. $Q^{\mathsf{T}}Q = I$) and U is upper triangular. Hence $Ax = b$ can be replaced by $Ux = Q^{\mathsf{T}}b$, which is easy to solve. Moreover, $\|Ax - b\| = \|Ux - Q^{\mathsf{T}}b\|$, which last form is easy to minimize. Gauss used the method in determining the orbit of the planetoid Pallas (Gauss [1811], cf. Goldstine [1977: Sec. 4.10]) and in geodesy (Gauss [1828]).

Note that the QU-decomposition differs from the Gaussian elimination process described in Section 3.3, where the matrix A is decomposed as $A = LU$, with L nonsingular and lower triangular, and U upper triangular, whence $Ax = b$ is equivalent to $Ux = L^{-1}b$.

In a letter of 26th December 1823 to C. L. Gerling, Gauss [1823] described an alternative method for solving linear equations of type $A^{\mathsf{T}}Ax = A^{\mathsf{T}}b$, namely an iterative method (called by Gauss an 'indirect' method). Starting with $x^* = 0$, repeat the following iteration. Select the index i for which the quotient

$$(2) \qquad q := (A^{\mathsf{T}}b - A^{\mathsf{T}}Ax^*)_i / (A^{\mathsf{T}}A)_{ii}$$

is as large as possible; increase the ith component of x^* by q, thus making the new x^*. 'So fahre ich fort, bis nichts mehr zu corrigiren ist' (up to the desired precision). The method 'lässt sich halb im Schlafe ausführen, oder man kann während desselben an andere Dinge denken'.

In Section 3.4 above we mentioned iterative methods designed by Gerling [1843], Jacobi [1845], and Seidel [1874]. The Gauss–Jordan elimination method seems to have been described first by Clasen [1888]; the name of W. Jordan is attached only as the method was publicized by the 5th edition of the *Handbuch der Vermessungskunde* (Jordan [1904]).

The theory of determinants was greatly extended by Cauchy [1815]. Among the several basics of determinants Cauchy derived are the product rule for determinants (det AB = det A.det B) and a description of the adjoint. (The product rule was also stated without satisfactory proof by Binet [1813].) Cauchy also introduced the term determinant and the notation a_{ij}. Also Binet [1813] and Scherk [1825] gave expositions of determinants.

Further important progress in matrix and determinant theory was made, *inter alia*, by Sylvester (who introduced in [1850] the word *matrix*), Cayley (e.g. [1855a, 1858], describing basic notions like identity matrix and the inverse of a matrix), Smith (e.g. [1861]), Kronecker, and Frobenius (e.g. [1879, 1880]).

Geometry and duality

The geometric idea of duality, saying that in many geometrical statements in projective planes the notions of points and lines can be interchanged, and similarly in projective 3-space the notions of points and planes, was detected and elaborated by Poncelet [1822, 1829], Gergonne [1825–6], Steiner [1832], and von Staudt [1847]. This phenomenon was explained analytically by Plücker [1829, 1830a, b, 1832], who observed that the coefficients of homogeneous linear equations can serve as coordinates of a point. This gives a duality between hyperplanes through the origin and lines through the origin.

According to Smith [1929: p. 524], 'all references to a geometry of more than three dimensions before 1827 are in the form of single sentences pointing out that we cannot go beyond a certain point in some process because there is no space of more than three dimensions, or mentioning something that would be true if there were such a space'. In Chapter 1 of Part 2 of *Der barycentrische Calcul*, Möbius [1827] studies the isomorphy of figures in two dimensions which can be transformed to each other by rotating through the third dimension. To do the same for figures in three dimensions one needs a fourth dimension, which is impossible ('Zur Coincidenz zweier sich gleichen und ähnlichen Systeme im Raume von drei Dimensionen... würde also, der Analogie nach zu schliessen, erforderlich seyn, dass man das eine System in einem Raume von vier Dimensionen eine halbe Umdrehung machen lassen könnte. Da aber ein solcher Raum nicht gedacht werden kann, so ist auch die Coincidenz in diesem Falle unmöglich').

The insight that geometry is possible also in more than three dimensions, and thus obtaining a spatial representation for linear algebra, was obtained through the work of Cayley [1843, 1846] (one can deal with 4-dimensional space 'sans recourir à aucune notion métaphysique à l'égard de la possibilité de l'espace à quatre dimensions'—Cayley [1846: §I]), Grassmann [1844, 1845] ('Dadurch geschieht es nun, dass die Sätze der Raumlehre eine Tendenz zur Allgemeinheit haben, die in ihr vermöge ihrer Beschränkung auf drei Dimensionen keine Befriedigung findet, sondern erst in der Ausdehnungslehre zur Ruhe kommt.'—Grassmann [1845: p. 338]), and Cauchy [1847] ('analytical points, lines,...'). See also Manning [1916: Introduction] and Wieleitner [1925].

FURTHER NOTES ON LINEAR ALGEBRA

Among the several introductory and advanced texts on linear algebra and matrix theory are: Bellman [1960], Eves [1966], Franklin [1968], Friedberg, Insel and Spence [1979], Gantmacher [1959], Gewirtz, Sitomer, and Tucker [1974] (introductory), Greub

[1975], Halmos [1974], Lancaster and Tismenetsky [1985], Lang [1966a], MacDuffee [1933] (historical), Marcus and Minc [1964], Nering [1963] (introductory), Birkhoff and Mac Lane [1977], Schwartz [1961] (introductory), and Strang [1980].

Algorithms in linear algebra are disscussed, among others, in: Atkinson [1978] (pp. 487–490 give a review of literature), Broyden [1975] (introductory), Conte and de Boor [1965], Faddeev and Faddeeva [1963], Faddeeva [1959], Forsythe and Moler [1967] (introductory), Fox [1964], Goult, Hoskins, Milner, and Pratt [1974], Golub and Van Loan [1983], Householder [1964], John [1966], Kellison [1975], Stewart [1973], Stoer and Bulirsch [1980], and Westlake [1968]. A compendium of linear algebra algorithms is given by Wilkinson and Reinsch [1971]. Borodin and Munro [1975] give a survey of computational complexity in linear algebra. See also Aho, Hopcroft, and Ullman [1974], Savage [1976], Strassen [1969], and Winograd [1980].

Iterative methods are discussed especially by Varga [1962] and Young [1971a]. Chvátal [1983: pp. 71–96] reviews the Gaussian elimination method and its numerical behaviour.

The standard work on the history of determinants is given by Muir [1890, 1898, 1911, 1920, 1923, 1930]. The history of numerical methods is described by Goldstine [1977]. The history of linear algebra in ancient civilizations is investigated by van der Waerden [1983]. Moreover, any book on the history of mathematics in general discusses the history of linear equations (e.g. Boyer [1968], Cajori [1895], Eves [1964], Kline [1972], Smith [1925]). Kloyda [1937] gives a bibliography on linear and quadratic equations over the period 1550–1660.

Part II

Lattices and linear
diophantine equations

Lattices and linear diophantine equations can be described, to a large extent, parallel to linear spaces and linear equations (Part I) and to polyhedra and linear inequalities (Part III).

In this part we first discuss in Chapter 4 the theoretical side of lattices and linear diophantine equations, basically due to Hermite and Minkowski. Fundamental is the *Hermite normal form* of a matrix. Next, in Chapters 5 and 6, we go into the algorithmic side.

In Chapter 5 we discuss first the classical *Euclidean algorithm* for finding the g.c.d., which also can be used for solving *one* linear diophantine equation in polynomial time. Next an extension to solving *systems* of linear diophantine equations in polynomial time is described, based on finding the Hermite normal form of a matrix.

Chapter 6 starts with a discussion of the classical *continued fraction method* for approximating a real number by rational numbers of low denominator (*diophantine approximation*). The continued fraction method is a variant of the Euclidean algorithm. Next we describe Lovász's important *basis reduction method* for lattices. This method applies to simultaneous diophantine approximation, to solving systems of linear diophantine equations, to linear programming (see Section 14.1), to integer linear programming (see Section 18.4), to factoring polynomials over $\mathbb{Q}$ in polynomial time, and to several other problems.

Part II concludes with historical and further notes on lattices and linear diophantine equations.

4

Theory of lattices and linear diophantine equations

In this chapter we describe some results from elementary number theory concerning lattices and linear diophantine equations. In Section 4.1 we show that each rational matrix of full row rank can be brought into the so-called *Hermite normal form*. We derive a characterization for the feasibility of a system of linear diophantine equations and derive the existence of a linearly independent basis for any rational lattice. In Section 4.2 we show uniqueness of the Hermite normal form. In Section 4.3 we study *unimodular* matrices, and in Section 4.4 we go into some further theoretical aspects.

4.1. THE HERMITE NORMAL FORM

A matrix of full row rank is said to be in *Hermite normal form* if it has the form $[B \ 0]$, where B is a nonsingular, lower triangular, nonnegative matrix, in which each row has a unique maximum entry, which is located on the main diagonal of B.

The following operations on a matrix are called *elementary* (*unimodular*) *column operations*:

(1)   (i) exchanging two columns;
          (ii) multiplying a column by -1;
          (iii) adding an integral multiple of one column to another column.

Theorem 4.1 (Hermite normal form theorem). *Each rational matrix of full row rank can be brought into Hermite normal form by a series of elementary column operations.*

Proof. Let A be a rational matrix of full row rank. Without loss of generality, A is integral. Suppose we have transformed A, by elementary column operations, to the form $\begin{bmatrix} B & 0 \\ C & D \end{bmatrix}$ where B is lower triangular and with positive diagonal.

Now with elementary column operations we can modify D so that its first row $(\delta_{11},\ldots,\delta_{1k})$ is nonnegative, and so that the sum $\delta_{11} + \cdots + \delta_{1k}$ is as small as possible. We may assume that $\delta_{11} \geqslant \delta_{12} \geqslant \cdots \geqslant \delta_{1k}$. Then $\delta_{11} > 0$, as A has full row rank. Moreover, if $\delta_{12} > 0$, by subtracting the second column of D from the first column of D, the first row will have smaller sum, contradicting our assumption. Hence $\delta_{12} = \cdots = \delta_{1k} = 0$, and we have obtained a larger lower triangular matrix (this last procedure corresponds to the Euclidean algorithm — cf. Section 5.1).

By repeating this procedure, the matrix A finally will be transformed into $[B\ 0]$ with $B = (\beta_{ij})$ lower triangular with positive diagonal. Next do the following:

(2)   for $i = 2,\ldots,n$ ($:=$ order of B), do the following: for $j = 1,\ldots,i-1$, add an integer multiple of the ith column of B to the jth column of B so that the (i,j)th entry of B will be nonnegative and less than β_{ii}.

(So the procedure is applied in the order: $(i,j) = (2,1),\ (3,1),\ (3,2),\ (4,1),\ (4,2),\ (4,3),\ldots$.) It is easy to see that after these elementary column operations the matrix is in Hermite normal form.   □

In Theorem 4.2 below we shall see that any rational matrix of full row rank has a *unique* Hermite normal form. So we can speak of *the* Hermite normal form of a matrix.

A first corollary gives necessary and sufficient conditions for the feasibility of systems of linear diophantine equations. This result is a corollary of *Kronecker's approximation theorem* [1884b] (cf. Koksma [1936: p. 83] and the historical notes at the end of Part II).

Corollary 4.1a. *Let A be a rational matrix and let b be a rational column vector. Then the system $Ax = b$ has an integral solution x, if and only if yb is an integer for each rational row vector y for which yA is integral.*

Proof. Necessity of the condition is trivial: if x and yA are integral vectors and $Ax = b$, then $yb = yAx$ is an integer.

To see sufficiency, suppose yb is an integer whenever yA is integral. Then $Ax = b$ has a (possibly fractional) solution, since otherwise $yA = 0$ and $yb = \frac{1}{2}$ for some rational vector y (by linear algebra — see Corollary 3.1b). So we may assume that the rows of A are linearly independent. Now both sides of the equivalence to be proved are invariant under elementary column operations (1). So by Theorem 4.1 we may assume that A is in Hermite normal form $[B\ 0]$. Since $B^{-1}[B\ 0] = [I\ 0]$ is an integral matrix, it follows from our assumption that also $B^{-1}b$ is an integral vector. Since

(3) $[B\ 0]\begin{pmatrix} B^{-1}b \\ 0 \end{pmatrix} = b$

the vector $x := \begin{pmatrix} B^{-1}b \\ 0 \end{pmatrix}$ is an integral solution for $Ax = b$.   □

Corollary 4.1a can be seen as the integer analogue of Corollary 3.1b and of Corollary 7.1d (Farkas' lemma). It can also be interpreted in terms of groups and lattices. A subset Λ of $\mathbb{R}^n$ is called an (*additive*) *group* if:

(4)   (i) $0 \in \Lambda$
        (ii) if $x, y \in \Lambda$ then $x + y \in \Lambda$ and $-x \in \Lambda$.

The group is said to be *generated by* $a_1, \ldots, a_m$ if

(5) $\Lambda = \{\lambda_1 a_1 + \cdots + \lambda_m a_m | \lambda_1, \ldots, \lambda_m \in \mathbb{Z}\}$.

The group is called a *lattice* if it can be generated by linearly independent vectors. These vectors then are called a *basis* for the lattice.

Note that if matrix B comes from matrix A by elementary column operations, then the columns of B generate the same group as the columns of A.

Theorem 4.1 has also the following consequence.

Corollary 4.1b. *If* $a_1, \ldots, a_m$ *are rational vectors, then the group generated by* $a_1, \ldots, a_m$ *is a lattice, i.e. is generated by linearly independent vectors.*

Proof. We may assume that $a_1, \ldots, a_m$ span all space. (Otherwise we could apply a linear transformation to a lower dimensional space.) Let A be the matrix with columns $a_1, \ldots, a_m$ (so A has full row rank). Let $[B\ 0]$ be the Hermite normal form of A. Then the columns of B are linearly independent vectors generating the same group as $a_1, \ldots, a_m$. $\qquad\square$

So if $a_1, \ldots, a_m$ are rational vectors we can speak of the lattice generated by $a_1, \ldots, a_m$. Given a rational matrix A, Corollary 4.1a gives necessary and sufficient conditions for being an element of the lattice Λ generated by the columns of A. The proof of Corollary 4.1a implies that if A has full row rank, with Hermite normal form $[B\ 0]$ (B lower triangular), then b belongs to Λ if and only if $B^{-1}b$ is integral.

Another consequence of Theorem 4.1 is:

Corollary 4.1c. *Let* A *be an integral* $m \times n$*-matrix of full row rank. Then the following are equivalent:*

(i) *the g.c.d. of the subdeterminants of* A *of order* m *is* 1;
(ii) *the system* $Ax = b$ *has an integral solution* x, *for each integral vector* b;
(iii) *for each vector* y, *if* yA *is integral, then* y *is integral.*

Proof. Since (i), (ii), and (iii) are invariant under elementary column operations on A, by Theorem 4.1 we may assume that A is in Hermite normal form $[B\ 0]$, with B triangular. Then each of (i), (ii), and (iii) is easily seen to be equivalent to: $B = I$. $\qquad\square$

It can be derived easily from the Hermite normal form theorem that for any rational system $Ax = b$ with at least one integral solution there exist integral vectors $x_0, x_1, \ldots, x_t$ such that

(6) $\{x \mid Ax = b; x \text{ integral}\} = \{x_0 + \lambda_1 x_1 + \cdots + \lambda_t x_t \mid \lambda_1, \ldots, \lambda_t \in \mathbb{Z}\}$

where $x_1, \ldots, x_t$ are linearly independent and $t = $ (number of columns of A) − rank (A). The existence of such a *system of fundamental solutions* was stated by Smith [1861].

4.2. UNIQUENESS OF THE HERMITE NORMAL FORM

The uniqueness of the Hermite normal form follows from the following theorem.

Theorem 4.2. *Let A and A' be rational matrices of full row rank, with Hermite normal forms $[B\ 0]$ and $[B'\ 0]$, respectively. Then the columns of A generate the same lattice as those of A', if and only if $B = B'$.*

Proof. Sufficiency is clear, as the columns of B and A generate the same lattice, and similarly for B' and A'.

To see necessity, suppose the columns of A and those of A' generate the same lattice Λ. Then the same holds for B and B', as they come by elementary column operations from A and A'. Write $B =: (\beta_{ij})$ and $B' =: (\beta'_{ij})$. Suppose $B \neq B'$, and choose $\beta_{ij} \neq \beta'_{ij}$ with i as small as possible. Without loss of generality, $\beta_{ii} \geqslant \beta'_{ii}$. Let b_j and b'_j be the jth columns of B and B'. Then $b_j \in \Lambda$ and $b'_j \in \Lambda$, and hence $b_j - b'_j \in \Lambda$. This implies that $b_j - b'_j$ is an integral linear combination of the columns of B. By our choice of i, the vector $b_j - b'_j$ has zeros in the first $i - 1$ positions. Hence, as B is lower triangular, $b_j - b'_j$ is an integral linear combination of the columns indexed $i, \ldots, n$. So $\beta_{ij} - \beta'_{ij}$ is an integral multiple of β_{ii}. However, this contradicts the fact that $0 < |\beta_{ij} - \beta'_{ij}| < \beta_{ii}$ (since if $j = i$, then $0 < \beta'_{ii} < \beta_{ii}$, and if $j < i$, then $0 \leqslant \beta_{ij} < \beta_{ii}$ and $0 < \beta'_{ij} < \beta'_{ii} \leqslant \beta_{ii}$).   □

Corollary 4.2a. *Every rational matrix of full row rank has a unique Hermite normal form.*

Proof. Take $A = A'$ in Theorem 4.2.   □

Note that if $\beta_{11}, \ldots, \beta_{mm}$ are the diagonal entries of the Hermite normal form $[B\ 0]$ of A, then for each $j = 1, \ldots, m$, the product $\beta_{11} \cdots \beta_{jj}$ is equal to the g.c.d. of the subdeterminants of order j of the first j rows of A (as this g.c.d. is invariant under elementary column operations). This is an alternative way of seeing that the main diagonal of the Hermite normal form is unique. (It also implies that the size of the Hermite normal form is polynomially bounded by the size of the original matrix—cf. Section 5.2.)

4.3. UNIMODULAR MATRICES

Series of elementary column operations can be described by so-called unimodular matrices. Let U be a nonsingular matrix. Then U is called *unimodular* if U is integral and has determinant ± 1. (Later we shall extend the notion of unimodularity to not-necessarily nonsingular matrices.) Unimodular matrices were studied by Smith [1861], Frobenius [1879, 1880], and Veblen and Franklin [1921-2].

Theorem 4.3. *The following are equivalent for a nonsingular rational matrix U of order n:*

  (i) *U is unimodular;*
 (ii) *U^{-1} is unimodular;*
(iii) *the lattice generated by the columns of U is $\mathbb{Z}^n$;*
(iv) *U has the identity matrix as its Hermite normal form;*
 (v) *U comes from the identity matrix by elementary column operations.*

Proof. (i) implies (ii), as $\det(U^{-1}) = (\det U)^{-1} = \pm 1$, and as each entry of U^{-1} is a subdeterminant of U, and hence is an integer. Similarly, (ii) implies (i).

The equivalence of (iii), (iv), and (v) follows directly from Theorem 4.2. The implication (v)$\Rightarrow$(i) is trivial: elementary column operations leave a matrix integral, and do not change its determinant (up to sign). Also the implication (i)$\Rightarrow$(iv) is easy: if B is the Hermite normal form of U, then B is integral and $|\det B| = |\det U| = 1$. Hence $B = I$. □

Corollary 4.3a. *Let A and A' be nonsingular matrices. Then the following are equivalent:*

  (i) *the columns of A and those of A' generate the same lattice;*
 (ii) *A' comes from A by elementary column operations;*
(iii) *$A' = AU$ for some unimodular matrix U (i.e. $A^{-1}A'$ is unimodular).*

Proof. Directly from Theorem 4.2 and 4.3. □

Corollary 4.3b. *For each rational matrix A of full row rank there is a unimodular matrix U such that AU is the Hermite normal form of A. If A is nonsingular, U is unique.*

Proof. Directly from Corollary 4.2a and Theorem 4.3. □

If A and B are monsingular matrices, and each column of B is in the lattice generated by the columns of A, then $\det B$ is an integral multiple of $\det A$. Furthermore, $|\det A| = |\det B|$ if and only if the lattice generated by the columns of A coincides with the lattice generated by the columns of B.

Proof. If each column of B is in the lattice generated by the columns of A then $B = AU$ for some integral matrix U. Therefore, $\det B = \det U \cdot \det A$ ($\det U$ is an integer), and $|\det B| = |\det A|$ if and only if U is unimodular. □

In Sections 19.1 and 21.4 we extend the notion of unimodularity to not-necessarily nonsingular matrices. Moreover, in Chapters 19-21 we shall consider *totally* unimodular matrices, in which by definition all subdeterminants are 0, $+1$, or -1.

4.4. FURTHER REMARKS

Dual lattices

If Λ is a full-dimensional lattice (i.e. Λ spans all space), define the *dual lattice* $\Lambda^{\perp}$ by:

(7) $\Lambda^{\perp} := \{z | zx \text{ is an integer for all } x \text{ in } \Lambda\}$.

If Λ is generated by the columns of the nonsingular matrix B, then $\Lambda^{\perp}$ is the lattice generated by the rows of B^{-1}.

[Indeed, the lattice generated by the rows of B^{-1} is contained in $\Lambda^{\perp}$ as each row of B^{-1} is contained in $\Lambda^{\perp}$ (since $B^{-1}B = I$). Conversely, if $z \in \Lambda^{\perp}$, then zB is integral. Hence $z = (zB)B^{-1}$ is an integral combination of the rows of B^{-1}.]

So $\Lambda^{\perp\perp} = \Lambda$.

It may be derived more generally from Theorem 4.1 that for all rational matrices A and B there are rational matrices C and D such that:

(8) $\{x | \exists y, \exists \text{integral } z : x = Ay + Bz\} = \{x | Cx = 0; Dx \text{ is integral}\}$.

Moreover, if A, B, C, D are related as in (8) then also:

(9) $\{u | \exists v, \exists \text{integral } w : u = vC + wD\} = \{u | uA = 0; uB \text{ is integral}\}$.

Smith normal form

Smith [1861] proved that any rational matrix A can be transformed by elementary row and column operations into the *Smith normal form* $\begin{bmatrix} D & 0 \\ 0 & 0 \end{bmatrix}$, where $D = \text{diag}(\delta_1, \ldots, \delta_k)$, with $\delta_1, \ldots, \delta_k$ positive rationals such that $\delta_1 | \delta_2 | \cdots | \delta_k$. (*Elementary row operations* are of course the operations (1) with column replaced by row.) This form is unique, as one easily checks that for each i, the product $\delta_1 \ldots \delta_i$ is equal to the g.c.d. of the subdeterminants of A of order i (as this g.c.d. is invariant under elementary row and column operations). The numbers $\delta_1, \ldots, \delta_k$ are called the *elementary divisors* of A.

To see that any rational $m \times n$ matrix can be brought into Smith normal form, suppose we have applied the elementary operations so that we obtained a matrix $B = (\beta_{ij})$ for which

(10) $\min \{|\beta_{ij}| \, | \, i = 1, \ldots, m; \, j = 1, \ldots, n; \, \beta_{ij} \neq 0\}$

is as small as possible. We may assume this minimum is attained by β_{11}. Moreover, we may assume that all other entries in the first row and column are 0, as by applying elementary operations we can have $0 \le \beta_{1j} < |\beta_{11}|$ for all $j \neq 1$ and $0 \le \beta_{i1} < |\beta_{11}|$ for all $i \neq 1$. Hence, by the minimality of $|\beta_{11}|$, $\beta_{1j} = 0$ and $\beta_{i1} = 0$ for $j \neq 1$ and $i \neq 1$. Now each entry of B is an integral multiple of β_{11}. Indeed, if β_{ij} is not a multiple of β_{11}, with elementary operations we can obtain the remainder of $\beta_{ij} \pmod{\beta_{11}}$ in the (i, j)th position, contradicting again the minimality of $|\beta_{11}|$. Write

(11) $B = \begin{bmatrix} \beta_{11} & 0 \\ 0 & C \end{bmatrix}$.

By induction, C can be transformed by elementary operations into Smith normal form

$$\begin{bmatrix} D' & 0 \\ 0 & 0 \end{bmatrix},$$ say with $D' = \mathrm{diag}(\delta_2, \ldots, \delta_k)$ and $\delta_2 | \delta_3 | \cdots | \delta_k$. Since δ_{11} divides all entries of C, β_{11} also divides δ_2, and hence also B can be brought into Smith normal form.

As for each j, the product $\delta_1 \cdots \delta_j$ is equal to the g.c.d. of the subdeterminants of A of order j, the size of the Smith normal form is polynomially bounded by the size of A. By the equivalence of (i) and (v) in Theorem 4.3, for any matrix A there are unimodular matrices U and V such that UAV is in Smith normal form.

A consequence of the existence of the Smith normal form is the famous theorem of Kronecker [1877] that each finite abelian group is the direct product of cyclic groups (see Cohn [1962]). The Smith normal form was also studied by Frobenius [1879, 1880].

It can be proved that if A and B are rational nonsingular matrices of order n, and g.c.d. $(\det A, \det B) = 1$, then $\mathrm{Snf}(A) \cdot \mathrm{Snf}(B) = \mathrm{Snf}(AB)$, where Snf denotes Smith normal form (see Newman [1972: pp. 33–34] and Marcus and Underwood [1972]).

Other characterizations of feasibility of linear diophantine equations

Corollary 4.1a extends the well-known result from elementary number theory that one linear equation $\alpha_1 \xi_1 + \cdots + \alpha_n \xi_n = \beta$ has a solution in integers $\xi_1, \ldots, \xi_n$, if and only if β is an integral multiple of the g.c.d. of $\alpha_1, \ldots, \alpha_n$. Another extension of this result to more dimensions is a characterization due to Heger [1858: p. 111] (cf. Smith [1861], Frobenius [1879], Veblen and Franklin [1921], Skolem [1938]):

(12)   let A be a rational matrix of full row rank, with m rows, and let b be a rational column m-vector. Then $Ax = b$ has an integral solution x, if and only if each subdeterminant of the matrix $[A\ b]$ of order m is an integral multiple of the g.c.d. of the subdeterminants of the matrix A of order m.

This may be proved in a similar way as Corollary 4.1a: both sides of the equivalence are invariant under elementary column operations on A, and the equivalence is easy for matrices in Hermite normal form.

Another characterization is:

(13)   the rational system $Ax = b$ has an integral solution x if and only if for each natural number M the congruence $Ax \equiv b \pmod{M}$ has an integral solution x

(cf. Mordell [1969]). This may be derived from Corollary 4.1a as follows. The necessity of the condition is trivial. To prove sufficiency, suppose $Ax = b$ has no integral solution. Then by Corollary 4.1a there exists a rational vector y with yA integral and yb not an integer. Let M be a natural number such that My is integral. Let x be an integral solution for $Ax \equiv b \pmod{M}$. So $M^{-1}(Ax - b)$ is integral. Hence $y(Ax - b)$ is an integer. Since yAx is an integer, it follows that yb is an integer. Contradiction.

For studies of the duality aspects of Corollary 4.1a, see Bachem and von Randow [1979] and Williams [1984].

5

Algorithms for linear diophantine equations

The classical Euclidean algorithm for finding the g.c.d. yields also a polynomial-time method for solving one rational linear diophantine equation. This is described in Section 5.1, as a preliminary to solving *systems* of rational linear diophantine equations, discussed in Sections 5.2 and 5.3. First in Section 5.2 we show that the size of the Hermite normal form of a rational matrix A is polynomially bounded by the size of A, and we derive that the solvability of systems of rational linear diophantine equations has a good characterization. Next in Section 5.3 we describe a polynomial-time algorithm for finding the Hermite normal form of a matrix, and we deduce that systems of rational linear diophantine equations can be solved in polynomial time.

5.1. THE EUCLIDEAN ALGORITHM

The Euclidean algorithm determines, in polynomial time, the g.c.d. of two positive rational numbers α and β. First α is replaced by $\alpha - \lfloor \alpha/\beta \rfloor \beta$, and (with the new α) β is replaced by $\beta - \lfloor \beta/\alpha \rfloor \alpha$. Next this is repeated until one of α and β is 0. In that case, the nonzero among α and β is the g.c.d. of the original α and β. This follows easily from the facts that g.c.d. $\{\alpha, \beta\} = $ g.c.d. $\{\alpha - \lfloor \alpha/\beta \rfloor \beta, \beta\}$ and g.c.d. $\{\alpha, 0\} = \alpha$.

With a little more administration one can find integers γ and ε satisfying $\gamma\alpha + \varepsilon\beta = $ g.c.d.$\{\alpha, \beta\}$, given rational numbers α and β, and one can solve one rational linear diophantine equation, as we shall describe now.

Let two positive rationals α and β be given. Determine a series of 3×2-matrices $A_0, A_1, A_2, \ldots$ as follows. The matrix A_0 is given by:

$$(1) \qquad A_0 := \begin{bmatrix} \alpha & \beta \\ 1 & 0 \\ 0 & 1 \end{bmatrix}.$$

If A_k has been found, say:

(2) $\qquad A_k = \begin{bmatrix} \alpha_k & \beta_k \\ \gamma_k & \delta_k \\ \varepsilon_k & \zeta_k \end{bmatrix},$

let A_{k+1} arise from A_k by the following rule:

(3)   (i) if k is even and $\beta_k > 0$, subtract $\lfloor \alpha_k/\beta_k \rfloor$ times the second column of A_k from the first column;

         (ii) if k is odd and $\alpha_k > 0$, subtract $\lfloor \beta_k/\alpha_k \rfloor$ times the first column of A_k from the second column;

This step is performed for $k = 0, 1, 2, \ldots$, until for $k = N$ (say) $\alpha_N = 0$ or $\beta_N = 0$. If $\beta_N = 0$ and $\alpha_N \neq 0$, then α_N is the g.c.d. of α and β, as during the iterations (3) the g.c.d. of the entries in the upper row of A_k does not change, and as $\alpha_N = $ g.c.d. $\{\alpha_N, 0\}$. Similarly, if $\alpha_N = 0$, $\beta_N \neq 0$, then $\beta_N = $ g.c.d. $\{\alpha, \beta\}$.

We also find integers γ and ε with $\gamma\alpha + \varepsilon\beta = $ g.c.d. $\{\alpha, \beta\}$. To see this, observe that $(1, -\alpha, -\beta)A_0 = (0, 0)$, and hence for each k, $(1, -\alpha, -\beta)A_k = (0, 0)$ as we did only elementary column operations. In the notation (2) this means:

(4) $\qquad \gamma_k\alpha + \varepsilon_k\beta = \alpha_k$

$\qquad\quad \delta_k\alpha + \zeta_k\beta = \beta_k.$

Therefore, if $\beta_N = 0$, $\alpha_N \neq 0$,

(5) $\qquad \gamma_N\alpha + \varepsilon_N\beta = \alpha_N = $ g.c.d. $\{\alpha, \beta\}$

$\qquad\quad \delta_N\alpha + \zeta_N\beta = 0.$

Similarly, if $\alpha_N = 0$, $\beta_N \neq 0$.

The Euclidean algorithm indeed terminates, within polynomial time. Before proving this, we observe that for each k:

(6) $\qquad -\alpha_k\delta_k + \beta_k\gamma_k = \beta$

$\qquad\quad \alpha_k\zeta_k - \beta_k\varepsilon_k = \alpha.$

The first line follows from the fact that the upper 2×2-determinants of A_0 and A_k are the same (as we did only elementary column operations), and the second line follows similarly by leaving out the middle rows of A_0 and A_k. By considering the lower 2×2-determinants of A_0 and A_k we obtain the equality:

(7) $\qquad \gamma_k\zeta_k - \delta_k\varepsilon_k = 1.$

Note moreover that for all k, $\gamma_k \geq 1$, $\zeta_k \geq 1$, and $\delta_k \leq 0$, $\varepsilon_k \leq 0$.

Theorem 5.1. *The Euclidean algorithm is a polynomial-time method.*

Proof. We may assume that α and β are natural numbers (otherwise, multiply throughout the procedure the first row of each matrix A_k by the product of the denominators of α and β; this modification does not change the complexity).

So all matrices A_k have nonnegative integers in their first row. Since at each iteration either α_k or β_k is reduced by a factor of at least 2, after at most $\lfloor \log_2\alpha \rfloor + \lfloor \log_2\beta \rfloor + 1$ iterations, one of α_k and β_k will be 0.

Now for each k: $\alpha_k \leqslant \alpha$ and $\beta_k \leqslant \beta$. From (6) it follows that $|\delta_k| \leqslant \beta, |\gamma_k| \leqslant \beta$, $|\zeta_k| \leqslant \alpha$, $|\varepsilon_k| \leqslant \alpha$, as long as $\alpha_k > 0$ and $\beta_k > 0$. This shows that throughout the method the sizes of the numbers are polynomially bounded by the sizes of α and β. $\square$

Corollary 5.1a. *A linear diophantine equation with rational coefficients can be solved in polynomial time.*

Proof. Let

$$(8) \qquad \alpha_1\xi_1 + \cdots + \alpha_n\xi_n = \beta$$

be a rational linear diophantine equation. The algorithm solving (8) is described recursively on n. The case $n = 1$ is trivial. Let $n \geqslant 2$. Find, with the Euclidean algorithm, α', γ, and ε satisfying

$$(9) \qquad \alpha' = \text{g.c.d.}\{\alpha_1, \alpha_2\} = \alpha_1\gamma + \alpha_2\varepsilon$$
$$\gamma, \varepsilon \text{ integers.}$$

Next solve the linear diophantine equation (in $n - 1$ variables)

$$(10) \qquad \alpha'\xi' + \alpha_3\xi_3 + \cdots + \alpha_n\xi_n = \beta.$$

If (10) has no integral solution, then neither has (8). If (10) has an integral solution $\xi', \xi_3, \ldots, \xi_n$, then $\xi_1 := \gamma\xi'$, $\xi_2 := \varepsilon\xi'$, $\xi_3, \ldots, \xi_n$ gives an integral solution to (8). This defines a polynomial algorithm. $\square$

In Section 5.3 we shall see that also *systems* of linear diophantine equations can be solved in polynomial time.

Notes for Section 5.1. The fact that the number of iterations in the Euclidean algorithm is bounded linearly in the logarithm of the smallest number was proved by Lamé [1844]. For more on the Euclidean algorithm, see Knuth [1969: pp. 316–364], Blankinship [1963], Bradley [1970a], Collins [1974], Dixon [1970, 1971], Harris [1970a], Heilbronn [1968], Rieger [1976], Schönhage [1971], and van Trigt [1978]. Heilbronn [1968] and Dixon [1970] showed that the number of iterations in the Euclidean algorithm applied to α, β $(\alpha \leqslant \beta)$ is 'almost always about' $12\pi^{-2}\log 2.\log\beta$. See also Harris [1970b], Merkes and Meyers [1973], Shea [1973], and Yao and Knuth [1975].

The solution of one linear diophantine equation was studied by Bond [1967], Kertzner [1981], Lehmer [1919], Levit [1956], Morito and Salkin [1979, 1980], and Weinstock [1960].

5.2. SIZES AND GOOD CHARACTERIZATIONS

We show that the Hermite normal form has polynomially bounded size, from which it follows that the characterizations given in Section 4.1 are good characterizations.

Theorem 5.2. *The Hermite normal form* $[B\ 0]$ *of a rational matrix* A *of full row rank has size polynomially bounded by the size of* A. *Moreover, there exists a unimodular matrix* U *with* $AU = [B\ 0]$, *such that the size of* U *is polynomially bounded by the size of* A.

Proof. We may assume that A is integral, as multiplying A by the product, say κ, of the denominators occurring in A also multiplies the Hermite normal form of A by κ.

The diagonal entries of B are divisors of subdeterminants of A (cf. Section 4.2). As each row of B has its maximum entry on the main diagonal of B, it follows that the size of $[B\ 0]$ is polynomially bounded by the size of A.

The second statement can be seen as follows. By permuting columns we may assume that $A = [A'\ A'']$, with A' nonsingular. The Hermite normal form of the matrix

(11) $\qquad \begin{bmatrix} A' & A'' \\ 0 & I \end{bmatrix}$ is $\begin{bmatrix} B & 0 \\ B' & B'' \end{bmatrix}$

for certain matrices B' and B''. As the sizes of B, B', and B'' are polynomially bounded by the size of A, also the size of the unimodular matrix

(12) $\qquad U := \begin{bmatrix} A' & A'' \\ 0 & I \end{bmatrix}^{-1} \begin{bmatrix} B & 0 \\ B' & B'' \end{bmatrix}$

is polynomially bounded by the size of A. Now $AU = [A'\ A'']U = [B\ 0]$. $\qquad \square$

Corollary 5.2a. *If a rational system* $Ax = b$ *has an integral solution, it has one of size polynomially bounded by the sizes of* A *and* b.

Proof. Without loss of generality, A has full row rank. Let $[B\ 0] = AU$ be the Hermite normal form of A, with U unimodular of size polynomially bounded by the size of A. Now $B^{-1}b$ is integral (as for any integral solution x of $Ax = b$ we have that $B^{-1}b = B^{-1}Ax = B^{-1}[B\ 0]U^{-1}x = [I\ 0]U^{-1}x$ is integral, since $U^{-1}x$ is integral). Therefore,

(13) $\qquad \tilde{x} := U \begin{pmatrix} B^{-1}b \\ 0 \end{pmatrix}$

is an integral solution of $Ax = b$, of size polynomially bounded by the sizes of A and b. $\qquad \square$

This implies that Corollary 4.1a in fact yields a good characterization (cf. von zur Gathen and Sieveking [1976, 1978]).

Corollary 5.2b. *The following problem has a good characterization: given a rational matrix* A *and a rational vector* b, *does the system* $Ax = b$ *have an integral solution?*

Proof. The general case can be easily reduced to the case where A has full row rank (using Corollary 3.3a). Let $[B \ 0] = AU$ be the Hermite normal form of A, with U unimodular of size polynomially bounded by the size of A.

If the answer is positive, there is a solution of polynomial size, by Corollary 5.2a.

If the answer is negative, by Corollary 4.1a there exists a rational row vector y with yA integral and yb not an integer. We can take y of polynomially bounded size: since $B^{-1}A$ is integral but $B^{-1}b$ not, we can take for y one of the rows of B^{-1}. $\square$

5.3. POLYNOMIAL ALGORITHMS FOR HERMITE NORMAL FORMS AND SYSTEMS OF LINEAR DIOPHANTINE EQUATIONS

The Euclidean algorithm gives the polynomial solvability of one linear diophantine equation—see Corollary 5.1a. It was shown by Frumkin [1976a, b] and von zur Gathen and Sieveking [1976] that also *systems* of linear diophantine equations can be solved in polynomial time.

Moreover, von zur Gathen and Sieveking [1976] and Votyakov and Frumkin [1976] showed that for any system $Ax = 0$ of homogeneous linear diophantine equations we can find a basis for the solution set in polynomial time. Frumkin [1976c] derived from this that also the Hermite normal form of a matrix can be determined in polynomial time. (Frumkin [1975, 1977] announces and reviews his and Votyakov's results.) A direct polynomial-time method bringing a matrix into Hermite normal form was given by Kannan and Bachem [1979]. They also gave a polynomial algorithm for the Smith normal form.

Several prescriptions for the order of the elementary column operations making the Hermite normal form had been proposed before (cf. Pace and Barnett [1974], Blankinship [1966a, b], Bodewig [1956], Bradley [1971a], Cheema [1966], Donaldson [1979], and Hu [1969: Appendix A]), but none of these methods could be shown to be polynomial-time.

The methods of Frumkin, Votyakov, von zur Gathen and Sieveking are based on an elimination process, at the same time reducing all numbers modulo a certain integer. Here we describe a variant of these methods (developed independently by Domich [1983] and Domich, Kannan and Trotter [1985]), which finds the Hermite normal form in polynomial time.

Polynomial algorithm to determine the Hermite normal form

Let an integral $m \times n$-matrix A, of full row rank, be given. Let M be the absolute value of the determinant of an (arbitrary) submatrix of A of rank m (such an M can be found in polynomial time with Gaussian elimination). Now the columns of A generate the same lattice as the columns of the matrix:

$$(14) \qquad A' := \begin{bmatrix} A & \begin{matrix} M & & & 0 \\ & \ddots & & \\ & & \ddots & \\ 0 & & & M \end{matrix} \end{bmatrix}$$

[*Proof.* A has a submatrix B of rank m with $|\det B| = M$. Then the matrix $(\det B)B^{-1}$ is integral, and $B((\det B)B^{-1}) = \pm M \cdot I$, so that each of the last m columns of A' is an integral combination of the columns of B.   □]

So the Hermite normal form of A is the same as that of A', except for the last m columns of the Hermite normal form of A', which are all-zero.

So it suffices to find the Hermite normal form of A'. The idea is that after each elementary column operation to bring A' to Hermite normal form we can do some further elementary column operations to bring all entries between 0 and M. More precisely, first we add integral multiples of the last m columns of A' to the first n columns of A' so that all components will be at least 0 and at most M. Next we apply the following procedure repeatedly:

Suppose we have obtained the matrix

$$
(15) \qquad
\begin{bmatrix}
B & 0 & 0 \\
\hline
 & & 0 \cdots\cdots 0 \\
C & D & M_{\ddots} \; 0 \\
 & & 0 \;_{\ddots} \\
 & & \quad\; M
\end{bmatrix}
$$

where B is a lower triangular $k \times k$-matrix, C is an $(m - k) \times k$-matrix, D is an $(m - k) \times (n + 1)$-matrix, such that the first row of D is nonzero. (In particular, (14) has form (15), with $k = 0$.)

Then apply the following repeatedly, writing $D =: (\delta_{ij})_{i=1, j=1}^{m-k, n+1}$:

(16)   if there are $i \neq j$ with $\delta_{1i} \geq \delta_{1j} > 0$, then:
  (i) subtract $\lfloor \delta_{1i}/\delta_{1j} \rfloor$ times the jth column of D from the ith column of D;
  (ii) next add integral multiples of the last $m - k - 1$ columns of (15) to the other columns to bring all components between 0 and M (i.e. reduce D modulo M).

(16) is repeated as long as the first row of D contains more than one nonzero entry. If it has exactly one nonzero, we start the procedure anew, with k replaced by $k + 1$.

If $k = m$, (15) is in the form $[B \; 0]$ with B lower triangular. Then finally with the following procedure it can be brought into Hermite normal form:

(17)   for $i = 2, \ldots, n$ ($:=$ order of B), do the following: for $j = 1, \ldots, i - 1$, add an integral multiple of the ith column of B to the jth column of B so that the (i, j)th entry of B will be nonnegative and less than β_{ii}.

Deleting the last m columns yields the Hermite normal form of A.

Theorem 5.3. *The above method finds the Hermite normal form in polynomial time.*

Proof. The execution time of each single application of (i) or (ii) in (16) can be easily seen to be polynomially bounded by n and $\log_2 M$. Since at each application of step (16), either the first row of D will contain one zero entry more, or the product of the nonzero entries in this row is reduced by a factor of at least 2 (as $(\delta_{1i} - \lfloor \delta_{1i}/\delta_{1j} \rfloor \delta_{1j}) \leqslant \frac{1}{2}\delta_{1i}$), and since all entries in the first row of D are at most M, after at most $n \cdot \log_2 M$ applications of step (16) the matrix D has at most one nonzero in the first row. In that case k is increased to $k + 1$.

So after at most $mn \cdot \log_2 M$ applications of (16) we have $k = m$. It is easy to see that during application of the final procedure (17) the numbers will not be larger than M^n in absolute value. Therefore, the running time of the algorithm is polynomially bounded.   □

Clearly, also, if A is not integral the Hermite normal form can be found in polynomial time: either we apply the above method, or we first multiply A by the product of the denominators of the entries of A.

Also a unimodular matrix bringing a matrix into Hermite normal form can be found in polynomial time.

Corollary 5.3a. *Given a rational matrix A of full row rank, we can find in polynomial time a unimodular matrix U such that AU is in Hermite normal form.*

Proof. Similar to the proof of Theorem 5.2.   □

Another consequence is a result of Frumkin [1976a, b] and von zur Gathen and Sieveking [1976].

Corollary 5.3b. *Given a system of rational linear equations, we can decide if it has an integral solution, and if so, find one, in polynomial time.*

Proof. Let the system $Ax = b$ be given. With the Gaussian elimination method we can decide if $Ax = b$ has any (possibly fractional) solution. If not, it certainly has no integral solution. If so, we can restrict $Ax = b$ to a subsystem with linearly independent left-hand sides. So we may assume that A has full row rank. Determine with the above method the Hermite normal form $[B \ 0]$ of A, and determine a unimodular matrix U with $AU = [B \ 0]$. If $B^{-1}b$ is integral, then

$$(18) \qquad x := U\begin{pmatrix} B^{-1}b \\ 0 \end{pmatrix}$$

is an integral solution for $Ax = b$. If $B^{-1}b$ is not integral, $Ax = b$ has no integral solution.   □

Von zur Gathen and Sieveking [1976] and Votyakov and Frumkin [1976] showed:

Corollary 5.3c. *Given a feasible system $Ax = b$ of rational linear diophantine equations, we can find in polynomial time integral vectors $x_0, x_1, \ldots, x_t$ such that*

$$(19) \qquad \{x \mid Ax = b; \ x \text{ integral}\} = \{x_0 + \lambda_1 x_1 + \cdots + \lambda_t x_t \mid \lambda_1, \ldots, \lambda_t \in \mathbb{Z}\}$$

with $x_1, \ldots, x_t$ linearly independent.

Proof. As in the proof of Corollary 5.3b we may assume that A has full row rank, and we can find in polynomial time a unimodular matrix U such that $AU = [B \ 0]$ is in Hermite normal form. Let $x_0, x_1, \ldots, x_t$ be given by

$$(20) \qquad [x_0, x_1, \ldots, x_t] = U \begin{bmatrix} B^{-1}b & 0 \\ 0 & I \end{bmatrix}.$$

Then $x_0, x_1, \ldots, x_t$ have the required properties. $\square$

Notes for Chapter 5. Further work on solving systems of linear diophantine equations is described by Chou and Collins [1982]. Also Fiorot and Gondran [1969] study solving systems of linear diophantine equations. Rosser [1942] studied extensions of the Euclidean algorithm to more than two variables. See also the references to extensions of the continued fraction method to more dimensions, given at the end of Chapter 6.

Computing the *Smith normal form* was studied by Bodewig [1956], Hu [1969: Appendix A], Bradley [1971a], Garfinkel and Nemhauser [1972a: Sec. 7.4], Pace and Barnett [1974], Hartley and Hawkes [1974], Rayward-Smith [1979], and Kannan and Bachem [1979]. The last reference gives a polynomial-time algorithm (see also Bachem and Kannan [1979] (applications)).

6

Diophantine approximation and basis reduction

We now study the problems of approximating real numbers by rational numbers of low denominator, i.e. *diophantine approximation*, and of finding a so-called *reduced basis* in a lattice. First, in Section 6.1, we discuss the *continued fraction method* for approximating *one* real number. Next, in Section 6.2, we describe Lovász's *basis reduction method* for lattices, finding a 'reduced' basis in a lattice. In Section 6.3 we give some of the several applications of this method, such as finding a short nonzero vector in a lattice and *simultaneous diophantine approximation* (approximating several reals by rationals with one common low denominator).

6.1. THE CONTINUED FRACTION METHOD

The continued fraction method, which may be seen as a variant of the Euclidean algorithm, yields approximations of a real number by rationals. It can be used to find integers p and q as described in the following classical result of Dirichlet [1842]:

Theorem 6.1 (Dirichlet's theorem). *Let α be a real number and let $0 < \varepsilon \leqslant 1$. Then there exist integers p and q such that:*

(1) $$\left| \alpha - \frac{p}{q} \right| < \frac{\varepsilon}{q} \quad \text{and} \quad 1 \leqslant q \leqslant \varepsilon^{-1}.$$

Proof. The proof is based on Dirichlet's *pigeon-hole principle*. Let $M := \lfloor \varepsilon^{-1} \rfloor$, and consider the numbers

(2) $$\{0\alpha\}, \{1\alpha\}, \{2\alpha\}, \{3\alpha\}, \ldots, \{M\alpha\}$$

where $\{\ \}$ denotes fractional part (i.e. $\{\beta\} = \beta - \lfloor \beta \rfloor$). Since $0 \leqslant \{i\alpha\} < 1$, there exist two different integers i and j with $0 \leqslant i, j \leqslant M$ and $\{(i-j)\alpha\} \leqslant 1/(M+1)$.

Let $q := i - j$ and $p := \lfloor q\alpha \rfloor$. Then

(3)
$$\left| \alpha - \frac{p}{q} \right| = \left| \frac{\{(i-j)\alpha\}}{q} \right| \leqslant \frac{1}{(M+1)|q|} < \frac{\varepsilon}{|q|}.$$

As $|q| \leqslant M \leqslant \varepsilon^{-1}$, Dirichlet's result (1) follows. $\qquad \square$

The *continued fraction method* for the real number $\alpha \geqslant 0$ can be described as follows. Determine $\gamma_1, \gamma_2, \gamma_3, \ldots$ successively by:

(4)
$$\gamma_1 := \alpha$$
$$\gamma_2 := (\gamma_1 - \lfloor \gamma_1 \rfloor)^{-1}$$
$$\gamma_3 := (\gamma_2 - \lfloor \gamma_2 \rfloor)^{-1}$$
$$\gamma_4 := (\gamma_3 - \lfloor \gamma_3 \rfloor)^{-1}$$
$$\vdots \qquad \vdots$$

The sequence stops if γ_N, say, is an integer, so that we cannot determine γ_{N+1}. Directly from (4) we see that $\gamma_1, \gamma_2, \gamma_3, \ldots$ satisfy:

(5)
$$\alpha = \gamma_1 = \lfloor \gamma_1 \rfloor + \frac{1}{\gamma_2} = \lfloor \gamma_1 \rfloor + \cfrac{1}{\lfloor \gamma_2 \rfloor + \cfrac{1}{\gamma_3}} = \lfloor \gamma_1 \rfloor + \cfrac{1}{\lfloor \gamma_2 \rfloor + \cfrac{1}{\lfloor \gamma_3 \rfloor + \cfrac{1}{\gamma_4}}} = \cdots$$

The associated rationals

(6)
$$\lfloor \gamma_1 \rfloor, \quad \lfloor \gamma_1 \rfloor + \frac{1}{\lfloor \gamma_2 \rfloor}, \quad \lfloor \gamma_1 \rfloor + \cfrac{1}{\lfloor \gamma_2 \rfloor + \cfrac{1}{\lfloor \gamma_3 \rfloor}},$$

$$\lfloor \gamma_1 \rfloor + \cfrac{1}{\lfloor \gamma_2 \rfloor + \cfrac{1}{\lfloor \gamma_3 \rfloor + \cfrac{1}{\lfloor \gamma_4 \rfloor}}}, \cdots$$

are called the *convergents* for α. If this sequence stops, its last term is equal to α, as one easily sees. It will be shown below that the sequence stops iff α is rational. For irrational α, the sequence converges to α.

The convergents can be described alternatively as follows. Let $\alpha \geqslant 0$ be a real number. Let p_1/q_1 and p_2/q_2 be the first two terms in (6), i.e.

(7)
$$\frac{p_1}{q_1} = \lfloor \alpha \rfloor, \quad \frac{p_2}{q_2} = \lfloor \alpha \rfloor + \frac{1}{\lfloor (\alpha - \lfloor \alpha \rfloor)^{-1} \rfloor}$$

where p_1, q_1, p_2, q_2 are nonnegative integers, with g.c.d.$(p_1, q_1) =$ g.c.d.$(p_2, q_2) = 1$ (if α is by chance an integer, we only determine p_1, q_1). Note that $p_1/q_1 \leqslant \alpha \leqslant p_2/q_2$ and $p_2 q_1 - p_1 q_2 = 1$.

Suppose we have found nonnegative integers $p_{k-1}, q_{k-1}, p_k, q_k$ such that

(8) $\dfrac{p_{k-1}}{q_{k-1}} < \alpha < \dfrac{p_k}{q_k}$ and $p_k q_{k-1} - p_{k-1} q_k = 1$

where k is even. ((8) implies g.c.d.(p_{k-1}, q_{k-1}) = g.c.d.$(p_k, q_k) = 1$.) Let t be the largest integer such that

(9) $\dfrac{p_{k-1} + t p_k}{q_{k-1} + t q_k} \leqslant \alpha.$

Define

(10) $p_{k+1} := p_{k-1} + t p_k$

         $q_{k+1} := q_{k-1} + t q_k.$

One easily checks $p_k q_{k+1} - p_{k+1} q_k = 1$, implying g.c.d.$(p_{k+1}, q_{k+1}) = 1$.

(9) and (10) give: $p_{k+1}/q_{k+1} \leqslant \alpha$. In the case of equality we stop. In the case of strict inequality, let u be the largest integer such that

(11) $\alpha \leqslant \dfrac{p_k + u p_{k+1}}{q_k + u q_{k+1}}.$

Define

(12) $p_{k+2} := p_k + u p_{k+1}$

         $q_{k+2} := q_k + u q_{k+1}.$

Again, one easily checks $p_{k+2} q_{k+1} - p_{k+1} q_{k+2} = 1$, implying g.c.d.$(p_{k+2}, q_{k+2}) = 1$.

(11) and (12) give $\alpha \leqslant p_{k+2}/q_{k+2}$. In the case of equality we stop. In the case of strict inequality, we know that (8) holds, with k increased by 2. So we can repeat the iteration, thus defining the sequence

(13) $\dfrac{p_1}{q_1}, \dfrac{p_2}{q_2}, \dfrac{p_3}{q_3}, \ldots$

If the sequence stops, its last term is α. It turns out that the sequence (13) is the same as the sequence (6) of convergents.

Proof. We first prove:

Claim. *If $0 < \alpha < 1$ and $p_1/q_1, p_2/q_2, p_3/q_3, \ldots$ is row (13) with respect to α, then q_2/p_2, $q_3/p_3, q_4/p_4, \ldots$ is row (13) with respect to $1/\alpha$.*

Proof of the claim. Let $p'_1/q'_1, p'_2/q'_2, p'_3/q'_3, \ldots$ be row (13) with respect to $1/\alpha$. Then the following holds:

(14) $\dfrac{q_2}{p_2} = \dfrac{p'_1}{q'_1} \leqslant \dfrac{1}{\alpha} \leqslant \dfrac{p'_2}{q'_2} = \dfrac{q_3}{p_3}.$

Indeed,

(15) $\dfrac{p'_1}{q'_1} = \left\lfloor \dfrac{1}{\alpha} \right\rfloor = \dfrac{q_2}{p_2},$

$$\frac{p'_2}{q'_2} = \left\lfloor \frac{1}{\alpha} \right\rfloor + \frac{1}{\lfloor (1/\alpha - \lfloor 1/\alpha \rfloor)^{-1} \rfloor} = \left\lfloor \frac{1}{\alpha} \right\rfloor + \frac{1}{t} = \frac{q_2}{p_2} + \frac{1}{t} = \frac{q_2 t + q_1}{p_2 t + p_1} = \frac{q_3}{p_3}$$

where t is the largest integer with

(16) $\quad \dfrac{p_2 t + p_1}{q_2 t + q_1} \leqslant \alpha, \quad$ i.e. $\quad t = \left\lfloor \dfrac{\alpha q_1 - p_1}{p_2 - \alpha q_2} \right\rfloor = \left\lfloor \left(\dfrac{1}{\alpha} - \left\lfloor \dfrac{1}{\alpha} \right\rfloor \right)^{-1} \right\rfloor,$

using $p_1 = 0$, $q_1 = 1$, $p_2 = 1$, $q_2 = \lfloor 1/\alpha \rfloor$. This proves (14).

By considering the way (13) is constructed, (14) implies that for each even k:

(17) $\quad \dfrac{q_k}{p_k} = \dfrac{p'_{k-1}}{q'_{k-1}} \leqslant \dfrac{1}{\alpha} \leqslant \dfrac{p'_k}{q'_k} = \dfrac{q_{k+1}}{p_{k+1}}.$

End of proof of the claim.

We next show that (13) is the same sequence as (6). Let $c_k(\alpha)$ denote the kth convergent of α, i.e. the kth term in (6). We show that $c_k(\alpha) = p_k/q_k$ by induction on k. We may assume that $0 < \alpha < 1$, since each term both in (6) and in (13) decreases by $\lfloor \alpha \rfloor$ if we replace α by $\alpha - \lfloor \alpha \rfloor$, and since the case $\alpha = 0$ is trivial.

If $0 < \alpha < 1$, then $c_k(\alpha) = (c_{k-1}(1/\alpha))^{-1}$, as one easily derives from (6). By induction and by the Claim, $c_{k-1}(1/\alpha) = q_k/p_k$. Hence $c_k(\alpha) = p_k/q_k$. $\qquad \square$

Note that for every k

(18) $\quad \left| \dfrac{p_{k+1}}{q_{k+1}} - \dfrac{p_k}{q_k} \right| = \dfrac{1}{q_k q_{k+1}}$

since $|p_{k+1} q_k - p_k q_{k+1}| = 1$. Moreover, the denominators $q_1, q_2, q_3, \ldots$ are monotonically increasing. So

(19) $\quad \left| \alpha - \dfrac{p_k}{q_k} \right| \leqslant \left| \dfrac{p_{k+1}}{q_{k+1}} - \dfrac{p_k}{q_k} \right| = \dfrac{1}{q_k q_{k+1}} < \dfrac{1}{q_k^2}.$

Therefore by calculating the convergents we find integers p, q for Dirichlet's result (1).

Theorem 6.2. *For any real numbers α and ε with $\alpha \geqslant 0$, $0 < \varepsilon \leqslant 1$, one of the convergents p_k/q_k for α satisfies (1).*

Proof. Let p_k/q_k be the last convergent for which $q_k \leqslant \varepsilon^{-1}$ holds. If $\alpha \neq p_k/q_k$, then p_{k+1}/q_{k+1} is defined, and by (19),

(20) $\quad \left| \alpha - \dfrac{p_k}{q_k} \right| \leqslant \dfrac{1}{q_k q_{k+1}} < \dfrac{\varepsilon}{q_k}.$

If $\alpha = p_k/q_k$, then trivially $|\alpha - p_k/q_k| < \varepsilon/q_k$. $\qquad \square$

In fact we can derive the following. As t respectively u are the largest integers satisfying (9) resp. (11), it follows that

(21) $\quad \dfrac{p_k + p_{k+1}}{q_k + q_{k+1}} > \alpha \quad$ if k is even, $\quad \dfrac{p_k + p_{k+1}}{q_k + q_{k+1}} < \alpha \quad$ if k is odd.

Therefore, u resp. t (in the next iteration) are at least 1. Hence $q_{k+2} \geq q_k + q_{k+1}$ for every k. Since $q_1 = 1$ and $q_2 \geq 1$ we know inductively that $q_{k+2} \geq (\frac{3}{2})^k$ (in fact, $q_{k+2} \geq F_k$, where F_k is the kth Fibonacci number). So with (19) we know that the convergents converge to α with 'exponential convergence speed'.

If α is rational, there will be a convergent equal to α, so that the sequence of convergents stops, the last term being α.

Suppose α is rational, and the sequence does not stop. Let $\alpha = p/q$, and let p_k/q_k be the first convergent with $q_k > q$. Then we obtain the contradiction:

$$(22) \qquad \frac{1}{qq_{k-1}} \leq \left| \alpha - \frac{p_{k-1}}{q_{k-1}} \right| \leq \frac{1}{q_k q_{k-1}} < \frac{1}{qq_{k-1}}.$$

Here the first inequality is based on the fact that $\alpha \neq p_{k-1}/q_{k-1}$.

Algorithmically, the convergents in the continued fraction method can be found in a way analogous to the Euclidean algorithm. The continued fraction method can be viewed as computing the g.c.d. of α and 1. For $\alpha \geq 0$, let

$$(23) \qquad A_0 := \begin{bmatrix} \alpha & 1 \\ 1 & 0 \\ 0 & 1 \end{bmatrix}.$$

If A_k has been found, say:

$$(24) \qquad A_k = \begin{bmatrix} \alpha_k & \beta_k \\ \gamma_k & \delta_k \\ \varepsilon_k & \zeta_k \end{bmatrix},$$

let A_{k+1} arise from A_k by the following rule:

25   (i) if k is even and $\beta_k > 0$, subtract $\lfloor \alpha_k/\beta_k \rfloor$ times the second column of A_k from the first column;

           (ii) if k is odd and $\alpha_k > 0$, subtract $\lfloor \beta_k/\alpha_k \rfloor$ times the first column of A_k from the second column.

This step is performed for $k = 0, 1, 2, \ldots$, until $\alpha_k = 0$ or $\beta_k = 0$.

Similarly as for the Euclidean algorithm, we know that for each k:

$$(26) \qquad \gamma_k \alpha + \varepsilon_k = \alpha_k$$
$$\delta_k \alpha + \zeta_k = \beta_k.$$

Now let the matrices $A_0, A_1, A_2, \ldots$ look like:

$$(27) \qquad \begin{bmatrix} \alpha & 1 \\ 1 & 0 \\ 0 & 1 \end{bmatrix}, \quad \begin{bmatrix} \alpha_1 & 1 \\ q_1 & 0 \\ -p_1 & 1 \end{bmatrix}, \quad \begin{bmatrix} \alpha_1 & \beta_2 \\ q_1 & -q_2 \\ -p_1 & p_2 \end{bmatrix},$$

$$\begin{bmatrix} \alpha_3 & \beta_2 \\ q_3 & -q_2 \\ -p_3 & p_2 \end{bmatrix}, \quad \begin{bmatrix} \alpha_3 & \beta_4 \\ q_3 & -q_4 \\ -p_3 & p_4 \end{bmatrix}, \ldots$$

Then the p_k, q_k found in this way are the same as found in the convergents (13).

Proof. Let $\tilde{p}_k$, $\tilde{q}_k$ denote the p_k, q_k obtained in (27). We show that $\tilde{p}_k = p_k$ and $\tilde{q}_k = q_k$ by induction on k, the cases $k = 1, 2$ being trivial.

First let k be even. By (25), $\tilde{p}_{k+1} = \tilde{p}_{k+1} + t\tilde{p}_k$ and $\tilde{q}_{k+1} = \tilde{q}_{k-1} + t\tilde{q}_k$, where $t = \lfloor \alpha_k/\beta_k \rfloor$. By (26), $\alpha_k = \tilde{q}_{k-1}\alpha - \tilde{p}_{k-1}$ and $\beta_k = -\tilde{q}_k\alpha + \tilde{p}_k$. Hence t is the largest integer such that

$$(28) \qquad t \leqslant \frac{\tilde{q}_{k-1}\alpha - \tilde{p}_{k-1}}{-\tilde{q}_k\alpha + \tilde{p}_k}, \quad \text{i.e.} \quad \frac{\tilde{p}_{k-1} + t\tilde{p}_k}{\tilde{q}_{k-1} + t\tilde{q}_k} \leqslant \alpha.$$

So the induction hypothesis: $\tilde{p}_{k-1} = p_{k-1}$, $\tilde{q}_{k-1} = q_{k-1}$, $\tilde{p}_k = p_k$, $\tilde{q}_k = q_k$, implies $\tilde{p}_{k+1} = p_{k+1}$ and $\tilde{q}_{k+1} = q_{k+1}$.

The case of k odd is shown similarly. $\qquad\square$

So from the Euclidean algorithm and Theorem 5.1 we know that if α is rational, the continued fraction method terminates, in polynomial time. Therefore, we have the following corollary.

Corollary 6.2a. *Given rationals α and ε ($0 < \varepsilon \leqslant 1$), the continued fraction method finds integers p and q as described in Dirichlet's theorem, in time polynomially bounded by the size of α.*

Proof. Directly from the above and Theorem 6.2 (if $\alpha < 0$, applied to $-\alpha$). $\qquad\square$

The following variant of Theorem 6.2 is due to Legendre [1798: pp. 27–29].

Theorem 6.3. *Let $\alpha \geqslant 0$ be a real number, and let p and q be natural numbers with $|\alpha - p/q| < 1/2q^2$. Then p/q occurs as convergent for α.*

Proof. The proof uses the fact that if $p'/q' < p''/q''$ are rational numbers ($p', p'' \in \mathbb{Z}$, $q', q'' \in \mathbb{N}$), then

$$(29) \qquad \frac{p''}{q''} - \frac{p'}{q'} = \frac{p''q' - p'q''}{q'q''} \geqslant \frac{1}{q'q''}.$$

To prove the theorem, first suppose that $p/q < \alpha$. Make the continued fraction expansion $p_1/q_1, p_2/q_2, \ldots$ of α, and let p_i/q_i be that convergent of α with $p_i/q_i \leqslant p/q$ and i as large as possible (so i is odd). We show that $p_i/q_i = p/q$. Suppose to the contrary that $p_i/q_i < p/q$. Then there exists an integer $u \geqslant 0$ such that

$$(30) \qquad \frac{p_i + up_{i+1}}{q_i + uq_{i+1}} \leqslant \frac{p}{q} < \frac{p_i + (u+1)p_{i+1}}{q_i + (u+1)q_{i+1}} \leqslant \alpha.$$

This is easy to see if $p_{i+1}/q_{i+1} = \alpha$. If $p_{i+1}/q_{i+1} > \alpha$, then by definition of p_{i+2}/q_{i+2} there exists an integer $t \geqslant 1$ with

$$(31) \qquad \alpha \geqslant \frac{p_i + tp_{i+1}}{q_i + tq_{i+1}} = \frac{p_{i+2}}{q_{i+2}} > \frac{p}{q}$$

which also implies the existence of an integer $u \geqslant 0$ satisfying (30).

Now $q_i + uq_{i+1} \leqslant q$, as

$$(32) \qquad \frac{1}{qq_{i+1}} \leqslant \frac{p_{i+1}}{q_{i+1}} - \frac{p}{q} \leqslant \frac{p_{i+1}}{q_{i+1}} - \frac{p_i + up_{i+1}}{q_i + uq_{i+1}} = \frac{1}{q_{i+1}(q_i + uq_{i+1})}$$

(by (29), (30), and since $p_{i+1}q_i - p_iq_{i+1} = 1$). Also $q_{i+1} \leqslant q$, as

$$(33) \qquad \frac{1}{qq_i} \leqslant \frac{p}{q} - \frac{p_i}{q_i} \leqslant \frac{p_{i+1}}{q_{i+1}} - \frac{p_i}{q_i} = \frac{1}{q_{i+1}q_i}$$

(by (29) and since $p_{i+1}q_i - p_iq_{i+1} = 1$).

Therefore, $q_i + (u+1)q_{i+1} \leqslant 2q$. Hence

$$(34) \qquad \alpha - \frac{p}{q} \geqslant \frac{p_i + (u+1)p_{i+1}}{q_i + (u+1)q_{i+1}} - \frac{p}{q} \geqslant \frac{1}{q(q_i + (u+1)q_{i+1})} \geqslant \frac{1}{2q^2},$$

which contradicts our assumption.

The case $p/q > \alpha$ is shown similarly.   □

Theorem 6.3 implies that if a rational number is near enough to a rational number of low denominator, this last rational can be found in polynomial time. More precisely:

Corollary 6.3a. *There exists a polynomial algorithm which, for given rational number α and natural number M, tests if there exists a rational number p/q with $1 \leqslant q \leqslant M$ and $|\alpha - p/q| < 1/2M^2$, and if so, finds this (unique) rational number.*

Proof. It is clear that if such a rational number p/q exists, it is unique, for if p'/q' is another such rational number, we have the contradiction:

$$(35) \qquad \frac{1}{M^2} \leqslant \frac{1}{qq'} \leqslant \left| \frac{p}{q} - \frac{p'}{q'} \right| \leqslant \left| \alpha - \frac{p}{q} \right| + \left| \alpha - \frac{p'}{q'} \right| < \frac{1}{2M^2} + \frac{1}{2M^2} = \frac{1}{M^2}.$$

First let $\alpha > 0$. Suppose p/q as required exists. As

$$(36) \qquad \left| \alpha - \frac{p}{q} \right| < \frac{1}{2M^2} \leqslant \frac{1}{2q^2}$$

Theorem 6.3 implies that p/q occurs as convergent of α. So it suffices to compute the convergents of α (as long as $q_i \leqslant M$), and to test if one of them satisfies (36).

If $\alpha < 0$ we apply the above to $-\alpha$, while for $\alpha = 0$ the algorithm is trivial.   □

The continued fraction method approximates one number. In Section 6.3 we shall see an extension to the approximation of vectors, i.e. to approximating several real numbers simultaneously, by rationals with one common low denominator: *simultaneous diophantine approximation*.

Notes for Section 6.1. Papadimitriou [1979] and Reiss [1979] (cf. Zemel [1981–2]) derived with continued fractions and the related Farey series that there exists an algorithm which identifies a number $s^* = p/q$, with $p, q \in \mathbb{N}$ and $p, q \leqslant M$, within $O(\log M)$ queries of the form: 'is $s^* \leqslant s$?' Indeed, it is clear that, by binary search, within $\lceil \log_2(2M^3) \rceil$ queries we can find $m \in \mathbb{N}$ such that $1 \leqslant m \leqslant 2M^3$ and

$$(37) \qquad \frac{m-1}{2M^2} < s^* \leqslant \frac{m}{2M^2}$$

(as $0 < s^* \leqslant M$). Now apply Corollary 6.3a to $\alpha := m/2M^2$.

Ursic and Patarra [1983] apply continued fractions to finding an exact solution for linear equations if an approximate solution is available.

An elementary exposition of continued fractions is given by Richards [1981].

The average length of the continued fraction method is studied by Heilbronn [1968], Dixon [1970], and Tonkov [1974]. For more on continued fractions, see Cheng and Pollington [1981] and Nathanson [1974].

Hurwitz [1891] showed that for each irrational number α, there exist infinitely many convergents p/q in the continued fraction approximation for α such that

$$(38) \qquad \left| \alpha - \frac{p}{q} \right| < \frac{1}{\sqrt{5}} \frac{1}{q^2}$$

and that the factor $1/\sqrt{5}$ here is best possible (as is shown by $\alpha = \frac{1}{2} + \frac{1}{2}\sqrt{5}$). See Cohn [1973] for a short proof.

6.2. BASIS REDUCTION IN LATTICES

Let A be a nonsingular matrix of order n, and let Λ be the lattice generated by the columns of A. From Chapter 4 we know that if B is another nonsingular matrix, whose columns also generate Λ, then $|\det A| = |\det B|$. So this number is independent of the choice of the basis, and is called the *determinant* of Λ, denoted by $\det \Lambda$. It is equal to the volume of the parallelepiped $\{\lambda_1 b_1 + \cdots + \lambda_n b_n | 0 \leqslant \lambda_i < 1$ for $i = 1, \ldots, n\}$, where $b_1, \ldots, b_n$ is any basis for Λ. This gives the well-known *Hadamard inequality*:

$$(39) \qquad \det \Lambda \leqslant \|b_1\| \cdots \|b_n\|$$

where $\|\cdot\|$ denotes Euclidean norm ($\|x\| = \sqrt{x^\mathsf{T} x}$).

It is trivial that equality in (39) occurs only if $b_1, \ldots, b_n$ are orthogonal, and that not every lattice has an orthogonal basis. A classical theorem of Hermite [1850a] states that for each n there exists a number $c(n)$ such that every n-dimensional lattice Λ has a basis $b_1, \ldots, b_n$ with

$$(40) \qquad \|b_1\| \cdots \|b_n\| \leqslant c(n) \det \Lambda.$$

Such a basis may be viewed as an approximation of an orthogonal basis.

Hermite showed that we can take

$$(41) \qquad c(n) = (\tfrac{4}{3})^{n(n-1)/4}.$$

Minkowski [1896] improved this by showing that even

$$(42) \qquad c(n) = 2^n/V_n \approx \left(\frac{2n}{\pi e} \right)^{n/2}$$

works, where V_n denotes the volume of the n-dimensional unit ball. However, for choice (42) of $c(n)$ no polynomial algorithm finding a basis satisfying (40) is known. But if we take $c(n) = 2^{n(n-1)/4}$, Lovász designed a polynomial algorithm finding such a basis (if the lattice is given by generators). In fact, Schnorr [1985] proved that for each fixed $\varepsilon > 0$ there exists a polynomial algorithm finding a basis satisfying (40) with $c(n) = (1 + \varepsilon)^{n(n-1)}$, if $n \geqslant n(\varepsilon)$.

We now first describe Lovász's basis reduction method (see Lenstra, Lenstra, and Lovász [1982]) in terms of positive definite matrices (the above formulation comes back in Corollary 6.4a).

Theorem 6.4 (Basis reduction method). *There exists a polynomial algorithm which, for given positive definite rational matrix D, finds a basis $b_1, \ldots, b_n$ (where $n := $ order (D)) for the lattice $\mathbb{Z}^n$ satisfying*

(43) $\qquad \| b_1 \| \cdots \| b_n \| \leqslant 2^{n(n-1)/4} \sqrt{\det D},$

where $\| x \| := \sqrt{x^\mathsf{T} D x}$.

Proof. I. We describe the algorithm. Let the positive definite $n \times n$-matrix D be given. Denote by $\perp$ orthogonality relative to the inner product defined by D. So:

(44) $\qquad x \perp y$ if and only if $x^\mathsf{T} D y = 0$.

Without loss of generality we assume that D is integral (otherwise, multiply D by an appropriate integer).

Construct a series of bases for $\mathbb{Z}^n$ as follows. The unit basis vectors form the first basis. Suppose inductively that we have found a basis $b_1, \ldots, b_n$ for $\mathbb{Z}^n$. Now apply the following Steps 1, 2, and 3.

Step 1. Denote by B_i the matrix with columns $b_1, \ldots, b_i$. Calculate

(45) $\qquad b_i^* = b_i - B_{i-1}(B_{i-1}^\mathsf{T} D B_{i-1})^{-1} B_{i-1}^\mathsf{T} D b_i.$

Comment 1. The vectors $b_1^*, \ldots, b_n^*$ form the *Gram–Schmidt orthogonalization* of $b_1, \ldots, b_n$ relative to $\perp$. That is, for $i = 1, \ldots, n$, b_i^* is the unique vector satisfying

(46) $\qquad b_i^* \perp b_1, \ldots, b_{i-1}, \quad$ and $\quad b_i^* - b_i \in \mathrm{lin.hull}\{b_1, \ldots, b_{i-1}\}.$

Geometrically, b_i^* is the perpendicular line (relative to $\perp$) from b_i to lin.hull $\{b_1, \ldots, b_{i-1}\} = \mathrm{lin.hull}\{b_1^*, \ldots, b_{i-1}^*\}$. The vectors $b_1^*, \ldots, b_n^*$ are pairwise orthogonal (relative to $\perp$), and for each i,

(47) $\qquad b_i = \lambda_1 b_1^* + \cdots + \lambda_{i-1} b_{i-1}^* + b_i^*$

for some $\lambda_1, \ldots, \lambda_{i-1}$.

Step 2. For $i = 2, \ldots, n$, do the following: for $j = i - 1, i - 2, \ldots, 1$, do: write $b_i = \lambda_1 b_1^* + \cdots + \lambda_{i-1} b_{i-1}^* + b_i^*$, and replace b_i by $b_i - \lfloor \lambda_j + \tfrac{1}{2} \rfloor b_j$.

Comment 2. In matrix language, Step 2 does the following. Let C_i be the matrix with columns $b_1^*, \ldots, b_i^*$. Then $B_n = C_n V$ for some upper-triangular matrix V, with 1's on the main diagonal (by (47)). By elementary column operations one can change V into a matrix W in upper-triangular form, with 1's on the diagonal, and all other entries at most $\frac{1}{2}$ in absolute value. Then the columns of $C_n W$ form the basis $b_1, \ldots, b_n$ as it is after applying Step 2.

Step 3. Choose, if possible, an index i such that $\| b_i^* \|^2 > 2 \| b_{i+1}^* \|^2$. Exchange b_i and b_{i+1}, and start with Step 1 anew.

If no such i exists, the algorithm stops.

Two further comments:

Comment 3. The replacements made in Step 2 do not change the Gram–Schmidt orthogonalization of $b_1, \ldots, b_n$. Note that at completion of Step 2, we have for $i = 1, \ldots, n$,

$$(48) \qquad b_i = \lambda_1 b_1^* + \cdots + \lambda_{i-1} b_{i-1}^* + b_i^*$$

with $|\lambda_j| \leqslant \frac{1}{2}$ for $j = 1, \ldots, i-1$, since $B_n = C_n W$.

Comment 4. In the notation of Step 1,

$$(49) \qquad \| b_1^* \|^2 \cdot \cdots \cdot \| b_i^* \|^2 = \det (B_i^T D B_i).$$

Indeed, let C_i be the matrix with columns $b_1^*, \ldots, b_i^*$. As in Comment 2, there is an upper triangular matrix V_i with 1's on the diagonal such that $C_i V_i = B_i$. Then, as $b_i^* \perp b_j^*$ if $i \neq j$,

$$(50) \qquad \| b_1^* \|^2 \cdot \cdots \cdot \| b_i^* \|^2 = \det (C_i^T D C_i) = \det (V_i^T C_i^T D C_i V_i) = \det (B_i^T D B_i).$$

In particular, as $|\det B_n| = 1$,

$$(51) \qquad \| b_1^* \|^2 \cdot \cdots \cdot \| b_n^* \|^2 = \det D.$$

II. *If* the algorithm stops (i.e. the choice of i in Step 3 is not possible), we have $\| b_i^* \|^2 \leqslant 2 \| b_{i+1}^* \|^2$ for $i = 1, \ldots, n-1$. Hence, inductively, $\| b_i^* \|^2 \leqslant 2^{k-i} \| b_k^* \|^2$ for $1 \leqslant i < k \leqslant n$. By Comments 1 and 3,

$$(52) \qquad \| b_k \|^2 \leqslant \| b_k^* \|^2 + \frac{1}{4} \sum_{i=1}^{k-1} \| b_i^* \|^2 \leqslant \| b_k^* \|^2 \left(1 + \frac{1}{4} \sum_{i=1}^{k-1} 2^{k-i} \right) \leqslant 2^{k-1} \| b_k^* \|^2.$$

Therefore, using (51),

$$(53) \qquad \| b_1 \|^2 \cdot \cdots \cdot \| b_n \|^2 \leqslant \prod_{k=1}^{n} (2^{k-1} \| b_k^* \|^2) = 2^{n(n-1)/2} \det D.$$

So the final basis $b_1, \ldots, b_n$ satisfies (43).

III. We still have to show *that* the algorithm terminates, in polynomial time. Consider the following function (using the notation of Step 1):

$$(54) \qquad F(b_1, \ldots, b_n) := \det (B_1^T D B_1) \det (B_2^T D B_2) \cdot \cdots \cdot \det (B_n^T D B_n).$$

Then, by Comment 4,

$$(55) \qquad F(b_1, \ldots, b_n) = \| b_1^* \|^{2n} \cdot \| b_2^* \|^{2n-2} \cdot \| b_3^* \|^{2n-4} \cdot \ldots \cdot \| b_n^* \|^2.$$

So at Step 2, $F(b_1, \ldots, b_n)$ does not change (cf. Comment 3).

If at Step 3 we interchange b_i and b_{i+1}, then only the term $\det(B_i^T D B_i)$ of (54) changes: the matrices $B_1, \ldots, B_{i-1}$ do not change at all, while in the matrices $B_{i+1}, \ldots, B_n$ two columns are interchanged, which leaves the corresponding terms of (54) invariant.

To avoid confusion, let $\bar{b}_1, \ldots, \bar{b}_n$ be the new basis (after application of Step 3), and let $\bar{b}_1^*, \ldots, \bar{b}_n^*$ be the new Gram–Schmidt orthogonalization. Then $\bar{b}_j = b_j$ for $j \neq i, i+1$, and $\bar{b}_i = b_{i+1}$ and $\bar{b}_{i+1} = b_i$. So $\bar{b}_j^* = b_j^*$ for $j \neq i, i+1$. Therefore,

$$(56) \qquad \bar{b}_i = b_{i+1} = \lambda_1 b_1^* + \cdots + \lambda_{i-1} b_{i-1}^* + \lambda_i b_i^* + b_{i+1}^*$$
$$= \lambda_1 \bar{b}_1^* + \cdots + \lambda_{i-1} \bar{b}_{i-1}^* + \lambda_i b_i^* + b_{i+1}^*$$

with the λ_j at most $\frac{1}{2}$ in absolute value. As $\lambda_i b_i^* + b_{i+1}^* \perp b_1, \ldots, b_{i-1}$, we know that $\bar{b}_i^* = \lambda_i b_i^* + b_{i+1}^*$. Therefore, by our choice of i in Step 3:

$$(57) \qquad \frac{\|\bar{b}_i^*\|^2}{\|b_i^*\|^2} = \frac{\lambda_i^2 \|b_i^*\|^2 + \|b_{i+1}^*\|^2}{\|b_i^*\|^2} = \lambda_i^2 + \frac{\|b_{i+1}^*\|^2}{\|b_i^*\|^2} < \frac{1}{4} + \frac{1}{2} = \frac{3}{4}.$$

So by interchanging b_i and b_{i+1}, $\det(B_i^T D B_i)$ is multiplied by a factor $\mu < 3/4$. Hence, at Step 3, the value of $F(b_1, \ldots, b_n)$ is multiplied by $\mu < 3/4$.

Let T be the maximum absolute value of the entries of D. Then initially, and hence by the above throughout the procedure, we have:

$$(58) \qquad \det(B_i^T D B_i) \leqslant (nT)^n$$

and hence

$$(59) \qquad F(b_1, \ldots, b_n) \leqslant (nT)^{n^2}.$$

Since $F(b_1, \ldots, b_n)$ is a positive integer (by (54), as $b_1, \ldots, b_n$ are linearly independent integral vectors, and D is integral), the algorithm terminates after at most

$$(60) \qquad \log_{4/3}((nT)^{n^2}) = n^2(\log_2 n + \log_2 T)\log_{4/3} 2.$$

iterations, which number is polynomially bounded by the size of D.

IV. We still must show that the sizes of the intermediate numbers also are polynomially bounded by the size of D. From (45) it follows that throughout the algorithm, the vectors $\det(B_{i-1}^T D B_{i-1}) \cdot b_i^*$ are integral. Hence the denominators in b_i^* are at most $\det(B_{i-1}^T D B_{i-1}) \leqslant (nT)^n$ (cf. (58)). This implies

$$(61) \qquad \|b_i^*\| \geqslant (nT)^{-n}$$

where $\|\cdot\|$ denotes the Euclidean norm ($\|x\| = \sqrt{x^T x}$). Let λ denote the smallest eigenvalue of D. Then, again denoting the maximum absolute value of the entries of D by T,

$$(62) \qquad \lambda^{-1} = \text{largest eigenvalue of } D^{-1} \leqslant \text{Tr}(D^{-1}) \leqslant n(nT)^n$$

(as the entries of D^{-1} are quotients of (sub)determinants of D, which are at most $(nT)^n$). Therefore, by (49), (58), and (61),

$$(63) \qquad \|b_i^*\|^2 \leqslant \lambda^{-1}\|b_i^*\|^2 = \lambda^{-1}\det(B_i^T D B_i)\|b_1^*\|^{-2} \cdot \ldots \cdot \|b_{i-1}^*\|^{-2}$$
$$\leqslant \lambda^{-n}\det(B_i^T D B_i)\|b_1^*\|^{-2} \cdot \ldots \cdot \|b_{i-1}^*\|^{-2}$$
$$\leqslant n^n(nT)^{n^2}(nT)^{2n^2} \leqslant (nT)^{3n^2 + 2n}.$$

Hence the sizes of the numerators and denominators in the b_i^* are polynomially bounded by the size of D.

After application of Step 2 we know by Comment 3:

(64) $\| b_i \| \leqslant \| b_1^* \| + \cdots + \| b_i^* \| \leqslant n(nT)^{3n^2 + 2n}$.

Since the b_i are integral vectors, it follows that also the sizes of the b_i are polynomially bounded by the size of D. $\square$

Corollary 6.4a. *There exists a polynomial algorithm which, for given nonsingular rational matrix A, finds a basis $b_1, \ldots, b_n$ (with $n := \mathrm{order}(A)$) for the lattice Λ generated by the columns of A satisfying*

(65) $\| b_1 \| \cdot \cdots \cdot \| b_n \| \leqslant 2^{n(n-1)/4} |\det A|$

where $\| x \| = \sqrt{x^T x}$.

Proof. Apply the basis reduction algorithm (Theorem 6.4) to the matrix $D := A^T A$. This gives a basis $b_1, \ldots, b_n$ for $\mathbb{Z}^n$ with

(66) $\| A b_1 \| \cdot \cdots \cdot \| A b_n \| = \sqrt{b_1^T D b_1} \ldots \sqrt{b_n^T D b_n} \leqslant 2^{n(n-1)/4} \sqrt{\det D}$
$\qquad\qquad = 2^{n(n-1)/4} |\det A|$.

The vectors $A b_1, \ldots, A b_n$ form a basis for Λ, and the corollary follows. $\square$

Notes for Section 6.2.   In fact, Hermite's factor $c(n) = (\frac{4}{3})^{n(n-1)/4}$ in (40) can be approached arbitrarily close by a slight adaptation of the above algorithm. Then one shows: *there exists an algorithm which, for given positive definite rational matrix D and given $\varepsilon > 0$, finds a basis $b_1, \ldots, b_n$ for the lattice $\mathbb{Z}^n$ satisfying*

(67) $\| b_1 \| \cdot \cdots \cdot \| b_n \| \leqslant (\frac{4}{3} + \varepsilon)^{n(n-1)/4} \sqrt{\det D}$

in time polynomially bounded by size(D) and by $1/\varepsilon$.

Kannan [1983b] gave an improved version of the basis reduction method, from which Schnorr [1985] derived that for each fixed $\varepsilon > 0$ one can find a basis $b_1, \ldots, b_n$ for $\mathbb{Z}^n$ with

(68) $\| b_1 \| \cdot \cdots \cdot \| b_n \| \leqslant (1 + \varepsilon)^{n(n-1)/4} \sqrt{\det D}$

in time polynomially bounded by size (D).

Lenstra [1983] showed with methods from the geometry of numbers that there exists a constant c such that for each fixed n, there is a polynomial-time algorithm finding, for given positive definite rational $n \times n$-matrix D, a basis $b_1, \ldots, n_n$ for $\mathbb{Z}^n$ such that

(69) $\| b_1 \| \cdot \cdots \cdot \| b_n \| \leqslant (cn)^n \sqrt{\det D}$.

The order of this bound is close to that of Minkowski's bound (42).

6.3 APPLICATIONS OF THE BASIS REDUCTION METHOD

Finding a short nonzero vector in a lattice

It is a classical result of Minkowski [1891] that any n-dimensional lattice Λ contains a nonzero vector b with

(70) $\| b \| \leqslant 2 \left(\dfrac{\det \Lambda}{V_n} \right)^{1/n}$,

where again V_n denotes the volume of the n-dimensional unit ball.

Sketch of proof. Let $r := \frac{1}{2} \min \{ \|b\| \, | \, b \in \Lambda; \; b \neq 0 \}$. Then for $x, y \in \Lambda$ with $x \neq y$ one has $B^\circ(x, r) \cap B^\circ(y, r) = \varnothing$ (where $B^\circ(x, r)$ denotes the open ball of radius r with centre x). Now $B^\circ(x, r)$ has volume $r^n V_n$. Hence the density of $\cup_{x \in \Lambda} B^\circ(x, r)$ is $r^n V_n (\det \Lambda)^{-1}$, which is at most 1. This gives (70).   □

To date, no polynomial algorithm finding a vector b satisfying (70) is known. It is conjectured that finding a shortest nonzero vector in Λ (given by a basis) is $\mathcal{NP}$-complete.

However, with the basis reduction method of Theorem 6.4 one can find a 'longer short vector' in a lattice, by taking the shortest vector in the basis constructed there.

Corollary 6.4b. *There exists a polynomial algorithm which, given a nonsingular rational matrix A, finds a nonzero vector b in the lattice Λ generated by the columns of A with*

(71) $\|b\| \leqslant 2^{n(n-1)/4} \cdot (\det \Lambda)^{1/n}$

where $n := \text{order} \, (A)$.

Proof. The shortest vector in the basis $b_1, \ldots, b_n$ described in Corollary 6.4a has length at most

(72) $(2^{n(n-1)/4} \det \Lambda)^{1/n} = 2^{(n-1)/4} (\det \Lambda)^{1/n}$.

[In fact, if we follow the algorithm as described in the proof of Theorem 6.4, and the notation of the proof of Corollary 6.4a, we can take Ab_1 for the vector b satisfying (71). Indeed, in part II of the proof of Theorem 6.4 we saw that $\|b_i^*\|^2 \leqslant 2^{k-i} \|b_k^*\|^2$ for $1 \leqslant i < k \leqslant n$. Moreover, by Comment 4 in that proof, $\|b_1^*\| \cdot \ldots \cdot \|b_n^*\| = \det \Lambda$. Hence,

(73) $\|Ab_1\| = \|b_1\| = \left(\prod_{k=1}^{n} \|b_1^*\| \right)^{1/n} \leqslant \left(\prod_{k=1}^{n} 2^{(k-1)/2} \|b_k^*\| \right)^{1/n}$

$= (2^{n(n-1)/4} \det \Lambda)^{1/n}.]$ □

So with the basis reduction method also a short nonzero vector can be found, albeit generally not the shortest.

Van Emde Boas [1981] proved that, for any fixed norm, the problem: 'given lattice Λ and vector a, find $b \in \Lambda$ with norm of $b - a$ as small as possible' is $\mathcal{NP}$-complete (it is the integer analogue of the, polynomially solvable, least squares problem). Algorithms for finding shortest vectors in lattices were described by Rosser [1942], Coveyou and MacPherson [1967], and Knuth [1969: pp. 89–98] (cf. also Dieter [1975]). Kannan [1983b] gave an algorithm for finding a shortest nonzero lattice vector using a variant of Lovász's basis reduction method.

Simultaneous diophantine approximation

Another application is to *simultaneous diophantine approximation*. Extending the 'pigeon-hole' argument of Theorem 6.1, Dirichlet [1842] showed that if

$\alpha_1, \ldots, \alpha_n, \varepsilon$ are real numbers with $0 < \varepsilon < 1$, then there exist integers $p_1, \ldots, p_n$ and q such that

(74) $\qquad \left| \alpha_i - \dfrac{p_i}{q} \right| < \dfrac{\varepsilon}{q}, \quad \text{for} \quad i = 1, \ldots, n, \quad \text{and} \quad 1 \leqslant q \leqslant \varepsilon^{-n}.$

One can find these integers in finite time (as there are at most ε^{-n} candidates for q), but no polynomial method is known (except for the case $n = 1$, where we can apply the technique of continued fractions—see Corollary 6.2a).

However, a consequence of Corollary 6.4b is that a weaker approximation can be found in polynomial time, which turns out to be strong enough, for example, for application in the ellipsoid method in Chapter 14.

Corollary 6.4c (Simultaneous diophantine approximation). *There exists a polynomial algorithm which for given vector $a \in \mathbb{Q}^n$ and rational ε with $0 < \varepsilon < 1$, finds an integral vector p and an integer q such that*

(75) $\qquad \left\| a - \dfrac{1}{q} \cdot p \right\| < \dfrac{\varepsilon}{q}, \quad \text{and} \quad 1 \leqslant q \leqslant 2^{n(n+1)/4} \varepsilon^{-n}.$

Proof. Let Λ be the lattice generated by the columns of the matrix:

(76) $\qquad A := \begin{bmatrix} 1 & & & & -\alpha_1 \\ & \ddots & & 0 & \vdots \\ & 0 & & \ddots & \vdots \\ & & & \cdot 1 & -\alpha_n \\ 0 & \cdots & \cdots & 0 & 2^{-n(n+1)/4}\varepsilon^{n+1} \end{bmatrix}$

where $a =: (\alpha_1, \ldots, \alpha_n)^{\mathrm{T}}$. So $\det \Lambda = \det A = 2^{-n(n+1)/4} \varepsilon^{n+1}$. Application of Corollary 6.4b to A gives a nonzero vector $b = (\beta_1, \ldots, \beta_{n+1})^{\mathrm{T}}$ in Λ with

(77) $\qquad \| b \| \leqslant 2^{n/4} (2^{-n(n+1)/4} \varepsilon^{n+1})^{1/(n+1)} = \varepsilon.$

Now we can calculate

(78) $\qquad \begin{bmatrix} p_1 \\ \vdots \\ p_n \\ q \end{bmatrix} := A^{-1} b$

which is an integral vector, since b is in Λ. So

(79) $\qquad b = A \begin{bmatrix} p_1 \\ \vdots \\ p_n \\ q \end{bmatrix}.$

Then $q \neq 0$, since otherwise $b = (p_1, \ldots, p_n, 0)^{\mathrm{T}}$, and hence $\| b \| \geqslant 1 > \varepsilon$ (as the p_i are integers and $b \neq 0$).

Without loss of generality, $q \geq 1$ (otherwise, replace b by $-b$). Then

(80) $\beta_i = p_i - q\alpha_i$,   for $i = 1, \ldots, n$,   and $\beta_{n+1} = q \cdot 2^{-n(n+1)/4} \varepsilon^{n+1}$.

Hence, defining $p := (p_1, \ldots, p_n)^{\mathsf{T}}$,

(81) $\| q \cdot a - p \| < \| b \| \leq \varepsilon$ and $q = \beta_{n+1} \cdot 2^{n(n+1)/4} \varepsilon^{-n-1} \leq 2^{n(n+1)/4} \varepsilon^{-n}$

(using $\beta_{n+1} \leq \| b \| \leq \varepsilon$). $\square$

Lagarias [1982] showed that the problem: 'Given rationals $\alpha_1, \ldots, \alpha_n$, a natural number M, and a rational $\varepsilon > 0$, decide if there exist integers $p_1, \ldots, p_n, q$ such that $|\alpha_i - (p_i/q)| \leq \varepsilon/q$ and $1 \leq q \leq M$' is $\mathcal{NP}$-complete. He showed moreover that, if we fix n, this problem is polynomially solvable. Lagarias also gave some sharpenings of Corollary 6.4c.

Finding the Hermite normal form

Another application of the basis reduction method is to finding the Hermite normal form of a matrix, in polynomial time. Let A be a nonsingular integral matrix of order n. Define

(82) $M := \lceil 2^{n(n-1)/4} |\det A| \rceil$

and let C arise from A by multiplying the ith row of A by M^{n-i}, for $i = 1, \ldots, n$. By Corollary 6.4a one can find in polynomial time a basis $b_1, \ldots, b_n$ for the lattice generated by the columns of C satisfying

(83) $\| b_1 \| \cdot \cdots \cdot \| b_n \| \leq 2^{n(n-1)/4} |\det C| = 2^{n(n-1)/4} M^{n(n-1)/2} |\det A|$.

Then the vectors $b_1, \ldots, b_n$ can be reordered in such a way that the matrix $[b_1, \ldots, b_n]$ is in lower triangular form. Indeed, by reordering we may assume that the jth coordinate of b_j is nonzero, for $j = 1, \ldots, n$. So the jth coordinate of b_j is at least M^{n-j} in absolute value, and hence $\| b_j \| \geq M^{n-j}$. Suppose that the ith coordinate of $b_{j'}$ is nonzero, for some $1 \leq i < j' \leq n$. Then $\| b_{j'} \| > M^{n-i} \geq M \cdot M^{n-j'}$, and

(84) $\| b_1 \| \cdot \cdots \cdot \| b_n \| > \left(\prod_{j=1}^{n} M^{n-j} \right) \cdot M \geq 2^{n(n-1)/4} M^{n(n-1)/2} |\det A|$

contradicting (83).

So $b_1, \ldots, b_n$ can be reordered so that the matrix $[b_1, \ldots, b_n]$ is in lower triangular form. This matrix can be easily modified by elementary column operations to Hermite normal form. Dividing the jth row by M^{n-j} (for each j) gives the Hermite normal form of A.

Other applications of the basis reduction method appear in Lenstra's integer linear programming algorithm (Section 18.4), in factoring polynomials (over the rationals) in polynomial time (see Lenstra, Lenstra, and Lovász [1982]), in breaking cryptographic codes (see Shamir [1982], Adleman [1983a, b], and Lagarias and Odlyzko [1983]), in disproving *Mertens' conjecture* (Odlyzko and te Riele [1984]), and in solving low density subset sum problems (Lagarias and Odlyzko [1983]).

Notes for Chapter 6. For more about more-dimensional continued fractions and simultaneous diophantine approximation, see Adams [1967, 1969a, b, 1971], Bernstein [1971], Brentjes [1981], Brun [1959, 1961], Bundschuh [1975], Cassels [1955], Cusick [1971, 1972a, b, 1974, 1977a, b, 1980], Davenport [1955], Davenport and Mahler [1946], Davenport and Schmidt [1967, 1969–70, 1970], Dubois and Rhin [1980, 1982], Ferguson and Forcade [1979], Fischer and Schweiger [1975], Furtwängler [1927, 1928], Güting [1975, 1976a, b], Jacobi [1868], Jurkat, Kratz and Peyerimhoff [1977, 1979], Jurkat and

Peyerimhoff [1976], Koksma and Meulenbeld [1941], Lagarias [1980a], Langmayr [1979a, b], Mack [1977], Mahler [1976], Mullender [1950], Nowak [1981], Paley and Ursell [1930], Peck [1961], Perron [1907], Pipping [1921, 1957], Podsypanin [1977], Poincaré [1884], Raisbeck [1950], Schmidt [1966], Schweiger [1973, 1977], Selmer [1961], Spohn [1968], Szekeres [1970], Thurnheer [1981], Tichy [1979], Vaughan [1978], and Wills [1968a,b, 1970].

For a survey on basis reduction algorithms, see Bachem and Kannan [1984].

Notes on lattices and linear diophantine equations

HISTORICAL NOTES

The Ancient Civilizations

The number theory discussed in this Part is to a large extent quite classic.

The ancient Greeks knew the Euclidean algorithm and the continued fraction method. The Euclidean algorithm was described in Book VII of Euclid's *Elements* (written $\pm$ 330–320 B.C.). Books V and X of the Elements describe the continued fraction method to test if two given lengths have a 'common measure'. Two ratios are defined to be equal if they have the same continued fraction expansion. The method was called ἀνθυφαιρεῖν ('reciprocal elimination') or ἀνταναιρεῖν ('cancellation of items on an account'). In the method, the larger number was not divided by the smaller, but the smaller was subtracted from the larger repeatedly.

These Books of the Elements are based on earlier work of Theaetetus (c. 415–c. 369 B.C.) (cf. Becker [1933]). Also Eudoxus (c. 408–c. 355 B.C.) and Aristotle (384–322 B.C.) knew the continued fraction method as a way of testing equality of two ratios. The method was used by Aristarchus of Samos (c. 310–c. 230 B.C.) and Archimedes (287–212 B.C.), the latter for determining square roots.

The Euclidean algorithm was also described in Book I of the *Chiu-Chang Suan-Shu* ('Nine Books on Arithmetic'). Book II of this collection discusses some special classes of linear diophantine equations. The *Sun-Tzu Suan-Ching* ('Master Sun's Arithmetic Manual'), written in China between A.D. 280 and 473, considers the 'Chinese remainder problem', which consists of a special type of system of linear diophantine equations (cf. Midonick [1968: pp. 211–218]).

Diophantos of Alexandria ($\pm$ 3rd century) has yielded the name *diophantine equation*. In his book *Arithmetika* he treats the problem of finding nonnegative rational solutions to equations (mainly quadratic), where, by making the equations homogeneous, the problem can be reduced to finding nonnegative integral solutions. He seems not to have considered solving linear equations in integers.

In fact, the solution of linear diophantine equations has not been found in any ancient Greek text. But, according to van der Waerden [1983], excellent Greek mathematicians like Archimedes or Apollonius would have had no difficulty in solving linear diophantine equations by means of the Euclidean algorithm.

Between 200 and 1200 several Indian mathematicians gave methods for solving linear diophantine equations in two unknowns. The earliest known systematic treatment of linear diophantine equations in two variables was given by Āryabhata (476–550), who proposed a continued-fraction-like method to find all integral solutions of $ax + by = c$. Āryabhata was an astronomer, interested in linear diophantine equations to determine periods of astronomical conjunctions. He and his followers, like Brahmagupta (b. 598), Bhāscara (629), Bhāscara Āchārya (1114–c. 1185) (cf. Midonick [1968: pp. 270–274]), and Parameshvara (1431), also considered special types of *systems* of two linear diophantine equations (cf. van der Waerden [1983: pp. 112–120], Kline [1972: pp. 184–188], Dickson [1920: pp. 41–44]).

16th to 18th century in Europe

In Europe, in the 17th and 18th century, solving linear diophantine equations in two or three variables, with reduction methods like the Euclidean algorithm, was studied, *inter alia*, by Bachet [1612], Kersey [1673], Rolle [1690], De Lagny [1697, 1733], Euler [1740; 1770a: 2.Theil, 2.Abschnitt, Capitel 1], Saunderson [1741], and Lagrange [1769a: Art. 7–8; 1769b; 1770: Lemme 1, §24; 1774: Section III; 1798] (cf. Dickson [1920: pp. 44–99]).

After some isolated work on *continued fractions* in the 16th and 17th century (Bombelli, Brouncker, Cataldi, Huygens), a systematic theory of continued fractions, as a main tool for approximation, was developed in the 18th century by Euler [1744], Lagrange [1769b, 1774, 1798], Lambert, and Legendre [1798]. Lagrange [1774: Section II] showed:

(1)   for any convergent p/q in the continued fraction method for a real number α one has $|p - \alpha q| < |p' - \alpha q'|$ whenever $p' < p$ and $q' < q$

(cf. Smith [1925: pp. 418–421], Jones and Thron [1980: pp. 1–11]).

Lagrange on binary quadratic forms

Lagrange [1775a] started the systematic study of *quadratic forms* in integer variables, following Fermat's work on special quadratic forms. Important is the concept of equivalence: two quadratic forms $x^{\mathsf{T}} D_1 x$ and $x^{\mathsf{T}} D_2 x$ (or two matrices D_1 and D_2) are *equivalent* if there exists a unimodular matrix U with $U^{\mathsf{T}} D_1 U = D_2$. So if D_1 and D_2 are equivalent, $x^{\mathsf{T}} D_1 x = t$ is solvable in integers if and only if $x^{\mathsf{T}} D_2 x = t$ is also.

Lagrange [1775a] studied the problem of *reducing* a binary quadratic form (*binary* means: in two variables). He showed that each positive definite 2×2-matrix D is equivalent to a matrix $\begin{bmatrix} a & b \\ b & c \end{bmatrix}$ with $|2b| \leqslant a \leqslant c$. A matrix with this last property was called by Lagrange *reduced*. It implies

(2) $ac \leqslant \frac{4}{3} \det D$

or, equivalently, that each lattice Λ in two dimensions has a basis b_1, b_2 with $\|b_1\| \cdot \|b_2\| \leqslant \sqrt{\frac{4}{3}} \det \Lambda$.

So each quadratic form $ax^2 + 2bxy + cy^2$ with $a > 0$, $ac - b^2 > 0$ is equivalent to a quadratic form $a'x^2 + 2b'xy + c'y^2$ with $|2b'| \leqslant a' \leqslant c'$. As for given $\Delta > 0$ there are only finitely many integral triples a, b, c with $ac - b^2 = \Delta$ and $|2b| \leqslant a \leqslant c$, this yields a classification of binary quadratic forms with integral coefficients of given negative discriminant.

Lagrange's result also directly implies that each lattice Λ in two dimensions contains a nonzero vector x such that

(3) $\|x\| \leqslant (\frac{4}{3})^{1/4} \sqrt{\det \Lambda}$.

It is not difficult to see that here the factor $(\frac{4}{3})^{1/4}$ is best possible.

Gauss and Seeber

In Art. 40 of his famous book *Disquisitiones Arithmeticae*, Gauss [1801] showed that a linear diophantine equation has a solution if and only if the g.c.d. of the coefficients divides the right-hand side.

The reduction problem in two variables was studied in Artt. 171–181 of the *Disquisitiones*, while Art. 213 contains a rudimentary version of the Hermite normal form for 2×2-matrices. In Artt. 266–285 Gauss starts the study of the reduction of *ternary quadratic forms* (i.e. in three variables). He proved that for each positive definite 3×3-matrix D there exists an integral vector $x \neq 0$ with $x^{\mathsf{T}} D x \leqslant \frac{4}{3} \sqrt[3]{\det D}$. Hence each lattice Λ in three dimensions contains a nonzero vector x with

$$(4) \qquad \| x \| \leqslant (\tfrac{4}{3})^{1/2} (\det \Lambda)^{1/3}.$$

Moreover, Gauss showed that there are only finitely many equivalence classes of integral positive definite 3×3-matrices of given determinant.

Gauss' study of ternary quadratic forms was continued by Seeber [1831], who modified Gauss' concept of reduced form, and showed that each lattice Λ in three dimensions has a basis b_1, b_2, b_3 with $\| b_1 \| \cdot \| b_2 \| \cdot \| b_3 \| \leqslant 3.\det \Lambda$. In an anonymous recension in the *Göttingische gelehrte Anzeigen*, Gauss [1831] simplified Seeber's arguments (however remarking: 'Wenn ein schwieriges Problem oder Theorem aufzulösen oder zu beweisen vorliegt, so ist allezeit der erste und mit gebührendem Danke zu erkennende Schritt, dass überhaupt eine Auflösung oder ein Beweis gefunden werde, und die Frage, ob dies nicht auf eine leichtere und einfachere Art hätte geschehen können, bleibt so lange eine müssige, als die Möglichkeit nicht zugleich durch die That entschieden wird.'). Moreover, Gauss reduced Seeber's factor 3 to the (best possible) 2. (Simple proofs of this were given by Dirichlet [1850] and Hermite [1850b].) Seeber [1831] also showed the existence of the Hermite normal form for 3×3-matrices.

Dirichlet on diophantine approximation

Fundamental results in diophantine analysis were obtained in the 1840's and 1850's by Dirichlet and Hermite.

Dirichlet [1842] first remarks that it has been long known from the theory of continued fractions that for each irrational number α there exist infinitely many pairs of integers p, q with $|p - \alpha q| < 1/q$. Next with the help of the following principle:

(5)   if $t + 1$ objects are put into t boxes, there exists a box containing at least two objects

(the *pigeon-hole principle*, or *drawer principle*, or *Schubfachprinzip*), he showed an extension:

(6)   *if* $\alpha_1, \ldots, \alpha_n, \varepsilon \in \mathbb{R}$ *with* $0 < \varepsilon < 1$, *there exist integers* $p_1, \ldots, p_n, q$ *such that* $|\alpha_1 p_1 + \cdots + \alpha_n p_n - q| < \varepsilon$ *and* $1 \leqslant |p_i| \leqslant \varepsilon^{-1/n}$ *for* $i = 1, \ldots, n$.

Moreover, Dirichlet says that analogous assertions hold for two or more simultaneous equations.

Hermite

Hermite [1849] showed that if $\alpha_1, \ldots, \alpha_n$ are relatively prime integers, then there exists an integral matrix A with $\det A = 1$ and with first row $(\alpha_1, \ldots, \alpha_n)$.

In a series of letters to Jacobi, published in the *Journal für die reine und angewandte Mathematik*, Hermite [1850a] stated some basic results on quadratic forms, approximation and diophantine equations. In his first letter Hermite showed:

(7)   *for each positive definite matrix D of order n there exists a nonzero integral vector x such that* $x^{\mathsf{T}} D x \leqslant (\tfrac{4}{3})^{(n-1)/2} (\det D)^{1/n}$.

This implies

(8)   each lattice Λ in n dimensions contains a nonzero vector x with $\|x\| \leq (\frac{4}{3})^{(n-1)/4}(\det \Lambda)^{1/n}$.

This extends the earlier results of Lagrange, Gauss, and Seeber for the cases $n = 2$ and 3. Hermite's proof is based on an interesting induction argument, which in fact yields a (non-polynomial-time) algorithm for finding x as in (8): given $x_0 \in \Lambda \setminus \{0\}$, let $\Omega := \{y \in \mathbb{R}^n \,|\, y^\mathsf{T}x \in \mathbb{Z}$ for each $x \in \Lambda; \; y^\mathsf{T}x_0 = 0\}$. So Ω is a lattice in $n - 1$ dimensions, and we can hence find $y_0 \in \Omega$ such that $\|y_0\| \leq (\frac{4}{3})^{(n-2)/4}(\det \Omega)^{1/(n-1)}$. Next find a nonzero vector x_1 in the lattice $K := \Lambda \cap \{x \,|\, y_0^\mathsf{T}x = 0\}$ with $\|x_1\| \leq (\frac{4}{3})^{(n-2)/4}(\det K)^{1/(n-1)}$ (which is again a problem in $n - 1$ dimensions). It is not difficult to show $\det \Omega \leq \|x_0\| \cdot (\det \Lambda)^{-1}$ and $\det K \leq \|y_0\| \det \Lambda$, and hence:

(9) $\|x_1\| \leq (\frac{4}{3})^{(n^2-2n)/(4n-4)}(\det \Lambda)^{(n-2)/(n-1)^2}\|x_0\|^{1/(n-1)^2}$.

Now replace x_0 by x_1, and start anew. (9) implies that by repeating this we finally find a vector x described in (8). (If $x_0 = x_1$ in (9) we obtain the inequality (8).) It is not difficult to see that this algorithm in fact finds a nonzero vector x in Λ with $\|x\| \leq (\frac{4}{3} + \varepsilon)^{(n-1)/4}(\det \Lambda)^{1/n}$, in time which is, *for each fixed* n, bounded by a polynomial in the size of the given generator for Λ and in $1/\varepsilon$.

Hermite mentions as an application of (8):

(10)   *if* $\alpha_1, \alpha_2, \varepsilon \in \mathbb{R}$ *with* $\varepsilon > 0$, *there exist integers* p_1, p_2, q *such that* $|p_1 - \alpha_1 q| < \varepsilon$, $|p_2 - \alpha_2 q| < \varepsilon$, *and* $1 \leq q \leq \frac{8}{9}\sqrt{3} \cdot \varepsilon^{-2}$.

as follows from (8) by taking the lattice generated by

(11) $$\begin{bmatrix} 1 \\ 0 \\ 0 \end{bmatrix}, \quad \begin{bmatrix} 0 \\ 1 \\ 0 \end{bmatrix}, \quad \begin{bmatrix} -\alpha_1 \\ -\alpha_2 \\ \frac{3}{8}\varepsilon^3 \sqrt{3} \end{bmatrix}.$$

Extending a definition given by Gauss, Hermite defines (recursively) a positive definite $n \times n$-matrix $D = (\delta_{ij})$ to be *reduced* if

(12)   (i) $\delta_{11} \leq (\frac{4}{3})^{(n-1)/2}(\det D)^{1/n}$ and $|\delta_{12}|, \ldots, |\delta_{1n}| < \frac{1}{2}\delta_{11}$;
         (ii) the positive definite $(n-1) \times (n-1)$-matrix D' obtained from D^{-1} by deleting the first row and first column is reduced.

Hermite showed that each positive definite matrix is equivalent to a reduced matrix, and that, if we fix Δ, there are only finitely many reduced integral positive definite matrices with determinant Δ. A consequence is that there are only finitely many equivalence classes of integral positive definite matrices of given determinant, a result proved earlier by Eisenstein [1847].

In the second letter, Hermite showed:

(13)   *each positive definite* $n \times n$-matrix D *is equivalent to a matrix whose diagonal entries have product at most* $(\frac{4}{3})^{n(n-1)} \det D$.

Equivalently:

(14)   *each full-dimensional lattice* Λ *in* $\mathbb{R}^n$ *has a basis* $b_1, \ldots, b_n$ *such that* $\|b_1\| \cdot \cdots \cdot \|b_n\| \leq (\frac{4}{3})^{n(n-1)/4} \det \Lambda$.

In the third letter, Hermite states without proof:

(15)   *in each full-dimensional lattice* Λ *in* $\mathbb{R}^n$, *any basis* $b_1, \ldots, b_n$ *for which* $(\|b_1\|, \ldots, \|b_n\|)$ *is lexicographically minimal satisfies* $\|b_1\| \cdot \cdots \cdot \|b_n\| \leq c(n) \det \Lambda$, *where* $c(n)$ *only depends on* n.

(For a proof, see Stouff [1902].)

Hermite [1851] states as an auxiliary remark, without proof, that each rational matrix can be transformed by elementary column operations to a form which is now known as the Hermite normal form. The main topic of the paper is again the reduction of quadratic forms.

As we saw in Corollary 4.1a, the Hermite normal form implies:

(16)   *A rational system $Ax = b$ of linear equations has an integral solution x if and only if yb is an integer for each rational row vector y with yA integral.*

Heger and Smith on systems of linear diophantine equations

Heger [1856] showed that a system of *two* rational linear diophantine equations $Ax = b$ has a solution if and only if the g.c.d. of the 2×2-subdeterminants of the matrix A divides the 2×2-subdeterminants of the matrix $[A \; b]$. He also gave a formula for the general solution of two linear diophantine equations, mentioning that the process is essentially the same for systems with more equations. This is elaborated in Heger [1858], where also the above statement is extended to systems of m linear diophantine equations.

Smith [1861] gives a review on integral matrices and linear diophantine equations, and he derived several new results. He studies unimodular matrices (called *prime matrices*). Unknown of Heger's results, he showed Heger's characterization again. (In a footnote, Smith admits priority to Heger.) Moreover, Smith showed that each rational matrix can be brought into the form now called Smith normal form—see Section 4.4. Smith also proved that the set of integral solutions of a system $Ax = b$ of rational linear equations can be described as $\{x_0 + \lambda_1 x_1 + \cdots + \lambda_t x_t | \lambda_1, \ldots, \lambda_t \in \mathbb{Z}\}$, where $x_1, \ldots, x_t$ are linearly independent and $t = $ (number of variables) $-$ rank (A) (assuming $Ax = b$ has at least one integral solution). Smith's work was extended by Frobenius [1879, 1880].

Chebyshev, Kronecker, and Hurwitz on approximation

Chebyshev [1866] showed that if α and β are real numbers such that α/β is irrational, then there exist infinitely many pairs p, q of integers such that $|\alpha q - p - \beta| < 2/|q|$, which upper bound was reduced to $1/2|q|$ by Hermite [1880].

Kronecker elaborated the applications of Dirichlet's pigeon-hole principle (5) to approximation. Kronecker [1883] showed that if $Ax = b$ is a system of real linear equations, and $t > 0$, then $Ax = b$ can be satisfied with an approximation 'of the order of' $t^{-n/m}$ (m and n being the numbers of rows and columns of A), by a vector x having numerators and denominators 'of the same order as' t. His method implies that if A is an $m \times n$-matrix with $m > n$, then for each $\varepsilon > 0$ there exists a nonzero integral vector x with $\| Ax \| < \varepsilon$. Another proof of this last statement is given in Kronecker [1884a]. In this paper also the elimination of variables and constraints in linear diophantine equations is studied.

Finally, Kronecker [1884b] combines many of the results above into the following famous theorem:

(17)   (Kronecker's approximation theorem) *Let A be a real $m \times n$-matrix and let $b \in \mathbb{R}^n$. Then the following are equivalent:*
         (i) *for each $\varepsilon > 0$ there is an $x \in \mathbb{Z}^n$ with $\| Ax - b \| < \varepsilon$;*
         (ii) *for each $y \in \mathbb{R}^m$, if $yA \in \mathbb{Z}^n$ then $yb \in \mathbb{Z}$.*

(Actually, Kronecker used the norm $\| \cdot \|_\infty$, but this is clearly irrelevant.) It is easy to see that (16) above is a corollary. (See also Perron [1921: Chapter V]. Several authors have re-proved (variants of) (17), e.g. Châtelet [1913: pp. 130–137; 1914], Giraud [1914], Kakeya [1913–4], and Riesz [1904].)

Hurwitz [1891] showed that for any irrational number α there exist infinitely many pairs of integers p, q with

$$(18) \qquad \left| \alpha - \frac{p}{q} \right| < \frac{1}{\sqrt{5}} \cdot \frac{1}{q^2}$$

and that the factor $1/\sqrt{5}$ is best possible (as is shown by $\alpha = \frac{1}{2} + \frac{1}{2}\sqrt{5}$). In fact, infinitely many convergents p/q in the continued fraction method satisfy (18).

Poincaré [1880] studied the elementary properties of lattices in two dimensions (extended by Châtelet [1913] to more dimensions).

Minkowski's geometry of numbers

At the end of the 19th century, Minkowski found several extensions and sharpenings of the results above, by his *geometry of numbers*. The following result, presented in Minkowski [1893], is basic:

(19)   *if $C \subseteq \mathbb{R}^n$ is a full-dimensional compact convex set, symmetric with respect to the origin, with volume at least 2^n, then C contains a nonzero integral vector.*

(*Symmetric with respect to the origin* means $-x \in C$ if $x \in C$.) The proof is based on measuring volumes, as an extension of the pigeon-hole principle: suppose C contains no integral vector other than the origin. Then the sets $x + \frac{1}{2}C$, for $x \in \mathbb{Z}^n$, are pairwise disjoint. However:

$$(20) \qquad \mathrm{vol}\left(\bigcup_{x \in \mathbb{Z}^n} (x + \tfrac{1}{2}C) \cap [0,1]^n \right) = \mathrm{vol}(\tfrac{1}{2}C) \geqslant 1 = \mathrm{vol}\,[0,1]^n$$

which is a contradiction (as $[0,1]^n$ is connected).

(19) is equivalent to:

(21)   *for any norm $\|\cdot\|$ on $\mathbb{R}^n$, there exists a nonzero integral vector x such that $\|x\| \leqslant 2 \cdot U_n^{-1/n}$, where U_n is the volume of $\{y \mid \|y\| \leqslant 1\}$.*

In particular, it implies a result of Minkowski [1891]:

(22)   *for any positive definite matrix D there exists a nonzero integral vector x such that $x^T D x \leqslant 4 \cdot V_n^{-2/n}(\det D)^{1/n}$,*

where V_n denotes the volume of the n-dimensional unit ball. As $4 \cdot V_n^{-2/n} \approx 2n/\pi e$, this sharpens Hermite's result (7) considerably.

[If we define γ_n as the smallest number which can replace the factor $4 \cdot V_n^{-2/n}$ in (22), then by Minkowski's result, $\gamma_n \lesssim 2n/\pi e$, while Minkowski [1893] showed $\gamma_n \gtrsim n/2\pi e$ (here $\gamma_n \lesssim 2n/\pi e$ means $\limsup_{n \to \infty}(\gamma_n/n) \leqslant 2/\pi e$; similarly for $\gtrsim$). So γ_n is of order linear in n. Blichfeldt [1914] showed that $\gamma_n \lesssim n/\pi e$, and recently Kabatyanskiĭ and Levenshteĭn [1978] that $\gamma_n \lesssim 0.872n/\pi e$. It is conjectured by several authors that $\lim_{n \to \infty}(\gamma_n/n) = 1/2\pi e$. At the moment the following values of γ_n are known exactly: $\gamma_1 = 1$, $\gamma_2 = 2/\sqrt{3}$, $\gamma_3 = \sqrt[3]{2}$, $\gamma_4 = \sqrt{2}$, $\gamma_5 = \sqrt[5]{8}$, $\gamma_6 = \sqrt[6]{64/3}$, $\gamma_7 = \sqrt[7]{64}$, $\gamma_8 = 2$ (as was shown by Korkine and Zolotareff [1872, 1973, 1877] and Blichfeldt [1935] (cf. Watson [1966])). See Chaundy [1946], Coxeter [1947], and Barnes [1959] for work on γ_9 and γ_{10}.]

In his famous book *Geometrie der Zahlen*, Minkowski [1896] gives a review of the above results and he derives some new theorems. In Section 37, he applies (21) to the norm $\|\cdot\|_\infty$:

(23)   *for each nonsingular matrix A there exists a nonzero integral vector x such that $\|Ax\|_\infty \leqslant (\det A)^{1/n}$.*

This *linear forms theorem* implies result (6) of Dirichlet.

Another corollary of (23) concerns simultaneous diophantine approximation, which was also considered by Dirichlet and Hermite (cf. 10)):

(24)   *if* $\alpha_1, \ldots, \alpha_n, \varepsilon \in \mathbb{R}, 0 < \varepsilon \leqslant 1$, *then there exist* $p_1, \ldots, p_n q \in \mathbb{Z}$ *with* $1 \leqslant q \leqslant \varepsilon^{-n}$ *and* $|\alpha_i q - p_i| < \varepsilon$ *for* $i = 1, \ldots, n$

(Minkowski [1896: pp. 110–112]). Finally, Minkowski [1896: Sec. 50] showed:

(25)   *each lattice* Λ *in* $\mathbb{R}^n$ *contains linearly independent vectors* $b_1, \ldots, b_n$ *such that* $\|b_1\| \cdots \|b_n\| \leqslant 2^n V_n^{-1} \det \Lambda$.

This closes our historical review. Summarizing, we saw two main proof methods: iterative methods (Euclidean algorithm, continued fractions, Hermite's method), and packing methods (pigeon-hole principle, geometry of numbers). Roughly speaking, the latter methods give better bounds, but the first yield better algorithms, which have culminated in Lovász's *basis reduction method* (see Section 6.2 above).

FURTHER NOTES ON LATTICES AND LINEAR DIOPHANTINE EQUATIONS

Classical books on the subjects treated in this part are: for the reduction theory of quadratic forms: Dirichlet [1879], Bachmann [1898, 1923]; for the geometry of numbers: Minkowski [1896]; for continued fractions: Perron [1913] (including historical notes); for diophantine approximation: Minkowski [1907], Koksma [1936]; for diophantine equations: Skolem [1938].

Other and more recent books are: Bachmann [1902–1910] (Euclidean algorithm, continued fractions), Bergström [1935] (lattices, diophantine approximation), Cassels [1957] (diophantine approximation), Cassels [1959] (lattices, reduction, geometry of numbers), Châtelet [1913] (lattices, linear diophantine equations, reduction), Cohn [1962] (lattices, quadratic forms), Davenport [1952] (continued fractions, quadratic forms), Dickson [1929, 1939] (quadratic forms—introductory), Dickson [1930] (quadratic forms), Hancock [1939] (geometry of numbers, diophantine approximation), Hardy and Wright [1979] (Euclidean algorithm, continued fractions, diophantine approximation, geometry of numbers), Henrici [1977] (continued fractions), Hua [1982] (continued fractions, unimodular and integral matrices, geometry of numbers, quadratic forms), Jones and Thron [1980] (continued fractions), Khintchine [1956] (continued fractions), Knuth [1969: pp. 316–364] (Euclidean algorithm), Lang [1966b] (diophantine approximation), Lekkerkerker [1969] (lattices, reduction, geometry of numbers), Mordell [1969] (diophantine equations), Newman [1972] (integral matrices, Hermite and Smith normal form, introduction to lattices and geometry of numbers), Niven [1967] (continued fractions, diophantine approximation—introductory), Nonweiler [1984: Ch. 2] (continued fractions—numerical applications), Olds [1963] (continued fractions), Perron [1921] (diophantine approximation), Schmidt [1980] (diophantine approximation), Stark [1970] (Euclidean algorithm, continued fractions—introductory), Uspensky and Heaslet [1939: Ch. III] (Euclidean algorithm—introductory), and Wall [1948] (continued fractions). For a survey on algorithmic aspects, see Frumkin [1982]. Van der Waerden [1956] surveys the reduction theory of positive quadratic forms. Lagarias [1980b] studies the worst-case complexity of algorithms for problems like reducing a quadratic form.

Hammer [1977] collected unsolved problems concerning lattice points.

For historical accounts, see Dickson [1920: Ch. II] (linear diophantine equations), Dickson [1923] (quadratic forms), Smith [1925: pp. 418–421] (continued fractions), Jones and Thron [1980: pp. 1–11] (continued fractions), Scharlau and Opolka [1980], Kloyda [1937] (linear and quadratic equations 1550–1660), and Brezinski [1981] (continued fractions).

Part III

Polyhedra, linear inequalities, and linear programming

The concepts mentioned in the title of this part belong to the central objects of this book. In a sense they form three sides of one and the same problem field. Polyhedra represent its geometric side, linear inequalities the algebraic side, and linear programming is the field in its optimizational form.

In Chapters 7–9 we discuss polyhedra, linear inequalities, and linear programming theoretically, while in Chapters 10–15 we go into the algorithmic aspects, including the theoretical complexity.

In Chapter 7 we discuss some fundamental concepts and results. As a basic result we give a theorem due to Farkas, Minkowski, Carathéodory, and Weyl. This theorem implies several other fundamental theorems, like Farkas' lemma, the Duality theorem of linear programming, the Decomposition theorem for polyhedra, and the Finite basis theorem for polytopes.

In Chapter 8 we discuss the geometric structure of polyhedra, studying objects like faces, facets, vertices, extremal rays, the characteristic cone, and the decomposition of polyhedra.

In Chapter 9 we deal with polarity and with blocking and anti-blocking polyhedra. Polarity is a classical geometric phenomenon, implying a certain duality relation between polyhedra. Related, and more recent, are the concepts of blocking and anti-blocking polyhedra, introduced by D. R. Fulkerson as a useful tool in polyhedral combinatorics.

In Chapter 10 we discuss the theoretical complexity of linear inequalities and linear programming. We make estimates on the sizes of solutions, and we derive that the characterizations given in Chapter 7 yield good characterizations, and that several problems on linear inequalities and linear programming are polynomially equivalent. These are prerequisites for Chapters 11–15 on solving linear inequalities and linear programming algorithmically.

In Chapter 11 we discuss the basic simplex method for linear programming, which is (for the present pivot rules) not a polynomial-time method, but which is very efficient in practice. In Chapter 12 we go into some further methods: primal–dual methods, the Fourier–Motzkin elimination method, and the relaxation method.

In Chapters 13 and 14 we discuss the ellipsoid method. In Chapter 13 we show

Khachiyan's result that the ellipsoid method solves linear programs in polynomial time. In Chapter 14 the ellipsoid method is extended to apply to more general problems, allowing application to combinatorial optimization problems.

In Chapter 15 we describe some further polynomiality results for linear programming: Karmarkar's polynomial-time algorithm, Tardos' work on strongly polynomial algorithms, Megiddo's linear-time algorithm for linear programming in fixed dimension, and algorithms for approximating polytopes by ellipsoids.

This part concludes with notes on history and literature.

7

Fundamental concepts and results on polyhedra, linear inequalities, and linear programming

In this chapter we first state a fundamental theorem on linear inequalities (Section 7.1), and next we derive as consequences some other important results, like the Finite basis theorem for cones and polytopes, the Decomposition theorem for polyhedra (Section 7.2), Farkas' lemma (Section 7.3), the Duality theorem of linear programming (Section 7.4), an affine form of Farkas' lemma (Section 7.6), Carathéodory's theorem (Section 7.7), and results for strict inequalities (Section 7.8). In Section 7.5 we give a geometrical interpretation of LP-duality. In Section 7.9 we study the phenomenon of *complementary slackness*.

Each of the results in this chapter holds both in real spaces and in rational spaces. In the latter case, all numbers occurring (like matrix and vector entries, variables) are restricted to the rationals.

7.1. THE FUNDAMENTAL THEOREM OF LINEAR INEQUALITIES

The fundamental theorem is due to Farkas [1894, 1898a] and Minkowski [1896], with sharpenings by Carathéodory [1911] and Weyl [1935]. Its geometric content is easily understood in three dimensions.

Theorem 7.1 (Fundamental theorem of linear inequalities). *Let $a_1, \ldots, a_m$, b be vectors in n-dimensional space. Then:*

either   I. *b is a nonnegative linear combination of linearly independent vectors from $a_1, \ldots, a_m$;*

or   II. *there exists a hyperplane* $\{x \mid cx = 0\}$, *containing* $t - 1$ *linearly independent vectors from* $a_1, \ldots, a_m$, *such that* $cb < 0$ *and* $ca_1, \ldots, ca_m \geqslant 0$, *where* $t := rank\{a_1, \ldots, a_m, b\}$.

Proof. We may assume that $a_1, \ldots, a_m$ span the n-dimensional space.

Clearly, I and II exclude each other, as otherwise, if $b = \lambda_1 a_1 + \cdots + \lambda_m a_m$ with $\lambda_1, \ldots, \lambda_m \geqslant 0$, we would have the contradiction

(1) $0 > cb = \lambda_1 \cdot ca_1 + \cdots + \lambda_m \cdot ca_m \geqslant 0.$

To see that at least one of I and II holds, choose linearly independent $a_{i_1}, \ldots, a_{i_n}$ from $a_1, \ldots, a_m$, and set $D := \{a_{i_1}, \ldots, a_{i_n}\}$. Next apply the following iteration:

(2)   (i) Write $b = \lambda_{i_1} a_{i_1} + \cdots + \lambda_{i_n} a_{i_n}$. If $\lambda_{i_1}, \ldots, \lambda_{i_n} \geqslant 0$, we are in case I.

    (ii) Otherwise, choose the smallest h among $i_1, \ldots, i_n$ with $\lambda_h < 0$. Let $\{x \mid cx = 0\}$ be the hyperplane spanned by $D \backslash \{a_h\}$. We normalize c so that $ca_h = 1$. [Hence $cb = \lambda_h < 0$.]

    (iii) If $ca_1, \ldots, ca_m \geqslant 0$ we are in case II.

    (iv) Otherwise, choose the smallest s such that $ca_s < 0$. Then replace D by $(D \backslash \{a_h\}) \cup \{a_s\}$, and start the iteration anew.

We are finished if we have shown that this process terminates. Let D_k denote the set D as it is in the kth iteration. If the process does not terminate, then $D_k = D_l$ for some $k < l$ (as there are only finitely many choices for D). Let r be the highest index for which a_r has been removed from D at the end of one of the iterations $k, k+1, \ldots, l-1$, say in iteration p. As $D_k = D_l$, we know that a_r also has been added to D in some iteration q with $k \leqslant q < l$. So

(3) $D_p \cap \{a_{r+1}, \ldots, a_m\} = D_q \cap \{a_{r+1}, \ldots, a_m\}.$

Let $D_p = \{a_{i_1}, \ldots, a_{i_n}\}$, $b = \lambda_{i_1} a_{i_1} + \cdots + \lambda_{i_n} a_{i_n}$, and let c' be the vector c found in (ii) of iteration q. Then we have the contradiction:

(4) $0 > c'b = c'(\lambda_{i_1} a_{i_1} + \cdots + \lambda_{i_n} a_{i_n}) = \lambda_{i_1} c' a_{i_1} + \cdots + \lambda_{i_n} c' a_{i_n} > 0.$

The first inequality was noted in (2) (ii) above. The last inequality follows from:

(5)   if $i_j < r$ then $\lambda_{i_j} \geqslant 0, c'a_{i_j} \geqslant 0$ $\left.\right\}$ $\left\{\begin{array}{l}\text{(as by (2) (ii), } r \text{ is the smallest index with } \lambda_r \\ < 0; \text{ similarly, by (2) (iv) } r \text{ is the smallest} \\ \text{index with } c'a_r < 0)\end{array}\right.$

    if $i_j = r$ then $\lambda_{i_j} < 0, c'a_{i_j} < 0$

    if $i_j > r$ then $c'a_{i_j} = 0$ (by (3) and (2) (ii)). $\square$

The above proof of this fundamental theorem also gives a fundamental algorithm: it is a disguised form of the famous *simplex method*, with *Bland's rule* incorporated—see Chapter 11 (see Debreu [1964] for a similar proof, but with a lexicographic rule).

7.2. CONES, POLYHEDRA, AND POLYTOPES

We first derive some geometrical consequences from the Fundamental theorem. A nonempty set C of points in Euclidean space is called a (*convex*) *cone* if $\lambda x + \mu y \in C$ whenever $x, y \in C$ and $\lambda, \mu \geq 0$. A cone C is *polyhedral* if

$$(6) \qquad C = \{x \mid Ax \leq 0\}$$

for some matrix A, i.e. if C is the intersection of finitely many linear half-spaces. Here a *linear half-space* is a set of the form $\{x \mid ax \leq 0\}$ for some nonzero row vector a. The cone *generated by* the vectors $x_1, \ldots, x_m$ is the set

$$(7) \qquad \text{cone}\{x_1, \ldots, x_m\} := \{\lambda_1 x_1 + \cdots + \lambda_m x_m \mid \lambda_1, \ldots, \lambda_m \geq 0\},$$

i.e. it is the smallest convex cone containing $x_1, \ldots, x_m$. A cone arising in this way is called *finitely generated*.

It now follows from the Fundamental theorem that for cones the concepts of 'polyhedral' and 'finitely generated' are equivalent (Farkas [1898a, 1902], Minkowski [1896], Weyl [1935]).

Corollary 7.1a (Farkas–Minkowski–Weyl theorem). *A convex cone is polyhedral if and only if it is finitely generated.*

Proof. I. To prove sufficiency, let $x_1, \ldots, x_m$ be vectors in $\mathbb{R}^n$. We show that cone$\{x_1, \ldots, x_m\}$ is polyhedral. We may assume that $x_1, \ldots, x_m$ span $\mathbb{R}^n$ (as we can extend any linear half-space H of lin.hull$\{x_1, \ldots, x_m\}$ to a linear half-space H' of $\mathbb{R}^n$ such that $H' \cap \text{lin.hull}\{x_1, \ldots, x_m\} = H$). Now consider all linear half-spaces $H = \{x \mid cx \leq 0\}$ of $\mathbb{R}^n$ such that $x_1, \ldots, x_m$ belong to H and such that $\{x \mid cx = 0\}$ is spanned by $n - 1$ linearly independent vectors from $x_1, \ldots, x_m$. By Theorem 7.1, cone $\{x_1, \ldots, x_m\}$ is the intersection of these half-spaces. Since there are only finitely many such half-spaces, the cone is polyhedral.

II. Part I above also yields the converse implication. Let C be a polyhedral cone, say $C = \{x \mid a_1^\mathsf{T} x \leq 0, \ldots, a_m^\mathsf{T} x \leq 0\}$ for certain column vectors $a_1, \ldots, a_m$. As by I above, each finitely generated cone is polyhedral, there exist column vectors $b_1, \ldots, b_t$ such that

$$(8) \qquad \text{cone}\{a_1, \ldots, a_m\} = \{x \mid b_1^\mathsf{T} x \leq 0, \ldots, b_t^\mathsf{T} x \leq 0\}.$$

We show that $C = \text{cone}\{b_1, \ldots, b_t\}$, implying that C is finitely generated.

Indeed, cone $\{b_1, \ldots, b_t\} \subseteq C$, as $b_1, \ldots, b_t \in C$, since $b_j^\mathsf{T} a_i \leq 0$ for $i = 1, \ldots, m$ and $j = 1, \ldots, t$, by (8). Suppose $y \notin \text{cone}\{b_1, \ldots, b_t\}$ for some $y \in C$. By Part I, cone$\{b_1, \ldots, b_t\}$ is polyhedral, and hence there exists a vector w such that $w^\mathsf{T} b_1, \ldots, w^\mathsf{T} b_t \leq 0$ and $w^\mathsf{T} y > 0$. Hence by (8), $w \in \text{cone}\{a_1, \ldots, a_m\}$, and hence $w^\mathsf{T} x \leq 0$ for all x in C. But this contradicts the facts that y is in C and $w^\mathsf{T} y > 0$. $\square$

A set P of vectors in $\mathbb{R}^n$ is called a (*convex*) *polyhedron* if

$$(9) \qquad P = \{x \mid Ax \leq b\}$$

for some matrix A and vector b, i.e. if P is the intersection of finitely many affine half-spaces. Here an *affine half-space* is a set of the form $\{x \mid wx \leqslant \delta\}$ for some nonzero row vector w and some number δ. If (9) holds, we say that $Ax \leqslant b$ *defines* or *determines* P. Trivially, each polyhedral cone is a polyhedron.

A set of vectors is a *(convex) polytope* if it is the convex hull of finitely many vectors.

It is intuitively obvious that the concepts of polyhedron and of polytope are related. This is made more precise in the Decomposition theorem for polyhedra (Motzkin [1936]) and in its direct corollary, the Finite basis theorem for polytopes (Minkowski [1896], Steinitz [1916], Weyl [1935]).

Corollary 7.1b (Decomposition theorem for polyhedra). *A set P of vectors in Euclidean space is a polyhedron, if and only if $P = Q + C$ for some polytope Q and some polyhedral cone C.*

Proof. I. First let $P = \{x \mid Ax \leqslant b\}$ be a polyhedron in $\mathbb{R}^n$. By Corollary 7.1a the polyhedral cone

$$(10) \qquad \left\{ \begin{pmatrix} x \\ \lambda \end{pmatrix} \middle| x \in \mathbb{R}^n; \lambda \in \mathbb{R}; \lambda \geqslant 0; Ax - \lambda b \leqslant 0 \right\}$$

is generated by finitely many vectors, say by $\begin{pmatrix} x_1 \\ \lambda_1 \end{pmatrix}, \ldots, \begin{pmatrix} x_m \\ \lambda_m \end{pmatrix}$. We may assume that each λ_i is 0 or 1. Let Q be the convex hull of the x_i with $\lambda_i = 1$, and let C be the cone generated by the x_i with $\lambda_i = 0$. Now $x \in P$, if and only if $\begin{pmatrix} x \\ 1 \end{pmatrix}$ belongs to (10), and hence, if and only if $\begin{pmatrix} x \\ 1 \end{pmatrix} \in \text{cone} \left\{ \begin{pmatrix} x_1 \\ \lambda_1 \end{pmatrix}, \ldots, \begin{pmatrix} x_m \\ \lambda_m \end{pmatrix} \right\}$. It follows directly that $P = Q + C$.

II. Let $P = Q + C$ for some polytope Q and some polyhedral cone C. Say $Q = \text{conv.hull} \{x_1, \ldots, x_m\}$ and $C = \text{cone} \{y_1, \ldots, y_t\}$. Then a vector x_0 belongs to P, if and only if

$$(11) \qquad \begin{pmatrix} x_0 \\ 1 \end{pmatrix} \in \text{cone} \left\{ \begin{pmatrix} x_1 \\ 1 \end{pmatrix}, \ldots, \begin{pmatrix} x_m \\ 1 \end{pmatrix}, \begin{pmatrix} y_1 \\ 0 \end{pmatrix}, \ldots, \begin{pmatrix} y_t \\ 0 \end{pmatrix} \right\}.$$

By Corollary 7.1a, the cone in (11) is equal to $\left\{ \begin{pmatrix} x \\ \lambda \end{pmatrix} \middle| Ax + \lambda b \leqslant 0 \right\}$ for some matrix A and vector b. Hence $x_0 \in P$, if and only if $Ax_0 \leqslant -b$, and therefore P is a polyhedron. $\square$

We shall say that P is *generated by the points* $x_1, \ldots, x_m$ and *by the directions* $y_1, \ldots, y_t$ if.

$$(12) \qquad P = \text{conv.hull} \{x_1, \ldots, x_m\} + \text{cone} \{y_1, \ldots, y_t\}.$$

This gives a 'parametric' description of the solution set of a system of linear inequalities. For more about decomposition of polyhedra, see Section 8.9 below.

The Finite basis theorem for polytopes can be derived from the Decomposition theorem. It is usually attributed to Minkowski [1896], Steinitz [1916] and Weyl [1935]. ('This classical result is an outstanding example of a fact which is completely obvious to geometric intuition, but which wields important algebraic content and is not trivial to prove'—R.T. Rockafellar.)

Corollary 7.1c (Finite basis theorem for polytopes). *A set P is a polytope if and only if P is a bounded polyhedron.*

Proof. Directly from Corollary 7.1b. $\square$

7.3. FARKAS' LEMMA AND VARIANTS

Another consequence of Theorem 7.1 is the well-known Farkas' lemma, proved first by Farkas [1894, 1898a] and Minkowski [1896].

Corollary 7.1d (Farkas' lemma). *Let A be a matrix and let b be a vector. Then there exists a vector $x \geq 0$ with $Ax = b$, if and only if $yb \geq 0$ for each row vector y with $yA \geq 0$.*

Proof. The necessity of the condition is trivial, as $yb = yAx \geq 0$ for all x and y with $x \geq 0$, $yA \geq 0$, and $Ax = b$. To prove sufficiency, suppose there is no $x \geq 0$ with $Ax = b$. Let $a_1, \ldots, a_m$ be the columns of A. Then $b \notin \text{cone}\{a_1, \ldots, a_m\}$, and hence, by Theorem 7.1, $yb < 0$ for some y with $yA \geq 0$. $\square$

Farkas' lemma is equivalent to: if the linear inequalities $a_1 x \leq 0, \ldots, a_n x \leq 0$ imply the linear inequality $wx \leq 0$, then w is a nonnegative linear combination of $a_1, \ldots, a_n$ (thus providing a 'proof' of the implication).

Geometrically, Farkas' lemma is obvious: the content is that if a vector b does not belong to the cone generated by the vectors $a_1, \ldots, a_n$, there exists a linear hyperplane separating b from $a_1, \ldots, a_n$.

There are several other, equivalent, forms of Farkas' lemma, like those described by the following two corollaries.

Corollary 7.1e (Farkas' lemma (variant)). *Let A be a matrix and let b be a vector. Then the system $Ax \leq b$ of linear inequalities has a solution x, if and only if $yb \geq 0$ for each row vector $y \geq 0$ with $yA = 0$.*

Proof. Let A' be the matrix $[I \ A \ -A]$. Then $Ax \leq b$ has a solution x, if and only if $A'x' = b$ has a nonnegative solution x'. Application of Corollary 7.1d to the latter system yields Corollary 7.1e. $\square$

[Kuhn [1956a] showed that this variant of Farkas' lemma can be proved in a nice short way with the 'Fourier–Motzkin elimination method'—see Section 12.2.]

Corollary 7.1f (Farkas' lemma (variant)). *Let A be a matrix and let b be a vector. Then the system $Ax \leqslant b$ has a solution $x \geqslant 0$, if and only if $yb \geqslant 0$ for each row vector $y \geqslant 0$ with $yA \geqslant 0$.*

Proof. Let A' be the matrix $[I \;\; A]$. Then $Ax \leqslant b$ has a solution $x \geqslant 0$, if and only if $A'x' = b$ has a solution $x' \geqslant 0$. Application of Corollary 7.1d to the latter system yields Corollary 7.1f.   □

7.4. LINEAR PROGRAMMING

Linear programming, LP for short, concerns the problem of maximizing or minimizing a linear functional over a polyhedron. Examples are

$$(13) \qquad \max \{cx \,|\, Ax \leqslant b\}, \quad \min \{cx \,|\, x \geqslant 0; Ax \leqslant b\}.$$

There are traces of the idea of linear programming in Fourier [1826b, 1827], but its creation as a discipline, together with the recognition of its importance, came in the 1940's by the work of Dantzig, Kantorovich, Koopmans, and von Neumann.

We first derive the important Duality theorem of linear programming, due to von Neumann [1947] and Gale, Kuhn, and Tucker [1951], as a corollary of the Fundamental theorem of linear inequalities.

Corollary 7.1g (Duality theorem of linear programming). *Let A be a matrix, and let b and c be vectors. Then*

$$(14) \qquad \max \{cx \,|\, Ax \leqslant b\} = \min \{yb \,|\, y \geqslant 0; yA = c\}$$

provided that both sets in (14) are nonempty.

Proof. Clearly, if $Ax \leqslant b$ and $y \geqslant 0$, $yA = c$, then $cx = yAx \leqslant yb$. So if the optima in (14) are finite, we have the inequality max $\leqslant$ min. So it suffices to show that there exist x, y such that $Ax \leqslant b$, $y \geqslant 0$, $yA = c$, $cx \geqslant yb$, i.e. that

$$(15) \qquad \text{there are } x, y \text{ such that } y \geqslant 0 \text{ and } \begin{bmatrix} A & 0 \\ -c & b^{\mathsf{T}} \\ 0 & A^{\mathsf{T}} \\ 0 & -A^{\mathsf{T}} \end{bmatrix} \begin{pmatrix} x \\ y^{\mathsf{T}} \end{pmatrix} \leqslant \begin{pmatrix} b \\ 0 \\ c^{\mathsf{T}} \\ -c^{\mathsf{T}} \end{pmatrix}.$$

By Corollary 7.1e, (15) is equivalent to:

$$(16) \qquad \text{if } u, \lambda, v, w \geqslant 0 \text{ with } uA - \lambda c = 0 \text{ and } \lambda b^{\mathsf{T}} + vA^{\mathsf{T}} - wA^{\mathsf{T}} \geqslant 0, \text{ then } \\ ub + vc^{\mathsf{T}} - wc^{\mathsf{T}} \geqslant 0.$$

To show (16), let u, λ, v, w satisfy the premise of (16). If $\lambda > 0$, then $ub = \lambda^{-1} \cdot \lambda b^{\mathsf{T}} u^{\mathsf{T}} \geqslant \lambda^{-1}(w - v)A^{\mathsf{T}}u^{\mathsf{T}} = \lambda^{-1}\lambda(w - v)c^{\mathsf{T}} = (w - v)c^{\mathsf{T}}$. If $\lambda = 0$, let $Ax_0 \leqslant b$ and $y_0 \geqslant 0$, $y_0 A = c$ (x_0, y_0 exist, as both sets in (14) are assumed to be nonempty). Then $ub \geqslant uAx_0 = 0 \geqslant (w - v)A^{\mathsf{T}}y_0^{\mathsf{T}} = (w - v)c^{\mathsf{T}}$.   □

Equivalent formulations. There are several equivalent forms for a linear programming problem. For example, all the following problem types are equivalent in the sense of being reducible to each other (A is a matrix, b a column vector, and c a row vector):

(17)   (i) $\max\{cx\,|\,Ax \leqslant b\}$
         (ii) $\max\{cx\,|\,x \geqslant 0, \, Ax \leqslant b\}$
         (iii) $\max\{cx\,|\,x \geqslant 0, \, Ax = b\}$
         (iv) $\min\{cx\,|\,Ax \geqslant b\}$
         (v) $\min\{cx\,|\,x \geqslant 0, \, Ax \geqslant b\}$
         (vi) $\min\{cx\,|\,x \geqslant 0, \, Ax = b\}$.

Thus (ii) is reducible to (i) as $x \geqslant 0$, $Ax \leqslant b$ is equivalent to $\begin{bmatrix} -I \\ A \end{bmatrix} x \leqslant \begin{pmatrix} 0 \\ b \end{pmatrix}$.

(iii) is reducible to (ii) by replacing $Ax = b$ by $Ax \leqslant b$, $-Ax \leqslant -b$. (i) is reducible to (iii) by replacing $Ax \leqslant b$ by $x', x'', \tilde{x} \geqslant 0$, $Ax' - Ax'' + \tilde{x} = b$, i.e. by

(18) $z \geqslant 0, [A \ -A \ I]z = b.$

Similarly, (iv), (v), and (vi) are equivalent. Moreover, (iii) and (vi) are equivalent by replacing c by $-c$. So any method solving one of the problems (17) can be easily modified to solve the other. We shall call each of the problems (17), and each other problem equivalent to (17) in a similar way, a *general* LP-problem.

Also the Duality theorem of linear programming has equivalent formulations, which are shown in a similar way. For example, one has:

(19) $\max\{cx\,|\,x \geqslant 0, \, Ax \leqslant b\} = \min\{yb\,|\,y \geqslant 0, \, yA \geqslant c\},$
         $\max\{cx\,|\,x \geqslant 0, \, Ax = b\} = \min\{yb\,|\,yA \geqslant c\}$

provided that, in each case, both sets are nonempty. These min–max relations follow by replacing A in (14) by the matrices

(20) $\begin{bmatrix} -I \\ A \end{bmatrix}$ and $\begin{bmatrix} -I \\ A \\ -A \end{bmatrix}$, respectively.

A general formulation is as follows. Let a matrix, a column vector, and a row vector be given:

(21) $\begin{bmatrix} A & B & C \\ D & E & F \\ G & H & K \end{bmatrix}, \quad \begin{bmatrix} a \\ b \\ c \end{bmatrix}, \quad (d, e, f),$

where $A, B, C, D, E, F, G, H, K$ are matrices, a, b, c are column vectors, and d, e, f are row vectors. Then

(22) $\max\{dx + ey + fz\,|\,x \geqslant 0; \, z \leqslant 0; \, Ax + By + Cz \leqslant a;$
         $Dx + Ey + Fz = b; \, Gx + Hy + Kz \geqslant c\}$
         $= \min\{au + bv + cw\,|\,u \geqslant 0; \, w \leqslant 0; \, uA + vD + wG \geqslant d;$
         $uB + vE + wH = e; \, uC + vF + wK \leqslant f\}$

provided that both sets are nonempty.

To derive (22) from (14), observe that the maximum in (22) is equal to:

(23) $\max \{ dx + ey_1 - ey_2 - fz \mid x, y_1, y_2, z \geqslant 0; Ax + By_1 - By_2 - Cz \leqslant a;$
$Dx + Ey_1 - Ey_2 - Fz \leqslant b; -Dx - Ey_1 + Ey_2 + Fz \leqslant -b;$
$-Gx - Hy_1 + Hy_2 + Kz \leqslant -c \}.$

Now (14) equates this maximum to a certain minimum, which can be seen to be equal to the minimum in (22).

Some LP-terminology. Some standard linear programming terminology is as follows. Equations like (14) and (19), and the most general (22), are called *linear programming duality equations*. They all state that the optimum value of a certain maximization problem is equal to that of a certain minimization problem. The minimization problem is called the *dual* problem of the maximization problem (which problem then is called *primal*), and conversely.

In case of (14), any vector x satisfying $Ax \leqslant b$ is called a *feasible solution* (for the maximum). The set of feasible solutions is called the *feasible region* (which is a polyhedron). If the feasible region is nonempty, the problem is *feasible*, otherwise *infeasible*.

The function $x \rightarrow cx$ is called the *objective function* or *cost function*, and cx is the *objective value* or *cost* of x. A feasible solution with objective value equal to the maximum is called an *optimum* (or *optimal*) solution.

The words '*for the maximum*' are added if confusion may arise, as similar terminology is used with respect to the minimum in the duality equation, where we may add '*for the minimum*', or alternatively use the adjective *dual*.

Infeasibility and unboundedness. The following can be proved similarly to the Duality theorem of linear programming. *If A is a matrix and b and c are vectors, then*

(24) $\sup \{ cx \mid Ax \leqslant b \} = \inf \{ yb \mid y \geqslant 0, \quad yA = c \}$

provided that at least one of these sets is nonempty (cf. the proof of Corollary 7.1h below). Obviously, (24) is not true if both sets are empty (which can be the case: e.g. take $A = (0)$, and $b = c = (-1)$ (there are less trivial examples)).

Summarizing, there are four possible cases:

(25)   (i) both optima in (24) are finite, in which case they are equal, and attained as maximum and minimum;
(ii) both optima in (24) are $+\infty$;
(iii) both optima in (24) are $-\infty$;
(iv) the supremum in (24) is $-\infty$, and the infimum in (24) is $+\infty$.

In particular, if one optimum in (24) is finite, then also the other is. If $\max \{ cx \mid Ax \leqslant b \}$ is unbounded from above, we shall also say that the maximum is $+\infty$. If it is infeasible, we shall say that it is $-\infty$. Similarly for the minimum and for other linear programming duality equations.

7.5. LP-DUALITY GEOMETRICALLY

We briefly interpret linear programming duality geometrically. Consider the LP-problem

(26) $\max\{cx|Ax \leqslant b\}$.

Let $P:= \{x|Ax \leqslant b\}$ be the feasible region, which is a polyhedron. Finding the maximum (26) can be seen as shifting the hyperplane orthogonal to the vector c, as long as it contains points in P.

Suppose the maximum is finite, say its value is δ, and attained by the element x^* of P. Let $a_1x \leqslant \beta_1, \ldots, a_kx \leqslant \beta_k$ be the inequalities from $Ax \leqslant b$ satisfied with equality by x^*.

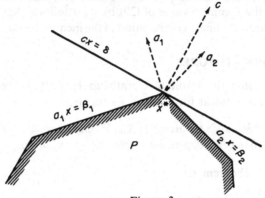

Figure 3

Now elementary geometric insight tells us that $cx = \delta$ is a *nonnegative* linear combination of $a_1x = \beta_1, \ldots, a_kx = \beta_k$ (cf. Figure 3). Say $c = \lambda_1a_1 + \cdots + \lambda_ka_k$ and $\delta = \lambda_1\beta_1 + \cdots + \lambda_k\beta_k$, with $\lambda_1 \ldots, \lambda_k \geqslant 0$. This implies:

(27) $\max\{cx|Ax \leqslant b\} = \delta = \lambda_1\beta_1 + \cdots + \lambda_k\beta_k \geqslant \min\{yb|y \geqslant 0, yA = c\}$

(the inequality follows since the λ_i give a feasible solution for the minimum). This yields the LP-duality equation

(28) $\max\{cx|Ax \leqslant b\} = \min\{yb|y \geqslant 0, yA = c\}$

as the inequality $\leqslant$ in (28) is trivial ($cx = yAx \leqslant yb$).

Note that also 'complementary slackness' (cf. Section 7.9) can be seen geometrically: if $ax^* < \beta$ for some inequality $ax \leqslant \beta$ from $Ax \leqslant b$, then the corresponding component in any optimal solution y^* for the minimum is zero.

We leave it to the reader to extend this insight to more general LP-problems.

7.6. AFFINE FORM OF FARKAS' LEMMA

Another consequence of the Fundamental theorem of linear inequalities is the following 'affine' form of Farkas' lemma (Haar [1918, 1924–6], Weyl [1935: Theorem 3]).

Corollary 7.1h. *Let the system $Ax \leqslant b$ of linear inequalities have at least one solution, and suppose that the linear inequality $cx \leqslant \delta$ holds for each x satisfying $Ax \leqslant b$. Then for some $\delta' \leqslant \delta$ the linear inequality $cx \leqslant \delta'$ is a nonnegative linear combination of the inequalities in the system $Ax \leqslant b$.*

Proof. Consider the linear programming duality equation

(29) $\max\{cx|Ax \leqslant b\} = \min\{yb|y \geqslant 0, yA = c\}$.

By assumption, the maximum has feasible solutions. Suppose the minimum has no feasible solutions. Then, by Farkas' lemma (Corollary 7.1d), there exists a vector z such that $Az \leqslant 0$ and $cz > 0$. So adding z to any feasible solution for the maximum increases its objective value by cz, contradicting the fact that $cx \leqslant \delta$ whenever $Ax \leqslant b$.

Hence both sets in (29) are nonempty, and therefore (29) holds by the Duality theorem. Denoting the common value of (29) by δ', it follows that $\delta' \leqslant \delta$, and that $cx \leqslant \delta'$ is a nonnegative linear combination of the inequalities in $Ax \leqslant b$.   □

7.7. CARATHÉODORY'S THEOREM

We have incorporated the result of Carathéodory [1911] already in the Fundamental theorem of linear inequalities. Explicitly, it states:

Corollary 7.1i (Carathéodory's theorem). *If* $X \subseteq \mathbb{R}^n$ *and* $x \in \mathrm{cone}(X)$, *then* $x \in \mathrm{cone}$ $\{x_1, \ldots, x_d\}$ *for some linearly independent vectors* $x_1, \ldots, x_d$ *in* X.

Proof. Directly from Theorem 7.1.   □

In fact, Carathéodory gave the following result:

Corollary 7.1j. *If* $X \subseteq \mathbb{R}^n$ *and* $x \in \mathrm{conv.hull}\ (X)$, *then* $x \in \mathrm{conv.hull}\ \{x_0, \ldots, x_d\}$ *for certain affinely independent vectors* $x_0, \ldots, x_d$ *in* X.

Proof. Let $Y := \left\{ \begin{pmatrix} z \\ 1 \end{pmatrix} \middle| z \in X \right\}$. Then $x \in \mathrm{conv.hull}(X)$ if and only if $\begin{pmatrix} x \\ 1 \end{pmatrix} \in \mathrm{cone}(Y)$.

Moreover, $x_0, \ldots, x_d$ are affinely independent if and only if $\begin{pmatrix} x_0 \\ 1 \end{pmatrix}, \ldots, \begin{pmatrix} x_d \\ 1 \end{pmatrix}$ are linearly independent. So Corollary 7.1i implies this corollary.   □

One similarly shows, with the help of Farkas' lemma (Corollary 7.1e): if a system $Ax \leqslant b$ of linear inequalities in n variables has no solution, then $Ax \leqslant b$ has a subsystem $A'x \leqslant b'$ of at most $n + 1$ inequalities having no solution.

7.8. STRICT INEQUALITIES

The Fundamental theorem also yields characterization for *strict* inequalities. The following result is due to Fourier [1826a], Kuhn [1956a], and Motzkin [1936] (cf. Stoer and Witzgall [1970: pp. 7–30]).

Corollary 7.1k ((Motzkin's) transposition theorem). *Let* A *and* B *be matrices and let* b *and* c *be column vectors. Then there exists a vector* x *with* $Ax < b$ *and* $Bx \leqslant c$, *if and only if*

(30)   *for all row vectors* $y, z \geqslant 0$:
   (i)   *if* $yA + zB = 0$ *then* $yb + zc \geqslant 0$; *and*
   (ii) *if* $yA + zb = 0$ *and* $y \neq 0$ *then* $yb + zc > 0$.

Proof. Conditions (i) and (ii) are obviously necessary. To prove sufficiency, suppose (30) is satisfied. By Corollary 7.1e and (30) (i), there exists an x with

$Ax \leqslant b$, $Bx \leqslant c$. Condition (30) (ii) implies that for each inequality $a_i x \leqslant \beta_i$ in the system $Ax \leqslant b$ there are no $y, z \geqslant 0$ with $yA + zB = -a_i$ and $yb + zc \leqslant -\beta_i$. By Farkas' lemma, this is equivalent to: for each inequality $a_i x \leqslant \beta_i$ in $Ax \leqslant b$ there is a vector x^i with: $Ax^i \leqslant b$, $Bx^i \leqslant c$, $a_i x^i < \beta_i$. Then the barycentre x of the x^i satisfies $Ax < b$, $Bx \leqslant c$. □

Special cases of Motzkin's transposition theorem are the Transposition theorem of Gordan [1873]:

(31)   *there is a vector x with $x \geqslant 0$, $x \neq 0$ and $Ax = 0$, if and only if there is no vector y with $yA > 0$;*

the Transposition theorem of Stiemke [1915]:

(32)   *there is a vector x with $x > 0$ and $Ax = 0$, if and only if $yA \geqslant 0$ implies $yA = 0$;*

and a result of Carver [1921–2]:

(33)   *$Ax < b$ is feasible, if and only if $y = 0$ is the only solution for $y \geqslant 0$, $yA = 0$, $yb \leqslant 0$.*

Motzkin [1951] and Slater [1951] derive the Duality theorem of linear programming from the Transposition theorem.

7.9. COMPLEMENTARY SLACKNESS

Let A be a matrix, let b be a column vector, and let c be a row vector. Consider the linear programming duality equation

(34) $\max\{cx \mid Ax \leqslant b\} = \min\{yb \mid y \geqslant 0, yA = c\}$.

Assume that both optima here are finite, and let x_0 and y_0 be feasible solutions (i.e. $Ax_0 \leqslant b$ and $y_0 \geqslant 0$, $y_0 A = c$). Then the following are equivalent:

(35)   (i) x_0 and y_0 are optimum solutions in (34);
      (ii) $cx_0 = y_0 b$;
      (iii) if a component of y_0 is positive, the corresponding inequality in $Ax \leqslant b$ is satisfied by x_0 with equality; i.e. $y_0(b - Ax_0) = 0$.

Phenomenon (iii) is called *complementary slackness*. (The equivalence of (i) and (ii) follows directly from the Duality theorem. The equivalence of (ii) and (iii) follows from the facts that $cx_0 = y_0 Ax_0 \leqslant y_0 b$, and that $y_0 Ax_0 = y_0 b$ if and only if (iii) is true.)

More generally, for each inequality $a_i x \leqslant \beta_i$ in $Ax \leqslant b$, exactly one of the following holds:

(36)   either   (i) the maximum in (34) has an optimum solution x_0 with $a_i x_0 < \beta_i$;
      or   (ii) the minimum in (34) has an optimum solution y_0 with positive ith component.

Proof. The fact that (i) and (ii) exclude each other follows from the equivalence of (35) (i) and (35) (iii). Suppose now (36) (i) is not satisfied. That is, there is no optimum solution x_0 for the maximum in (34) with $a_i x_0 < \beta_i$. Let δ be the common value of (34). Then $Ax \leqslant b, cx \geqslant \delta$ implies $a_i x \geqslant \beta_i$. So, by Corollary 7.1h, $yA - \lambda c = -a_i$ and $yb - \lambda \delta \leqslant -\beta_i$ for some $y, \lambda \geqslant 0$. Let y' arise from y by increasing its ith component by one. If $\lambda = 0$, then $yA + a_i = 0$ and $yb + \beta_i \leqslant 0$. Hence if y_0 attains the minimum in (34), then also $y_0 + y'$ attains the minimum, and hence (36) (ii) is fulfilled. If $\lambda > 0$, then $\lambda^{-1} y'$ attains the minimum (as $\lambda^{-1} y'A = \lambda^{-1} yA + \lambda^{-1} a_i = c$ and $\lambda^{-1} y'b = \lambda^{-1} yb + \lambda^{-1} \beta \leqslant \delta$), and has positive ith component. This shows (36). $\qquad\square$

From Carathéodory's theorem (Corollary 7.1i) we derive:

Corollary 7.1l. *Let A be a matrix, let b be a column vector, and let c be a row vector. If the optimum in the LP-problems*

$$(37) \qquad \max\{cx | Ax \leqslant b\} = \min\{yb | y \geqslant 0; yA = c\}$$

is finite, then the minimum is attained by a vector $y \geqslant 0$ whose positive components correspond to linearly independent rows of A.

Proof. Let the maximum be attained by x^*, with optimal value δ. Since the minimum is also δ, we know that the vector $(c \ \delta)$ is a nonnegative combination of the rows of the matrix $[A \ b]$. Hence, by Carathéodory's theorem, $(c \ \delta) = y[A \ b]$ for a $y \geqslant 0$ whose positive components correspond to linearly independent rows of $[A \ b]$. Let y' be the positive part of y, and $[A' \ b']$ the corresponding part of $[A \ b]$. So $[A' \ b']$ has full row rank. As $A'x^* = b'$ (by complementary slackness), A' has full row rank, and the theorem follows. $\qquad\square$

Note that (36) implies that the system $Ax \leqslant b$ can be partitioned into two subsystems $A_1 x \leqslant b_1$ and $A_2 x \leqslant b_2$ such that

$$(38) \qquad \max\{cx | A_1 x < b_1, A_2 x = b_2\} = \min\{y_2 b_2 | y_2 > 0, y_2 A_2 = c\}$$

(by taking convex combinations of the optimum solutions as described in (36)).

So the following extension of Motzkin's transposition theorem (Corollary 7.1k) follows: if $A_1 x \leqslant b_1$ and $A_2 x \leqslant b_2$ are systems of linear inequalities, and c is a vector, then either $\max\{cx | A_1 x \leqslant b_1, A_2 x \leqslant b_2\}$ is attained by a vector x with $A_1 x < b_1$, or $\min\{y_1 b_1 + y_2 b_2 | y_1, y_2 \geqslant 0, y_1 A_1 + y_2 A_2 = c\}$ is attained by vectors y_1, y_2 with $y_1 \neq 0$.

Similar considerations can be made with respect to the other linear programming duality equations.

Exercise (Robert Freund). Given a system $Ax \leqslant b$ of linear inequalities, describe a linear programming problem whose optimal solution immediately tells us which inequalities among $Ax \leqslant b$ are always satisfied with equality.

7.10. APPLICATION: MAX-FLOW MIN-CUT

We will derive the 'max-flow min-cut' theorem from LP-duality and complementary slackness.

Let $D = (V, A)$ be a directed graph, let $r, s \in V$, and $c: A \to \mathbb{R}_+$. As usual, let $\delta^+(W)$ and $\delta^-(W)$ denote the sets of arcs leaving W and entering W, respectively. Let $\delta^+(v) := \delta^+(\{v\})$ and $\delta^-(v) := \delta^-(\{v\})$, for $v \in V$.

The Duality theorem of linear programming yields the following equality:

$$(39) \qquad \max(x(\delta^+(r)) - x(\delta^-(r))) \qquad\qquad = \min \sum_{a \in A} c_a y_a$$

subject to

$$x(\delta^-(v)) = x(\delta^+(v)) \qquad (v \in V, v \neq r, s)$$
$$0 \leqslant x_a \leqslant c_a \qquad\qquad (a \in A)$$

subject to

$$y_a \geqslant 0 \qquad\qquad (a \in A)$$
$$z_v \in \mathbb{R} \qquad\qquad (v \in V)$$
$$z_w - z_v + y_a \geqslant 0 \qquad (a = (v, w) \in A)$$
$$z_r = 1, z_s = 0.$$

Here $x(A') := \sum_{a \in A'} x_a$ for $A' \subseteq A$. (One may see that the conditions $z_r = 1$, $z_s = 0$ in the minimum are equivalent to $z_r - z_s = 1$; with this last condition, (39) is easily seen to follow from (14).)

The maximum in (39) can be seen as the maximum amount of r–s-flow subject to the capacity constraint c. We show that the minimum is equal to the minimum capacity of an r–s-cut, i.e. to

$$(40) \qquad \min_{\substack{W \subseteq V \\ r \in W, s \notin W}} c(\delta^+(W)).$$

Let x^*, y^*, z^* form an optimum solution for the problems in (39). Define

$$(41) \qquad W := \{v \in V \mid z_v^* \geqslant 1\}.$$

Then $r \in W$, $s \notin W$.

If $a = (v, w) \in \delta^+(W)$, then $z_w^* - z_v^* + y_a^* \geqslant z_w^* - z_v^* > 0$, and hence, by complementary slackness, $x_a^* = c_a$. If $a = (v, w) \in \delta^-(W)$, then $y_a^* \geqslant z_v^* - z_w^* > 0$, and hence, again by complementary slackness, $x_a^* = 0$. Therefore,

$$(42) \qquad x^*(\delta^+(r)) - x^*(\delta^-(r)) = \sum_{v \in W} (x^*(\delta^+(v)) - x^*(\delta^-(v)))$$

$$= x^*(\delta^+(W)) - x^*(\delta^-(W)) = c(\delta^+(W)).$$

The first equality here follows from the primal conditions in (39). The second equality follows from the fact that if $a = (u, w) \in A$ and $u, w \in W$, then x_a^* occurs as term both in $x^*(\delta^+(u))$ and in $x^*(\delta^-(w))$, so that it is cancelled if we sum $x^*(\delta^+(v)) - x^*(\delta^-(v))$ over all $v \in W$. What is left is $x^*(\delta^+(W)) - x^*(\delta^-(W))$. The third equality follows from our remarks above.

So the maximum amount of flow is equal to the capacity of the r–s-cut $\delta^+(W)$. Since no r–s-cut can have lower capacity than the maximum amount of flow, we have obtained the max-flow min-cut theorem of Ford and Fulkerson [1956] and Elias, Feinstein, and Shannon [1956]:

Theorem 7.2 (Max-flow min-cut theorem). *The maximum value of r–s-flow subject to capacity c, is equal to the minimum capacity of an r–s-cut.*

Proof. See above. □

This result can also be interpreted as the minimum in (39) having an integral optimum solution: indeed, replacing the optimum y^*, z^* by $\tilde{y}, \tilde{z}$ with

$$(43) \qquad \tilde{y}_a := 1 \quad \text{if} \quad a \in \delta^+(W)$$
$$\tilde{y}_a := 0 \quad \text{otherwise}$$
$$\tilde{z}_v := 1 \quad \text{if} \quad v \in W$$
$$\tilde{z}_v := 0 \quad \text{otherwise}$$

gives another optimum solution, as it is feasible, and its objective value is equal to the capacity of cut $\delta^+(W)$.

If c is integral, also the maximum in (39) has an integral optimum solution, which is a result of Dantzig [1951b]—see Application 19.2.

Notes for Chapter 7. Variants of the theory discussed in this chapter are studied by Ben-Israel [1969], Berman [1973], Berman and Ben-Israel [1971a,b, 1973] Camion [1968], Fulkerson [1968], Kortanek and Rom [1971], Kortanek and Soyster [1972], Mangasarian [1981a], Minty [1974], and Rockafellar [1970: pp. 203–209].

8

The structure of polyhedra

The concepts and results treated in Chapter 7 lead us to a more geometric approach to polyhedra. In this chapter we review results on the geometric structure of polyhedra, which are to a large extent obvious, and which are generally also not difficult to prove formally. The reader wishing to trust his or her geometric intuition may omit the parts of this chapter printed in smaller type, where strictly mathematical proofs are hinted at.

For more extensive treatments see, for example, Grünbaum [1967: Ch. 3], Rockafellar [1970: Section 19], Stoer and Witzgall [1970: Chapter 2], Kuhn and Tucker [1956] (especially Goldman [1956], Goldman and Tucker [1956], and Hoffman and Kruskal [1956]).

Again we note: *Each of the results in this chapter holds both in real spaces and in rational spaces.* In the latter case, all numbers occurring (like matrix and vector entries, variables) are restricted to the rationals.

In Section 7.2 we gave the definitions of (polyhedral) cone, polyhedron, and polytope, and we showed:

(1)   (i) *a cone is polyhedral, if and only if it is finitely generated;*

         (ii) *P is a polyhedron, if and only if $P = Q + C$ for some polytope Q and some polyhedral cone C;*

        (iii) *P is a polytope, if and only if P is a bounded polyhedron;*

(cf. Corollaries 7.1a, b, c).

Throughout this chapter, let $Ax \leqslant b$ be a feasible system of linear inequalities, in n variables, and let $P := \{x \mid Ax \leqslant b\}$ be the corresponding nonempty polyhedron in n dimensions.

8.1. IMPLICIT EQUALITIES AND REDUNDANT CONSTRAINTS

An inequality $ax \leqslant \beta$ from $Ax \leqslant b$ is called an *implicit equality* (*in $Ax \leqslant b$*) if $ax = \beta$ for all x satisfying $Ax \leqslant b$. We use the following notation:

(2) $A^= x \leqslant b^=$ is the system of implicit equalities in $Ax \leqslant b$

      $A^+ x \leqslant b^+$ is the system of all other inequalities in $Ax \leqslant b$.

It is easy to see that there is an x in P satisfying

(3) $A^= x = b^=, A^+ x < b^+$.

A constraint in a constraint system is called *redundant* (*in* the system) if it is implied by the other constraints in the system. (So redundant constraints can be removed—however, deleting one redundant constraint can make other redundant constraints irredundant, so that generally not all the redundant constraints can be deleted at the same time.) The system is *irredundant* if it has no redundant constraints.

8.2. CHARACTERISTIC CONE, LINEALITY SPACE, AFFINE HULL, DIMENSION

The *characteristic cone* of P, denoted by char.cone$P$, is the polyhedral cone

(4)   char.cone $P := \{y \mid x + y \in P \text{ for all } x \text{ in } P\} = \{y \mid Ay \leqslant 0\}$

(the last equality is easy, using the fact that $P \neq \varnothing$). [Some authors call char.cone$P$ the *recession cone of P*.] It follows straightforwardly that:

(5)   (i) $y \in$ char.cone$P$ *if and only if there is an x in P such that $x + \lambda y \in P$ for all $\lambda \geqslant 0$;*

           (ii) $P +$ char.cone$P = P$;

           (iii) P *is bounded if and only if* char.cone$P = \{0\}$;

           (iv) *if $P = Q + C$, with Q a polytope and C a polyhedral cone, then $C =$ char.cone$P$.*

The nonzero vectors in char.cone$P$ are called the *infinite directions* of P. The *lineality space* of P, denoted by lin.space$P$, is the linear space

(6)   lin.space$P :=$ char.cone$P \cap -$char.cone$P = \{y \mid Ay = 0\}$.

If the lineality space has dimension zero, P is called *pointed*.

It is easy to prove that each nonempty polyhedron P can be uniquely represented as

(7) $P = H + Q$

where H is a linear space, and Q is a nonempty pointed polyhedron with $Q \subseteq H^{\perp}$ (the only possibility is: $H :=$ lin.space$P$, and $Q := H^{\perp} \cap P$).

The affine hull of P satisfies (using notation (2)):

(8)   aff.hull $P = \{x \mid A^{=}x = b^{=}\} = \{x \mid A^{=}x \leqslant b^{=}\}$.

Here the inclusions $\subseteq$ are trivial. To show that $\{x \mid A^{=}x \leqslant b^{=}\} \subseteq$ aff.hull$P$, suppose $A^{=}x_0 \leqslant b^{=}$. Let x_1 be such that $A^{=}x_1 = b^{=}, A^{+}x_1 < b^{+}$ (cf. (3)). If $x_0 = x_1$ then $x_0 \in P \subseteq$ aff.hull$P$. If $x_0 \neq x_1$ the line segment connecting x_0 and x_1 contains more than one point in P, and hence $x_0 \in$ aff.hull$P$.

In particular, if $ax \leqslant \beta$ is an implicit equality in $Ax \leqslant b$, the equality $ax = \beta$ is already implied by $A^{=}x \leqslant b^{=}$.

The *dimension* of P is the dimension of its affine hull. So by (8),

(9)   *the dimension of P is equal to n minus the rank of the matrix $A^{=}$.*

P is *full-dimensional* if its dimension is n. Hence, by (8), P is full-dimensional if and only if there are no implicit equalities.

By convention, $\dim(\emptyset) = -1$.

8.3. FACES

If c is a nonzero vector, and $\delta = \max\{cx\,|\,Ax \leqslant b\}$, the affine hyperplane $\{x\,|\,cx = \delta\}$ is called a *supporting hyperplane* of P. A subset F of P is called a *face* of P if $F = P$ or if F is the intersection of P with a supporting hyperplane of P. Directly we have:

(10)   *F is a face of P if and only if there is a vector c for which F is the set of vectors attaining* $\max\{cx\,|\,x \in P\}$ *provided that this maximum is finite (possibly c = 0).*

Alternatively, F is a face of P if and only if F is nonempty and

(11) $F = \{x \in P\,|\,A'x = b'\}$

for some subsystem $A'x \leqslant b'$ of $Ax \leqslant b$.

To see that this is an equivalent characterization, let F be a face of P, say $F = \{x \in P\,|\,cx = \delta\}$ for some vector c with $\delta = \max\{cx\,|\,x \in P\}$. By the Duality theorem of linear programming, $\delta = \min\{yb\,|\,y \geqslant 0, yA = c\}$. Let y_0 attain this minimum, and let $A'x \leqslant b'$ be the subsystem of $Ax \leqslant b$ consisting of those inequalities in $Ax \leqslant b$ which correspond to positive components of y_0. Then $F = \{x \in P\,|\,A'x = b'\}$, since if $Ax \leqslant b$, then: $cx = \delta \Leftrightarrow y_0 Ax = y_0 b \Leftrightarrow A'x = b'$. So (11) is satisfied. Conversely, if $\{x \in P\,|\,A'x = b'\}$ is nonempty, it is the set of points of P attaining $\max\{cx\,|\,x \in P\}$, where c is the sum of the rows of A'.

It follows that

(12)   (i)   *P has only finitely many faces*;
        (ii) *each face is a nonempty polyhedron*;
        (iii) *if F is a face of P and F' ⊆ F, then: F' is a face of P if and only if F' is a face of F.*

8.4. FACETS

A *facet* of P is a maximal face distinct from P (maximal relative to inclusion).

Theorem 8.1. *If no inequality in $A^+x \leqslant b^+$ (cf. (2)) is redundant in $Ax \leqslant b$, then there exists a one-to-one correspondence between the facets of P and the inequalities in $A^+x \leqslant b^+$, given by*

(13) $F = \{x \in P\,|\,a_i x = \beta_i\}$

for any facet F of P and any inequality $a_i x \leqslant \beta_i$ from $A^+x \leqslant b^+$.

Proof. This one-to-one correspondence follows directly from (i) and (ii) below:

(i) We show that each facet has a representation (13). Let $F = \{x \in P\,|\,A'x = b'\}$ be a facet of P, with $A'x \leqslant b'$ a subsystem of $A^+x \leqslant b^+$. Let $a_1 x \leqslant \beta_1$ be an inequality from $A'x \leqslant b'$. Then $F' := \{x \in P \mid a_1 x = \beta_1\}$ is a face of P with $F \subseteq F' \subseteq P$. Since $F' \neq P$ (as $a_1 x \leqslant \beta_1$ is not an implicit equality), we know that $F = \{x \in P \mid a_1 x = \beta_1\}$.

(ii) Let $ax \leqslant \beta$ be an inequality from $A^+ x \leqslant b^+$, and let $A'x \leqslant b'$ be the other inequalities from $A^+ x \leqslant b^+$. Then there exists an x_0 with $A^= x_0 = b^=$, $ax_0 = \beta$, $A'x_0 < b'$. Indeed, by (3), there exists an x_1 with $A^= x_1 = b^=$, $A^+ x_1 < b^+$. As $ax \leqslant \beta$ is not redundant in $Ax \leqslant b$, there is an x_2 with $A^= x_2 = b^=$, $A'x_2 \leqslant b'$, $ax_2 > \beta$. Hence some convex combination x_0 of x_1 and x_2 has the required properties. $\square$

In the case where (13) holds we say that $a_i x \leqslant \beta_i$ *defines* or *determines* the facet F.

Representations (11) and (13) for faces and facets directly imply:

(14)   *each face of P, except for P itself, is the intersection of facets of P.*

It also follows that

(15)   *P has no faces different from P if and only if P is an affine subspace.*

[Indeed, P is an affine subspace if and only if $A^+ x \leqslant b^+$ can be taken to be empty.]

Moreover,

(16)   *the dimension of any facet of P is one less than the dimension of P.*

To prove (16), we may assume that $Ax \leqslant b$ is irredundant. Let $F = \{x \in P \mid a_i x = \beta_i\}$, where $a_i x \leqslant \beta_i$ is an inequality from $A^+ x \leqslant b^+$. As $Ax \leqslant b$ is irredundant, the only implicit equalities in the system $Ax \leqslant b, a_i x \geqslant \beta_i$ are the inequalities in $A^= x = b^=$, $a_i x \leqslant \beta_i$, $a_i x \geqslant \beta_i$. (Since if $Ax \leqslant b$, $a_i x \geqslant \beta_i$ implies that $a_j x = \beta_j$ for some $a_j x \leqslant \beta_j$ in $A^+ x \leqslant b^+$, then $F \subseteq \{x \in P \mid a_j x = \beta_j\}$. As these sets are facets, it follows from Theorem 8.1 that $a_i x \leqslant \beta_i$ is the same as $a_j x \leqslant \beta_j$.) Hence, by (9) the dimension of F is equal to n minus the rank of the matrix $\begin{bmatrix} A^= \\ a_i \end{bmatrix}$. Since $a_i x \leqslant \beta_i$ is not an implicit equality, it follows that this rank is $1 +$ the rank of $A^=$. Hence the dimension of F is one less than the dimension of P.

In the case of full-dimensional P, if $Ax \leqslant b$ is irredundant, there is a one-to-one correspondence between the inequalities in $Ax \leqslant b$ and the facets of P. Combined with (16) this implies:

(17)   *if P is full-dimensional, and $Ax \leqslant b$ is irredundant, then $Ax \leqslant b$ is the unique minimal representation of P, up to multiplication of inequalities by positive scalars.*

As a theorem we state:

Theorem 8.2. *The inequalities in $A^+ x \leqslant b^+$ are irredundant in $Ax \leqslant b$ if and only if for all distinct $a_1 x \leqslant \beta_1$ and $a_2 x \leqslant \beta_2$ from $A^+ x \leqslant b^+$ there is a vector x' satisfying $Ax' \leqslant b, a_1 x' = \beta_1, a_2 x' < \beta_2$.*

Proof. First suppose each inequality in $A^+ x \leqslant b^+$ is irredundant in $Ax \leqslant b$ and let $a_1 x \leqslant \beta_1$ and $a_2 x \leqslant \beta_2$ be distinct inequalities from $A^+ x \leqslant b^+$. Then $\{x \in P | a_1 x = \beta_1\} \not\subseteq \{x \in P | a_2 x = \beta_2\}$, as these sets are distinct facets. This proves necessity.

Conversely, suppose the inequality $a_1 x \leqslant \beta_1$ from $A^+ x \leqslant b^+$ is redundant in $Ax \leqslant b$. Then the set $\{x \in P | a_1 x = \beta_1\}$ is contained in some facet of P, say in $\{x \in P | a_2 x = \beta_2\}$, where $a_2 x \leqslant \beta_2$ is from $A^+ x \leqslant b^+$. This contradicts the existence of a vector x' in P with $a_1 x' = \beta_1$ and $a_2 x' < \beta_2$, proving sufficiency. $\square$

In particular, we have:

Corollary 8.2a. *If $P = \{x | Ax \leqslant b\}$ is full-dimensional, then the system $Ax \leqslant b$ is irredundant if and only if for all distinct $a_1 x \leqslant \beta_1$ and $a_2 x \leqslant \beta_2$ from $Ax \leqslant b$ there is a vector x' in P satisfying $a_1 x' = \beta_1$, $a_2 x' < \beta_2$.*

Proof. Directly from Theorem 8.2. $\square$

Another useful characterization is as follows.

Theorem 8.3. *Let S be a finite set of vectors such that $P := $ conv.hull S is full-dimensional. Then any inequality $cx \leqslant \delta$ determines a facet of P if and only if:*

(18) $cx \leqslant \delta$ *is valid for all vectors in S, and for each row vector a which is not a nonnegative multiple of c there exist x', x'' in S such that $ax' < ax''$ and $cx' = \delta$.*

Proof. To see necessity, let $cx \leqslant \delta$ determine a facet of P. Then clearly $cx \leqslant \delta$ is valid for all x in S. Let a be a row vector with $a \neq \lambda c$ for all $\lambda \geqslant 0$. Let $\beta := \max\{ax | x \in S\}$ be attained by $x'' \in S$. As, $F := \{x \in P | cx = \delta\}$ is a facet of P, $F \not\subseteq \{x \in P | ax = \beta\}$. Hence there exists $x \in F$ with $ax < \beta$. Therefore, $ax' < ax''$ for some $x' \in S \cap F$.

To see sufficiency, suppose (18) holds and $F := \{x \in P | cx = \delta\}$ is not a facet. One easily derives from (18) that F is nonempty (by taking an arbitrary $a \notin \{\lambda c | \lambda \geqslant 0\}$). So F is a face. Let $F' := \{x \in P | ax = \beta\}$ be a facet of P with $F \subset F'$, where $ax \leqslant \beta$ is a valid inequality for P. Then $a \neq \lambda c$ for all $\lambda \geqslant 0$. By (18) there are $x', x'' \in S$ with $ax' < ax''$ and $cx' = \delta$. It follows that $\beta \geqslant ax'' > ax'$, and hence $x' \in F \setminus F'$, contradicting the fact that $F \subseteq F'$. $\square$

As an extension of (17), the representation

(19) $P = $ aff.hull $P \cap \{x | A^+ x \leqslant b^+\}$

with each row of A^+ being in the linear space parallel to aff.hull$P$, and with $A^+ x \leqslant b^+$ irredundant, is unique up to multiplication of inequalities by positive scalars.

8.5. MINIMAL FACES AND VERTICES

A minimal face of P is a face not containing any other face. By (12) (iii) and (15),

(20)   *a face F of P is a minimal face if and only if F is an affine subspace.*

In fact, there is the following result of Hoffman and Kruskal [1956].

Theorem 8.4 *A set F is a minimal face of P, if and only if $\varnothing \neq F \subseteq P$ and*

(21) $F = \{x \mid A'x = b'\}$

for some subsystem $A'x \leqslant b'$ of $Ax \leqslant b$.

Proof. If F has a representation (21), it is an affine subspace, and hence, by (20), a minimal face. Conversely, let F be a minimal face of P. We can write $F = \{x \mid A''x \leqslant b'', A'x = b'\}$, where $A''x \leqslant b''$ and $A'x \leqslant b'$ are subsystems of $Ax \leqslant b$. We can choose A'' as small as possible. In particular, all inequalities in $A''x \leqslant b''$ are irredundant in the system $A''x \leqslant b''$, $A'x = b'$. As F has no facets, $A''x \leqslant b''$ is empty, i.e. (21) holds.   □

Moreover,

(22)   *each minimal face of P is a translate of the lineality space of P.*

If the minimal face is represented as in (21), then $\{x \mid A'x = 0\} \supseteq \{x \mid Ax = 0\} = \text{lin.space} P$. Moreover, the rank of A' is equal to the rank of A, since otherwise there would exist an inequality $a_1 x \leqslant \beta_1$ in $Ax \leqslant b$ with a_1 not in the linear hull of the rows of A'. In that case we have the contradiction: $F \subseteq \{x \mid A'x = b', a_1 x \leqslant \beta_1\} \subset \{x \mid A'x = b'\} = F$.

In particular, all minimal faces of P have the same dimension, namely n minus the rank of A. Moreover, if P is pointed, each minimal face consists of just one point. These points (or these minimal faces) are called the *vertices* of P. So

(23)   *each vertex is determined by n linearly independent equations from the systems $Ax = b$.*

A vertex of $\{x \mid Ax \leqslant b\}$ is called also a *basic (feasible) solution* of $Ax \leqslant b$. If it attains max $\{cx \mid Ax \leqslant b\}$ for some objective vector c, it is called a *basic optimum solution*.

8.6. THE FACE-LATTICE

By (11), the intersection of two faces is empty or a face again. Hence, the faces, together with $\varnothing$, form a lattice under inclusion, which is called the *face-lattice* of P.

If P has dimension k, and the lineality space of P has dimension t, then any maximal chain in the face-lattice contains $\varnothing$ and exactly one face of each of the dimensions $t, \ldots, k$ (this follows from (16) and (22)).

Note that *each face F of dimension $k - 2$ is contained in exactly two facets. The intersection of these facets is F.*

To see this we may assume that $Ax \leqslant b$ is irredundant. Let $F_1 = \{x \in P \mid a_1 x = \beta_1\}$ be a facet of P, with $a_1 x \leqslant \beta_1$ from $A^+ x \leqslant b^+$. If F is a facet of F_1, then $F = \{x \in F_1 \mid a_2 x = \beta_2\}$ for some inequality $a_2 x \leqslant \beta_2$ from $A^+ x \leqslant b^+$. So if F_2 is the facet $\{x \in P \mid a_2 x = \beta_2\}$ of P, then $F = F_1 \cap F_2$. Now suppose $F \subset F_3 = \{x \in P \mid a_3 x = \beta_3\}$ for some $a_3 x \leqslant \beta_3$ from $A^+ x \leqslant b^+$, with $F_3 \neq F_1, F_2$. Hence $F = \{x \in P \mid a_1 x = \beta_1, a_2 x = \beta_2, a_3 x = \beta_3\}$. Since the dimension of F is $k - 2$, the rank of the matrix

(24)
$$\begin{bmatrix} A^= \\ a_1 \\ a_2 \\ a_3 \end{bmatrix}$$

is two more than the rank of $A^=$. Hence the system $A^= x = b^=$, $a_1 x \leqslant \beta_1$, $a_2 x \leqslant \beta_2$, $a_3 x \leqslant \beta_3$ is redundant, contradicting our assumption that $Ax \leqslant b$ is irredundant.

8.7. EDGES AND EXTREMAL RAYS

Again, let t be the dimension of the lineality space of P. Let G be a face of P of dimension $t + 1$. So the facets of G are minimal faces. Then there exists a subsystem $A'x \leqslant b'$ of $Ax \leqslant b$, with $\mathrm{rank}(A') = \mathrm{rank}(A) - 1$, and there exist inequalities $a_1 x \leqslant \beta_1$ and $a_2 x \leqslant \beta_2$ (not necessarily distinct) from $Ax \leqslant b$ such that

(25) $G = \{x \mid a_1 x \leqslant \beta_1, a_2 x \leqslant \beta_2, A'x = b'\}.$

Indeed, there is a representation $G = \{x \mid A''x \leqslant b'', A'x = b'\}$, with $A''x \leqslant b''$ and $A'x \leqslant b'$ subsystems of $Ax \leqslant b$, with A'' as small as possible. So $A'x = b'$ contains all implicit equalities of the system $A''x \leqslant b''$, $A'x \leqslant b'$, $A'x \geqslant b'$, and hence $\mathrm{rank}\,(A') = n - \dim(G) = n - t - 1$. Since the rank of A is $n - t$, the sets $\{x \mid a_j x = \beta_j, A'x = b'\}$, for $a_j x \leqslant \beta_j$ in $A''x \leqslant b''$, form parallel hyperplanes in the space $\{x \mid A'x = b'\}$. Since no inequality of $A''x \leqslant b''$ is redundant in $A''x \leqslant b''$, $A'x = b'$, A'' has at most two rows.

It follows that $G = \mathrm{lin.space}\ P + l$, where l is a line-segment or a half-line. If P is pointed (i.e. lin.space $P = \{0\}$), then G is called an *edge* if l is a line-segment, and G is called an *(extremal) ray* if l is a half-line. It also follows that G has at most two facets, being minimal faces of P. Two minimal faces of P are called *adjacent* or *neighbouring* if they are contained in one face of dimension $t + 1$.

8.8. EXTREMAL RAYS OF CONES

For the moment, we restrict ourselves to polyhedral cones. Let C be the cone $\{x \mid Ax \leqslant 0\}$. Clearly, the only minimal face of C is its lineality space lin.space C. Let t be the dimension of the lineality space. A face of C of dimension $t + 1$ is called a *minimal proper face*. So if $t = 0$, the minimal proper faces are just the extremal rays of C.

Let $G_1, \ldots, G_s$ be the minimal proper faces of C. Each of the G_i can be represented as

(26) $G_i = \{x \mid a_i x \leqslant 0, A'x = 0\}$

(cf. (25)), where A' is a submatrix of A, and a_i is a row of A, such that rank $\begin{bmatrix} A' \\ a_i \end{bmatrix} = n - t$, and lin.space $C = \{x \mid a_i x = 0, A'x = 0\}$.

Now choose for each $i = 1, \ldots, s$, a vector y_i from $G_i \backslash$ lin.space C, and choose $z_0, \ldots, z_t$ in lin.space C such that lin.space $C = \text{cone}\{z_0, \ldots, z_t\}$. Then

(27) $C = \text{cone}\{y_1, \ldots, y_s, z_0, \ldots, z_t\}$.

This can be shown by induction on $k - t$, where $k = \dim(C)$. If $k = t$, the assertion is trivial. Suppose $k \geqslant t + 1$. Write $C = \{x \mid A'''x \leqslant 0, A''x = 0\}$, where $A''x = 0$ contains all implicit equalities, and where A''' is as small as possible. By induction we know that each facet of C is contained in cone $\{y_1, \ldots, y_s, z_0, \ldots, z_t\}$. Now let x be a vector in C. Let λ be the largest number such that $x - \lambda y_1$ belong to C. Since $a_1 x \leqslant 0$ and $a_1 y_1 < 0$, this λ exists and is nonnegative. Moreover, for at least one row a of A''' we have $a(x - \lambda y_1) = 0$. Hence $x - \lambda y_1$ belongs to a facet of C, and therefore cone$\{y_1, \ldots, y_s, z_0, \ldots, z_t\}$ contains $x - \lambda y_1$, and hence also x.

So if C is pointed, C is generated by nonzero representatives from its extremal rays.

8.9. DECOMPOSITION OF POLYHEDRA

Let $F_1, \ldots, F_r$ be the minimal faces of the polyhedron $P = \{x \mid Ax \leqslant b\}$, and choose an element x_i from F_i, for $i = 1, \ldots, r$. Then

(28) $P = \text{conv.hull}\{x_1, \ldots, x_r\} + \text{char.cone } P$.

Clearly, the inclusion $\supseteq$ holds. The converse inclusion can be proved by induction on the dimension k of P. Let $t = \dim(\text{lin.space } P)$. If $k = t$, the inclusion is trivial. Suppose $k > t$. By induction we know that each facet of P is contained in the right-hand side of (28). Now take a point x in P. If $x - x_1 \in \text{char.cone } P$, then $x \in \{x_1\} + \text{char.cone } P$, and hence x belongs to the right-hand side of (28). If $x - x_1 \notin \text{char.cone} P$, there exists a largest number $\lambda \geqslant 0$ such that $x + \lambda(x - x_1)$ belongs to P. Then $x + \lambda(x - x_1)$ attains at least one of the inequalities in $A^+ x \leqslant b^+$ with equality, and hence it belongs to a facet of P. Therefore, the right-hand side of (28) contains $x + \lambda(x - x_1)$, and hence also x (as x is on the line segment connecting x_1 and $x + \lambda(x - x_1)$).

So any polyhedron P has a unique minimal representation as

(29) $P = \text{conv.hull}\{x_1, \ldots, x_r\} + \text{cone}\{y_1, \ldots, y_t\} + \text{lin.space } P$

where $x_1, \ldots, x_r, y_1, \ldots, y_t$ are othogonal to the lineality space of P, and where the y_i are unique up to multiplication by a positive scalar.

The existence of a representation (29) follows from (27) and (28). To show uniqueness, observe that if P is represented by (29), each face of P contains some element from

$x_1, \ldots, x_r$, and each face of char.cone P, except lin.space P, contains some element from $y_1, \ldots, y_t$.

Representation (29) is in a sense an 'internal' representation of a polyhedron, as compared with the 'external' representation (19).

Let F be a face of P. Since the minimal faces of F are also minimal faces of P, and since the minimal proper faces of char.cone F are also minimal proper faces of char.cone P, it follows that F has a representation as

$$(30) \qquad F = \text{conv.hull}(X') + \text{cone}(Y') + \text{lin.space } P$$

with $X' \subseteq \{x_1, \ldots, x_r\}$ and $Y' \subseteq \{y_1, \ldots, y_l\}$.

If P is pointed, then lin.space $P = \{0\}$, and the x_i are exactly the vertices of P, and the y_i are nonzero representatives of the extremal rays of the characteristic cone of P. It follows that if P is a polytope, say,

$$(31) \qquad P = \text{conv.hull}\{x_1, \ldots, x_r\},$$

then all vertices of P occur among $x_1, \ldots, x_r$. So if $\{x_1, \ldots, x_r\}$ is minimal, it is the set of vertices of P.

Summarizing, we have the following.

Theorem 8.5. *Let $P = \{x \mid Ax \leqslant b\}$ be a nonempty polyhedron.*

(i) Choose for each minimal face F of P, an arbitrary vector $x_F \in F$ (which is an arbitrary solution of $A'x = b'$ for some subsystem $A'x \leqslant b'$ of $Ax \leqslant b$);

(ii) Choose for each minimal proper face F of char.cone P, an arbitrary vector y_F in $F \setminus \text{lin.space } P$ (which is an arbitrary solution of $A'x = 0$; $ax < 0$ for some row submatrix $\begin{bmatrix} A' \\ a \end{bmatrix}$ of A);

(iii) Choose an arbitrary collection $z_1, \ldots, z_t$ generating lin.space P (which can be an arbitrary fundamental system of solutions of $Ax = 0$). Then:

$$(32) \qquad P = \text{conv.hull}\{x_F \mid F \text{ minimal face of } P\} + \text{cone}\{y_F \mid F \text{ minimal proper}$$
$$\text{face of char.cone } P\} + \text{lin.hull}\{z_1, \ldots, z_t\}.$$

Proof. See above. $\qquad\qquad\qquad\qquad\qquad\qquad\qquad\qquad\qquad\qquad\qquad\qquad\square$

8.10. APPLICATION: DOUBLY STOCHASTIC MATRICES

We close this chapter with two illustrations of how polyhedra can be used in proving combinatorial theorems. In this section we consider doubly stochastic matrices and perfect matchings in bipartite graphs, while in the next section we pass to general undirected graphs.

A square matrix $A = (\alpha_{ij})_{i,j=1}^n$ is called *doubly stochastic* if

$$(33) \qquad \sum_{i=1}^n \alpha_{ij} = 1 \quad (j = 1, \ldots, n);$$

$$\sum_{j=1}^{n} \alpha_{ij} = 1 \quad (i = 1, \ldots, n);$$

$$\alpha_{ij} \geqslant 0 \quad (i, j = 1, \ldots, n).$$

A *permutation matrix* is a $\{0, 1\}$-matrix with exactly one 1 in each row and in each column. Birkhoff [1946] and von Neumann [1953] showed:

Theorem 8.6. *Matrix A is doubly stochastic if and only if A is a convex combination of permutation matrices.*

Proof. Sufficiency of the condition is direct, as each permutation matrix is doubly stochastic.

Necessity is proved by induction on the order n of A, the case $n = 1$ being trivial. Consider the polytope P (in n^2 dimensions) of all doubly stochastic matrices of order n. So P is defined by (33). To prove the theorem, it suffices to show that each vertex of P is a convex combination of permutation matrices. To this end, let A be a vertex of P. Then n^2 linearly independent constraints among (33) are satisfied by A with equality. As the first $2n$ constraints in (33) are linearly dependent, it follows that at least $n^2 - 2n + 1$ of the α_{ij} are 0. So A has a row with $n - 1$ 0's and one 1. Without loss of generality, $\alpha_{11} = 1$. So all other entries in the first row and in the first column are 0. Then deleting the first row and first column of A gives a doubly stochastic matrix of order $n - 1$, which by our induction hypothesis is a convex combination of permutation matrices. Therefore, A itself is a convex combination of permutation matrices.   □

A corollary is that each regular bipartite graph G of degree $r \geqslant 1$ has a perfect matching (proved implicitly by Frobenius [1912, 1917] and explicitly by König [1916]). To see this, let $\{1, \ldots, n\}$ and $\{n + 1, \ldots, 2n\}$ be the colour classes of G, and consider the $n \times n$-matrix $A = (\alpha_{ij})$ with

(34) $\alpha_{ij} := \dfrac{1}{r} \cdot$ (number of edges connecting i and $n + j$).

Then A is doubly stochastic, and hence, by Theorem 8.6, there is a permutation matrix $B = (\beta_{ij})$ such that $\alpha_{ij} > 0$ if $\beta_{ij} = 1$. This permutation matrix thus gives a perfect matching in G.

Moreover, Theorem 8.6 is equivalent to the following characterization of the perfect matching polytope for bipartite graphs. Let $G = (V, E)$ be an undirected graph. The *perfect matching polytope* of G is the convex hull of the characteristic vectors of perfect matchings of G. So P is a polytope in $\mathbb{R}^E$. Clearly, each vector x in the perfect matching polytope satisfies:

(35) $x_e \geqslant 0 \qquad (e \in E)$

$\quad x(\delta(v)) = 1 \qquad (v \in V)$

where $\delta(v)$ denotes the set of edges incident with v, and $x(E') := \sum_{e \in E'} x_e$ for $E' \subseteq E$. Theorem 8.6 implies that if G is bipartite, then the perfect matching polytope is completely determined by (35):

Corollary 8.6a. *The perfect matching polytope of a bipartite graph is determined by (35).*

Proof. We have to show that each vector x satisfying (35) is a convex combination of characteristic vectors of perfect matching in G, if G is bipartite.

Let x satisfy (35). Let V_1 and V_2 be the two colour classes of G. Then

(36) $|V_1| = \sum_{v \in V_1} x(\delta(v)) = \sum_{v \in V_2} x(\delta(v)) = |V_2|$

(the middle equality follows as both sides are equal to $x(E)$). Without loss of generality, $V_1 = \{1, \ldots, n\}$ and $V_2 = \{n + 1, \ldots, 2n\}$. Let $A = (\alpha_{ij})$ be the $n \times n$-matrix defined by:

(37) $\alpha_{ij} := 0, \quad$ if $\{i, n + j\} \notin E$

$\alpha_{ij} := x_e, \quad$ if $e = \{i, n + j\} \in E$.

Then A is doubly stochastic, by (35), and hence, by Theorem 8.6, A is a convex combination of permutation matrices. Each of these permutation matrices corresponds to a perfect matching in G, and thus we have decomposed x as a convex combination of characteristic vectors of perfect matchings in G.   □

8.11. APPLICATION: THE MATCHING POLYTOPE

As a second application we discuss the matching polytope. It generalizes the previous application to general, not necessarily bipartite, graphs.

Let $G = (V, E)$ be an undirected graph, with $|V|$ even, and let P be again the associated *perfect matching polytope*, i.e. P is the convex hull of the characteristic vectors (in $\mathbb{R}^E$) of the perfect matchings in G. So P is a polytope in $|E|$-dimensional space.

For non-bipartite graphs, the constraints (35) generally are not enough to determine P: let G be the graph of Figure 4 and let $x_e = \frac{1}{2}$ for each edge of G. Then x satisfies (35) but x is not in P ($P = \varnothing$, as G has no perfect matchings). So in general we need more than (35).

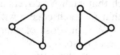

Figure 4

Edmonds' *matching polyhedron theorem* [1965c] states that P is exactly the polytope defined by the following linear inequalities:

(38) $x_e \geq 0 \qquad (e \in E)$

$x(\delta(v)) = 1 \qquad (v \in V)$

$x(\delta(W)) \geq 1 \qquad (W \subseteq V, |W| \text{ odd})$.

Here $\delta(W)$ is the set of edges of G intersecting W in exactly one vertex, $\delta(v) := \delta(\{v\})$, and $x(E') := \sum_{e \in E'} x_e$ for $E' \subseteq E$.

Let Q be the set of vectors in $\mathbb{R}^E$ satisfying (38). As the characteristic vector of any perfect matching in G satisfies (38), it trivially follows that $P \subseteq Q$—the content of Edmonds' theorem is the converse inclusion, which we show now.

Theorem 8.7 (Edmonds' matching polyhedron theorem). *The perfect matching polytope is determined by* (38).

Proof. Suppose $Q \not\subseteq P$, and let G be a smallest graph with $Q \not\subseteq P$ (that is, with $|V| + |E|$ as small as possible). Let x be a vertex of Q not contained in P. Then for each edge e, $x_e > 0$—otherwise we could delete e from E to obtain a smaller counterexample. Also $x_e < 1$—otherwise we could replace E by $E \setminus \{e\}$ and V by $V \setminus e$. Moreover, $|E| > |V|$—otherwise, either G is disconnected (in which case one of the components of G will be a smaller counterexample), or G has a vertex v of degree 1 (in which case the edge e incident with v has $x_e = 1$), or G is an even circuit (for which the theorem is trivial).

Since x is a vertex of Q, there are $|E|$ linearly independent constraints among (38) satisfied by x with equality. As $x_e > 0$ for all e, and $|E| > |V|$, it implies that there exists a $W \subseteq V$ with $|W|$ odd, $|W| \geq 3$, $|V \setminus W| \geq 3$, and $x(\delta(W)) = 1$.

Let G_1 and G_2 arise from G by contracting $V \setminus W$ and W, respectively. That is, G_1 has vertex set $W \cup \{w^*\}$, where w^* is some new vertex, and edge set $\{e \in E \mid e \subseteq W\} \cup \{\{v, w^*\} \mid v \in W, \text{and } \{v, w\} \in E \text{ for some } w \in V \setminus W\}$. Similarly, G_2 has vertex set $(V \setminus W) \cup \{w_*\}$, where w_* is some new vertex, and edge set $\{e \in E \mid e \subseteq V \setminus W\} \cup \{\{w_*, v\} \mid v \in V \setminus W, \text{and } \{w, v\} \in E \text{ for some } w \in W\}$.

Let x_1 and x_2 be the corresponding 'projections' of x onto the edge sets of G_1 and G_2, respectively. That is $x_1(e) := x(e)$ if $e \subseteq W$, and $x_1(\{v, w^*\}) := \Sigma_{w \in V \setminus W} x(\{v, w\})$. Similarly, $x_2(e) = x(e)$ if $e \subseteq V \setminus W$, and $x_2(\{w_*, v\}) := \Sigma_{w \in W} x(\{w, v\})$.

Now x_1 and x_2 satisfy the inequalities (38) with respect to the graphs G_1 and G_2, as can be checked easily. Since each of G_1 and G_2 is smaller than G, it follows that x_1 and x_2 can be decomposed as convex combinations of characteristic vectors of perfect matchings in G_1 and G_2, respectively. These decompositions can be easily glued together to form a decomposition of x as a convex combination of characteristic vectors of perfect matchings in G, contradicting our assumption.

[This glueing can be done, for example, as follows. By the rationality of x (as it is a vertex of Q), there exists a natural number K such that, for $i = 1, 2$, Kx_i is the sum of the characteristic vectors of K perfect matchings, say $M_1^i, \ldots, M_K^i$, in G_i (possibly with repetitions).

Each of the matchings M_j^1 corresponds to a matching N_j^1 in G such that N_j^1 covers all vertices in W and exactly one vertex in $V \setminus W$, in such a way that for each edge e with $|W \cap e| \geqslant 1$, the number of j with $e \in N_j^1$ is equal to $Kx(e)$. Similarly, from the M_j^2 we obtain matchings N_j^2 in G such that each N_j^2 covers all vertices in $V \setminus W$ and exactly one vertex in W, in such a way that for each edge e with $|(V \setminus W) \cap e| \geqslant 1$, the number of j with $e \in N_j^2$ is equal to $Kx(e)$. Without loss of generality, for each j, N_j^1 and N_j^2 have an edge, viz. in $\delta(W)$, in common; so $N_j^1 \cup N_j^2$ is a perfect matching in G. Then each edge e of G is in exactly $Kx(e)$ of the perfect matching $N_j^1 \cup N_j^2$. Hence x is a convex combination of characteristic vectors of matchings.] $\square$

By a standard construction one can derive Edmonds' characterization of the *matching polytope*, being the convex hull of characteristic vectors of (not-necessarily perfect) matchings in G. Again, $G = (V, E)$ is an undirected graph, but now $|V|$ may be odd. Edmonds showed that the matching polytope is determined by the following inequalities:

(39)
      (i) $x(e) \geqslant 0$ $(e \in E)$

      (ii) $x(\delta(v)) \leqslant 1$ $(v \in V)$

      (iii) $x(\langle W \rangle) \leqslant \frac{1}{2}(|W| - 1)$ $(W \subseteq V, |W| \text{ odd})$.

[Here $\langle W \rangle$ denotes the set of edges contained in W.] Again it is clear that each vector in the matching polytope satisfies (39), as the characteristic vector of each matching satisfies (39).

Corollary 8.7a. *The matching polytope is determined by the inequalities* (39).

Proof. Let $x \in \mathbb{R}^E$ satisfy (39). Let $G^* = (V^*, E^*)$ be a disjoint copy of G, where the copy of vertex v will be denoted by v^*, and the copy of edge $e (= \{v, w\})$ will be denoted by $e^* (= \{v^*, w^*\})$. Let $\tilde{G}$ be the graph with vertex set $V \cup V^*$ and edge set $E \cup E^* \cup \{\{v, v^*\} \mid v \in V\}$. Define $\tilde{x}(e) := \tilde{x}(e^*) := x(e)$ for e in E, and $\tilde{x}(\{v, v^*\}) := 1 - x(\delta(v))$, for v in V. Now condition (38) is easily derived for $\tilde{x}$ with respect to $\tilde{G}$. (i) and (ii) are trivial. To prove (iii) we have to show, for $V_1, V_2 \subseteq V$ with $|V_1| + |V_2|$ odd, that $x(\delta(V_1 \cup V_2^*)) \geqslant 1$. Indeed, we may assume by symmetry that $|V_1 \setminus V_2|$ is odd. Hence

(40) $\tilde{x}(\delta(V_1 \cup V_2^*)) \geqslant \tilde{x}(\delta(V_1 \setminus V_2)) + \tilde{x}(\delta(V_2^* \setminus V_1^*)) \geqslant \tilde{x}(\delta(V_1 \setminus V_2))$

          $= |V_1 \setminus V_2| - 2x(\langle V_1 \setminus V_2 \rangle) \geqslant 1$

by elementary counting and (39) (iii).

Hence $\bar{x}$ is a convex combination of perfect matchings in $\bar{G}$. By restriction to x and G it follows that x is a convex combination of matchings in G. $\square$

Not always all inequalities (39) give facets of the matching polytope (see Pulleyblank and Edmonds [1974] for a characterization). However:

Theorem 8.8. *If $G = K_n$ (the complete graph on n vertices) with $n \geqslant 4$, then each inequality in (39) determines a facet of the matching polytope of G.*

Proof. We apply Theorems 8.1 and 8.2. As the matching polytope is full-dimensional (as it contains the all-zero vector, and all unit basis vectors), by Theorem 8.1 it suffices to show that the system (39) is irredundant. Simple case-checking shows that for each two distinct inequalities $a_1 x \leqslant \beta_1$ and $a_2 x \leqslant \beta_2$ in (39), there exists a matching M whose characteristic vector x satisfies $a_1 x = \beta_1, a_2 x < \beta_2$. Hence by Theorem 8.2 the system (39) is irredundant. $\square$

9

Polarity, and blocking and anti-blocking polyhedra

An important role in polyhedral theory, and also in polyhedral combinatorics, is played by *polarity*, in several forms. Besides the classical polarity (treated in Section 9.1), there are the *blocking* and *anti-blocking* relations between polyhedra (Sections 9.2 and 9.3). Blocking and anti-blocking were introduced by Fulkerson, as a tool in combinatorial optimization.

Again: *Each of the results in this chapter holds both in real spaces and in rational spaces*. In the latter case, all numbers occurring (like matrix or vector entries, variables) are restricted to the rationals.

9.1. POLARITY

If $X \subseteq \mathbb{R}^n$, the *polar* X^* of X is the set:

$$(1) \qquad X^* := \{z \in \mathbb{R}^n | z^T x \leqslant 1 \text{ for all } x \text{ in } X\}.$$

Theorem 9.1. *Let P be a polyhedron in $\mathbb{R}^n$, containing the origin. Then:*

  (i) *P^* is a polyhedron again;*
 (ii) *$P^{**} = P$;*
(iii) *$x \in P$ if and only if $\forall z \in P^*: z^T x \leqslant 1$;*
 (iv) *if $P = \text{conv.hull} \{0, x_1, \ldots, x_m\} + \text{cone} \{y_1, \ldots, y_t\}$ then $P^* = \{z \in \mathbb{R}^n | z^T x_i \leqslant 1$ for $i = 1, \ldots, m; z^T y_j \leqslant 0$ for $j = 1, \ldots, t\}$, and conversely.*

Proof. First suppose P is as described in (iv). Then it follows directly from definition (1) that P^* is as described in (iv). In particular, P^* is a polyhedron. This shows (i) and the first part of (iv).

We next show (ii), where the inclusion $\supseteq$ is easy. To prove the reverse inclusion, suppose $y \in P^{**} \backslash P$. Then there is an inequality $ax \leqslant \beta$ satisfied by all x in P, but not by y. Since $0 \in P$, $\beta \geqslant 0$. If $\beta > 0$, then $\beta^{-1} \cdot a^T \in P^*$, contradicting the fact that y is in P^{**} and $\beta^{-1} \cdot ay > 1$. If $\beta = 0$, then $\lambda a^T \in P^*$ for all $\lambda \geqslant 0$.

However, $ay > 0$, and hence $\lambda ay > 1$ for some $\lambda > 0$, contradicting the fact that y is in P^{**}. This shows (ii).

(iii) directly follows from (ii): $x \in P \Leftrightarrow x \in P^{**} \Leftrightarrow \forall z \in P^{*} : z^{\mathsf{T}} x \leqslant 1$.

Finally, to see the 'converse' in (iv), suppose that P^{*} is as described in (iv), and define $Q := \text{conv.hull}\{0, x_1, \ldots, x_m\} + \text{cone}\{y_1, \ldots, y_t\}$. Then by the first part of (iv), $Q^{*} = P^{*}$, and hence by (ii), $P = P^{**} = Q^{**} = Q$. $\square$

Several characteristics of polyhedra have a 'polar' characteristic:

Corollary 9.1a. *Let P be a polyhedron in $\mathbb{R}^n$, containing the origin. Then:*

(i) *P is full-dimensional $\Leftrightarrow P^{*}$ is pointed;*
(ii) *P has dimension $k \Leftrightarrow \text{lin.space}(P^{*})$ has dimension $n - k$;*
(iii) *0 is an internal point of $P \Leftrightarrow P^{*}$ is bounded.*

Proof. Let P and P^{*} be as described in (iv) of Theorem 9.1. To see (ii), note that the rank of the set of vectors $x_1, \ldots, x_m, y_1, \ldots, y_t$ is equal to the dimension of P, and is also equal to n minus the dimension of the lineality space of P^{*} (cf. the definition of lineality space in Section 8.2). This shows (ii). Clearly, (ii) implies (i).

To see (iii), 0 is an internal point of $P^{*} \Leftrightarrow$ we can take $t = 0$ in (iv) of Theorem 9.1 $\Leftrightarrow P$ is bounded. This implies (iii). $\square$

Another consequence of (iv) of Theorem 9.1 is that, if P is full-dimensional and the origin is in its interior, there is a one–one correspondence between the facets of P and the vertices of P^{*}. One can prove more generally that the face-lattice of P^{*} is the reverse of the face-lattice of P.

If $C \subseteq \mathbb{R}^n$ is a polyhedral cone, then C^{*} is a polyhedral cone again, with:

(2) $\qquad C^{*} = \{z \in \mathbb{R}^n \mid z^{\mathsf{T}} x \leqslant 0 \text{ for all } x \text{ in } C\}$.

Again, there is a one–one correspondence between the faces of C and those of C^{*}, in such a way that a face of C of dimension k is associated to a face of C^{*} of dimension $n - k$. The face-lattices (now not including $\varnothing$) are opposite.

If the origin does not belong to the polyhedron P, then $P^{**} = \text{conv.hull}$ $(P \cup \{0\})$. We shall sometimes take polars relative to another vector than the origin. That is, we consider $(P - x_0)^{*}$ for some vector x_0 in P, which makes it possible also to apply the above to polyhedra not containing the origin.

9.2. BLOCKING POLYHEDRA

Two other types of polarity were introduced by Fulkerson [1970a, 1971, 1972]: the blocking and anti-blocking polyhedra. Fulkerson noticed the importance of these notions in polyhedral combinatorics (the field treating combinatorial optimization problems with polyhedra and linear programming).

Often with the theory of blocking and anti-blocking polyhedra, one min–max relation can be seen to follow from another, and conversely. This yields a certain polarity between min–max results (see the examples below).

We first consider blocking. We say that a polyhedron P in $\mathbb{R}^n$ is of *blocking type* if $P \subseteq \mathbb{R}^n_+$ and if $y \geq x \in P$ implies $y \in P$. It follows directly that a polyhedron P in $\mathbb{R}^n$ is of blocking type if and only if there exist vectors $c_1, \ldots, c_m$ in $\mathbb{R}^n_+$ such that

(3) $P = \text{conv.hull}\{c_1, \ldots, c_m\} + \mathbb{R}^n_+ =: \{c_1, \ldots, c_m\}^{\uparrow}$.

Similarly, P is of blocking type if and only if there are vectors $d_1, \ldots, d_t$ in $\mathbb{R}^n_+$ such that

(4) $P = \{x \in \mathbb{R}^n_+ \mid d_j^\mathsf{T} x \geq 1 \text{ for } j = 1, \ldots, t\}$.

For any polyhedron P in $\mathbb{R}^n$, define its *blocking polyhedron* $B(P)$ by:

(5) $B(P) := \{z \in \mathbb{R}^n_+ \mid z^\mathsf{T} x \geq 1 \text{ for all } x \text{ in } P\}$.

The following theorem of Fulkerson [1970a, 1971] is analogous to Theorem 9.1 (including the proof).

Theorem 9.2. *Let $P \subseteq \mathbb{R}^n$ be a polyhedron of blocking type. Then:*

(i) $B(P)$ *is again a polyhedron of blocking type;*
(ii) $B(B(P)) = P$;
(iii) $x \in P$ *if and only if: $x \geq 0$ and $\forall z \in B(P): z^\mathsf{T} x \geq 1$;*
(iv) *if* $P = \{c_1, \ldots, c_m\}^{\uparrow}$ *then* $B(P) = \{z \in \mathbb{R}^n_+ \mid z^\mathsf{T} c_i \geq 1$ *for* $i = 1, \ldots, m\}$, *and conversely.*

Proof. First suppose $P = \{c_1, \ldots, c_m\}^{\uparrow}$. Then it follows directly from definition (5) that $B(P)$ is as described in (iv). In particular, $B(P)$ is a polyhedron of blocking type. This shows (i) and the first part of (iv).

We next show (ii), where the inclusion $\supseteq$ is easy. To prove the reverse inclusion, suppose $y \in B(B(P)) \setminus P$. As P is of blocking type, it is of form (4). Hence there exists an inequality $ax \geq 1$ satisfied by all x in P, but not by y, with $a \geq 0$. Then $a^\mathsf{T} \in B(P)$, contradicting the facts that $y \in B(B(P))$ and $ay < 1$. This shows (ii).

(iii) directly follows from (ii): $x \in P \Leftrightarrow x \in B(B(P)) \Leftrightarrow x \geq 0$ and $\forall z \in B(P): z^\mathsf{T} x \geq 1$.

Finally, to see the 'converse' in (iv), suppose that $B(P)$ is as described in (iv), and define $Q := \{c_1, \ldots, c_m\}^{\uparrow}$. Then $B(Q) = B(P)$, and hence by (ii), $P = B(B(P)) = B(B(Q)) = Q$. $\square$

If P is a polyhedron of blocking type, and $R := B(P)$, then the pair P, R is called a *blocking pair of polyhedra*.

The theorem implies that, if $c_1, \ldots, c_m, d_1, \ldots, d_t \in \mathbb{R}^n_+$, then:

(6) $\{c_1, \ldots, c_m\}^{\uparrow} = \{x \in \mathbb{R}^n_+ \mid d_j^\mathsf{T} x \geq 1 \quad \text{for} \quad j = 1, \ldots, t\}$

if and only if

(7) $\{d_1, \ldots, d_t\}^\uparrow = \{z \in \mathbb{R}^n_+ \mid z^\mathsf{T} c_i \geqslant 1 \quad \text{for } i = 1, \ldots, m\}.$

Proof. (6) is equivalent to: $\{c_1, \ldots, c_m\}^\uparrow = B(\{d_1, \ldots, d_t\}^\uparrow)$. This implies $\{d_1, \ldots, d_t\}^\uparrow = B(B(\{d_1, \ldots, d_t\}^\uparrow)) = B(\{c_1, \ldots, c_m\}^\uparrow)$, which is equivalent to (7).   □

Similarly, Theorem 9.2 implies that if matrix C has rows $c_1^\mathsf{T}, \ldots, c_m^\mathsf{T} \geqslant 0$ and matrix D has rows $d_1^\mathsf{T}, \ldots, d_t^\mathsf{T} \geqslant 0$, then:

(8)   *for all* $w \in \mathbb{R}^n_+$: $\min\{wc_1, \ldots, wc_m\} = \max\{y\mathbf{1} \mid y \geqslant 0, yD \leqslant w\}$

if and only if

(9)   *for all* $w \in \mathbb{R}^n_+$: $\min\{wd_1, \ldots, wd_t\} = \max\{y\mathbf{1} \mid y \geqslant 0, yC \leqslant w\}.$

[Here **1** stands for all-one column vectors.]

Proof. By LP-duality, $\max\{y\mathbf{1} \mid y \geqslant 0, yD \leqslant w\} = \min\{wz \mid z \geqslant 0, Dz \geqslant \mathbf{1}\}$. So (8) is equivalent to: $\min\{wc_1, \ldots, wc_m\} = \min\{wz \mid z \geqslant 0, Dz \geqslant \mathbf{1}\}$ for all $w \in \mathbb{R}^n_+$. This last is equivalent to (6). Similarly, (9) is equivalent to (7). So the equivalence of (8) and (9) follows from the equivalence of (6) and (7).   □

Note that, by continuity, in (8) and (9) we may restrict w to rational, and hence to integral vectors, without changing these conditions. This will allow us to prove them by induction.

The equivalence of (8) and (9) yields that one min–max relation (viz. (8)) implies another (viz. (9)), and conversely. This has several applications in combinatorics.

Application 9.1. *Max-flow min-cut.* We describe one illustrative example here. Let $D = (V, A)$ be a directed graph, and let r and s be vertices of D. Let $c_1, \ldots, c_m \in \mathbb{R}^A_+$ be the incidence vectors of the r–s-paths in D (so they are $\{0, 1\}$-vectors, with 1's in coordinates corresponding to arcs in the path). Similarly, let $d_1, \ldots, d_t \in \mathbb{R}^A_+$ be the incidence vectors of the r–s-cuts (i.e. arc sets intersecting all r–s-paths).

Considering a given function $l: A \to \mathbb{Z}_+$ as a 'length' function, one easily verifies: the minimum length of an r–s-path is equal to the maximum number of r–s-cuts (repetition allowed) such that no arc a is in more than $l(a)$ of these r–s-cuts. [Indeed, the inequality $\min \geqslant \max$ is easy. To see the reverse inequality, let p be the minimum length of an r–s-path. For $i = 1, \ldots, p$, let V_i be the set

(10) $V_i := \{v \in V \mid \text{the shortest } r\text{–}v\text{-path has length at least } i\}.$

Then $\delta^-(V_1), \ldots, \delta^-(V_p)$ are r–s-cuts as required.] This directly implies (8) (with w replacing l). By the equivalence of (8) and (9), we know that (9) holds. This is equivalent to the *max-flow min-cut theorem* of Ford and Fulkerson [1956] and Elias, Feinstein, and Shannon [1956] (Theorem 7.2):

(11)   the maximum amount of r–s-flow subject to the 'capacity' function w is equal to the minimum capacity of an r–s-cut.

Thus y in (9) is considered as a function describing a flow through r–s-paths.

(In fact, the flow function y may be taken to be integral if w is integral, but this does not seem to follow from the theory of blocking polyhedra.)

So the polyhedra $\{c_1,\ldots,c_m\}^\uparrow$ ('everything above the convex hull of r–s-paths') and $\{d_1,\ldots,d_t\}^\uparrow$ ('everything above the convex hull of r–s-cuts') form a blocking pair of polyhedra.

Another, symmetric, characterization of the blocking relation is the 'length–width inequality' given by Lehman [1965].

Theorem 9.3 (Lehman's length–width inequality). *Let P and R be polyhedra of blocking type in $\mathbb{R}^n$. Then P, R form a blocking pair of polyhedra if and only if:*

(12)   (i) $z^\mathsf{T}x \geqslant 1$ for all $x\in P$ and $z\in R$;
          (ii) $\min\{lx\,|\,x\in P\}\cdot\min\{wz\,|\,z\in R\} \leqslant lw^\mathsf{T}$ for all $l, w\in\mathbb{R}^n_+$.

Proof. If P, R form a blocking pair of polyhedra, then (12) (i) is clearly satisfied. To see (12) (ii), choose $l, w\in\mathbb{R}^n_+$. Let $\alpha:=\min\{lx\,|\,x\in P\}$ and $\beta:=\min\{wz\,|\,z\in R\}$. If $\alpha=0$ or $\beta=0$, the inequality in (12) (ii) is trivial. If $\alpha>0$ and $\beta>0$, by rescaling l and w we may assume that $\alpha=\beta=1$. So $l^\mathsf{T}\in B(P)=R$, and $w^\mathsf{T}\in B(R)$. Therefore, $lw^\mathsf{T}\geqslant 1=\alpha\beta$, proving the inequality in (12) (ii). This shows necessity of (12).

To see sufficiency, suppose (12) holds. By (12) (i), $R\subseteq B(P)$. To show $B(P)\subseteq R$, take $y\in B(P)$ (hence $\min\{y^\mathsf{T}x\,|\,x\in P\}\geqslant 1$). To show that y is in R, take $v\in B(R)$ (hence $\min\{v^\mathsf{T}z\,|\,z\in R\}\geqslant 1$). Then by (12) (ii), $y^\mathsf{T}v\geqslant 1$. As this is true for all v in $B(R)$, it follows that y is in $B(B(R))=R$.   □

Again, in (12) (ii) we may restrict l and w to $\mathbb{Z}^n_+$, which again may be useful in proving the length–width inequality inductively. Note that, again by continuity, Theorem 9.3 is equivalent to: if $c_1,\ldots,c_m,\ d_1,\ldots,d_t\in\mathbb{R}^n_+$, then $\{c_1,\ldots,c_m\}^\uparrow$ and $\{d_1,\ldots,d_t\}^\uparrow$ form a blocking pair of polyhedra if and only if

(13)   (i) $d_j^\mathsf{T}c_i\geqslant 1$ for $i=1,\ldots,m;\ j=1,\ldots,t$;
          (ii) $\min\{lc_1,\ldots,lc_m\}\cdot\min\{wd_1,\ldots,wd_t\}\leqslant lw^\mathsf{T}$ for all $l, w\in\mathbb{Z}^n_+$.

9.3. ANTI-BLOCKING POLYHEDRA

The anti-blocking case is almost completely analogous to the blocking case dealt with above, and arises mostly by reversing inequality signs and by interchanging 'minimum' and 'maximum'.

A polyhedron P in $\mathbb{R}^n$ is said to be of *anti-blocking type* if $\varnothing\neq P\subseteq\mathbb{R}^n_+$ and if $x\in P$ and $0\leqslant y\leqslant x$ implies $y\in P$. It follows directly that a polyhedron P in $\mathbb{R}^n$ is of anti-blocking type if and only if $P\neq\varnothing$ and there are $c_1,\ldots,c_m$ in $\mathbb{R}^n_+$ and $M\subseteq\{1,\ldots,n\}$ such that

(14) $P=\{c_1,\ldots,c_m\}^\downarrow_M:=\mathbb{R}^n_+\cap(\text{conv.hull}\{c_1,\ldots,c_m\}+\{(\xi_1,\ldots,\xi_n)^\mathsf{T}\in$
          $\mathbb{R}^n\,|\,\xi_i\leqslant 0$ for $i\notin M\})$.

Similarly, P is of anti-blocking type if and only if there are $d_1,\ldots,d_t$ in $\mathbb{R}^n_+$ and $L\subseteq\{1,\ldots,n\}$ such that

(15) $P=\{x=(\xi_1,\ldots,\xi_n)^\mathsf{T}\in\mathbb{R}^n_+\,|\,d_jx\leqslant 1$ for $j=1,\ldots,t;\ \xi_i=0$ for $i\in L\}$.

For any polyhedron P in $\mathbb{R}^n$ define its *anti-blocking polyhedron* $A(P)$ by:

(16) $A(P) := \{z \in \mathbb{R}^n_+ \,|\, z^T x \leqslant 1 \text{ for all } x \text{ in } P\}$.

The following theorem of Fulkerson [1971, 1972] is analogous to Theorems 9.1 and 9.2.

Theorem 9.4. *Let* $P \subseteq \mathbb{R}^n$ *be a polyhedron of anti-blocking type. Then:*

  (i) $A(P)$ *is again a polyhedron of anti-blocking type*;
 (ii) $A(A(P)) = P$;
(iii) $x \in P$ *if and only if:* $x \geqslant 0$ *and* $\forall z \in A(P)$: $z^T x \leqslant 1$;
(iv) *if* $P = \{c_1, \ldots, c_m\}_M^\downarrow$ *then* $A(P) = \{z = (\zeta_1, \ldots, \zeta_n)^T \in \mathbb{R}^n_+ \,|\, z^T c_j \leqslant 1 \text{ for } j = 1, \ldots, m;$
      $\zeta_i = 0 \text{ for } i \in M\}$, *and conversely.*

Proof. Similar to the proof of Theorem 9.2. $\square$

If P is a polyhedron of anti-blocking type, and $R := A(P)$, then the pair P, R is called an *anti-blocking pair of polyhedra*.

Let us denote

(17) $\{c_1, \ldots, c_m\}^\downarrow := \{c_1, \ldots, c_m\}_\varnothing^\downarrow$.

Theorem 9.4 implies that, if $c_1, \ldots, c_m, d_1, \ldots, d_t \in \mathbb{R}^n_+$, then:

(18) $\{c_1, \ldots, c_m\}^\downarrow = \{x \in \mathbb{R}^n_+ \,|\, d_j^T x \leqslant 1 \text{ for } j = 1, \ldots, t\}$

if and only if

(19) $\{d_1, \ldots, d_t\}^\downarrow = \{z \in \mathbb{R}^n_+ \,|\, c_i^T z \leqslant 1 \text{ for } i = 1, \ldots, m\}$

(cf. (6) and (7)). Similarly, Theorem 9.4 implies that if matrix C has rows $c_1^T, \ldots, c_m^T \geqslant 0$ and matrix D has rows $d_1^T, \ldots, d_t^T \geqslant 0$, then:

(20)   *for all* $w \in \mathbb{R}^n_+$: $\max \{w c_1, \ldots, w c_m\} = \min \{y \mathbf{1} \,|\, y \geqslant 0, yD \geqslant w\}$

if and only if

(21)   *for all* $w \in \mathbb{R}^n_+$: $\max \{w d_1, \ldots, w d_t\} = \min \{y \mathbf{1} \,|\, y \geqslant 0, yC \geqslant w\}$

(cf. (8) and (9)). Again, the equivalence of (20) and (21) yields that one min–max relation (viz. (20)) implies another (viz. (21)), and conversely. This again has its application in combinatorics—see below.

Again, in (20) and (21) we may restrict w to integral vectors, without changing these conditions. This will allow us to prove them by induction.

Also the 'length–width inequality' (Theorem 9.3) has its anti-blocking analogue, which is again due to Lehman [1965].

Theorem 9.5 (Lehman's length–width inequality). *Let* P *and* R *be polyhedra of anti-blocking type in* $\mathbb{R}^n$. *Then* P, R *form an anti-blocking pair of polyhedra if and only if*:

(22)   (i) $z^T x \leqslant 1$ for all $x \in P$ and $z \in R$;
          (ii) $\max\{lx \mid x \in P\} \cdot \max\{wz \mid z \in R\} \geqslant lw^T$ for all $l, w \in \mathbb{R}^n_+$.

Proof. Similar to the proof of Theorem 9.3.   □

Again, in (22) (ii) we may restrict l and w to $\mathbb{Z}^n_+$. So Theorem 9.5 is equivalent to: *if* $c_1, \ldots, c_m, d_1, \ldots, d_t \in \mathbb{R}^n_+$, *then* $\{c_1, \ldots, c_m\}^\downarrow$ *and* $\{d_1, \ldots, d_t\}^\downarrow$ *form an anti-blocking pair of polyhedra if and only if*

(23)   (i) $d_j^T c_i \leqslant 1$ for $i = 1, \ldots, m; j = 1, \ldots, t$;
          (ii) $\max\{lc_1, \ldots, lc_m\} \cdot \max\{wd_1, \ldots, wd_t\} \geqslant lw^T$ for all $l, w \in \mathbb{Z}^n_+$.

Application 9.2. *Clique and coclique polytopes.* Let $G = (V, E)$ be an undirected graph. Let $P_{cl}(G)$ be the *clique polytope* of G, which is defined as the convex hull of the incidence vectors (in $\mathbb{R}^V$) of all cliques of G. Similarly, let $P_{cocl}(G)$ be the *coclique polytope* of G, being the convex hull of the incidence vectors of all cocliques of G. So

(24) $P_{cocl}(G) = P_{cl}(\bar{G})$

where $\bar{G}$ denotes the complementary graph of G.

Clearly, both $P_{cl}(G)$ and $P_{cocl}(G)$ are of anti-blocking type. If x and y are the incidence vectors of a clique and of a coclique, respectively, then $x^T y \leqslant 1$. This implies:

(25) $P_{cl}(G) \subseteq A(P_{cocl}(G))$ and $P_{cocl}(G) \subseteq A(P_{cl}(G))$.

Note that $P_{cl}(G) = A(P_{cocl}(G))$ means that $P_{cl}(G)$ is determined by the linear inequalities:

(26) $x(v) \geqslant 0$ $(v \in V)$;
          $\sum_{v \in I} x(v) \leqslant 1$ $(I \subseteq V, I \text{ coclique})$.

It follows directly from Theorem 9.4 and (24) that

(27) $P_{cl}(G) = A(P_{cocl}(G)) \Leftrightarrow P_{cocl}(G) = A(P_{cl}(G)) \Leftrightarrow P_{cl}(\bar{G}) = A(P_{cocl}(\bar{G}))$.

So the property that $P_{cl}(G)$ is determined by (26) is closed under taking the complementary graph of G. Lovász [1972], Fulkerson [1970b, 1973] and Chvátal [1975a] showed that graphs with this property are exactly the so-called perfect graphs.

Application 9.3. *Matchings and edge-colourings.* Let $G = (V, E)$ be an undirected graph. Let $P_{mat}(G)$ be the *matching polytope* of G, being the convex hull of the incidence vectors (in $\mathbb{R}^E$) of all matchings in G. Edmonds [1965c] showed that the matching polytope is determined by the following linear inequalities:

(28) $x(e) \leqslant 0$ $(e \in E)$
          $\sum_{e \ni v} x(e) \leqslant 1$ $(v \in V)$
          $\sum_{e \subseteq U} x(e) \leqslant \lfloor \tfrac{1}{2} |U| \rfloor$ $(U \subseteq V)$

(Corollary 8.7a).

By scalar multiplication, we can normalize system (28) to: $x \geqslant 0$, $Cx \leqslant 1$, for a certain matrix C (deleting the inequalities corresponding to $U \subseteq V$ with $|U| \leqslant 1$). As the matching polytope is of anti-blocking type, Theorem 9.4 applies, which gives that the anti-blocking polyhedron $A(P_{mat}(G))$ of $P_{mat}(G)$ is equal to $\{z \in \mathbb{R}^E \mid z \geqslant 0; Dz \leqslant 1\}$, where the rows of D are the incidence vectors of all matchings in G. So by (20):

(29) $\max\left\{\Delta(G), \max_{U \subseteq V, |U| \geq 2} \dfrac{|\langle U \rangle|}{\lfloor \frac{1}{2}|U| \rfloor}\right\} = \min\{y\mathbf{1} \mid y \geq 0; yD \geq 1\}.$

Here $\Delta(G)$ denotes the maximum degree of G, and $\langle U \rangle$ is the collection of all edges contained in U.

The minimum in (29) can be interpreted as the *fractional edge-colouring number* $\chi^*(G)$ of G. If the minimum is attained by an integral optimum solution y, it is equal to the edge-colouring number $\chi(G)$ of G (which is the minimum number of colours needed to colour the edges of G such that no two edges of the same colour intersect), since

(30) $\chi(G) = \min\{y\mathbf{1} \mid y \geq 0; yD \geq 1; y \text{ integral}\}.$

It is a theorem of Vizing [1964] that $\chi(G) = \Delta(G)$ or $\chi(G) = \Delta(G) + 1$, if G is a simple graph. If G is the so-called Petersen graph, then $\Delta(G) = \chi^*(G) = 3$ and $\chi(G) = 4$. Seymour [1979] conjectures that for each, possibly nonsimple, graph one has $\chi(G) \leq \max\{\Delta(G) + 1, \lceil \chi^*(G) \rceil\}$.

Further notes on Chapter 9. The theory of blocking and anti-blocking polyhedra has been studied by Aráoz [1973, 1979], Aráoz, Edmonds, and Griffin [1983], Bland [1978], Griffin [1977], Griffin, Aráoz, and Edmonds [1982], Huang and Trotter [1980], Johnson [1978], Tind [1974, 1977, 1979, 1980], and Tind and Wolsey [1982].

10

Sizes and the theoretical complexity of linear inequalities and linear programming

In this chapter we investigate the sizes of solutions of linear inequalities and linear programming. From the estimates we derive that linear inequality and linear programming problems have good characterizations—see Section 10.1. In Section 10.2 we define the notions of *vertex complexity* and *facet complexity* of a polyhedron, and we make some further estimates. In Section 10.3 we show that the problem of solving linear inequality systems is polynomially equivalent to the problem of solving LP-problems, and to several other problems. Finally, in Section 10.4 we very briefly discuss *sensitivity analysis* in linear programming, describing how optimum solutions change if we modify the right-hand sides of the constraints.

10.1. SIZES AND GOOD CHARACTERIZATIONS

The following is a direct consequence of the theory of polyhedra and the estimates on sizes in linear algebra (Section 3.2).

Theorem 10.1. *If the system $Ax \leqslant b$ of rational linear inequalities has a solution, it has one of size polynomially bounded by the sizes of A and b.*

Proof. Let $\{x \mid A'x = b'\}$ be a minimal face of the polyhedron $\{x \mid Ax \leqslant b\}$, where $A'x \leqslant b'$ is a subsystem of $Ax \leqslant b$. Then by Corollary 3.2b this minimal face contains a vector of size polynomially bounded by the sizes of A' and b', and hence by the sizes of A and b. □

Theorem 10.1 implies that Farkas' lemma provides us with a good characterization for each of the following problems:

(1)   Given a rational matrix A and a rational vector b, does $Ax \leqslant b$ have a solution?

(2)   Given a rational matrix A and a rational vector b, does $Ax = b$ have a
           nonnegative solution?
(3)   (LP-problem) Given a rational matrix A, a rational column vector b, a
           rational row vector c, and a rational δ, does $Ax \leqslant b$, $cx > \delta$ have a
           solution?

Corollary 10.1a. *Each of the problems* (1), (2), (3) *has a good characterization.*

Proof. If the answer in (1) is positive, this can be proved by specifying a solution
x_0. By Theorem 10.1 we can take this x_0 of size polynomially bounded by the
sizes of A and b. If the answer in (1) is negative, by Farkas' lemma (Corollary
7.1e) we can prove this by specifying a vector y and $y \geqslant 0$, $yA = 0$, $yb = -1$.
Again by Theorem 10.1, there exists such a y of size polynomially bounded by
the sizes of A and b.

Good characterization for (2) and (3) follow similarly from Farkas' lemma
(in the form of Corollary 7.1d), and from the affine form of Farkas' lemma
(Corollary 7.1h).   □

So each of the problems (1), (2), and (3) belongs to the complexity class
$\mathcal{NP} \cap \mathrm{co} - \mathcal{NP}$. It is easy to see that if one of them is polynomially solvable,
then so too are the others. In Section 10.3 below we shall see that in that case
also explicit solutions can be found in polynomial time. In fact, in Chapter 13
we shall see that these problems *are* polynomially solvable, with the ellipsoid
method.

10.2. VERTEX AND FACET COMPLEXITY

In this section we make some more precise estimates. Let $P \subseteq \mathbb{R}^n$ be a rational
polyhedron. Define the *facet complexity* of P to be the smallest number φ such
that: $\varphi \geqslant n$ and there exists a system $Ax \leqslant b$ of rational linear inequalities defining
P, where each inequality in $Ax \leqslant b$ has size at most φ. The *vertex complexity*
of P is the smallest number v such that: $v \geqslant n$ and there exist rational vectors
$x_1, \ldots, x_k, y_1, \ldots, y_t$ with

(4) $P = \mathrm{conv.hull}\{x_1, \ldots, x_k\} + \mathrm{cone}\{y_1, \ldots, y_t\}$

where each of $x_1, \ldots, x_k, y_1, \ldots, y_t$ has size at most v. (Note that facet and vertex
complexity are also defined if P has no facets or vertices.)

The following theorem shows that facet and vertex complexity are polynomially related. It extends Theorem 10.1.

Theorem 10.2. *Let P be a rational polyhedron in $\mathbb{R}^n$ of facet complexity φ and
vertex complexity v. Then $v \leqslant 4n^2\varphi$ and $\varphi \leqslant 4n^2v$.*

Proof. We first show that $v \leqslant 4n^2\varphi$. Let $P = \{x \mid Ax \leqslant b\}$, where each inequality
in $Ax \leqslant b$ has size at most φ. By Theorem 8.5 it suffices to show:

(i) that each minimal face of P contains a vector of size at most $4n^2\varphi$—this follows directly from Corollary 3.2d (each minimal face is of form $\{x \mid A'x = b'\}$, with $A'x \leqslant b'$ being a subsystem of $Ax \leqslant b$);

(ii) that the lineality space of P is generated by vectors of size at most $4n^2\varphi$—this again follows from Corollary 3.2d (the lineality space is $\{x \mid Ax = 0\}$);

(iii) that each minimal proper face F of char.cone P contains a vector y, not in lin.space P, of size at most $4n^2\varphi$—also this follows from Corollary 3.2d (as $F = \{x \mid A'x = 0, ax \leqslant 0\}$ for some submatrix A' of A and some row a of A, where a is not in the linear hull of the rows of A').

We next show $\varphi \leqslant 4n^2v$. The case $n = 1$ being easy, we assume $n \geqslant 2$. Let $P = \text{conv.hull } X + \text{cone } Y$, where X and Y are sets of rational vectors each of size at most v. First suppose that P is full-dimensional. Then every facet is determined by a linear equation of form

$$(5) \qquad \det \begin{bmatrix} 1 & 1 & \cdots 1 & 0 & \cdots 0 \\ x & x_1 \cdots x_k & y_1 & \cdots & y_{n-k} \end{bmatrix} = 0$$

where $x_1, \ldots, x_k \in X$ and $y_1, \ldots, y_{n-k} \in Y$. Expanding this determinant by its first column we obtain (with $x = (\xi_1, \ldots, \xi_n)^T$):

$$(6) \qquad \sum_{i=1}^{n} (-1)^i (\det D_i) \xi_i = -\det D_0$$

where D_i denotes the $(i+1, 1)$th minor of the matrix in (5). By Theorem 3.2, $\det D_0$ has size less than $2nv$, and for $i \neq 0$, $\det D_i$ has size less than $2n(v+1)$. Hence the equation, and the corresponding inequality for the facet, have size at most $4n^2v$.

If P is not full-dimensional, we can find similarly inequalities of size at most $4n^2v$, defining aff.hull P. Moreover, there exist $n - \dim(P)$ coordinates so that their deletion makes P full-dimensional. This projected polyhedron Q has vertex complexity at most v, and hence, by the above, it can be defined by linear inequalities of size at most $4(n - \dim(P))^2v$. By adding zero coefficients at the deleted coordinates, we can extend these inequalities to inequalities valid for P. Together with the inequalities defining aff.hull P, they define P. $\qquad \square$

So if $Ax \leqslant b$ has a solution, it has one of size polynomially bounded by the maximum row size of the matrix $[A\ b]$ (which can be much smaller than size $([A\ b])$).

A consequence of Theorem 10.2 concerns sizes of LP-solutions.

Corollary 10.2a. *Let A be a rational $m \times n$-matrix and let b and c be rational vectors, such that the optima in*

$$(7) \qquad \max\{cx \mid Ax \leqslant b\} = \min\{yb \mid y \geqslant 0, yA = c\}$$

are finite. Let σ be the maximum size of the entries in A, b, and c. Then:

(i) *the maximum in* (7) *has an optimum solution of size polynomially bounded by n and σ;*

(ii) *the minimum in* (7) *has an optimum solution with at most n positive entries, each having size polynomially bounded by n and σ;*

(iii) *the optimum value of* (7) *has size polynomially bounded by n and σ.*

Proof. As the facet complexity of $\{x \mid Ax \leqslant b\}$ is at most $(n+1)(\sigma+1)$, and as the maximum (7) is attained by some vector x_i in (4), we know by Theorem 10.2 that it is attained by some vector of size at most $4n^2(n+1)(\sigma+1)$, proving (i).

The minimum in (7) has an optimum solution y, whose positive entries correspond to linearly independent rows of A. So by Corollary 3.2d, the size of this y is polynomially bounded by n and σ, which gives (ii).

Assertion (iii) directly follows from (i). $\qquad\square$

Note that the number m of (primal) constraints is not used in these size estimates.

Another consequence of Theorem 10.2 is the following, which we need later.

Corollary 10.2b. *Let P be a rational polyhedron in $\mathbb{R}^n$ of facet complexity φ. Define*

$$(8) \qquad Q := P \cap \{x = (\xi_1, \ldots, \xi_n)^\mathsf{T} \mid -2^{5n^2\varphi} \leqslant \xi_i \leqslant 2^{5n^2\varphi} \text{ for } i = 1, \ldots, n\}.$$

Then $\dim Q = \dim P$.

Proof. As P has vertex complexity at most $4n^2\varphi$ (by Theorem 10.2), there are vectors $x_1, \ldots, x_k,\ y_1, \ldots, y_t$ of size at most $4n^2\varphi$ such that $P = \text{conv.hull} \{x_1, \ldots, x_k\} + \text{cone} \{y_1, \ldots, y_t\}$. Now as each component of each x_i, y_j is at most $2^{4n^2\varphi}$ in absolute value, we know:

$$(9) \qquad \text{conv.hull} \{x_1, \ldots, x_k, x_1 + y_1, \ldots, x_1 + y_t\} \subseteq Q \subseteq P.$$

As the first polyhedron in (9) has the same dimension as P, the conclusion $\dim Q = \dim P$ follows. $\qquad\square$

Another estimate is:

Theorem 10.3. *If $\max \{cx \mid Ax \leqslant b\}$ is finite $(A, b, c$ rational$)$ then the optimum value has size at most $4(\text{size } A + \text{size } b + \text{size } c)$.*

Proof. Let λ^* be the optimum value, and let x^* be an optimum solution. Equivalently, $\begin{pmatrix} x^* \\ \lambda^* \end{pmatrix}$ is an optimum solution of the LP-problem

$$(10) \qquad \max \lambda$$
$$\text{such that } \lambda - cx \leqslant 0,$$
$$Ax \leqslant b.$$

We may assume that $\begin{pmatrix} x^* \\ \lambda^* \end{pmatrix}$ is determined by setting $n + 1$ inequalities from (10) to equality. Cramer's rule then gives that the numerator and the denominator of λ^* are at most Δ in absolute value, where Δ is the largest absolute value of the subdeterminants of proper submatrices of the matrix

(11) $\qquad \begin{bmatrix} 1 & -c & 0 \\ 0 & A & b \end{bmatrix}.$

By Theorem 3.2, $\text{size}(\Delta) \leqslant 2(\text{size}\,A + \text{size}\,b + \text{size}\,c)$. Hence, the size of λ^* is at most $4(\text{size}\,A + \text{size}\,b + \text{size}\,c)$. $\qquad\qquad\qquad\qquad\qquad\qquad\qquad\square$

10.3. POLYNOMIAL EQUIVALENCE OF LINEAR INEQUALITIES AND LINEAR PROGRAMMING

We now show that if one of the problems (1), (2), (3) is polynomially solvable, then also each of the other two is polynomially solvable, and explicit solutions can be found in polynomial time (cf., for example, Papadimitriou [1979]). To this end we formulate the following further problems:

(12)   Given a rational matrix A and a rational vector b, test if $Ax \leqslant b$ has a solution, and if so, find one.

(13)   Given a rational matrix A and a rational vector b, test if $Ax = b$ has a nonnegative solution, and if so, find one.

(14)   (LP-problem) Given a rational matrix A, a rational column vector b, and a rational row vector c, test if $\max\{cx \mid Ax \leqslant b\}$ is infeasible, finite, or unbounded. If it is finite, find an optimal solution. If it is unbounded, find a feasible solution x_0, and find a vector z with $Az \leqslant 0$ and $cz > 0$.

Theorem 10.4. *If one of the problems* (1), (2), (3), (12), (13), (14) *is polynomially solvable, then so is each of the other problems.*

[In Chapter 13 we shall see that these problems indeed are polynomially solvable.]

Proof. Clearly, the polynomial solvability of (12), (13), (14), respectively, implies that of (1), (2), (3), respectively.

The equivalence of (1) and (2) is standard: testing $Ax \leqslant b$ for feasibility is equivalent to testing $Ax' - Ax'' + \tilde{x} = b$ for having a nonnegative solution x', x'', $\tilde{x}$. Testing $Ax = b$ for having a nonnegative solution is equivalent to testing $x \geqslant 0$, $Ax \leqslant b$, $-Ax \leqslant -b$ for feasibility. Similarly, the equivalence of (12) and (13) follows.

If (14) is polynomially solvable, then also (12) is polynomially solvable, as we can take $c = 0$ in (14). Similarly, the polynomial solvability of (3) implies that of (1) (taking $c = 0$, $\delta = -1$ in (3)).

If (1) is polynomially solvable, then also (12) is polynomially solvable. Suppose (1) is polynomially solvable. We describe a polynomial-time algorithm for (12), i.e., one finding a solution of

(15) $\qquad a_1 x \leqslant \beta_1, \ldots, a_t x \leqslant \beta_t, a_{t+1} x = \beta_{t+1}, \ldots, a_m x = \beta_m$

if there exists one (where the a_i and β_i are rational). The algorithm is described recursively on t. If $t = 0$, we can find a solution with the Gaussian elimination method. If $t > 0$, first test if (15) has a solution (with the polynomial-time method for (1)). If it has a solution, find a solution of

(16) $\qquad a_1 x \leqslant \beta_1, \ldots, a_{t-1} x \leqslant \beta_{t-1}, a_t x = \beta_t, a_{t+1} x = \beta_{t+1}, \ldots, a_m x = \beta_m,$

if it exists. If we find one, we have a solution of (15). If (16) has no solution, then $a_t x \leqslant \beta_t$ is redundant in (15). So recursively we can find a solution of

(17) $\qquad a_1 x \leqslant \beta_1, \ldots, a_{t-1} x \leqslant \beta_{t-1}, a_{t+1} x = \beta_{t+1}, \ldots, a_m x = \beta_m,$

which solution is also a solution of (15).

It is clear that this forms a polynomial-time algorithm for (12).

We are finished after showing that the polynomial solvability of (12) implies that of (14). Let (12) be polynomially solvable. Let the input A, b, c for (14) be given. With the polynomial-time method for (12) we test if $Ax \leqslant b$ is feasible, and if so, find a solution x_0. Next, again with the method for (12), we test if $y \geqslant 0$, $yA = c$ is feasible. If it is feasible, also the system

(18) $\qquad Ax \leqslant b, y \geqslant 0, yA = c, cx \geqslant yb$

has a solution x^*, y^*, which we can find with the method for (12). Then x^* is an optimum solution for $\max\{cx \mid Ax \leqslant b\}$, by the Duality theorem of linear programming. If $y \geqslant 0$, $yA = c$ is infeasible, then by Farkas' lemma there exists a vector z such that $Az \leqslant 0, cz = 1$. Again such a z can be found with the method for (12). This solves (14) in polynomial time. $\qquad\qquad\qquad\qquad\square$

For a long time it was an open question whether there is a polynomial algorithm for linear programming, or for the equivalent problems (1)–(3). This question was answered by Khachiyan [1979], who showed that the *ellipsoid method* for nonlinear programming can be adapted to solve linear inequalities and linear programming in polynomial time—see Chapter 13.

10.4. SENSITIVITY ANALYSIS

An important purpose of determining an optimum solution of the dual of an LP-problem is that it makes possible the following *sensitivity analysis*.

Consider the following primal (max) LP-problem and its dual (min):

(19) $\max \{cx | Ax \leqslant b\} = \min \{yb | y \geqslant 0; yA = c\}$.

Let δ be the optimum value, and let y^* be an optimum dual solution. Now if we replace b by $b - e$, for some vector e, we know for the new optimum value δ':

(20) $\delta' \leqslant \delta - y^*e$.

Indeed, $\delta' \leqslant y^*(b - e) = \delta - y^*e$, as y^* is a feasible solution of $\min \{y(b - e) | y \geqslant 0;$ $yA = c\}$, and $\delta = y^*b$.

So y^* yields a bound for the change of the optimum value in terms of the change in the RHS. Moreover, if the problem is nondegenerate (more precisely, if the rows of A which have equality in $Ax \leqslant b$ for each optimum solution are linearly independent), then

(21) $\delta' = \delta - y^*e$

as long as $\| e \| \leqslant \varepsilon$ for some $\varepsilon > 0$.

So y^* gives a measure for the sensitivity of the optimum value of the maximum (19) for perturbation of b.

Inequality (20) implies that if both $\max \{cx | Ax \leqslant b'\}$ and $\max \{cx | Ax \leqslant b''\}$ are finite, then

(22) $|\max \{cx | Ax \leqslant b''\} - \max \{cx | Ax \leqslant b'\}| \leqslant n\Delta \| c \|_1 \cdot \| b'' - b' \|_\infty$

where Δ is an upper bound on the absolute values of the entries of B^{-1} for all nonsingular submatrices B of A. As usual,

(23) $\| x \|_1 = |\xi_1| + \cdots + |\xi_n|$
              $\| x \|_\infty = \max \{|\xi_1|, \ldots, |\xi_n|\}$

for $x = (\xi_1, \ldots, \xi_n)$ or its transpose.

To show (22), assume without loss of generality $\max \{cx | Ax \leqslant b'\} \leqslant \max \{cx | Ax \leqslant b''\}$. Let y^* be a basic optimum solution of $\min \{yb' | y \geqslant 0, yA = c\}$. As y^* is basic, $\| y^* \|_1 \leqslant n\Delta \| c \|_1$. Hence, by (20),

(24) $\max \{cx | Ax \leqslant b''\} - \max \{cx | Ax \leqslant b'\} \leqslant - y^*(b' - b'')$
                        $\leqslant \| y^* \|_1 \| b' - b'' \|_\infty \leqslant n\Delta \| c \|_1 \| b' - b'' \|_\infty$.

Inequality (22) can also be derived from the following theorem due to Cook, Gerards, Schrijver, and Tardos [1986], extending results of Hoffman [1952] and Mangasarian [1981b].

Theorem 10.5. *Let A be an $m \times n$-matrix, and let Δ be such that for each nonsingular submatrix B of A all entries of B^{-1} are at most Δ in absolute value. Let c be a row n-vector, and let b' and b'' be column m-vectors such that* max $\{cx\,|\,Ax \leqslant b'\}$ and max $\{cx\,|\,Ax \leqslant b''\}$ are finite. Then for each optimum solution x' of the first maximum there exists an optimum solution x'' of the second maximum with

(25) $\|x' - x''\|_\infty \leqslant n\Delta \|b' - b''\|_\infty.$

Proof. I. We first show the case $c = 0$, i.e. if x' satisfies $Ax' \leqslant b'$, and if $Ax \leqslant b''$ is feasible, then $Ax'' \leqslant b''$ for some x'' satisfying (25). Indeed, if such an x'' does not exist, then the system

(26) $Ax \leqslant b''$

$\qquad x \leqslant x' + \mathbf{1}\varepsilon$

$\qquad -x \leqslant -x' + \mathbf{1}\varepsilon$

has no solution (where $\varepsilon := n\Delta \|b' - b''\|_\infty$, and $\mathbf{1}$ is an all-one vector). By Farkas' lemma, it means that $yA + u - v = 0$, $yb'' + u(x' + \mathbf{1}\varepsilon) + v(-x' + \mathbf{1}\varepsilon) < 0$ for some vectors $y, u, v \geqslant 0$. As $Ax \leqslant b''$ is feasible, we know that $u + v \neq 0$. We may assume $\|u + v\|_1 = 1$. Moreover, we may assume by Carathéodory's theorem that the positive part $\tilde{y}$ of y corresponds to linearly independent rows of A. So $\tilde{y} = B^{-1} (\tilde{v} - \tilde{u})$ for some nonsingular submatrix B of A and some part $\tilde{u}, \tilde{v}$ of u, v. Hence $\|y\|_1 \leqslant n\Delta \|u + v\|_1 = n\Delta$. This gives the contradiction:

(27) $0 > yb'' + u(x' + \mathbf{1}\varepsilon) + v(-x' + \mathbf{1}\varepsilon) = yb'' - yAx' + \|u + v\|_1\varepsilon$

$\qquad \geqslant y(b'' - b') + \varepsilon \geqslant -\|y\|_1 \|b' - b''\|_\infty + \varepsilon \geqslant 0.$

II. We next show the result in general. Let x' be an optimum solution of the first maximum, and let x^* be any optimum solution of the second maximum. Let $A_1 x \leqslant b'_1$ be those inequalities from $Ax \leqslant b'$ satisfied by x' with equality. Then $yA_1 = c$ for some $y \geqslant 0$ (by LP-duality). Since x' satisfies $Ax' \leqslant b'$, $A_1 x' \geqslant A_1 x^* - \mathbf{1}\|b' - b''\|_\infty$ (as $A_1 x' = b'_1 \geqslant b''_1 - \mathbf{1}\|b'_1 - b''_1\|_\infty \geqslant A_1 x^* - \mathbf{1}\|b' - b''\|_\infty$). and since the system $Ax \leqslant b''$, $A_1 x \geqslant A_1 x^*$ is feasible (as x^* is a solution), we know from Part I above that $Ax'' \leqslant b''$, $A_1 x'' \geqslant A_1 x^*$ for some x'' with $\|x' - x''\|_\infty \leqslant n\Delta \|b' - b''\|_\infty$. As $cx'' = yA_1 x'' \geqslant yA_1 x^* = cx^*$, we know that x'' is an optimum solution of max $\{cx\,|\,Ax \leqslant b''\}$. $\square$

Note that the factor $n\Delta$ in (25) is tight, as is shown by taking

(28) $A = \begin{bmatrix} +1 & & & & \\ -1 & +1 & & & 0 \\ & -1 & \ddots & & \\ & & & \ddots & \\ 0 & & & & \ddots \\ & & & & -1 & +1 \end{bmatrix}, \quad \Delta = 1, \quad b' = \begin{bmatrix} 1 \\ 1 \\ \vdots \\ 1 \end{bmatrix},$

$$b'' = \begin{bmatrix} 0 \\ 0 \\ \vdots \\ \vdots \\ 0 \end{bmatrix}, \quad x' = \begin{bmatrix} 1 \\ 2 \\ \vdots \\ \vdots \\ n \end{bmatrix}, \quad c = (0, \ldots, 0).$$

[For related results, see Robinson [1973].]

11

The simplex method

Chapters 11–15 treat the algorithmic side of polyhedra and linear programming. In the present chapter we discuss the most prominent algorithm, the *simplex method*.

The simplex method was designed by Dantzig [1951a], and is, at the moment, *the* method for linear programming. Although some artificial examples show exponential running time, in practice and on the average the method is very efficient.

We do not aim at teaching the simplex method here—for this we refer to the text-books on linear programming mentioned at the end of Part III.

In Section 11.1 we describe the simplex method, and show that it terminates. In Section 11.2 we discuss how the method can be performed in practice, with simplex tableaux and the pivoting operation. In Section 11.3 we make some remarks on pivot selection and cycling, and on the theoretical and practical complexity. Next, in Sections 11.4 and 11.5 we discuss the worst-case and the average running time of the simplex method. Finally, in Sections 11.6 and 11.7 we give two variants of the simplex method: the *revised simplex method* and the *dual simplex method*.

11.1. THE SIMPLEX METHOD

The idea of the simplex method is to make a trip on the polyhedron underlying a linear program, from vertex to vertex along edges, until an optimal vertex is reached. This idea is due to Fourier [1826b], and was mechanized algebraically by Dantzig [1951a]. We describe the simplex method, including 'Bland's pivoting rule' [1977a].

The simplex method if a vertex is known. First suppose that we wish to solve

$$(1) \qquad \max \{cx \,|\, Ax \leqslant b\}$$

and that we know a vertex x_0 of the feasible region $P := \{x \,|\, Ax \leqslant b\}$, which we assume to be pointed. We assume the inequalities in $Ax \leqslant b$ to be ordered:

$$(2) \qquad a_1 x \leqslant \beta_1, \ldots, a_m x \leqslant \beta_m.$$

Choose a subsystem $A_0 x \leqslant b_0$ of $Ax \leqslant b$ such that $A_0 x_0 = b_0$ and A_0 is nonsingular. Determine u so that $c = uA$ and u is 0 at components outside A_0 (so cA_0^{-1} is calculated, and 0's are added).

Case 1. $u \geqslant 0$. Then x_0 is optimal, since

(3) $cx_0 = uAx_0 = ub \geqslant \min\{yb \mid y \geqslant 0; yA = c\} = \max\{cx \mid Ax \leqslant b\}$.

So at the same time, u is an optimal solution for the dual problem of (1).

Case 2. $u \not\geqslant 0$. Choose the smallest index i^* for which u has negative component v_{i^*}. Let y be the vector with $ay = 0$ for each row a of A_0 if $a \neq a_{i^*}$, and $a_{i^*}y = -1$ (i.e. y is the appropriate column of $-A_0^{-1}$). [Note that, for $\lambda \geqslant 0$, $x_0 + \lambda y$ traverses an edge or ray of P, or is outside P for all $\lambda > 0$. Moreover,

(4) $cy = uAy = -v_{i^*} > 0$.]

Case 2 splits into two cases:

Case 2a. $ay \leqslant 0$ for each row a of A. Then $x_0 + \lambda y$ is in P for all $\lambda \geqslant 0$, and hence the maximum (1) is unbounded (using (4)).

Case 2b. $ay > 0$ for some row a of A. Let λ_0 be the largest λ such that $x_0 + \lambda y$ belongs to P, i.e.

(5) $\lambda_0 := \min\left\{ \left. \dfrac{\beta_j - a_j x_0}{a_j y} \right| j = 1, \ldots, m; a_j y > 0 \right\}$.

Let j^* be the smallest index attaining this minimum. Let A_1 arise from A_0 by replacing row a_{i^*} by a_{j^*}, and let $x_1 := x_0 + \lambda_0 y$. So $A_1 x_1 = b_1$, where b_1 is the part of b corresponding to A_1. Start the process anew with A_0, x_0 replaced by A_1, x_1.

Repeating this we find A_0, x_0; A_1, x_1; A_2, x_2;...

Theorem 11.1. *The above method terminates.*

Proof. Denote by $A_k x \leqslant b_k$, x_k, u_k, y_k the subsystem of $Ax \leqslant b$, the vertex, the vectors u, y as they are in the kth iteration. From (4) we know

(6) $cx_0 \leqslant cx_1 \leqslant cx_2 \leqslant \ldots$

where $cx_k = cx_{k+1}$ only if $x_{k+1} = x_k$ (as if $x_{k+1} \neq x_k$, then $\lambda_0 > 0$, and hence $cx_{k+1} > cx_k$).

Suppose the method does not terminate. Then there exist k, l such that $k < l$ and $A_k = A_l$ (since there are only finitely many choices for A_k). Hence $x_k = x_l$, and therefore $x_k = x_{k+1} = \cdots = x_l$. Let r be the highest index for which a_r has been removed from A_t in one of the iterations $t = k, k+1, \ldots, l$, say in iteration p. As $A_k = A_l$, we know that a_r also has been added to A_q in some iteration q with $k \leqslant q < l$. It follows that

(7)   for $j > r$: a_j occurs in $A_p \Leftrightarrow a_j$ occurs in A_q.

By (4), $u_p A y_q = cy_q > 0$. So $v_{pj}(a_j y_q) > 0$ for at least one j (denoting $u_p = (v_{p1}, \ldots, v_{pm})$. However:

(8)   if a_j does not occur in A_p: $v_{pj} = 0$;

$$\left.\begin{array}{l} \text{if } a_j \text{ occurs in } A_p \text{ and } j < r: v_{pj} \geqslant 0,\ a_j y^q \leqslant 0 \\ \text{if } a_j \text{ occurs in } A_p \text{ and } j = r: v_{pj} < 0,\ a_j y^q > 0 \end{array}\right\} \left\{\begin{array}{l} \text{as } r \text{ is the smallest index} \\ j \text{ with } v_{pj} < 0, \text{ and } r \text{ is} \\ \text{the smallest index } j \text{ with} \\ a_j x_q = \beta_j \text{ and } a_j y^q > 0 \end{array}\right.$$

if a_j occurs in A_p and $j > r$: $a_j y^q = 0$ (by (7)).

So the method terminates. $\square$

The general case. What to do if no initial vertex is known? Consider the LP-problem

(9) $\max \{cx | x \geqslant 0;\ Ax \leqslant b\}$

which is a 'general' linear program: each LP-problem can be reduced to (9). We can write $Ax \leqslant b$ as $A_1 x \leqslant b_1$, $A_2 x \geqslant b_2$ with $b_1 \geqslant 0$ and $b_2 > 0$. Consider

(10) $\max \{1(A_2 x - \tilde{x}) | x, \tilde{x} \geqslant 0;\ A_1 x \leqslant b_1;\ A_2 x - \tilde{x} \leqslant b_2\}$

where $\tilde{x}$ is a new variable vector, and 1 denotes an all-one row vector. Then $x = 0$, $\tilde{x} = 0$ defines a vertex of the feasible region of (10). Therefore, we can solve (10) with the method described above. If the maximum value for (10) is $1 b_2$, say with optimal vertex x^*, $\tilde{x}^*$, then x^* is a vertex of the feasible region of (9). Hence we can continue with the method as described above.

If the maximum value of (10) is less than $1 b_2$, then (9) is infeasible.

This is a complete description of the simplex method. The method is in practice very fast. Its success also comes from a compact representation of the successive x_k, A_k, b_k, u_k, y_k in a so-called *simplex tableau*, and by a direct method of obtaining x_{k+1}, A_{k+1}, b_{k+1}, u_{k+1}, y_{k+1} by updating x_k, A_k, b_k, u_k, y_k, the *pivoting* operation. This is described in the next section.

Note. The proof of the Fundamental theorem of linear inequalities (Theorem 7.1) consists of the simplex method for the special case of maximizing cx over $Ax \leqslant 0$. Actually, the fact that the algorithm in the proof of Theorem 7.1 terminates also implies that the simplex method above terminates (so the termination proof above could be deleted): suppose, from iteration s on, we do not leave a certain vertex, which we may assume to be the origin 0; then we can forget all inequalities in $Ax \leqslant b$ with nonzero right-hand side, as they will never be included in A_k; but then, from iteration s on, the simplex method described above is parallel to maximizing cx over $Ax \leqslant 0$, which must terminate by the proof of Theorem 7.1.

11.2. THE SIMPLEX METHOD IN TABLEAU FORM

In the previous section we gave a complete description of the simplex method. We did not discuss, however, how the algorithm is organized in practice. Basic ingredients here are the *simplex tableaux* and the *pivoting* operation.

In tableau form, the simplex method is usually applied to the 'general' LP-problem

(11) $\max \{cx \mid x \geqslant 0; \ Ax \leqslant b\}$.

So compared with the description in Section 7.1, the system $Ax \leqslant b$ is replaced by

(12) $\begin{bmatrix} -I \\ A \end{bmatrix} x \leqslant \begin{pmatrix} 0 \\ b \end{pmatrix}$.

Let A have order $m \times n$. As in Section 7.1, let a vertex x_k of (12) be given, together with a nonsingular submatrix C_k of $\begin{bmatrix} -I \\ A \end{bmatrix}$ of order n, such that C_k corresponds to constraints in (12) satisfied with equality by x_k. Let $\tilde{x}_k := b - Ax_k$ (the *slack*).

We say that, for each $i = 1, \ldots, n + m$, the following *correspond* to each other:

(13)   (i) the ith constraint in (12);

(ii) the ith row of $\begin{bmatrix} -I \\ A \end{bmatrix}$;

(iii) the ith component in x_k, $\tilde{x}_k$;

(iv) the ith column of $[A \ I]$.

Using this terminology, we know:

(14)   If a row of $\begin{bmatrix} -I \\ A \end{bmatrix}$ occurs in C_k, the corresponding component of x_k, $\tilde{x}_k$ is 0.

Let B_k be the submatrix of $[A \ I]$ consisting of those columns of $[A \ I]$ whose corresponding row of $\begin{bmatrix} -I \\ A \end{bmatrix}$ is *not* in C_k. The nonsingularity of C_k immediately gives the nonsingularity of B_k. Moreover, we know:

(15)   (i) the part of x_k, $\tilde{x}_k$ corresponding to columns outside B_k is zero;

(ii) the part of x_k, $\tilde{x}_k$ corresponding to columns in B_k is $B_k^{-1}b$.

$\left[\text{Here (i) follows from (14), while (ii) follows from (i) and from } [A \ I] \begin{pmatrix} x_k \\ \tilde{x}_k \end{pmatrix} = b. \right]$

Now define

(16) $(u'_k \ u''_k) := -(c, 0) + c_B B_k^{-1} [A \ I]$

where c_B is the part of the vector $(c, 0)$ corresponding to B_k. Then one can show that (u'_k, u''_k) is equal to the vector u_k as calculated in iteration k of the simplex method.

Indeed, u'_k, u''_k is 0 in components corresponding to columns in B_k, i.e. corresponding to rows outside C_k. Moreover,

(17) $$(u'_k, u''_k) \begin{bmatrix} -I \\ A \end{bmatrix} = -(c, 0) \begin{bmatrix} -I \\ A \end{bmatrix} + c_B B_k^{-1} [A \ I] \begin{bmatrix} -I \\ A \end{bmatrix} = c.$$

As u_k is the unique vector which is 0 in components corresponding to rows outside C_k and with $u_k \begin{bmatrix} -I \\ A \end{bmatrix} = c$, we know $u_k = (u'_k, u''_k)$.

The simplex tableau. The *simplex tableau associated with B_k* now is, by definition, the following matrix:

(18) $$\begin{bmatrix} u'_k & u''_k & cx_k \\ B_k^{-1}A & B_k^{-1} & B_k^{-1}b \end{bmatrix}.$$

All information necessary for carrying out the simplex method is contained in this tableau. By (15), the vector $(x_k, \tilde{x}_k)$ can be read off from $B_k^{-1}b$.

The following observation follows with (16) from linear algebra:

(19)   (18) is the unique matrix obtainable from $\begin{bmatrix} -c & 0 & 0 \\ A & I & b \end{bmatrix}$ by repeatedly:

  − multiplying a row by a nonzero real number;
  − adding a real multiple of a row (except for the top row) to another row (including the top row);
  in such a way that finally the submatrix B_k of the part $[A \ I]$ has become the identity matrix, with 0's above it in the top row.

The pivoting operation. Now if u'_k, u''_k is nonnegative, x_k is optimal (*Case 1* in Section 11.1). If not all components of u'_k, u''_k are nonnegative, select the lowest index i^* for which the i^*th component of u'_k, u''_k is negative (*Case 2*). The i^*th row of $\begin{bmatrix} -I \\ A \end{bmatrix}$ is *in* C_k, and hence the i^*th column of $[A \ I]$ is *out* of B_k.

What is now the vector y determined in *Case 2* of Section 11.1? By definition, y is the unique vector satisfying $dy = 0$ for each row d of C_k, except for the i^*th row of $\begin{bmatrix} -I \\ A \end{bmatrix}$, for which $dy = -1$. Putting $\tilde{y} := -Ay$, it means that $\begin{pmatrix} y \\ \tilde{y} \end{pmatrix}$ is the unique vector satisfying $[A \ I] \begin{pmatrix} y \\ \tilde{y} \end{pmatrix} = 0$ such that each component of $\begin{pmatrix} y \\ \tilde{y} \end{pmatrix}$ corresponding to a row in C_k is 0, except for the i^*th component, which is 1. Therefore, denoting the i^*th column of $[A \ I]$ by g,

(20) $\begin{pmatrix} -B_k^{-1}g \\ 1 \end{pmatrix}$ is the part of $\begin{pmatrix} y \\ \tilde{y} \end{pmatrix}$ corresponding to B_k, g,

where y, $\tilde{y}$ have 0's outside B_k, g (as rows in C_k correspond to columns out of B_k). This determines y, $\tilde{y}$.

Next, in *Case 2b*, we must replace $(x_k, \tilde{x}_k)$ by $(x_{k+1}, \tilde{x}_{k+1}) = (x_k, \tilde{x}_k) + \lambda_0(y, \tilde{y})$, where λ_0 is the largest number for which x_{k+1} is feasible. As $B_k^{-1}b$ is the part of x_k, $\tilde{x}_k$ corresponding to B_k, while x_k, $\tilde{x}_k$ is 0 outside B_k, this implies that λ_0 is the largest number for which

$$(21) \qquad \begin{pmatrix} B_k^{-1}b \\ 0 \end{pmatrix} + \lambda_0 \begin{pmatrix} -B_k^{-1}g \\ 1 \end{pmatrix} \geqslant \begin{pmatrix} 0 \\ 0 \end{pmatrix}$$

i.e.

$$(22) \qquad \lambda_0 = \min \{(B_k^{-1}b)_t/(B_k^{-1}g)_t \mid t = 1, \ldots, m; (B_k^{-1}g)_t > 0\}$$

$((\cdots)_t$ denoting tth component of $\cdots$). If t^* attains this minimum, let the t^*th column of B_k have index j^* as column of $[A\ I]$. In case there are several t^* attaining the minimum, we choose the t^* with j^* as small as possible.

So now we know that B_{k+1} arises from B_k by replacing the t^*th column by the i^*th column of $[A\ I]$. The simplex tableau associated with B_{k+1} is easily obtained from the simplex tableau associated with B_k: add real multiples of the t^*th row of the tableau (18) to the other rows, and multiply the t^*th row by a real, in such a way that finally the j^*th column has become a unit basis vector, with its 1 in position t^*. This *pivoting* does not change the $m - 1$ unit basis vectors e_s with $s \neq t^*$. So by observation (19) we have found the $(k + 1)$th simplex tableau.

If the minimum (22) does not exist, i.e. $B_k^{-1}g \leqslant 0$, then we saw in *Case 2a* that the maximum is unbounded, and y is a ray of the feasible region with $cy > 0$.

This finishes the algebraic description of the simplex method.

Note. It is not necessary that the order of the columns of B_k is the same as in $[A\ I]$. This means that the part corresponding to B_k in $[B_k^{-1}A\ B_k^{-1}]$ can be an identity matrix with its columns permuted (i.e. a permutation matrix). This does not conflict with the above. To keep track of the positions of the unit basis vectors, one sometimes writes the variables ξ_i to the right of the tableau, in such a way that if the unit basis vector e_s occurs as the jth column, we put ξ_j to the right of the sth row of the tableau.

Some simplex terminology. The variables ξ_i in x, $\tilde{x}$ for which the corresponding column of $[A\ I]$ belongs to B_k are the *(current) basic variables*. The other variables are the *(current) nonbasic variables*. The set of basic variables is the *(current) basis*. $(x_k, \tilde{x}_k)$ or just x_k is the *(current) basic (feasible) solution*. When applying a pivot step, the variable ξ_{j^*} is said to *enter the basis*, while ξ_{i^*} is said to *leave the basis*.

Summarizing the simplex method: Suppose we wish to solve the LP-problem

$$(23) \qquad \max \{cx \mid x \geqslant 0; Ax \leqslant b\}$$

where A is an $m \times n$-matrix, b is a column m-vector and c is a row n-vector.

('*Phase II*') $b \geqslant 0$. Start with the tableau

(24)
$$\begin{array}{cc|c} -c & 0 & 0 \\ \hline A & I & b \end{array}$$

Next apply the following *pivoting* operation as often as possible. Let a tableau of form

(25)
$$\begin{array}{c|c} u & \delta \\ \hline D & f \end{array}$$

be given where $D = (\delta_{ij})$ is an $m \times (n+m)$-matrix, $f = (\varphi_1, \ldots, \varphi_m)^{\mathsf{T}}$ is a column m-vector, $u = (v_1, \ldots, v_{n+m})$ is a row $(n+m)$-vector, and δ is a real number, satisfying

(26) (i) $f \geqslant 0$;

   (ii) for each $t = 1, \ldots, m$: there is a column $\sigma(t)$ of the matrix $\begin{bmatrix} u \\ D \end{bmatrix}$ which is a unit basis vector with its 1 in the tth row of D.

Clearly, (24) satisfies (26). Make a new tableau satisfying (26) as follows.

(27) (i) Choose the smallest j^* such that $v_{j^*} < 0$;
   (ii) Next choose $t^* = 1, \ldots, m$ such that $\delta_{t^* j^*} > 0$ and

   $$\varphi_{t^*}/\delta_{t^* j^*} = \min \{\varphi_t / \delta_{t j^*} | t = 1, \ldots, m; \delta_{t j^*} > 0\}$$

   and such that $\sigma(t^*)$ is as small as possible.

   (iii) Next divide the t^*th row of the tableau by $\delta_{t^* j^*}$. Moreover, add real multiples of the t^*th row to the other rows of the tableau so as to obtain in the j^*th column of the tableau only zeros except for the 1 in the t^*th position.

If no choice of j^* in (i) is possible, the current basic solution is optimal. If no choice of t^* in (ii) is possible, the maximum (23) is unbounded.

('*Phase I*') $b \not\geqslant 0$. Write $Ax \leqslant b$ as $A_1 x \leqslant b_1$, $A_2 x \geqslant b_2$, with $b_1 \geqslant 0$, $b_2 > 0$. Let A_1 and A_2 have order $m_1 \times n$ and $m_2 \times n$, respectively. Solve with the method described above the LP-problem

(28) $\max \{1(A_2 x - \tilde{x}) | x, \tilde{x} \geqslant 0; A_1 x \leqslant b_1; A_2 x - \tilde{x} \leqslant b_2\}$

where 1 denotes the all-one row m_2-vector. So we make as starting tableau:

(29)
$$\begin{array}{cccc|c} 1A_2 & 1 & 0 & 0 & 0 \\ \hline A_1 & 0 & I & 0 & b_1 \\ A_2 & -I & 0 & I & b_2 \end{array}$$

and next we apply the pivoting operation (27) as often as possible.

If in the final tableau the number δ (notation (25)) is strictly less than $1b_2$, then (23) is infeasible.

Otherwise, $\delta = 1b_2$. Let the slack vector variables be denoted by x', x'' with $A_1x + x' = b_1$ and $A_2x - \tilde{x} + x'' = b_2$. If one of the variables in x'' is still among the final basic variables, the corresponding basic feasible solution yet has 0 in this position. The corresponding unit basis column vector in the final tableau occurs among the $(m + n + 1)$th to $(m + n + m_2)$th. Now for each such slack variable in x'' which is still in the final basis, we multiply the corresponding row (i.e. the row in which the corresponding basic column vector has its 1) by -1. Next we leave out the $(m + n + 1)$th to $(m + n + m_2)$th columns from the tableau, and we leave out the top row. [Note that still all unit basis vectors occur among the columns, and that the tableau left could be made from

$$(30) \qquad \begin{array}{ccc|c} A_1 & 0 & I & b_1 \\ A_2 & -I & 0 & b_2 \end{array}$$

by multiplying rows by nonzero reals, and by subtracting multiples of one row from another row—so the new tableau represents a system of linear equations equivalent to the original system $A_1x + x' = b_1, A_2x - \tilde{x} = b_2$.]

We add as top row the row $(-c, 0, 0)$, where 0 is the all-zero row m-vector. We subtract multiples of the other rows from the top row in a such a way that 'above' the n unit basis vectors there are only 0's. After having done this, the tableau is in form (25), and comes from $\dfrac{-c \quad 0 \mid 0}{A \quad I \mid b}$ by the operations described in (19). We thus have obtained a good starting tableau for Phase II.

The big-M method. What we described above is the so-called *two-phases* method: Phase I consists of finding a feasible solution, Phase II of finding an optimal solution. Often, the two phases are combined in the so-called *big-M method*. In this variant, to solve (23), again $Ax \leqslant b$ is written as $A_1x \leqslant b_1, A_2x \geqslant b_2$ with $b_1 \geqslant 0$ and $b_2 > 0$. Then (23) is equivalent to

$$(31) \qquad \max\{cx \mid x, \tilde{x} \geqslant 0; \; A_1x \leqslant b_1; \; A_2x - \tilde{x} = b_2\}.$$

This LP-problem can be solved by solving the LP-problem:

$$(32) \qquad \max\{cx + M1(A_2x - \tilde{x}) \mid x, \tilde{x} \geqslant 0; \; A_1x \leqslant b_1; \; A_2x - \tilde{x} \leqslant b_2\}.$$

This is a problem with nonnegative right-hand sides and $\leqslant$-constraints, so that 'Phase II' can be applied directly. Here 1 denotes an all-one row vector, and M is a number that is chosen so large that it forces equality in $A_2x - \tilde{x} \leqslant b_2$, implying $A_2x \geqslant b_2$. Hence we would obtain an optimal solution for (31). Such an M indeed exists.

[To see this, let P and Q denote the feasible regions of (23) and (32), respectively. Let M satisfy:

$$(33) \qquad M > \max\left\{\left.\frac{cx'' - cx'}{1(b_2 - A_2x'' + \tilde{x}'')}\right| x' \text{ is a vertex of } P, \text{ and } \begin{pmatrix} x'' \\ \tilde{x}'' \end{pmatrix} \text{ is} \right.$$
$$\left. \text{a vertex of } Q \text{ with } A_2x'' - \tilde{x}'' \neq b_2 \right\},$$

$$\text{and } M > \max\left\{\left.\frac{cx}{1(\tilde{x} - A_2x)}\right| \begin{pmatrix} x \\ \tilde{x} \end{pmatrix} \text{ is an extremal ray of the characteristic cone of } Q \right.$$
$$\left. \text{with } A_2x - \tilde{x} \neq 0 \right\}.$$

Now after applying Phase II of the simplex method to problem (32) we are in one of the following three cases:

Case 1. We have an optimum solution $x^*, \tilde{x}^*$ for (32) satisfying $A_2 x^* - \tilde{x}^* = b_2$. Then $x^*, \tilde{x}^*$ is an optimum solution for (31), and hence x^* itself is an optimum solution for (23).

Case 2. We have an optimum solution $x^*, \tilde{x}^*$ for (32) with $A_2 x^* - \tilde{x}^* \neq b_2$. Then the problems (23) and (31) have no feasible solution. For suppose to the contrary that P is nonempty. Then P has a vertex, say, x'. As $\begin{pmatrix} x^* \\ \tilde{x}^* \end{pmatrix}$ is a vertex of Q with $A_2 x^* - \tilde{x}^* \neq b_2$, it follows from (33) that $cx' + M \cdot 1 b_2 > cx^* + M \cdot 1(A_2 x^* - \tilde{x}^*)$. So $x^*, \tilde{x}^*$ would not be an optimum solution of (32), as $(x', A_2 x' - b_2)$ is a better solution.

Case 3. The maximum (32) is unbounded, and the simplex method gives an extremal ray $\begin{pmatrix} x^* \\ \tilde{x}^* \end{pmatrix}$ of the characteristic cone $\left\{ \begin{pmatrix} x \\ \tilde{x} \end{pmatrix} \middle| x, \tilde{x} \geqslant 0; \ A_1 x \leqslant 0; \ A_2 x - \tilde{x} \leqslant 0 \right\}$ of Q with $cx^* + M \cdot 1(A_2 x^* - \tilde{x}^*) > 0$. By (33), this implies that $1(A_2 x^* - \tilde{x}^*) = 0$. Hence x^* is in the characteristic cone of P, with $cx^* > 0$. So also problem (23) is unbounded.]

It is not difficult to compute a number M satisfying (33) from the input data (23), in polynomial time (using, for example, the techniques of Chapter 10). However, much simpler than plugging in an explicit value of M is to operate with M as just a formal entity. That is, in the upper row of the simplex-tableau, instead of numbers, we now have linear forms $\alpha M + \beta$ in M, with α and β real numbers. If during the simplex iterations we have to add a multiple of another row to this upper row, we now may multiply this other row by such a linear form, and then add it to the upper row. Note that we never need to divide by a linear form. If, in order to select a pivot column, we have to find a negative entry in the upper row, we now have to find there a form $\alpha M + \beta$ with $\alpha < 0$, or with $\alpha = 0$ and $\beta < 0$ (so we use here a lexicographic ordering). This again gives a completely deterministic algorithm, and the above proof that the method terminates (including Bland's anti-cycling rule) will hold also for the more general case.

11.3. PIVOT SELECTION, CYCLING, AND COMPLEXITY

At the pivot step of the simplex method (cf. (27)):

(34)   (i) we must choose an index j_0 such that $v_{j_0} < 0$;
   (ii) we must choose an index k_0 such that $\delta_{k_0 j_0} > 0$, and

$$\varphi_{k_0}/\delta_{k_0 j_0} = \min \{\varphi_k/\delta_{k j_0} | k = 1, \ldots, m; \ \delta_{k j_0} > 0\}.$$

What to choose in case of ties? In our description above we have incorporated Bland's pivoting rule, which says that in (34) (i) we must choose the smallest possible j_0, and in (34) (ii) we must choose k_0 with $\sigma(k_0)$ as small as possible. That is, under the conditions given in (34), the variable with smallest possible index is chosen to enter the basis, and next the variable with smallest possible index is chosen to leave the basis.

Several other selection rules to break ties in (34) (i) have been proposed:

(35)   (i) choose j_0 with $v_{j_0} < 0$ and v_{j_0} as small as possible (Dantzig's original rule [1951a] or *nonbasic gradient method*);
   (ii) choose j_0 which gives the best improvement in the current objective

value δ, i.e. with $-v_{j_0}\varphi_{k_0}/\delta_{k_0j_0}$ as large as possible (*best improvement rule* (Dantzig [1951a]));

(iii) choose j_0 for which the edge from the old to the new basic solution has most acute angle with the objective vector c, i.e. with $v_{j_0}^2/(1 + \Sigma_{k=1}^m \delta_{kj_0}^2)$ as large as possible (*steepest edge rule* or *all-variable gradient method* (Dickson and Frederick [1960], Kuhn and Quandt [1963])).

These three rules (including (ii)) do not prescribe a choice of the k_0 in the case of ties in (34) (ii).

However, these rules do not prevent *cycling*: after a finite number of iterations the method can find a simplex tableau which has already occurred earlier, so that the method is in a loop and will not terminate (see Beale [1955], Hoffman [1953]). Dantzig, Orden, and Wolfe [1955] showed that the following lexicographic rule will avoid cycling:

(36)   (i) in (34) (i), choose an arbitrary j_0;
          (ii) in (34) (ii), choose k_0 such that the vector $(\delta_{k_01}, \ldots, \delta_{k_0n+m})/\delta_{k_0j_0}$ is lexicographically minimal.

Here we take the *backward* lexicographic ordering: $(\lambda_1, \ldots, \lambda_t)$ is *lexicographically less than* $(\mu_1, \ldots, \mu_t)$ if the two vectors differ, and for the highest index i with $\lambda_i \neq \mu_i$ one has $\lambda_i < \mu_i$.

Note that, if i_0 has been chosen, the choice of k_0 is unique, as the last m columns of the matrix D form a nonsingular matrix (notation of (25)).

Rule (36) will avoid cycling, even if we choose j_0 arbitrarily: at each pivot step the upper row (u, δ) of the simplex tableau is lexicographically increased (again in the backward lexicographic ordering).

Earlier, Charnes [1952] proposed the following method: perturb the problem slightly, by replacing the feasible region $\{x \geqslant 0 | Ax \leqslant b\}$ by $\{x \geqslant 0 | Ax \leqslant b + e\}$, where $e = (\varepsilon, \varepsilon^2, \ldots, \varepsilon^m)^T$, and where ε is a small positive number. If we choose ε small enough, the final tableau in the perturbed problem must have the same basis as the original problem. An appropriate number ε could be found explicitly with methods as described in Chapter 10. However, just as at the big-M method, we could consider ε as just a formal entity, and operate with polynomials in ε. Then the perturbance technique can be seen to coincide with the lexicographic rule described above.

Complexity of the simplex method. From a practical point of view most of the above considerations turn out to be irrelevant. In real-world problems, where usually Dantzig's pivot rule is used, the bad luck of having ties in the choices (34) and of cycling hardly occurs (Hoffman, Mannos, Sokolowsky, and Wiegmann [1953]). It is generally believed that in practice LP-problems are disinclined to be degenerate. [*Degenerate* means that there exist two bases with the same basic feasible solution. So cycling can occur only if the problem is degenerate.]

According to Kotiah and Steinberg [1978] however cycling occurs more often in practical problems than is generally believed (they mention problems from queueing theory), but at computations it is countered by the randomizing effect of rounding (they impute the general belief to the fact that instructors are generally not experienced practitioners) (see also Wolfe [1963]).

In practice the simplex method is very fast: experience suggest that the number of pivot steps is about linear in the problem dimensions. Also in theory the average number of steps can be shown to be linear, assuming a certain natural probabilistic model—see Section 11.5. First in Section 11.4 we discuss the worst-case behaviour of the simplex method.

Notes. Computational experiments with the simplex method are described by Hoffman, Mannos, Sokolowsky, and Wiegmann [1953], Dantzig [1963], Kuhn and Quandt [1963], Wolfe [1963], Wolfe and Cutler [1963], Mueller and Cooper [1965], Ravindran [1973], Crowder and Hattingh [1975], Orden [1976], Benichou, Gauthier, Hentges, and Ribière [1977], Borgwardt [1977b, 1982b], Dunham, Kelly, and Tolle [1977], Goldfarb and Reid [1997], Avis and Chvátal [1978], Ho and Loute [1980, 1983], Haimovich [1983], and Garcia, Gould, and Guler [1984]. For a survey, see Shamir [1987].

11.4. THE WORST-CASE BEHAVIOUR OF THE SIMPLEX METHOD

Klee and Minty [1972] showed that the simplex method, with Dantzig's pivoting rule (35) (i), is not a polynomial-time method. They constructed an artificial class of LP-problems for which the number of pivot steps grows exponentially fast. These problems have as feasible region a deformation of the n-dimensional cube, which makes the simplex method pass all 2^n vertices:

$$(37) \quad \text{LP}_n \quad \begin{cases} \text{maximize } 2^{n-1}\xi_1 + 2^{n-2}\xi_2 + \cdots + 2\xi_{n-1} + \xi_n \\ \text{subject to} \quad \xi_1 \qquad\qquad\qquad\qquad\qquad \leqslant 5 \\ \qquad\qquad 4\xi_1 + \xi_2 \qquad\qquad\qquad\qquad \leqslant 25 \\ \qquad\qquad 8\xi_1 + 4\xi_2 + \xi_3 \qquad\qquad\qquad \leqslant 125 \\ \qquad\qquad \vdots \qquad\qquad\qquad\qquad\qquad\qquad\quad \vdots \\ \qquad 2^n\xi_1 + 2^{n-1}\xi_2 + 2^{n-2}\xi_3 + \cdots + 4\xi_{n-1} + \xi_n \leqslant 5^n \\ \qquad\qquad \xi_1, \ldots, \xi_n \geqslant 0. \end{cases}$$

Theorem 11.2. *Applying the simplex method to* LP_n, *with Dantzig's pivot selection rule, requires* 2^n *tableaux.*

Proof. Let $\eta_1, \ldots, \eta_n$ be the slack variables added to LP_n. We first make two observations.

(38) (i) $\xi_1 = \xi_2 = \cdots = \xi_{n-1} = 0$, $\xi_n = 5^n$ is the unique optimal solution for LP_n. So LP_n has as basic variables in the final tableau: $\eta_1, \ldots, \eta_{n-1}, \xi_n$.

(ii) In each tableau and for each i, exactly one of ξ_i and η_i is a basic variable: otherwise (as there are n basic and n nonbasic variables), for some i, both ξ_i and η_i are nonbasic; then $\eta_i = 0$, and hence the ith constraint of LP_n has equality; moreover, $\xi_i = 0$, but then equality in the ith constraint contradicts the $(i-1)$th constraint.

We now show by induction on n:

(39) LP_n requires 2^n simplex tableaux, and in each of these tableaux the top row is integral.

For $n = 1$ this is trivial to check. Suppose we have shown this for LP_n. Consider the first and 2^nth tableau for LP_n:

(40)

$-c$	-1	0	0	0
A	0	I	0	b
$2c$	1	0	1	5^n

Tableau 1

c	0	0	1	5^n
A	0	I	0	b
$2c$	1	0	1	5^n

Tableau 2^n

Here we have written the objective function of LP_n as $cx + \xi_n$ (with $x = (\xi_1, \ldots, \xi_{n-1})^T$), the first $n-1$ constraints as $Ax \le b$, and the nth constraint as $2cx + \xi_n \le 5^n$. We also used observation (38) (i) to obtain Tableau 2^n.

We next claim that the following are tableaux for LP_{n+1} (ignore the shadows for a moment):

(41)

$-2c$	-2	-1	0	0	0	0
A	0	0	I	0	0	b
$2c$	1	0	0	1	0	5^n
$4c$	4	1	0	0	1	5^{n+1}

Tableau 1

$2c$	0	-1	0	2	0	$2 \cdot 5^n$
A	0	0	I	0	0	b
$2c$	1	0	0	1	0	5^n
$-4c$	0	1	0	-4	1	5^n

Tableau 2^n

$-2c$	0	0	0	-2	1	$3 \cdot 5^n$
A	0	0	I	0	0	b
$2c$	1	0	0	1	0	5^n
$-4c$	0	1	0	-4	1	5^n

Tableau 2^n+1

$2c$	2	0	0	0	1	5^{n+1}
A	0	0	I	0	0	b
$2c$	1	0	0	1	0	5^n
$4c$	4	1	0	0	1	5^{n+1}

Tableau 2^{n+1}

Clearly, Tableau 1 here is the first tableau for LP_{n+1}. If we delete the shadowed part in Tableau 1 we obtain Tableau 1 of LP_n with top row multiplied by 2. Hence the first 2^n tableaux for LP_{n+1} will be similar to those for LP_n: as long as the top row in the tableaux for LP_{n+1} has some negative components in the nonshadowed part, these components will be at most -2 (as by our induction hypothesis the top row in each tableau for LP_n is integral). Hence our pivot column will not be the $(n+1)$th column. Moreover, our pivot row will not be the bottom row, as this would yield both ξ_{n+1} and η_{n+1} nonbasic, contradicting (38) (ii).

Therefore, after $2^n - 1$ pivot steps, the basic variables are $\eta_1, \ldots, \eta_{n-1}, \xi_n,$ η_{n+1}. So Tableau 2^n for LP_{n+1} is as in (41). The 2^nth pivot step is unique and clear. Forgetting again the shadowed part in Tableau $2^n + 1$ we have again a tableau similar to Tableau 1 of LP_n: the difference is that the top row is multiplied by 2, there is a $3 \cdot 5^n$ in the right upper corner, and two columns are interchanged. But these differences are irrelevant: again from the 2^n tableaux necessary for LP_n we can derive that $2^n - 1$ pivot steps are necessary for LP_{n+1} to obtain the final Tableau 2^{n+1}. $\square$

Corollary 11.2a. *The simplex method with Dantzig's pivot selection rule is not a polynomial-time method.*

Proof. See Theorem 11.2 (note that the size of problem (37) is $O(n^3)$). $\square$

The example can be easily modified to obtain that also with Bland's pivot selection rule the method is not polynomial, which is a result of Avis and Chvátal [1978]:

Corollary 11.2b. *The simplex method with Bland's pivot selection rule is not a polynomial-time method.*

Proof. Reorder the variables in LP_n as $\xi_1, \eta_1, \xi_2, \eta_2, \ldots, \xi_n, \eta_n$. It can be checked that for leaving the basis we have chosen always the candidate first occurring in this new order (at choosing a pivot row there is never a tie). This can be obtained directly by the induction of the proof of Theorem 11.2. (If we wish to keep slack variables at the end of the tableau, we can just add some new slack variables $\zeta_1, \ldots, \zeta_n$ with cost 0.) $\square$

Similarly, Jeroslow [1973], Goldfarb and Sit [1979], and Goldfarb [1983] showed that the best improvement rule, the steepest edge rule, and Borgwardt's *Schatteneckenalgorithmus* (see Section 11.5), respectively, do not give a polynomial-time version of the simplex method. The 'bad' linear programs arise by a further deformation of the cube, like abrasion of edges, so that the simplex method overlooks a short way to the optimum vertex. The main problem seems to be that the simplex method is 'myopic', and cannot see 'cut-off paths'. It does not find a path which is locally bad but globally good.

Besides, experiments have indicated that if the constraints are of a combinatorial or structural nature (e.g. all coefficients are 0,1 or coming from 'Krawchuk polynomials'), the simplex method often is slow, and with the nonbasic gradient method there is a tendency to cycling (A. E. Brouwer, K. Truemper). Apparently, this is not of much influence on the average behaviour: it has been shown by Borgwardt [1982a, b] that the average running time of the simplex method is polynomially bounded—see Section 11.5.

It is still an open question, however, whether there exists a pivot rule which makes the simplex method polynomial-time for each linear program. (Note that if there is a pivot rule which can be shown to need only polynomially many iterations in solving any linear program, then it yields a polynomial-time method: in each simplex tableau all numbers have size polynomially bounded by the size of the original input data, as one easily derives with Cramer's rule from (16) and (18).)

Note 1. Related is the following problem. For $m, n \in \mathbb{N}$, let $f(m, n)$ be the smallest number such that if P is an n-dimensional polyhedron with m facets, then any two vertices of P are connected by a path of length at most $f(m, n)$ (where a *path, connecting* the vertices v_0 and v_t, is a series of vertices $v_0, \ldots, v_t$ of P such that v_{i-1} and v_i are adjacent in P, for $i = 1, \ldots, t$; the *length* of this path is t). Larman [1970] found $f(m, n) \leqslant 2^{n-2} m$ (in particular, $f(m, n)$ is defined). A prominent open question is: is $f(m, n)$ bounded by a polynomial in m and n? In 1957, W. M. Hirsch (cf. Dantzig [1963]: p. 160]) conjectured that, if we restrict P to *bounded* polyhedra, then $f(m, n) \leqslant m - n$. Klee and Walkup [1967] showed that we may not delete boundedness here, and they proved the conjecture if $m - n \leqslant 5$. See also Saaty [1955, 1963], Grünbaum and Motzkin [1962], Goldman and Kleinman [1964], Klee [1964a, b, 1965a, b, 1966a, b], Quandt and Kuhn [1964], Saigal [1969], Adler [1974], Adler and Dantzig [1974], Adler, Dantzig, and Murty [1974], Barnette [1974], Adler and Saigal [1976], Lawrence [1978], Walkup [1978], Mani and Walkup [1980], Todd [1980], and Altshuler [1985].

McMullen [1970] showed that an n-dimensional polytope with m facets has at least

$$(42) \qquad \binom{m - \left\lfloor \dfrac{n+1}{2} \right\rfloor}{m - n} + \binom{m - \left\lfloor \dfrac{n+2}{2} \right\rfloor}{m - n}$$

vertices (cf. Klee [1974], Brøndsted [1983: Ch. 3], McMullen and Shephard [1971]). Gale [1963, 1964] showed that (42) is best possible.

Note 2. Dantzig [1951b] and Orden [1955-6] have specialized the simplex method to so-called *transportation* or *network problems*, i.e. linear programs with $\{0, \pm 1\}$-constraint matrix, with at most one $+1$ and at most one -1 in each column. The simplex tableaux then are replaced by 'transportation tableaux'.

Zadeh [1973a, b] and Cunningham [1979] have given network problems for which certain pivot selection rules imply exponentially many iterations in the worst case (cf. also Niedringhaus and Steiglitz [1978], Gassner [1964], Cunningham and Klincewicz [1983]).

Cunningham [1976, 1979] and Barr, Glover, and Klingman [1977] designed an anti-cycling rule for network problems. Ikura and Nemhauser [1983] gave a pivot selection rule for network problems which, combined with a scaling technique of Edmonds and Karp [1972], gives a polynomial-time method for network problems. Orlin [1985a, b] gives a pivot selection rule implying a strongly polynomial upper bound on the number of iterations, i.e. the number of iterations is bounded by a polynomial in the problem dimensions (and not in the sizes of the input numbers). See also Roohy-Laleh [1981], Balinski [1983, 1984, 1985], Hung [1983], Hung and Rom [1980], Bertsekas [1981], and Goldfarb [1985].

For the practical implementation of the simplex method for network problems, see Ali, Helgason, Kennington, and Lall [1978] and Glover and Klingman [1976].

11.5. THE AVERAGE RUNNING TIME OF THE SIMPLEX METHOD

Recently, Borgwardt [1982a, b] was able to show that on the average the simplex method is a polynomial-time method, in a certain natural probabilistic model, and using an interesting pivot selection rule, the *Schatteneckenal-*

gorithmus. Borgwardt showed that the average number of (Phase II) pivot steps at solving

(43) $\max\{cx\,|\,Ax \leqslant b\}$

is $O(n^3 m^{1/(n-1)})$. Here Borgwardt assumed that b is positive. (Throughout this section, m and n denote the number of rows and of columns, respectively, of A.)

Using a different probabilistic model, and a different pivot rule, Smale [1983a, b] independently showed, for every fixed m, an upper bound of $O((\log n)^{m^2 + m})$ at solving

(44) $\max\{cx\,|\,x \geqslant 0;\ Ax \leqslant b\}$.

Smale's bound is not polynomially bounded, but sometimes is better than Borgwardt's (e.g. if m is fixed and $n \to \infty$).

Haimovich [1983] combined Borgwardt's pivoting rule with a more general version of Smale's probabilistic model, and showed that the average number of pivot steps is in fact linear: it can be bounded by

(45) $\min\{\tfrac{1}{2}n, \tfrac{1}{2}(m - n + 1), \tfrac{1}{8}(m + 1)\}$,   for problems of type (43)

$\min\{\tfrac{1}{2}n, \tfrac{1}{2}(m + 1), \tfrac{1}{8}(m + n + 1)\}$,   for problems of type (44).

[Here Borgwardt's condition $b > 0$ is no longer assumed.] This agrees with the practical experience that the number of pivot steps is about linear.

We restrict ourselves to deriving the bound $\tfrac{1}{2}n$ for problems of type (43). The model is quite natural. The LP-problems are given as (43), together with an initial vertex x_0. Next we choose *at random* a row vector $\bar{c}$ such that

(46) $\bar{c}x_0 = \max\{\bar{c}x\,|\,Ax \leqslant b\}$.

[It is easy to find for any vertex x_0 a vector $\bar{c}$ satisfying (46) (e.g. we could add up the rows of A which have equality in $Ax_0 \leqslant b$), but this is not a random choice.] The function $x \to \bar{c}x$ is called the *co-objective function*. Now we assume that our probabilistic measure and our random choice satisfy the following conditions:

(47)   (i)   for each fixed $\bar{c}$, reversing some inequality signs in $Ax \leqslant b$ (so that the maximum of $\bar{c}x$ over the modified system is finite), or replacing c by $-c$, does not change the probability of A, b, c, $\bar{c}$;

(ii)   the class of choices A, b, c, $\bar{c}$ for which there exist n or less linearly dependent rows in

$$\begin{bmatrix} c \\ \bar{c} \\ A \end{bmatrix}$$

or $n + 1$ or less linearly dependent rows in $[A,\ b]$, has probability 0.

In particular, an LP-problem is nondegenerate with probability 1.

The pivoting rule for deriving the upper bound of $\tfrac{1}{2}n$ is Borgwardt's *Schatteneckenalgorithmus:*

(48)   Given a vertex x_k of P maximizing $(\lambda c + \bar{c})x$ over P, for some $\lambda \geqslant 0$, but
          not maximizing cx over P, find a vertex x_{k+1} of P adjacent to x_k which
          maximizes $(\lambda'c + \bar{c})x$ over P for some $\lambda' > \lambda$, or find a ray y starting in x_k
          for which $(\lambda'c + \bar{c})y > 0$ for some $\lambda' > \lambda$.

So starting with $x_k = x_0$ and $\lambda = 0$, this describes a pivoting rule for the simplex
method. Below we shall see that rule (48) is a most natural and elegant tool in
studying the average speed of the simplex method.

It is not difficult to implement rule (48) in the simplex tableau framework: if for
$c_k := \lambda c + \bar{c}$, vertex x_k maximizes $c_k x$ over P, we can make the corresponding simplex
tableau T with nonnegative top row. Next replace c_k by $c_k + \mu c$, for $\mu > 0$, making μ
just large enough that the top row in the tableau corresponding to $c_k + \mu c$ is nonnegative,
but with a 0 in at least one position which was positive in T. Then we can pivot, and find
x_{k+1}. For details, see Borgwardt [1982a].

Note that, with probability 1, we have:

(49) $\begin{aligned} cx_{k+1} &> cx_k \\ \bar{c}x_{k+1} &< \bar{c}x_k \end{aligned}$

since $cx_{k+1} \neq cx_k$, $\bar{c}x_{k+1} \neq \bar{c}x_k$ (by (47) (ii)), and since (writing $c_k := \bar{c} + \lambda c$ and
$c_{k+1} := \bar{c} + \lambda'c$)

(50) $(\lambda' - \lambda)(cx_{k+1} - cx_k) = (c_{k+1}x_{k+1} - c_{k+1}x_k) + (c_k x_k - c_k x_{k+1}) \geqslant 0$
          $(\lambda' - \lambda)(\bar{c}x_k - \bar{c}x_{k+1}) = \lambda'(c_k x_k - c_k x_{k+1}) + \lambda(c_{k+1}x_{k+1} - c_{k+1}x_k) \geqslant 0$

which imply (49).
One similarly shows that if one finds in (48) a ray y of P, then

(51) $cy > 0, \bar{c}y < 0$

giving that (43) is unbounded.
Geometrically, step (48) means the following. Project P onto the plane spanned
by c^T and $\bar{c}^T$. We obtain a polyhedron Q in two dimensions. The vertices of Q
are exactly the projections of those vertices x of P which maximize $(\lambda c + \mu\bar{c})x$
for some λ, μ. The step from x_k to x_{k+1} in (48) means in Q that we go from the
projection of x_k to the adjacent vertex in the direction of c. This explains the
name *Schatteneckenalgorithmus* ('shadow vertices algorithm').
We now prove Haimovich's bound.

Theorem 11.3. *The class of LP-problems, given as in* (43), *can be solved with the
simplex method, using the Schatteneckenalgorithmus, in at most* $\frac{1}{2}n$ *pivot iterations
on the average, under any probability measure satisfying* (47).

Proof. Let the LP-problem max $\{cx \mid Ax \leqslant b\}$ be given, together with a vertex
x_0 of $\{x \mid Ax \leqslant b\}$ and a vector $\bar{c}$ such that x_0 maximizes $\bar{c}x$ over $Ax \leqslant b$. Let
m and n be the numbers of rows and columns of A.
Let $z_1, \ldots, z_t$ be those vectors in $\mathbb{R}^n$ which can be determined by some choice

of n linearly independent equations from $Ax = b$. So $t = \binom{m}{n}$ with probability 1.

Consider the class $\mathcal{L}$ of LP-problems $\max\{\tilde{c}x \mid \tilde{A}x \leqslant \tilde{b}\}$ which arise from $\max\{cx \mid Ax \leqslant b\}$ by reversing some (or none) of the inequality signs and/or replacing c by $-c$, and for which

(52) $\max\{\tilde{c}x \mid \tilde{A}x \leqslant \tilde{b}\}$

is finite. We claim $|\mathcal{L}| = 2\binom{m}{n}$. Indeed, by (47), with probability 1, each $\max\{\tilde{c}x \mid \tilde{A}x \leqslant \tilde{b}\}$ is uniquely attained by a vertex of $\tilde{A}x \leqslant \tilde{b}$, and hence by one of the vectors z_i. Moreover, for each z_i there is exactly one choice of $\tilde{A}x \leqslant \tilde{b}$ for which $\max\{\tilde{c}x \mid \tilde{A}x \leqslant \tilde{b}\}$ is attained by z_i. [Since, with probability 1, there are exactly n inequalities in $\tilde{A}x \leqslant \tilde{b}$, say $\tilde{A}'x \leqslant \tilde{b}'$, which are satisfied by z_i with equality. Hence the inequality sign is strict in the other $m - n$ inequalities, and therefore we cannot reverse the sign. Moreover, $\tilde{c}$ must be in the cone generated by the rows of $\tilde{A}'$, and hence also here the signs are uniquely determined.] So to z_i there correspond the programs $\max\{cx \mid \tilde{A}x \leqslant \tilde{b}\}$ and $\max\{-cx \mid \tilde{A}x \leqslant \tilde{b}\}$. Therefore, $|\mathcal{L}| = 2\binom{m}{n}$.

Next, consider the class of edges and rays formed by the systems $\tilde{A}x \leqslant \tilde{b}$. Each set of $n - 1$ equations from $Ax = b$ gives a line (1-dimensional affine subspace) in $\mathbb{R}^n$, containing $m - n + 1$ of the z_i. The z_i subdivide this line into $m - n$ edges and two rays.

Let l be such a line, and let z_i and z_j be two neighbouring points on l (i.e. the line segment $\overline{z_i z_j}$ contains no other points from $z_1, \ldots, z_t$). Then the edge $\overline{z_i z_j}$ is traversed in at most (in fact, exactly) one of the programs in $\mathcal{L}$. For suppose $\overline{z_i z_j}$ is traversed in more than one of these programs $\max\{\tilde{c}x \mid \tilde{A}x \leqslant \tilde{b}\}$. By (48), there is a λ such that

(53) $\overline{z_i z_j} = \{x \mid \tilde{A}x \leqslant \tilde{b}\} \cap \{x \mid (\bar{c} + \lambda \tilde{c})x = (\bar{c} + \lambda \tilde{c})z_i\}$

and such that $(\bar{c} + \lambda \tilde{c})x \leqslant (\bar{c} + \lambda \tilde{c})z_i$ is valid for $\{x \mid \tilde{A}x \leqslant \tilde{b}\}$. By (47) we know that with probability 1 there are exactly $n - 1$ of the inequalities $\tilde{A}x \leqslant \tilde{b}$, which are tight on $\overline{z_i z_j}$, say $\tilde{A}'x \leqslant \tilde{b}'$. Hence the inequality signs in the other inequalities cannot be reversed. Moreover, there is only one vector u and only one λ with $uA' = \bar{c} + \lambda \tilde{c}$ (with probability 1). Since $(\bar{c} + \lambda \tilde{c})x \leqslant (\bar{c} + \lambda \tilde{c})z_i$ is a supporting half-space, we know $u > 0$. The uniqueness of u implies the uniqueness of the inequality signs in $\tilde{A}x \leqslant \tilde{b}$. Finally, let without loss of generality $\bar{c}z_i < \bar{c}z_j$. If $cz_i > cz_j$, then (49) implies $\tilde{c} = c$. If $cz_i < cz_j$, (49) implies $\tilde{c} = -c$. So there is also only one choice for $\tilde{c}$. Concluding, the edge $\overline{z_i z_j}$ is traversed in only one program in $\mathcal{L}$.

Similarly, each ray $z_i + y\mathbb{R}_+$ is traversed only if $\bar{c}y > 0$ (by (51)), and then in only one program in $\mathcal{L}$.

Now on each line determined by $n - 1$ equations from $Ax = b$ there are $m - n$

edges and one ray $z_i + y\mathbb{R}_+$ with $\bar{c}y > 0$. So the total number of such edges and such rays is $(m - n + 1)\binom{m}{n-1} = \binom{m}{n}n$. Therefore, the average number of pivot steps is at most:

(54) $$\frac{\binom{m}{n}n}{|\mathscr{L}|} = \tfrac{1}{2}n.$$ $\square$

The theorem also applies to programs of type $\max\{cx\,|\,x \geqslant 0;\ Ax \leqslant b\}$. Then flipping inequalities in the system $x \geqslant 0$ is equivalent to replacing columns of A by their opposites. Consider any probability measure which is invariant under multiplying rows or columns of $\begin{bmatrix} c & 0 \\ A & b \end{bmatrix}$ by -1, and such that the class of LP-programs for which

(55) $$\begin{bmatrix} c \\ \bar{c} \\ A \end{bmatrix} \quad \text{or} \quad [A\ b]$$

has a singular submatrix, has probability 0. Then the class of LP-problems $\max\{cx\,|\,x \geqslant 0;\ Ax \leqslant b\}$, with given initial vertex x_0 and vector $\bar{c}$ such that x_0 maximizes $\bar{c}x$ over $x \geqslant 0;\ Ax \leqslant b$, can be solved on the average in $\tfrac{1}{2}n$ pivot steps (n = number of columns of A). To derive this from Theorem 11.3, we can put probability measure 0 on the set of all problems which are not of type $\max\{cx\,|\,x \geqslant 0;\ Ax \leqslant b\}$.

So on the average the simplex method is very good: the average number of pivot steps is linear in the number of variables (note that the size of the numbers occurring in the simplex tableaux is polynomially bounded by the size of the original linear program).

Haimovich [1983] obtained some better bounds (always $m :=$ number of rows of A, $n :=$ number of columns of A):

(56) (i) $\left(\dfrac{2}{n} + \dfrac{2}{m+1}\right)^{-1}\ [\leqslant \tfrac{1}{2}\min\{n, m+1, \tfrac{1}{4}(m+n+1)\}]$ for problems of type $\max\{cx\,|\,x \geqslant 0;\ Ax \leqslant b\}$;

(ii) $\left(\dfrac{2}{n} + \dfrac{2}{m-n+1}\right)^{-1}\ [\leqslant \tfrac{1}{2}\min\{n, m-n+1, \tfrac{1}{4}(m+1)\}]$ for problems of type $\max\{cx\,|\,Ax \leqslant b\}$;

(iii) $\left(\dfrac{2}{n-m} + \dfrac{2}{m+1}\right)^{-1}\ [\leqslant \tfrac{1}{2}\min\{n-m, m+1, \tfrac{1}{4}(n+1)\}]$ for problems of type $\max\{cx\,|\,x \geqslant 0;\ Ax = b\}$.

Haimovich describes some more bounds, and he gives a comparative analysis of probabilistic models designed by him and other authors.

Earlier work on the average behaviour of the simplex method is described in Orden [1971, 1976, 1979, 1980], Liebling [1973], Borgwardt [1977a, b, 1978, 1979], Dantzig [1980], Körner [1980], Kelly [1981], Ross [1981], and May and Smith [1982].

Adler [1983], Adler, Karp, and Shamir [1986, 1987], Adler and Megiddo [1983, 1984], Blair [1986], Haimovich [1984a, b], Megiddo [1986a], Saigal [1983, Todd [1986] and Vershik and Sporyshev [1983] made a further analysis of the average speed of the simplex method and of the related Lemke's algorithm for linear complementarity problems.

Schmidt [1968] showed that, in a certain probabilistic model, for each dimension n there exists a constant c_n such that the average number of vertices of systems $Ax \leqslant b$ is at most c_n. Berenguer [1978] (cf. Adler and Berenguer [1981]), and independently May and Smith [1982], showed $c_n = 2^n$. See also Rényi and Sulanke [1963, 1968], Cover and Efrom [1967], Carnal [1970], Prékopa [1972], Kelly and Tolle [1979, 1981], Adler and Berenguer [1983], and Berenguer and Smith [1986]. See Rényi and Sulanke [1963, 1968], Shamos [1978], and May and Smith [1982] for further surveys and references.

See Shamir [1987] for a survey on the average speed of the simplex method.

11.6. THE REVISED SIMPLEX METHOD

When executing the simplex method it is not essential to write down the tableaux completely: each tableau is determined already by the choice of the basic variables, and we could restrict our administration. This suggests a reduction in the complexity of the method, especially if the number of columns is high relative to the number of rows. It has led to the *revised simplex method*. This method calculates and stores only the last $m + 1$ columns of the tableau (using the notation of Section 11.2) together with the basic variables. That is, the following *reduced tableau* is stored, where B^{-1} is a matrix with m columns:

$$
(57) \quad
\begin{array}{c|cc}
v & \delta & \xi_0 \\
 & & \xi_{\sigma(1)} \\
\hline
B^{-1} & f & \vdots \\
 & & \xi_{\sigma(m)}
\end{array}
$$

So if we have started with (24), and we have stored (57), then the corresponding tableau in the original nonrevised simplex method is:

$$
(58) \quad
\begin{array}{cc|cc}
vA - c & v & \delta & \xi_0 \\
 & & & \xi_{\sigma(1)} \\
\hline
B^{-1}A & B^{-1} & f & \vdots \\
 & & & \xi_{\sigma(m)}
\end{array}
$$

In order to make a pivot step we must choose a negative entry in v or in $vA - c$. After that the next reduced tableau can be determined from the reduced tableau (57) and from the column in (58) corresponding to the chosen negative component of $(vA - c, v)$. Termination of the method can be guaranteed by using the lexicographic rule (36).

So we only call for a column of A when we need it as a pivot column. This *column generation technique* can be useful if n is large compared with m. As an example, in combinatorial applications we sometimes must solve an LP-problem of form

(59) $\min \{yb \mid y \geqslant 0; yA \geqslant c\}$

where all constraints in $yA \geqslant c$ are known, but not explicitly written down. Although writing down all inequalities might be time-consuming, there sometimes exists a fast subroutine which tests if a given vector y satisfies $y \geqslant 0$, $yA \geqslant c$, and which finds a violated inequality, if it exists. In that case we can solve the dual problem

(60) $\max \{cx \mid x \geqslant 0; Ax \leqslant b\}$

with the revised simplex method. If we have the reduced tableau (57), we can test with the fast subroutine whether there is a negative component in v or in $vA - c$. If so, this negative component determines the pivot column, and we can calculate the next reduced tableau. If not, we know that (57) corresponds to an optimal tableau, and we can find the optimum solutions for (59) and (60).

   This column generation technique connects to the discussion of the separation and the optimization problem, and their polynomial equivalence by the ellipsoid method, in Chapter 14.

Example. Let $D = (V, A)$ be a directed graph, and let $r, s \in V$. Let $\mathscr{P}$ be the collection of $r - s$-paths in D. Let M be the $A \times \mathscr{P}$-matrix with

(61) $M_{a,P} = 1$ if a occurs in P,
                  $= 0$ otherwise,

for $a \in A$, $P \in \mathscr{P}$. Let $b:A \to \mathbb{R}_+$. One version of the maximum flow problem is to solve

(62) $\max \{1x \mid x \geqslant 0; Mx \leqslant b\}$.

So b is the capacity function here. [Note that, for each $z \in \mathbb{R}^A$, z is an $r - s$-flow, iff $z = Mx$ for some $x \in \mathbb{R}_+^{\mathscr{P}}$. The value of z then is equal to $1x$. 1 denotes an all-one row vector.]
   The dual of (62) is:

(63) $\min \{yb \mid y \geqslant 0; yM \geqslant 1\}$.

We can solve this problem with the revised simplex method, without writing down the whole matrix M (which can have very many columns). We only must be able to test if a given vector y satisfies: $y \geqslant 0$; $yM \geqslant 1$. This is easy: if $y \geqslant 0$, then $yM \geqslant 1$ holds iff each $r - s$-path in D has length at least 1, considering y as the length function. So $y \geqslant 0$; $yM \geqslant 1$ can be tested with Dijkstra's algorithm. If $yM \geqslant 1$ does not hold, Dijkstra's method gives a path of length less than 1, i.e. a violated inequality in $yM \geqslant 1$.

Note. The revised simplex method was developed by Dantzig [1953], Orchard-Hays [1954], and Dantzig and Orchard-Hays [1954]. For an analysis, see Smith and Orchard-Hays [1963] and Wagner [1957a]. Column generation techniques were described by Eisemann [1957], Ford and Fulkerson [1958-9], Gilmore and Gomory [1961], and Manne [1957-8].

11.7. THE DUAL SIMPLEX METHOD

   Lemke [1954] and Beale [1954] designed a dual version of the simplex method, as follows. Let a simplex tableau:

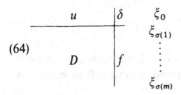

(64)

be given, satisfying:

(65)   (i) $u \geqslant 0$;
        (ii) for all $i = 1, \ldots, m$: column $\sigma(i)$ of the matrix $\begin{bmatrix} u \\ D \end{bmatrix}$ is a unit basis
        vector, with its 1 in the ith row of D

(so now f may have negative entries, but u not). Then we can apply the following *dual pivot step*:

(66)   (i)   Choose $t_0 = 1, \ldots, m$ such that $\varphi_{t_0} < 0$ and such that $\sigma(t_0)$ is as small as possible.
        (ii) Next choose the smallest j_0 such that $\delta_{t_0 j_0} < 0$ and

$$v_{j_0}/\delta_{t_0 j_0} = \max \{v_j/\delta_{t_0 j} | j = 1, \ldots, n + m; \delta_{t_0 j} < 0\}.$$

        (iii) Apply (iii) of the (primal) pivot step (27).

[Denoting $u = (v_1, \ldots, v_{n+m})$ and $f = (\varphi_1, \ldots, \varphi_m)^T$.] After application of a dual pivot step, we have again a tableau of form (64) satisfying (65). Similarly as in Section 11.2 one shows that after a finite number of repetitions of the dual pivot step one will end up with a tableau where no choices as described in (66) (i) and (ii) are possible. If no choice as in (i) is possible then $f \geqslant 0$, and the basic solution is feasible, and optimal, as $u \geqslant 0$, thus solving the corresponding LP-problem. If the choice in (i) is possible, but not the choice in (ii), then the problem is infeasible: row t_0 of the tableau corresponds to a linear equation, valid for the feasible region, which equation has nonnegative coefficients and a negative right-hand side. This equation has no nonnegative solution.

This method is called the *dual simplex method*, as opposed to the *primal* simplex method described before. The dual simplex method can be considered as the primal simplex method applied to the dual problem, in an alternative administration. Note that the δ in the dual simplex method is monotonically non-increasing.

As an application, suppose we wish to solve the LP-problem min $\{cx | x \geqslant 0;$ $Ax \leqslant b\}$, with c nonnegative. This can be done by making up the initial tableau (24) for the equivalent problem max $\{-cx | x \geqslant 0; Ax \leqslant b\}$. This tableau satisfies (65). So we can apply the dual simplex method, and the final tableau will give the optimum solution. So in this case the problem can be solved without the 'big-M' method, even if not all components of b are nonnegative.

Adding extra constraints. Another application of the dual simplex method is as follows. Suppose we have solved, with the primal or the dual simplex method, the LP-problem max $\{cx | Ax \leqslant b\}$, where the final tableau is as in (64), with $u \geqslant 0$, $f \geqslant 0$. Now we want to add an extra constraint, say $a'x \leqslant \beta'$, to the LP-problem.

That is, we want to solve

(67) $\max \{cx \,|\, Ax \leqslant b; a'x \leqslant \beta'\}$.

Instead of applying the simplex method anew to solve this LP-problem, we can use the final tableau for $\max \{cx \,|\, Ax \leqslant b\}$. To this end, add to (64) an extra row and column as follows:

(68)

	u		0	δ	ξ_0
			0		$\xi_{\sigma(1)}$
	D		$\vdots$	f	$\vdots$
			$\vdots$		$\vdots$
			0		$\xi_{\sigma(m)}$
	a'	$0\cdots\cdots 0$	1	β'	ξ_{n+m+1}

Here ξ_{n+m+1} is a new 'slack' variable. That is, we added the constraint $a'x + \xi_{n+m+1} = \beta'$. Since $u \geqslant 0$, (68) satisfies (65)(i). In order to fulfil (65)(ii), for $i = 1, \ldots, m$, subtract a real multiple of the ith row in (68) from the bottom row, so that the bottom row will have zeros in positions corresponding to the basic variables $\xi_{\sigma(1)}, \ldots, \xi_{\sigma(m)}$. After this, (65) is satisfied, and we can apply the dual simplex method. The final tableau solves (67).

Changing right-hand sides. The dual simplex method can also be applied if after solving $\max \{cx \,|\, x \geqslant 0, Ax \leqslant b\}$ with, say, the primal simplex method, we replace the initial RHS b by a slight perturbation b'. We may hope that a new optimum solution is not far from the old. Thus we replace the column $B^{-1}b$ of the final tableau by $B^{-1}b'$. If $B^{-1}b'$ is nonnegative we immediately have an optimum tableau, and an optimum solution, for the new problem. If some components of $B^{-1}b'$ are negative, we can carry out some dual pivot steps to obtain a new optimum solution.

Notes for Chapter 11. The combinatorial aspects of the simplex method are abstracted in the setting of *oriented matroids*—see Bland [1977b] (cf. Bland [1978], Cottle and Dantzig [1968], Rockafellar [1969], Todd [1976]).

Crowder and Hattingh [1975] study the 'partially normalized pivot selection rule'.

12

Primal–dual, elimination, and relaxation methods

In this chapter we collect three methods for linear programming and linear inequalities: the primal–dual method (Section 12.1), the Fourier–Motzkin elimination method (Section 12.2), and the relaxation method (Section 12.3).

12.1. THE PRIMAL–DUAL METHOD

As a generalization of similar methods for network flow and transportation problems, Dantzig, Ford, and Fulkerson [1956] designed the *primal–dual method* for linear programming. The general idea is as follows. Starting with a dual feasible solution y, the method searches for a primal feasible solution x satisfying the complementary slackness condition with respect to y. If such a primal solution has been found, x and y are optimum primal and dual solutions. If no such primal solution is found, the method prescribes a modification of the dual feasible solution y, after which we start anew.

The problem now is how to find a primal feasible solution satisfying the complementary slackness condition, and how to modify the dual solution if no such primal solution is found. For general linear programs this problem can be seen to amount to another linear program, which is generally simpler than the original linear program. To solve this simpler problem, we could insert any LP-method, e.g. the simplex method. In many combinatorial applications, however, this simpler LP-problem is a simpler combinatorial optimization problem, for which direct methods are available. Thus, if we can describe a combinatorial optimization problem as a linear program (which is a main issue in *polyhedral combinatorics*) the primal–dual method provides us with a scheme for reducing one combinatorial problem to an easier combinatorial problem.

We describe the primal–dual method more precisely. Suppose we wish to solve the LP-problem

(1) $\min \{ cx \mid x \geqslant 0; Ax = b \}$

which is one form of a general LP-problem, where $A = [a_1, \ldots, a_n]$ is an $m \times n$-matrix, b is an m-vector, and $c = (\gamma_1, \ldots, \gamma_n)$ is an n-vector. The dual LP-problem is:

(2) $\max \{yb \mid yA \leqslant c\}$.

The primal–dual method consists of repeating the following iteration.

Primal–dual iteration. Suppose we have a feasible solution y_0 for the dual problem (2). Let A' be the submatrix of A consisting of those columns a_j of A for which $y_0 a_j = \gamma_j$. To find a feasible primal solution for which the complementary slackness condition holds, solve the *restricted linear program*:

(3) $\min \{\lambda \mid x', \lambda \geqslant 0; A'x' + b\lambda = b\} = \max \{yb \mid yA' \leqslant 0; yb \leqslant 1\}$.

If the optimum value in (3) is 0, let x'_0, λ be an optimum solution for the minimum. So $x'_0 \geqslant 0$, $A'x'_0 = b$ and $\lambda = 0$. Hence by adding zero-components, we obtain a vector $x_0 \geqslant 0$ such that $Ax_0 = b$ and $(x_0)_j = 0$ if $y_0 a_j < \gamma_j$. By complementary slackness it follows that x_0 and y_0 are optimum solutions for (1) and (2).

If the optimum value in (3) is strictly positive, then it is 1. Let y be an optimum solution for the maximum. Let θ be the largest rational number such that

(4) $(y_0 + \theta y)A \leqslant c$.

[Note that $\theta > 0$, since $yA' \leqslant 0$, $y_0 A \leqslant c$ and $y_0 a_j < \gamma_j$ if a_j is not in A'. Note also that θ is easy to calculate. In the case $\theta = \infty$, the maximum (2) is unbounded, as then $(y_0 + \theta y)A \leqslant c$ and $(y_0 + \theta y)b \to \infty$ if $\theta \to \infty$]. Reset $y_0 := y_0 + \theta y$, and start the iteration anew.

Further discussion. The method reduces problem (1) to problem (3), which is often an easier problem. It actually consists of testing feasibility of: $x' \geqslant 0$; $A'x' = b$. In several combinatorial applications this turned out to be a successful approach (see Application 12.1).

We have yet to discuss some points. The method can only work if the minimum (1) is not $-\infty$, as otherwise there are no dual solutions. Often an initial dual feasible solution can be derived easily. If not, we could add to the constraints in (1) the extra constraints:

(5) $\xi_0 + \xi_1 + \cdots + \xi_n = M, \quad \xi_0 \geqslant 0$

where $x = (\xi_1, \ldots, \xi_n)^{\mathsf{T}}$ and ξ_0 is a new variable, and where M is an upper bound on $\xi_1 + \cdots + \xi_n$ for some optimum solution for (1). Such an M could be found, for example, with the methods described in Section 10.1. Then adding (5) to (1) does not change the problem. Moreover, the extended problem has a direct dual feasible solution. (The idea of adding (5) is due to Beale [1954] and others.)

Next observe that if, in a certain iteration, we have found a primal optimum solution x'_0, λ for (3) with $\lambda = 1$, this gives a primal feasible solution in the next iteration. Indeed, if y is an optimum dual solution in (3), and $(x'_0)_j$ is positive (so a_j occurs in A'), then by complementary slackness, $ya_j = 0$. Hence $(y_0 + \theta y)a_j = y_0 a_j = \gamma_j$. This gives that column a_j occurs in A' in the next iteration again.

More strongly, the last argument holds also if we replace '$(x'_0)_j$ is positive' by 'variable $(x')_j$ is basic in the final tableau', when solving (3) with the simplex method. This implies that the optimum tableau in one iteration can be used as starting tableau in the next iteration, after replacing some nonbasic columns by some other columns, corresponding to the change in A'. At the first iteration we could start with the basic feasible solution $x' = 0$, $\lambda = 1$.

Since any basic feasible solution in (3) is also a basic feasible solution for the problem $\min \{\lambda \,|\, x, \lambda \geq 0; Ax + b\lambda = b\}$, and since the optima (3) in the subsequent iterations do not increase (as the optimum solution is feasible in the next iteration), the primal–dual algorithm terminates if we can avoid repetition of bases. This can be done, for example, by using the lexicographic rule (cf. Section 11.3).

In many combinatorial applications one uses other methods than the simplex method to solve (3), which usually directly imply termination of the method. Then often the primal–dual method only serves as a manual to design a direct algorithm, without any explicit reference to linear programming (cf. Papadimitriou and Steiglitz [1982]).

Remark. Note that the primal–dual method can equally be considered as a *gradient method*. Suppose we wish to solve $\max \{yb \,|\, yA \leq c\}$, and we have a feasible solution y_0. This y_0 is not optimal iff we can find a vector u such that $ub > 0$ and u is a *feasible direction* (i.e. $(y_0 + \lambda u)A \leq c$ for some $\lambda > 0$). If we let A' consist of those columns of A in which $yA \leq c$ has equality, then u is a feasible direction iff $uA' \leq 0$. So u is found by solving

$$(6) \qquad \max \{ub \,|\, uA' \leq 0; ub \leq 1\},$$

which is the RHS in (3).

Application 12.1. Max-flow min-cut. Let $D = (V, A)$ be a directed graph, and let $r, s \in V$. Let furthermore a 'capacity' function $c: A \to \mathbb{Q}_+$ be given. The *maximum flow problem* is to find the maximum amount of flow from r to s, subject to the capacity function c:

$$(7) \qquad \max \sum_{a \in \delta^+(r)} x(a) - \sum_{a \in \delta^-(r)} x(a)$$

$$\text{subject to:} \sum_{a \in \delta^+(v)} x(a) - \sum_{a \in \delta^-(v)} x(a) = 0 \quad (v \in V, v \neq r, s)$$

$$0 \leq x(a) \leq c(a) \qquad\qquad (a \in A).$$

[Here $\delta^+(v)(\delta^-(v))$ denotes the set of arcs leaving (entering) v.] By our remark above, if we have a feasible solution x_0, we have to find a function $u: A \to \mathbb{R}$ such that

$$(8) \qquad \sum_{a \in \delta^+(r)} u(a) - \sum_{a \in \delta^-(r)} u(a) = 1,$$

$$\sum_{a \in \delta^+(v)} u(a) - \sum_{a \in \delta^-(v)} u(a) = 0 \quad (v \in V, v \neq r, s),$$

$$u(a) \geq 0 \qquad\qquad \text{if } a \in A, x_0(a) = 0,$$

$$u(a) \leq 0 \qquad\qquad \text{if } a \in A, x_0(a) = c(a).$$

One easily checks that this problem is equivalent to the problem of finding an undirected path from r to s in $D = (V, A)$, such that for any arc a in the path:

(9)   if $x_0(a) = 0$,   arc a is traversed forwardly,

           if $x_0(a) = c(a)$,   arc a is traversed backwardly,

           if $0 < x_0(a) < c(a)$, arc a is traversed forwardly or backwardly.

If we have found such a path, we find u as in (8) (by taking $u(a) = +1$, resp. -1, if a occurs in the path forwardly, resp. backwardly, and $u(a) = 0$ if a does not occur in the path). Next, for some $\lambda > 0$, $x_0 + \lambda u$ is a feasible solution for (7), with objective value higher than that of x_0. Taking the highest λ for which $x_0 + \lambda u$ is feasible gives us the next feasible solution. The path is called a *flow-augmenting path*. Note that this highest λ satisfies

(10) $\lambda = \min \{ \min \{ c(a) - x_0(a) | a \text{ occurs in the path forwardly} \}$,
           $\min \{ x_0(a) | a \text{ occurs in the path backwardly} \} \}$.

So for at least one arc a in the flow-augmenting path we have for the new flow x'_0 that $x'_0(a) = c(a)$ while $x_0(a) < c(a)$, or that $x'_0(a) = 0$ while $x'_0(a) > 0$.

Iterating this process gives finally an optimal flow. This is exactly Ford and Fulkerson's algorithm [1957] for finding a maximum flow, which therefore is an example of a primal–dual method.

In fact, a version of this algorithm can be shown to be a polynomial-time method, as was proved by Dinits [1970] and Edmonds and Karp [1972]. Note that at each iteration one searches for a *directed r–s-path* in the 'auxiliary' digraph $\tilde{D} = (V, \tilde{A})$, where

(11) $(v, w) \in \tilde{A}$ iff $\begin{cases} (v, w) \in A \text{ and } x_0(v, w) < c(v, w) \\ \text{or } (w, v) \in A \text{ and } x_0(w, v) > 0. \end{cases}$

By (10), $\tilde{D}$ changes with each iteration.

Now Dinits and Edmonds and Karp showed that if we choose, at each iteration, a *shortest r–s-path* in $\tilde{D}$ (i.e. one with a minimum number of arcs) as our flow-augmenting path, then we obtain a polynomial-time method.

To prove this, it suffices to show that the number of iterations is polynomially bounded. Let $\tilde{D}_1, \tilde{D}_2, \tilde{D}_3, \ldots$ denote the sequence of auxiliary digraphs $\tilde{D}$. Let $\varphi(\tilde{D}_i)$ denote the length (= number of arcs) of a shortest r–s-path in $\tilde{D}_i$ and let $\psi(\tilde{D}_i)$ denote the number of arcs of $\tilde{D}_i$ which are contained in the union of all shortest r–s-paths in $\tilde{D}_i$. So $\varphi(\tilde{D}_i) \le |V|$ (except if we have reached optimality, when $\varphi(\tilde{D}_i) = \infty$) and $\psi(\tilde{D}_i) \le |A|$ (as never both (v, w) and (w, v) belong to the union of shortest paths). We show that for each i:

(12)   (i) $\varphi(\tilde{D}_{i+1}) \ge \varphi(\tilde{D}_i)$
           (ii) if $\varphi(\tilde{D}_{i+1}) = \varphi(\tilde{D}_i)$ then $\psi(\tilde{D}_{i+1}) < \psi(\tilde{D}_i)$.

This immediately implies that after at most $|V| \cdot |A|$ iterations we reach optimality.

Proof of (12). We first show the following. For each $v \in V$, let ρ_v denote the distance (= number of arcs in a shortest directed path) from r to v in $\tilde{D}_i$. Then we have for each arc (v, w) of $\tilde{D}_{i+1}$:

(13)   (i) $\rho_w \le \rho_v + 1$ if (v, w) is also an arc of $\tilde{D}_i$

           (ii) $\rho_w \le \rho_v - 1$ if (v, w) is not an arc of $\tilde{D}_i$.

Indeed, if (v, w) is an arc of $\tilde{D}_i$ then (13)(i) follows from the definition of ρ_v and ρ_w. If (v, w) is not an arc of $\tilde{D}_i$ then by our definition of changing flows and obtaining auxiliary digraphs, we know that (w, v) was an arc of $\tilde{D}_i$, and that (w, v) belongs to a shortest r–s-path in $\tilde{D}_i$. Hence $\rho_v = \rho_w + 1$ and (13)(ii) follows.

Now (12)(i) is easy: if $(r, v_1), (v_1, v_2), \ldots, (v_{k-1}, s)$ form a shortest r–s-path in $\tilde{D}_{i+1}$, then

(13) gives

$$(14) \qquad \varphi(\tilde{D}_{i+1}) = k \geqslant (\rho_s - \rho_{v_{k-1}}) + \cdots + (\rho_{v_2} - \rho_{v_1}) + (\rho_{v_1} - \rho_r)$$
$$= \rho_s - \rho_r = \rho_s = \varphi(\tilde{D}_i).$$

Moreover, by (13)(ii), we can only have equality here if each arc of the path is also an arc of $\tilde{D}_i$. So if $\varphi(\tilde{D}_{i+1}) = \varphi(\tilde{D}_i)$ then $\psi(\tilde{D}_{i+1}) \leqslant \psi(\tilde{D}_i)$. But by (10), at least one arc (v, w) in the flow-augmenting path chosen in $\tilde{D}_i$ is not an arc of $\tilde{D}_{i+1}$ any more. Hence if $\varphi(\tilde{D}_{i+1}) = \varphi(\tilde{D}_i)$ then $\psi(\tilde{D}_{i+1}) < \psi(\tilde{D}_i)$.   □

The method also provides us with an r-s-cut of minimum capacity: in the auxiliary digraph $\tilde{D} = (V, \tilde{A})$ made after attaining an optimum flow x^*, there does not exist any r-s-path. Hence the set $W := \{v \in V | \text{ there exists an } r\text{-}v\text{-path in } \tilde{D}\}$ contains r but not s. Moreover, if $(v, w) \in \delta_A^+(W) (= \text{set of arcs in } A \text{ leaving } W)$ then $x^*(v, w) = c(v, w)$ (otherwise $(v, w) \in \tilde{A}$ and hence $w \in W$). If $(v, w) \in \delta_A^-(W)$ ($= \text{set of arcs in } A \text{ entering } W$) then $x^*(v, w) = 0$ (otherwise $(w, v) \in \tilde{A}$, and hence $v \in W$). So the capacity of the r-s-cut $\delta_A^+(W)$ is exactly equal to the value of x^*. So by Theorem 7.2, $\delta_A^+(W)$ is a minimum capacitated r-s-cut.

12.2. THE FOURIER–MOTZKIN ELIMINATION METHOD

Since elimination works well for solving linear equations, it is natural to investigate a similar method for linear inequalities, as was done by Fourier [1827], Dines [1918-9], and Motzkin [1936]. This *Fourier–Motzkin elimination method* is well illustrated by application to the problem:

(15)   given matrix A and vector b, test if $Ax \leqslant b$ has a solution, and if so find one.

Let A have m rows and n columns. As we may multiply each inequality by a positive scalar, we may assume that all entries in the first column of A are 0 or ± 1. So the problem is to solve (maybe after a reordering of the inequalities):

$$(16) \qquad \begin{aligned} \xi_1 + a_i x' &\leqslant \beta_i & (i = 1, \ldots, m') \\ -\xi_1 + a_i x' &\leqslant \beta_i & (i = m' + 1, \ldots, m'') \\ a_i x' &\leqslant \beta_i & (i = m'' + 1, \ldots, m). \end{aligned}$$

where $x = (\xi_1, \ldots, \xi_n)^\mathsf{T}$ and $x' = (\xi_2, \ldots, \xi_n)^\mathsf{T}$, and where $a_1, \ldots, a_m$ are the rows of A with the first entry deleted. Since the first two lines of (16) are equivalent to:

$$(17) \qquad \max_{m'+1 \leqslant j \leqslant m''} (a_j x' - \beta_j) \leqslant \xi_1 \leqslant \min_{1 \leqslant i \leqslant m'} (\beta_i - a_i x')$$

the unknown ξ_1 can be eliminated. This gives that (16) is equivalent to the system of linear inequalities:

$$(18) \qquad \begin{aligned} a_j x' - \beta_j &\leqslant \beta_i - a_i x' & (i = 1, \ldots, m'; j = m' + 1, \ldots, m'') \\ a_i x' &\leqslant \beta_i & (i = m'' + 1, \ldots, m), \end{aligned}$$

i.e. to

$$(19) \qquad \begin{aligned} (a_i + a_j) x' &\leqslant \beta_i + \beta_j & (i = 1, \ldots, m'; j = m' + 1, \ldots, m'') \\ a_i x' &\leqslant \beta_i & (i = m'' + 1, \ldots, m). \end{aligned}$$

This system has $m'(m'' - m') + m - m''$ constraints, and $n - 1$ unknowns. Any solution x' of (19) can be extended to a solution (ξ_1, x') of (16), by choosing ξ_1 such that (17) is satisfied.

By repeating this procedure, we can successively eliminate the first $n - 1$ components of x, and we end up with an equivalent problem in one unknown, which is trivial. From any solution for this final system, a solution for the original system can be derived.

Geometrically, the method consists of successive projections: the polyhedron (19) is the projection of (16) along the ξ_1-axis.

The method can be applied by hand if n is small, but can be quite time-consuming for problems in many variables. The method is not polynomial.

Examples showing that the Fourier–Motzkin elimination method is not polynomial are as follows. Let p be a natural number, and let $n := 2^p + p + 2$. Consider any system of linear inequalities in the unknowns $\xi_1, \ldots, \xi_n$ which contains as left-hand sides all possible forms $\pm \xi_i \pm \xi_j \pm \xi_k$, for $1 \leqslant i < j < k \leqslant n$ (so there are $8\dbinom{n}{3}$ constraints). By induction on t one easily shows that after eliminating the first t variables, we have the following forms among the left-hand sides:

$$(20) \qquad \pm \xi_{j_1} \pm \xi_{j_2} \pm \cdots \pm \xi_{j_s} \qquad (t + 1 \leqslant j_1 < j_2 < \cdots < j_s \leqslant n)$$

where $s := 2^t + 2$. However, the size of the initial system can be $O(n^4)$, while the system after p eliminations contains at least

$$(21) \qquad 2^{2^p + 2} \geqslant 2^{n/2}$$

inequalities.

To obtain a method for finding a nonnegative solution of a system $Ax = b$ of linear equations, we can first transform the system $Ax = b$ into an equivalent system $[I \ A']\begin{pmatrix} y \\ z \end{pmatrix} = b'$ (by Gaussian elimination). Next we can apply the above method to the system $z \geqslant 0$, $A'z \leqslant b'$.

The method can also be easily modified to a method for linear programming. Indeed, if we wish to find an optimum solution for $\max \{cx \mid Ax \leqslant b\}$, apply the Fourier–Motzkin elimination method to the system $Ax \leqslant b$, $\lambda - cx \leqslant 0$, where λ is a new unknown. So we want to find a solution $\begin{pmatrix} x \\ \lambda \end{pmatrix}$ with λ as large as possible.

Now the components of x can be eliminated successively, so that λ is the only variable left, which next can be chosen as large as possible.

Note that the Fourier–Motzkin elimination method can be extended in an obvious way to include also strict inequalities.

Notes for Section 12.2. Kuhn [1956a] derived from the Fourier–Motzkin elimination method a nice short proof of Farkas' lemma, in the form of Corollary 7.1e, by induction on the dimension of x. Also Farkas' [1896] proof followed these lines.

[Indeed, suppose the system $Ax \leqslant b$ has no solution. Denote the equivalent system (19) by $A'x' \leqslant b'$. Since also $A'x' \leqslant b'$ has no solution, by induction we know that the vector $(0, \ldots, 0, -1)$ is a nonnegative combination of the rows of the matrix $[A' \ b']$. Now each row of the matrix $[0 \ A' \ b']$ (where 0 stands for an all-zero column) is a sum of two rows of $[A \ b]$. Hence the vector $(0, 0, \ldots, 0, -1)$ is a nonnegative combination of the rows of $[A \ b]$, thus proving Farkas' lemma.]

Kuhn [1956a] also gave a survey of the Fourier–Motzkin elimination method. Historically, it is the 'pre-linear programming' method for solving linear inequalities, see Fourier [1827], Dines [1917, 1918–9, 1925, 1927a], Dines and McCoy [1933], and Motzkin [1936]. (See also Chernikov [1961, 1965], Duffin [1967, 1974], Klee [1973], Kohler [1967, 1973].)

The *dual* of the Fourier–Motzkin elimination method consists of eliminating one inequality and introducing many new variables. This was used by Farkas [1898a] to show that each polyhedral cone is finitely generated, which implies Farkas' lemma. This dual was also studied by Abadie [1964] and Dantzig and Eaves [1973].

Nelson [1982] studied the Fourier–Motzkin elimination method for systems in which each inequality contains at most two variables (i.e. all other variables have coefficient 0). Note that this property is maintained under elimination of variables. Nelson found for this special case a running time of order $O(m \cdot n^{\lceil 2 \log n \rceil + 3} . \log n)$ (where m = number of inequalities, n = number of variables).

12.3. THE RELAXATION METHOD

The relaxation method was introduced by Agmon [1954] and Motzkin and Schoenberg [1954]. It solves systems of linear inequalities, and by the equivalence discussed in Section 10.3, it can be adapted to solve systems of equations in nonnegative numbers, and linear programming. We here give a brief description and analysis. For an extensive treatment, see Goffin [1980].

Description of the relaxation method. Let a system $Ax \leq b$ of linear inequalities be given. Choose a vector z^0, and a number $\lambda > 0$. Determine vectors $z^1, z^2, \ldots$ inductively as follows. Suppose z^i has been found ($i \geq 0$). Check whether $Az^i \leq b$ is satisfied. If so, we have found a solution, and the method stops. If not, choose an inequality $a_k x \leq \beta_k$ from $Ax \leq b$ violated by z^i. Let z^{i+1} be given by:

$$(22) \qquad z^{i+1} := z^i + \lambda \frac{(\beta_k - a_k z^i)}{\|a_k\|^2} \cdot a_k^\mathsf{T}.$$

(So z^{i+1} arises from z^i by going along the projection vector of z^i onto the hyperplane $a_k x = \beta_k$. For $\lambda = 2$ we have that z^{i+1} is the reflection of z^i into this hyperplane. If $\lambda = 1$, z^{i+1} is the projection of z^i onto this hyperplane.)

Further discussion. The above description leaves a number of questions open: What to choose for λ and z^0? Which violated inequality should one choose if there are several? Does the method terminate? What is the running time of the method?

As violated inequality one can take the 'most violated' inequality, i.e. inequality $a_k x \leq \beta_k$ from $Ax \leq b$ which maximizes

$$(23) \qquad \mu_i := \frac{a_k z^i - \beta_k}{\|a_k\|}.$$

(which is the Euclidean distance of z^i to the hyperplane $a_k x = \beta_k$). In the proofs below we shall assume that we take such a most violated inequality.

Motzkin and Schoenberg [1954] showed:

Theorem 12.1. *If* $P := \{x \mid Ax \leqslant b\}$ *is full-dimensional, and we take* $\lambda = 2$ *and* z^0 *arbitrarily, the relaxation method terminates.*

Proof. Suppose the method does not terminate. As P is full-dimensional, there exists a vector x^0 and a real number $r > 0$ such that $B(x^0, r) \subseteq P$. If $a_k x \leqslant \beta_k$ is the inequality in which we reflect z^i to obtain z^{i+1}, let μ_i be as given by (23), and let ρ_i be the Euclidean distance of x^0 to the hyperplane $a_k x = \beta_k$. Then $\rho_i \geqslant r$, and hence, by elementary geometry (using the fact that $z^{i+1} - z^i$ is orthogonal to $\{x \mid a_k x = \beta_k\}$),

$$(24) \qquad \|z^{i+1} - x^0\|^2 = \|z^i - x^0\|^2 - 4\mu_i \rho_i \leqslant \|z^i - x^0\|^2 - 4\mu_i r.$$

Hence, by induction,

$$(25) \qquad 0 \leqslant \|z^i - x^0\|^2 \leqslant \|z^0 - x^0\|^2 - 4r \sum_{j=0}^{i-1} \mu_j.$$

Therefore, $\sum_{j=0}^{\infty} \mu_j$ converges. As $\|z^{i+1} - z^i\| = 2\mu_i$, it follows that also $\sum_i \|z^{i+1} - z^i\|$ converges, and hence also $\sum_i (z^{i+1} - z^i)$. So $\lim_{i \to \infty} z^i$ exists, say it is z^*. Then $Az^* \leqslant b$, as $\lim_{i \to \infty} \mu_i = 0$.

Let $A'x \leqslant b'$ be the system of those inequalities from $Ax \leqslant b$ which are satisfied with equality by z^*, and let $A''x \leqslant b''$ be the other inequalities. As $A''z^* < b''$, there exists an index t such that $A''z^i < b''$ for $i \geqslant t$. Hence, if $i \geqslant t$, z^{i+1} is obtained from z^i by reflecting in an inequality $a_k x \leqslant \beta_k$ with $a_k z^* = \beta_k$. This immediately gives $\|z^{i+1} - z^*\| = \|z^i - z^*\|$ for $i \geqslant t$, and hence $\|z^i - z^*\| = \|z^t - z^*\|$ for $i \geqslant t$. As $\lim_{i \to \infty} z^i = z^*$, it follows that $z^t = z^*$, contradicting our assumption of non-termination. $\qquad\qquad\square$

For $\lambda = 2$, full-dimensionality is essential for termination, even for convergence: for the system $1 \leqslant \xi \leqslant 1$, starting with $z^0 = 0$, the method does not converge. Agmon [1954] and Motzkin and Schoenberg [1954] showed that if we take $0 < \lambda < 2$, the method terminates or converges to a solution, if the system is feasible (cf. also Hoffman [1952], Goffin [1980], Jeroslow [1979d], Todd [1979]).

Theorem 12.2. *If* P *is nonempty, and we choose* $0 < \lambda < 2$ *and* z^0 *arbitrarily, the relaxation method either terminates, or the sequence* $z^0, z^1, z^2, \ldots$ *converges to a point* z^* *on the boundary of* P, *with geometric convergence speed, i.e. there exist* v, θ $(0 < \theta < 1)$ *such that for all* i,

$$(26) \qquad \|z^i - z^*\| \leqslant v\theta^i.$$

Proof. Suppose the method does not terminate. Let $\tilde{A}x \leqslant \tilde{b}$ be the system arising from $Ax \leqslant b$ by scaling each inequality so that each row of $\tilde{A}$ has Euclidean norm 1. Let

(27) $\Delta =$ an upper bound for the absolute values of the entries of B^{-1}, for all nonsingular submatrices B of $\tilde{A}$.

We first show that, for each i,

(28) $$\frac{d(z^{i+1}, P)}{d(z^i, P)} \leqslant \sqrt{1 - \frac{\lambda(2 - \lambda)}{n^3 \Delta^2}} =: \theta.$$

Again, let $a_k x \leqslant \beta_k$ be the inequality from $Ax \leqslant b$ used to obtain z^{i+1} from z^i, and let μ_i be as in (23). Let x_0 be the point in P nearest to z^i. As z^i satisfies $\tilde{A}x \leqslant \tilde{b} + \mu_i \mathbf{1}$, we know by Theorem 10.5 that

(29) $\|z^i - x\|_\infty \leqslant n\Delta\mu_i$

for some x in P. Hence

(30) $\|z^i - x_0\| = d(z^i, P) \leqslant \sqrt{n}\|z^i - x\|_\infty \leqslant n^{3/2}\Delta\mu_i$

for some x_0 in P. Note moreover

(31) $\|z^i - x_0\|^2 - \|z^{i+1} - x_0\|^2 = \mu_i^2 \lambda(2 - \lambda) + 2\mu_i\lambda\frac{\beta_k - a_k x_0}{\|a_k\|} \geqslant \mu_i^2 \lambda(2 - \lambda)$

(using that $a_k x_0 \leqslant \beta_k$ and that $z^i - z^{i+1}$ is orthogonal to $\{x \mid a_k x = \beta_k\}$). Therefore,

(32) $$\frac{d(z^{i+1}, P)}{d(z^i, P)} \leqslant \frac{\|z^{i+1} - x_0\|}{\|z^i - x_0\|} = \sqrt{1 - \frac{\|z^i - x_0\|^2 - \|z^{i+1} - x_0\|^2}{\|z^i - x_0\|^2}}$$

$$\leqslant \sqrt{1 - \frac{\mu_i^2 \lambda(2 - \lambda)}{n^3 \Delta^2 \mu_i^2}} = \sqrt{1 - \frac{\lambda(2 - \lambda)}{n^3 \Delta^2}}.$$

This proves (28). By induction it follows that, if we define $\sigma := d(z^0, P)$, we have $d(z^i, P) \leqslant \theta^i \sigma$. Now $\mu_i \leqslant d(z^i, P) \leqslant \theta^i \sigma$. Therefore,

(33) $\|z^{i+1} - z^i\| \leqslant 2\mu_i \leqslant 2\theta^i \sigma$.

So $z^0, z^1, z^2, \dots$ is a Cauchy-sequence, say with limit z^*. Then z^* belongs to P (as $d(z^*, P) = d(\lim_{i \to \infty} z^i, P) = \lim_{i \to \infty} d(z^i, P) = 0$). Moreover, with elementary calculus one derives from (33) that

(34) $\|z^i - z^*\| \leqslant \theta^i \left(\dfrac{2\sigma}{1 - \theta}\right)$.

This shows (26).   □

Motzkin and Schoenberg also showed that for any $\lambda < 2$ there is a full-dimensional polyhedron in two dimensions where the method converges, but does not terminate.

If $0 < \lambda < 1$, this is easy: consider the system $\xi \leqslant 0$ in one variable, and start with $z^0 = 1$. Then $z^k = (1 - \lambda)^k$, and we never obtain a solution.

If $1 \leqslant \lambda < 2$, choose ψ so that $0 < \psi < \frac{1}{4}\pi$ and so that

(35) $\dfrac{\tan(\frac{1}{4}\pi - \psi)}{\tan(\frac{1}{4}\pi + \psi)} > \lambda - 1,$

and consider the system in the two variables ξ and η:

(36) $-(\tan\psi)\xi \leqslant \eta \leqslant (\tan\psi)\xi$.

If we choose our starting vector z^0 in the region

(37) $S:=\left\{\begin{pmatrix}\xi\\\eta\end{pmatrix}\,\middle|\,\xi < \eta < -\xi\right\}$,

then $z^0, z^1, z^2, \ldots$ never leave S. Indeed, let $z^k = \begin{pmatrix}\rho\cos\varphi\\\rho\sin\varphi\end{pmatrix}$ belong to S, i.e. $\rho > 0$, $\frac{3}{4}\pi <$ $\varphi < \frac{5}{4}\pi$. Without loss of generality, $\varphi \leqslant \pi$, and we use the inequality $\eta \leqslant (\tan\psi)\xi$ as violated inequality. One easily checks that $z^{k+1} = \begin{pmatrix}\rho'\cos\varphi'\\\rho'\sin\varphi'\end{pmatrix}$ with $\rho' > 0$ and $\varphi' = \pi + \psi + \arctan$ $((\lambda - 1)\tan(\pi - \varphi + \psi))$. Because of (35) we know $\frac{3}{4}\pi < \varphi' < \frac{5}{4}\pi$, and hence z^{k+1} belongs to S. So we never hit a solution of (36).

However, the method can be made to terminate and to yield exact solutions by adding the simultaneous diophantine approximation algorithm (Section 6.3).

Theorem 12.3. *A combination of the relaxation method and the simultaneous approximation method solves a system $Ax \leqslant b$ of rational linear inequalities in time polynomially bounded by size(A,b) and by Δ, where Δ satisfies (27).*

Proof. Apply the relaxation method to $Ax \leqslant b$, with $\lambda = 1$ and starting with $z^0 = 0$, up to z^N, where

(38) $N := n^3\Delta^2\lceil(2\log_2(n^3\Delta^2) + 14n^2\varphi)\rceil$

where n is the number of columns of A and φ is the maximum row size of the matrix $[A\ b]$. (Note that the size of vector z^i is polynomially bounded by n, φ, Δ, and i, as follows inductively from (22).)

By Theorem 10.2, if $Ax \leqslant b$ has a solution, it has one of size $4n^2\varphi$. So the σ defined in the proof of Theorem 12.2 satisfies

(39) $\sigma \leqslant 2^{4n^2\varphi}$.

Now if we take as in (28),

(40) $\theta = \sqrt{1 - \dfrac{1}{n^3\Delta^2}}$

then, by (34), vector z^N satisfies

(41) $d(z^N, P) \leqslant \theta^N\left(\dfrac{2\sigma}{1-\theta}\right) \leqslant 2^{-8n\varphi}$

if P is nonempty. Let

(42) $\varepsilon := 2^{-3\varphi}$.

By Corollary 6.4c we can find in polynomial time an integral vector p and an integer q such that

(43) $\qquad \left\| z^N - \dfrac{1}{q} \cdot p \right\| < \dfrac{\varepsilon}{q}$ and $\quad 1 \leqslant q \leqslant 2^{n(n+1)/4} \varepsilon^{-n}$.

We claim that, if $P \neq \varnothing$, then $(1/q) \cdot p \in P$ (which proves the theorem). Indeed, let $ax \leqslant \beta$ be an inequality from $Ax \leqslant b$. Then

(44) $\quad ap - q\beta = a(p - qz^N) + q(az^N - \beta) \leqslant q\|a\| \cdot \left\|\dfrac{1}{q}p - z^N\right\| + q\|a\| \cdot d(z^N, P)$

$\qquad \leqslant \|a\|(\varepsilon + 2^{n(n+1)/4}\varepsilon^{-n}d(z^N, P)) < 2^{-\varphi}$.

As the denominators of β and of the components of a have least common multiple at most 2^φ, it follows that $ap - q\beta \leqslant 0$, and hence $a((1/q)p) \leqslant \beta$. $\qquad \square$

Goffin [1982] and Todd [1979] showed that for any λ there is a series of full-dimensional polyhedra for which the relaxation method requires exponentially increasing running time to get within distance 1 from the feasible region.

To see this, consider, for each fixed natural number k, the system of linear inequalities:

(45) $\qquad \begin{aligned} \xi + k\eta &\geqslant k \\ \xi - k\eta &\geqslant k \end{aligned}$

in the variables ξ, η (cf. Figure 5). Apply the relaxation method, with some $0 < \lambda \leqslant 2$, starting with $z^0 = \begin{pmatrix} 0 \\ 0 \end{pmatrix}$. Let $z^n =: \begin{pmatrix} \xi_n \\ \eta_n \end{pmatrix}$. By inspection of the figure we know that the series $z^0, z^1, z^2, \ldots$ will start with oscillating around the x-axis. Moreover, one easily shows that $|\eta_{n+1}| \leqslant |\eta_n| + 2$. Hence, by induction, $|\eta_n| \leqslant 2n$. Also, $\xi_{n+1} \leqslant \xi_n + \lambda(|\eta_n| + 1) \leqslant \xi_n + 4n + 2$. Therefore, by induction, $\xi_n \leqslant 2n^2$. Now if $\begin{pmatrix} \xi \\ \eta \end{pmatrix}$ satisfies (45), then $\xi \geqslant k$. So we do not attain a solution as long as $2n^2 < k$, i.e. $n < \sqrt{\tfrac{1}{2}k}$. Hence, while the size of system (45) is $O(\log k)$, we have to make at least $\sqrt{\tfrac{1}{2}k}$ steps.

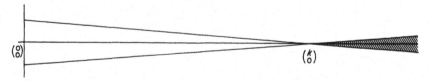

Figure 5

Note. The following interesting observation is due to Goffin [1984]. It is a result of K. Löwner and of John [1948] that for each system $Ax \leqslant b$ in n unknowns, with bounded full-dimensional feasible region P, there exists an affine transformation τ such that

(46) $\qquad B(0,1) \subseteq \tau P \subseteq B(0,n)$.

Then for the transformed polytope τP we know that $d(z, \tau P) \leqslant n\mu$, where μ is the maximum distance of z to the hyperplanes defining the facets of τP. To see this, let v be the largest real number so that $vz \in \tau P$. Let $ax \leqslant \beta$ be an inequality defining a facet of τP so that $a(vz) = \beta$ (this exists). From (46) we know that $\|vz\| \leqslant n$ and $\|a\| \geqslant \beta$ (as

$a/\|a\| \in \tau P)$. Hence

(47) $d(z, \tau P) \leqslant d(z, vz) = (1 - v)\|z\| = \|vz\| \dfrac{az - \beta}{\beta} \leqslant n \dfrac{az - \beta}{\|a\|} \leqslant n\mu.$

So we can replace $n^{3/2}\Delta\mu_i$ in (30) by $n\mu_i$, and hence $n^3\Delta^2$ in (28) by n^2. However, for using this algorithmically we should be able to find τ as in (46).

In Section 15.4 we shall see that there is a polynomial-time algorithm which, for each system $Ax \leqslant b$, with bounded full-dimensional feasible region P (say) finds an affine transformation τ satisfying a somewhat weaker condition, viz.

(48) $B(0, 1) \subseteq \tau P \subseteq B(0, n^{3/2}).$

This means that, in the transformed space, we can replace $n^3\Delta^2$ by n^3 in (28), and hence, after transforming the space, Theorem 12.3 yields a polynomial-time algorithm for solving a system $Ax \leqslant b$ with bounded full-dimensional feasible region. In Section 13.4 we shall see that this implies that any system $Ax \leqslant b$ can be solved in polynomial time. So the relaxation method can be modified to a polynomial algorithm. *However*, in finding τ we at the same time find a solution, namely $\tau^{-1}0$, for $Ax \leqslant b$. So adding the relaxation method is superfluous!

If A is totally unimodular (i.e. each submatrix of A has determinant $+1$, -1, or 0 — cf. Chapter 19), then we can take $\Delta = \sqrt{n}$ in Theorem 12.3. Hence we have a result of Maurras, Truemper, and Akgül [1981]:

(49)   *if A is totally unimodular, then $Ax \leqslant b$ can be solved with the relaxation method in time polynomially bounded by the sizes of A and b.*

In fact, Maurras, Truemper, and Akgül showed that any *linear program* with a totally unimodular constraint matrix can be solved in polynomial time with a modification of the relaxation method.

For a variant of the relaxation method, see Censor and Elfving [1982]. See also Drezner [1983].

13

Khachiyan's method for linear programming

In Chapters 13 and 14 we discuss the ellipsoid method for linear programming and polyhedra. The ellipsoid method was developed for nonlinear programming by Shor [1970a, b, 1977] and Yudin and Nemirovskiĭ [1976a, b]. Khachiyan [1979, 1980] showed that an extension solves linear programming problems in polynomial time (cf. Gács and Lovász [1980]). A further extension is to linear programming over 'implicitly given polyhedra', which yields the polynomial solvability of several combinatorial optimization problems, if they can be described suitably by polyhedra (Karp and Papadimitriou [1980], Padberg and Rao [1980], Grötschel, Lovász, and Schrijver [1981]).

In Chapter 13 we consider Khachiyan's method for linear programming, and in Chapter 14 we deal with the extension to implicitly given polyhedra.

The ellipsoid method is based on finding a series of ellipsoids of decreasing volume. Therefore, before discussing the method, we first give in Section 13.1 some elementary facts on ellipsoids and positive definite matrices. Next in Section 13.2 we give an outline of Khachiyan's method for solving a system of linear inequalities, under the assumptions that the feasible region is bounded and full-dimensional, and that we can work with infinite precision. To allow rational arithmetic, we need two approximation lemmas for ellipsoids, discussed in Section 13.3. Then in Section 13.4 we give a precise description of Khachiyan's method. Finally, in Section 13.5 we go briefly into the practical complexity of the method, and in Section 13.6 we make some historical and further notes.

For more extensive discussions of the ellipsoid method, see Bland, Goldfarb, and Todd [1981], Grötschel, Lovász, and Schrijver [1986], and Schrader [1982, 1983].

13.1. ELLIPSOIDS

A real-valued symmetric matrix D is called *positive definite* if all its eigenvalues are positive. There are the following equivalences: D is positive definite $\Leftrightarrow$ $D = B^T B$ for some nonsingular matrix $B \Leftrightarrow x^T D x > 0$ for each nonzero vector x.

A set E of vectors in $\mathbb{R}^n$ is called an *ellipsoid* if there exists a vector $z \in \mathbb{R}^n$ and a positive definite matrix D of order n such that

$$(1) \qquad E = \text{ell}(z, D) := \{x \mid (x - z)^T D^{-1} (x - z) \leqslant 1\}.$$

Here the parameters z and D are uniquely determined by the ellipsoid E. The vector z is the *centre* of E. It follows from the characterization of positive definite matrices given above that a set E is an ellipsoid if and only if E is an affine transformation of the unit ball $\{x \mid x^\mathsf{T} x \leqslant 1\}$.

A basic technique in the ellipsoid method is that of halving ellipsoids, as described in the following theorem.

Theorem 13.1. *Let $E = \mathrm{ell}\,(z, D)$ be an ellipsoid in $\mathbb{R}^n$, and let a be a row n-vector. Let E' be an ellipsoid containing $\mathrm{ell}\,(z, D) \cap \{x \mid ax \leqslant az\}$, such that E' has smallest volume. Then E' is unique, and $E' = \mathrm{ell}\,(z', D')$, where*

(2) $$z' := z - \frac{1}{n+1} \cdot \frac{Da^\mathsf{T}}{\sqrt{aDa^\mathsf{T}}}$$

$$D' := \frac{n^2}{n^2 - 1}\left(D - \frac{2}{n+1} \cdot \frac{Da^\mathsf{T} aD}{aDa^\mathsf{T}}\right).$$

Moreover

(3) $$\frac{\mathrm{vol}\,E'}{\mathrm{vol}\,E} < e^{-1/(2n+2)}.$$

Proof. It is elementary geometry to see that (2) indeed defines the ellipsoid of smallest volume in the case that $\mathrm{ell}\,(z, D)$ is the unit ball, i.e. if $z = 0$ and $D = I$.

Since any other ellipsoid is an affine transformation of the unit ball, and since, at an affine transformation, the volumes of all ellipsoids are multiplied by a constant factor, also for arbitrary ellipsoids this smallest ellipsoid is unique, and the parameters (2) follow from those for the unit ball.

To see (3), use

(4) $$\frac{\mathrm{vol}\,(\mathrm{ell}\,(z', D'))}{\mathrm{vol}\,(\mathrm{ell}\,(z, D))} = \sqrt{\frac{\det D'}{\det D}} = \left(\frac{n^2}{n^2 - 1}\right)^{(n-1)/2} \cdot \left(\frac{n}{n+1}\right) < e^{-1/2(n+1)}.$$

The first equality is well-known; the last inequality follows from the fact that $1 + x < e^x$ if $x \neq 0$. To see the second equality, we may assume $D = I$, since we may assume that $\mathrm{ell}\,(z, D)$ is the unit ball. Then:

(5) $$\sqrt{\frac{\det D'}{\det D}} = \sqrt{\det D'} = \left(\frac{n^2}{n^2 - 1}\right)^{n/2}\left(\det\left(I - \frac{2}{n+1} \cdot \frac{a^\mathsf{T} a}{aa^\mathsf{T}}\right)\right)^{1/2}$$

$$= \left(\frac{n^2}{n^2 - 1}\right)^{n/2} \cdot \left(1 - \frac{2}{n+1}\right)^{1/2}$$

$$= \left(\frac{n^2}{n^2 - 1}\right)^{(n-1)/2}\left(\frac{n}{n+1}\right).$$

The third equality here follows from the fact that the only nonzero eigenvalue of the matrix $(aa^\mathsf{T})^{-1} \cdot a^\mathsf{T} a$ is 1, with multiplicity 1. $\square$

Note that if D has smallest eigenvalue s^2 and largest eigenvalue S^2, and $z \in \mathbb{R}^n$, then

(6) $B(z,s) \subseteq \text{ell}(z,D) \subseteq B(z,S)$.

This implies that for $\varepsilon \geq 0$:

(7) $\text{ell}(z,(1 + \varepsilon/S)^2 D) \subseteq B(\text{ell}(z,D),\varepsilon) \subseteq \text{ell}(z,(1 + \varepsilon/s)^2 D)$.

13.2. KHACHIYAN'S METHOD: OUTLINE

We outline Khachiyan's method by applying it to the problem of finding a solution of a rational linear inequality system $Ax \leq b$, where A is an $m \times n$-matrix ($n \geq 2$). In this outline we make two assumptions:

(8)   (i) the polyhedron $P := \{x \mid Ax \leq b\}$ is bounded and full-dimensional;
             (ii) we can make our calculations with infinite precision.

In the next sections we shall abandon these assumptions.

Let $v := 4n^2 \varphi$, where φ is the maximum row size of the matrix $[A \; b]$. So by Theorem 10.2, each vertex of P has size at most v. Let $R := 2^v$. Then $P \subseteq \{x \mid \|x\| \leq R\}$.

Khachiyan's method consists of determining recursively a sequence of ellipsoids $E_0, E_1, E_2, \ldots$ of decreasing volume, such that $P \subseteq E_i$ for every i. So a sequence of centres $z^0, z^1, z^2, \ldots$ and a sequence of symmetric positive definite matrices $D_0, D_1, D_2, \ldots$ is determined, where

(9) $E_i = \text{ell}(z^i, D_i)$.

The first ellipsoid E_0 is the ball $\{x \mid \|x\| \leq R\}$, that is:

(10) $z^0 := 0, \quad D_0 := R^2 \cdot I$.

Suppose now the ellipsoid $E_i \supseteq P$ has been determined, in terms of the parameters z^i and D_i ($i \geq 0$). If z^i satisfies $Ax \leq b$, we have found a solution and the method stops. If z^i does not satisfy $Ax \leq b$, let $a_k x \leq \beta_k$ be an inequality from $Ax \leq b$ violated by z^i. Let E_{i+1} be the ellipsoid of smallest volume containing the half $E_i \cap \{x \mid a_k x \leq a_k z^i\}$ of E_i. The parameters z^{i+1} and D_{i+1} follow from z^i, D_i, and a_k by the formulas (2) above. The ellipsoid E_{i+1} again contains P, as $P \subseteq E_i$ and $P \subseteq \{x \mid a_k x \leq \beta_k\} \subseteq \{x \mid a_k x \leq a_k z^i\}$.

By Theorem 13.1,

(11) $\dfrac{\text{vol } E_{i+1}}{\text{vol } E_i} < e^{-1/(2n+2)}$.

Since $\text{vol } E_0 \leq (2R)^n$, inductively with (11) we know

(12) $\text{vol } E_i \leq e^{-i/(2n+2)} \cdot (2R)^n$.

On the other hand, P is assumed to be full-dimensional, and hence its volume is bounded from below as follows. Let $x_0, \ldots, x_n$ be affinely independent vertices of P. Then

(13) $\text{vol } P \geqslant \text{vol(conv.hull} \{x_0, \ldots, x_n\})$

$$= \frac{1}{n!} \left| \det \begin{bmatrix} 1 & \cdots & 1 \\ x_0 & & x_n \end{bmatrix} \right| \geqslant n^{-n} 2^{-nv} \geqslant 2^{-2nv}$$

(the lower bound on the determinant follows from the fact that each of $x_0, \ldots, x_n$ has size at most v, and that hence the determinant has denominator at most 2^{nv}). Hence, if we reach the ellipsoid E_N with $N := 16n^2 v$, we have the contradiction

(14) $2^{-2nv} \leqslant \text{vol } P \leqslant \text{vol } E_N < e^{-N/(2n+2)} \cdot (2R)^n \leqslant 2^{-2nv}$

(using $P \subseteq E_N$ and $R = 2^v$). So if P is nonempty, before reaching E_N we find a solution z^i for $Ax \leqslant b$. As N is polynomially bounded by the sizes of A and b, this solves the problem in polynomial time.

In the description above we have passed over some problems: we restricted ourselves to bounded, full-dimensional polyhedra, and we assumed we can make our calculations with infinite precision. If we were working in rational arithmetic, the square root in (2) would prevent the calculations from being made exactly. Moreover, we did not deal with the question of whether the numbers involved do not become too large. Although we need to evaluate only a polynomial number of ellipsoids, we still must check that the defining parameters do not grow exponentially in size.

In Section 13.4 we will show that these problems can be overcome. Before, we prove two auxiliary lemmas.

13.3. TWO APPROXIMATION LEMMAS

In this section we prove two auxiliary theorems on the approximation of ellipsoids.

Theorem 13.2 (Ellipsoid approximation lemma). *Let* $E = \text{ell}(z, D)$ *be an ellipsoid, where* $z \in \mathbb{R}^n$ *and* D *is a positive definite* $n \times n$*-matrix with all eigenvalues at least* s^2 *and at most* S^2. *Let* $\varepsilon > 0$ *and let* $p \geqslant 3n + |\log_2 S| + 2|\log_2 s| + |\log_2 \varepsilon|$. *Let* $\tilde{z}$ *and* $\tilde{D}$ *arise from* z *and* D *by rounding all entries up to* p *binary digits behind the point (taking care that* $\tilde{D}$ *be symmetric). Let* $\tilde{E} := \text{ell}(\tilde{z}, \tilde{D})$. *Then:*

(i) *all eigenvalues of* $\tilde{D}$ *are at least* $\frac{1}{2}s^2$ *and at most* $2S^2$;
(ii) $E \subseteq B(\tilde{E}, \varepsilon)$;
(iii) $\text{vol } \tilde{E} / \text{vol } E \leqslant 1 + \varepsilon$.

Proof. (i) Let $x \in \mathbb{R}^n$ with $\|x\| = 1$. As each entry of $\tilde{D} - D$ is at most 2^{-p} in absolute value,

(15) $\|x^\mathsf{T}(\tilde{D} - D)x\| \leqslant n^2 2^{-p} \leqslant \frac{1}{2}s^2$.

So from $s^2 \leqslant x^\mathsf{T} D x \leqslant S^2$, we know $\frac{1}{2}s^2 \leqslant x^\mathsf{T} \tilde{D} x \leqslant 2S^2$, proving (i).

(ii) Write $D = B^2$, where B is a symmetric matrix. Then

(16)   all eigenvalues of $B^{-1}\tilde{D}B^{-1}$ differ by at most $s^{-2}n^2 2^{-p}$ from 1.

Indeed, if $\|x\| = 1$, then $\|x^T B^{-1}\|^2 = x^T D^{-1} x \leqslant s^{-2}$, and hence

(17) $|x^T B^{-1}\tilde{D}B^{-1}x - 1| = |x^T B^{-1}(\tilde{D} - D)B^{-1}x| \leqslant n^2 2^{-p}s^{-2}$

which implies (16).

From (16) it immediately follows that all eigenvalues of $B\tilde{D}^{-1}B$ differ by at most $2s^{-2}n^2 2^{-p}$ from 1. Therefore, if $x^T D^{-1} x \leqslant 1$ then

(18) $x^T \tilde{D}^{-1} x = x^T B^{-1}(B\tilde{D}^{-1}B)B^{-1}x \leqslant 1 + 2s^{-2}n^2 2^{-p} \leqslant (1 + s^{-2}n^2 2^{-p})^2.$

Hence, as $\|\tilde{z} - z\| \leqslant n2^{-p}$ and $\tilde{D}$ has smallest eigenvalue at least $\frac{1}{2}s^2$, we have, using (7),

(19) $\mathrm{ell}(z, D) \subseteq \mathrm{ell}(z, (1 + s^{-2}n^2 2^{-p})^2 \tilde{D}) \subseteq B(\mathrm{ell}(z, \tilde{D}), \sqrt{2}\, Ss^{-2}n^2 2^{-p})$

$\subseteq B(\mathrm{ell}(\tilde{z}, \tilde{D}), \sqrt{2}\, Ss^{-2}n^2 2^{-p} + n2^{-p}) \subseteq B(\mathrm{ell}(\tilde{z}, \tilde{D}), \varepsilon)$

proving (ii).

(iii) By (16),

(20) $\dfrac{\mathrm{vol}\,\tilde{E}}{\mathrm{vol}\,E} = \sqrt{\dfrac{\det \tilde{D}}{\det D}} = \sqrt{\det B^{-1}\tilde{D}B^{-1}} \leqslant (1 + s^{-2}n^2 2^{-p})^{n/2} \leqslant 1 + \varepsilon$

showing (iii). $\square$

Theorem 13.3. *Let $E = \mathrm{ell}(z, D)$ be an ellipsoid, where all eigenvalues of D are at least s^2 and at most S^2. Let H be an affine half-space containing z, and let $\varepsilon \geqslant 0$. Then $B(E, \varepsilon) \cap H \subseteq B(E \cap H, \varepsilon S/s)$.*

Proof. Without loss of generality, $z = 0$. Let $x \in B(E, \varepsilon) \cap H$. If $x \in E$, then $x \in E \cap H$, and hence $x \in B(E \cap H, \varepsilon S/s)$. If $x \notin E$, let y be the point in E nearest to x. So $\|x - y\| \leqslant \varepsilon$. Then

(21) $\sqrt{x^T D^{-1} x} \leqslant \sqrt{(x - y)^T D^{-1}(x - y)} + \sqrt{y^T D^{-1} y} \leqslant \varepsilon s^{-1} + 1.$

Let $w := (x^T D^{-1} x)^{-1/2} \cdot x$. Then $w \in E \cap H$ (since x and 0 belong to H and $x^T D^{-1} x > 1$), and

(22) $\|x - w\| = (1 - (x^T D^{-1} x)^{-1/2})\|x\|$

$= ((x^T D^{-1} x)^{1/2} - 1)(x^T D^{-1} x)^{-1/2}\|x\| \leqslant \dfrac{S}{s}\varepsilon$

since D^{-1} has smallest eigenvalue at least S^{-2}. This shows $x \in B(E \cap H, \varepsilon S/s)$. $\square$

13.4. KHACHIYAN'S METHOD MORE PRECISELY

We now give a precise description of Khachiyan's method. Suppose we wish to test feasibility of a given system $Ax \leq b$ of rational linear inequalities, in n variables ($n \geq 2$).

Let φ be the maximum row size of the matrix $[A \ b]$. Define

$$(23) \qquad v := 4n^2 \varphi$$
$$R := 2^v$$
$$N := 32n^2 v$$
$$p := 5N^2.$$

The algorithm determines a series of ellipsoids $E_0, E_1, E_2, \ldots$, given as $E_i = \text{ell}(z^i, D_i)$. Let $E_0 := \{x \mid \|x\| \leq R\}$, i.e. $z^0 := 0$ and $D_0 := R^2 \cdot I$.

If z^i, D_i have been found, check if z^i satisfies $Ax \leq b$. If so, we have found a solution. If not, let $a_k x \leq \beta_k$ be an inequality from $Ax \leq b$ violated by z^i, and let z^{i+1} and D_{i+1} arise from z' and D' in (2) (with $z = z^i$, $D = D_i$, $a = a_k$), by rounding all entries up to p binary digits behind the decimal point (taking care that D_{i+1} be symmetric). Inductively it follows that, for $i \leq N$,

$$(24) \qquad \text{the eigenvalues of } D_i \text{ are at least } 8^{-i}R^2 \text{ and at most } 8^i R^2.$$

This is clear for $i = 0$. Suppose we have proved (24) for a certain i. As the ellipsoid $\text{ell}(z', D')$ contains half of $\text{ell}(z^i, D_i)$, we know that the smallest eigenvalue of D' is at least $\frac{1}{4} 8^{-i} R^2$. Moreover, from (2) it follows that the largest eigenvalue of D' is at most $4 \cdot 8^i \cdot R^2$. Now the Ellipsoid approximation lemma (Theorem 13.2) gives that all eigenvalues of D_{i+1} are between $8^{-(i+1)}R^2$ and $8^{i+1}R^2$.

In particular, the matrices D_i are positive definite, for $i \leq N$. Also for $i \leq N$,

$$(25) \qquad \|z^i\| \leq 8^i R.$$

Indeed, this is true for $i = 0$. Suppose we know (25) for a certain $i < N$. As z' is in $\text{ell}(z^i, D_i)$, we know with (24) that $\|z' - z^i\| \leq 8^{i/2} R$. Hence

$$(26) \qquad \|z^{i+1}\| \leq \|z^{i+1} - z'\| + \|z' - z^i\| + \|z^i\| \leq n2^{-p} + 8^{i/2}R + 8^i R \leq 8^{i+1}R.$$

Claim. *If we reach E_N (so no z^i did satisfy $Ax \leq b$), then $P = \{x \mid Ax \leq b\}$ is not full-dimensional or not bounded.*

Proof of the Claim. By Theorems 13.1 and 13.2,

$$(27) \qquad \frac{\text{vol } E_{i+1}}{\text{vol } E_i} = \frac{\text{vol } E_{i+1}}{\text{vol } E'} \frac{\text{vol } E'}{\text{vol } E_i} \leq \left(1 + \frac{1}{8n}\right) e^{-1/(2n+2)} < e^{-1/8n}$$

(where $E' := \text{ell}(z', D')$). Therefore, as $\text{vol } E_0 \leq (2R)^n$,

$$(28) \qquad \text{vol } E_N < (2R)^n e^{-N/8n}.$$

Moreover, if P is bounded, then inductively, for $i \leq N$,

$$(29) \qquad P \subseteq B(E_i, 2^{-2N - 4N^2 + 4Ni}).$$

For $i = 0$ this is immediate, as all vertices of P have size at most v (Theorem 10.2). Suppose we know (29) for a certain $i < N$. Then by Theorems 13.2 and 13.3,

(30) $\quad P \subseteq B(E_i, 2^{-2N-4N^2+4Ni}) \cap \{x \mid ax \leqslant az^i\} \subseteq B(E_i \cap \{x \mid ax \leqslant az^i\}, 8^i 2^{-2N-4N^2+4Ni})$

$\qquad \subseteq B(E', 2^{-2N-4N^2+4Ni+3i}) \subseteq B(E_{i+1}, 2^{-2N-4N^2+4Ni+3i} + 2^{-2N-4N^2})$

$\qquad \subseteq B(E_{i+1}, 2^{-2N-4N^2+4N(i+1)}).$

Hence (29) gives

(31) $\quad P \subseteq B(E_N, 2^{-2N}) \subseteq (1 + 2^{-2N} 8^{N/2} R^{-1}) * E_N \subseteq 2 * E_N$

(where $\lambda * E$ denotes the blow-up of the ellipsoid E from its centre by a factor λ—so $\lambda * \text{ell}(z, D) := \text{ell}(z, \lambda^2 \cdot D)$). If P is full-dimensional, then $\text{vol} P \geqslant 2^{-2nv}$ (see (13)). So if P is full-dimensional and we could reach E_N, we would have the contradiction

(32) $\quad 2^{-2nv} \leqslant \text{vol} P \leqslant \text{vol}(2 * E_N) = 2^n \text{vol} E_N < 2^n (2R)^n e^{-N/8n} \leqslant 2^{-2nv}.$

End of proof of the Claim.

So if P is full-dimensional and bounded, we can find in polynomial time a solution of $Ax \leqslant b$. Note that (24) and (25) imply that the components of z^i and D_i are not larger than $8^N R$ in absolute value, and hence we can operate with them and store them in polynomial time and space.

The general case. Suppose now we do not require $P = \{x \mid Ax \leqslant b\}$ to be full-dimensional and bounded.

First, the general case can be reduced to the bounded case as follows. By Theorem 10.2, if $Ax \leqslant b$ has a solution, it has a solution of size at most $v := 4n^2 \varphi$, where φ is the maximum row size of the matrix $[A \; b]$. So $Ax \leqslant b$ has a solution, if and only if the system

(33) $\quad Ax \leqslant b, \; -2^v \leqslant \xi_1 \leqslant 2^v, \ldots, -2^v \leqslant \xi_n \leqslant 2^v$

has a solution (where $x = (\xi_1, \ldots, \xi_n)^{\mathsf{T}}$). As the size of system (33) is polynomially bounded by the size of the original system $Ax \leqslant b$, this reduces the general case to the bounded case.

To reduce the bounded case to the full-dimensional and bounded case, let $P = \{x \mid Ax \leqslant b\}$ be a bounded polyhedron. Let $\varepsilon := \frac{1}{2} n^{-1} 2^{-4n^2 \sigma}$, where σ is the size of the system $Ax \leqslant b$. Let

(34) $\quad P^\varepsilon := \{x \mid Ax \leqslant b + \varepsilon\}$

where ε stands for an all-ε column vector. Then P is empty if and only if P^ε is empty.

Proof. As $P \subseteq P^\varepsilon$, if $P^\varepsilon = \varnothing$ then $P = \varnothing$. To see the converse, suppose $P = \varnothing$. By Farkas' lemma (Corollary 7.1e), there is a vector $y \geqslant 0$ with $yA = 0$ and $yb = -1$. By Theorem 10.2 we can take y with size at most $4n^2 \sigma$. Therefore, $y(b + \varepsilon) < 0$, which implies that $Ax \leqslant b + \varepsilon$ has no solution, i.e. $P^\varepsilon = \varnothing$. $\qquad \square$

Since P is bounded, also P^ε is bounded (as if P and P^ε are nonempty, both have the same characteristic cone $\{x \mid Ax \leqslant 0\}$). Moreover, if P is nonempty, P^ε is full-dimensional (as any vector in P is an interior point of P^ε). The size of the system $Ax \leqslant b + \varepsilon$ is polynomially bounded by the size of $Ax \leqslant b$. Since, by the claim, P^{ε} can be tested for nonemptiness in polynomial time, also $Ax \leqslant b$ can be tested for feasibility in polynomial time.

In Theorem 10.4 we saw that this last implies that we can also find in polynomial time an explicit solution of $Ax \leqslant b$, if it exists. Similarly, a nonnegative solution for a system of linear equations can be found in polynomial time, if it exists. And—what caused the big excitement—linear programming problems can be solved in polynomial time.

Theorem 13.4 (Khachiyan's theorem). *Systems of rational linear inequalities, and linear programming problems with rational data, can be solved in polynomial time.*

Proof. See above.   □

One can derive that, if T is the maximum absolute value of the entries in A, b (where A is an integral $m \times n$-matrix and $b \in \mathbb{Z}^m$), then Khachiyan's method solves $Ax \leqslant b$ in $O(n^5 \cdot \log T)$ arithmetric operations on numbers with $O(n^3 \cdot \log T)$ digits—altogether $O(n^8 \cdot \log^2 T)$ bit operations.

Remark. It follows easily from Theorem 13.4 that for any given system $Ax \leqslant b$ of linear inequalities, the implicit equalities and the redundant constraints can be identified in polynomial time. So the affine hull and the facets of the polyhedron $P := \{x | Ax \leqslant b\}$ can be described in polynomial time. Moreover, if P is nonempty, a minimal face (i.e. a vertex if P is pointed) can be found by solving the LP-problem $\max \{cx | Ax \leqslant b\}$, for $c := (1, \varepsilon, \varepsilon^2, \ldots, \varepsilon^{m-1})A$, where ε is a small positive number (m being the number of rows of A). If ε is chosen small enough, any optimal solution x^* belongs to a minimal face of P. Such an ε can be computed in polynomial time, e.g. with the methods of Section 10.2. The inequalities in $Ax \leqslant b$ which are satisfied by x^* with equality determine, if they are put to equality, a minimal face of P.

Also systems with strict inequalities: $A_1 x < b_1$, $A_2 x \leqslant b_2$, can be solved in polynomial time. This follows from the polynomial solvability of systems of nonstrict inequalities, e.g. with the estimates made in Section 10.2.

13.5. THE PRACTICAL COMPLEXITY OF KHACHIYAN'S METHOD

Rather than revolutionizing the solving of linear programs, Khachiyan's publication has brought to the fore the gap between complexity theory and practice, and has caused a reconsideration of the value of 'polynomiality' of algorithms. Although the ellipsoid method is theoretically efficient (i.e. polynomial-time), computational experiments with the method are very discouraging, and, at the moment of writing, it is in practice by no means a competitor of the, theoretically inefficient, simplex method. Also if we could economize the rough estimates above, the number of ellipsoids we have to evaluate, together with the precision of the calculations, still seem much too high to be feasible for any nontrivial problem. If we abandon the precision during the calculations, the ellipsoid method practically can solve linear inequality systems in up to ± 15 variables. Halfin [1983] reports on an implementation of Khachiyan's method solving linear programs with up to 50 variables, some of which required more than 24000 iterations.

Some improvements have been proposed, like taking 'deep cuts' (Shor and Gershovich [1979]), or representing the positive definite matrices by their 'Cholesky factorization', to reduce the numerical instability (cf. Goldfarb and Todd [1982]), but they have not yet led to a breakthrough in the practical application of the ellipsoid method (see also Ursic [1982] for an interesting modification).

One reason for the bad performance may be that the number of iterations depends on the sizes of the numbers in the input. Although this does not conflict with the definition of polynomiality, one might have hoped that the sizes of these numbers influence only the sizes of the numbers occurring when executing the method. Algorithms with this characteristic are called *strongly polynomial*—see Section 15.2.

13.6. FURTHER REMARKS

Historically. Nice surveys on the history of the ellipsoid method and its predecessors are given by Bland, Goldfarb, and Todd [1981] and Schrader [1982, 1983]. They sketch the following line from the relaxation method (Section 12.3) to the ellipsoid method.

The relaxation method finds a series of points $z^0, z^1, z^2, \ldots$, where each z^{i+1} comes from projecting z^i into a violated inequality (which can be seen as a special case of the *subgradient method* introduced by Shor [1964]—see Note in Section 24.3). Shor [1970a,b] realized that the relaxation method can be improved by transforming the space at each step, so that the meaning of 'projecting' (and more generally of subgradient) varies at each step.

On the other hand, Levin [1965] and Newman [1965] gave the following *method of central sections*, approximating the minimum of a convex function f over a polytope P as follows. Let $P_0 := P$. If P_i has been found, let z^i be the centre of gravity of P_i, let a be the gradient of f in z^i (or a 'subgradient' in the case where f is not differentiable), and let $P_{i+1} := \{x \in P_i | a^\mathsf{T} x \leqslant a^\mathsf{T} z^i\}$. Since f is convex, its minimum is attained in P_{i+1}. As vol $P_{i+1} \leqslant (1 - e^{-1})volP_i$, the z^i converge to the minimum (see also Yudin and Nemirovskiĭ [1976a]).

However, calculating these centres of gravity can be difficult in practice. Yudin and Nemirovskiĭ [1976b] (cf. Shor [1977]) noticed that if we replace the polytopes by ellipsoids (E_0 is an ellipsoid containing P; E_{i+1} is the smallest ellipsoid containing $\{x \in E_i | a^\mathsf{T} x \leqslant a^\mathsf{T} z^i\}$), the method is more feasible, and applies to any convex programming problem. This is the ellipsoid method. They also observed that this method can be seen as a special case of Shor's method of projections with transformation of the space: the positive definite matrix defining the ellipsoid gives the transformation.

Khachiyan [1979, 1980] (cf. Gács and Lovász [1981]) next showed that for linear programming problems the ellipsoid method can be modified to give exact solutions, in polynomial time.

Both the 'polyhedron method' of Levin and Newman, and the ellipsoid method can be seen as a method of 'central sections' of a certain class $\mathscr{C}$ of geometric bodies: let a point z^i in a given set $C_i \in \mathscr{C}$ be given; find an affine half-space H, with z^i on its boundary, such that the points we are searching for are in H; next let C_{i+1} be a set in $\mathscr{C}$ with small volume such that $C_{i+1} \supseteq C_i \cap H$, and let z^{i+1} be the centre of gravity of C_{i+1}.

In this way the relaxation method can be considered as a 'ball method', where $\mathscr{C}$ is the class of balls. Yamnitsky [1982] (cf. Yamnitsky and Levin [1982]) showed that if we take for $\mathscr{C}$ the class of all simplices, we obtain a polynomial-time 'simplex method' for linear programming!

For a survey on the ellipsoid and related methods, see also the book by Shor [1979].

14

The ellipsoid method for polyhedra more generally

In Chapter 13 we described Khachiyan's method for solving explicitly given systems of linear inequalities, and explicitly given linear programs. In this chapter, we describe an extension to the solution of problems which are, in a sense, implicitly given.

A look at Khachiyan's method shows that it is not necessary to have an explicit list of all inequalities ready at hand. It suffices to be able to test if a given vector (in particular, the centre of the ellipsoid) is a solution of the inequality system, and if not, to find one violated inequality. This was observed by Karp and Papadimitriou, Padberg and Rao, and Grötschel, Lovász, and Schrijver.

In Section 14.1 we show the result of Grötschel, Lovász, and Schrijver that if we have a polynomial-time 'separation algorithm' for a polyhedron, we can find a vector in the polyhedron in polynomial time. In Section 14.2 we derive from this the polynomial-time equivalence of the 'separation problem' and the 'optimization problem'. In Section 14.3 we discuss some further implications, like finding a vertex or determining the dimension of a polyhedron in polynomial time.

14.1. FINDING A SOLUTION WITH A SEPARATION ALGORITHM

Suppose we know that the polyhedron $P \subseteq \mathbb{R}^n$ is defined by rational linear inequalities of size at most φ. We do not know these inequalities explicitly. Suppose, however, that we can solve the following problem in time polynomially bounded by n and φ:

(1)   *Separation problem.* Given a vector $y \in \mathbb{Q}^n$, decide whether y belongs to P or not, and, in the latter case, find a vector $a \in \mathbb{Q}^n$ such that $ax < ay$ for all x in P.

Then, with ellipsoids, one can find a vector in P or decide that P is empty, in time polynomially bounded by n and φ.

It is not difficult to derive this implication from Section 13.4 in the case where P is full-dimensional, as was shown by Grötschel, Lovász, and Schrijver [1981], Karp and Papadimitriou [1980], and Padberg and Rao [1980]. Non-full-dimensional polyhedra give more problems, as we cannot apply directly the reduction method of Section 13.4

(i.e. replacing $Ax \leqslant b$ by $Ax \leqslant b + \varepsilon$). Karp and Papadimitriou [1980] showed that the implication also holds for general, not-necessarily full-dimensional, polyhedra, provided that the output in (1) is a violated inequality $ax \leqslant \beta$ of size at most φ. In Grötschel, Lovász, and Schrijver [1984, 1988] it is shown that this condition can be deleted.

It will follow also that if we know a polynomial algorithm for the separation problem (1), then we can optimize a linear functional cx over P (i.e. solve an LP-problem over P), in polynomial time. This finds its application in combinatorial optimization, where the polyhedra generally have too many facets to be listed explicitly. Yet we know sometimes a polynomial method solving (1), and then with the ellipsoid method the polynomial solvability of the combinatorial problem follows.

In the following theorem we state and prove the above more precisely. A *separation algorithm* for a polyhedron P is an algorithm solving the separation problem (1). Note that such an algorithm can be considered formally as just a string of symbols.

In this chapter we assume $n \geqslant 2$.

Theorem 14.1. *There exists an algorithm* FEAS *and a polynomial* $\Psi(\xi, \eta)$, *such that if* FEAS *is given the input* (n, φ, SEP), *where*

(2)   *n and φ are natural numbers and* SEP *is a separation algorithm for some rational polyhedron P in $\mathbb{R}^n$ defined by linear inequalities of size at most φ,*

then FEAS *finds a vector in P or concludes that P is empty, in time bounded by $T\Psi(n, \varphi)$, where T is the maximum time required by* SEP *for inputs of size at most $\Psi(n, \varphi)$.*

[So n, φ, and SEP are the only data of P we need to know *a priori*.]

Proof. I. We first show the existence of a subroutine:

(3)   There exists an algorithm FEAS″ and a polynomial $\Psi''(\xi, \eta)$, such that if FEAS″ is given the input (n, v, SEP), where n and v are natural numbers and SEP is a separation algorithm for a bounded rational polyhedron P in $\mathbb{R}^n$, with vertices of size at most v, then:

either   (i)   FEAS″ concludes that P is empty;
or   (ii) FEAS″ gives a vector x in P;
or   (iii) FEAS″ gives a nonzero vector $c \in \mathbb{Z}^n$ and an integer δ such that $P \subseteq \{x \in \mathbb{R}^n | cx = \delta\}$, and $\|c\| \leqslant 2^{8nv}$ and $|\delta| \leqslant 2^{9nv}$;

in time bounded by $T\Psi''(n, v)$, where T is the maximum time needed by SEP for inputs of size at most $\Psi''(n, v)$.

We describe FEAS″. Let (n, v, SEP) as above be given. Define:

(4) $R := 2^v$
          $N := 125n^3 v$
          $p := 5N^2$.

As P is a polytope, we know $P \subseteq \{x \mid \|x\| \leqslant R\}$. The method determines inductively a series of ellipsoids $E_0, E_1, E_2, \ldots$, each 'almost' containing P. So a series of centres $z^0, z^1, z^2, \ldots$ and positive definite matrices $D_0, D_1, D_2, \ldots$ is determined, where $E_i = \text{ell}(z^i, D_i)$.

Again, $E_0 := \{x \mid \|x\| \leqslant R\}$, i.e. $z^0 := 0$ and $D_0 := R^2 \cdot I$. Suppose we have determined $E_i = \text{ell}(z^i, D_i)$. Check with SEP if z^i belongs to P. If so, we have found a vector in P, and conclude to (ii) in (3). If not, SEP gives a vector a such that $ax < az^i$ for all x in P. If $a = 0$ we conclude that P is empty ((i) in (3)). If $a \neq 0$, let z^{i+1}, D_{i+1} arise from z', D' in Theorem 13.1 (with $z := z^i, D := D_i$) by rounding their entries up to p binary digits behind the decimal point (so that D_{i+1} is symmetric). By the Ellipsoid approximation lemma (Theorem 13.2) we know that D_{i+1} again is positive definite. Let $E_{i+1} := \text{ell}(z^{i+1}, D_{i+1})$.

If we reach E_N, we can compute a hyperplane containing P as follows. As in (28) and (31) of Section 13.4:

(5) $\text{vol } E_N < (2R)^n e^{-N/8n}$

$P \subseteq 2 * E_N$.

So for each x in P,

(6) $(x - z^N)^{\mathsf{T}} D_N^{-1} (x - z^N) \leqslant 4$.

Moreover,

(7) $\det D_N = \dfrac{(\text{vol } E_N)^2}{V_n^2} \leqslant (2R)^{2n} e^{-N/4n} n^{2n} \leqslant 2^{-24n^2 v}$

(where V_n is the volume of the n-dimensional unit ball; so $V_n \geqslant n^{-n}$).

We now find a vector u with

(8) $\|u\| \geqslant 1, u^{\mathsf{T}} D_N u \leqslant 2^{-22nv}$

as follows. (*Comment*: we could find such a vector by computing an eigenvector belonging to the smallest eigenvalue of D_N. However, to avoid discussions on polynomiality and rounding, we describe a more direct method.) By (7),

(9) $\det D_N^{-1} \geqslant 2^{24n^2 v} \geqslant n^n 2^{22n^2 v}$

and hence D_N^{-1} has largest eigenvalue at least $n \cdot 2^{22nv}$. Hence $\text{Tr} D_N^{-1} \geqslant n \cdot 2^{22nv}$ (as D_N^{-1} is positive definite). We may assume that the diagonal entry $(D_N^{-1})_{11}$ of D_N^{-1} is the largest diagonal entry (if not, permute coordinates), so that

(10) $\gamma := (D_N^{-1})_{11} \geqslant \dfrac{1}{n} \cdot \text{Tr}(D_N^{-1}) \geqslant 2^{22nv}$.

Let v be the first column of D_N^{-1}, and let $u := \gamma^{-1} \cdot v$. Then

(11) $u^{\mathsf{T}} D_N u = \gamma^{-2} \cdot v^{\mathsf{T}} D_N v = \gamma^{-2} \cdot (D_N^{-1})_{11} = \gamma^{-1} \leqslant 2^{-22nv}$.

Moreover, the first component v_1 of u is $\gamma^{-1}(D_N^{-1})_{11} = 1$, so that $\|u\| \geqslant 1$, and hence we have (8).

Let $\beta := u^\mathsf{T} z^N$. Then for each x in P (using the Cauchy–Schwarz inequality (cf. Section 1.3) and (6)):

$$(12) \qquad |u^\mathsf{T} x - \beta| = |u^\mathsf{T}(x - z^N)| \leqslant (u^\mathsf{T} D_N u)^{1/2}((x - z^N)^\mathsf{T} D_N^{-1}(x - z^N))^{1/2}$$
$$\leqslant 2 \cdot 2^{-11nv} \leqslant 2^{-10nv}.$$

Now with the Simultaneous diophantine approximation algorithm (Corollary 6.4c) we can find an integral vector $c = (\gamma_1, \ldots, \gamma_n)$ and integers δ and q such that

$$(13) \qquad \|qu^\mathsf{T} - c\| \leqslant 2^{-4v}$$
$$|q\beta - \delta| \leqslant 2^{-4v}$$
$$1 \leqslant q \leqslant 2^{(n+1)(n+2)/4} 2^{4(n+1)v} \leqslant 2^{7nv}.$$

Then $\gamma_1 \neq 0$, since

$$(14) \qquad |\gamma_1| \geqslant |qv_1| - |\gamma_1 - qv_1| \geqslant q - 2^{-4v} > 0$$

(as $v_1 = 1$—see above). For each vertex x of P (cf. (12)):

$$(15) \qquad |cx - \delta| \leqslant |(c - qu^\mathsf{T})x| + |q(u^\mathsf{T} x - \beta)| + |q\beta - \delta|$$
$$\leqslant \|c - qu^\mathsf{T}\| \cdot \|x\| + q|u^\mathsf{T} x - \beta| + |q\beta - \delta|$$
$$\leqslant 2^{-4v} 2^v + 2^{7nv} 2^{-10nv} + 2^{-4v} < 2^{-v}.$$

However, as c and δ are integral, and x has size at most v, $cx - \delta$ is a rational number with denominator at most 2^v. Hence $cx = \delta$ for all vertices x of P, and therefore for all vectors x in P.

Note that

$$(16) \qquad \|c\| \leqslant \|qu^\mathsf{T}\| + \|c - qu^\mathsf{T}\| \leqslant qn + 2^{-4v} \leqslant qn + 1 \leqslant n2^{7nv} + 1 \leqslant 2^{8nv}$$

(as $\|u\| \leqslant n$, since $(D_N^{-1})_{11}$ is the maximum element in absolute value in the first column of D_N^{-1}, as D_N^{-1} is positive definite).

If $|\delta| > 2^{9nv}$ we conclude to (3)(i), as P is empty, since if x is a vertex of P we have the contradiction:

$$(17) \qquad 2^{9nv} < |\delta| = |cx| \leqslant \|c\| \cdot \|x\| \leqslant 2^{8nv} \cdot 2^v \leqslant 2^{9nv}.$$

If $|\delta| \leqslant 2^{9nv}$, we have (3)(iii).

This describes the algorithm FEAS". We finally estimate its running time. Similarly as in Section 13.4, we know that for each $i \leqslant N$, $\|z^i\| \leqslant 8^N R$ and all eigenvalues of D_i are at most $8^N R^2$. Therefore, all components of z^i and of D_i are at most $8^N R^2$ in absolute value. Moreover, they all have at most p binary digits behind the decimal point. As beside at most N applications of SEP to the input z^i, only elementary arithmetic operations are performed on the z^i, D_i, and the output of SEP, there exists a polynomial $\Psi''(\xi, \eta)$ such that the algorithm needs time at most $T\Psi''(n, v)$, where T is the maximum time needed by SEP if it is given an input of size at most $\Psi''(n, v)$.

II. The algorithm FEAS" is used as a subroutine in the following, which is almost the same as required by the theorem, except for the restriction to bounded polyhedra and for the fact that we consider the sizes of vertices instead of sizes of linear inequalities.

(18)   There exists an algorithm FEAS' and a polynomial $\Psi'(\xi, \eta)$, such that if FEAS' is given the input (n, v, SEP), where n and v are natural numbers and SEP is a separation algorithm for a bounded rational polyhedron P in $\mathbb{R}^n$, with vertices of size at most v, then:

either   (i)   FEAS' concludes that P is empty;
or   (ii)   FEAS' finds a vector x in P;

in time bounded by $T \cdot \Psi'(n, v)$, where T is the maximum time needed by SEP for inputs of size at most $\Psi'(n, v)$.

We describe FEAS'. Let (n, v, SEP) as above be given. We apply the following repeatedly.

Suppose we have found linearly independent integral row vectors $c_1, \ldots, c_k$ and integers $\delta_1, \ldots, \delta_k$ such that

(19) $P \subseteq \{x \mid c_1 x = \delta_1, \ldots, c_k x = \delta_k\} =: H;$

$\|c_1\|, \ldots, \|c_k\| \leq 2^{8nv}; |\delta_1|, \ldots, |\delta_k| \leq 2^{9nv}.$

(Initially, $k = 0$.) Without loss of generality we can write

(20) $\begin{bmatrix} c_1 \\ \vdots \\ c_k \end{bmatrix} = [C_1 \quad C_2]$

where C_1 is a nonsingular matrix of order k (possibly after permuting coordinates). Let $d := (\delta_1, \ldots, \delta_k)^{\mathsf{T}}$. Let the polyhedron P_k arise from P by projecting onto the last $n - k$ coordinates, i.e.

(21) $P_k := \left\{ y \in \mathbb{R}^{n-k} \, \middle| \, \begin{pmatrix} z \\ y \end{pmatrix} \in P \text{ for some } z \in \mathbb{R}^k \right\}.$

By (19) and (20),

(22) $y \in P_k$ if and only if $\begin{pmatrix} C_1^{-1} d - C_1^{-1} C_2 y \\ y \end{pmatrix} \in P.$

Moreover, all vertices of P_k have again size at most v, as they arise from vertices of P by deleting the first k coordinates.

We describe a separation algorithm SEP_k for P_k. If $w \in \mathbb{Q}^{n-k}$ is given, apply SEP to the vector

(23) $v := \begin{pmatrix} C_1^{-1} d - C_1^{-1} C_2 w \\ w \end{pmatrix}.$

If SEP asserts that v is in P then w belongs to P_k. If SEP gives a vector a such that $ax < av$ for all x in P, let $a = (a_1, a_2)$, with a_1 and a_2 vectors of length k

and $n - k$, respectively. Then

(24) $(a_2 - a_1 C_1^{-1} C_2)y < (a_2 - a_1 C_1^{-1} C_2)w$

for all y in P_k, and SEP_k gives $a_2 - a_1 C_1^{-1} C_2$ as output. This defines SEP_k.

Now apply the algorithm FEAS'' of (3) to $(n - k, v, SEP_k)$. Then there are three possible outcomes.

(i) FEAS'' concludes that P_k is empty. Then FEAS' concludes that P is empty, i.e. we have (i) in (18).

(ii) FEAS'' gives a vector y in P_k. Then FEAS' finds with (22) a vector in P, which yields (ii) in (18).

(iii) FEAS'' gives a nonzero vector $c \in \mathbb{Z}^{n-k}$ and an integer δ such that $P_k \subseteq \{y \in \mathbb{R}^{n-k} | cy = \delta\}$, $\|c\| \leqslant 2^{8(n-k)v}$, and $|\delta| \leqslant 2^{9nv}$. Taking $c_{k+1} := (0, c) \in \mathbb{Z}^n$ and $\delta_{k+1} := \delta$, we have $P \subseteq \{x \in \mathbb{R}^n | c_{k+1} x = \delta_{k+1}\}$, and therefore we can start anew with k replaced by $k + 1$.

This describes the algorithm FEAS'. To analyse its running time, by (3) an application of FEAS'' to $(n - k, v, SEP_k)$ needs time at most $T_k \Psi''(n - k, v)$, where T_k is the maximum time needed by SEP_k for inputs of size at most $\Psi''(n - k, v)$. By (23) and the estimates in (19), there is a fixed polynomial $\Phi(\xi, \eta, \zeta)$ such that $T_k \leqslant T$, where T is the maximum time needed by SEP for inputs of size at most $\Phi(n, v, \Psi''(n, v))$. Since we do further only elementary arithmetic operations, (18) follows.

III. We finally consider the general, not-necessarily bounded, case, which can be reduced to the bounded case as follows. If (n, φ, SEP) as in the theorem is given, then $P \neq \varnothing$ if and only if the polyhedron

(25) $Q := P \cap \{x = (\xi_1, \ldots, \xi_n)^T \in \mathbb{R}^n | -2^{4n^2 \varphi} \leqslant \xi_i \leqslant 2^{4n^2 \varphi} \text{ for } i = 1, \ldots, n\}$

is nonempty, by Theorem 10.2. So Q is defined by linear inequalities of size at most $5n^2 \varphi$, and hence, by Theorem 10.2 again, the vertices of Q have size at most $20n^4 \varphi$. Moreover, SEP is easily transformed to a separation algorithm SEP' for Q. Therefore applying FEAS' to $(n, 20n^4 \varphi, SEP')$ has the same effect as required for FEAS applied to (n, φ, SEP). The estimate on the running time follows easily. This finishes the proof of Theorem 14.1. $\square$

14.2. EQUIVALENCE OF SEPARATION AND OPTIMIZATION

Theorem 14.1 has several corollaries. One of them, especially important for optimization, concerns the following problem for a rational polyhedron P in $\mathbb{R}^n$:

(26)   **Optimization problem.** Given the input $c \in \mathbb{Q}^n$, conclude with one of the following:

   (i) give a vector x_0 in P with $cx_0 = \max\{cx | x \in P\}$;
   (ii) give a vector y_0 in char.cone P with $cy_0 > 0$;
   (iii) assert that P is empty.

An algorithm solving the optimization problem is called an *optimization algorithm* for *P*. Now, roughly speaking, if the separation problem is polynomially solvable, then also the optimization problem is polynomially solvable. More generally and more precisely:

Corollary 14.1a. *There exists an algorithm ELL such that if ELL is given the input* $(n, \varphi, \text{SEP}, c)$, *where*

(27)   *n and φ are natural numbers, and SEP is a separation algorithm for some rational polyhedron P in $\mathbb{R}^n$, defined by linear inequalities of size at most φ, and $c \in \mathbb{Q}^n$,*

then ELL solves the optimization problem for P, for the input c, in time polynomially bounded by n, φ, the size of c, and the running time of SEP.

[The definition of *polynomially bounded by* occurs in Section 2.4.]

Proof. We describe ELL. Let $(n, \varphi, \text{SEP}, c)$ satisfy (27). Let σ be the size of c. If $\max \{cx | x \in P\}$ is finite, it is attained by a vector of size at most $4n^2\varphi$ (Theorem 10.2). Hence

(28) $\text{size}(\max \{cx | x \in P\}) \leqslant 2(\sigma + 4n^2\varphi).$

Let

(29) $\tau := 3\sigma + 12n^2\varphi.$

Apply the following 'binary search' procedure. Let $m_0 := -2^\tau$, $M_0 := 2^\tau$. If m_i and M_i have been found, test with the algorithm FEAS of Theorem 14.1 whether the polyhedron

(30) $P \cap \{x | cx \geqslant \tfrac{1}{2}(m_i + M_i)\}$

is empty (one easily derives from SEP a separation algorithm for the polyhedron (30)). If so, let $m_{i+1} := m_i$ and $M_{i+1} := \tfrac{1}{2}(m_i + M_i)$. If not, let $m_{i+1} := \tfrac{1}{2}(m_i + M_i)$ and $M_{i+1} := M_i$. Stop if $i = K := 3\tau + 2$. There are three possibilities (note that inductively $M_i - m_i = 2^{\tau + 1 - i}$, so that $M_K - m_K = 2^{-2\tau - 1}$):

Case 1. $-2^\tau < m_K < M_K < 2^\tau$. In this case $M := \max \{cx | x \in P\}$ is finite, and $m_K \leqslant M < M_K$. Since M has size at most τ, its denominator is at most 2^τ, and hence M can be calculated with the continued fraction method (Corollary 6.3a) from m_K, as $|M - m_K| < (M_K - m_K) = 2^{-2\tau - 1}$. A vector in P maximizing cx can be found by applying the algorithm FEAS of Theorem 14.1 to the polyhedron $P \cap \{x | cx = M\}$ (one easily derives from SEP a separation algorithm for this polyhedron). So we have output (i) in (26).

Case 2. $M_K = 2^\tau$. Then $\max \{cx | x \in P\} = \infty$. Indeed, as $m_K = \tfrac{1}{2}(m_{K-1} + M_{K-1})$, we know that there is a vector x in P with $cx \geqslant m_K = 2^\tau - 2^{-2\tau - 1} > 2^{2\sigma + 8n^2\varphi}$. If the maximum is finite, by (28) it will be at most $2^{2\sigma + 8n^2\varphi}$.

We can find a vector y_0 in char.cone P with $cy_0 > 0$ as follows. First find with FEAS a vector x_0 in P. Next we define a separation algorithm SEP′ for char.cone P. Let $y \in \mathbb{Q}^n$ be given. Define $\lambda := 2^{3\varphi + \text{size}(x_0) + \text{size}(y)}$. Test with SEP if $x_0 + \lambda y$ belongs to P. If SEP asserts that $x_0 + \lambda y$ is not in P, and gives a vector a such that $ax < a(x_0 + \lambda y)$ for all x in P, then y is not in char.cone P and $az < ay$ for all z in char.cone P (since $\lambda \cdot az = a(x_0 + \lambda z) - ax_0 < a(x_0 + \lambda y) - ax_0 = \lambda \cdot ay$). If SEP asserts that $x_0 + \lambda y$ is in P, then y is in char.cone P. For suppose not. Then there would be a linear inequality $ax \leqslant \beta$, of size at most φ, valid for P, but with $ay > 0$. Then

(31) $\qquad ay \geqslant 2^{-\text{size}(a) - \text{size}(y)} \geqslant 2^{-\varphi - \text{size}(y)}$.

Therefore,

(32)
$$a(x_0 + \lambda y) = ax_0 + \lambda \cdot ay \geqslant - \|a\| \cdot \|x_0\| + \lambda \cdot ay$$
$$\geqslant -2^\varphi 2^{\text{size}(x_0)} + 2^{3\varphi + \text{size}(x_0) + \text{size}(y)} 2^{-\varphi - \text{size}(y)} > 2^\varphi \geqslant \beta$$

contradicting the fact that $x_0 + \lambda y$ is in P.

This defines the separation algorithm SEP′ for char.cone P, which is easily modified to a separation algorithm for char.cone $P \cap \{y | cy = 1\}$. With FEAS one finds a vector y_0 in this last polyhedron, which yields the output (ii) in (26).

Case 3. $m_K = -2^\tau$. Then P is empty. For suppose $P \neq \varnothing$. Then P contains a vector x of size at most $4n^2\varphi$ (Theorem 10.2), so that $cx \geqslant -2^{\sigma + 4n^2\varphi} \geqslant -2^\tau + 2^{-2\tau - 1} = M_K = \frac{1}{2}(m_{K-1} + M_{K-1})$, contradicting the definition of M_K. This yields the output (iii) in (26).

We leave it to the reader to become convinced of the polynomiality of the above method. $\qquad\qquad\qquad\qquad\qquad\qquad\qquad\qquad\qquad\qquad\qquad\square$

A particular case of Corollary 14.1a is Khachiyan's result that linear programming is polynomially solvable: an explicitly given system $Ax \leqslant b$ yields directly a separation algorithm for the polyhedron $P := \{x | Ax \leqslant b\}$, with running time polynomially bounded by the sizes of A and b: one can test the inequalities in $Ax \leqslant b$ one by one. Hence Corollary 14.1a gives that any linear program over P can be solved in polynomial time.

Corollary 14.1a states that the complexity of the optimization problem is polynomially bounded by the complexity of the separation problem. Interestingly, by polarity this also holds the other way around.

Corollary 14.1b. *There exists an algorithm* ELL* *such that if* ELL* *is given the input* $(n, \varphi, \text{OPT}, y)$, *where*

(33) $\qquad n$ *and* φ *are natural numbers,* OPT *is an optimization algorithm for some rational polyhedron* P *in* $\mathbb{R}^n$, *defined by linear inequalities of size at most* φ, *and* $y \in \mathbb{Q}^n$,

then ELL* *solves the separation problem for* P, *for the input* y, *in time polynomially bounded by* n, φ, *the size of* y *and the running time of* OPT.

Proof. We describe ELL*. Let n, φ, OPT, y satisfying (33) be given. We solve the separation problem for P for the input y.

First apply OPT to the input 0. If OPT answers that P is empty, we decide that $y \notin P$ and that $0x < 0y$ for all x in P.

Suppose OPT gives a vector x_0 in P (trivially maximizing $0x$ over P). Then the size of x_0 is polynomially bounded by n, φ, and the running time of OPT. Define the 'polar' $(P - x_0)^*$ of P with respect to x_0 by:

(34) $(P - x_0)^* := \{z \in \mathbb{R}^n | z^T(x - x_0) \leqslant 1 \text{ for all } x \text{ in } P\}$.

By Theorem 10.2 there are vectors $x_1, \ldots, x_t, y_1, \ldots, y_s \in \mathbb{Q}^n$ of size at most $4n^2\varphi$ such that

(35) $P = \text{conv.hull}\{x_1, \ldots, x_t\} + \text{cone}\{y_1, \ldots, y_s\}$.

Hence

(36) $(P - x_0)^* = \{z \in \mathbb{R}^n | z^T(x_i - x_0) \leqslant 1 \ (i = 1, \ldots, t); \ z^T y_j \leqslant 0 \ (j = 1, \ldots, s)\}$.

So $(P - x_0)^*$ is a polyhedron defined by linear inequalities of size at most φ^*, where φ^* is polynomially bounded by n, φ, and the running time of OPT.

Now we define a separation algorithm SEP* for $(P - x_0)^*$ as follows.

(37)   Let $w \in \mathbb{Q}^n$ be given. Apply OPT to the input w^T.

> *Case 1.* OPT gives a vector x^* in P with $w^T x^* = \max\{w^T x | x \in P\}$. If $w^T(x^* - x_0) \leqslant 1$ then SEP* will assert that w is in $(P - x_0)^*$. If $w^T(x^* - x_0) > 1$ then SEP* asserts that w is not in $(P - x_0)^*$, and that $(x^* - x_0)^T z < (x^* - x_0)^T w$ for all z in $(P - x_0)^*$.

> *Case 2.* OPT gives a vector y_0 in char.cone P with $w^T y_0 > 0$. Then SEP* asserts that w is not in $(P - x_0)^*$, and $y_0^T z < y_0^T w$ for all z in $(P - x_0)^*$ (this is justified, as $z^T y_0 \leqslant 0$ for all z in $(P - x_0)^*$).

This describes SEP*. Note that the running time of SEP* is polynomially bounded by n, φ, and the running time of OPT.

Now apply the algorithm ELL of Corollary 14.1a to the input $(n, \varphi^*, \text{SEP*}, y^T - x_0^T)$. So this solves the optimization problem for $(P - x_0)^*$ for input $y^T - x_0^T$.

If ELL gives that $\max\{(y - x_0)^T z | z \in (P - x_0)^*\}$ is at most 1, then ELL* will assert that y is in P (this is justified, as by polarity (Theorem 9.1), $y \in P$ if and only if $z^T(y - x_0) \leqslant 1$ for all $z \in (P - x_0)^*$).

If ELL gives that $\max\{(y - x_0)^T z | z \in (P - x_0)^*\}$ is attained by z_0 with $(y - x_0)^T z_0 > 1$, then ELL* will assert that y is not in P, and that $z_0^T x < z_0^T y$ for all x in P (as $z_0^T x = z_0^T(x - x_0) + z_0^T x_0 \leqslant 1 + z_0^T x_0 < z_0^T(y - x_0) + z_0^T x_0 = z_0^T y$).

If ELL gives a vector z_0 in char.cone $((P - x_0)^*)$ such that $(y - x_0)^T z_0 > 0$, then ELL* tells that y is not in P and $z_0^T x < z_0^T y$ for all x in P (indeed: for each $\lambda > 0$: $\lambda z_0 \in (P - x_0)^*$, and therefore $\lambda z_0^T(x - x_0) \leqslant 1$ for each $\lambda \geqslant 0$, implying $z_0^T x \leqslant z_0^T x_0 < z_0^T y$ for all x in P).

This finishes the description of ELL*. It is easy to see that the running time of ELL* is bounded as required in the Corollary.   □

Corollaries 14.1a and 14.1b show that the separation problem and the optimization problem are in a sense 'polynomially equivalent'. More precisely, we have the following.

Let for each $i \in \mathbb{N}$, P_i be a rational polyhedron, and suppose:

(38)   we can compute, for each $i \in \mathbb{N}$, in time polynomially bounded by $\log i$, natural numbers n_i and φ_i such that $P_i \subseteq \mathbb{R}^{n_i}$ and such that P_i has facet complexity at most φ_i.

[Often the index set $\mathbb{N}$ is replaced by other collections of strings of symbols, like collections of matrices, of graphs, or of inequality systems. Then $\log i$ is replaced by the size of the string. For example, we could take, for each undirected graph G, P_G to be the 'matching polytope' of G (being the convex hull of the characteristic vectors of matchings in G—so $P_G \subseteq \mathbb{R}^E$, where E is the edge set of G). See also Application 14.1 below.]

We say that *the separation problem is polynomially solvable* for the class $(P_i | i \in \mathbb{N})$ if it satisfies (38) and if there exists an algorithm, which for any input (i, y), with $i \in \mathbb{N}$, $y \in \mathbb{Q}^{n_i}$, solves the separation problem for P_i for the input y, in time polynomially bounded by $\log i$ and size(y).

The optimization problem is polynomially solvable for $(P_i | i \in \mathbb{N})$ is defined similarly, after replacing 'separation' by 'optimization'. The two concepts turn out to be equivalent:

Corollary 14.1c. *For any class $(P_i | i \in \mathbb{N})$, the separation problem is polynomially solvable if and only if the optimization problem is polynomially solvable.*

Proof. Directly from Corollaries 14.1a and 14.1b.   □

Therefore, we call a class $(P_i | i \in \mathbb{N})$ of rational polyhedra *polynomially solvable* if the separation problem (equivalently, the optimization problem) is polynomially solvable for this class.

It follows from Corollary 14.1c that polynomially solvable classes of polyhedra are closed under taking element-wise intersections and unions.

Corollary 14.1d. *If $(P_i | i \in \mathbb{N})$ and $(Q_i | i \in \mathbb{N})$ are polynomially solvable classes of polyhedra (such that for each i, P_i and Q_i are in the same space), then also the classes $(P_i \cap Q_i | i \in \mathbb{N})$ and $(\text{conv.hull}(P_i \cup Q_i) | i \in \mathbb{N})$ are polynomially solvable.*

Proof. Use that separation algorithms for P_i and Q_i immediately yield a separation algorithm for $P_i \cap Q_i$, and that optimization algorithms for P_i and Q_i immediately yield an optimization algorithm for conv.hull$(P_i \cup Q_i)$, with polynomially related running times.   □

Moreover, polynomially solvable classes of polyhedra are closed under taking several forms of polars (cf. Chapter 9):

Corollary 14.1e. *Let* $(P_i | i \in \mathbb{N})$ *be a class of polyhedra.*

(i) *If each* P_i *contains the origin, then the class* $(P_i | i \in \mathbb{N})$ *is polynomially solvable, if and only if the class* $(P_i^* | i \in \mathbb{N})$ *of polars is polynomially solvable.*

(ii) *If each* P_i *is of 'blocking type', then the class* $(P_i | i \in \mathbb{N})$ *is polynomially solvable, if and only if the class* $(B(P_i) | i \in \mathbb{N})$ *of blocking polyhedra is polynomially solvable.*

(iii) *If each* P_i *is of 'anti-blocking type', then the class* $(P_i | i \in \mathbb{N})$ *is polynomially solvable, if and only if the class* $(A(P_i) | i \in \mathbb{N})$ *of anti-blocking polyhedra is polynomially solvable.*

Proof. The assertions easily follow from the above, by observing that the separation problem for a polyhedron containing the origin is a special case of the optimization problem for its polar. Similarly for polyhedra of blocking and anti-blocking type. $\qquad \square$

Assertions (ii) and (iii) in particular will be the basis for deriving the polynomial solvability of one class of combinatorial optimization problems from that of another class, and conversely. For example, in networks, a polynomial algorithm for finding minimum capacity cuts can be derived from one for finding shortest paths, and conversely. Another application is as follows.

Application 14.1. Arborescences. Let $D = (V, A)$ be a directed graph, and let r be a vertex of D. A collection A' of arcs is called an r-arborescence if it forms a rooted spanning tree, with root r. So for each vertex v there exists a unique directed r–v-path in A'. A collection A' of arcs is an r-cut if $A' = \delta^-(U)$ for some nonempty subset U of $V \setminus \{r\}$. So each r-arborescence intersects each r-cut. In fact, the r-arborescences are the (inclusion-wise) minimal sets intersecting all r-cuts. Conversely, the (inclusion-wise) minimal r-cuts are the minimal sets intersecting all r-arborescences.

Recall the notation of Section 9.2: if $c_1, \ldots, c_m$ are vectors in $\mathbb{R}^A$, then

(39) $\qquad \{c_1, \ldots, c_m\}^\uparrow := \text{conv.hull}\,\{c_1, \ldots, c_m\} + \mathbb{R}^A_+$.

Now let $c_1, \ldots, c_m$ be the incidence vectors (in $\mathbb{R}^A$) of the r-cuts. Let $d_1, \ldots, d_t$ be the incidence vectors (again in $\mathbb{R}^A$) of the r-arborescences. It follows from a theorem of Edmonds [1967a] (which we will prove in Application 22.1 below) that the polyhedra

(40) $\qquad P_{D,r} := \{c_1, \ldots, c_m\}^\uparrow$ and $\quad Q_{D,r} := \{d_1, \ldots, d_t\}^\uparrow$

form a blocking pair of polyhedra, i.e.

(41) $\qquad P_{D,r} = B(Q_{D,r})$.

So we can apply (ii) of Corollary 14.1e to $P_{D,r}$ and $Q_{D,r}$.

First we show that the optimization problem for the class $(P_{D,r} | D$ digraph, r vertex$)$ is polynomially solvable. Let digraph $D = (V, A)$, $r \in V$ and $w \in \mathbb{Q}^A$ be given. Solving the optimization problem is equivalent to solving

(42) $\qquad \min \{wx | x \in P_{D,r}\}$.

If w has some negative component, then we easily find a vector x in char.cone$(P_{D,r}) = \mathbb{R}^A_+$ with $wx < 0$. If all components of w are nonnegative, (42) amounts to finding an r-cut of minimum 'capacity', considering w as a capacity function. We can do this by finding for each $s \neq r$ a minimum capacitated r–s-cut $C_{r,s}$ (in polynomial-time, by Dinits' variant of Ford and Fulkerson's method—see Application 12.1). The cut among the $C_{r,s}$ of minimum capacity is a minimum capacitated r-cut.

So the class $(P_{D,r}|D$ digraph, r vertex) is polynomially solvable. Hence by Corollary 14.1e, also the class of blocking polyhedra $(B(P_{D,r})|D$ digraph, r vertex) $= (Q_{D,r}|D$ digraph, r vertex) is polynomially solvable. It follows that for any digraph $D = (V, A)$, $r \in V$ and 'length' function $l: A \to \mathbb{Q}_+$, we can find a minimum length r-arborescence in polynomial time.

In fact, by Corollary 14.1e, the polynomial solvability of the minimum capacitated r-cut problem is *equivalent* to the polynomial solvability of the minimum length r-arborescence problem.

14.3. FURTHER IMPLICATIONS

With the ellipsoid method, several other problems on a polyhedron can be solved in polynomial time, as soon as we can solve the separation or the optimization problem in polynomial time. Among these problems are those of determining the affine hull or the lineality space, finding a minimal face, or a vector in the relative interior, writing a vector as a convex linear combination of vertices, and determining an optimal 'dual solution' in an LP-problem over the polyhedron. In this section we discuss such problems.

We first describe how to find the lineality space, the affine hull and the characteristic cone of a polyhedron (determining the affine hull is due to Edmonds, Lovász, and Pulleyblank [1982]).

Corollary 14.1f. *There exist algorithms which for given input (n, φ, SEP), where SEP is a separation algorithm of a nonempty rational polyhedron P in $\mathbb{R}^n$, defined by linear inequalities of size at most φ, solve the following problems:*

(i) *find linearly independent linear equations $c_1 x = \delta_1, \ldots, c_t x = \delta_t$ such that* aff.hull $P = \{x | c_1 x = \delta_1, \ldots, c_t x = \delta_t\}$;

(ii) *find linearly independent vectors $a_1, \ldots, a_k$ such that* lin.space $P = \{x | a_1 x = \cdots = a_k x = 0\}$;

(iii) *solve the separation problem for* char.cone P;

each in time polynomially bounded by n, φ, and the running time of SEP.

Proof. I. We first show that if (n, φ, SEP) as above is given, and $P = L$ is a *linear subspace* of $\mathbb{R}^n$, we can find linearly independent vectors $c_1, \ldots, c_t, x_1, \ldots, x_d$ such that

(43) $L = \{x | c_1 x = \cdots = c_t x = 0\} = \text{lin.hull} \{x_1, \ldots, x_d\}$,

in time polynomially bounded by n, φ, and the running time of SEP. Indeed, suppose we have found certain linearly independent vectors $x_1, \ldots, x_d$ in L, such that we can write, possibly after permuting coordinates,

(44) $[x_1, \ldots, x_d] = \begin{bmatrix} X_1 \\ X_2 \end{bmatrix}$

where X_1 is a nonsingular matrix of order d (at start $d = 0$). Let

(45) $L_d := L \cap \left\{ \begin{pmatrix} \mathbf{0} \\ y \end{pmatrix} \middle| y \in \mathbb{R}^{n-d} \right\}$

(0 here denoting the origin in $\mathbb{R}^d$), i.e. L_d consists of all vectors in L with zeros in the first d coordinates. We try to find a nonzero vector in L_d by searching for a vector in one of the polyhedra

(46) $P_{d,i} := L_d \cap \{x = (\xi_1, \ldots, \xi_n)^\mathsf{T} \in \mathbb{R}^n \mid \xi_i \geqslant 1\}$

for $i = d + 1, \ldots, n$. It is easy to derive from SEP a separation algorithm for $P_{d,i}$. (Note that $P_{d,i}$ is defined by linear inequalities of size at most φ.) Hence, with Theorem 14.1 we can decide that $P_{d,i}$ is empty or find a vector in it. If we find a vector in some $P_{d,i}$, by permuting coordinates we may assume that this vector is in $P_{d,d+1}$. Calling this vector x_{d+1} we apply the procedure anew with d replaced by $d + 1$.

If each of the $P_{d,i}$ turns out to be empty, then $L_d = \{0\}$, and hence $L = \text{lin.hull}\{x_1, \ldots, x_d\}$. Taking X_1 and X_2 as in (44) we know that

(47) $L = \{x \mid [X_2 X_1^{-1} - I]x = 0\}$

and so we find $c_1, \ldots, c_t$ as in (43).

II. Next, if (n, φ, SEP) as in Corollary 14.1f is given, we can find the affine hull of P as follows. First, with Theorem 14.1 we can find a vector x_0 in P (or decide that P is empty, in which case aff.hull $P = \varnothing$). Let L be the linear space

(48) $L := \text{aff.hull } P - x_0$.

It is easy to derive an optimization algorithm OPT_L for L from an optimization algorithm OPT_P for P (and hence from SEP, by Corollary 14.1a):

(49)   If $c \in \mathbb{Q}^n$ is given, give the inputs c and $-c$ to OPT_P. If OPT_P answers with a y in char.cone P such that $cy > 0$ or $-cy > 0$, then OPT_L will answer with y or $-y$, respectively (as $y, -y \in \text{char.cone } L = L$). If OPT_P answers with vectors x_1 and x_2 in P such that $cx_1 = \max\{cx \mid x \in P\}$ and $-cx_2 = \max\{-cx \mid x \in P\}$, then OPT_L will answer with $(x_1 - x_2) \in \text{char.cone } L$, $c(x_1 - x_2) > 0$ if $cx_1 > cx_2$, and with $0 \in L$, $c0 = \max\{cx \mid x \in L\}$ if $cx_1 = cx_2$.

So we have an optimization algorithm for L, and hence by Corollary 14.1b we have also a separation algorithm for L. With Part I of this proof we find linearly independent vectors $c_1, \ldots, c_t$ such that $L = \{x \mid c_1 x = \cdots = c_t x = 0\}$. Then aff.hull $P = \{x \mid c_1 x = c_1 x_0, \ldots, c_t x = c_t x_0\}$. This shows (i).

III. From (n, φ, SEP) as in Corollary 14.1f we can find a separation algorithm SEP' for char.cone P as follows. With Corollary 14.1a we can derive an optimization algorithm for P. Now an optimization algorithm OPT' for char.cone P is easily derived from OPT: for input $c \in \mathbb{Q}^n$, OPT' will give the same output as OPT, except when OPT answers with $x_0 \in \mathbb{Q}^n$ such that $cx_0 = \max\{cx \mid x \in P\}$; in that case OPT' will answer with the origin $0 \in \mathbb{Q}^n$, as then

$c0 = \max\{cx|x \in \text{char.cone } P\}$. From OPT' we derive SEP' with Corollary 14.1 b. This shows (iii).

IV. Finally, for (n, φ, SEP) as in Corollary 14.1f we can find vectors as required in (ii) by combining I and III above: to this end observe that

$$(50) \qquad \text{lin.space } P = \text{char.cone } P \cap - \text{char.cone } P$$

and hence that from a separation algorithm for char.cone P one easily derives a separation algorithm for lin.space P. This shows (ii).

We leave it to the reader to verify that each of the derivations above is polynomial relative n, φ, and the running time of SEP.   □

As by Corollary 14.1f we are able to identify the affine hull and the lineality space of P in polynomial time, we can restrict our discussion now to full-dimensional pointed polyhedra, as the general case can be reduced to the latter case by some simple linear transformations.

Corollary 14.1g. *There exist algorithms which for given input (n, φ, SEP), where SEP is a separation algorithm of a full-dimensional pointed rational polyhedron P in $\mathbb{R}^n$, defined by linear inequalities of size at most φ, solve the following problems:*

  (i) *find an internal point of P;*
 (ii) *find a vertex of P;*
(iii) *find a facet of P (i.e. find c, δ such that $P \cap \{x|cx = \delta\}$ is a facet of P);*
(iv) *for any rational vector x in P, find vertices $x_0, \ldots, x_t$ of P, extremal rays $y_1, \ldots, y_{n-t}$ of char.cone P and $\lambda_0, \ldots, \lambda_n \geqslant 0$ such that $x = \lambda_0 x_0 + \cdots + \lambda_t x_t + \lambda_{t+1} y_1 + \cdots + \lambda_n y_{n-t}$, $\lambda_0 + \cdots + \lambda_t = 1$ and $x_1 - x_0, \ldots, x_t - x_0, y_1, \ldots, y_{n-t}$ are linearly independent;*
 (v) *for any c in $\mathbb{Q}^n$ with $\max\{cx|x \in P\}$ bounded, find linearly independent facets $c_1 x = \delta_1, \ldots, c_n x = \delta_n$ and rationals $\pi_1, \ldots, \pi_n \geqslant 0$ such that $c_1, \ldots, c_n$ are linearly independent, $c = \pi_1 c_1 + \cdots + \pi_n c_n$, and $\max\{cx|x \in P\} = \pi_1 \delta_1 + \cdots + \pi_n \delta_n$ (i.e. find an optimal dual solution for any linear program over P);*

in time polynomially bounded by n, φ, the size of the further input data, and the running time of SEP.

Proof. Let (n, φ, SEP) be given, where SEP is a separation algorithm of a rational polyhedron $P \subseteq \mathbb{R}^n$, defined by linear inequalities of size at most φ. With Corollary 14.1a we can find an optimization algorithm OPT for P, with polynomially related running time. With this algorithm we check whether P is empty or not. If $P = \varnothing$, the problems above become trivial. So we may assume that P is nonempty. The theorem then is implied by the following results, where we leave it to the reader to verify polynomiality.

I. *If P is a polytope, we can find a vertex of P.* To show this, determine successively (where $x = (\xi_1, \ldots, \xi_n)^\mathsf{T}$)

(51) $\gamma_1 := \max \{\xi_1 \mid x \in P\}$

$\gamma_2 := \max \{\xi_2 \mid x \in P; \xi_1 = \gamma_1\}$

$\vdots \qquad \vdots$

$\gamma_n := \max \{\xi_n \mid x \in P; \xi_1 = \gamma_1, \ldots, \xi_{n-1} = \gamma_{n-1}\}.$

Then $(\gamma_1, \ldots, \gamma_n)^\mathsf{T}$ is a vertex of P.

II. *If P is a polytope, and a vector c in $\mathbb{Q}^n$ is given, we can find a vertex of P maximizing cx over P.* First we find with OPT $\delta := \max \{cx \mid x \in P\}$. Next with I above we find a vertex of the polytope $\{x \in P \mid cx = \delta\}$, which is a vertex of P maximizing cx.

III. *If P is a polytope, we can find affinely independent vertices $x_0, \ldots, x_d$ of P such that $d = \dim(P)$* (cf. Edmonds, Lovász, and Pulleyblank [1982]). We start with finding one vertex x_0 of P (cf. I). The next vertices are found inductively. If affinely independent vertices $x_0, \ldots, x_i$ and linearly independent equations $c_1 x = \delta_1, \ldots, c_j x = \delta_j$ have been found ($i \geqslant 0, j \geqslant 0$), such that

(52) $x_0, \ldots, x_i \in P \subseteq \{x \mid c_1 x = \delta_1, \ldots, c_j x = \delta_j\}$

and if $i + j < n$, choose c orthogonal to the affine hull of $x_0, \ldots, x_i$ and to $c_1, \ldots, c_j$. With II we can find vertices x' and x'' of P maximizing cx and $-cx$, respectively. If $cx' = cx''$, let $c_{j+1} := c$ and $\delta_{j+1} := cx'$. If $cx' > cx''$, then $cx' > cx_0$ or $cx'' < cx_0$. In the first case, let $x_{i+1} := x'$. Otherwise, let $x_{i+1} := x''$.

So either i is replaced by $i + 1$, or j by $j + 1$, and we can repeat the iteration. We stop if $i + j = n$, in which case we have that $i = \dim(P)$. (Note that throughout the iterations the size of c can be polynomially bounded by n and φ.)

IV. *We can find a vector in the relative interior of P* (the *relative interior* of P is $\{x \in \mathbb{R}^n \mid \exists \varepsilon > 0: B(x, \varepsilon) \cap \text{aff.hull } P \subseteq P\}$). Indeed, let

(53) $Q := \{x = (\xi_1, \ldots, \xi_n)^\mathsf{T} \in P \mid -2^{5n^2\varphi} \leqslant \xi_i \leqslant 2^{5n^2\varphi}, \text{ for } i = 1, \ldots, n\}.$

By Corollary 10.2b, Q is a polytope of the same dimension as P. Let d be this dimension. With III above we can find affinely independent vertices $x_0, \ldots, x_d$ of Q. The barycentre $(x_0 + \cdots + x_d)/(d + 1)$ will be in the relative interior of Q, and hence also of P. This shows (i).

V. *If P is pointed, we can find a vertex of P.* Indeed, consider the polar of the characteristic cone of P:

(54) $(\text{char.cone } P)^* = \{c \in \mathbb{R}^n \mid c^\mathsf{T} y \leqslant 0 \text{ for all } y \text{ in char.cone} P\}$

$= \{c \in \mathbb{R}^n \mid \max \{c^\mathsf{T} x \mid x \in P\} \text{ is finite}\}.$

Note that as P is pointed, $(\text{char.cone } P)^*$ is full-dimensional. It is easy to derive from OPT a separation algorithm for $(\text{char.cone } P)^*$. So with IV above we can find a vector c in the interior of (char.cone P)*. Then the face of vectors x in P attaining max $\{c^T x | x \in P\}$ is a polytope (as max $\{c'^T x | x \in P\}$ is finite if c' is in some neighbourhood of c). With I above we can find a vertex of this face, and hence of P. This shows (ii).

VI. *If P is full-dimensional, we can find a facet of P.* Indeed, with IV above we can find a vector x_0 in the interior of P. Let

$$(55) \qquad (P - x_0)^* = \{z | z^T(x - x_0) \leqslant 1 \text{ for all } x \text{ in } P\}.$$

It is easy to derive from OPT a separation algorithm SEP* for $(P - x_0)^*$ (cf. (37) in the proof of Corollary 14.1b). As P is full-dimensional, $(P - x_0)^*$ is pointed, and hence with V we can find a vertex z_0 of $(P - x_0)^*$. Then $P \cap \{x | z_0^T x = z_0^T x_0 + 1\}$ is a facet of P. This shows (iii).

VII. *If P is a polytope, then for any given rational vector y in P we can find affinely independent vertices $x_0, \ldots, x_d$ of P, and $\lambda_0, \ldots, \lambda_d \geqslant 0$ such that $\lambda_0 + \cdots + \lambda_d = 1$ and $y = \lambda_0 x_0 + \cdots + \lambda_d x_d$.* This can be seen by induction on the dimension of P. First find an arbitrary vertex x_0 of P. If $y = x_0$ we are done. If $y \neq x_0$, consider the half-line starting in x_0 through y:

$$(56) \qquad L := \{x_0 + \lambda(y - x_0) | \lambda \geqslant 0\}.$$

Let y' be the last point of this half-line in P. That is, let $\lambda' := \max\{\lambda | x_0 + \lambda(y - x_0) \in P\}$ (which can be found by binary search, or alternatively by applying an optimization algorithm for $\{\lambda | \lambda \geqslant 0, x_0 + \lambda(y - x_0) \in P\}$, which algorithm can be derived from OPT), and let $y' := x_0 + \lambda'(y - x_0)$. Then y is on the line-segment connecting x_0 and y'. Moreover, y' is on a proper face of P. To describe this, again let

$$(57) \qquad (P - x_0)^* := \{z \in \mathbb{R}^n | z^T(x - x_0) \leqslant 1 \text{ for all } x \text{ in } P\}.$$

It is easy to derive from SEP an optimization algorithm OPT* for $(P - x_0)^*$. So we can find a vector $z_0 \in \mathbb{Q}^n$ with $z_0^T(y' - x_0)$ maximizing $z^T(y' - x_0)$ over $z \in (P - x_0)^*$. Then

$$(58) \qquad F := \{x \in P | z_0^T(x - x_0) = 1\}$$

is a proper face of P, containing y'. It is easy to derive from SEP a separation algorithm SEP_F for F. Now, by induction, we can write y' as a convex combination of vertices of F. Hence, as y is on the line-segment connecting x_0 and y', we can write y as a convex combination of vertices of P.

VIII. *If P is pointed cone, and y in P is given, then we can find linearly independent extremal rays $y_1, \ldots, y_d$ of P and $\lambda_1, \ldots, \lambda_d \geqslant 0$ such that $y = \lambda_1 y_1 + \cdots + \lambda_d y_d$.* To see this, let c_0 be an internal vector of the polar $P^* = \{c | c^T x \leqslant 0 \text{ for all } x \in P\}$ of P, which can be found by IV (as P is pointed, P^* is full-dimensional). Then

$$(59) \qquad P \cap \{x | c_0^T x = 0\} = \{0\}.$$

For suppose there is a vector x in P with $c_0^\mathsf{T} x = 0$ and $x \neq 0$. Then there is a slight perturbation $\tilde{c}_0$ of c_0 such that $\tilde{c}_0 \in P^*$ and $\tilde{c}_0^\mathsf{T} x > 0$, which is a contradiction. Hence

(60) $Q := P \cap \{x \,|\, c_0^\mathsf{T} x = -1\}$

is a polytope (as its characteristic cone is (59)). It is easy to derive from SEP a separation algorithm SEP_Q for Q.

Now if $y = 0$ we can choose $d = 0$. If $y \neq 0$, then by (59) $c_0^\mathsf{T} y < 0$, and hence $-(c_0^\mathsf{T} y)^{-1} y \in Q$. By VII we can find affinely independent vertices $y_1, \ldots, y_d$ of Q and $\mu_1, \ldots, \mu_d \geqslant 0$ such that $\mu_1 + \cdots + \mu_d = 1$ and $-(c_0^\mathsf{T} y)^{-1} y = \mu_1 y_1 + \cdots + \mu_d y_d$. Since $y_1, \ldots, y_d$ are extremal rays of P, this finishes our proof.

IX. *If P is a pointed polyhedron, and a rational vector x in P is given, then we can find vertices $x_0, \ldots, x_d$ of P, extremal rays $y_1, \ldots, y_t$ of char.cone P, and $\lambda_0, \ldots, \lambda_d$, $\mu_1, \ldots, \mu_t \geqslant 0$ such that $x = \lambda_0 x_0 + \cdots + \lambda_d x_d + \mu_1 y_1 + \cdots + \mu_t y_t, \lambda_0 + \cdots + \lambda_d = 1$, and $x_1 - x_0, \ldots, x_d - x_0, y_1, \ldots, y_t$ are linearly independent.* Indeed, consider the cone in $\mathbb{R}^{n+1}$:

(61) $C := \left\{ \binom{x}{\lambda} \,\middle|\, \lambda > 0 \text{ and } \lambda^{-1} \cdot x \in P, \text{ or } \lambda = 0 \text{ and } x \in \text{char.cone } P \right\}.$

So if $P = \{x \,|\, Ax \leqslant b\}$, then $C = \left\{ \binom{x}{\lambda} \,\middle|\, \lambda \geqslant 0; \ Ax - \lambda \cdot b \leqslant 0 \right\}$. One easily derives from SEP a separation algorithm SEP_C for C. Now given x in P, by VIII we can describe $\binom{x}{1}$ as a nonnegative linear combination of extremal rays of C. This easily yields a decomposition of x as required. This shows (iv).

X. *If P is full-dimensional, then given $c_0 \in \mathbb{Q}^n$ with $\max \{c_0 x \,|\, x \in P\}$ finite, we can find facets $c_1 x = \delta_1, \ldots, c_t x = \delta_t$ of P and numbers $\pi_1, \ldots, \pi_t \geqslant 0$ such that $c_1, \ldots, c_t$ are linearly independent, $c_0 = \pi_1 c_1 + \cdots + \pi_t c_t$, and $\max \{c_0 x \,|\, x \in P\} = \pi_1 \delta_1 + \cdots + \pi_t \delta_t$.* Indeed, if $\max \{c_0 x \,|\, x \in P\}$ is attained by $x_0 \in P$ (which can be found with OPT), let C be the cone:

(62) $C := \{c^\mathsf{T} \in \mathbb{R}^n \,|\, \max \{cx \,|\, x \in P\} = cx_0\}.$

One easily derives from SEP an optimization algorithm for C, and hence with Corollary 14.1b also a separation algorithm for C. Then with VIII we can write c_0^T as a nonnegative linear combination of linearly independent extremal rays $c_1^\mathsf{T}, \ldots, c_t^\mathsf{T}$ of C:

(63) $c_0^\mathsf{T} = \pi_1 c_1^\mathsf{T} + \cdots + \pi_t c_t^\mathsf{T}, \quad \pi_1, \ldots, \pi_t \geqslant 0.$

If we define $\delta_1 := c_1 x_0, \ldots, \delta_t := c_t x_0$, then $c_1 x = \delta_1, \ldots, c_t x = \delta_t$ determine facets of P, and $\max \{c_0 x \,|\, x \in P\} = \pi_1 \delta_1 + \cdots + \pi_t \delta_t$, as required. This shows (v).   □

Various other problems on a polyhedron can be solved if the string (n, φ, SEP), or (n, φ, OPT), is given, in time polynomially bounded by n, φ, and the running time of the given separation or optimization algorithm, like:

- determining the dimension of P;
- deciding whether P is bounded ($\Leftrightarrow \dim$ (char.cone $P) = 0$);
- solving the *violation problem*: given a linear inequality $cx \leqslant \delta$, conclude that $cx \leqslant \delta$ is valid for all vectors in P, or find a vector in P violating $cx \leqslant \delta$;
- given $x \in P$, solving the separation problem for the smallest face of P containing x (or: finding linearly independent equations determining facets of P with intersection this smallest face);
- given $x \in P$, deciding whether x is a vertex of P;
- given two vertices of P, deciding whether they are adjacent vertices of P.

Conversely, given the input $(n, \varphi, \text{VIOL})$, where VIOL is an algorithm for the violation problem for a polyhedron $P \subseteq \mathbb{R}^n$ defined by linear inequalities of size at most φ, one can derive a separation algorithm for P, with running time polynomially bounded by n, φ, and the running time of VIOL. So also the violation problem and the separation problem are polynomially equivalent.

Moreover, any optimization algorithm OPT for a nonempty pointed polyhedron P can be 'polynomially' modified to another optimization algorithm OPT' for P which always gives as output a vertex of P or an extremal ray of the characteristic cone of P. Similarly, any separation algorithm SEP for a polyhedron can be 'polynomially' modified to another separation algorithm SEP' for P which always gives facets of P as separating hyperplanes.

Finally, Yudin and Nemirovskiĭ [1976b] (cf. Grötschel, Lovász and Schrijver [1988]) showed that there is an algorithm such that, if it is given the input $(n, \varphi, x_0, \text{MEM})$, where MEM is an algorithm solving the following problem for a polyhedron $P \subseteq \mathbb{R}^n$:

(64)   (*membership problem*) given $x \in \mathbb{Q}^n$, decide whether x is in P or not;

where P is defined by linear inequalities of size at most φ, and where $x_0 \in P$, then the algorithm solves the separation problem for P, in time polynomially bounded by n, φ, size (x_0), and the running time of MEM.

So assuming that we know an element x_0 of P in advance, we can leave out that the input algorithm gives separating hyperplanes. The algorithm is based on an extension of the ellipsoid method, the *shallow cut ellipsoid method*. Another application of this method is discussed in Section 15.4.

15

Further polynomiality results in linear programming

In this chapter we collect four further polynomiality results in linear programming.

In Section 15.1 we describe the recent algorithm of Karmarkar, which is a polynomial-time method for linear programming, which also in practice seems to work well (at the moment of writing).

In Section 15.2 we discuss *strongly polynomial* algorithms for linear programming, i.e. algorithms which consist of a number of arithmetic operations (on numbers of size polynomially bounded by the input size), which number is polynomially bounded by the *dimensions* of the linear program only (i.e. by the number n of variables and m of constraints—the number of arithmetic operations does not depend on the sizes of the input numbers, unlike the ellipsoid method). No strongly polynomial algorithm for linear programming is known (at the moment of writing), but some interesting positive results have been obtained by Megiddo and by Tardos—see Section 15.2.

Megiddo also designed an algorithm for linear programming which is, if we fix the number n of variables, a linear-time method: the algorithm consists of at most $C_n \cdot m$ arithmetic operations (on numbers of size polynomially bounded by the input size). This algorithm is described in Section 15.3.

Finally, in Section 15.4 we describe a polynomial-time method of Goffin and of Grötschel, Lovász, and Schrijver which finds, within a given full-dimensional polytope, an ellipsoid whose blow-up by a factor $n + 1$ *contains* the polytope (n is the dimension of the space).

15.1. KARMARKAR'S POLYNOMIAL-TIME ALGORITHM FOR LINEAR PROGRAMMING

Karmarkar [1984] presented a new algorithm for linear programming, which is, like the ellipsoid method, a polynomial-time method, but which also is claimed to be fast in practice. The method consists of $O(n^5 \cdot \log T)$ arithmetic operations on numbers with $O(n^2 \cdot \log T)$ digits, where T is the maximum absolute value of the input numbers (which are assumed to be integers—the ellipsoid method requires $O(n^5 \cdot \log T)$ arithmetic operations on numbers with $O(n^3 \cdot \log T)$ digits). At the moment of writing, no conclusive computational results were available to us. We here restrict ourselves to a theoretical description of the algorithm.

The algorithm applies to problems of the following form:

(1)   given an integral $m \times n$ — matrix A satisfying $A1 = 0$, and given an integral row vector $c \in \mathbb{Z}^n$, find a vector $x \geqslant 0$ such that $Ax = 0$, $1^T x = 1$, $cx \leqslant 0$.

[Here **1** denotes an all-one column vector.] At the end of this section we shall see that such a polynomial-time algorithm yields a polynomial-time algorithm for the general linear programming problem.

Assume without loss of generality that the rows of A are linearly independent, and that $n \geqslant 2$. Throughout we use

(2) $\alpha := \frac{1}{2}, \quad r := \sqrt{\dfrac{n}{n-1}}.$

Starting with $x^0 := (1/n, \ldots, 1/n)^T$, the algorithm determines a sequence of vectors $x^0, x^1, x^2, \ldots$ such that $Ax^k = 0$, $1^T x^k = 1$, $x^k > 0$, with the following recursion: denote $x^k := (\xi_1^{(k)}, \ldots, \xi_n^{(k)})^T$, and let D be the diagonal matrix:

(3) $D := \operatorname{diag}(\xi_1^{(k)}, \ldots, \xi_n^{(k)}).$

Define z^{k+1} and x^{k+1} as:

(4) z^{k+1} is the vector attaining $\min \{(cD)z \,|\, (AD)z = 0;\ 1^T z = n;\ z \in B(1, \alpha r)\}$, $x^{k+1} := (1^T D z^{k+1})^{-1} D z^{k+1}.$

Note that z^{k+1} minimizes $(cD)z$ over the intersection of a ball with an affine space, implying

(5) $z^{k+1} = 1 - \alpha r \dfrac{(I - DA^T(AD^2 A^T)^{-1}AD - n^{-1} \cdot 11^T)Dc^T}{\|(I - DA^T(AD^2 A^T)^{-1}AD - n^{-1} \cdot 11^T)Dc^T\|}$

[$\|\cdot\|$ is the Euclidean norm.]

Proof of (5). The minimum in (4) can be determined by projecting $(cD)^T$ onto the space $\{z \,|\, (AD)z = 0;\ 1^T z = 0\}$, thus obtaining the vector

(6) $p := (I - DA^T(AD^2 A^T)^{-1} AD - n^{-1} \cdot 11^T)Dc^T.$

[Indeed, $ADp = 0$, $1^T p = 0$, as one easily checks (using $AD1 = Ax^k = 0$), and $cD - p^T$ is a linear combination of rows of AD and of the row vector 1^T.]

Then z^{k+1} is the vector reached from **1** by going over a distance αr in the direction $-p$, i.e.

(7) $z^{k+1} = 1 - \alpha r \dfrac{p}{\|p\|}.$

This is equivalent to (5). $\square$

Since $1^T z = n$, $z \in B(1, \alpha r)$ implies $z > 0$ (cf. (i) in the Lemma below), we know that for each k: $Ax^k = 0$, $1^T x^k = 1$, $x^k > 0$.

To show correctness and convergence of the algorithm, we use the following lemma in elementary calculus. Notation, for $x = (\xi_1 \dots, \xi_n)^{\mathsf{T}}$

(8) $\Pi x := \xi_1 \cdot \dots \cdot \xi_n$.

Lemma. *Let* $n \in \mathbb{N}$, $H := \{x \in \mathbb{R}^n \mid 1^{\mathsf{T}} x = n\}$, $0 \leqslant \alpha < 1$. *Then*

(i) $H \cap B(1, r) \subseteq H \cap \mathbb{R}^n_+ \subseteq H \cap B(1, (n-1)r)$;
(ii) *if* $x \in H \cap B(1, \alpha r)$, *then* $\Pi x \geqslant (1 - \alpha)(1 + \alpha/(n-1))^{n-1}$.

Proof. (i) Let $x = (\xi_1, \dots, \xi_n)^{\mathsf{T}} \in H \cap B(1, r)$. To show $x \in \mathbb{R}^n_+$, suppose without loss of generality $\xi_1 < 0$. Since $x \in B(1, r)$, we know $(\xi_2 - 1)^2 + \dots + (\xi_n - 1)^2 \leqslant r^2 - (\xi_1 - 1)^2 < r^2 - 1 = 1/(n-1)$. Hence, with the Cauchy–Schwarz inequality: $(\xi_2 - 1) + \dots + (\xi_n - 1) \leqslant \sqrt{n-1} \cdot \sqrt{(\xi_2 - 1)^2 + \dots + (\xi_n - 1)^2} < 1$.   Therefore, $\xi_1 + \dots + \xi_n < \xi_2 + \dots + \xi_n < n$, contradicting the fact that x belongs to H.

Next let $x \in H \cap \mathbb{R}^n_+$. Then $(\xi_1 - 1)^2 + \dots + (\xi_n - 1)^2 = (\xi_1^2 + \dots + \xi_n^2) - 2(\xi_1 + \dots + \xi_n) + n \leqslant (\xi_1 + \dots + \xi_n)^2 - 2(\xi_1 + \dots + \xi_n) + n = n^2 - 2n + n = n(n-1) = (n-1)^2 r^2$. (The last inequality follows from the fact that $x \geqslant 0$.) So $x \in B(1, (n-1)r)$.

(ii) We first show the following:

(9)   let $\lambda, \mu \in \mathbb{R}$; if ξ^*, η^*, ζ^* achieve $\min \{\xi \eta \zeta \mid \xi + \eta + \zeta = \lambda, \xi^2 + \eta^2 + \zeta^2 = \mu\}$,
        and $\xi^* \leqslant \eta^* \leqslant \zeta^*$, then $\eta^* = \zeta^*$.

By replacing ξ, η, ζ by $\xi - \tfrac{1}{3}\lambda, \eta - \tfrac{1}{3}\lambda, \zeta - \tfrac{1}{3}\lambda$, we may assume that $\lambda = 0$. Then it is clear that the minimum is nonpositive, and hence $\xi^* \leqslant 0 \leqslant \eta^* \leqslant \zeta^*$. Therefore,

(10) $\xi^* \eta^* \zeta^* \geqslant \xi^* \left(\dfrac{\eta^* + \zeta^*}{2} \right)^2 = \tfrac{1}{4} \xi^{*3} \geqslant -\dfrac{\mu}{18}\sqrt{6\mu}$.

The first inequality here follows from the geometric–arithmetic mean inequality (note that $\xi^* \leqslant 0$), and the second inequality follows from the fact that if $\xi + \eta + \zeta = 0$, $\xi^2 + \eta^2 + \zeta^2 = \mu$, then $\xi \geqslant -\tfrac{1}{3}\sqrt{6\mu}$.

On the other hand, $(\xi, \eta, \zeta) := (-\tfrac{1}{3}\sqrt{6\mu}, \tfrac{1}{6}\sqrt{6\mu}, \tfrac{1}{6}\sqrt{6\mu})$ satisfies $\xi + \eta + \zeta = 0$, $\xi^2 + \eta^2 + \zeta^2 = \mu$, $\xi \eta \zeta = -(\mu/18)\sqrt{6\mu}$. Hence we have equality throughout in (10). Therefore, $\eta^* = \zeta^*$, proving (9).

We now prove (ii) of the lemma. The case $n = 2$ being easy, assume $n \geqslant 3$. Let x attain

(11) $\min \{\Pi x \mid x \in H \cap B(1, \alpha r)\}$.

Without loss of generality, $\xi_1 \leqslant \xi_2 \leqslant \dots \leqslant \xi_n$. Then for all $1 \leqslant i < j < k \leqslant n$, the vector (ξ_i, ξ_j, ξ_k) attains

(12) $\min \{\xi \eta \zeta \mid \xi + \eta + \zeta = \xi_i + \xi_j + \xi_k, \xi^2 + \eta^2 + \zeta^2 = \xi_i^2 + \xi_j^2 + \xi_k^2\}$

(otherwise we could replace the components ξ_i, ξ_j, ξ_k of x by better values). Hence by (9), $\xi_j = \xi_k$. Therefore $\xi_2 = \xi_3 = \dots = \xi_n$. As $x \in H \cap B(1, \alpha r)$, this implies $\xi_1 = 1 - \alpha$, and $\xi_2 = \dots = \xi_n = (1 + \alpha/(n-1))$. This shows (ii). $\square$

The operativeness of the algorithm now follows from the following theorem.

Theorem 15.1. *Suppose* (1) *has a solution. Then for all* k:

$$(13) \qquad \frac{(cx^{k+1})^n}{\Pi x^{k+1}} \leqslant \frac{e^{-2\alpha}}{1-\alpha} \frac{(cx^k)^n}{\Pi x^k}.$$

Proof. First note

$$(14) \qquad \frac{(cx^{k+1})^n}{\Pi x^{k+1}} \cdot \frac{\Pi x^k}{(cx^k)^n} = \frac{(cDz^{k+1})^n}{\Pi(Dz^{k+1})} \cdot \frac{\Pi x^k}{(cx^k)^n} = \left(\frac{cDz^{k+1}}{cD1}\right)^n \cdot \frac{1}{\Pi z^{k+1}},$$

using (4) and the facts that $\Pi(Dz^{k+1}) = (\Pi x^k)(\Pi z^{k+1})$ and $x^k = D1$.

We next show that, if (1) has a solution, then

$$(15) \qquad \frac{(cD)z^{k+1}}{(cD)1} \leqslant 1 - \frac{\alpha}{n-1}.$$

Indeed, $Ax = 0$, $x \geqslant 0$, $cx \leqslant 0$ for some $x \neq 0$. Hence $ADz = 0$, $z \geqslant 0$, $cDz \leqslant 0$ for some $z \neq 0$. We may assume, $1^T z = n$. Hence,

$$(16) \qquad 0 \geqslant \min\{(cD)z \mid z \in \mathbb{R}^n_+, ADz = 0, 1^T z = n\}$$
$$\geqslant \min\{(cD)z \mid z \in B(1,(n-1)r), ADz = 0, 1^T z = n\}$$

(the last inequality follows from (i) of the Lemma).

The last minimum in (16) is attained by the vector $1 - (n-1)r(p/\|p\|)$, as $1 - \alpha r(p/\|p\|)$ attains the minimum in (4). Therefore, $cD(1 - (n-1)r(p/\|p\|)) \leqslant 0$. This implies

$$(17) \qquad cDz^{k+1} = cD\left(1 - \alpha r \frac{p}{\|p\|}\right) = \left(1 - \frac{\alpha}{n-1}\right)cD1$$
$$+ \frac{\alpha}{n-1} cD\left(1 - (n-1)r\frac{p}{\|p\|}\right) \leqslant \left(1 - \frac{\alpha}{n-1}\right)(cD)1$$

proving (15).

Therefore, as $\Pi z^{k+1} \geqslant (1-\alpha)(1 + \alpha/(n-1))^{n-1}$, by (ii) of the Lemma,

$$(18) \qquad \left(\frac{cDz^{k+1}}{cD1}\right)^n \frac{1}{\Pi z^{k+1}} \leqslant \left(1 - \frac{\alpha}{n-1}\right)^n \cdot \frac{1}{(1-\alpha)\left(1+\dfrac{\alpha}{n-1}\right)^{n-1}} \leqslant \frac{e^{-2\alpha}}{1-\alpha}$$

(as $(1-x)/(1+x) \leqslant e^{-2x}$ for $x \geqslant 0$, since the function $e^{-2x} - (1-x)/(1+x)$ is 0 for $x = 0$, and has nonnegative derivative if $x \geqslant 0$). (14) and (18) combined gives (13). $\qquad\qquad\square$

Now for $\alpha = \frac{1}{2}$ (which minimizes $e^{-2\alpha}/(1-\alpha)$), we have

$$(19) \qquad \frac{(cx^{k+1})^n}{\Pi x^{k+1}} < \frac{2}{e} \cdot \frac{(cx^k)^n}{\Pi x^k}$$

if (1) has a solution. So by induction we have, as long as $cx^0 \geqslant 0$, $cx^1 \geqslant 0, \ldots, cx^k \geqslant 0$:

$$(20) \qquad cx^k \leqslant \frac{cx^k}{(\Pi x^k)^{1/n}} < \left(\frac{2}{e}\right)^{k/n} \cdot \frac{cx^0}{(\Pi x^0)^{1/n}} \leqslant \left(\frac{2}{e}\right)^{k/n} \cdot nT,$$

where T is the maximum absolute value of the entries in A and c. This gives, if we take

$$(21) \qquad N := \left\lceil \left(\frac{2}{1 - \ln 2} \right) \cdot n^2 \cdot \log_2(nT) + 1 \right\rceil,$$

the following theorem.

Theorem 15.2. *If* (1) *has a solution, then* $cx^k < n^{-n}T^{-n}$ *for some* $k = 0, \ldots, N$.

Proof. Suppose $cx^k \geqslant n^{-n}T^{-n}$ for all $k = 0, \ldots, N$. Then (20) holds for $k = N$, implying $cx^N < n^{-n}T^{-n}$. Contradiction. $\qquad\qquad\square$

So if (1) has a solution, by Theorem 15.2 the method yields a vector x^k with $cx^k < n^{-n}T^{-n}$. With elementary linear algebra we can find a vertex x^* of $\{x \mid x \geqslant 0, Ax = 0, \mathbf{1}^T x = 1\}$ such that $cx^* \leqslant cx^k$. As, by Cramer's rule, the entries in x^* have common denominator at most $n^n T^n$, and as c is integral, it follows that $cx^* \leqslant 0$. So x^* is a solution of (1).

So we have a polynomial-time algorithm for solving problem (1), if we can show that it suffices to make all calculations (especially calculating the Euclidean norm in (5)) with precision of p binary digits behind the point, where p is polynomially bounded by the sizes of A and c. This last is left to the reader—it can be done in a way similar to Khachiyan's method (see Section 13.4). Karmarkar claims that we can take $p = O(n^2 \cdot \log T)$. Since updating x^k takes $O(n^3)$ arithmetic operations (cf. (5)), this makes an algorithm consisting of $O(n^5 \cdot \log T)$ arithmetic operations on numbers of size $O(n^2 \cdot \log T)$.

We finally show that solving (1) in polynomial time suffices to show that the general linear programming problem is polynomially solvable. Indeed, by Theorem 10.4, we know that it suffices to be able to decide if a system $x \geqslant 0$, $Ax = b$ is feasible, in polynomial time. To this end, we can find a number M such that if $x \geqslant 0$, $Ax = b$ has a solution, it has one with $\mathbf{1}^T x \leqslant M$. By scaling b, we may assume $M = 1$. By adding a slack variable, we may assume that we ask for the feasibility of $x \geqslant 0$, $\mathbf{1}^T x = 1$, $Ax = b$. By subtracting multiples of $\mathbf{1}^T x = 1$ from the equations in $Ax = b$, we may assume $b = 0$. Thus it suffices to check the feasibility of

$$(22) \qquad x \geqslant 0, \ \lambda \geqslant 0, \ \mathbf{1}^T x + \lambda = 1, \ Ax - \lambda A \mathbf{1} = 0$$

such that $\lambda \leqslant 0$. As for $x := \mathbf{1}$, $\lambda := 1$ we have $Ax - \lambda \cdot A \mathbf{1} = 0$, this is a special case of (1), and we can apply Karmarkar's algorithm.

15.2. STRONGLY POLYNOMIAL ALGORITHMS

A drawback of the LP-methods of Khachiyan and Karmarkar is that the number of iterations, and hence the number of arithmetic operations performed, depends on the number of bits of the numbers in input data. It leaves open the question whether there exists an algorithm solving LP-problems with n variables and m constraints in at most $p(m, n)$ elementary arithmetic operations, where $p(m, n)$ is a polynomial. Here elementary arithmetic operations are: addition, subtraction, multiplication, division, comparison; they are supposed to operate on rationals of size polynomially bounded by the size of the input. Such an algorithm is called *strongly polynomial* or *genuinely polynomial*.

Megiddo [1983] showed that any linear program with n variables and m constraints, such that each of the constraints, and the objective function, has at most two nonzero coefficients, can be solved in $O(mn^3 \log n)$ elementary arithmetic operations. (This result extends earlier work of Nelson [1982] (who used Fourier–Motzkin elimination), Shostak [1981], Aspvall and Shiloach [1980a], and Aspvall [1980].)

Another interesting result was obtained by Tardos [1985, 1986], who showed that any linear program max $\{cx \,|\, Ax \leqslant b\}$ can be solved in at most $p(\text{size } A)$ elementary arithmetic operations on numbers of size polynomially bounded by size (A, b, c), where p is a polynomial. Thus the sizes of b and c do not contribute in the number of arithmetic steps. In particular the class of LP-problems where A is a $\{0, \pm 1\}$-matrix has a strongly polynomial algorithm. This includes the network flow problems, and therefore Tardos' result answers a long-standing open question asked by Edmonds and Karp [1972]. It also includes 'multi-commodity' flow problems.

We now describe Tardos' method. First three lemmas are proved.

Lemma A. *Suppose that we can solve the rational LP-problem* max $\{cx \,|\, Ax \leqslant b\}$ *in at most* $f(\text{size}(A), \text{size}(b), \text{size}(c))$ *elementary arithmetic operations on numbers of size polynomially bounded by* size(A, b, c). *Then for each such LP-problem we can*:

(23)   *either*   (i) *find* z^0 *satisfying* $Az^0 = b$;

      or   (ii) *find an inequality* $ax \leqslant \beta$ *in* $Ax \leqslant b$ *such that, if the maximum is finite, then* $ax^* < \beta$ *for some optimum solution* x^*;

in at most $f(\text{size}(A)), Q(\text{size}(A)), \text{size}(c)) + Q(\text{size}(A))$ *elementary arithmetic operations on numbers of size polynomially bounded by* size(A, b, c), *for some fixed polynomial* Q.

Proof. Let the rational LP-problem max $\{cx \,|\, Ax \leqslant b\}$ be given, where A has order $m \times n$. Without loss of generality, A is integral. Let T be the maximum absolute value of the entries of A. Let

(24) $\Delta := (nT)^n$.

So Δ is an upper bound on the absolute values of the subdeterminants of A. Choose z^0 so that $\|b - Az^0\|_2$ is as small as possible. [This can be done with the least square method, i.e. by solving $A^T Az = A^T b$ by Gaussian elimination— see Section 3.3.] Let

(25) $b' := b - Az^0$.

If $b' = 0$, we have found z^0 as in (23) (i). If $b' \neq 0$ let

(26) $b'' := \dfrac{mn\Delta^2}{\|b'\|_\infty} \cdot b'$

where $\|\cdot\|_\infty$ denotes the l_∞-norm ($\|x\|_\infty = $ maximum absolute value of the components of x). Note that $Ax \leqslant b''$ arises from $Ax \leqslant b$ by a translation and a multiplication: $x \mapsto (mn\Delta^2 / \|b'\|_\infty)(x - z^0)$. Hence maximizing cx over $Ax \leq b$ is equivalent to maximizing cx over $Ax \leq b''$: x^* is an optimum solution of $\max \{cx | Ax \leq b\}$ if and only if $(mn\Delta^2 / \|b'\|_\infty) (x^* - z^0)$ is an optimum solution of $\max \{cx | Ax \leq b''\}$.

Now solve

(27) $\max \{cx | Ax \leq \lceil b'' \rceil\}$

with the algorithm mentioned in the premise of the lemma.

If (27) is $\pm \infty$, then also $\max \{cx | Ax \leq b''\}$ is $\pm \infty$, as one easily checks.

If (27) is finite, let x^0 be an optimum solution of (27). Then there exists an inequality $ax \leq \beta''$ in $Ax \leq b''$ for which

(28) $ax^0 \leq \beta'' - n\Delta^2.$

This follows from:

(29) $\|b'' - Ax^0\|_\infty \geq \dfrac{1}{m} \|b'' - Ax^0\|_2$

$$= \frac{1}{m} \cdot \frac{mn\Delta^2}{\|b'\|_\infty} \left\| b - A \left(z^0 + \frac{\|b'\|_\infty}{mn\Delta^2} x^0 \right) \right\|_2 \geq \frac{n\Delta^2}{\|b'\|_\infty} \|b - Az^0\|_2$$

$$= \frac{n\Delta^2}{\|b'\|_\infty} \|b'\|_2 \geq n\Delta^2.$$

We claim that if $\max \{cx | Ax \leq b''\}$ is finite, then $ax^* < \beta''$ for some optimum solution x^* (and hence for the corresponding inequality $ax \leq \beta$ in $Ax \leq b$ we have $ax^* < \beta$ for some optimum solution x^* of $\max \{cx | Ax \leq b\}$, i.e. we have (23) (ii)). Indeed, by Theorem 10.5,

(30) $\|x^* - x^0\|_\infty \leq n\Delta \|b'' - \lceil b'' \rceil\|_\infty < n\Delta$

for some optimum solution x^* of $\max \{cx | Ax \leq b''\}$. Therefore,

(31) $ax^* \leq ax^0 + \|a\|_1 \|x^* - x^0\|_\infty < ax^0 + (nT)n\Delta \leq \beta'' - n\Delta^2$
         $+ (nT)n\Delta \leq \beta''.$

Now solving the linear program (27) takes, by the premise of the lemma, at most

(32) $f(\text{size}(A), \text{size}(\lceil b'' \rceil), \text{size}(c)) \leq f(\text{size}(A), Q'(\text{size}(A)), \text{size}(c))$

elementary arithmetic operations on numbers of size polynomially bounded by size (A, b, c) for some polynomial Q' (as $\|\lceil b'' \rceil\|_\infty = mn\Delta^2$). Moreover, we did at most Q'' (size(A)) further elementary arithmetic operations, for some polynomial Q''. This proves Lemma A.   □

Lemma B. We can find a solution of a given rational system $Ax \leq b$ of linear inequalities (or decide to its infeasibility) in at most $p(\text{size}(A))$ elementary arithmetic operations on numbers of size polynomially bounded by size(A, b), for some fixed polynomial p.

Proof. Let A have order $m \times n$. Without loss of generality, A is integral. Let T be the maximum absolute value of the entries in A. Let $\Delta := (nT)^n$, and let $c := (\Delta + 1, (\Delta + 1)^2, \ldots, (\Delta + 1)^m)A$. Then, if $Ax \leqslant b$ is feasible, $\max \{cx \mid Ax \leqslant b\}$ is finite. Moreover, it is attained at a unique minimal face. This follows from the fact that, if $r := \text{rank}(A)$, then c is not a linear combination of less than r rows of A. To see this, assume without loss of generality that the first r columns of A are linearly independent. Let them form the $m \times r$-submatrix A' of A. Let c' be the first r components of c. If c is a linear combination of less than r rows of A, then there exists an $(r - 1) \times r$-submatrix B of A' of rank $r - 1$ so that the matrix $\begin{bmatrix} B \\ c' \end{bmatrix}$ is singular. Hence

$$(33) \qquad 0 = \det \begin{bmatrix} B \\ c' \end{bmatrix} = (\Delta + 1) \det \begin{bmatrix} B \\ a_1' \end{bmatrix} + (\Delta + 1)^2 \det \begin{bmatrix} B \\ a_2' \end{bmatrix}$$
$$+ \cdots + (\Delta + 1)^m \det \begin{bmatrix} B \\ a_m' \end{bmatrix}$$

where a_i' denotes the ith row of A'. As each $\det \begin{bmatrix} B \\ a_i' \end{bmatrix}$ is at most Δ in absolute value, it follows that each $\det \begin{bmatrix} B \\ a_i' \end{bmatrix}$ is 0. This contradicts the fact that A' has rank r.

Now we know, from Khachiyan's method, that the premise of Lemma A holds for some polynomial f. Repeated application of Lemma A yields a splitting of $Ax \leqslant b$ into subsystems $A_1 x \leqslant b_1$ and $A_2 x \leqslant b_2$ and a vector z^0 such that $A_2 z^0 = b_2$ and such that, if $\max \{cx \mid Ax \leqslant b\}$ is finite, then $A_1 x^* < b_1$ for some, and hence for each, optimum solution x^*. (We use here that the maximum is attained at a unique minimal face.) Now if $A_1 z^0 \leqslant b_1$ then clearly z^0 is a solution of $Ax \leqslant b$. If $A_1 z^0 \leqslant b_1$ does not hold, then $Ax \leqslant b$ is infeasible: otherwise, z^0 would belong to the (minimal) face of optimum solutions of $\max \{cx \mid Ax \leqslant b\}$, implying $A_1 z^0 \leqslant b_1$.

As size(c) is polynomially bounded by size(a), and since we apply the algorithm of Lemma A at most m times, we get the time bound as required. $\qquad \square$

Lemma C. *Suppose we can solve the rational LP-problems*

$$(34) \qquad \max \{cx \mid Ax \leqslant b\} = \min \{yb \mid y \geqslant 0, yA = c\}$$

in at most $p(\text{size}(A), \text{size}(b)).\text{size}(c)^k$ elementary arithmetic operations on numbers of size polynomially bounded by size(A, b, c), for some polynomial p and some natural number k. Then we can solve the problems (34) also in at most $q(\text{size}(A)) \cdot \text{size}(c)^k$ elementary arithmetic operations on numbers of size polynomially bounded by size(A, b, c), for some polynomial q.

Proof. We first test, with the algorithm of Lemma B, if $Ax \leqslant b$ is feasible, and if $y \geqslant 0$, $yA = c$ is feasible. If one of them is infeasible, we know that the optima

(34) are not finite. Therefore, we may assume that the optima are feasible, and hence finite.

According to the premise of Lemma C, we can take in the premise of Lemma A $f(\kappa, \lambda, \mu) = p(\kappa, \lambda)\mu^k$. Repeating the algorithm of Lemma A we can split $Ax \leq b$ into subsystems $A_1 x \leq b_1$ and $A_2 x \leq b_2$ and find a vector z^0 such that $A_2 z^0 = b_2$ and such that $A_1 x^* < b_1$ for some optimum solution x^* of $\max\{cx|Ax \leq b\}$. Hence the components corresponding to A_1 in each optimum dual solution y in (34) are 0, so that

$$(35) \qquad \min\{yb|y \geq 0, yA = c\} = \min\{y_2 b_2 | y_2 \geq 0, y_2 A_2 = c\}.$$

By the algorithm of Lemma B we can find a feasible solution y_2^* for the second minimum here, which is automatically an optimum solution, as $y_2 b_2 = y_2 A_2 z^0 = cz^0$ for each feasible solution y_2.

Let $A_2' x \leq b_2'$ be the subsystem of $A_2 x \leq b_2$ corresponding to positive components of y_2^*. By complementary slackness, it follows that

$$(36) \qquad \{x|Ax \leq b, A_2' x = b_2'\}$$

is the set of optimum solutions of $\max\{cx|Ax \leq b\}$. With the algorithm of Lemma B we can find such a solution.   □

We now can prove Tardos' theorem.

Theorem 15.3. *There exists an algorithm which solves a given rational LP-problem* $\max\{cx|Ax \leq b\}$ *in at most* $P(\text{size}(A))$ *elementary arithmetic operations on numbers of size polynomially bounded by* $\text{size}(A, b, c)$, *for some polynomial P.*

Proof. By Khachiyan's theorem (Theorem 13.4), there exists a polynomial p_1 and an algorithm which solves the linear programs

$$(37) \qquad \max\{cx|Ax \leq b\} = \min\{yb|y \geq 0, yA = c\}$$

in at most $p_1(\text{size}(A), \text{size}(b), \text{size}(c))$ elementary arithmetic operations on numbers of size polynomially bounded by $\text{size}(A, b, c)$. Hence by Lemma C we can solve (37) in at most $p_2(\text{size}(A), \text{size}(b))$ elementary arithmetic operations on numbers of polynomial size, for some polynomial p_2 (this follows, by writing $\max\{cx|Ax \leq b\} = -\min\{-cx' + cx''|x', x'', \tilde{x} \geq 0, \quad Ax' - Ax'' + \tilde{x} = b\} = -\max\{-yb|y \geq 0, yA \leq c, yA \geq c\} = \min\{yb|y \geq 0, yA = c\}$).

Applying Lemma C again (now with $k = 0$), we can solve (37) in at most $P(\text{size}(A))$ elementary arithmetic operations on numbers of polynomial size, for some polynomial P.   □

Corollary 15.3a. *There exists a strongly polynomial algorithm for rational LP-problems with* $\{0, \pm 1\}$-*constraint matrix.*

Proof. Directly from Theorem 15.3.   □

An interesting related result was shown by Frank and Tardos [1985]. Let $(P_i|i \in \mathbb{N})$ be a polynomially solvable class of polyhedra, as defined in Section 14.2.

Then there exists a polynomial algorithm for the optimization problem for $(P_i | i \in \mathbb{N})$ (input: $i \in \mathbb{N}$, $c \in \mathbb{Q}^{n_i}$), which consists of $p(\log i, n_i, \varphi_i)$ elementary arithmetic operations, for some polynomial p (using n_i and φ_i as in Section 14.2). So the size of c does not influence the number of elementary arithmetic operations to be performed. Similarly for the separation problem. Frank and Tardos' method is based on simultaneous diophantine approximation (Section 6.3).

15.3. MEGIDDO'S LINEAR-TIME LP-ALGORITHM IN FIXED DIMENSION

It is not difficult to see that for any fixed dimension n there is a polynomial-time algorithm for linear programming in n variables (since if $Ax \leqslant b$ has m constraints in n variables, there are at most $\binom{m}{n}$ minimal faces, and they can be enumerated in time $O(m^n)$). Interestingly, Megiddo [1984] showed that for each fixed n there is an algorithm solving linear programs with n variables and m constraints in $O(m)$ elementary arithmetic operations on numbers of polynomial size. (It extends earlier results of Megiddo [1982] and Dyer [1984a, b] for $n = 2, 3$.)

A fundamental subroutine in Megiddo's algorithm is that for finding the median of a number of rationals in linear time, due to Blum, Floyd, Pratt, Rivest, and Tarjan [1973] (cf. Aho, Hopcroft, and Ullman [1974: Sec. 3.6] and Schönhage, Paterson, and Pippenger [1976]). The *median* of a sequence of m real numbers $q_1, \ldots, q_m$ is the number q such that $q_i \leqslant q$ for at least $\lceil \frac{1}{2} m \rceil$ indices i, and $q_i \geqslant q$ for at least $\lceil \frac{1}{2} (m + 1) \rceil$ indices i.

Lemma (Linear-time median-finding). *There exists a constant C such that we can find the median of given rationals $q_1, \ldots, q_m$ in at most $C.m$ arithmetic operations (viz. comparisons).*

Proof. It is shown more generally that if we are given rationals $q_1, \ldots, q_m$ and a natural number $k \leqslant m$, then we can find the kth smallest entry among $q_1, \ldots, q_m$ (i.e. the rational r such that $q_i \leqslant r$ for at least k indices i, and $q_i \geqslant r$ for at least $m - k + 1$ indices i) in at most $C.m$ arithmetic operations. The algorithm is described recursively. Without loss of generality, m is a multiple of 10 (otherwise, add some large dummy elements) and the q_i are different (otherwise replace each q_i by $q_i + i\varepsilon$, where ε is a small rational).

Let $t := m/5$. For $j = 1, \ldots, t$, determine the median q_j' of $q_{5j-4}, q_{5j-3}, q_{5j-2}, q_{5j-1}, q_{5j}$. Determine the median r' of $q_1', \ldots, q_t'$ (this, of course, can be done by determining, with the recursion, the $\frac{1}{2} t$th smallest entry in $q_1', \ldots, q_t'$). Now split the indices $i = 1, \ldots, m$ as

$$(38) \qquad I := \{i \mid q_i \leqslant r'\}, \quad J := \{i \mid q_i > r'\}.$$

If $|I| \geqslant k$, determine (recursively) the kth smallest entry in $\{q_i | i \in I\}$. If $|I| < k$, determine (recursively) the $(k - |I|)$th smallest entry in $\{q_i | i \in J\}$. In both cases, the answer is the kth smallest entry in $q_1, \ldots, q_m$.

Running time. The basis for deriving a linear time bound is the observation that $|I| \leqslant \frac{7}{10}m$ and $|J| \leqslant \frac{7}{10}m$. Indeed, as r' is the median of $q'_1, \ldots, q'_t$, we know that $q'_j \leqslant r'$ for at least $\frac{1}{2}t$ of the indices j. As each q'_j is the median of $q_{5j-4}, \ldots, q_{5j}$, it follows that $q_i \leqslant r'$ for at least $3 \cdot \frac{1}{2}t = \frac{3}{10}m$ of the indices i. Hence $|J| \leqslant \frac{7}{10}m$. Similarly, $|I| \leqslant \frac{7}{10}m$.

Clearly, there is a constant c_1 so that the sequence of medians q'_j of $q_{5j-4}, \ldots, q_{5j}$ can be found in at most $c_1 m$ comparisons. Now suppose that the median r' of $q'_1, \ldots, q'_t$ can be found in at most $Dt = D \cdot m/5$ comparisons. Moreover, suppose that the kth smallest entry in $\{q_i | i \in I\}$, respectively the $(k-|I|)$th smallest entry in $\{q_i | i \in J\}$, can be found in at most $D|I| \leqslant \frac{7}{10}Dm$, respectively in at most $D|J| \leqslant \frac{7}{10}Dm$, comparisons. Then, in order to prove that the whole algorithm requires linear time only, it suffices to show that there exists a number D such that for all m

(39) $c_1 m + \frac{1}{5}Dm + \frac{7}{10}Dm \leqslant Dm.$

Indeed, $D = 10c_1$ does so.   □

Now we show Megiddo's result.

Theorem 15.4. *For each natural number n there exists a constant K_n and a polynomial p such that: each given rational LP-problem $\max\{cx | Ax \leqslant b\}$, where A has order $m \times n$, can be solved by at most $K_n m$ elementary arithmetic operations on rationals of size at most $p(\text{size}(A, b, c))$.*

Proof. For recursive purposes, the theorem will be proved with the following concept of 'solving $\max\{cx | Ax \leqslant b\}$':

(40)   either   (i)   finding an optimum solution, if the maximum is finite;
         or   (ii) finding a solution of $Ax \leqslant b$, together with a vector z such that $Az \leqslant 0$, $cz > 0$, if the maximum is $+\infty$;
         or   (iii) finding a vector $y \geqslant 0$ such that $yA = 0$, $yb < 0$, if the maximum is $-\infty$ (i.e. if $Ax \leqslant b$ is infeasible).

The theorem is proved by induction on n, the case $n = 1$ being easy. Let $n \geqslant 2$ and suppose we know the theorem for $n - 1$.

For any linear program $\max\{cx | Ax \leqslant b\}$ and any linear equation $fx = \delta$ ($f \neq 0$) we say that *we know that $fx < \delta$ holds for the optimum,* when

(41)   either   (i)   we know that $Ax \leqslant b$ is feasible, and that $\max\{cx | Ax \leqslant b, fx \leqslant \delta\} > \max\{cx | Ax \leqslant b, fx = \delta\}$
         or   (ii) we know a vector $y \geqslant 0$ such that $yA = f$ and $yb < \delta$.

[Here we take $\infty > \gamma$ if γ is finite, but $\infty \not> \infty$. So (i) implies that $fx^* < \delta$ for each optimum solution (if the maximum is finite) or $fz < 0$ for each z with

$Az \leqslant 0$, $cz > 0$ (if the maximum is $+\infty$). Knowledge (ii) is also allowed if $Ax \leqslant b$ is feasible. Note that (ii) implies that $fx < \delta$ for each feasible solution of $Ax \leqslant b$.]

Similarly, we say that *we know that $fx > \delta$ holds for the optimum*, when

(42) either (i) we know that $Ax \leqslant b$ is feasible, and that $\max\{cx | Ax \leqslant b, fx \geqslant \delta\} > \max\{cx | Ax \leqslant b, fx = \delta\}$

   or (ii) we know a vector $y \geqslant 0$ such that $yA = -f$ and $yb < -\delta$.

We say *we know the position of the optimum relative to $fx = \delta$* if either we know that $fx < \delta$ holds for the optimum or we know that $fx > \delta$ holds for the optimum. To curtail arguments, we say, if $f = 0$, that we know the position of the optimum relative to $fx = \delta$ (irrespective of whether $\delta > 0$, $\delta = 0$, or $\delta < 0$).

The proof is based on the following claim.

Claim. *There exists a constant L_n such that if we are given a rational LP-problem*

(43) $\max\{cx | Ax \leqslant b\}$

(where A has order $m \times n$), and we are given a system of p linear equations $Fx = d$ in n variables, then we can either solve (43) or find $\lceil p/2^{2^{r-1}} \rceil$ equations from $Fx = d$ relative to each of which we know the position of the optimum of (43), in at most $L_n(p + m)$ elementary arithmetic operations on numbers of size polynomially bounded by $\mathrm{size}(A, b, c)$, where $r := \mathrm{rank}$ (F).

Proof of the Claim. The algorithm meant is defined recursively on $\mathrm{rank}(F)$.

I. First suppose $\mathrm{rank}(F) = 1$ (this can be checked in $O(p)$ elementary arithmetic operations). We may suppose that the first column of F is not all-zero (otherwise, permute coordinates). Since we trivially know the position of the optimum relative to equations with all coefficients 0, we may assume that F has no all-zero row. By scaling, we may assume that all entries in the first column of F are 1. It follows that all rows of F are the same, say f. Let q be the median of the components of d (found in linear time—see the Lemma above). By applying an affine transformation to $\mathbb{R}^n$, we may assume that $f = (1, 0, \ldots, 0)$ and $q = 0$. Next solve

(44) $\max\{cx | Ax \leqslant b, fx = q\}$.

As this is a problem in $n - 1$ dimension (as $fx = q$ means that the first component of x is 0), by our induction hypothesis we can solve (44) in at most $K_{n-1}m$ elementary arithmetic operations on numbers of polynomial size. We split into three cases.

Case 1. (44) is finite. Let it be attained by x^*. Let $A_1 x \leqslant b_1$ be the part of $Ax \leqslant b$ satisfied by x^* with equality. Next solve

(45)   (a) $\max\{cz | A_1 z \leqslant 0, fz = 1\}$

  and (b) $\max\{cz | A_1 z \leqslant 0, fz = -1\}$

which are again problems in $n-1$ dimensions, so that they can be solved in at most $K_{n-1}m$ elementary arithmetic operations on numbers of polynomial size. Then we have the following three subcases:

(46)   (α) (45)(a) is positive. Then $\max\{cx\,|\,Ax\leqslant b, fx\geqslant q\} > \max\{cx\,|\,Ax\leqslant b, fx = q\}$, and hence we know the position of the optimum relative to each equation $fx = \delta$ from $Fx = d$ with $\delta\leqslant q$; these are at least half of the equations in $Fx = d$.

  (β) (45)(b) is positive. Then, similarly, $\max\{cx\,|\,Ax\leqslant b, fx\leqslant q\} > \max\{cx\,|\,Ax\leqslant b, fx = q\}$, and hence we know the position of the optimum relative to at least half of the equations in $Fx = d$.

  (γ) Both optimum (45)(a) and (b) are infeasible or nonpositive. Then x^* is an optimum solution of $\max\{cx\,|\,Ax\leqslant b\}$, thus solving (43).

Case 2. (44) is $+\infty$. Then we obtain, by our definition (40) of 'solving', vectors x', z such that $Ax'\leqslant b$, $fx' = q$, $Az\leqslant 0$, $fz = 0$, $cz > 0$. Hence $\max\{cx\,|\,Ax\leqslant b\} = +\infty$, and x', z form a solution of (43) (in the sense of (40)).

Case 3. (44) is $-\infty$. Again by definition (40), we obtain y, λ such that $y\geqslant 0$, $yA + \lambda f = 0$, $yb + \lambda q < 0$. If $\lambda = 0$, then $Ax\leqslant b$ is infeasible, and we have solved $\max\{cx\,|\,Ax\leqslant b\}$ (in the sense of (40)). If $\lambda > 0$, then without loss of generality $\lambda = 1$. Then $yA = -f$ and $yb < -q$. Hence we conclude to (42)(ii) for each of the equations $fx = \delta$ with $\delta\leqslant q$. Similarly, if $\lambda < 0$, then without loss of generality $\lambda = -1$. Then $yA = f$ and $yb < q$. Hence we conclude to (41)(ii) for each of the equations $fx = \delta$ with $\delta\geqslant q$.

II. Next suppose $r = \operatorname{rank}(F)\geqslant 2$. Then we can find two columns of F which are linearly independent (this can be done easily in $O(p)$ elementary arithmetic operations). We may suppose these are the first two columns of F. Without loss of generality each entry in the first column of F is 0 or 1. Let $F'x = d'$ ($F''x = d''$, respectively) be the equations in $Fx = d$ with first coefficient 0 (1, respectively). Let F' and F'' have p' and p'' rows, respectively.

As $\operatorname{rank}(F') < \operatorname{rank}(F)$, by our recursion we can either solve $\max\{cx\,|\,Ax\leqslant b\}$ or find

(47) $s' := \lceil p'/2^{2^{r-2}}\rceil$

of the equations in $F'x = d'$ relative to which we know the position of the optimum.

In the latter case, consider next $F''x = d''$. Let q be the median of the entries in the second column of F''. Split $F''x = d''$ into two subsystems $G'x = e'$ and $G''x = e''$ of $\lceil\frac{1}{2}p''\rceil$ equations each, so that each entry in the second column of G' (G'', respectively) is at most q (at least q, respectively). (If p'' is odd, repeat one of the equations in $Fx = d$ which has first coefficient 1 and second coefficient q, and take this equation as first equation both in $G'x = e'$ and in $G''x = e''$.)

As the system $(G' - G'')x = e' - e''$ has $\operatorname{rank}(G' - G'') < \operatorname{rank}(F)$ (since $G' - G''$

has all-zero first column, but F not), by recursion we can either solve $\max\{cx\,|\,Ax \leqslant b\}$ or find

(48) $t := \lceil p''/2^{2^{r-2}} \rceil$

of the equations in $(G' - G'')x = e' - e''$ relative to each of which we know the position of the optimum. In the latter case, we have equations $g'_1 x = \varepsilon'_1, \ldots,$ $g'_t x = \varepsilon'_t$ in $G'x = e'$, and equations $g''_1 x = \varepsilon''_1, \ldots, g''_t x = \varepsilon''_t$ in $G''x = e''$, so that we know the position of the optimum relative to the equations $(g'_1 - g''_1)x = \varepsilon'_1 - \varepsilon''_1, \ldots, (g'_t - g''_t)x = \varepsilon'_t - \varepsilon''_t$.

Now, for $k = 1, \ldots, t$, we can make a convex combination

(49) $(\lambda_k g'_k + \mu_k g''_k)x = \lambda_k \varepsilon'_k + \mu_k \varepsilon''_k$

$(\lambda_k, \mu_k \geqslant 0, \lambda_k + \mu_k = 1)$, so that the first and second coefficients of (49) are 1 and q, respectively. So

(50) $\text{rank} \begin{bmatrix} \lambda_1 g'_1 + \mu_1 g''_1 \\ \vdots \\ \lambda_t g'_t + \mu_t g''_t \end{bmatrix} < \text{rank}(F)$

as the first matrix in (50) has its first two columns linearly dependent, but F not. Hence, by recursion we can either solve $\max\{cx\,|\,Ax \leqslant b\}$ or find

(51) $s := \lceil t/2^{2^{r-2}} \rceil = \lceil p''/2^{2^{r-1}} \rceil$

equations among (49) relative to which we know the position of the optimum. Without loss of generality, let it be the equations from (49) with $k = 1, \ldots, s$.

Now we show that for each $k = 1, \ldots, s$: we know the position of the optimum relative to *either* $g'_k x = \varepsilon'_k$ or $g''_k x = \varepsilon''_k$, as we know it relative to *both* $(g'_k - g''_k)x = \varepsilon'_k - \varepsilon''_k$ and $(\lambda_k g'_k + \mu_k g''_k)x = \lambda_k \varepsilon'_k + \mu_k \varepsilon''_k$ (**this is the heart of the proof**).

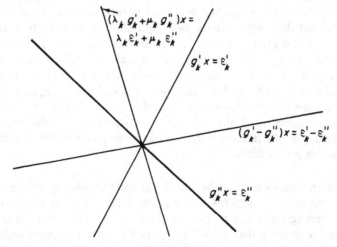

Figure 6

This follows from the fact that:

$$(52) \quad g'_k x < \varepsilon'_k \equiv \mu_k \times [(g'_k - g''_k)x < \varepsilon'_k - \varepsilon''_k] + 1 \times [(\lambda_k g'_k + \mu_k g''_k)x < \lambda_k \varepsilon'_k + \mu_k \varepsilon''_k]$$

$$g'_k x > \varepsilon'_k \equiv \mu_k \times [(g'_k - g''_k)x > \varepsilon'_k - \varepsilon''_k] + 1 \times [(\lambda_k g'_k + \mu_k g''_k)x > \lambda_k \varepsilon'_k + \mu_k \varepsilon''_k]$$

$$g''_k x < \varepsilon''_k \equiv \lambda_k \times [(g'_k - g''_k)x > \varepsilon'_k - \varepsilon''_k] + 1 \times [(\lambda_k g'_k + \mu_k g''_k)x < \lambda_k \varepsilon'_k + \mu_k \varepsilon''_k]$$

$$g''_k x > \varepsilon''_k \equiv \lambda_k \times [(g'_k - g''_k)x > \varepsilon'_k - \varepsilon''_k] + 1 \times [(\lambda_k g'_k + \mu_k g''_k)x > \lambda_k \varepsilon'_k + \mu_k \varepsilon''_k].$$

So, altogether, we know the position of the optimum relative to $s' + s \geq \lceil p/2^{2^{r-1}} \rceil$ of the equations in $Fx = d$.

End of proof of the Claim.

Apply the Claim to the linear program $\max \{cx \mid Ax \leq b\}$ and the system of equations $Ax = b$. We then either solve the linear program, or we find $\lceil m/2^{2^{n-1}} \rceil$ equations among $Ax = b$ relative to which we know the position of $\max \{cx \mid Ax \leq b\}$. This requires $L_n m$ elementary arithmetic operations. Next leave out these equations from $Ax = b$, thus giving $m_1 := m - \lceil m/2^{2^{n-1}} \rceil$ equations $A_1 x = b_1$. Apply the Claim to the problem $\max \{cx \mid Ax \leq b\}$ and the system $A_1 x = b_1$. This requires $L_n m_1$ elementary arithmetic operations. We either solve the linear program, or we find $\lceil m_1/2^{2^{n-r}} \rceil$ equations among $A_1 x = b_1$ relative to which we know the position of the optimum. Continuing, we either solve $\max \{cx \mid Ax \leq b\}$, or we find the position of the optimum with respect to each of the equations in $Ax = b$, in at most

$$(53) \quad L_n(m + m_1 + m_2 + \cdots) \leq L_n m \sum_{k=0}^{\infty} \left(1 - \frac{1}{2^{2^{n-1}}}\right)^k = L_n m \cdot 2^{2^{n-1}}$$

elementary arithmetic operations, where $m_{k+1} := m_k - \lceil m_k/2^{2^{n-1}} \rceil$ $(k = 1, 2, \ldots)$.

However, suppose we know the position of the optimum relative to each of the equations in $Ax = b$: if the maximum is finite, this would imply that for each optimum solution no inequality in $Ax \leq b$ is satisfied with equality—contradicting the fact that there exists a basic optimum solution. If the maximum is $+\infty$, there would exist vectors x', z and an inequality $ax \leq \beta$ in $Ax \leq b$ such that $Ax' \leq b$, $Az \leq 0$, $cz > 0$, $ax' = \beta$, $az = 0$. This contradicts the fact that we know the position of the optimum relative to $ax = \beta$. If $Ax \leq b$ is infeasible, this would imply that for each inequality $ax \leq \beta$ we know y_a, λ_a such that $y_a \geq 0$ and $y_a A + \lambda_a a = 0$, $y_a b + \lambda_a \beta < 0$. If one of the λ_a is nonnegative, we have a solution as required in (40)(iii). If each λ_a is negative, it would yield that, for each inequality $ax \leq \beta$ in $Ax \leq b$, the strict inequality $ax < \beta$ is implied by $Ax \leq b$. This clearly is not possible.   □

Megiddo mentions as applications of this method linear programs with few variables and many constraints, which occur, for example, in the linear separability problem (given many points in $\mathbb{R}^n$, organized in two disjoint sets, find a hyperplane separating the two sets), and in Chebyshev approximation (given matrix A and vector b, find vector x minimizing $\|b - Ax\|_\infty$; special case: given many points $(x_1, \lambda_1), \ldots, (x_m, \lambda_m)$ in $\mathbb{R}^{n+1}$, find a linear form $ax + \beta$ minimizing $\max_{i=1,\ldots,m} |\lambda_i - ax_i - \beta|$.

15.4. SHALLOW CUTS AND ROUNDING OF POLYTOPES

Several authors have observed that the efficiency of the ellipsoid method can be increased by taking 'deeper cuts': e.g. for the hyperplane separating the centre of the ellipsoid from the polytope we can choose a hyperplane *supporting* the polytope, rather than passing through the centre of the ellipsoid. Then generally a smaller piece than half of the ellipsoid is to be included in the next ellipsoid. Thus this next ellipsoid can be chosen smaller, which will speed up the method correspondingly.

Yudin and Nemirovskiĭ [1976b] noticed that, on the other hand, one can allow less deep, more shallow cuts, which yet will keep the method polynomial-time. If the next ellipsoid is the smallest ellipsoid containing a piece of the current ellipsoid which is a little larger than just half of it, we still can derive a polynomial method, with some interesting applications. One application, due to Yudin and Nemirovskiĭ, was mentioned in Section 14.3: from a polynomial membership algorithm for a polyhedron, together with the knowledge of an upper bound on its face-complexity and of one vector in it, one can derive a polynomial separation algorithm. So if we know an element of the polyhedron in advance, then from a membership test we can derive separating hyperplanes. For a description, see Grötschel, Lovász, and Schrijver [1988].

Another application is an approximation to the following result of K. Löwner (as reported by Danzer, Grünbaum, and Klee [1963: p. 139]) and John [1948] (see also Danzer, Laugwitz, and Lenz [1957]):

(54)   For each full-dimensional compact convex set K in $\mathbb{R}^n$ there exists a vector $z \in \mathbb{R}^n$ and a positive definite matrix D of order n such that $\mathrm{ell}(z, n^{-2} D) \subseteq K \subseteq \mathrm{ell}(z, D)$.

Note that $\mathrm{ell}(z, n^{-2}D)$ arises by multiplying $\mathrm{ell}(z, D)$ from its centre by a factor $1/n$. Löwner and John's result reads in other words: for each full-dimensional compact convex set K in $\mathbb{R}^n$ there exists an affine transformation τ of $\mathbb{R}^n$ such that $B(0, 1/n) \subseteq \tau K \subseteq B(0, 1)$. So τ gives a 'rounding' of the convex set K. The factor $1/n$ is best possible, as is shown if K is a simplex.

Goffin [1984], extending a subroutine of Lenstra [1983], showed that one can approximate the ellipsoids in (54) in polynomial time, in a sense made more precise in Theorem 15.6 below. This result will be used in Section 18.4 at the discussion of Lenstra's algorithm for integer linear programming.

We now first give an extension of Theorem 13.1.

Theorem 15.5. *Let* $E = \mathrm{ell}(z, D)$ *be an ellipsoid in* $\mathbb{R}^n$, *and let* $a \in \mathbb{R}^n$. *Let* $-1 < \beta \leqslant 1/n$, *and let* E' *be an ellipsoid containing* $E \cap \{x \mid ax \leqslant az + \beta\sqrt{aDa^{\mathsf{T}}}\}$, *such that E' has smallest volume. Then E' is unique, and* $E' = \text{ell}(z', D')$, *where*

(55) $\qquad z' := z - \dfrac{1 - n\beta}{n + 1} \cdot \dfrac{Da^{\mathsf{T}}}{\sqrt{aDa^{\mathsf{T}}}}$

$\qquad D' := \dfrac{n^2(1 - \beta^2)}{n^2 - 1} \cdot \left(D - \dfrac{2}{n + 1} \cdot \dfrac{1 - n\beta}{1 - \beta} \cdot \dfrac{Da^{\mathsf{T}}aD}{\sqrt{aDa^{\mathsf{T}}}} \right).$

Moreover,

(56) $\qquad \dfrac{\text{vol } E'}{\text{vol } E} < e^{-n(\beta - 1/n)^2/3}.$

Proof. Similarly as in the proof of Theorem 13.1, we need to check this theorem only for $D = I$, $z = 0$, and $a = (1, 0, \ldots, 0)$. The general case then follows by applying affine transformations.

The proof of (55) is straightforward with elementary geometry. As for (56) we have:

(57) $\qquad \dfrac{\text{vol } E'}{\text{vol } E} = \sqrt{\dfrac{\det D'}{\det D}} = \left(\dfrac{n^2(1 - \beta^2)}{n^2 - 1} \right)^{n/2} \cdot \left(\dfrac{(n - 1)(1 + \beta)}{(n + 1)(1 - \beta)} \right)^{1/2}$

$\qquad = \left(\dfrac{n^2}{n^2 - 1} \right)^{(n-1)/2} (1 - \beta^2)^{(n-1)/2} \cdot \dfrac{n}{n + 1}(1 + \beta)$

$\qquad = \left(1 + \dfrac{1}{n^2 - 1} - \beta^2 - \dfrac{\beta^2}{n^2 - 1} \right)^{(n-1)/2} \left(1 - \dfrac{1}{n + 1} + \beta - \dfrac{\beta}{n + 1} \right)$

$\qquad \leqslant e^{-(1 - \beta n)^2/2(n + 1)} < e^{-n(\beta - 1/n)^2/3}.$

(using $1 + x \leqslant e^x$). $\qquad\qquad\qquad\qquad\qquad\qquad\qquad\qquad\qquad\qquad\qquad\qquad\square$

Note that if we take $\beta \geqslant 1/n$ in Theorem 15.5 then $E' = E$, and we have no shrinking of volume.

Theorem 15.6. *There exists an algorithm such that if a rational system $Ax \leqslant b$ is given, where A has n columns and $P := \{x \mid Ax \leqslant b\}$ is full-dimensional and bounded, and if a rational γ with $0 < \gamma < 1/n$ is given, then an ellipsoid $\text{ell}(z, D)$ is found such that $\text{ell}(z, \gamma^2 D) \subseteq P \subseteq \text{ell}(z, D)$, in time polynomially bounded by n, $1/(1 - \gamma n)$, and the sizes of A, b, and γ.*

Proof. We describe the algorithm. Let n, γ, $Ax \leqslant b$, and P as above be given. We can find an upper bound v on the vertex-complexity of P, where v is polynomially bounded by the sizes of A and b (cf. Theorem 10.2). Construct a series $E_0, E_1, E_2, \ldots$ of ellipsoids as follows. Let

(58) $\qquad R := 2^v$

$\qquad N := 144n^2 v(1 - \gamma n)^{-2}$

$\qquad p := 5N^2$

$\qquad \beta := \dfrac{1}{2}\left(\gamma + \dfrac{1}{n} \right).$

Let $E_0 = \text{ell}(z^0, D_0) := B(0, R)$, i.e. $z^0 := 0$, $D_0 := R^2 \cdot I$. So $P \subseteq E_0$. If $E_i = \text{ell}(z^i, D_i)$ has been found, look for an inequality $a_k x \leqslant \beta_k$ from $Ax \leqslant b$ such that

(59) $\qquad a_k z^i > \beta_k - \beta \cdot \sqrt{a_k D_i a_k^{\mathsf{T}}}.$

(Note that, despite the square root, this can be done in polynomial time, as (59) is equivalent to: $a_k z^i > \beta_k$ or $(a_k z^i - \beta_k)^2 < \beta^2 (a_k D_i a_k^{\mathsf{T}})$.) If we find such an inequality, let $E' = \text{ell}(z', D')$ arise from E_i as in Theorem 15.5 (with $z := z^i$, $D := D_i$, $a := a_k$). Let z^{i+1} and D_{i+1} arise from z' and D' by rounding all components up to p binary digits behind the decimal point (such that D_{i+1} is symmetric).

Inductively, as in Section 13.4, one shows that the eigenvalues of D_i are at least $8^{-i} R^2$ and at most $8^i R^2$, and that $\|z^i\| \leqslant 8^i R$. Moreover, if $i \leqslant N$,

(60) $\qquad P \subseteq B(E_i, 2^{-2N - 4N^2 + 4Ni}).$

If we reached E_N, we would have:

(61) $\qquad P \subseteq B(E_N, 2^{-2N}) \subseteq (1 + 2^{-2N} 8^{N/2} R^{-1}) * E_N \subseteq 2 * E_N$

(where $\lambda * E$ denotes the blow-up of the ellipsoid E with a factor λ from its centre). Moreover, by Theorems 13.2 and 15.5,

(62) $\qquad \dfrac{\text{vol } E_{i+1}}{\text{vol } E_i} = \dfrac{\text{vol } E_{i+1}}{\text{vol } E'} \cdot \dfrac{\text{vol } E'}{\text{vol } E_i} < e^{-n(\beta - 1/n)^2/3} \left(1 + \tfrac{1}{6} n \left(\dfrac{1}{n} - \beta \right)^2 \right)$

$$\leqslant e^{-n(\beta - 1/n)^2/3} e^{n(\beta - 1/n)^2/6} = e^{-n(\beta - 1/n)^2/6}.$$

Since $\text{vol } E_0 \leqslant (2R)^n$, it follows with induction that

(63) $\qquad \text{vol } E_N < (2R)^n e^{-nN(\beta - 1/n)^2/6}.$

Since all vertices of P have size at most v, we know that $\text{vol } P \geqslant n^{-n} 2^{-nv}$ (cf. (13) in Section 13.2). This gives the contradiction

(64) $\qquad n^{-n} 2^{-nv} \leqslant \text{vol } P \leqslant 2^n \text{vol } E_N < 2^n (2R)^n e^{-nN(\beta - 1/n)^2/6} \leqslant n^{-n} 2^{-nv}.$

Therefore, for some $i < N$, there is no inequality $a_k x \leqslant \beta_k$ in $Ax \leqslant b$ satisfying (59). This implies

(65) $\qquad \text{ell}(z^i, \beta^2 D_i) \subseteq P.$

Indeed, if $y \in \text{ell}(z^i, \beta^2 D_i)$, then for any inequality $a_k x \leqslant \beta_k$ from $Ax \leqslant b$:

(66) $\qquad a_k y = a_k(y - z^i) + a_k z^i \leqslant \sqrt{a_k D_i a_k^{\mathsf{T}}} \sqrt{(y - z^i)^{\mathsf{T}} D_i^{-1}(y - z^i)}$

$$+ \beta_k - \beta \sqrt{a_k D_i a_k^{\mathsf{T}}} \leqslant \beta_k.$$

Hence y belongs to P.

On the other hand, from (60) we know:

(67) $\qquad P \subseteq B(E_i, 2^{-2N}) \subseteq (1 + 2^{-2N} 8^{N/2} R^{-1}) * E_i \subseteq (1 + 2^{-N/2}) * E_i$

$$= \text{ell}(z^i, (1 + 2^{-N/2})^2 D_i) \subseteq \text{ell}(z^i, (\beta/\gamma)^2 D_i).$$

Taking $z := z^i$ and $D := (\beta/\gamma)^2 D_i$ gives the ellipsoid as described by the theorem. We leave checking the running time bound to the reader. $\qquad\square$

So if γ tends to $1/n$, the ellipsoid of Theorem 15.6 tends to the ellipsoid described in (54).

One of the consequences was mentioned in discussing the relaxation method (Section 12.3).

Corollary 15.6a. *There exists an algorithm such that if a rational system $Ax \leqslant b$ is given, where $P := \{x \mid Ax \leqslant b\}$ is full-dimensional and bounded, a linear transformation τ (i.e. a nonsingular matrix) and a vector x_0 are found such that*

$$(68) \qquad B(x_0, 1) \subseteq \tau P \subseteq B(x_0, n^{3/2})$$

in time polynomially bounded by the sizes of A and b.

Proof. Take $\gamma = n^{-5/4}$ in Theorem 15.6. We find a vector z and a positive definite matrix D such that

$$(69) \qquad \mathrm{ell}(z, \gamma^2 D) \subseteq P \subseteq \mathrm{ell}(z, D).$$

Next find a nonsingular matrix U such that $U^{\mathsf{T}} U$ is near enough to D^{-1} so that for $\tau : x \to n^{11/8} U x$ we have:

$$
\begin{aligned}
(70) \qquad B(\tau z, 1) &= \tau\{x \mid (x - z)^{\mathsf{T}} U^{\mathsf{T}} U (x - z) \leqslant n^{-11/4}\} \\
&\subseteq \tau\{x \mid (x - z)^{\mathsf{T}} D^{-1} (x - z) \leqslant n^{-5/2}\} \\
&= \tau(\mathrm{ell}(z, \gamma^2 D)) \subseteq \tau P \subseteq \tau(\mathrm{ell}(z, D)) \\
&= \tau\{x \mid (x - z)^{\mathsf{T}} D^{-1} (x - z) \leqslant 1\} \subseteq \tau\{x \mid (x - z)^{\mathsf{T}} U^{\mathsf{T}} U (x - z) \leqslant n^{1/4}\} \\
&= B(\tau z, n^{3/2}).
\end{aligned}
$$

The first and the last inclusion here give the conditions for the approximation $D^{-1} \approx U^{\mathsf{T}} U$. The details are left to the reader.   □

Grötschel, Lovász, and Schrijver [1988] showed a similar result in the case where the polytope P is not given by an explicit system of inequalities but by a separation algorithm. In this case we cannot simply test (59) for all inequalities. Then, in order to determine the next ellipsoid, first an affine transformation is applied, transforming the current ellipsoid to the unit ball. Next look around from the origin in $2n$ orthogonal directions. If in one of these directions the vector at distance β is outside the transformed polytope, we can cut the current ellipsoid with a shallow cut (the separation algorithm provides us with this cut). If in each of these directions the vector at distance β belongs to the transformed polytope, an appropriate shrinking of the unit ball will be included in the transformed polytope—in fact, a shrinking by a factor $n^{-1/2}\beta$. Back to the original space this means that a shrinking by a factor $n^{-1/2}\beta$ of the current ellipsoid is contained in the original polytope P. We find an ellipsoid $\mathrm{ell}(z, D)$ with $\mathrm{ell}(z, \gamma^2 D) \subseteq P \subseteq \mathrm{ell}(z, D)$, for $0 < \gamma < n^{-3/2}$.

Notes for Section 15.4. For formulas for the minimum volume ellipsoid containing the part of an ellipsoid between two parallel hyperplanes, see König and Pallaschke [1981], Krol and Mirman [1980], Shor and Gershovich [1979], and Todd [1982].

Notes on polyhedra, linear inequalities, and linear programming

HISTORICAL NOTES

Polyhedra, linear inequalities, and linear programming represent three faces of one and the same problem field: a geometric, an algebraic, and an optimizational face. Their interrelation seems to have been observed first by Fourier in the 1820's, and after that again only in the 20th century.

Polyhedra up to the 18th century

It is known that the Egyptians of the Middle Kingdom (c. 2050–1800 B.C.) studied the problem of measuring the volume of certain polyhedra, like truncated pyramids. Polyhedra, mainly two- and three-dimensional, were also studied in the 5th–2nd centuries B.C. by the Greeks, among which are the Pythagoreans, Plato, Euclid (Book XII of the Elements), and Archimedes. The Chinese *Chiu-Chang Suan-Shu* (Nine Books of Arithmetic) written *c.* 100 B.C., also considered the problem of the volume of certain polyhedra.

In the 17th century, Descartes discovered the formula $v + f = e + 2$ for three-dimensional simple polyhedra, where $v, e,$ and f are the number of vertices, edges, and facets, respectively. This formula, rediscovered in the 18th century by Euler, is fundamental for the combinatorial part of polyhedral theory.

Mechanics and inequalities

The first interest in inequalities arose from mechanics. The following is a special case of the *principle of virtual velocity*, set up by Johann Bernoulli in 1717. Let the position x of a mass point be restricted to some region R. A vector y is called a *virtual velocity* in the point $x^* \in R$, if

(1)   there exists a curve C in R starting in x^* such that the half-line $\{x^* + \lambda y | \lambda \geqslant 0\}$ is tangent to C.

Let the mass point be in position x^* subject to the force b (i.e. a force $\|b\|$ in the direction b). The principle of virtual velocity says, under the following assumption:

(2)   with any virtual velocity y also $-y$ is a virtual velocity,

that the mass point is in equilibrium if and only if for each virtual velocity y, the 'power' $b^\mathsf{T} y$ is zero.

In his famous book *Méchanique analitique*, Lagrange [1788] observes that if R is the set of vectors x satisfying

(3) $f_1(x) = \cdots = f_m(x) = 0$

for certain differentiable functions $f_i : \mathbb{R}^n \to \mathbb{R}$, where the gradients $\nabla f_1(x^*), \ldots, \nabla f_m(x^*)$ are linearly independent, then the assumption (2) is satisfied, and (1) is equivalent to

(4) $\nabla f_1(x^*)^\mathsf{T} y = \cdots = \nabla f_m(x^*)^\mathsf{T} y = 0.$

Hence ('par la théorie de l'élimination des équations linéaires'), (1) implying $b^\mathsf{T} y = 0$ means:

(5) $b + \lambda_1 \nabla f_1(x^*) + \cdots + \lambda_m \nabla f_m(x^*) = 0$

for certain real numbers $\lambda_1, \ldots, \lambda_m$ (Lagrange [1788: Part I, Sec. 4, Art. 2]). The λ_i are the well-known *Lagrange multipliers*.

Lagrange remarks that (5) together with (3) form as many equations as there are unknowns $(\lambda_1, \ldots, \lambda_m$ and the components of $x^*)$, from which x^* can be found. Lagrange [1788: Part I, Sec. 4, Art. 7] explains that (5) may be understood as that each 'obstacle' $f_i(x) = 0$ could be replaced by an appropriate force $\lambda_i \nabla f_i(x^*)$, so that together they neutralize force b. In Part II, Section 4 of his book, Lagrange [1788] investigates the analogue for the 'dynamical' equilibrium.

In his book *Théorie des fonctions analytiques*, Lagrange [1797: Part II, Ch. XI, Secs. 57–58] studies minimizing or maximizing a function $f : \mathbb{R}^n \to \mathbb{R}$ subject to the constraint $g(x) = 0$ $(g : \mathbb{R}^n \to \mathbb{R})$. He concludes that if f and g are differentiable, a necessary condition for the minimum or maximum is: $\nabla f(x^*) + \lambda \nabla g(x^*) = 0$ for some real number λ.

More generally, Lagrange remarks that if we want to minimize or maximize $f : \mathbb{R}^n \to \mathbb{R}$ subject to $g_1(x) = \cdots = g_m(x) = 0$ (all functions differentiable), then it suffices to consider the stationary points of the function $L : \mathbb{R}^{n+m} \to \mathbb{R}$ given by: $L(x, \lambda_1, \ldots, \lambda_m) := f(x) + \lambda_1 g_1(x) + \cdots + \lambda_m g_m(x)$, the famous *Lagrange function*.

The above all bears upon optimizing subject to *equality* constraints. In his paper *Mémoire sur la statique*, Fourier [1798] aims at providing a solid base to the principle of virtual velocity and to Lagrange's theory. He also extends the principle to the case where the feasible region R is not-necessarily given by equations like (3): he allows inequalities, since often constraints are only one-sided ('il arrive souvent que les points du système s'appuient seulement sur les obstacles fixes, sans y êtres attachés' (Fourier [1798: Section 11])). Fourier then arrives at an extension of the principle of virtual velocity:

(6)   A mass point in x^*, whose position is restricted to R, and which is subject to a force b, is in equilibrium, if and only if $b^\mathsf{T} y \leqslant 0$ for each virtual velocity y.

This principle is sometimes called *Fourier's principle*.

Apparently not knowing Fourier's paper, the mathematical economist Cournot [1827] also considered the problem of inequality constraints in mechanics. He argued as follows. Let the position x of a mass point be restricted by

(7) $f_1(x) \geqslant 0, \ldots, f_m(x) \geqslant 0$

each f_i being differentiable. Let the point be subject to a force b, and in equilibrium in position x^*. Without loss of generality, $f_1(x^*) = \cdots = f_t(x^*) = 0, f_{t+1}(x^*) > 0, \ldots, f_m(x^*) > 0$. Then we can replace the first t 'obstacles' in (7) by forces $\lambda_1 \nabla f_1(x^*), \ldots, \lambda_t \nabla f_t(x^*)$, with appropriately chosen $\lambda_i \geqslant 0$, so that we still have equilibrium. Hence

(8) $b + \lambda_1 \nabla f_1(x^*) + \cdots + \lambda_t \nabla f_t(x^*) = 0.$

So this gives a 'mechanical proof' of Farkas' lemma (which is the special case where each f_i

is a linear function). Next Cournot argues that if y is a virtual velocity in x^*, then we must have:

(9) $\nabla f_1(x^*)^\mathsf{T} y \geqslant 0, \ldots, \nabla f_t(x^*)^\mathsf{T} y \geqslant 0.$

Combining (8) and (9) gives

(10) $b^\mathsf{T} y \leqslant 0.$

Thus a special case of Fourier's principle is shown. It seems that for Cournot this result came from just an excursion to mechanics, and that he did not observe any economical interpretation. (Related research is described in Cournot [1830a, b, 1832]. For Cournot's work in economics, see Roy [1939], Nichol [1938], and Fisher [1938].)

Gauss [1829] also considered the problem of mechanical equilibrium. He described a 'new general principle', the *principle of least constraint*: the movement of a system of mass points, whose position is restricted, always agrees as much as possible with the free movement. Gauss describes what 'as much as possible' should mean. The new principle contains Fourier's principle (6) as special case. A statement similar to (6) also occurs in Section 1 of Gauss' *Principia generalia theoriae figurae fluidorum in statu aequilibrii* [1832].

Independently of Fourier, Cournot, and Gauss, the Russian physicist Ostrogradsky [1838a] considered inequality constraints in mechanics. He first states the principle (6) (which therefore is called *Ostrogradsky's principle* in the Russian literature). Next he says that the virtual velocities y in a position x of a mass point which is subject to restrictions often can be characterized by a system of linear inequalities

(11) $a_1^\mathsf{T} y \geqslant 0, \ldots, a_t^\mathsf{T} y \geqslant 0.$

He assumes that $a_1, \ldots, a_t$ *are linearly independent.* In that case one can write, if the mass point is subject to force b, and in equilibrium:

(12) $b + \lambda_1 a_1 + \cdots + \lambda_t a_t + \lambda_{t+1} a_{t+1} + \cdots + \lambda_m a_m = 0$

for certain $a_{t+1}, \ldots, a_m$, with $a_1, \ldots, a_m$ linearly independent, and $\lambda_1, \ldots, \lambda_m \in \mathbb{R}$. Now according to the principle (6), any y satisfying (11) must satisfy $b^\mathsf{T} y \leqslant 0$. As $a_1, \ldots, a_m$ are linearly independent, this directly implies $\lambda_1, \ldots, \lambda_t \geqslant 0$ and $\lambda_{t+1} = \cdots = \lambda_m = 0$, as one easily checks. Ostrogradsky also remarks that if we replace any $\geqslant$ in (11) by $=$, the corresponding multiplier λ_i will not be restricted in sign any more (see also Ostrogradsky [1838b]).

Concluding, both Cournot and Ostrogradsky made attempts to obtain Farkas' lemma. Cournot's proof, however, is based on mechanical principles, while Ostrogradsky only considers the simple case where the linear inequalities are linearly independent. The first strictly mathematical proofs for the general case were given in the 1890's by Farkas and Minkowski—see below.

Fourier's further work on linear inequalities

Fourier was one of the first to recognize the importance of inequalities for applied mathematics. Besides mechanics, he mentioned applications to least-square problems, to the support of weight (where the inequalities represent the limits of safe loading), to elections (where the inequalities represent majorities), and to probability theory. (See Grattan-Guinness [1970], who checked Fourier's unpublished manuscripts in the Bibliothèque Nationale in Paris.)

Fourier is also a pioneer in observing the link between linear inequalities, polyhedra, and optimization. Fourier [1826b] (repeated in [1827]) describes a rudimentary version of the simplex method. It solves $\min\{\zeta | (\xi, \eta, \zeta) \in P\}$, where P is a polyhedron in three dimensions (a 'vase'), defined by linear inequalities of form

(13) $\qquad \left. \begin{array}{l} \zeta \geqslant \alpha_i \xi + \beta_i \eta + \gamma_i \\ \zeta \geqslant -\alpha_i \xi - \beta_i \eta - \gamma_i \end{array} \right\} \qquad (i = 1, \ldots, m).$

So the minimum is equal to the minimum of $\| Ax - b \|_\infty$ for some $m \times 2$ matrix A and m-vector b.

The method now is:

> Pour atteindre promptement le point inférieur du vase, on élève en un point quelconque du plan horizontal, par example à l'origine des x et y, une ordonnée verticale jusqu'à la rencontre du plan le plus élevé, c'est-à-dire que parmi tous les points d'intersection que l'on trouve sur cette verticale, on choisit le plus distant du plan des x et y. Soit m_1 ce point d'intersection placé sur le plan extrême. On descend sur ce même plan depuis le point m_1 jusqu'à un point m_2 d'une arête du polyèdre, et en suivant cette arête, on descend depuis le point m_2 jusqu'au sommet m_3 commun à trois plans extrêmes. A partir du point m_3 on continue de descendre suivant une seconde arête jusqu'à un nouveau sommet m_4, et l'on continue l'application du même procédé, en suivant toujours celle des deux arêtes qui conduit à un sommet moins élevé. On arrive ainsi très-prochainement au point le plus bas du polyèdre.

According to Fourier [1826b, 1827], this description suffices to understand the method in more dimensions. ('Le calcul des inégalités fait connaître que le même procédé convient à un nombre quelconque d'inconnues, parce que les fonctions extrêmes ont dans tous les cas des propriétés analogues à celles des faces du polyèdre qui sert de limite aux plans inclinés. En général les propriétés des faces, des arêtes, des sommets et des limites de tous les ordres, subsistent dans l'analyse générale, quel que soit le nombre des inconnues.' ([1827: p. 1]))

The problem of minimizing $\| Ax - b \|_\infty$ was studied earlier by Laplace [1978: Livre III, No. 39] and Legendre [1805: Appendix], and later on again by de la Vallée Poussin (see below).

In [1827] Fourier proposed a method for solving an arbitrary system of linear inequalities, consisting of successive elimination of variables. The method is now known as the Fourier–Motzkin elimination method (see Section 12.2). It has been discussed by Navier [1825] and Cournot [1826] (see Kohler [1973] for a translation of Fourier [1827]). Also Fourier [1826a] reports on linear inequalities.

Geometry of polyhedra and duality in the 19th century

During the 19th century, also the geometry of polyhedra was developed by A. L. Cauchy, J. Steiner, J. J. Sylvester, A. Cayley, A. F. Möbius, T. P. Kirkman, H. Poincaré, L. Schläfli, and P. G. Tait. The combinatorial classification of polyhedra was studied in particular, mainly in two and three dimensions. Here the Finite basis theorem for polytopes was implicitly assumed.

We have mentioned already in the historical notes on linear algebra that in the 1820's and 1830's the ideas of duality and polarity took shape. Gergonne [1820-1, 1824-5, 1825-6], Poncelet [1822, 1829], and von Staudt [1847] studied polar combinatorial properties of polyhedra. The duality of vertices and facets of 3-dimensional polytopes was observed by several authors (e.g. Poncelet [1829], von Staudt [1847]): if $c_1, \ldots, c_t$ are the vertices of a 3-dimensional polytope containing the origin, and $a_1^T x = 1, \ldots, a_m^T x = 1$ represent its facets, then $a_1, \ldots, a_m$ are the vertices of a polytope with facets $c_1^T x = 1, \ldots, c_t^T x = 1$.

In particular, in von Staudt's book *Geometrie der Lage* [1847] the polarity of polyhedra and convex sets is studied (Sections 179–181, 192–194). It is stated that if we have a polarity in 3-dimensional projective space, then the set of planes touching the surface of a convex set forms itself (under the polarity) the surface of a convex set. We quote Section 193 of von Staudt's book:

193. Wird die Oberfläche k eines Körpers K von keiner Geraden in mehr als zwei Punkten geschnitten, so soll der Inbegriff s von allen Ebenen, welche die Fläche k berühren, der diese Fläche oder auch der den Körper K umhüllende Ebenenbündel genannt werden. Durch den Ebenenbündel k wird das System von allen im Raume denkbaren Ebenen in zwei Systeme getheilt, von welchen das eine S von der Fläche ausgeschlossen und (185) das Reciproke von einem Körper ist, das andere aber unendlich viele Ebenenbündel erster Ordnung in sich enthält. Eine Ebene, welche nicht in der gemeinschaftlichen Grenze dieser beiden Systeme liegt, gehört dem erstern oder dem letztern an, je nachdem sie die Fläche k nicht schneidet oder schneidet. Ist K ein ebenflächiger Körper, welcker n Ecken hat, so ist der Ebenenbündel s aus n Ebenen-bündeln, deren jeder vom Mantel einer Strahlenpyramide (von einer Raumecke) ausgeschlossen ist, zusammengesetzt, so dass aber auch noch die Grenzen dieser Ebenenbündel ihm angehören. Wird die Fläche k in jedem ihrer Punkte nur von einer Ebene berührt, so ist s der ihr sich anschmiegende Ebenenbündel. Bezeichnet man das System von Geraden, welche die Fläche k schneiden durch V und das System von Geraden, welche mit ihr keinen Punkt gemein haben, durch W, so gilt Folgendes:

Jede Gerade des Systems V wird durch die beiden Punkte, in welchen sie die Fläche k schneidet, in zwei Strecken getheilt, in welchen die eine in den Körper K be-schrieben ist, die andere aber ausserhalb desselben liegt. Alle Ebenen, welche durch die Gerade gehen, liegen ausserhalb des Systems S.	Jeder Ebenenbüschel I. Ordnung, dessen Axe eine Gerade des Systems W ist, wird durch die beiden Ebenen, welche er mit dem Ebenenbündel s gemein hat, in zwei Winkel getheilt, von welchen der eine dem Systeme S angehört, der andere aber um den Körper K beschrieben ist. Alle Punkte der Geraden liegen ausserhalb des Körpers K.
Liegt eine Gerade h in der gemeinschaftlichen Grenze der beiden Systeme V, W, so hat das gerade Gebilde h mit der Fläche k entweder α) nur einen Punkt oder β) eine Strecke gemein.	der Ebenenbüschel h mit dem Ebenen-bündel s entweder α_1) nur eine Ebene oder β_1) einen Winkel gemein.

Wenn K ein ebenflächiger Körper ist, so findet der Fall $\alpha\alpha_1$ oder $\alpha\beta_1$ oder $\beta\alpha_1$ oder $\beta\beta_1$ statt, je nachdem die Gerade h die Fläche k in einem in einer Kante befindlichen Punkte oder in einem Eckpunkte oder in einer Strecke, welche keine Kante oder eine Kante ist, berührt.

Die gemeinschaftliche Grenze der beiden Systeme V, W hat mit jeder Ebene, welche ausserhalb des Systems S liegt, einen Strahlenbüschel gemein, der die Linie umhüllt, in welcher die Ebene die Fläche k schneidet.	jedem Strahlenbündel, dessen Mittelpunkt ausserhalb des Körpers K liegt, eine Kegel-fläche gemein, welche der dem Strahlen-bündel und dem Ebenenbündel s gemein-schaftliche Ebenenbüschel umhüllt.

(I particularly thank Professor Werner Fenchel for pointing out to me von Staudt's book together with the relevant passages.)

1873–1902: Gordan, Farkas and Minkowski

Gordan [1873] studied linear inequalities algebraically. He seems to be the first to give necessary and sufficient conditions for linear inequalities, showing that

(15)   *a system of linear equations $Ax = 0$ has a nonzero nonnegative solution if and only if there is no linear combination yA of the rows of A such that $yA < 0$.*

Gordan's proof is by induction on the number of rows of A. By interchanging x and y, Gordan's result is equivalent to:

(16) $Ax < 0$ has a solution if and only if $y = 0$ is the only solution of $yA = 0$; $y \geq 0$.

At the end of the 19th century, Farkas and Minkowski independently studied linear inequalities, and laid the foundation for the algebraic theory of polyhedra.

The interest of the physicist Farkas in linear inequalities came again from mechanics, especially from the lacunary proof of Ostrogradsky (see above). Farkas [1894] (German translation: [1895]) showed what is now known as Farkas' lemma:

(17) $Ax \leq 0$ implies $cx \leq 0$, if and only if $yA = c$ for some $y \geq 0$

thus extending Ostrogradsky's proof to the general case. The heart of Farkas' proof is elimination of columns of A, which is correct but stated rather cryptically by Farkas ('(da durch Eliminationen positiver Glieder mittels entsprechender negativen ... nach und nach aequivalenten Teil-Systeme erhalten werden können.)' is added in the German translation).

In Farkas [1896] (German translation: [1897–9]) a simpler proof of (17) is given ('Der Beweis den ich schon früher mittheilte, enthält einige Hilfssätze ohne Beweis, und ist davon abgesehen auch schwerfällig.'). The new proof is by elimination of variables. In Farkas [1898a] (German translation: [1897–9]) it is shown moreover that

(18)   each polyhedral cone is finitely generated.

This fact is proved with a dual form of the Fourier–Motzkin elimination method, viz. by elimination of constraints. Farkas [1898b] (German translation: [1898c]) gives an even simpler proof of (17), again by elimination of variables ('ich habe die Absicht, den möglichst einfachsten, ganz strengen und vollkommenen Beweis dieses Satzes zu geben.').

In Farkas [1900a] and [1902] surveys of the above results are given. In [1900a] the mechanical applications are discussed, while in [1902] the mathematical implications are described. In the latter paper (Section VII, §1), Farkas observes:

(19)   if $\{x \mid Ax \leq 0\} = \text{cone}\{y_1, \ldots, y_m\}$,
          then $\{c \mid Ax \leq 0 \Rightarrow cx \leq 0\} = \{c \mid cy_1, \ldots, cy_m \leq 0\}$.

This is of course easy, but it shows that Farkas was aware of the exchangeability of coefficients and variables. With this duality principle in mind, (17), (18) and (19) together immediately imply that each finitely generated cone is polyhedral, a result stated explicitly by Weyl [1935]—see below. Also in several other publications Farkas discussed linear inequalities in relation to mechanics: [1900b, 1906, 1914, 1917a, b, 1918a, b, 1923, 1926] (cf. Prékopa [1980]).

The interest of Minkowski for convexity and linear inequalities came from number theory. He derived several important results on convex sets and their volume (see Minkowski's Gesammelte Abhandlungen [1911a]). Section 19 (Anhang über lineare Ungleichungen) of Minkowski's book Geometrie der Zahlen [1896] is devoted to linear inequalities, where first the following is shown:

(20)   each polyhedral cone $\{x \mid Ax \leq 0\}$ is finitely generated; if A has full column rank then the cone is generated by rays each of which is determined by $n - 1$ linearly independent equations from $Ax = 0$.

From this, Minkowski derives:

(21)   If $Ax \leq 0$ implies $cx \leq 0$ then $yA = c$ for some $y \geq 0$

(i.e. Farkas' lemma), under the assumption that A has full column rank. Minkowski's proof runs as follows. First it is shown that without loss of generality we may assume that $\{x \mid Ax \leq 0\}$ is full-dimensional (i.e. that $Ax < 0$ is feasible). Next, let $y_1, \ldots, y_t$ be generators for $\{x|Ax \leqslant 0\}$. Then clearly

(22) $\{c|Ax \leqslant 0 \Rightarrow cx \leqslant 0\} = \{c|cy_1 \leqslant 0, \ldots, cy_t \leqslant 0\}$.

So (22) is a polyhedral cone, and hence by (20) it is finitely generated, namely by rays as described in (20). Minkowski next showed that each such ray is a nonnegative multiple of a row of A. Hence each c in (22) is a nonnegative combination of rows of A, thus proving (21).

At this last argument, Minkowski explicitly mentioned that the cone (22) is the cone generated by the rows of A. So Minkowski knew the duality relation between rays and inequalities.

Minkowski observed that the inhomogeneous case can be reduced to the homogeneous case by introduction of a new variable (i.e. $Ax \leqslant b$ becomes $Ax - b\lambda \leqslant 0$).

Minkowski also introduced several geometric concepts like supporting plane, extremal point, polar body. In Minkowski [1911b: §8] (posthumously published in the *Gesammelte Abhandlungen*), it is shown that if K is a compact convex set in three dimensions, with the origin 0 in its interior, and $K^* := \{y|y^T x \leqslant 1$ for all x in $K\}$ is the polar body of K, then K^* again contains the origin in its interior, and $(K^*)^* = K$. See also Minkowski [1897].

1908–1918: Voronoï, Carathéodory, Steinitz, Stiemke, Haar

Voronoï [1908] studied linear inequalities in relation to number theory (viz. the reduction of quadratic forms). After having derived Gordan's characterization of the full-dimensionality of $\{x|Ax \leqslant 0\}$ (i.e. of the feasibility of $Ax < 0$—cf. (16)), and Farkas' and Minkowski's result that each polyhedral cone is finitely generated, Voronoï introduced (Section 14) the polar cone C^* (called *domaine corrélatif*) of a polyhedral cone C, and he showed:

(23)   (i) $(C^*)^* = C$;
         (ii) C is full-dimensional if and only if C^* is pointed;
         (iii) the face-lattice of C^* is the converse of the face-lattice of C.

Next, Carathéodory [1911] needed convexity in studying coefficients of Fourier series. As 'eine freie Bearbeitung' of Minkowski's investigations, he showed:

(24)   (i) if $x \notin \text{conv.hull}\{a_1, \ldots, a_m\}$ then there is a separating hyperplane;
         (ii) if $x \in \text{conv.hull}\{a_1, \ldots, a_m\}$ then x is a convex combination of some affinely independent vectors from $a_1, \ldots, a_m$.

[Schmidt [1913] applied Minkowski's and Carathéodory's results to Hilbert's proof of Waring's theorem.]

In a series of papers, Steinitz [1913, 1914, 1916] studied relatively convergent sequences with the help of convex sets. In [1913: §10] he showed again Carathéodory's result (24) (ii), and in [1914: §19] he re-introduced the polar cone C^* (called *komplementäre Strahlsystem*), and he re-proved (23) (i). Moreover, in [1914: §20] Steinitz showed a theorem amounting to: for any matrix A with n columns,

(25) $\{x|Ax \leqslant 0\} = \{0\}$ if and only if $yA = 0$ for some $y \geqslant 0$, where the rows of A corresponding to positive components of y span $\mathbb{R}^n$.

In [1916] Steinitz especially studied convex sets, with results:

(26)   (i) if C is a closed convex set, then
             $\{c|\max\{cx|x \in C\}$ is finite$\} = (\text{char.cone } C)^*$ (§26);
         (ii) a set is a polytope if and only if it is a bounded polyhedron (§29).

Moreover, in §§30–32 Steinitz investigated the polarity of convex sets (in projective and in Euclidean space), with the result that, if K is a closed convex set containing the origin, then $K^{**} = K$.

Stiemke [1915] proved a further characterization for strict linear inequalities:

(27) $\exists x > 0: Ax = 0$ *if and only if* $\forall y: yA \leqslant 0 \Rightarrow yA = 0$

and he re-proved Gordan's characterization (16).

Haar [1918] extended Farkas' lemma to the inhomogeneous case:

(28)   *if $Ax \leqslant b$ is feasible, and $Ax \leqslant b$ implies $cx \leqslant \delta$, then $yA = c$, $yb \leqslant \delta$ for some $y \geqslant 0$.*

De la Vallée Poussin's prototype of the simplex method

Motivated by Goedseels' book *Théorie des erreurs d'observation* [1907], de la Vallée Poussin [1910–1] designed a method for Chebyshev approximation: given matrix A and vector b, minimize $\|Ax - b\|_\infty$. The method can be seen as a precursor of the simplex method, and as an algebraic analogue to Fourier's method described above.

Let A have order $m \times n$. Throughout, de la Vallée Poussin assumes that each n rows of A are linearly independent. He first shows that if the minimum of $\|Ax - b\|_\infty$ is M, then for the minimizing x^* we know that $Ax^* - b$ is equal to $\pm M$ in at least $n + 1$ constraints. (Today we know this as

(29) $M = \min\{\lambda| -\lambda \cdot 1 \leqslant Ax - b \leqslant \lambda \cdot 1\}$

represents M as a linear program in $n + 1$ variables and hence there exists a basic optimum solution (1 denotes an all-one column vector).) De la Vallée Poussin derives from this that M can be calculated by minimizing $\|\tilde{A}x - \tilde{b}\|_\infty$ for each choice $[\tilde{A} \; \tilde{b}]$ of $n + 1$ rows from $[A \; b]$, and then taking the maximum of these minima. Minimizing $\|\tilde{A}x - \tilde{b}\|_\infty$ can be done by observing that we want to have the minimum positive number M for which there exists a column $\{\pm 1\}$-vector ε such that $\tilde{A}x + \varepsilon M = \tilde{b}$. By Cramer's rule, $M = \det[\tilde{A} \; \tilde{b}]/\det[\tilde{A} \; \varepsilon]$. So we must maximize $|\det[\tilde{A} \; \varepsilon]|$ over all $\{\pm 1\}$-vectors ε, which is easy by developing $\det[\tilde{A} \; \varepsilon]$ to the last column.

De la Vallée Poussin next describes the following shorter procedure for minimizing $\|Ax - b\|_\infty$ (shorter than considering all submatrices $[\tilde{A} \; \tilde{b}]$ of $[A \; b]$):

(30)   First minimize $\|\tilde{A}x - \tilde{b}\|_\infty$ for some arbitrary choice $[\tilde{A} \; \tilde{b}]$ of $n + 1$ rows from $[A \; b]$, say the minimum is M_1, attained by x_1. Next calculate $\|Ax_1 - b\|_\infty$. If $\|Ax_1 - b\|_\infty \leqslant M_1$, then we know $\min_x \|Ax - b\|_\infty = M_1$, and we are finished. If $\|Ax_1 - b\|_\infty > M_1$, let (a_i, β_i) be a row of $[A \; b]$ with $|a_i x - \beta_i| > M_1$. Then

$$\min \left\| \begin{bmatrix} \tilde{A} \\ a_i \end{bmatrix} x - \begin{bmatrix} \tilde{b} \\ \beta_i \end{bmatrix} \right\|_\infty > M_1$$

and so there exist $n + 1$ rows from

$$\begin{bmatrix} \tilde{A} & \tilde{b} \\ a_i & \beta_i \end{bmatrix}$$

say $[\tilde{\tilde{A}} \tilde{\tilde{b}}]$ for which the minimum of $\|\tilde{\tilde{A}}x - \tilde{\tilde{b}}\|_\infty$ is larger than M_1. Let this minimum be M_2, attained by x_2. Repeat with $\tilde{A}, \tilde{b}, M_1, x_1$ replaced by $\tilde{\tilde{A}}, \tilde{\tilde{b}}, M_2, x_2$.

It is clear that this procedure terminates, as there are only finitely many choices $[\tilde{A}\tilde{b}]$, and as the sequence $M_1, M_2, \ldots$ is monotonically increasing. The procedure is the simplex method applied to (29).

1916–1936

With respect purely to the theory of linear inequalities, the period 1916–1936 appears to be one of consolidation. The work consists to a large extent of variations, extensions, applications, and re-provings of the fundamental results obtained between 1873 and 1916. We shall discuss the papers written between 1916 and 1936 briefly.

An article of Lovitt [1916], on determining weightings in preferential voting using linear inequalities in 3 variables, was a motivation for Dines to study linear inequalities. In [1917] he gave an algebraic elaboration of Lovitt's geometric arguments. Dines [1918–9] gave a characterization (different from Gordan's (16)), of the feasibility of $Ax < 0$, in terms of the *inequality-rank*, and he derived a characterization for $Ax < b$. The proof is by elimination of variables, now known as the Fourier–Motzkin elimination method. Also Carver [1921–2] gave a characterization of the feasibility of $Ax < b$:

(31) $Ax < b$ *is feasible if and only if* $y = 0$ *is the only solution of* $y \geqslant 0; yA = 0; yb \leqslant 0$.

Dines [1925] compared his own characterization with Carver's, while La Menza [1930] and Dines [1930] made variations on this theme.

Dines [1927a] described an algorithm for finding positive solutions for a system of linear equations, based on elimination of equations. Dines [1926–7] showed how to adapt his proof (Dines [1925]) of Gordan's characterization to obtain a proof of Stiemke's characterization (27). See also Gummer [1926].

Huber [1930: pp. 59–60] gave a proof due to Furtwängler of:

(32) $x > 0; Ax > 0$ *is feasible if and only if* $y = 0$ *is the only solution of* $y \geqslant 0; yA \leqslant 0$.

(Fujiwara [1930] observed the necessity of this last condition; cf. Takahashi [1932].)

Stokes [1931] extended the theory of linear inequalities further, studying mixed systems of linear equations and inequalities, strict and nonstrict, and deriving geometrical characterizations. Also Schlauch [1932] studied mixed systems. Van der Corput [1931a] characterized the linear hull of a cone $\{x \mid Ax \leqslant 0\}$. The theses of Stoelinga [1932] and Bunt [1934] develop further theory on convex sets and linear inequalities. Also La Menza [1930, 1932a, b] studied linear inequalities.

Among the generalizations published are those to infinitely many constraints (Haar [1918, 1924–6]), to infinitely many variables (Dines [1927b, c, 1930], Schoenberg [1932]), and to integration (Kakeya [1913–4], Haar [1918, 1924–6], Fujiwara [1928], Dines [1936]).

Dines and McCoy [1933], Weyl [1935], and Motzkin [1936] gave systematic surveys of the theory of linear inequalities and its relations to polyhedra. Weyl stated the Fundamental theorem of linear inequalities (Theorem 7.1 above), and he derived several duality results for polyhedral cones and polyhedra, like the Finite basis theorem for polytopes (Corollary 7.1c above).

Motzkin stated the Decomposition theorem for polyhedra (Corollary 7.1b above), he characterized feasibility of mixed systems $A_1 x < 0$, $A_2 x \leqslant 0$ (cf. Section 7.8), and he described the Fourier–Motzkin elimination method (Section 12.2), which Motzkin also used in an alternative proof of the feasibility characterization of $A_1 x < 0$, $A_2 x \leqslant 0$.

On the other hand, the study of *convexity* flourished during the first half of the 20th century, due to the work of, *inter alia*, H. Brunn, H. Minkowski [1896, 1911b], E. Helly, C. Carathéodory, E. Steinitz, J. Radon, H. Hahn, S. Banach, M. G. Krein, D. P. Milman on extremal and separation properties, and Brunn, Minkowski, T. Bonnesen, W. Fenchel, and A. D. Aleksandrov on mixed volumes of convex bodies. Many of the results are described in Bonnesen and Fenchel's book *Theorie der konvexen Körper* [1934]. Another important result is L. E. J. Brouwer's fixed point theorem, and its extension by S. Kakutani.

The geometric theory of polyhedra was described in Steinitz's *Vorlesungen über die Theorie der Polyeder* [1934].

Game theory

Certain two-person games can be formulated mathematically as the problem to solve, for given $m \times n$-matrix A,

$$(33) \qquad \max_x \min_y xAy$$

where $x = (\xi_1, \ldots, \xi_m)$ and $y = (\eta_1, \ldots, \eta_n)^\mathsf{T}$ range over all nonnegative vectors with $\xi_1 + \cdots + \xi_m = 1$ and $\eta_1 + \cdots + \eta_n = 1$. Trivially one has

$$(34) \qquad \max_x \min_y xAy \leqslant \min_y \max_x xAy.$$

This problem was studied first by Borel [1921, 1924, 1927], especially for skew-symmetric matrices A (i.e. $A^\mathsf{T} = -A$). Borel [1924: Note IV; 1927] conjectured that there exist skew-symmetric matrices A for which one has strict inequality in (34), which means that the LHS is negative and the RHS is positive (as for skew-symmetric A, $\max_x \min_y xAy = - \min_y \max_x xAy$).

Von Neumann [1928] however proved that for *each* matrix A one has equality in (34):

$$(35) \qquad \max_x \min_y xAy = \min_y \max_x xAy.$$

This equality can be seen to follow from the LP-duality theorem, using the fact that for any fixed x^*, $\min_y x^*Ay$ is equal to the smallest component of the vector x^*A, so that (33) is equal to (denoting $A = (\alpha_{ij})$):

$$(36) \qquad \max \lambda$$
$$\text{subject to} \quad \sum_{i=1}^{m} \alpha_{ij}\xi_i \geqslant \lambda \qquad\qquad (j = 1, \ldots, n)$$
$$\sum_{i=1}^{n} \xi_i = 1$$
$$\xi_i \geqslant 0 \qquad\qquad (i = 1, \ldots, m).$$

Similarly, $\min_y \max_x xAy$ is equal to the minimum value of the dual of (36). Thus we have (35).

Von Neumann gave an *ad hoc* proof of (35), while Ville [1938] showed how it follows from results on linear inequalities.

In fact it can be seen that (35) is equivalent to the Duality theorem of linear programming—see Dantzig [1951c] and Gale, Kuhn, and Tucker [1951]. (See von Neumann [1945] for a historical comment.)

Mathematical economics: the Walras equations and extensions

In the 1930s, several mathematical methods for economics were developed which are related to linear programming.

Classical are the *Walras equations* [1874], which were simplified by Cassel [1918: Kap. IV] to:

$$(37) \qquad b = Ax; \ z = yA; \ x = f(z)$$

for given $m \times n$-matrix $A = (\alpha_{ij})$, column m-vector b, and function $f : \mathbb{R}^n \to \mathbb{R}^n$.

The problem of solving (37) can be interpreted as follows. There are m materials and n products. The ith component β_i of b stands for the quantity available of material i, entry α_{ij} for the quantity of material i necessary for producing one unit of product j, while the jth component ξ_j of the unknown vector x denotes the number of units of product j to be produced. So $Ax = b$ says that all available material is used for production. Component η_i

of y gives the (unknown) price of one unit of material i, while component ζ_j of z gives the (unknown) price of one unit of product j. So $yA = z$ means that the price of a product is the sum of the prices of the materials processed. Finally, for each $j = 1, \ldots, n$, the demand ξ_j of product j depends by f on the prices of the products. (Sometimes this last is inverted, when the prices are depending on the demands.)

Walras claims that there is equilibrium if (37) is satisfied. These represent $m + 2n$ equations in $m + 2n$ unknowns, so one might hope that x, y, z can be determined from (37).

However, often (37) has no solution (e.g. if $m > n$, so that $Ax = b$ is overdetermined), or has a solution which is not economically meaningful, viz. if some of the variables are negative (cf. Neisser [1932], von Stackelberg [1933]). Therefore, Zeuthen [1933] and Schlesinger [1935] proposed the following modification of (37):

(38) $\quad Ax \leqslant b; \; yA = z; \; x, y, z \geqslant 0; \; x = f(y, z); \; y(b - Ax) = 0.$

So the supply of material might be more than necessary for production. The condition $y(b - Ax) = 0$ then states that if there is more available of material i than necessary for production, the price η_i of this material is 0—a form of complementary slackness. See Wald [1935, 1936a, b] for a study of the solvability of (38).

[The book of Dorfman, Samuelson, and Solow [1958: Ch. 13] gives a further extension of (38):

(39) $\quad Ax \leqslant w; \; yA \geqslant z; \; x, y, z, w \geqslant 0; \; x = f(y, z);$

$\quad\quad w = g(y, z); \; y(w - Ax) = 0; \; (yA - z)x = 0.$

Here A, f, and g are given, and x, y, z, w are variable. So now also the availability of goods is variable, and depends on the prices. Moreover, the price of a product j may be less than the sum of the prices of the materials needed ($yA \geqslant z$), in which case product j is not produced (by $(yA - z)x = 0$). We refer to the book mentioned and to Kuhn [1956c] for a discussion of the relation of (39) with linear programming and fixed point theorems.]

Mathematical economics: von Neumann's growth model

Another interesting model is the *growth model* proposed by von Neumann [1937], which also gives rise to a system of linear inequalities, where again duality (of prices and production intensities) and complementary slackness show up.

Von Neumann showed that, given two nonnegative $m \times n$-matrices $A = (\alpha_{ij})$ and $B = (\beta_{ij})$, the following system in the variables x, y, γ, δ has a solution, provided that A has no all-zero column:

(40) $\quad \gamma \cdot Ax \leqslant Bx$

$\quad\quad \delta \cdot yA \geqslant yB$

$\quad\quad y(Bx - \gamma \cdot Ax) = 0$

$\quad\quad (\delta \cdot yA - yB)x = 0$

$\quad\quad \gamma, \delta, x, y \geqslant 0, \; x \neq 0, \; y \neq 0.$

Moreover, γ and δ are uniquely determined by A and B and $\gamma = \delta$, if we know that $A + B > 0$.

Interpretation: Each row-index i stands for a 'good', while each column index j stands for a 'process'. Process j can convert (in one time unit, say a year) α_{ij} units of good i ($i = 1, \ldots, m$) into β_{ij} units of good i ($i = 1, \ldots, m$). So a process uses goods as materials, and gives goods as products. Component ξ_j of x denotes the 'intensity' by which we let process j work, while component η_i of y is the price of one unit of good i. The number γ is the factor by which all intensities are multiplied each year. So $\gamma \cdot Ax \leqslant Bx$ requires that, for each good i, the amount of good i produced in year k is at least the amount of good i required in year $k + 1$. Moreover, if there is strict surplus here, the equation $y(\gamma \cdot Ax - Bx) = 0$ says that the price of good i is 0. The number δ denotes the interest factor $(:= 1 + z/100$, where z is the rate of interest). Then $\delta \cdot yA \geqslant yB$ says that, for each process j, the price of the goods produced at the end of year k is at most the price of the goods necessary for the process at the beginning of year k, added with interest. Moreover, if there is strict inequality here, the equation $(\delta \cdot yA - yB)x = 0$ says that the intensity of process j is 0.

[We prove the feasibility of (40) with the theory of linear inequalities. Let δ^* be the smallest real number $\delta \geqslant 0$ for which

(41) $\exists y \geqslant 0: y \neq 0, \delta \cdot yA \geqslant yB$.

By Motzkin's transposition theorem (Corollary 7.1k) this implies that for each δ with $0 \leqslant \delta < \delta^*$ one has:

(42) $\exists x \geqslant 0: x \neq 0, \delta \cdot Ax < Bx$

          i.e. $\exists x \geqslant 0, \|x\| = 1, \delta \cdot Ax < Bx$.

Hence, by continuity and compactness arguments, we know:

(43) $\exists x \geqslant 0: \|x\| = 1, \delta^* \cdot Ax \leqslant Bx$.

Now let $x^* \geqslant 0, y^* \geqslant 0$ be such that $x^* \neq 0, y^* \neq 0, \delta^* Ax^* \leqslant Bx^*$ and $\delta^* \cdot y^*A \geqslant y^*B$. Then this gives a solution of (40), with $\gamma := \delta^*$ (note that $y^*Bx^* \leqslant \delta^* \cdot y^*Ax^* \leqslant y^*Bx^*$, proving the third and fourth line in (40)).]

Von Neumann's proof of the solvability of (40) shows that his model always has an equilibrium, with a unique growth and interest factor $\gamma = \delta$.

In fact, if $A + B$ is positive, von Neumann's result is easily seen to be equivalent to:

(44) $\max \{\gamma | \exists x \neq 0: x \geqslant 0, \gamma \cdot Ax \leqslant Bx\} = \min \{\delta | \exists y \neq 0: y \geqslant 0, \delta \cdot yA \geqslant yB\}$.

So 'the maximum growth factor is equal to the minimum interest factor'. (44) is equivalent to:

(45) $\displaystyle \max_{x} \min_{y} \frac{yBx}{yAx} = \min_{y} \max_{x} \frac{yBx}{yAx}$

where x and y range over $x \geqslant 0, x \neq 0$ and $y \geqslant 0, y \neq 0$. By taking A an all-one matrix, von Neumann's min–max result in game theory follows (see above). If we take $B = bc$ in (45), where b is a positive column vector and c is a positive row vector, then (45) becomes the LP-duality equation

(46) $\max \{cx | x \geqslant 0, Ax \leqslant b\} = \min \{yb | y \geqslant 0, yA \geqslant c\}$.

So von Neumann's result contains LP-duality for the case where all data is positive.

Karush's work in nonlinear programming

In his M.Sc. thesis, Karush [1939] studied the problem of minimizing a function f over a domain given by inequality constraints. He derives necessary conditions and sufficient conditions, involving first and second derivatives. One of his results is as follows.

Suppose differentiable functions $f, g_1, \ldots, g_m : \mathbb{R}^n \to \mathbb{R}$, with continuous derivatives, are given, and we wish to minimize $f(x)$ subject to

(47) $g_1(x) \geqslant 0, \ldots, g_m(x) \geqslant 0$.

If the minimum is attained by x^0 we may assume that $g_1(x^0) = \cdots = g_m(x^0) = 0$, since if 'say, $g_1(x^0) > 0$ then by continuity $g_1(x) \geqslant 0$ for all x sufficiently close to x^0 and hence the condition $g_1(x) \geqslant 0$ puts no restriction on the problem so far as the theory of relative minima is concerned'. Now Karush's Theorem 3:2 states:

(48)   'Suppose that for each admissible direction λ there is an admissible arc issuing from x^0 in the direction λ. Then a first necessary condition for $f(x^0)$ to be a minimum is that there exist multipliers $\chi_\alpha \leqslant 0$ such that the derivatives F_{x_i} of the function $F = f + \chi_\alpha g_\alpha$ all vanish at x^0.'

Here an *admissible direction* is a nonzero vector y such that $\nabla g_i(x^0)^\mathsf{T} y \geqslant 0$ for all $i = 1, \ldots, m$. A regular arc $x(t)(0 \leqslant t \leqslant t_0)$ is *admissible* if $g_i(x(t)) \geqslant 0$ for all $t \in [0, t_0]$. *Issuing from x^0 in the direction λ* means $x(0) = x^0$ and $x'(0) = \lambda$. Moreover, Karush uses $\chi_\alpha g_\alpha$ for $\chi_1 g_1 + \cdots + \chi_m g_m$.

It is not difficult to derive the Duality theorem of linear programming from (48), by taking all functions linear.

Karush's result is fundamental for nonlinear programming. (Previously, one generally reduced inequality constraints to equality constraints by introducing slack variables: thus $g_1(x) \geqslant 0$ becomes $g_1(x) - \eta_1^2 = 0$ (cf. Hancock [1917])—however, Lagrange's condition (5) thus obtained is weaker than Karush's). Extensions were given by John [1948] and Kuhn and Tucker [1951]. The premise in (48), which links back to the mechanical conditions (9) and (1), was called by Kuhn and Tucker the *constraint qualification*. The use of (48) is that when minimizing f over (47) we only have to consider those vectors x satisfying (47) which:

(49)   (i) do *not* satisfy the constraint qualification;
        or (ii) do satisfy the necessary condition described in (48).

Often, there are only finitely many such x, so that the one minimizing $f(x)$ is easily selected.

Transportation and more general problems: Kantorovich, Hitchcock, Koopmans

Transportation problems belong to the oldest operations research problems studied— see Monge [1784] (cf. Taton [1951] and the historical notes on integer linear programming). They also led to linear programming, through the work of Kantorovich, Hitchcock, and Koopmans.

Kantorovich [1939] described the relevance of certain linear programming problems for organizing and planning production:

'I want to emphasize again, that the greater part of the problems of which I shall speak, relating to the organization and planning of production, are connected specifically with the Soviet system of economy and in the majority of cases do not arise in the economy of a capitalist society. There the choice of output is determined not by the plan but by the interests and profits of individual capitalists. The owner of the enterprise chooses for production those goods which at a given moment have the highest price, can most easily be sold, and therefore give the largest profit. The raw material used is not that of which there are huge supplies in the country, but that which the entrepreneur can buy most cheaply. The question of the maximum utilization of equipment is not raised; in any case, the majority of enterprises work at half capacity.

In the USSR the situation is different. Everything is subordinated not to the interests and advantage of the individual enterprise, but to the task of fulfilling the state plan. The basic task of an enterprise is the fulfillment and overfulfillment of its plan, which is a part of the general state plan. Moreover, this not only means fulfillment of the plan in aggregate terms (i.e. total value of output, total tonnage, and so on), but the certain fulfillment of the plan for all kinds of output; that is, the fulfillment of the assortment plan (the fulfillment of the plan for each kind of output, the completeness of individual items of output, and so on).'

(From Kantorovich, L. V. (1960), *Management Science* 6, 366–422. Reproduced by permission of Professor L. V. Kantorovich and The Institute of Management Sciences.)

Kantorovich gave a dual-simplex-method type method for solving the following transportation problem:

(50)   maximize λ
          subject to: $\sum_{i=1}^{m} \xi_{ij} = 1$ $(j = 1, \ldots, n)$

                           $\sum_{j=1}^{n} \alpha_{ij}\xi_{ij} = \lambda$ $(i = 1, \ldots, m)$

                           $\xi_{ij} \geqslant 0$ $(i = 1, \ldots, m; j = 1, \ldots, n)$

and for, in Kantorovich's terminology, 'Problem C':

(51)   maximize λ
          subject to: $\sum_{i=1}^{m} \xi_{ij} = 1$ $(j = 1, \ldots, n)$

                           $\sum_{j=1}^{m} \sum_{i=1}^{n} \alpha_{ijk}\xi_{ij} = \lambda$ $(k = 1, \ldots, t)$

                           $\xi_{ij} \geqslant 0$ $(i = 1, \ldots, m; j = 1, \ldots, n)$.

[*Interpretation*: Let there be n machines, which can do m jobs. Let there be one final product consisting of t parts. When machine i does job j, α_{ijk} units of part k are produced $(k = 1, \ldots, t)$. Now ξ_{ij} denotes the fraction of time machine i does job j. The number λ will be the amount of the final product produced.]

Kantorovich's method consists of determining dual variables ('resolving multipliers') and finding the corresponding primal solution. If the primal solution is not feasible, the dual solution is modified following prescribed rules. Kantorovich also indicates the role of the dual variables in sensitivity analysis, and he shows that a feasible solution for (51) can be shown to be optimal by specifying optimal dual variables.

In fact, as described in footnote 2 of Koopmans [1959–60], Problem C ((51)) is equivalent to the general linear programming problem, so that Kantorovich actually gave a method for general linear programming.

[To see this, suppose we want to solve $\max\{cx \,|\, x \geqslant 0, Ax = b\}$. We may suppose we know an upper bound M on the variables in the optimal solution. Then, if A has order $m \times n$,

(52) $\max\{cx \,|\, x \geqslant 0; \; Ax = b\} = \max\{cx \,|\, x \geqslant 0; \; \mu \geqslant 0; \; \mathbf{1}^{\mathsf{T}}x + \mu = nM; \; Ax = b\}$
          $= nM \cdot \max\{cx \,|\, x \geqslant 0; \; \mu \geqslant 0; \; \mathbf{1}^{\mathsf{T}}x + \mu = 1; \; Ax = (nM)^{-1}b\}$

where $\mathbf{1}$ is an all-one column vector. It follows that also $\max\{cx \,|\, x \geqslant 0; \mathbf{1}^{\mathsf{T}}x = 1; Ax = b\}$ is a general LP-problem. Now

(53) $\max\{cx \,|\, x \geqslant 0; \; \mathbf{1}^{\mathsf{T}}x = 1; \; Ax = b\} = \max\{\lambda \,|\, x \geqslant 0; \; \mathbf{1}^{\mathsf{T}}x = 1; \; cx = \lambda; \; Ax = b\}$
          $= \max\{\lambda \,|\, x \geqslant 0; \; \mathbf{1}^{\mathsf{T}}x = 1; \; cx = \lambda; \; (A - b\mathbf{1}^{\mathsf{T}} + \mathbf{1}c)x = \mathbf{1}\lambda\}$.

The last problem here is a special case of Problem C (taking $n = 1$ in (51)).]

Hitchcock [1941] considers the problem

(54) $\min \sum_{i=1}^{m} \sum_{j=1}^{n} \gamma_{ij}\xi_{ij}$

          subject to $\sum_{i=1}^{m} \xi_{ij} = \delta_j$ $(j = 1, \ldots, n)$

                     $\sum_{j=1}^{n} \xi_{ij} = \beta_i$ $(i = 1, \ldots, m)$

                     $\xi_{ij} \geqslant 0$ $(i = 1, \ldots, m; \; j = 1, \ldots, n)$.

He shows that the minimum is attained at a vertex of the feasible region, and he describes a scheme to solve (54) which has much in common with the (primal) simplex method, including pivoting (eliminating and introducing basic variables) and the fact that nonnegativity of certain dual variables implies optimality.

Koopmans [1948] describes 'transshipment problems' and their relevance, and gives a 'local search' method to solve these problems, stating that this procedure leads to optimality.

For comments, see Koopmans [1959–60, 1961–2], Charnes and Copper [1961–2], and Vajda [1962–3].

Linear programming

The study of transportation problems formed an important prelude to the creation of the field of linear programming in the 1940s, by the work of Dantzig, Kantorovich, Koopmans and von Neumann.

The mathematical fundamentals were laid by von Neumann [1947] in a working paper (published first in von Neumann's Collected Works), where it is shown that solving max $\{cx | x \geqslant 0, Ax \leqslant b\}$ is the same as solving the system of linear inequalities

(55) $x \geqslant 0, Ax \leqslant b, yA \geqslant c, y \geqslant 0, cx \geqslant yb.$

Gale, Kuhn, and Tucker [1951] next stated and proved the Duality theorem of linear programming:

(56) $\max \{cx | x \geqslant 0, Ax \leqslant b\} = \min \{yb | y \geqslant 0, yA \geqslant c\}.$

Both von Neumann and Gale, Kuhn, and Tucker used Farkas' lemma.

Moreover, Dantzig [1951a] designed the simplex method, which especially has decided the success of linear programming in practice. Further important contributions to linear programming and linear inequalities were made by, *inter alia*, E. M. L. Beale, A. Charnes, W. W. Cooper, G. B. Dantzig, A. J. Goldman, A. J. Hoffman, H. W. Kuhn, W. Orchard-Hays, A. Orden, A. W. Tucker, H. M. Wagner, and P. Wolfe. We refer to Dantzig [1963, 1982] and Dorfman [1984] for surveys of origins and development of linear programming.

FURTHER NOTES ON POLYHEDRA, LINEAR INEQUALITIES, AND LINEAR PROGRAMMING

Literature. We first collect some references to further literature.

Linear inequalities. Kuhn [1956a], Tucker [1956], Good [1959], Chernikov [1971], and Solodovnikov [1977] (elementary).

Dines and McCoy [1933] and Kuhn and Tucker [1956] gave surveys and lists of references. Prékopa [1980] describes the history of Farkas' lemma.

Polyhedra. Wiener [1864], Eberhard [1891], Brückner [1900], Schläfli [1901] (historical); Steinitz [1934], Grünbaum [1967] (standard references); Aleksandrov [1958], Barnette [1979], Bartels [1973], Brøndsted [1983], Coxeter [1948], Grünbaum and Shephard [1969], Klee [1959, 1964a, 1969], and McMullen and Shephard [1971]. Most of these books and papers have emphasis on the combinatorial properties of polyhedra (like the structure of the face-lattice and the number of vertices and facets).

The origins of polyhedral theory are described in Coxeter [1948], Grünbaum [1967], and Steinitz [1934].

Applications of polyhedral theory in combinatorics are described in Yemelichev, Kovalev, and Kravtsov [1981].

Linear programming. Dantzig [1963] (a standard text, with a thorough historical survey and a comprehensive bibliography), Bazaraa and Jarvis [1977], Charnes and Cooper [1961], Chvátal [1983] (a recommended recent book), Driebeek [1969], Ferguson and Sargent [1958], Gale [1960], Gass [1975], Hadley [1962], Yudin and Gol'shtein [1965], Kim [1971], Murtagh [1981], Murty [1976,1983], van de Panne [1971], Simonnard [1966], Spivey and Thrall [1970], Thompson [1971] (introductory), and Zionts [1974].

Specialized in the practical aspects of linear programming (large-scale) are: Beale [1968], Dantzig, Dempster, and Kallio [1981], Gill, Murray, and Wright [1981], Lasdon [1970], and Orchard-Hays [1968].

For bibliographies, see Kuhn and Tucker [1956], Riley and Gass [1958], Rohde [1957], and Wagner [1957b]. A historical collection of papers was edited by Koopmans [1951].

An important collection of papers was edited by Kuhn and Tucker [1956], in which polyhedra, linear inequalities, and linear programming are brought together.

Convexity. Bonnesen and Fenchel [1934] (historical); Rockafellar [1970], Stoer and Witzgall [1970] (two standard references); Eggleston [1958], Gruber and Wills [1983], Hadwiger [1955], Klee [1963], Lay [1982], Leichtweiss [1980], Marti [1977], Valentine [1964], and Wets [1976].

For historical notes and a bibliography, see Fenchel [1953]. More historical notes are given in Fenchel [1983].

The books of Shor [1979], Nemirovsky and Yudin [1983], and Grötschel, Lovász, and Schrijver [1988] survey the ellipsoid method and related methods for optimization.

Further remarks

Algorithms for finding *all* vertices of a polytope, or all extreme rays of a polyhedral cone, were given by Altherr [1975, 1978], Balinski [1961], Burdet [1974], Burger [1956], Charnes, Cooper and Henderson [1953], Chernikova [1964, 1965, 1968], Dyer [1983], Dyer and Proll [1977], Galperin [1976], Greenberg [1975], Mañas and Nedoma [1968], Mattheiss [1973], Mattheiss and Schmidt [1980], McRae and Davidson [1973], Motzkin, Raiffa, Thompson, and Thrall [1953], Rubin [1975a], Shachtman [1974]. For surveys, see Chvátal [1983:Ch.18] and Mattheiss and Rubin [1980].

Finding the convex hull of a set of given points in the plane, or in higher dimensional space, was studied by Akl [1979], Akl and Toussaint [1978], Andrew [1979], Avis [1980b, 1982], Bykat [1978], Dévai and Szendrényi [1979], Devroye [1980], Eddy [1977a, b], Fournier [1979], Graham [1972], Green and Silverman [1979], Jaromczyk [1981], Jarvis [1973], Koplowitz and Jouppi [1978], McCallum and Avis [1979], Preparata and Hong [1977], Shamos and Hoey [1975], Swart [1985], van Emde Boas [1980], and Yao [1981]—see Preparata and Shamos [1985].

Burger [1956], Gerstenhaber [1951], Gale [1951], and Wets and Witzgall [1968] studied polyhedral cones. Barker [1973, 1978] discussed faces and duality in convex cones. Murty [1971] and Bachem and Grötschel [1981] studied adjacency on polyhedra. Klee [1978] studied the faces of a polyhedron relative to its polar. LP- and other dualities, and the analogues of Farkas' lemma, are discussed by Bachem and von Randow [1979] and Dixon [1981].

Wolfe [1976] discussed the problem of finding the point of smallest Euclidean norm in a given polyhedron.

A negative result was obtained by Chandrasekaran, Kabadi and Murty [1981–2]: testing degeneracy of a given LP-problem is $\mathcal{NP}$-complete. Also $\mathcal{NP}$-complete is the problem of finding a basic feasible solution of given objective value. Murty [1972] considered the problem of finding a basic feasible solution of an LP-problem such that a given set of variables belongs to the basic variables.

Dobkin, Lipton and Reiss [1979] showed that linear programming is "log-space hard for $\mathscr{P}$", which means that if linear programming is solvable in poly-log space (i.e., the space, except for writing down the problem, is $O(\log \sigma)^k$ for some k, where σ is the size of the problem), then each problem in $\mathscr{P}$ is poly-log space solvable.

Traub and Woźniakowski [1981–2] comment on the complexity of linear programming in an infinite precision model.

Robinson [1975, 1977] and Mangasarian [1979, 1981a] studied stability and uniqueness of LP-solutions.

Part IV

Integer linear programming

Integer linear programming (abbreviated: *ILP*) investigates linear programming problems in which the variables are restricted to integers: the general problem is to determine $\max\{cx|Ax \leqslant b; x \text{ integral}\}$. ILP can be considered equivalently as the search for lattice points in polyhedra, or as solving systems of linear equations in nonnegative integers. So it combines the quests of nonnegativity and of integrality.

This combination is not as well-solved as its constituents. For example, no satisfactory analogues of Farkas' lemma and of LP-duality have been found. If we restrict in the LP-duality equation the variables to integers, in general we lose equality and obtain a strict inequality. It seems difficult to close this 'duality gap' generally—no satisfactory *integer* linear programming duality equation has been found.

Moreover, no polynomial algorithm for ILP has been discovered. In fact the ILP-problem is $\mathcal{NP}$-complete, and hence it is generally believed not to be polynomially solvable. This conforms to the practical experience: solving ILP-problems with present-day methods is hard and time-consuming. Large-scale integer linear programs seem as yet practically unsolvable.

Despite this intractability, or thanks to it, integer linear programming has led to some interesting results, and the remainder of this book is devoted to them.

In Chapter 16 we give a general introduction to integer linear programming, and we study some polyhedral aspects of integer linear programming, like the fact that the *integer hull* P_I of a rational polyhedron P (i.e. the convex hull of the integral vectors in P) is again a polyhedron.

In Chapter 17 we make some estimates on solutions of ILP-problems, yielding that integer linear programming belongs to $\mathcal{NP}$. We also give some sensitivity and proximity results.

In Chapter 18 we study the complexity of integer linear programming. We show that the integer linear programming problem is $\mathcal{NP}$-complete, and we derive that several other problems are $\mathcal{NP}$-complete. We also describe Lenstra's algorithm for ILP, which is, for each fixed number of variables, a polynomial-time method, and we discuss a 'dynamic programming' approach to ILP, yielding 'pseudo-polynomial' algorithms for a fixed number of constraints.

In Chapters 19–21 we study *totally unimodular* matrices, which are matrices with all subdeterminants equal to 0, $+1$ or -1. They are important in integer linear programming, as each linear program with integral input data and with totally unimodular constraint matrix has an integral optimum solution. After having given in Chapter 19 fundamental properties and examples of totally unimodular matrices, we describe in Chapter 20 a polynomial-time algorithm for recognizing total unimodularity (based on

Seymour's fundamental decomposition theorem for totally unimodular matrices). In Chapter 21 we deal with some further theory related to total unimodularity.

Chapter 22 discusses the theory of *total dual integrality*, a clarifying concept in integer linear programming and combinatorial optimization, introduced by Edmonds and Giles. It yields that certain linear programs have integral optimum solutions, if their duals have integral optimum solutions.

In Chapter 23 we go into the theory of *cutting planes*, based on work of Gomory and Chvátal. It gives a systematic procedure for finding the inequalities determining the integer hull P_I from those determining a polyhedron P.

Finally, in Chapter 24 we survey some further methods in integer linear programming: branch and bound methods, Gomory's method of 'corner polyhedra', Lagrangean relaxation, and Benders decomposition.

Part IV concludes with historical and further notes.

We remark here that most of the results in this part only hold in rational spaces, with all data rational.

16

Introduction to integer linear programming

In this chapter we describe some introductory theory for integer linear programming. After the introductory Section 16.1, we show in Section 16.2 that the integer hull P_I of a rational polyhedron P (i.e. the convex hull of the integral vectors in P) is a polyhedron again. In Section 16.3 we consider *integral* polyhedra, i.e. polyhedra P satisfying $P = P_I$. This corresponds to linear programs which have, for each objective function, an integral optimum solution. In Section 16.4 we discuss so-called *Hilbert bases* for rational polyhedral cones. These Hilbert bases will be used in Chapter 21 on total dual integrality. In Section 16.5 we describe a theorem of Doignon, an analogue of Carathéodory's theorem. Finally, in Section 16.6 we make some notes on the *knapsack problem* and *aggregation* and in Section 16.7 on *mixed integer linear programming*.

16.1. INTRODUCTION

One form of the *integer linear programming problem* is:

(1)   given rational matrix A, and rational vectors b and c, determine $\max\{cx\,|\,Ax \leq b;\ x\ \text{integral}\}$.

Another form is:

(2)   given rational matrix A, and rational vectors b and c, determine $\max\{cx\,|\,x \geq 0;\ Ax = b;\ x\ \text{integral}\}$.

In fact, (1) and (2) are polynomially equivalent: in (1) we may assume that A, b are integral (by multiplying A, b by the product of the denominators occurring in A and b), and then $\max\{cx\,|\,Ax \leq b\}$ is equivalent to $\max\{cx' - cx''\,|\,x', x'', \tilde{x} \geq 0,\ Ax' - Ax'' + \tilde{x} = b;\ x', x'', \tilde{x}\ \text{integral}\}$, a special case of (2). The reverse reduction is even more direct: $x \geq 0$, $Ax = b$ is equivalent to $-x \leq 0$, $Ax \leq b$, $-Ax \leq -b$. These reductions are Karp-reductions.

In general we will have strict inequality in a corresponding duality relation:

(3) $\max\{cx\,|\,Ax \leq b;\ x\ \text{integral}\} \leq \min\{yb\,|\,y \geq 0;\ yA = c;\ y\ \text{integral}\}$.

For example, take $A = (2)$, $b = (1)$, $c = (1)$, where the maximum is 0 and the minimum is infeasible, while the corresponding LP-optima both are $\frac{1}{2}$.

It is easy to see that (1) and (2) are also polynomially equivalent to the problems:

(4)   given rational matrix A and rational vector b, decide if $Ax \leqslant b$ for some integral vector x;

(5)   given rational matrix A and rational vector b, decide if $Ax = b$ for some nonnegative integral vector x.

Farkas' lemma (Corollary 7.1d) and the analogous Corollaries 3.1b and 4.1a suggest the following: the rational system $Ax = b$ has a nonnegative integral solution x if and only if yb is a nonnegative integer whenever yA is a nonnegative integral vector. This is however not true: e.g. take $A = (2 \;\; 3)$, $b = (1)$. It is easy to see that the first half of the statement implies the second half, but generally not conversely. Similarly, we have only 'weak duality' in (3), i.e. only the given inequality holds.

Closing the duality gap in a satisfactory way would possibly imply that the integer linear programming problem has a good characterization, i.e. belongs to $\mathcal{NP} \cap \text{co-}\mathcal{NP}$ (as in Section 10.1 for LP). But this is unknown, and seems unlikely: the ILP-problem is $\mathcal{NP}$-complete, as we shall show in Chapter 18. We first discuss in this chapter some polyhedral and further theory.

We mainly restrict ourselves to integer linear programming with rational input data. If we do not make this restriction it is not always clear that optima exist (e.g. $\sup\{\xi + \eta\sqrt{2}|\xi + \eta\sqrt{2} \leqslant \tfrac{1}{2}; \xi, \eta \text{ integer}\} = \tfrac{1}{2}$, but no ξ, η attain the supremum).

Finally, a definition: the *LP-relaxation* of the integer linear programming problem

(6) $\max\{cx | Ax \leqslant b; x \text{ integral}\}$

is the LP-problem

(7) $\max\{cx | Ax \leqslant b\}$.

Clearly, (7) gives an upper bound for (6).

16.2. THE INTEGER HULL OF A POLYHEDRON

Define, for any polyhedron P, the *integer hull* P_I of P by:

(8) $P_I :=$ the convex hull of the integral vectors in P.

So the ILP-problem (1) is equivalent to determining $\max\{cx | x \in P_I\}$ for $P := \{x | Ax \leqslant b\}$.

We first show the result of Meyer [1974] that if P is rational, then P_I is a polyhedron again. (This is trivial if P is bounded.) Note that for any rational polyhedral cone C,

(9) $C_I = C$

(as C is generated by rational, and hence by integral, vectors).

Theorem 16.1. *For any rational polyhedron P, the set P_1 is a polyhedron again. If P_1 is nonempty, then char.cone $(P_1) = $ char.cone(P).*

Proof. Let $P = Q + C$, where Q is a polytope and C is the characteristic cone of P (cf. Corollary 7.1b). Let C be generated by the integral vectors $y_1, \ldots, y_s$, and let B be the polytope

(10) $\qquad B := \left\{ \sum_{i=1}^{s} \mu_i y_i \mid 0 \leqslant \mu_i \leqslant 1 \quad \text{for} \quad i = 1, \ldots, s \right\}.$

We show that $P_1 = (Q + B)_1 + C$, which implies the theorem, as $Q + B$ is bounded, and hence $(Q + B)_1$ is a polytope, and as if $P_1 \neq \varnothing$, then C is its characteristic cone.

In order to show that $P_1 \subseteq (Q + B)_1 + C$, let p be an integral vector in P. Then $p = q + c$ for some $q \in Q$ and $c \in C$. Then $c = b + c'$ for some $b \in B$ and some integral vector $c' \in C$ (indeed, write $c = \sum \mu_i y_i$ $(\mu_i \geqslant 0)$; then take $c' := \sum \lfloor \mu_i \rfloor y_i$ and $b := c - c'$). Hence $p = (q + b) + c'$, and $q + b \in (Q + B)_1$ (as $q + b = p - c'$ and p and c' are integral). So $p \in (Q + B)_1 + C$.

The reverse inclusion follows from: $(Q + B)_1 + C \subseteq P_1 + C = P_1 + C_1 \subseteq (P + C)_1 = P_1$ (using (9)). $\square$

A direct consequence looks at first sight surprising: any integer linear programming problem can be written as $\max\{cx \mid x \in Q\}$ for some polyhedron Q, and hence is nothing other than a linear programming problem! But the harm is in 'can be written': to apply LP-techniques we must represent Q, i.e. P_1, by linear inequalities. This generally is a difficult task and, for example, blocks polynomiality from being transferred in this way from LP to ILP. In Chapter 23 we shall see the (nonpolynomial) 'cutting plane method' for finding the inequalities defining P_1.

As we shall see in Section 17.4, Theorem 16.1 can be extended to:

(11)   *for each rational matrix A there exists a rational matrix M such that: for each column vector b there exists a column vector d such that $\{x \mid Ax \leqslant b\}_1 = \{x \mid Mx \leqslant d\}$.*

So the coefficients of the inequalities defining P_1 can be described by the coefficients of the inequalities defining P.

16.3. INTEGRAL POLYHEDRA

A crucial phenomenon is that the polyhedra occurring in polyhedral combinatorics frequently have $P = P_1$. A rational polyhedron with this property is called an *integral* polyhedron. One easily shows that for a rational polyhedron P, the following are equivalent:

(12)   (i) P is integral, i.e. $P = P_1$, i.e. P is the convex hull of the integral vectors in P;

    (ii) each face of P contains integral vectors;

    (iii) each minimal face of P contains integral vectors;

    (iv) $\max\{cx|x\in P\}$ is attained by an integral vector, for each c for which the maximum is finite.

So if P is pointed, then P is integral if and only if each vertex is integral.

If P is integral, we can solve any integer linear program over P in polynomial time:

Theorem 16.2. *There is a polynomial algorithm which, given a rational system $Ax \leqslant b$ defining an integral polyhedron, and given a rational vector c, finds an optimum solution for the ILP-problem $\max\{cx|Ax \leqslant b; x$ integral$\}$ (if it is finite).*

Proof. With Khachiyan's method (Chapter 13) one can find $\delta := \max\{cx|Ax \leqslant b\}$, and a system of linear equations $A'x = b'$ determining a minimal face F of $\{x|Ax \leqslant b\}$ so that each x in F satisfies $cx = \delta$. If $Ax \leqslant b$ determines an integral polyhedron, F contains an integral vector, which can be found in polynomial time (see Corollary 5.3b). This integral vector is an optimum solution for the ILP-problem. $\square$

More generally one has, in the terminology of Section 14.2:

Theorem 16.3. *Let $(P_i|i\in\mathbb{N})$ be a class of integral polyhedra for which the separation problem is polynomially solvable. Then there is a polynomial algorithm which, for given $i\in\mathbb{N}$ and $c\in\mathbb{Q}^{n_i}$, finds an integral vector x_0 in P_i attaining $\max\{cx|x\in P_i; x$ integral$\}$.*

Proof. Like the above. $\square$

Note that an algorithm which,

(13)   given a rational matrix A and rational vectors b and c such that $\max\{cx|Ax \leqslant b\}$ is attained by an integral vector x,

finds an integral optimum solution for $\max\{cx|Ax \leqslant b\}$, is as strong as one finding an integral solution for a given system $Ax \leqslant b$. We could take $c = 0$ in (13): if $Ax \leqslant b$ has an integral solution, the algorithm finds it; if $Ax \leqslant b$ has no integral solution, the algorithm will get stuck in polynomial time or will go beyond its polynomial time limit (as the input is not 'valid').

16.4. HILBERT BASES

Related is the concept of a Hilbert basis for a polyhedral cone, introduced by Giles and Pulleyblank [1979]. A finite set of vectors $a_1,\ldots,a_t$ is a *Hilbert basis* if each integral vector b in cone $\{a_1,\ldots,a_t\}$ is a nonnegative integral combination of $a_1,\ldots,a_t$. Especially, we are interested in *integral* Hilbert bases, which are Hilbert bases consisting of integral vectors only. The following was observed first essentially by Gordan [1873] (cf. Hilbert [1890]). Uniqueness for pointed cones was shown by van der Corput [1931b, c].

Theorem 16.4. *Each rational polyhedral cone C is generated by an integral Hilbert basis. If C is pointed there is a unique minimal integral Hilbert basis generating C (minimal relative to taking subsets).*

Proof. Let C be a rational polyhedral cone, generated by, say, $b_1, \ldots, b_k$. Without loss of generality, $b_1, \ldots, b_k$ are integral vectors. Let $a_1, \ldots, a_t$ be all integral vectors in the polytope

$$(14) \qquad \{\lambda_1 b_1 + \cdots + \lambda_k b_k | 0 \leqslant \lambda_i \leqslant 1 \ (i = 1, \ldots, k)\}.$$

Then $a_1, \ldots, a_t$ form an integral Hilbert basis, generating C. Indeed, they generate C, as $b_1, \ldots, b_k$ occur among $a_1, \ldots, a_t$, and as (14) is contained in C. To see that $a_1, \ldots, a_t$ form a Hilbert basis, let b be an integral vector in C. Then there are $\mu_1, \ldots, \mu_k \geqslant 0$ such that

$$(15) \qquad b = \mu_1 b_1 + \cdots + \mu_k b_k.$$

Then

$$(16) \qquad b = \lfloor \mu_1 \rfloor b_1 + \cdots + \lfloor \mu_k \rfloor b_k + ((\mu_1 - \lfloor \mu_1 \rfloor) b_1 + \cdots + (\mu_k - \lfloor \mu_k \rfloor) b_k).$$

Now the vector

$$(17) \qquad b - \lfloor \mu_1 \rfloor b_1 - \cdots - \lfloor \mu_k \rfloor b_k = (\mu_1 - \lfloor \mu_1 \rfloor) b_1 + \cdots + (\mu_k - \lfloor \mu_k \rfloor) b_k$$

occurs among $a_1, \ldots, a_t$, as the left-hand side of (17) clearly is integral, and the right-hand side of (17) clearly belongs to (14). Since also $b_1, \ldots, b_k$ occur among $a_1, \ldots, a_t$, it follows that (16) decomposes b as a nonnegative integral combination of $a_1, \ldots, a_t$. So $a_1, \ldots, a_t$ form a Hilbert basis.

Next, suppose C is pointed. Define

$$(18) \qquad H := \{a \in C | a \neq 0; a \text{ integral}; a \text{ is not the sum of two other integral vectors in } C\}.$$

It is immediate that any integral Hilbert basis generating C must contain H as a subset. So H is finite, as H is contained in (14). To see that H itself is a Hilbert basis generating C, let b be a vector such that $bx > 0$ if $x \in C \setminus \{0\}$ (b exists as C is pointed). Suppose not every integral vector in C is a nonnegative integral combination of vectors in H. Let c be such a vector, with bc as small as possible (this exists, as c must be in the set (14)). Then c is not in H. Hence $c = c_1 + c_2$ for certain nonzero integral vectors c_1 and c_2 in C. Then $bc_1 < bc$ and $bc_2 < bc$. Hence both c_1 and c_2 are nonnegative integral combinations of vectors in H, and therefore c is also. $\qquad \square$

It is easy to see that if the cone is not pointed there is no unique minimal integral Hilbert basis generating the cone. Note that if a vector c belongs to a minimal integral Hilbert basis generating a pointed cone, then the components of c are relatively prime integers.

Combining the methods of Theorems 16.1 and 16.4 one easily deduces that for any rational polyhedron P there exist integral vectors $x_1, \ldots, x_t, y_1, \ldots, y_s$ such that

(19) $\{x \mid x \in P; \; x \text{ integral}\} = \{\lambda_1 x_1 + \cdots + \lambda_t x_t + \mu_1 y_1 + \cdots + \mu_s y_s \mid \lambda_1, \ldots, \lambda_t,$
         $\mu_1, \ldots, \mu_s \text{ nonnegative integers, with } \lambda_1 + \cdots + \lambda_t = 1\}.$

So this yields a parametric solution for linear diophantine inequalities.

Hilbert bases are studied by Bachem [1978], Graver [1975], Huet [1978], Jeroslow [1978b]. For other finite basis and periodicity results, see Weidner [1976].

16.5. A THEOREM OF DOIGNON

Doignon [1973] (cf. Bell [1977] and Scarf [1977]) showed the following interesting analogue of Carathéodory's theorem (cf. Section 7.7). Let a system

(20) $a_1 x \leqslant \beta_1, \ldots, a_m x \leqslant \beta_m$

of linear inequalities in n variables be given.

Theorem 16.5. *If (20) has no integral solution, then there are 2^n or less constraints among (20) which already have no integral solution.*

Proof. Suppose (20) has no integral solution. We may assume that if we delete one of the constraints in (20), the remaining system has an integral solution. This means: there exist integral vectors $x^1, \ldots, x^m$ so that, for $j = 1, \ldots, m$, x^j violates $a_j x \leqslant \beta_j$ but satisfies all other inequalities in (20).

We must show $m \leqslant 2^n$. So assume $m > 2^n$. Let

(21) $Z := \mathbb{Z}^n \cap \text{conv.hull} \{x^1, \ldots, x^m\}.$

Choose $\gamma_1, \ldots, \gamma_m$ so that:

(22)   (i) $\gamma_j \geqslant \min \{a_j z \mid z \in Z; \; a_j z > \beta_j\}$,
        (ii) the system $a_1 x < \gamma_1, \ldots, a_m x < \gamma_m$ has no solution in Z,

and so that $\gamma_1 + \cdots + \gamma_m$ is as large as possible. Such $\gamma_1, \ldots, \gamma_m$ exist, since the set of $(\gamma_1, \ldots, \gamma_m)$ satisfying (22) is nonempty (as is shown by taking equality in (22)(i)), bounded (since $\gamma_j \leqslant a_j x^j$ for all j, as otherwise $x^j \in Z$ would be a solution of $a_1 x < \gamma_1, \ldots, a_m x < \gamma_m$), and closed (since the negation of (21)(ii): $\exists z \in Z$: $\forall j = 1, \ldots, m$: $\gamma_j > a_j z$, defines an open set).

Since $\gamma_1 + \cdots + \gamma_m$ is as large as possible, for each $j = 1, \ldots, m$ there exists $y^j \in Z$ so that $a_j y^j = \gamma_j$ and $a_i y^j < \gamma_i$ $(i \neq j)$. As $m > 2^n$, there exist k, l $(k \neq l)$ so that $y^k \equiv y^l \pmod{2}$. Hence $\frac{1}{2}(y^k + y^l)$ belongs to Z and satisfies $a_1 x < \gamma_1, \ldots,$ $a_m x < \gamma_m$. This contradicts (22)(ii). $\square$

A direct corollary is a theorem of Scarf [1977] (cf. Todd [1977]):

Corollary 16.5a. *Let $Ax \leqslant b$ be a system of linear inequalities in n variables, and* *let $c \in \mathbb{R}^n$. If* $\max \{cx \,|\, Ax \leqslant b; \, x \text{ integral}\}$ *is finite, then*

(23) $\max \{cx \,|\, Ax \leqslant b; \, x \text{ integral}\} = \max \{cx \,|\, A'x \leqslant b'; \, x \text{ integral}\}$

for some subsystem $A'x \leqslant b'$ *of* $Ax \leqslant b$ *with at most* $2^n - 1$ *inequalities.*

Proof. Let $\mu := \max \{cx \,|\, Ax \leqslant b; \, x \text{ integral}\}$. Hence for each $t \in \mathbb{N}$, the system $Ax \leqslant b$, $cx \geqslant \mu + 1/t$ has no integral solution. Therefore, by Theorem 16.5, for each $t \in \mathbb{N}$ there is a subsystem of $Ax \leqslant b$, $cx \geqslant \mu + 1/t$ of at most 2^n constraints having no integral solution. Since $Ax \leqslant b$ does have an integral solution (as μ is finite), each such subsystem contains the constraint $cx \geqslant \mu + 1/t$. Hence there is a subsystem $A'x \leqslant b'$ of $Ax \leqslant b$ with at most $2^n - 1$ constraints so that the system $A'x \leqslant b'$, $cx \geqslant \mu + 1/t$ has no integral solution for infinitely many values of $t \in \mathbb{N}$. Therefore, $A'x \leqslant b'$, $cx > \mu$ has no integral solution. This implies (23). $\square$

Note that the bound 2^n in Theorem 16.5 is best possible, as is shown by the system

(24) $$\sum_{i \in I} \xi_i - \sum_{i \notin I} \xi_i \leqslant |I| - 1 \quad (I \subseteq \{1, \ldots, n\})$$

of 2^n constraints in the n variables $\xi_1, \ldots, \xi_n$.

16.6. THE KNAPSACK PROBLEM AND AGGREGATION

The *knapsack problem* is

(25)   given rational vectors a and c, and a rational β, determine $\max \{cx \,|\, ax \leqslant \beta; \, x \, \{0,1\}\text{-vector}\}$.

This special case of the integer linear programming problem turns out to be also $\mathcal{NP}$-complete. Also the following related problems are $\mathcal{NP}$-complete:

(26)   given rational vector a and rational β decide if $ax = \beta$ for some $\{0,1\}$-vector x;

(27)   given rational vector a and rational β decide if $ax = \beta$ for some nonnegative integral vector x.

The $\mathcal{NP}$-completeness of these problems will be shown in Section 18.2. The reduction there is based on the idea of *aggregating* linear equations, which we now discuss briefly from a theoretical point of view.

First note that, if a system of rational linear equations has no (i) nonnegative, (ii) integral, (iii) arbitrary solution, respectively, then it is possible to make a linear combination of these equations which also has no such solutions. It is easy to see that this is equivalent to Corollaries 7.1d (Farkas' lemma), 4.1a, and 3.1b.

Geometrically cases (i) and (ii) can be interpreted as: each affine subspace S of $\mathbb{Q}^n$ with $S \cap \mathbb{Q}_+^n = \varnothing$ ($S \cap \mathbb{Z}^n = \varnothing$, respectively) is contained in an affine hyperplane H with $H \cap \mathbb{Q}_+^n = \varnothing$ ($H \cap \mathbb{Z}^n = \varnothing$, respectively).

Concerning $\mathbb{Z}_+^n$, Shevchenko [1976] (following work of Mathews [1897],

Bradley [1971b], and Glover and Woolsey [1972]) proved the following: let $S = \{x \mid Ax = b\}$ be an affine subspace of $\mathbb{Q}^n$; then S is contained in an affine hyperplane H with $H \cap \mathbb{Z}^n_+ = S \cap \mathbb{Z}^n_+$ if and only if S itself is an affine hyperplane or the cone generated by the columns of A is pointed. That is, if A has rank at least two, then there exists a linear combination of the equation in $Ax = b$ having the same nonnegative integral solutions as $Ax = b$, if and only if the cone generated by the columns of A is pointed.

The problem of finding this linear combination is called the *aggregation problem*. Aggregation reduces the problem of solving a system of linear equations in nonnegative integers to the problem of solving one such equation.

See also Babaev and Mamedov [1978], Babayev and Glover [1984], Chvátal and Hammer [1977], Fayard and Plateau [1973], Glover [1975a], Kannan [1983a], Kendall and Zionts [1977], Padberg [1972], Rosenberg [1974], Veselov and Shevchenko [1978], and Weinberg [1977, 1979a, b]. For more references, see Leveque [1974: Vol. 2, pp. 7–16], Kastning [1976: pp. 251–252], Hausmann [1978: p. 174], and von Randow [1982: p. 182; 1985: p. 206]. Cf. also Section 18.2.

16.7. MIXED INTEGER LINEAR PROGRAMMING

The *mixed integer linear programming problem* is:

(28)   given rational matrices A, B and rational vectors b, c, d, determine
          $\max \{cx + dy \mid Ax + By \leqslant b; x \text{ integral}\}$.

So it combines the linear and the integer linear programming problem; hence it is a difficult problem.

Again one can prove, by methods similar to those used for Theorem 16.1, that the set

(29) $\text{conv.hull} \left\{ \begin{pmatrix} x \\ y \end{pmatrix} \middle| Ax + By \leqslant b; x \text{ integral} \right\}$

is a polyhedron.

In this book we do not treat the mixed integer linear programming problem extensively. We occasionally discuss mixed integer linear programming results, viz. in Sections 18.4 (Lenstra's algorithm), 23.1 (cutting planes), and 24.5 (Benders' decomposition).

17

Estimates in integer linear programming

This chapter collects estimates and proximity and sensitivity results in integer linear programming.

In Section 17.1 we show that if an integer linear programming problem has a finite optimum, then it has an optimum solution of polynomially bounded size. It means that the facet complexity of P_1 can be bounded by a polynomial in that of P itself. It implies that the integer linear programming problem belongs to the class $\mathcal{NP}$. Next in Section 17.2 we discuss how close ILP-solutions are to a solution of the corresponding LP-relaxation, and conversely. It implies estimates on the sensitivity of ILP-solutions under changing right-hand sides. In Section 17.3 we show that if a feasible solution of an ILP-problem is not optimal, then there exists a better feasible solution 'nearby'. Finally, in Section 17.4 we show that, roughly speaking, the left-hand sides (i.e. coefficients) of the facet equations for the integer hull P_1 only depend on the left-hand sides of the facet equations for P.

17.1. SIZES OF SOLUTIONS

In this section we make some estimates on the sizes of solutions of ILP-problems. In particular, we show that if an ILP-problem has a finite optimum, then it has an optimum solution of size polynomially bounded by the size of the input. This implies that the problem belongs to the complexity class $\mathcal{NP}$ — see also Section 18.1.

First, similarly as in the proof of Theorem 16.1, one shows that the facet complexity of the integer hull P_1 is polynomially bounded by the facet complexity of P (cf. Karp and Papadimitriou [1980]). This will follow from the following theorem, which we need also for other purposes.

Theorem 17.1. *Let* $P = \{x \mid Ax \leqslant b\}$, *where* A *is an integral* $m \times n$*-matrix and* b *is an integral vector. Then*

(1) $P_1 = \text{conv.hull} \{x_1, \ldots, x_t\} + \text{cone} \{y_1, \ldots, y_s\}$

where $x_1, \ldots, x_t, y_1, \ldots, y_s$ *are integral vectors with all components at most* $(n + 1)\Delta$ *in absolute value, where* Δ *is the maximum absolute value of the subdeterminants of the matrix* $[A \ b]$.

Proof. The proof is parallel to that of Theorem 16.1. The case $P_1 = \emptyset$ being trivial, assume $P_1 \neq \emptyset$. By Theorem 16.1 and Cramer's rule we can write char.cone $P_1 =$ char.cone $P =$ cone $\{y_1, \ldots, y_s\}$, where $y_1, \ldots, y_s$ are integral vectors with each component being a subdeterminant of A—in particular each component is at most Δ in absolute value.

Moreover

(2) $\qquad P = \text{conv.hull} \{z_1, \ldots, z_k\} + \text{cone} \{y_1, \ldots, y_s\}$

where $z_1, \ldots, z_k$ are vectors with each component being a quotient of subdeterminants of the matrix $[A \ b]$—in particular, each component is at most Δ in absolute value. Now let $x_1, \ldots, x_t$ be the integral vectors contained in

(3) $\qquad$ conv.hull $\{z_1, \ldots, z_k\} + \{\mu_1 y_1 + \cdots + \mu_s y_s | 0 \leqslant \mu_j \leqslant 1 \ (j = 1, \ldots, s);$
$\qquad$ at most n of the μ_j are nonzero$\}$.

Then each component of each x_i is at most $(n + 1)\Delta$ in absolute value. Moreover, (1) holds. To see this, we have to show that each minimal face F of P_1 contains at least one x_i. Any such F certainly contains an integral vector, say x^*. As x^* belongs to P, we can write

(4) $\qquad x^* = \lambda_1 z_1 + \cdots + \lambda_k z_k + \mu_1 y_1 + \cdots + \mu_s y_s$

where $\lambda_1, \ldots, \lambda_k, \mu_1, \ldots, \mu_s \geqslant 0$, $\lambda_1 + \cdots + \lambda_k = 1$, and at most n of the μ_j nonzero (by Carathéodory's theorem (Corollary 7.1i)). Then

(5) $\qquad \tilde{x} := \lambda_1 z_1 + \cdots + \lambda_k z_k + (\mu_1 - \lfloor \mu_1 \rfloor) y_1 + \cdots + (\mu_s - \lfloor \mu_s \rfloor) y_s$

is again an integral vector in F, as one easily checks. Moreover, it is one of the x_i, which proves our claim. $\qquad\qquad\qquad\qquad\qquad\qquad\qquad\qquad\qquad\qquad\quad \square$

Corollary 17.1a. *Let P be a rational polyhedron in $\mathbb{R}^n$ of facet complexity φ. Then P_1 has facet complexity at most $24n^5\varphi$.*

Proof. Let $P = \{x | Ax \leqslant b\}$, where each row of $[A \ b]$ has size at most φ. Each inequality $ax \leqslant \beta$ in $Ax \leqslant b$ can be multiplied by the product of the denominators in $ax \leqslant \beta$, which is at most 2^φ, to obtain an equivalent integral inequality of size at most $(n + 1)\varphi$. In this way we obtain the inequality system $\tilde{A}x \leqslant \tilde{b}$. Each submatrix of $[\tilde{A} \ \tilde{b}]$ has size at most $(n + 1)^2\varphi$, and hence, by Theorem 3.2, the size of the largest absolute value Δ of subdeterminants of $[\tilde{A} \ \tilde{b}]$ is at most $2(n + 1)^2\varphi$. So $\Delta \leqslant 2^{2(n + 1)^2\varphi}$. Hence, by Theorem 17.1, there exist integral vectors $x_1, \ldots, x_t, y_1, \ldots, y_s$ such that

(6) $\qquad P_1 = \text{conv.hull} \{x_1, \ldots, x_t\} + \text{cone} \{y_1, \ldots, y_s\}$

where each component of each x_i, y_j is at most $(n + 1)\Delta$ in absolute value, and hence has size at most $2(n + 1)^2 \varphi + \lceil \log_2 (n + 1) \rceil + 3$. Therefore, for each i

$$(7) \qquad \text{size}(x_i) \leqslant n + n(2(n + 1)^2 \varphi + \lceil \log_2 (n + 1) \rceil + 3) \leqslant 6n^3 \varphi.$$

Similarly for the y_j. So the vertex complexity of P_1 is at most $6n^3 \varphi$, and hence, by Theorem 10.2, the facet complexity of P_1 is at most $24n^5 \varphi$. $\square$

Another consequence of Theorem 17.1 is that if a polyhedron contains integral vectors, it contains integral vectors of small size (cf. Borosh and Treybig [1976, 1979], Cook [1976], von zur Gathen and Sieveking [1978], Kannan [1976], Kannan and Monma [1978], and Papadimitriou [1981]).

Corollary 17.1b. *Let P be a rational polyhedron in $\mathbb{R}^n$ of facet complexity φ. If P contains an integral vector, it contains one of size at most $6n^3 \varphi$.*

Proof. Cf. the proof of Corollary 17.1a. $\square$

So by enumeration we can find an integral vector x satisfying $Ax \leqslant b$, or decide that no such vector exists, in finite time.

Corollary 17.1c. *Let P be a rational polyhedron in $\mathbb{R}^n$ of facet complexity φ, and let $c \in \mathbb{Q}^n$. If $\max \{cx \mid x \in P;\ x \text{ integral}\}$ is finite, it is attained by a vector of size at most $6n^3 \varphi$.*

Proof. Cf. the proof of Theorem 17.1: the maximum is attained by one of the vectors x_i. $\square$

Corollary 17.1d. *The integer linear programming problem belongs to $\mathcal{NP}$.*

Proof. It follows from Corollary 17.1b that if a system of rational linear inequalities $Ax \leqslant b$ has an integral solution, it has one of size polynomially bounded by size(A, b). [Similarly, if $Ax = b$ has a nonnegative integral solution, it has one of size polynomially bounded by size(A, b).] $\square$

17.2. DISTANCES OF OPTIMUM SOLUTIONS

We first give an estimate on the proximity of ILP-solutions and solutions of the corresponding LP-relaxation. It is due to Cook, Gerards, Schrijver, and Tardos [1986], and strengthens a result of Blair and Jeroslow [1979: Thm 1.2].

Theorem 17.2. *Let A be an integral $m \times n$-matrix, such that each subdeterminant is at most Δ in absolute value, and let b and c be vectors. Suppose that*

$$(8) \qquad \text{both} \quad \text{(i) } \max \{cx \mid Ax \leqslant b\}$$
$$\text{and} \quad \text{(ii) } \max \{cx \mid Ax \leqslant b;\ x \text{ integral}\}$$

are finite. Then:

I. *For each optimum solution y of* (i) *there exists an optimum solution z of* (ii) *with* $\|y - z\|_\infty \leqslant n\Delta$;

II. *For each optimum solution z of* (ii) *there exists an optimum solution y of* (i) *with* $\|y - z\|_\infty \leqslant n\Delta$.

[As usual, $\|\cdots\|_\infty$ is the maximum absolute value of the components of $\cdots$.]

Proof. Let y be any optimum solution of (i) and let z be any optimum solution of (ii). Split $Ax \leqslant b$ as $A_1 x \leqslant b_1$, $A_2 x \leqslant b_2$ such that

$$(9) \qquad A_1 y < A_1 z, \; A_2 y \geqslant A_2 z.$$

Consider the cone $C := \{u \mid A_1 u \leqslant 0; \; A_2 u \geqslant 0\}$. As $A_1 y < A_1 z \leqslant b_1$, we know, by complementary slackness, that $vA_2 = c$ for some $v \geqslant 0$. Hence $cu \geqslant 0$ for each u in C. Moreover, $y - z \in C$, and hence

$$(10) \qquad y - z = \lambda_1 u_1 + \cdots + \lambda_t u_t$$

for some $\lambda_1, \ldots, \lambda_t \geqslant 0$ and linearly independent vectors $u_1, \ldots, u_t$ in C, such that for each i, u_i is integral and $\|u_i\|_\infty \leqslant \Delta$ (as C is generated by integral vectors u with $\|u\|_\infty \leqslant \Delta$, by Cramer's rule—cf. Section 3.2).

Now if $0 \leqslant \mu_i \leqslant \lambda_i$ for all i, then

$$(11) \qquad z + \mu_1 u_1 + \cdots + \mu_t u_t$$

satisfies $Ax \leqslant b$, as

$$(12) \qquad A_1(z + \mu_1 u_1 + \cdots + \mu_t u_t) \leqslant A_1 z \leqslant b_1 \text{ and}$$
$$A_2(z + \mu_1 u_1 + \cdots + \mu_t u_t) = A_2(y - (\lambda_1 - \mu_1)u_1 - \cdots - (\lambda_t - \mu_t)u_t)$$
$$\leqslant A_2 y \leqslant b_2$$

(as $A_1 u_i \leqslant 0$, $A_2 u_i \geqslant 0$ for all i).

Now let

$$(13) \qquad z' := z + \lfloor \lambda_1 \rfloor u_1 + \cdots + \lfloor \lambda_t \rfloor u_t = y - \{\lambda_1\}u_1 - \cdots - \{\lambda_t\}u_t$$

(where $\{\lambda_i\} = \lambda_i - \lfloor \lambda_i \rfloor$). Then z' is an integer solution of $Ax \leqslant b$ with $cz' \geqslant cz$ (as $cu_i \geqslant 0$ for all i). So z' is an optimum solution of (ii). Moreover,

$$(14) \qquad \|y - z'\|_\infty = \|\{\lambda_1\}u_1 + \cdots + \{\lambda_t\}u_t\|_\infty \leqslant \sum_{i=1}^{t} \|u_i\|_\infty \leqslant n\Delta.$$

This shows I.

Similarly, let

$$(15) \qquad y' := y - \lfloor \lambda_1 \rfloor u_1 - \cdots - \lfloor \lambda_t \rfloor u_t = z + \{\lambda_1\}u_1 + \cdots + \{\lambda_t\}u_t.$$

Then y' satisfies $Ax \leqslant b$. Also, $cy' \geqslant cy$. For otherwise $\lfloor \lambda_i \rfloor > 0$ and $cu_i > 0$ for some i; then (by (12)) $z + \lfloor \lambda_i \rfloor u_i$ is an integral solution of $Ax \leqslant b$, and $c(z + \lfloor \lambda_i \rfloor u_i) > cz$, contradicting the optimality of z. Moreover,

$$(16) \qquad \|z - y'\|_\infty = \|\{\lambda_1\}u_1 + \cdots + \{\lambda_t\}u_t\|_\infty \leqslant \sum_{i=1}^{t} \|u_i\|_\infty \leqslant n\Delta.$$

This shows II. $\qquad \square$

This theorem extends results of Gomory [1965] and Wolsey [1981b] that for any rational matrix A there exist finitely many vectors $x^1, \ldots, x^N$ such that, for any row vector c and any integral column vector b, if x^* is an optimum basic (i.e. vertex) solution of the linear program $\max \{cx \mid x \geqslant 0; Ax = b\}$, then the integer linear program $\max \{cx \mid x \geqslant 0; Ax = b; x \text{ integral}\}$ is attained by one of the vectors $x^* - x^k$ $(k = 1, \ldots, N)$.

The following example shows that the bound $n\Delta$ in Theorem 17.1 is best possible:

(17) $\qquad A := \begin{bmatrix} +1 & 0 & 0 \cdots \cdots \cdots \cdots \cdots 0 \\ -1 & +1 & 0 & & \\ 0 & -1 & +1 & & \\ & & & \ddots & 0 \\ 0 \cdots \cdots \cdots \cdots \cdots \cdots 0 & -1 & +1 \end{bmatrix}, \quad \Delta := 1,$

$\qquad b := \begin{bmatrix} \beta \\ \beta \\ \beta \\ \vdots \\ \beta \end{bmatrix}, \quad c := (1, \ldots, 1).$

For each β with $0 \leqslant \beta < 1$ the following are the unique optimum solutions for $\max \{cx \mid Ax \leqslant b\}$ and $\max \{cx \mid Ax \leqslant b; x \text{ integral}\}$:

(18) $\qquad y := \begin{bmatrix} \beta \\ 2\beta \\ \vdots \\ n\beta \end{bmatrix}, \quad z = \begin{bmatrix} 0 \\ 0 \\ \vdots \\ 0 \end{bmatrix}.$

Note that $\|y - z\|_\infty = n\Delta\beta$. (This example is due to L. Lovász.)

Theorem 17.2 combined with the sensitivity result for linear programming (Theorem 10.5) yields the following corollary, due to Cook, Gerards, Schrijver, and Tardos [1986], which is a sharpened form of a theorem of Blair and Jeroslow [1977: Thm 2.1].

Corollary 17.2a. *Let A be an integral $m \times n$-matrix, such that each subdeterminant of A is at most Δ in absolute value, let b' and b'' be column m-vectors, and let c be a row n-vector. Suppose $\max \{cx \mid Ax \leqslant b'; x \text{ integral}\}$ and $\max \{cx \mid Ax \leqslant b''; x \text{ integral}\}$ are finite. Then for each optimum solution z' of the first maximum there exists an optimum solution z'' of the second maximum such that*

(19) $\qquad \|z' - z''\|_\infty \leqslant n\Delta(\|b' - b''\|_\infty + 2).$

Proof. Let z' be an optimum solution of $\max \{cx \mid Ax \leqslant b'; x \text{ integral}\}$. By II of Theorem 17.2 there exists an optimum solution x' of $\max \{cx \mid Ax \leqslant b'\}$ with $\|z' - x'\|_\infty \leqslant n\Delta$. By Theorem 10.5 there exists an optimum solution x'' of

$\max \{cx | Ax \leqslant b''\}$ with $\| x' - x'' \|_{\infty} \leqslant n\Delta \| b' - b'' \|_{\infty}$. By I of Theorem 17.2 there exists an optimum solution z'' of $\max \{cx | Ax \leqslant b''; x \text{ integral}\}$ with $\| x'' - z'' \|_{\infty} \leqslant n\Delta$. Combining the three inequalities yields (19). $\qquad \square$

References to other sensitivity results in integer linear programming are given at the end of Section 23.7.

17.3. FINITE TEST SETS FOR INTEGER LINEAR PROGRAMMING

Another estimate is given in the following result of Cook, Gerards, Schrijver, and Tardos [1986]. It implies a theorem of Graver [1975] (cf. Blair and Jeroslow [1982: Lemma 4.3]) that for any integer linear programming problem $\max \{cx | Ax \leqslant b; x \text{ integral}\}$ there exists a finite set $x^1, \ldots, x^N$ (a 'test set') of integral vectors, depending only on the constraint matrix, such that: a feasible solution x^* is optimal, if and only if $c(x^* + x^k) \leqslant cx^*$ whenever $x^* + x^k$ is feasible $(k = 1, \ldots, N)$.

Theorem 17.3. *Let A be an integral $m \times n$-matrix, with all subdeterminants at most Δ in absolute value, let b be a column m-vector and let c be a row n-vector. Let z be a feasible, but not optimal, solution of $\max \{cx | Ax \leqslant b; x \text{ integral}\}$. Then there exists a feasible solution z' such that $cz' > cz$ and $\| z - z' \|_{\infty} \leqslant n\Delta$.*

Proof. As z is not optimal, we know $cz'' > cz$ for some integral solution z'' of $Ax \leqslant b$. We can split $Ax \leqslant b$ into subsystems $A_1 x \leqslant b_1$ and $A_2 x \leqslant b_2$ such that

$$(20) \qquad A_1 z \leqslant A_1 z'', \quad A_2 z \geqslant A_2 z''.$$

Let C be the cone

$$(21) \qquad C := \{u | A_1 u \geqslant 0; A_2 u \leqslant 0\}$$

Then $z'' - z$ belongs to C, and hence

$$(22) \qquad z'' - z = \lambda_1 u_1 + \cdots + \lambda_t u_t$$

for certain $\lambda_1, \ldots, \lambda_t \geqslant 0$ and linearly independent integral vectors $u_1, \ldots, u_t$ in C with $\| u_i \|_{\infty} \leqslant \Delta$ (by Cramer's rule—cf. Section 3.2).

Now note that, if $0 \leqslant \mu_i \leqslant \lambda_i (i = 1, \ldots, t)$, then

$$(23) \qquad z + \mu_1 u_1 + \cdots + \mu_t u_t = z'' - (\lambda_1 - \mu_1) u_1 - \cdots - (\lambda_t - \mu_t) u_t$$

satisfies $Ax \leqslant b$, as

$$(24) \qquad A_1(z'' - (\lambda_1 - \mu_1) u_1 - \cdots - (\lambda_t - \mu_t) u_t) \leqslant A_1 z'' \leqslant b_1 \text{ and}$$
$$A_2(z + \mu_1 u_1 + \cdots + \mu_t u_t) \leqslant A_2 z \leqslant b_2.$$

Now there are two cases.

Case 1. $\lambda_i \geqslant 1$ and $cu_i > 0$ for some i. Then $z' := z + u_i$ is a feasible solution of $\max \{cx | Ax \leqslant b; x \text{ integral}\}$ (by (24)) such that $cz' > cz$ and $\| z' - z \|_{\infty} \leqslant \Delta \leqslant n\Delta$.

Case 2. For all i, $\lambda_i \geqslant 1$ implies $cu_i \leqslant 0$. Then

$$(25) \qquad z' := z + \{\lambda_1\}u_1 + \cdots + \{\lambda_t\}u_t = z'' - \lfloor \lambda_1 \rfloor u_1 - \cdots - \lfloor \lambda_t \rfloor u_t$$

(where $\{\lambda_i\} = \lambda_i - \lfloor \lambda_i \rfloor$) is a feasible solution of $\max\{cx | Ax \leqslant b; x \text{ integral}\}$ (by (24)) such that $cz' \geqslant cz'' > cz$ and $\|z' - z\|_\infty \leqslant \sum_{i=1}^{t} \|u_i\|_\infty \leqslant n\Delta$. $\qquad \square$

17.4. THE FACETS OF P_1

In Section 17.1 we saw that the facet complexity of the integer hull P_1 is polynomially bounded by the facet complexity of the polyhedron P itself. We shall now see that this is also true if we do not take into account the right-hand sides of the defining inequalities. This was shown by Wolsey [1981b: Thm.2'] and Cook, Gerards, Schrijver, and Tardos [1986] (cf. Blair and Jeroslow [1982: Thm 6.2]).

Theorem 17.4. *For each rational matrix A there exists an integral matrix M such that: for each column vector b there exists a column vector d such that*

$$(26) \qquad \{x | Ax \leqslant b\}_1 = \{x | Mx \leqslant d\}.$$

In fact, if A is integral, and each subdeterminant of A is at most Δ in absolute value, then we can take all entries of M at most $n^{2n}\Delta^n$ in absolute value (n being the number of columns of A).

Proof. Without loss of generality, A is integral. Let Δ be the maximum absolute value of the subdeterminants of A. Let M be the matrix with rows all integral vectors in the set

$$(27) \qquad L := \{w | \exists y \geqslant 0 : yA = w; \ w \text{ integral}; \ \|w\|_\infty \leqslant n^{2n}\Delta^n\}.$$

We claim that this is the matrix as required. To show this, let b be a column vector.

If $Ax \leqslant b$ has no solution, then it is clear that we can find a vector d such that $Mx \leqslant d$ has no solution (as the rows of A occur among L).

If $Ax \leqslant b$ has a solution, but no integral solution, then the cone $\{x | Ax \leqslant 0\}$ is not full-dimensional, and hence $Ax \leqslant 0$ has an implicit equality, say $a_i x \leqslant 0$. Hence both a_i and $-a_i$ belong to L, and hence we can choose d so that $\{x | Mx \leqslant d\} = \varnothing$.

So we may assume that $Ax \leqslant b$ has an integral solution. For each row vector c let

$$(28) \qquad \delta_c := \max\{cx | Ax \leqslant b; x \text{ integral}\}.$$

It suffices to show that

$$(29) \qquad \{x | Ax \leqslant b\}_1 = \{x | wx \leqslant \delta_w \text{ for each } w \text{ in } L\}.$$

Here $\subseteq$ follows directly from (28). To show $\supseteq$, let $cx \leqslant \delta_c$ be a valid inequality for $\{x | Ax \leqslant b\}_1$. Let $\max\{cx | Ax \leqslant b; x \text{ integral}\}$ be attained by z. Then by Theorem 17.3,

(30) $C := \text{cone}\{x - z \,|\, Ax \leqslant b;\ x \text{ integral}\}$

         $= \text{cone}\{x - z \,|\, Ax \leqslant b;\ x \text{ integral};\ \|x - z\|_\infty \leqslant n\Delta\}.$

Hence, by Cramer's rule, this cone can be described as

(31) $C = \{u \,|\, wu \leqslant 0 \text{ for all } w \text{ in } L_1\}$

for some subset L_1 of L. For each w in L_1 we have $\delta_w = wz$. Moreover, $cu \leqslant 0$ for each u in C (as z is an optimum solution of max$\{cx \,|\, Ax \leqslant b;\ x \text{ integral}\}$). Therefore, $c = \lambda_1 w_1 + \cdots + \lambda_t w_t$ for some $\lambda_1, \ldots, \lambda_t \geqslant 0$ and some $w_1, \ldots, w_t \in L_1$. Moreover, $\delta_c = cz = \lambda_1 w_1 z + \cdots + \lambda_t w_t z = \lambda_1 \delta_{w_1} + \cdots + \lambda_t \delta_{w_t}$. It follows that $cx \leqslant \delta_c$ is valid for the RHS of (29).   □

For more on the facets of P_1, see Section 18.3.

18

The complexity of integer linear programming

In this chapter we study the complexity of integer linear programming, with as main results the $\mathcal{NP}$-completeness of the ILP-problem, and its polynomial solvability if we fix the number of variables.

In Section 18.1 we show that the ILP-problem is $\mathcal{NP}$-complete; it therefore belongs to the hardest problems in the complexity class $\mathcal{NP}$. Several related problems are also $\mathcal{NP}$-complete: the knapsack problem, solving one linear equation in nonnegative integers, splitting a set of rationals into two sets with equal sum (see Section 18.2), testing membership of the integer hull P_1 of a polyhedron P, testing if an integral vector is a vertex of P_1, testing adjacency on P_1 (see Section 18.3). Moreover, if $\mathcal{NP} \neq$ co-$\mathcal{NP}$, P_1 generally has 'difficult' facets, as we shall also discuss in Section 18.3.

Next we go over to more positive results. In Section 18.4 we show the result of Lenstra, that for each fixed number of variables there exists a polynomial-time algorithm for integer linear programming. In Section 18.5 it is shown that, if we fix an upper bound on the coefficients in the objective function, then the knapsack problem can be solved in polynomial time. The method, due to Bellman and Dantzig, is based on 'dynamic programming', and, as Ibarra and Kim showed, yields a 'fully polynomial approximation scheme' for the knapsack problem. Finally, in Section 18.6 we show the more general result of Papadimitriou that, for each fixed number of equality constraints, there is a 'pseudo-polynomial' algorithm for the ILP-problem $\max\{cx \mid x \geqslant 0; Ax = b\}$. Also this method is based on dynamic programming.

18.1. ILP IS $\mathcal{NP}$-COMPLETE

We show that the integer linear programming problem, in the following form:

(1)   given rational matrix A and rational vector b, does $Ax \leqslant b$ have an integral solution x?

is $\mathcal{NP}$-complete. This result, essentially due to Cook [1971], implies that ILP belongs to the most difficult problems in the class $\mathcal{NP}$.

Theorem 18.1. *The integer linear programming problem* (1) *is $\mathcal{NP}$-complete.*

Proof. By Corollary 17.1d, (1) belongs to $\mathcal{NP}$.

To show that (1) is $\mathcal{NP}$-complete, we have to show that any decision problem $\mathcal{L}$ in $\mathcal{NP}$ is Karp-reducible to (1). Therefore, let $\mathcal{L} \subseteq \{0, 1\}^*$ be a problem in $\mathcal{NP}$. So there exists a polynomial ϕ and a problem $\mathcal{L}'$ in $\mathcal{P}$ so that for each string z in $\{0, 1\}^*$:

(2) $z \in \mathcal{L} \Leftrightarrow \exists y \in \{0, 1\}^*:(z; y) \in \mathcal{L}'$ and $\text{size}(y) = \phi(\text{size}(z))$.

We use the following as our computer model M (*k-tape Turing machine*—see Aho, Hopcroft, and Ullman [1974]):

(3)   (i) M is, at each *moment*, in one of a finite number of *states* $q_1, \ldots, q_h$;
       (ii) M has k, two-sided infinite, *tapes*, each consisting of a countable number of *cells* (numbered $\ldots, -2, -1, 0, 1, 2, \ldots$), which can contain one of the symbols 0 and 1;
       (iii) at each moment, M *reads*, on each of the tapes, exactly one cell;
       (iv) suppose, at a certain moment t, M reads cell c_i of tape i, for $i = 1, \ldots, k$; then the symbols read, together with the state of M at moment t, uniquely determine the following:
             (a) the state of M at moment $t + 1$;
             (b) the symbol in cell c_i of tape i ($i = 1, \ldots, k$) at moment $t + 1$ (the symbols in all other cells remain unchanged);
             (c) which of the cells $c_i - 1, c_i, c_i + 1$ of tape i will be read at moment $t + 1$ ($i = 1, \ldots, k$).

So (a), (b), and (c) only depend on the state of M and what is read by M at moment t.

In fact, we can do without states. We can imitate states by tapes: we can replace the h states by h extra tapes, of which always cell numbered 0 is read, so that the symbol in cell 0 of the ith extra tape being 1 corresponds to the fact that the machine is in state q_i (so at each moment exactly one of these extra tapes has a 1 in cell 0, all other extra tapes having 0 in cell 0). So we may assume $h = 0$.

Hence mathematically the model can be described completely by a function

(4) $F := \{0, 1\}^k \rightarrow \{0, 1\}^k \times \{-1, 0, +1\}^k$.

$F(\sigma_1, \ldots, \sigma_k) = ((\sigma'_1, \ldots, \sigma'_k), (v_1, \ldots, v_k))$ should be interpreted as: if, at any moment t, M reads symbol σ_i in cell c_i of tape i ($i = 1, \ldots, k$), then symbol σ_i in cell c_i of tape i is replaced by σ'_i, and at moment $t + 1$ the cell read on tape i is $c_i + v_i$ ($i = 1, \ldots, k$).

Now $\mathcal{L}'$ belonging to $\mathcal{P}$ means that there exists a computer model M as above (i.e. a function F as in (4)) and a polynomial ψ such that:

(5)   a string $(z; y) = (\zeta_1, \ldots, \zeta_n; \eta_1, \ldots, \eta_{\phi(n)})$ belongs to $\mathcal{L}'$ if and only if the following holds:
       if: at moment 0, on tape 1 the symbols in cells $1, \ldots, 2n$ are $1, \zeta_1, 1$, $\zeta_2, \ldots, 1, \zeta_n$, and on tape 2 the symbols in cells $1, \ldots, \phi(n)$ are η_1,

$\eta_2, \ldots, \eta_{\phi(n)}$, while all other cells contain symbol 0, and on each tape cell number 0 is read,

then: at moment $\psi(n)$, M reads symbol 0 on tape 1.

Without loss of generality, $\psi(n) \geqslant 2n$ and $\psi(n) \geqslant \phi(n)$ for all n.

We show that the above implies that $\mathcal{L}$ is Karp-reducible to problem (1). Suppose we want to know whether a given string $z = (\zeta_1, \ldots, \zeta_n)$ belongs to $\mathcal{L}$. So we must decide if $(z; y)$ belongs to $\mathcal{L}'$ for some string $y = (\eta_1, \ldots, \eta_{\phi(n)})$. We shall now consider the symbols on the tapes of our machine M as variables. If c_{it} denotes the cell of tape i read at moment t, let $\delta_{i,t,j}$ stand for the symbol in cell $c_{it} + j$ of tape i at moment t. So $\delta_{i,t,0}$ is the symbol read at moment t on tape i. Now consider the following entries:

$$(6) \qquad \delta_{i,0,-\psi(n)}, \ldots\ldots\ldots\ldots, \delta_{i,0,0}, \ldots\ldots\ldots\ldots, \delta_{i,0,\psi(n)} \qquad (i = 1, \ldots, k)$$

$$\delta_{i,1,-\psi(n)+1}, \ldots\ldots\ldots, \delta_{i,1,0}, \ldots\ldots\ldots, \delta_{i,1,\psi(n)-1} \qquad (i = 1, \ldots, k)$$

$$\delta_{i,\psi(n)-1,-1}, \delta_{i,\psi(n)-1,0}, \delta_{i,\psi(n)-1,1} \qquad\qquad (i = 1, \ldots, k)$$

$$\delta_{i,\psi(n),0} \qquad\qquad\qquad (i = 1, \ldots, k).$$

Each entry $\delta_{i,t+1,j}$ on the $(t+1)$th line is determined uniquely by at most $k+3$ entries on the tth line, namely by

$$(7) \qquad \begin{array}{l} \delta_{1,t,0}, \ldots, \delta_{k,t,0} \\ \delta_{i,t,j-1}, \delta_{i,t,j}, \delta_{i,t,j+1} \end{array}.$$

Moreover, there exists a finite algorithm to find out how $\delta_{i,t+1,j}$ depends on the variables (7) (as there exists a function F as in (4)). We can describe this dependence by linear inequalities in $\delta_{i,t+1,j}$ and the variables (7), so that the $\{0,1\}$-solutions uniquely determine $\delta_{i,t+1,j}$ as a function of the variables (7) (that is, we can find linear inequalities in the variables (7) and in $\delta_{i,t+1,j}$ so that: for each $(\delta_{1,t,0}, \ldots, \delta_{k,t,0}, \delta_{i,t,j-1}, \delta_{i,t,j}, \delta_{i,t,j+1}) \in \{0,1\}^{k+3}$ there is a unique $\delta_{i,t+1,j} \in \{0,1\}$ so that this setting of variables is feasible).

It follows from (5) that z belongs to $\mathcal{L}$ if and only if the system Σ in the variables (6) consisting of all these inequalities, together with the following constraints:

$$(8) \qquad \delta_{1,0,2j} = \zeta_j \qquad (j = 1, \ldots, n),$$
$$\delta_{1,0,2j-1} = 1 \qquad (j = 1, \ldots, n),$$
$$\delta_{i,0,j} = 0 \qquad (i = 1 \text{ and } j \leqslant 0 \text{ or } j > 2n; \text{ or } i = 2$$
$$\text{and } j \leqslant 0 \text{ or } j > \phi(n); \text{ or } i \geqslant 3),$$
$$\delta_{1,\psi(n),0} = 0,$$
$$0 \leqslant \delta_{i,t,j} \leqslant 1 \qquad (\text{all } i, t, j)$$

has an integral solution. As there exists a polynomial-time algorithm to construct the system Σ from z, this reduces problem $\mathcal{L}$ to the integer linear programming problem. $\qquad\qquad\qquad\qquad\qquad\qquad\qquad\qquad\qquad\qquad\qquad\qquad\qquad\square$

18.2. $\mathcal{NP}$-COMPLETENESS OF RELATED PROBLEMS

From Theorem 18.1 the $\mathcal{NP}$-completeness of several related problems can be derived. First consider the problem:

(9)   given rational matrix A and rational vector b, does $Ax = b$ have a nonnegative integral solution x?

Corollary 18.1a. *Problem* (9) *is* $\mathcal{NP}$-*complete.*

Proof. (1) can be reduced to (9). To see this, observe that in (1) we may assume A and b to be integral. Then $Ax \leqslant b$ has an integral solution if and only if $Ax' - Ax'' + \tilde{x} = b$ has a nonnegative integral solution $x', x'', \tilde{x}$.   □

In this and in subsequent proofs, the reductions are Karp-reductions (cf. Section 2.6). Moreover, we leave it to the reader to verify that the problems in question belong to $\mathcal{NP}$ (this follows easily with Corollary 17.1b).

Next we restrict (9) to $\{0, 1\}$-solutions:

(10)   given rational matrix A and rational vector b, does $Ax = b$ have a $\{0, 1\}$-solution x?

Corollary 18.1b. *Problem* (10) *is* $\mathcal{NP}$-*complete.*

Proof. (9) can be reduced to (10): suppose we wish to solve $Ax = b$ in nonnegative integers. Let φ be the maximum row size of the matrix $[A\ b]$. By Corollary 17.1b, if $Ax = b$ has a nonnegative integral solution, it has one of size at most $M := 6n^3 \varphi$ (n being the number of variables). Now replace each component ξ_i of x by $\sum_{j=0}^{M} 2^j \xi_{ij}$, where the ξ_{ij} are variables restricted to 0 and 1 ($i = 1, \dots, n$; $j = 0, \dots, M$). Then $Ax = b$ has a nonnegative solution if and only if the system in the variables ξ_{ij} has a $\{0, 1\}$-solution.   □

Problem (10) remains $\mathcal{NP}$-complete if we restrict the system to only one equation:

(11)   given a rational vector a and a rational β, does $ax = \beta$ have a $\{0, 1\}$-solution?

Corollary 18.1c. *Problem* (11) *is* $\mathcal{NP}$-*complete.*

Proof. (10) can be reduced to (11): Suppose we wish to find a $\{0, 1\}$-vector x satisfying $a_1 x = \beta_1, \dots, a_m x = \beta_m$. Without loss of generality, $a_1, \dots, a_m, \beta_1, \dots, \beta_m$ are integral. Let n be the dimension of the a_i and let T be the maximum absolute value of the components of the a_i and β_i. Then for any $\{0, 1\}$-vector x: $a_1 x = \beta_1, \dots, a_m x = \beta_m$ if and only if $(a_1 + (2nT)a_2 + \cdots + (2nT)^{m-1}a_m)x = \beta_1 + (2nT)\beta_2 + \cdots + (2nT)^{m-1}\beta_m$.   □

This implies that also the knapsack problem:

(12)   given rational vectors a, c and rationals β and δ, is there a $\{0, 1\}$-vector x
        with $ax \leqslant \beta,\ cx \geqslant \delta$?

is $\mathcal{N}\mathcal{P}$-complete, a result proved by Karp [1972].

Corollary 18.1d. *The knapsack problem* (12) *is* $\mathcal{N}\mathcal{P}$-*complete.*

Proof. Directly from Corollary 18.1c, by taking $a = c$ and $\beta = \delta$.   □

In fact, we may assume in (11) that a is nonnegative:

(13)   given rational vector $a \geqslant 0$ and rational β, does $ax = \beta$ have a $\{0, 1\}$-
        solution x?

Corollary 18.1e. *Problem* (13) *is* $\mathcal{N}\mathcal{P}$-*complete.*

Proof. (11) can be reduced to (13). Suppose we wish to find a $\{0, 1\}$-vector x
satisfying $ax = \beta$. By replacing component ξ_i of x by $(1 - \xi_i)$ whenever its
coefficient in ax is negative, we obtain an equivalent instance of problem (13).
    □

[For an approach to problem (13) with the basis reduction method, see Lagarias
and Odlyzko [1983].]

In fact we can take β equal to half of the sum of the components of a:

(14)   given rationals $\alpha_1, \ldots, \alpha_n \in \mathbb{Q}_+$, is there a subset S of $\{1, \ldots, n\}$ with
        $\sum_{i \in S} \alpha_i = \frac{1}{2}(\alpha_1 + \cdots + \alpha_n)$?

Corollary 18.1f. *Problem* (14) *is* $\mathcal{N}\mathcal{P}$-*complete.*

Proof. (13) can be reduced to (14). Let $a = (\alpha_1, \ldots, \alpha_n) \geqslant 0$ and $\beta \in \mathbb{Q}$ as in (13)
be given. We may assume that $\beta \leqslant \frac{1}{2}(\alpha_1 + \cdots + \alpha_n)$ (otherwise, replace β by
$(\alpha_1 + \cdots + \alpha_n) - \beta$, and any solution x by $1 - x$). Let $\alpha_0 := (\alpha_1 + \cdots + \alpha_n) - 2\beta$.
Then

(15) $\exists \{0, 1\}$-vector x with $ax = \beta \Leftrightarrow \exists S \subseteq \{0, \ldots, n\}$ with $\sum_{i \in S} \alpha_i = \frac{1}{2} \sum_{i=0}^{n} \alpha_i$.

To see $\Rightarrow$, let $x = (\xi_1, \ldots, \xi_n)^{\mathsf{T}}$ be a $\{0, 1\}$-vector x with $ax = \beta$. Let $S := \{0\} \cup$
$\{i | \xi_i = 1\}$. Then $\sum_{i \in S} \alpha_i = \alpha_0 + ax = \alpha_0 + \beta = \frac{1}{2}(\alpha_0 + \cdots + \alpha_n)$.

To see $\Leftarrow$, let $S \subseteq \{0, \ldots, n\}$ with $\sum_{i \in S} \alpha_i = \frac{1}{2} \sum_{i=0}^{n} \alpha_i$. We may assume $0 \in S$
(otherwise replace S by $\{0, \ldots, n\} \setminus S$). Let $x = (\xi_1, \ldots, \xi_n)^{\mathsf{T}}$ be the $\{0, 1\}$-vector with
$\xi_i = 1$ if and only if $i \in S$. Then

(16) $ax = \sum_{i \in S \setminus \{0\}} \alpha_i = -\alpha_0 + \sum_{i \in S} \alpha_i = -\alpha_0 + \frac{1}{2} \sum_{i=0}^{n} \alpha_i = -\frac{1}{2}\alpha_0 + \frac{1}{2} \sum_{i=1}^{n} \alpha_i = \beta.$

This shows (15), which reduces problem (13) to problem (14).   □

Also the following problem is $\mathcal{NP}$-complete (Lueker [1975]):

(17)   given rational vector a and rational β, does $ax = \beta$ have a nonnegative integral solution x?

Corollary 18.1g. *Problem* (17) *is* $\mathcal{NP}$-*complete.*

Proof. (13) can be reduced to (17). Suppose we wish to find a $\{0,1\}$-vector $x = (\xi_1, \ldots, \xi_n)^\mathsf{T}$ such that

(18) $\alpha_1 \xi_1 + \cdots + \alpha_n \xi_n = \beta$

where $\alpha_1, \ldots, \alpha_n, \beta$ are positive natural numbers (any instance of (13) can be easily reduced to this). We may assume $\beta \leqslant \alpha_1 + \cdots + \alpha_n$. Let σ be the smallest natural number such that

(19) $\Sigma := \alpha_1 + \cdots + \alpha_n < 2^\sigma$.

Consider the equation

$$
\begin{aligned}
(20) \quad & (2^0 + 2^{3n\sigma}\alpha_1)\xi_1 + (2^{3\sigma} + 2^{3n\sigma}\alpha_2)\xi_2 + \cdots + (2^{3(i-1)\sigma} + 2^{3n\sigma}\alpha_i)\xi_i + \cdots \\
& + (2^{3(n-1)\sigma} + 2^{3n\sigma}\alpha_n)\xi_n + (2^0 + 2^{(3n+1)\sigma}\alpha_1)\eta_1 + (2^{3\sigma} + 2^{(3n+1)\sigma}\alpha_2)\eta_2 + \cdots \\
& + (2^{3(i-1)\sigma} + 2^{(3n+1)\sigma}\alpha_i)\eta_i + \cdots + (2^{3(n-1)\sigma} + 2^{(3n+1)\sigma}\alpha_n)\eta_n \\
& = 2^0 + 2^{3\sigma} + \cdots + 2^{3(i-1)\sigma} + \cdots + 2^{3(n-1)\sigma} + 2^{3n\sigma}\beta + 2^{(3n+1)\sigma}(\Sigma - \beta).
\end{aligned}
$$

We show:

(21)   equation (18) has a $\{0,1\}$-solution $(\xi_1, \ldots, \xi_n)^\mathsf{T} \Leftrightarrow$ equation (20) has a nonnegative integral solution $(\xi_1, \ldots, \xi_n, \eta_1, \ldots, \eta_n)^\mathsf{T}$.

As 2^σ has size polynomially bounded by the size of (18), this reduces (13) to (17).

To see $\Rightarrow$ in (21), one easily checks that if $(\xi_1, \ldots, \xi_n)^\mathsf{T}$ is a $\{0,1\}$-solution of (18), then $(\xi_1, \ldots, \xi_n, 1 - \xi_1, \ldots, 1 - \xi_n)^\mathsf{T}$ is a nonnegative integral solution of (20).

To see $\Leftarrow$, let $(\xi_1, \ldots, \xi_n, \eta_1, \ldots, \eta_n)^\mathsf{T}$ be a nonnegative integral solution of (20). We first show $\xi_1 + \eta_1 = 1$. Indeed, reducing (20) modulo $2^{3\sigma}$ gives:

(22) $\xi_1 + \eta_1 \equiv 1 \pmod{2^{3\sigma}}$.

So if $\xi_1 + \eta_1 \neq 1$ then $\xi_1 + \eta_1 > 2^{3\sigma}$. Then we have for the LHS and the RHS of (20):

(23) $\text{LHS} \geqslant 2^{3n\sigma}\alpha_1\xi_1 + 2^{(3n+1)\sigma}\alpha_1\eta_1 \geqslant 2^{3n\sigma}\alpha_1(\xi_1 + \eta_1)$

$> 2^{3n\sigma}\alpha_1 2^{3\sigma} \geqslant 2^{(3n+3)\sigma} \geqslant \text{RHS}$

which contradicts equality of LHS and RHS.

If we have shown $\xi_1 + \eta_1 = \cdots = \xi_{i-1} + \eta_{i-1} = 1$, we can derive similarly that $\xi_i + \eta_i = 1$, now reducing modulo $2^{3i\sigma}$:

(24) $2^{3(i-1)\sigma}(\xi_i + \eta_i) \equiv 2^{3(i-1)\sigma} \pmod{2^{3i\sigma}}$.

If $\xi_i + \eta_i \neq 1$ then $\xi_i + \eta_i > 2^{3\sigma}$, and we derive the same contradiction as in (23) (now with subscripts 1 replaced by i).

So, by induction, $(\xi_1, \ldots, \xi_n, \eta_1, \ldots, \eta_n)^{\mathsf{T}}$ is a $\{0,1\}$-solution of (20) with $\xi_1 + \eta_1 = \cdots = \xi_n + \eta_n = 1$. Hence,

$$(25) \qquad 2^{3n\sigma}(\alpha_1 \xi_1 + \cdots + \alpha_n \xi_n) + 2^{(3n+1)\sigma}(\alpha_1 \eta_1 + \cdots + \alpha_n \eta_n)$$
$$= 2^{3n\sigma}\beta + 2^{(3n+1)\sigma}(\Sigma - \beta).$$

Replacing η_i by $1 - \xi_i$ gives

$$(26) \qquad 2^{3n\sigma}(1 - 2^{\sigma})(\alpha_1 \xi_1 + \cdots + \alpha_n \xi_n) = 2^{3n\sigma}(1 - 2^{\sigma})\beta$$

and hence $(\xi_1, \ldots, \xi_n)^{\mathsf{T}}$ is a $\{0, 1\}$-solution of (18). $\qquad\square$

Lagarias [1982] showed that already the following problem is $\mathcal{NP}$-complete: decide if a given system of rational linear inequalities, where each inequality has at most two nonzero coefficients, has an integral solution.

18.3. COMPLEXITY OF FACETS, VERTICES, AND ADJACENCY ON THE INTEGER HULL

The polyhedron P_{I} can be quite intractable compared with the polyhedron P. First of all, the number of facets of P_{I} can be exponentially large relative to the size of the system defining P. This is shown by a class of examples studied by Edmonds [1965c].

Theorem 18.2. *There is no polynomial ϕ such that for each rational polyhedron $P = \{x \mid Ax \leqslant b\}$, the integer hull P_{I} of P has at most $\phi(\mathrm{size}(A, b))$ facets.*

Proof. Choose a natural number $n \geqslant 4$. Let A be the $V \times E$-incidence matrix of the complete undirected graph $K_n = (V, E)$. So A is the $n \times \binom{n}{2}$-matrix, with as its columns all possible $\{0, 1\}$-vectors with exactly two 1's. Let $P := \{x \geqslant 0 \mid Ax \leqslant 1\}$, where $\mathbf{1}$ denotes an all-one vector. Then P_{I} (which is the matching polytope of K_n) has at least $\binom{n}{2} + 2^{n-1}$ facets, as according to Theorem 8.8, each of the following inequalities determines a facet of P_{I}:

$$(27) \qquad \text{(i)} \ x(e) \geqslant 0 \qquad\qquad (e \in E)$$
$$\text{(ii)} \ \sum_{e \ni v} x(e) \leqslant 1 \qquad (v \in V)$$
$$\text{(iii)} \ \sum_{e \subseteq U} x(e) \leqslant \lfloor \tfrac{1}{2} |U| \rfloor \qquad (U \subseteq V, |U| \text{ odd}, |U| \geqslant 3).$$

As $\binom{n}{2} + 2^{n-1}$ cannot be bounded by a polynomial in $\mathrm{size}(A, b) = O(n^3)$, the theorem follows. $\qquad\square$

In fact, Edmonds showed that (27) gives all facets of P_{I} for the polyhedra P mentioned—see Corollary 8.7a. This implies that the given class of polyhedra is not too bad: although there are exponentially many facets, we have at least an explicit description of them. Given an inequality, we can quickly test whether it determines a facet of such a P_1. Moreover, as was shown by Padberg and Rao [1982], the system (27) can be tested in polynomial time, i.e. the separation problem for these P_1 is polynomially solvable.

For general polyhedra, however, the situation seems to be worse, as follows from studies by Karp and Papadimitriou [1980], Papadimitriou [1978], and Papadimitriou and Yannakakis [1982]. Karp and Papadimitriou showed that, in general, P_1 has 'difficult' facets, under the generally accepted assumption that $\mathcal{NP} \neq$ co-$\mathcal{NP}$. More precisely, there is the following theorem.

Theorem 18.3. *The following are equivalent:*

(i) $\mathcal{NP} \neq$ co-$\mathcal{NP}$;

(ii) *there exists no polynomial ϕ such that for each rational matrix A and each rational vector b, and for each inequality $cx \leqslant \delta$ defining a facet of the integer hull P_1 (where $P := \{x \mid Ax \leqslant b\}$ and $P_1 \neq \emptyset$), the fact that $cx \leqslant \delta$ is valid for P_1 has a proof of length at most $\phi(\text{size}(A, b, c, \delta))$.*

(ii) can be described more formally as follows: there exists no set $\mathcal{L}$ in $\mathcal{NP}$ such that:

(28) $\{(A, b, c, \delta) \mid cx \leqslant \delta$ determines a facet of $\{x \mid Ax \leqslant b\}_1\} \subseteq \mathcal{L}$

          $\subseteq \{(A, b, c, \delta) \mid cx \leqslant \delta$ is valid for $\{x \mid Ax \leqslant b\}_1\} =: \mathcal{N}.$

Proof. We prove the theorem by considering the problem:

(29)   given matrix A, column vector b, row vector c, and rational δ, is there an integral vector x such that $Ax \leqslant b$, $cx > \delta$ (knowing that $Ax \leqslant b$ has an integral solution)?

A negative answer to (29) has a polynomial-length proof if and only if (29) belongs to co-$\mathcal{NP}$, that is (as (29) is $\mathcal{NP}$-complete), if and only if $\mathcal{NP} =$ co-$\mathcal{NP}$.

So $\mathcal{NP} =$ co-$\mathcal{NP}$ if and only if problem (29) belongs to co-$\mathcal{NP}$, i.e. if and only if the set $\mathcal{N}$ in (28) belongs to $\mathcal{NP}$.

We first show (i) $\Rightarrow$ (ii). Suppose there exists a polynomial ϕ as in (ii). Assume (29) has a negative answer, that is, if x is an integral vector with $Ax \leqslant b$ then $cx \leqslant \delta$. Let $P := \{x \mid Ax \leqslant b\}$. Then there are facets $c_1x \leqslant \delta_1, \ldots, c_tx \leqslant \delta_t$ of P_1 and nonnegative numbers $\lambda_1, \ldots, \lambda_t$ such that:

(30) $\lambda_1 c_1 + \cdots + \lambda_t c_t = c,$

          $\lambda_1 \delta_1 + \cdots + \lambda_t \delta_t \leqslant \delta.$

We can take $t \leqslant n$ (n being the number of columns of A), and by the estimates made in Section 17.1, we can take $c_1, \ldots, c_t, \delta_1, \ldots, \delta_t, \lambda_1, \ldots, \lambda_t$ with size polynomially bounded by the sizes of A, b, and c. By the assumption in (ii), the fact that $c_ix \leqslant \delta_i$ is a valid inequality for P_1 has a polynomial-length proof. So also the fact that $cx \leqslant \delta$ holds for P_1 has a polynomial-length proof, as

(31) $\qquad cx = (\lambda_1 c_1 + \cdots + \lambda_t c_t)x \leqslant \lambda_1 \delta_1 + \cdots + \lambda_t \delta_t \leqslant \delta.$

So a negative answer to (29) has a proof of polynomial length.

More formally, if $\mathscr{L}$ satisfies (28), then define

(32) $\qquad \mathscr{M} := \{(A, b, c, \delta, c_1, \ldots, c_n, \delta_1, \ldots, \delta_n, \lambda_1, \ldots, \lambda_n) | \lambda_1, \ldots, \lambda_n \geqslant 0;$
$\qquad\qquad \lambda_1 c_1 + \cdots + \lambda_n c_n = c; \lambda_1 \delta_1 + \cdots + \lambda_n \delta_n \leqslant \delta; (A, b, c_i, \delta_i) \in \mathscr{L} \text{ for } i = 1, \ldots, n\}.$

As $\mathscr{L}$ is in $\mathscr{N}\mathscr{P}$, also $\mathscr{M}$ is in $\mathscr{N}\mathscr{P}$ (as membership of $\mathscr{M}$ is easily reduced to membership of $\mathscr{L}$). Now $\mathscr{N}$ in (28) satisfies:

(33) $\qquad (A, b, c, \delta) \in \mathscr{N} \Leftrightarrow$ there is a string y such that $(A, b, c, \delta, y) \in \mathscr{M}$.

This implies that $\mathscr{N}$ belongs to $\mathscr{N}\mathscr{P}$, i.e. that problem (29) belongs to co-$\mathscr{N}\mathscr{P}$, which implies $\mathscr{N}\mathscr{P} = \text{co-}\mathscr{N}\mathscr{P}$.

To see (ii)$\Rightarrow$(i), assume $\mathscr{N}\mathscr{P} = \text{co-}\mathscr{N}\mathscr{P}$. Then, in particular, problem (29) belongs to co-$\mathscr{N}\mathscr{P}$. That means that a negative answer to (29) has a polynomial-length proof. This is the negation of (ii).

More formally, if $\mathscr{N}\mathscr{P} = \text{co-}\mathscr{N}\mathscr{P}$, then the set $\mathscr{N}$ in (28) is in $\mathscr{N}\mathscr{P}$, and hence we can take $\mathscr{L} = \mathscr{N}$. $\qquad\qquad\qquad\qquad\qquad\qquad\qquad\qquad\qquad\qquad\qquad\qquad\square$

So if we believe that $\mathscr{N}\mathscr{P} \neq \text{co-}\mathscr{N}\mathscr{P}$, we may not hope for a 'concise' description of P_I in terms of linear inequalities (like (27)), for general polyhedra P. Note that Edmonds' theorem mentioned above implies that such a description does exist if $P = \{x | x \geqslant 0; Ax \leqslant 1\}$ where A is a $\{0, 1\}$-matrix with in each column exactly two 1's. By Edmonds' description (27) of the matching polytope, (ii) of Theorem 18.3 does not hold if we restrict P_I to matching polytopes.

It follows from the proof of Theorem 18.3 that we may restrict the polyhedra $P = \{x | Ax \leqslant b\}$ to polyhedra corresponding with any $\mathscr{N}\mathscr{P}$-complete problem, like the knapsack problem (13) or (17). That is, if $\mathscr{N}\mathscr{P} \neq \text{co-}\mathscr{N}\mathscr{P}$, then for $a \in \mathbb{Q}^n_+$, $\beta \in \mathbb{Q}_+$ also the following polyhedra have 'difficult' facets, in the sense of Theorem 18.3:

(34) $\qquad \{x | x \geqslant 0; ax \leqslant \beta\}_I$
$\qquad\quad \{x | 0 \leqslant x \leqslant 1; ax \leqslant \beta\}_I$.

The second polytope is called a *knapsack polytope*, and was studied by Balas [1975a] and Wolsey [1975].

Stronger than Theorem 18.3 is the following result of Papadimitriou and Yannakakis [1982].

Theorem 18.4. *The following problem is $\mathscr{N}\mathscr{P}$-complete:*

(35) $\qquad$ given a rational system $Ax \leqslant b$ of linear inequalities and a column vector y, does y belong to P_I (where $P := \{x | Ax \leqslant b\}$)?

Proof. First, problem (35) belongs to the class $\mathscr{N}\mathscr{P}$. A positive answer can be proved by specifying integral vectors $x_1, \ldots, x_n$ satisfying $Ax \leqslant b$, and nonnegative numbers $\lambda_1, \ldots, \lambda_n$ such that $\lambda_1 + \cdots + \lambda_n = 1$ and $\lambda_1 x_1 + \cdots + \lambda_n x_n = y$. By Section 17.1, $x_1, \ldots, x_n, \lambda_1, \ldots, \lambda_n$ can be taken with size polynomially bounded by the sizes of A, b, and y.

Next, to show that (35) is $\mathcal{NP}$-complete, we show that the $\mathcal{NP}$-complete problem (14) is Karp-reducible to (35). Let $\alpha_1, \ldots, \alpha_n \in \mathbb{Q}_+$. Let P be the polyhedron defined by the system:

$$(36) \qquad \alpha_1 \xi_1 + \cdots + \alpha_n \xi_n = \tfrac{1}{2}(\alpha_1 + \cdots + \alpha_n),$$
$$0 \leqslant \xi_i \leqslant 1 \qquad\qquad (i = 1, \ldots, n).$$

Let $y := (\tfrac{1}{2}, \ldots, \tfrac{1}{2})^\mathsf{T}$. Then y belongs to P_1 if and only if (36) has an integral solution, i.e. if and only if the answer to problem (14) is positive. This directly follows from the fact that if $(\xi_1, \ldots, \xi_n)^\mathsf{T}$ is an integral solution to (36), then also $(1 - \xi_1, \ldots, 1 - \xi_n)^\mathsf{T}$ is an integral solution to (36), and hence $y = \tfrac{1}{2}(\xi_1, \ldots, \xi_n)^\mathsf{T} + \tfrac{1}{2}(1 - \xi_1, \ldots, 1 - \xi_n)^\mathsf{T}$ is in P_1. $\square$

Papadimitriou and Yannakakis gave a different proof, showing that testing membership of the 'traveling salesman polytope' is $\mathcal{NP}$-complete.

Theorem 18.4 indeed generalizes Theorem 18.3: if (ii) of Theorem 18.3 holds, then any negative answer to (35) has a polynomial-length proof, consisting of an inequality, defining a facet of P_1, violated by y, together with a polynomial-length proof that the inequality is valid for P_1. Hence (35) belongs to co-$\mathcal{NP}$, which implies with Theorem 18.4 that $\mathcal{NP} = \text{co-}\mathcal{NP}$.

Also testing a vector for being not a vertex of P_1 is $\mathcal{NP}$-complete.

Theorem 18.5. *The following problem is $\mathcal{NP}$-complete:*

(37) given a rational system $Ax \leqslant b$ of linear inequalities and an integral solution y of $Ax \leqslant b$, is y not a vertex of P_1 (where $P := \{x \mid Ax \leqslant b\}$)?

Proof. First, (37) belongs to $\mathcal{NP}$, since if y is not a vertex of P_1, there exist linearly independent integral solutions $x_1, \ldots, x_n$ of $Ax \leqslant b$ such that $y = \lambda_1 x_1 + \cdots + \lambda_n x_n$ for some $\lambda_1, \ldots, \lambda_n > 0$ with $\lambda_1 + \cdots + \lambda_n = 1$ and $n \geqslant 2$. These $x_1, \ldots, x_n$ can be chosen with size polynomially bounded by the sizes of A, b and y.

Next, to show that (37) is $\mathcal{NP}$-complete, we show that the $\mathcal{NP}$-complete problem (13) is Karp-reducible to (37). Let $\alpha_1, \ldots, \alpha_n, \beta \in \mathbb{Q}_+$ be a given input for (13). Let $a := (\alpha_1, \ldots, \alpha_n)$. Our task is to decide whether there exists a $\{0, 1\}$-vector x with $ax = \beta$. Consider the polyhedron $P \subseteq \mathbb{R}^{n+1}$ consisting of all vectors $(\xi_0, \xi_1, \ldots, \xi_n)^\mathsf{T}$ satisfying:

$$(38) \qquad \beta \xi_0 - \alpha_1 \xi_1 - \cdots - \alpha_n \xi_n = 0,$$
$$-1 \leqslant \xi_0 \leqslant +1,$$
$$-1 \leqslant 2\xi_i - \xi_0 \leqslant +1 \qquad (i = 1, \ldots, n).$$

Suppose $(\xi_0, \ldots, \xi_n)^\mathsf{T}$ is an integral vector in P. If $\xi_0 = 0$ then $\xi_1 = \cdots = \xi_n = 0$. If $\xi_0 = 1$ then $\xi_1, \ldots, \xi_n \in \{0, 1\}$. If $\xi_0 = -1$ then $\xi_1, \ldots, \xi_n \in \{0, -1\}$. Hence:

(39) $ax = \beta$ has a $\{0, 1\}$-solution $\Leftrightarrow (0, 0, \ldots, 0)^\mathsf{T}$ is not the only integral solution of (38) $\Leftrightarrow (0, 0, \ldots, 0)^\mathsf{T}$ is not a vertex of P_1.

The last equivalence follows from the fact that if x is an integral solution of (38), then also $-x$ is an integral solution of (38). $\square$

Moreover, testing non-adjacency of vertices of P_1 turns out to be $\mathcal{NP}$-complete, as was shown by Papadimitriou [1978].

Theorem 18.6. *The following problem is $\mathcal{NP}$-complete:*

(40)   given a rational system $Ax \leqslant b$ of linear inequalities, with $P := \{x \mid Ax \leqslant b\} \subseteq [0,1]^n$, and given integral solutions x_1 and x_2 of $Ax \leqslant b$, are x_1 and x_2 non-adjacent vertices of P_1?

Proof. First, problem (40) belongs to $\mathcal{NP}$, as x_1 and x_2 are non-adjacent vertices if and only if $\frac{1}{2}(x_1 + x_2)$ is a proper convex combination of x_1, x_2 and other integer solutions of $Ax \leqslant b$.

Next, to show that (40) is $\mathcal{NP}$-complete, we show that the $\mathcal{NP}$-complete problem (14) is Karp-reducible to (40). Let $\alpha_1, \ldots, \alpha_n \in \mathbb{Q}_+$ be an input for problem (14). Without loss of generality, $\alpha_1, \ldots, \alpha_n > 0$. Define

(41) $\alpha_{n+1} := \alpha_{n+2} := \frac{1}{2}(\alpha_1 + \cdots + \alpha_n)$.

Define the polyhedron $P \subseteq [0,1]^{n+2}$ by:

(42) $\alpha_1 \xi_1 + \cdots + \alpha_n \xi_n + \alpha_{n+1} \xi_{n+1} + \alpha_{n+2} \xi_{n+2} = (\alpha_1 + \cdots + \alpha_n)$,

          $0 \leqslant \xi_i \leqslant 1$ $(i = 1, \ldots, n+2)$.

Then $x_1 := (0, \ldots, 0, 1, 1)^T$ and $x_2 := (1, \ldots, 1, 0, 0)^T$ belong to P, and are hence vertices of P_1. Any other integral solution for (42) has $\xi_{n+1} = 1$ and $\xi_{n+2} = 0$, or $\xi_{n+1} = 0$ and $\xi_{n+2} = 1$. Moreover, $\alpha_1 \xi_1 + \cdots + \alpha_n \xi_n = \frac{1}{2}(\alpha_1 + \cdots + \alpha_n)$, i.e. it gives a solution for (14). Now if $(\xi_1, \ldots, \xi_{n+2})^T$ is an integral solution for (42), then also $(1 - \xi_1, \ldots, 1 - \xi_{n+2})^T$ is one. Therefore we have:

(43) x_1 and x_2 are nonadjacent in $P_1 \Leftrightarrow x_1$ and x_2 are not the only integral solutions of $(42) \Leftrightarrow \alpha_1 \xi_1 + \cdots + \alpha_n \xi_n = \frac{1}{2}(\alpha_1 + \cdots + \alpha_n)$ has a $\{0,1\}$-solution.

This reduces problem (14) to (40). $\square$

This proof differs from that of Papadimitriou, which in fact shows that testing non-adjacency on the 'traveling salesman polytope' is $\mathcal{NP}$-complete.

The $\mathcal{NP}$-completeness of testing non-adjacency appears not to go parallel with each $\mathcal{NP}$-complete problem. Chvátal [1975a] showed that there is a polynomial-time test for adjacency of vertices of the polytope P_1 for $P := \{x \geqslant 0 \mid x \leqslant 1; Ax \leqslant 1\}$, where A is a $\{0,1\}$-matrix. Note that P_1 is the 'clique polytope' of a certain graph, and that optimizing over P_1 (i.e. finding a maximum clique) is $\mathcal{NP}$-complete.

Remark 18.1. Rubin [1970] (cf. Jeroslow [1969], Jeroslow and Kortanek [1971], Halfin [1972]) showed that if φ_k denotes that kth Fibonacci number (defined recursively by:

$\varphi_1 = \varphi_2 = 1$, $\varphi_k = \varphi_{k-1} + \varphi_{k-2}$ $(k \geqslant 3)$, and the polytope P_k in $\mathbb{R}^2$ is defined by the inequalities

$$(44) \qquad \varphi_{2k}\xi + \varphi_{2k+1}\eta \leqslant \varphi_{2k+1}^2 - 1, \xi \geqslant 0, \eta \geqslant 0$$

then the convex hull $(P_k)_I$ of the integral vectors in P_k has $k + 3$ vertices and $k + 3$ facets. So we cannot bound the number of constraints defining P_I in terms of the dimension and the number of constraints defining P. (Jeroslow [1971] gave a similar example in two dimensions with only two defining constraints.)

Hayes and Larman [1983] showed that if $P \subseteq \mathbb{R}^n$ is defined by

$$(45) \qquad x \geqslant 0, ax \leqslant \beta,$$

where $a \in \mathbb{Z}^n$, $a > 0$, $\beta \in \mathbb{Z}$, $\beta > 0$, then the number of vertices of P_I is at most $(\log_2(2 + 2\beta/\alpha))^n$, where α is the smallest component of a.

[*Proof.* P_I has no two vertices $x = (\xi_1, \ldots, \xi_n)^T$ and $y = (\eta_1, \ldots, \eta_n)^T$ with: $\lfloor \log_2(\xi_1 + 1) \rfloor = \lfloor \log_2(\eta_1 + 1) \rfloor, \ldots, \lfloor \log_2(\xi_n + 1) \rfloor = \lfloor \log_2(\eta_n + 1) \rfloor$. [Otherwise $2x - y \geqslant 0$, $2y - x \geqslant 0$ as one easily checks. Moreover, $a(2x - y) + a(2y - x) = ax + ay \leqslant 2\beta$. Hence $a(2x - y) \leqslant \beta$ or $a(2y - x) \leqslant \beta$. Therefore, $2x - y \in P$ or $y - 2x \in P$. This contradicts the fact that both x and y are vertices of P_I.]

So the number of vertices of P_I is at most $(1 + \log_2(M + 1))^n$, where M is the largest possible component of any vertex of P_I. Since $M \leqslant \beta/\alpha$, the upper bound follows.]

Extension of these arguments yields that if we fix m (but not the dimension), for any polyhedron P defined by m linear inequalities, the number of minimal faces of P_I is polynomially bounded by the size of the system defining P.

18.4. LENSTRA'S ALGORITHM FOR INTEGER LINEAR PROGRAMMING

Lenstra [1983] showed that for each fixed natural number n, there is a polynomial algorithm for integer linear programming problems in n variables. Equivalently, for each fixed natural number n, there is a polynomial algorithm solving systems of linear inequalities in n integer variables.

Lenstra's result extends earlier results of Scarf [1981a, b] for the case of two variables (cf. Hirschberg and Wong [1976] and Kannan [1980]).

We have met the two basic ingredients of Lenstra's method before: the basis reduction method (Section 6.2) and the rounding method for polyhedra (Section 15.4). In fact these methods were developed, by Lovász, Goffin, and Grötschel, Lovász, and Schrijver [1984, 1988], in consequence of Lenstra's work. Lovász noticed that these two methods imply the following theorem, from which Lenstra's result can be derived.

Theorem 18.7. *There exists a polynomial algorithm which finds, for any system $Ax \leqslant b$ of rational linear inequalities, either an integral vector y satisfying $Ay \leqslant b$, or a nonzero integral vector c such that*

$$(46) \qquad \max\{cx \mid Ax \leqslant b\} - \min\{cx \mid Ax \leqslant b\} \leqslant 2n(n + 1)2^{n(n-1)/4},$$

where n is the number of columns of A.

Proof. We describe the algorithm. Let $Ax \le b$ be given, where A has n columns, and define $P := \{x \mid Ax \le b\}$. With the ellipsoid method we decide if P is full-dimensional and if P is bounded. Thus we are in one of the following three cases.

Case 1. P is not full-dimensional. With the ellipsoid method we can find a nonzero vector c and a rational δ such that $P \subseteq \{x \mid cx = \delta\}$, which yields an answer as required.

Case 2. P is bounded and full-dimensional. We can find in polynomial time, with the 'rounding algorithm' of Theorem 15.6 (taking $\gamma = 1/(n+1)$), a vector $z \in \mathbb{Q}^n$ and a positive definite matrix D such that

$$(47) \qquad \text{ell}\left(z, \frac{1}{(n+1)^2} D \right) \subseteq P \subseteq \text{ell}(z, D).$$

Next, with the basis reduction method (Theorem 6.4) we can find a basis $b_1, \ldots, b_n$ for the lattice $\mathbb{Z}^n$ such that

$$(48) \qquad \| b_1 \| \cdot \ldots \cdot \| b_n \| \le 2^{n(n-1)/4} \sqrt{\det D^{-1}}$$

where for $x \in \mathbb{R}^n$,

$$(49) \qquad \| x \| := \sqrt{x^{\mathsf{T}} D^{-1} x}.$$

Without loss of generality $\| b_1 \| \le \cdots \le \| b_n \|$. Write $z = \lambda_1 b_1 + \cdots + \lambda_n b_n$, and let $y := \lfloor \lambda_1 \rfloor b_1 + \cdots + \lfloor \lambda_n \rfloor b_n$.

If y satisfies $Ay \le b$ we are done. If not, by (47), $y \notin \text{ell}(z, (n+1)^{-2} D)$. Hence

$$(50) \qquad \frac{1}{(n+1)^2} < (y-z)^{\mathsf{T}} D^{-1} (y-z)$$

$$= \| y - z \|^2 = \| (\lambda_1 - \lfloor \lambda_1 \rfloor) b_1 + \cdots + (\lambda_n - \lfloor \lambda_n \rfloor) b_n \|^2 .$$

Therefore,

$$(51) \qquad \frac{1}{n+1} < \| (\lambda_1 - \lfloor \lambda_1 \rfloor) b_1 + \cdots + (\lambda_n - \lfloor \lambda_n \rfloor) b_n \| \le$$

$$(\lambda_1 - \lfloor \lambda_1 \rfloor) \| b_1 \| + \cdots + (\lambda_n - \lfloor \lambda_n \rfloor) \| b_n \|.$$

So we know

$$(52) \qquad \| b_n \| \ge \frac{1}{n(n+1)}.$$

Now let c be a nonzero vector with $cb_1 = \cdots = cb_{n-1} = 0$, such that the components of c are relatively prime integers. We show that c is a valid output, i.e. c satisfies (46). Let B_i be the matrix with columns $b_1, \ldots, b_i$. Then (cf. Section 1.3)

$$(53) \qquad (cDc^{\mathsf{T}}) \det (B_n^{\mathsf{T}} D^{-1} B_n) = (cb_n)^2 \det (B_{n-1}^{\mathsf{T}} D^{-1} B_{n-1}).$$

Since $\mathbb{Z}^n$ is the lattice generated by $b_1, \ldots, b_n$ we know that $cb_n = \pm 1$, and that $\det B_n = \pm 1$. So (53) gives:

$$(54) \qquad (cDc^\mathsf{T}) \det D^{-1} = \det(B_{n-1}^\mathsf{T} D^{-1} B_{n-1}).$$

Moreover, by Hadamard's inequality (cf. Section 1.3):

$$(55) \qquad \det(B_{n-1}^\mathsf{T} D^{-1} B_{n-1}) \leqslant (b_1^\mathsf{T} D^{-1} b_1) \cdot \ldots \cdot (b_{n-1}^\mathsf{T} D^{-1} b_{n-1})$$
$$= \|b_1\|^2 \cdot \ldots \cdot \|b_{n-1}\|^2.$$

Therefore, if $x^\mathsf{T} D^{-1} x \leqslant 1$, we have

$$(56) \qquad |cx| \leqslant \sqrt{cDc^\mathsf{T}} \cdot \sqrt{x^\mathsf{T} D^{-1} x} \leqslant \sqrt{cDc^\mathsf{T}} = \sqrt{\det D} \cdot \sqrt{\det(B_{n-1}^\mathsf{T} D^{-1} B_{n-1})}$$
$$\leqslant \sqrt{\det D} \cdot \|b_1\| \cdot \ldots \cdot \|b_{n-1}\| \leqslant 2^{n(n-1)/4} \|b_n\|^{-1} \leqslant n(n+1) 2^{n(n-1)/4}$$

(using the Cauchy–Schwarz inequality (cf. Section 1.3), and (48), (52), (54), (55)). So if $y \in \text{ell}(z, D)$ then $|c(y - z)| \leqslant n(n+1) 2^{n(n-1)/4}$, implying

$$(57) \qquad \max\{cx \mid x \in P\} - \min\{cx \mid x \in P\}$$
$$\leqslant \max\{cx \mid x \in \text{ell}(z, D)\} - \min\{cx \mid x \in \text{ell}(z, D)\} \leqslant 2n(n+1) 2^{n(n-1)/4}$$

as required.

Case 3. P is unbounded. Let φ be the maximum row size of the matrix $[A \; b]$. Define

$$(58) \qquad Q := P \cap \{x = (\xi_1, \ldots, \xi_n)^\mathsf{T} \mid |\xi_i| \leqslant 2^{13n^2\varphi} \quad \text{for} \quad i = 1, \ldots, n\}.$$

Then Q is bounded, and Case 1 or Case 2 above applies to Q. If we find an integral vector y in Q, then y is in P and we are done. Suppose we find a nonzero integral vector c such that

$$(59) \qquad \max\{cx \mid x \in Q\} - \min\{cx \mid x \in Q\} \leqslant 2n(n+1) 2^{n(n-1)/4}.$$

We show that the same holds for P, i.e. that (46) holds. By Theorem 10.2,

$$(60) \qquad P = \text{conv.hull}\{v_1, \ldots, v_k\} + \text{cone}\{e_1, \ldots, e_l\}$$

where each of $v_1, \ldots, v_k, e_1, \ldots, e_l$ has size at most $4n^2\varphi$. In particular, $v_1, \ldots, v_k$ belong to Q. So if

$$(61) \qquad \max\{cx \mid x \in P\} - \min\{cx \mid x \in P\} > \max\{cx \mid x \in Q\} - \min\{cx \mid x \in Q\}$$

then $ce_j \neq 0$ for some j. As e_j has size at most $4n^2\varphi$ and c is integral, we know that $|ce_j| \geqslant 2^{-4n^2\varphi}$. Moreover, $z := v_1 + 2^{8n^2\varphi} e_j$ belongs to Q, as all components of v_1 and of e_j are at most $2^{4n^2\varphi}$ in absolute value, and hence all components of z are at most $2^{13n^2\varphi}$ in absolute value. However,

$$(62) \qquad \max\{cx \mid x \in Q\} - \min\{cx \mid x \in Q\} \geqslant |cz - cv_1| = 2^{8n^2\varphi} |ce_j| \geqslant 2^{4n^2\varphi}$$
$$> 2n(n+1) 2^{n(n-1)/4}$$

contradicting (59). So (61) is not true, and (59) implies (46). $\qquad \square$

Corollary 18.7a (Lenstra's algorithm). *For each fixed natural number n, there exists a polynomial algorithm which finds an integral solution for a given rational system $Ax \leqslant b$, in n variables, or decides that no such solution exists.*

Proof. The proof is by induction on n, the case $n = 1$ being easy. If $Ax \leqslant b$ is given, where A has n columns, apply the algorithm of Theorem 18.7. If it gives an integral solution for $Ax \leqslant b$ we are done. Suppose it gives a nonzero integral vector c with

$$(63) \qquad \max\{cx \,|\, x \in P\} - \min\{cx \,|\, x \in P\} \leqslant 2n(n+1)2^{n(n-1)/4}$$

where $P := \{x \,|\, Ax \leqslant b\}$. Without loss of generality, the components of c are relatively prime integers. Determine $\mu := \min\{cx \,|\, x \in P\}$, and consider the polyhedra

$$(64) \qquad P_t := \{x \in P \,|\, cx = t\},$$

for $t = \lceil \mu \rceil, \ldots, \lceil \mu + 2n(n+1)2^{n(n-1)/4} \rceil$. Then each integral solution of $Ax \leqslant b$ is in one of these P_t. So testing each of the P_t for containing integral vectors suffices to solve $Ax \leqslant b$ in integers.

Let U be an $(n-1) \times n$-matrix such that the matrix $\begin{bmatrix} c \\ U \end{bmatrix}$ is nonsingular, integral, and unimodular (i.e. has determinant ± 1; U can be found with the methods of Section 5.3). Then the linear transformation $\Phi: \mathbb{R}^n \to \mathbb{R}^{n-1}$ with $\Phi(x) := Ux$ brings the polyhedron P_t to

$$(65) \qquad Q_t := \left\{ y \in \mathbb{R}^{n-1} \,\middle|\, A\begin{bmatrix} c \\ U \end{bmatrix}^{-1} \begin{pmatrix} t \\ y \end{pmatrix} \leqslant b \right\}.$$

Moreover, if $x \in P_t \cap \mathbb{Z}^n$ then $Ux \in Q_t \cap \mathbb{Z}^{n-1}$, and if $y \in Q_t \cap \mathbb{Z}^{n-1}$ then $\begin{bmatrix} c \\ U \end{bmatrix}^{-1} \begin{pmatrix} t \\ y \end{pmatrix} \in P_t \cap \mathbb{Z}^n$. So the problem is reduced to finding an integral vector in one of the polyhedra Q_t. This means that we have to solve at most $2n(n+1)2^{n(n-1)/4} + 1$ $(n-1)$-dimensional problems. By the induction hypothesis this can be done in polynomial time (note that the sizes of c and U can be polynomially bounded by the sizes of A and b, as they are found with polynomial algorithms). $\qquad \square$

Corollary 18.7b. *For each fixed natural number n, there exists a polynomial algorithm which solves the integer linear programming problem $\max\{cx \,|\, Ax \leqslant b;\ x \text{ integral}\}$, where A has n columns, and where all input data are rational.*

Proof. Using the estimates on the sizes of optimal solutions of ILP-problems (cf. Corollary 17.1c), this corollary easily follows from the previous corollary by applying a binary search technique. $\qquad \square$

Corollary 18.7b extends Corollary 18.7a (the case $c = 0$). Lenstra observed that the following more general result holds.

Corollary 18.7c. *For each fixed natural number n, there exists a polynomial algorithm which solves the integer linear programming problem* $\max\{cx\,|\,Ax \le b;$ x integral$\}$*, where A has rank at most n, and where all input data are rational.*

Proof. If A, b, c are given, where A has rank $r \le n$, let without loss of generality

$$(66) \qquad A = \begin{bmatrix} A_1 \\ A_2 \end{bmatrix}$$

where A_1 consists of r linearly independent rows (found in polynomial time with the Gaussian elimination method—see Section 3.3). Let U be a nonsingular, integral, unimodular matrix such that $A_1 U$ is in Hermite normal form (U can be found in polynomial time—see Section 5.3). Then, if $A' := AU$ and $c' := cU$,

$$(67) \qquad \max\{cx\,|\,Ax \le b; x \text{ integral}\} = \max\{c'y\,|\,A'y \le b; y \text{ integral}\}.$$

As only r columns of A' contain nonzeros, by deleting the all-zero columns of A' we can reduce the original problem to an ILP-problem with at most n columns. Corollary 18.7b thus implies Corollary 18.7c. $\qquad\square$

Note that Corollary 18.7b implies that also ILP-problems of form $\max\{cx\,|\,x \ge 0; Ax = b\}$ can be solved in polynomial time, for a fixed number of variables. It also follows that for each fixed k, there is a polynomial-time algorithm to solve a system $Ax = b$ of linear equations in nonnegative integer variables, where $\{x\,|\,Ax = b\}$ has dimension k (as we may assume A has full row rank, we can find a unimodular matrix U so that AU is in Hermite normal form; the problem then becomes to find an integral vector y such that $AUy = b$, $Uy \ge 0$, which is essentially a problem in k variables (cf. Rubin [1984-5])).

Theorem 18.7 can be extended to the case where the polyhedron is not given by an explicit system $Ax \le b$, but by a separation algorithm (cf. Grötschel, Lovász, and Schrijver [1988]). In that case, the bound $2n(n+1)2^{n(n-1)/4}$ is replaced by $2n^3 2^{n(n-1)/4}$. It follows that also in Corollaries 18.7a, b, c the system $Ax \le b$ may be replaced by a separation algorithm. Lenstra [1983] noticed that this implies the following for the *mixed integer linear programming problem*:

(68) *for each natural number n, there exists a polynomial algorithm which solves* $\max\{cx + dy\,|\,Ax + By \le b; x \text{ integral}\}$*, where A has rank at most n, and where all input data are rational.*

Sketch of proof. The separation problem for the polyhedron $Q := \{x\,|\,\exists y: Ax + By \le b\}$ is polynomially solvable, as by Farkas' lemma:

$$(69) \qquad x \in Q \Leftrightarrow \exists y: By \le b - Ax \Leftrightarrow (\forall z \ge 0: zB = 0 \Rightarrow zAx \le zb).$$

As the last expression can be tested in polynomial time, yielding a separating hyperplane if $x \notin Q$, we have a polynomial separation algorithm for Q. Note that the constraint matrix defining Q has rank at most n. So by the extension of Corollary 18.7c we can find a vector in $\{x\,|\,\exists y: Ax + By \le b; x \text{ integral}\}$, or decide that it is empty, in polynomial time. Hence the same is true for $\{(x,y)\,|\,Ax + By \le b; x \text{ integral}\}$. By applying binary search it follows that also the mixed integer linear programming is polynomially solvable. $\qquad\square$

[This idea relates to Benders' decomposition method—see Section 24.5.]

Notes for Section 18.4. Improved versions of Lenstra's algorithm were given by Kannan [1983b] and Babai [1984]. Babai showed that the upper bound in (46) may be improved to $n^{3/2}2^{n/2}$. Feit [1984] describes an algorithm for solving ILP-problems with two variables in time $O(m(L+\log m))$, where m is the number of constraints and L is the maximum size of the numbers occurring in the input.

18.5. DYNAMIC PROGRAMMING APPLIED TO THE KNAPSACK PROBLEM

Bellman [1956, 1957a, b] and Dantzig [1957] introduced a 'dynamic programming' approach for ILP-problems, especially for knapsack problems. Essentially it consists of finding a shortest path in a certain large graph. This gives a *pseudo-polynomial* algorithm for the knapsack problem (i.e. an algorithm whose running time is polynomially bounded by the size of the input and by the maximum absolute value of the numerators and denominators of the numbers occurring in the input). It was shown by Papadimitriou [1981] that it yields also a pseudo-polynomial algorithm for integer linear programming, if the number of constraints is fixed.

In this section we describe the method for solution of the knapsack problem, and in the next section we go over to the general integer linear programming problem. Note that these problems are $\mathcal{NP}$-complete, so that polynomial-time methods—measuring size in binary notation—are generally believed not to exist.

Let the knapsack problem be given in the following form:

(70) $\min\{cx\,|\,ax \geqslant \beta;\, x\,\{0,1\}\text{-vector}\}$

where $a = (\alpha_1,\ldots,\alpha_n)$ and $c = (\gamma_1,\ldots,\gamma_n)$ are integral vectors, and β is an integer. Let S be the maximum absolute value of the entries of a.

Make a directed graph $D = (V, A)$ with vertex set

(71) $V := \{0,\ldots,n\} \times \{-nS,\ldots,+nS\}$

and arc set A given by:

(72) $((j,\delta),(i,\delta')) \in A \Leftrightarrow j = i - 1$ and $\delta' - \delta \in \{0, \alpha_i\}$.

Define the *length* $l((i-1,\delta),(i,\delta'))$ of arc $((i-1,\delta),(i,\delta'))$ by

(73) $l((i-1,\delta),(i,\delta')) := \gamma_i$ if $\delta' = \delta + \alpha_i$

$\qquad\qquad\qquad\qquad\quad := 0$ if $\delta' = \delta$.

Now it is easy to see that any directed path Π in D from $(0,0)$ to (n,β') for some $\beta' \geqslant \beta$ yields a feasible solution $x = (\xi_1,\ldots,\xi_n)^{\mathsf{T}}$ of (70), viz. we take:

(74) $\xi_i := 0$ if $((i-1,\delta),(i,\delta))$ belongs to Π for some δ;

$\quad\ \ \xi_i := 1$ if $((i-1,\delta),(i,\delta+\alpha_i))$ belongs to Π for some δ.

Moreover, the length of Π (:= the sum of the lengths of the arcs in Π) is equal to cx. So we can solve (70) by finding a shortest path from $(0,0)$ to (n,β') for some $\beta' \geqslant \beta$.

It is easy to see that such a shortest path can be found in time $O(|V|\cdot\log(nT))$, where T is the maximum absolute value of the entries of c.

[Indeed, let $l_{i,\delta}$ be the length of the shortest path from $(0,0)$ to (i,δ) (with $l_{i,\delta} = \infty$ if no path from $(0,0)$ to (i,δ) exists). Then $l_{i,\delta}$ can be computed recursively, using the formula

(75) $\qquad l_{i,\delta} = \min\{l_{i-1,\delta}, l_{i-1,\delta-a_i} + \gamma_i\}.$

So for each i, δ, $l_{i,\delta}$ can be computed in time $O(\log(nTS))$. Thus, the total amount of time to find $l_{n,\beta}, l_{n,\beta+1}, \ldots, l_{n,nS}$ is $O(|V| \cdot \log(nT))$.]

So (70) can be solved in time $O(n^2 \cdot S \cdot \log(nTS))$ (assuming without loss of generality $|\beta| \leqslant nS$).

With the same digraph and length function as above, we can, for given integral row vectors a and c, and given natural number δ, determine the highest integer β for which there exists a $\{0,1\}$-vector x satisfying $cx \leqslant \delta$ and $ax \geqslant \beta$ (as β is the highest number such that $l_{n,\beta} \leqslant \delta$—cf. (75)). Again this takes time $O(n^2 S \cdot \log(nTS))$. The number β satisfies:

(76) $\qquad \beta = \max\{ax \mid cx \leqslant \delta; x \ \{0,1\}\text{-vector}\}.$

So also this knapsack problem can be solved in time $O(n^2 S \cdot \log(nTS))$ (assuming, without loss of generality, $|\delta| \leqslant nT$). Hence, by interchanging the roles of a and c,

(77) $\qquad \max\{cx \mid ax \leqslant \beta; x \ \{0,1\}\text{-vector}\}$

can be solved in time $O(n^2 T \cdot \log(nTS))$. Concluding we have:

Theorem 18.8. *Given integral row vectors* $a, c \in \mathbb{Q}^n$ *and an integer* β, *the problem*

(78) $\qquad \max\{cx \mid ax \leqslant \beta; x \ \{0,1\}\text{-vector}\}$

can be solved in time $O(n^2 S \cdot \log(nST))$ *and also in time* $O(n^2 T \cdot \log(nTS))$, *where* S *is the maximum absolute value of the components of* a *(assuming without loss of generality* $|\beta| \leqslant nS$), *and* T *is the maximum absolute value of the components of* c.

Proof. See above. Use that if we replace a, c, and β by their opposites, problem (78) passes into problem (70). $\qquad\qquad\qquad\qquad\qquad\qquad\qquad\qquad \square$

It is clear that this algorithm is not polynomial-time: by taking $a = (1, 2, 4, \ldots, 2^{n-1})$, the digraph D has $O(2^n)$ vertices (even if we leave out all vertices which cannot be reached from $(0,0)$ by a directed path). See also Chvátal [1980].

Ibarra and Kim [1975] showed that Theorem 18.8 implies that knapsack problems can be approximated in polynomial time, in the following sense.

Corollary 18.8a (Ibarra–Kim theorem). *There exists an algorithm which, given nonnegative integral vectors* a, $c \in \mathbb{Q}^n$, *an integer* β, *and a rational* $\varepsilon > 0$, *finds a* $\{0,1\}$-*vector* x^* *such that* $ax^* \leqslant \beta$ *and*

(79) $\qquad cx^* \geqslant (1 - \varepsilon) \cdot \max\{cx \mid ax \leqslant \beta; x \ \{0,1\}\text{-vector}\}$

in time polynomially bounded by $1/\varepsilon$ *and by the sizes of* a, c, *and* β.

Proof. Let a, c, β, ε be given, and let S and T be as in Theorem 18.8. We may assume that $S \leqslant \beta$ (otherwise we can delete variables), and that hence

(80) $\qquad \max\{cx \mid ax \leqslant \beta; x\ \{0, 1\}\text{-vector}\} \geqslant T$.

Define vector c' as

(81) $\qquad c' := \left\lfloor \dfrac{n}{\varepsilon T} \cdot c \right\rfloor$

(where $\lfloor \ \ \rfloor$ denotes component-wise lower integer part). The maximum absolute value of the components of c' is $\lfloor n/\varepsilon \rfloor$. Hence, by Theorem 18.8, we can find in time $O(n^3 \cdot \varepsilon^{-1} \cdot \log(\varepsilon^{-1} nS))$ a $\{0, 1\}$-vector x^* attaining

(82) $\qquad \max\{c'x \mid ax \leqslant \beta; x\ \{0, 1\}\text{-vector}\}$.

We show that this x^* satisfies (79). Let $\tilde{x}$ attain the maximum in (79). Define $\theta := \varepsilon T/n$. Then

(83) $\qquad (1 - \varepsilon) \cdot c\tilde{x} \leqslant c\tilde{x} - \theta n \leqslant \theta \cdot c'\tilde{x} \leqslant \theta \cdot c'x^* \leqslant cx^*$.

[*Proof.* First $\leqslant$: by (80), $\varepsilon \cdot c\tilde{x} \geqslant \varepsilon T = \theta n$. Second $\leqslant$: each component of $c - \theta c' = c - \theta \lfloor \theta^{-1}c \rfloor$ is at most θ. Third $\leqslant$: x^* attains maximum (82). Last $\leqslant$: $\theta c' = \theta \lfloor \theta^{-1}c \rfloor \leqslant c$.]
(83) implies (79). $\qquad\qquad\qquad\qquad\qquad\qquad\qquad\qquad\qquad\qquad\qquad\qquad\square$

This proof is due to Garey and Johnson [1979].

The Ibarra–Kim theorem asserts that the knapsack problem has a so-called *fully polynomial approximation scheme*. Note that it is polynomial in $1/\varepsilon$ and not in $\log(1/\varepsilon)$ ($\approx \text{size}(\varepsilon)$). This last would be better, and in fact would imply the polynomial solvability of the ($\mathcal{NP}$-complete) knapsack problem: let $c = (\gamma_1, \ldots, \gamma_n)$, and choose $\varepsilon := \frac{1}{4}(\gamma_1 + \cdots + \gamma_n)^{-1}$. Applying Ibarra and Kim's method we find x^* with $ax^* \leqslant \beta$ satisfying (79). Then $cx^* \leqslant \gamma_1 + \cdots + \gamma_n$. Therefore:

(84) $\qquad cx^* \leqslant \max\{cx \mid ax \leqslant \beta; x\ \{0, 1\}\text{-vector}\} \leqslant \dfrac{1}{1 - \varepsilon} \cdot cx^* \leqslant cx^* + \tfrac{1}{2}$.

As cx^* and the maximum are integers, x^* attains the maximum. So if the running time of the method is polynomially bounded also by $\log(1/\varepsilon)$ we have a polynomial algorithm for the knapsack problem.

Notes for Section 18.5. In fact, Ibarra and Kim showed that, for any fixed $\varepsilon > 0$, there is an algorithm as in Corollary 18.8a consisting of $O(n \cdot \log n)$ elementary arithmetic operations.

Lawler [1977, 1979] gives some sharpenings and extensions. In particular, he showed that Corollary 18.8a also holds after replacing '$x\{0, 1\}$-vector' in (79) by 'x nonnegative integral'. Kannan [1983a] showed that, if $\mathcal{NP} \neq \mathcal{P}$, we may not replace furthermore $ax \leqslant \beta$ by $ax = \beta$ in (79).

See also Gilmore and Gomory [1966], Martello and Toth [1984]. Nemhauser and Ullmann [1968–9], Sahni [1975], Weingartner and Ness [1967], and Johnson [1974a].

**18.6. DYNAMIC PROGRAMMING APPLIED TO INTEGER
LINEAR PROGRAMMING**

The general ILP-problem can be approached similarly by dynamic programming (shortest path).

Let M be an integral $m \times n$-matrix, and let b and c be integral vectors. Suppose we wish to solve the integer linear programming problem

(85) $\max \{cx | x \geqslant 0; Mx = b; x \text{ integral}\}$.

Let S be an upper bound on the absolute values of the components of M and b. We know that if (85) is finite, it has an optimal solution x with components at most $(n + 1)(mS)^m$ (by Theorem 17.1, as for the system $x \geqslant 0$, $Mx = b$ we have $\Delta \leqslant m!S^m \leqslant (mS)^m$). Let

(86) $U := (n + 1)S(mS)^m$.

Consider the directed graph $D = (V, A)$ with vertex set

(87) $V := \{0, \dots, n\} \times \{-U, \dots, +U\}^m$

and with arc set A given by:

(88) $((j, b'), (i, b'')) \in A \Leftrightarrow j = i - 1$, and $b'' - b' = k \cdot m_i$ for some $k \in \mathbb{Z}_+$

where m_i denotes the ith column of M. Let the arc described in (88) have 'length' $-k\gamma_i$, where γ_i denotes the ith component of c.

Again it is easy to see that any directed path in D from $(0, 0)$ to (n, b) yields a feasible solution $x = (\xi_1, \dots, \xi_n)^T$ of (85): if $((i - 1, b'), (i, b' + km_i))$ is in this path, let $\xi_i := k$. Any feasible solution x of (85) arises in this way from a path from $(0, 0)$ to (n, b). Moreover, the length of the path ($:=$ the sum of the lengths of its arcs) is equal to $-cx$. So we can solve (85) by finding a shortest path in D from $(0, 0)$ to (n, b).

More precisely, if no directed path from $(0, 0)$ to (n, b) exists, (85) is infeasible (since if (85) is feasible, it has a feasible solution x with all components at most $(n + 1)(mS)^m$). If such a path does exist, and if the 'LP-relaxation' $\max \{cx | x \geqslant 0; Mx = b\}$ is unbounded, also (85) is unbounded (cf. Theorem 16.1). If the LP-relaxation is finite, the shortest directed path from $(0, 0)$ to (n, b) gives the optimal solution for (85).

Since a shortest path can be found in time polynomially bounded by $|V|$ and the size of the length function (cf. Section 18.5), we can solve (85) in time polynomially bounded by $n, (2U + 1)^m$, and the size of c. This implies that, for any fixed number of constraints, ILP-problems can be solved in pseudo-polynomial time. That is, we have a result of Papadimitriou [1981]:

Theorem 18.9. *For any fixed number m, there exists an algorithm which solves the integer linear programming problem*

(89) $\max \{cx | x \geqslant 0; Ax = b; x \text{ integral}\}$

where A is an integral $m \times n$-matrix and b and c are integral vectors, in time *polynomially bounded by* n, size(c) *and* S ($:=$ *maximum absolute value of the components of* A *and* b).

Proof. See the above algorithm. $\square$

Corollary 18.9a. *For any fixed natural number* m, *there exists an algorithm which solves a system* $Ax = b$ *of linear equations in nonnegative integers, where* A *is an integral* $m \times n$-*matrix, in time polynomially bounded by* n *and* S ($:=$ *the maximum absolute value of the components of* A *and* b).

Proof. Take $c = 0$ in Theorem 18.9. $\square$

Corollary 18.9b. *For any fixed natural number* m, *there exists an algorithm which solves a system* $Ax \leqslant b$ *of linear inequalities in integers, where* A *is an integral* $m \times n$-*matrix, in time polynomially bounded by* n *and* S ($:=$ *the maximum absolute value of the components of* A *and* b).

Proof. We may assume b is integral. Replace $Ax \leqslant b$ by $Ax' - Ax'' + \tilde{x} = b$, x', x'', $\tilde{x} \geqslant 0$, and apply Corollary 18.9a. $\square$

Corollary 18.9c. *For any fixed natural number* m, *there exists an algorithm which solves the integer linear programming problems*

(90) $\max \{cx \,|\, Ax \leqslant b; x \text{ integral}\}$,

          $\max \{cx \,|\, x \geqslant 0; Ax \leqslant b; x \text{ integral}\}$

where A *is an integral* $m \times n$-*matrix and* b *and* c *are integral vectors, in time polynomially bounded by* n, size(c) *and* S ($:=$ *maximum absolute value of the components of* A *and* b).

Proof. Directly from Theorem 18.9. $\square$

The algorithm is easily modified to a pseudo-polynomial algorithm for the problems (85), (90) with 'x integral' replaced by '$x\{0, 1\}$-vector'. A special case then is Theorem 18.8.

Wolsey [1974a] gives a *view* of dynamic programming methods in integer linear programming. See also Beckmann [1970].

19

Totally unimodular matrices: fundamental properties and examples

In Chapters 19 to 21 we consider *totally unimodular* matrices, which are matrices with all subdeterminants equal to 1, -1, or 0 (in particular, each entry is 1, -1, or 0). Totally unimodular matrices yield a prime class of linear programming problems with integral optimum solutions.

In the present chapter we discuss the basic theory and examples of totally unimodular matrices. In Section 19.1 we describe Hoffman and Kruskal's characterization of totally unimodular matrices, providing the link between total unimodularity and integer linear programming. In Section 19.2 we give some further characterizations. In Section 19.3 we discuss some classes of examples of totally unimodular matrices, with the *network matrices* as prominent class. These matrices turn out to form the basis for all totally unimodular matrices, as follows from a deep theorem of Seymour. It states that each totally unimodular matrix arises by certain compositions from network matrices and from certain 5×5-matrices. Seymour's theorem will be discussed, but not proved, in Section 19.4.

One of the consequences of Seymour's theorem is a polynomial-time method for testing total unimodularity, which we shall discuss in Chapter 20. In Chapter 21 we deal with some further theory related to total unimodularity.

19.1. TOTAL UNIMODULARITY AND OPTIMIZATION

A matrix A is *totally unimodular* if each subdeterminant of A is 0, $+1$, or -1. In particular, each entry in a totally unimodular matrix is 0, $+1$, or -1.

A link between total unimodularity and integer linear programming is given by the following fundamental result.

Theorem 19.1. *Let A be a totally unimodular matrix and let b be an integral vector. Then the polyhedron $P := \{x \mid Ax \leqslant b\}$ is integral.*

Proof. Let $F = \{x \mid A'x = b'\}$ be a minimal face of P, where $A'x \leqslant b'$ is a subsystem of $Ax \leqslant b$, with A' having full row rank. Then $A' = [U \ V]$ for some matrix U with det $U = \pm 1$ (possibly after permuting coordinates). Hence

(1) $x := \begin{pmatrix} U^{-1}b' \\ 0 \end{pmatrix}$

is an integral vector in F.   □

It follows that each linear program with integer data and totally unimodular constraint matrix has an integral optimum solution:

Corollary 19.1a. *Let A be a totally unimodular matrix, and let b and c be integral vectors. Then both problems in the LP-duality equation*

(2) $\max \{cx \mid Ax \leqslant b\} = \min \{yb \mid y \geqslant 0; yA = c\}$

have integral optimum solutions.

Proof. Directly from Theorem 19.1, using the fact that with A also the matrix

(3) $\begin{bmatrix} I \\ A^{\mathsf{T}} \\ -A^{\mathsf{T}} \end{bmatrix}$

is totally unimodular.   □

Hoffman and Kruskal [1956] showed, as we shall see in Corollary 19.2a, that the above property more or less characterizes total unimodularity. This is a main result of this section.

Veinott and Dantzig [1968] gave a short proof of Hoffman and Kruskal's result, which uses an extension of the notion of unimodularity defined in Chapter 4. There a nonsingular matrix was defined to be unimodular if it is integral and has determinant ± 1. Let, for any $m \times n$-matrix A of full row rank, a *basis* of A be a nonsingular submatrix of A of order m. Now, as an extension, a matrix A of full row rank is *unimodular* if A is integral, and each basis of A has determinant ± 1. It is easy to see that a matrix A is totally unimodular if and only if the matrix $[I \ A]$ is unimodular.

The proof of Veinott and Dantzig is based on the following interesting auxiliary result.

Theorem 19.2. *Let A be an integral matrix of full row rank. Then the polyhedron $\{x \mid x \geqslant 0; Ax = b\}$ is integral for each integral vector b, if and only if A is unimodular.*

Proof. Let A have m rows and n columns. First suppose that A is unimodular. Let b be an integral vector, and let x' be a vertex of the polyhedron $\{x \mid x \geqslant 0; Ax = b\}$. Then there are n linearly independent constraints satisfied by x' with equality. Therefore the columns of A corresponding to nonzero components of x' are linearly independent. We can extend these columns to a basis B of A. Then x'

restricted to the coordinates corresponding to B is equal to $B^{-1}b$, which is integral (as det $B = \pm 1$). Since outside B, x' is zero, it follows that x' is integral.

To see the converse, suppose that $\{x \mid x \geqslant 0; Ax = b\}$ is integral for each integral vector b. Let B be a basis of A. To prove that B is unimodular, it suffices to show that $B^{-1}t$ is integral for each integral vector t. So let t be integral. Then there exists an integral vector y such that $z := y + B^{-1}t \geqslant 0$. Then $b := Bz$ is integral. Let z' arise from z by adding zero-components to z so as to obtain $Az' = Bz = b$. Then z' is a vertex of $\{x \mid x \geqslant 0; Ax = b\}$ (as it is in the polyhedron, and satisfies n linearly independent constraints with equality). Then z' is integral. Therefore z and $B^{-1}t = z - y$ also are integral. $\qquad\square$

As a direct corollary we obtain the following important theorem of Hoffman and Kruskal [1956].

Corollary 19.2a (Hoffman and Kruskal's theorem). *Let A be an integral matrix. Then A is totally unimodular if and only if for each integral vector b the polyhedron $\{x \mid x \geqslant 0; Ax \leqslant b\}$ is integral.*

Proof. Recall that A is totally unimodular if and only if the matrix $[I \ A]$ is unimodular. Furthermore, for any integral vector b, the vertices of the polyhedron $\{x \mid x \geqslant 0; Ax \leqslant b\}$ are integral if and only if the vertices of the polyhedron $\{z \mid z \geqslant 0; [I \ A]z = b\}$ are integral. So Theorem 19.2 directly gives the corollary. $\qquad\square$

Since for any totally unimodular matrix A also the matrix

$$(4) \qquad \begin{bmatrix} I \\ -I \\ A \\ -A \end{bmatrix}$$

is totally unimodular, it follows that an integral matrix A is totally unimodular if and only if for all integral vectors a, b, c, d the vertices of the polytope $\{x \mid c \leqslant x \leqslant d; a \leqslant Ax \leqslant b\}$ are integral. Similarly one may derive from Hoffman and Kruskal's theorem that an integral matrix A is totally unimodular if and only if one of the following polyhedra has all vertices integral, for *each* integral vector b and for *some* integral vector c:

$$(5) \qquad \{x \mid x \leqslant c; Ax \leqslant b\}, \ \{x \mid x \leqslant c; Ax \geqslant b\}, \ \{x \mid x \geqslant c; Ax \leqslant b\},$$
$$\{x \mid x \geqslant c; Ax \geqslant b\}.$$

Formulated in terms of linear programming Hoffman and Kruskal's theorem says the following.

Corollary 19.2b. *An integral matrix A is totally unimodular if and only if for all integral vectors b and c both sides of the linear programming duality equation*

(6) $\max \{cx | x \geqslant 0, Ax \leqslant b\} = \min \{yb | y \geqslant 0, yA \geqslant c\}$

are achieved by integral vectors x and y (if they are finite).

Proof. This follows from Corollary 19.2a (using that the transpose of a totally unimodular matrix is totally unimodular again). $\qquad\square$

19.2. MORE CHARACTERIZATIONS OF TOTAL UNIMODULARITY

There are other characterizations of totally unimodular matrices. We collect some of them, together with the above characterization, in the following theorem. Characterizations (ii) and (iii) are due to Hoffman and Kruskal [1956], characterization (iv) to Ghouila-Houri [1962] (cf. Padberg [1976]), characterizations (v) and (vi) to Camion [1963a, b, 1965], and characterization (vii) to R. E. Gomory (see Camion [1965]).

Theorem 19.3. *Let A be a matrix with entries* 0, $+1$, *or* -1. *Then the following are equivalent:*

(i) *A is totally unimodular, i.e. each square submatrix of A has determinant* 0, $+1$, *or* -1;

(ii) *for each integral vector b the polyhedron* $\{x | x \geqslant 0, Ax \leqslant b\}$ *has only integral vertices;*

(iii) *for all integral vectors a, b, c, d the polyhedron* $\{x | c \leqslant x \leqslant d, a \leqslant Ax \leqslant b\}$ *has only integral vertices;*

(iv) *each collection of columns of A can be split into two parts so that the sum of the columns in one part minus the sum of the columns in the other part is a vector with entries only* 0, $+1$, *and* -1;

(v) *each nonsingular submatrix of A has a row with an odd number of nonzero components;*

(vi) *the sum of the entries in any square submatrix with even row and column sums is divisible by four;*

(vii) *no square submatrix of A has determinant* $+2$ *or* -2.

Proof. Let A be an $m \times n$-matrix. The equivalence of (i), (ii), and (iii) has been shown above (Corollary 19.2a).

(iii) $\rightarrow$ (iv). Let A be totally unimodular, and choose a collection of columns of A. Consider the polyhedron

(7) $P := \{x | 0 \leqslant x \leqslant d, \lfloor \frac{1}{2} \cdot Ad \rfloor \leqslant Ax \leqslant \lceil \frac{1}{2} \cdot Ad \rceil\}$

where d is the characteristic vector of the collection of chosen columns, and $\lfloor \ \rfloor$ and $\lceil \ \rceil$ denote component-wise lower and upper integer parts of vectors. Since P is nonempty (as $\frac{1}{2}d \in P$), P has at least one vertex, say x, which is, by (iii), a $\{0, 1\}$-vector. Now $y := d - 2x$ has components only 0, $+1$, and -1, and $y \equiv d \pmod 2$. Furthermore, Ay has components only $+1$, -1, and 0. So y yields a partition of the columns as required.

(iv) → (v). Let B be a square submatrix of A with each row sum even. By (iv) there exists a $\{+1, -1\}$-vector x such that Bx is a $\{0, +1, -1\}$-vector. Since B has even row sums we know that $Bx = 0$. Since $x \neq 0$, B is singular.

(iv) → (vi). Let B be a square submatrix of A with each row sum and each column sum even. By (iv), the columns of B can be split into two classes B_1 and B_2 so that the sum of the columns in B_1 is the same vector as the sum of the columns in B_2 (as each row sum of B is even). Let $\sigma(B_i)$ denote the sum of the entries in B_i. Then $\sigma(B_1) = \sigma(B_2)$, and $\sigma(B_1)$ is even (as each column sum of B is even). Hence $\sigma(B_1) + \sigma(B_2)$ is divisible by 4.

(v) → (vii) and (vi) → (vii). Suppose the equivalence of (i)–(vii) has been proved for all proper submatrices of A. Assume furthermore that (v) or (vi) is true for A. If (vii) is not true, then det $A = \pm 2$, and each square proper submatrix of A has determinant 0 or ± 1. Since det $A \equiv 0 \pmod 2$ the columns of A are linearly dependent over $GF(2)$, the field with two elements. Now for each proper submatrix of A, linear dependence over $\mathbb{Q}$ coincides with linear dependence over $GF(2)$ (since each square proper submatrix of A has determinant 0 or ± 1). Since the columns of A are linearly independent over $\mathbb{Q}$, it follows that the sum of *all* columns of A is a vector having even components only. But this contradicts (v).

One similarly shows that the sum of the rows of A has even components only. Let B arise from A by deleting the first row of A. By our hypothesis there exists a $\{+1, -1\}$-vector x such that Bx is a $\{0, +1, -1\}$-vector. The matrix B has even row sums, which implies that $Bx = 0$. So

$$(8) \qquad Ax = \begin{bmatrix} \alpha \\ 0 \\ \vdots \\ 0 \end{bmatrix}$$

for some integer α. Now $|\alpha| = |\det A| = 2$, since

$$(9) \qquad A \cdot \left[x \; \begin{array}{|c} 0 \cdots 0 \\ \hline \\ I \\ \\ \end{array} \right] = \left[\begin{array}{c|c} \alpha & \\ 0 & \\ \vdots & A' \\ 0 & \end{array} \right]$$

(where A' arises from A by deleting the first column of A), and the determinants of the three matrices in this identity are equal to ± 2, ± 1, and $\pm \alpha$, respectively. As $1 - x$ has even components only, $1^{\mathsf T} A(1 - x) \equiv 0 \pmod 4$. As $1^{\mathsf T} Ax = \alpha \equiv 2 \pmod 4$ we arrive at $1^{\mathsf T} A 1 \equiv 2 \pmod 4$, contradicting (vi).

(vii) → (i). It suffices to show that each square $\{0, \pm 1\}$-matrix B with $|\det B| > 2$ has a square submatrix with determinant ± 2. Let B be such a matrix, of order, say, n, and consider the matrix $C := [B \; I]$. Let C' arise from C by adding or subtracting rows to or from other rows, and by multiplication of columns by

-1, such that (I) C' is a $\{0, \pm 1\}$-matrix, (II) C' contains among its columns the n unit basis column vectors, and (III) C' contains among its first n columns as many unit basis column vectors as possible. Let k be the number of unit basis vectors in the first n columns of C'. We may suppose, without loss of generality, that

(10) $\qquad C' = \left[\begin{array}{c|c|c} I_k & B' & 0 \\ \hline 0 & & I_{n-k} \end{array}\right]$

for a certain square matrix B' of order n (where I_k and I_{n-k} denote the identity matrices of order k and $n-k$). Since the first n columns of C, and hence also of C', form a matrix with determinant not equal to ± 1, we know that $k < n$. So there is a 1 (without loss of generality) occurring in some position (i, j) of C' with $k + 1 \leqslant i, j \leqslant n$. By our assumption (III) we cannot transfer column j to a unit vector by elementary row operations without violating condition (I). Hence there is a pair i', j' such that the 2×2-submatrix with row indices i and i' and column indices j and j' has the form $\begin{bmatrix} 1 & 1 \\ 1 & -1 \end{bmatrix}$ or $\begin{bmatrix} 1 & -1 \\ -1 & -1 \end{bmatrix}$. Now the submatrix of C' formed by the columns j, j' and the unit column vectors, except the ith and i'th unit vectors, has determinant ± 2. So also the corresponding columns of C form a matrix with determinant ± 2. This implies that B has a submatrix with determinant ± 2. $\qquad\square$

Baum and Trotter [1977] gave the following characterization:

Theorem 19.4. *An integral matrix A is totally unimodular if and only if for all integral vectors b and y, and for each natural number $k \geqslant 1$, with $y \geqslant 0$, $Ay \leqslant kb$, there are integral vectors $x_1, \ldots, x_k$ in $\{x \geqslant 0 \,|\, Ax \leqslant b\}$ such that $y = x_1 + \cdots + x_k$.*

Proof. **Sufficiency.** To show that A is totally unimodular, it is enough to show that for each integral vector b, all vertices of the polyhedron $P := \{x \geqslant 0 \,|\, Ax \leqslant b\}$ are integral. Suppose x_0 is a non-integral vertex. Let k be the l.c.m. of the denominators occurring in x_0. Then $y := kx_0$ satisfies $y \geqslant 0$, $Ay \leqslant kb$. Therefore, $y = x_1 + \cdots + x_k$ for certain integral vectors $x_1, \ldots, x_k$ in P. Hence $x_0 = (x_1 + \cdots + x_k)/k$ is a convex combination of integral vectors in P, contradicting the fact that x_0 is a non-integral vertex of P.

Necessity. Let A be totally unimodular. Choose integral vectors b and y, and a natural number $k \geqslant 1$, such that $y \geqslant 0$, $Ay \leqslant kb$. We show that there are integral vectors $x_1, \ldots, x_k$ in $\{x \geqslant 0 \,|\, Ax \leqslant b\}$ with $y = x_1 + \cdots + x_k$, by induction on k. The case $k = 1$ is trivial. If $k \geqslant 2$, we know that the polyhedron

(11) $\qquad \{x \,|\, 0 \leqslant x \leqslant y;\ Ay - kb + b \leqslant Ax \leqslant b\}$

is nonempty, as $k^{-1}y$ is in it. Since A is totally unimodular, (11) has an integral vertex, call it x_k. Let $y' := y - x_k$. Then y' is integral, and $y' \geqslant 0$, $Ay' \leqslant (k-1)b$.

Hence, by induction, $y' = x_1 + \cdots + x_{k-1}$ for integral vectors $x_1, \ldots, x_{k-1}$ in $\{x \geqslant 0 | Ax \leqslant b\}$. So $y = x_1 + \cdots + x_k$ is a decomposition as required. $\square$

Another characterization we need later is:

Theorem 19.5. *Let* A *be a matrix of full row rank. Then the following are equivalent*:

(i) *for each basis* B *of* A, *the matrix* $B^{-1} A$ *is integral*;

(ii) *for each basis* B *of* A, *the matrix* $B^{-1} A$ *is totally unimodular*;

(iii) *there exists a basis* B *of* A *for which* $B^{-1} A$ *is totally unimodular*.

Proof. We may assume that $A = [I \ C]$ for some matrix C, as (i), (ii), and (iii) are invariant under premultiplying A by a nonsingular matrix.

Now one easily checks that each of (i), (ii), and (iii) is equivalent to each basis of $[I \ C]$ being unimodular. $\square$

The following characterization is due to Chandrasekaran [1969]: *a matrix* A *is totally unimodular if and only if for each nonsingular submatrix* B *of* A *and for each nonzero* $\{0, \pm 1\}$-*vector* y, *the g.c.d. of the entries in* yB *is* 1.

Proof. Necessity. We know that B^{-1} is integral. Let $k := $ g.c.d. (yB) (meaning the g.c.d. of the components of yB). Then $k^{-1} yB$ is integral, and hence $k^{-1} y = k^{-1} yBB^{-1}$ is integral. As y is a $\{0, \pm 1\}$-vector, we know $k = 1$.

Sufficiency. We derive (v) of Theorem 19.3. Let B be a nonsingular submatrix of A. Then the components of $1B$ have g.c.d. 1. Then one of the columns of B must have an odd number of nonzero entries. $\square$

This proof, due to Tamir [1976], shows that in fact A is totally unimodular if and only if for each nonsingular submatrix B of A the g.c.d. of the entries in $1B$ is 1.

It is easy to see that a nonsingular matrix B is unimodular if and only if g.c.d. $(yB) = $ g.c.d.(y) for each vector y. Chandrasekaran [1969] showed that here one may not restrict y to $\{0, \pm 1\}$-vectors. This is shown by taking

$$(12) \qquad B = \begin{bmatrix} 1 & 0 & 1 & 0 \\ 1 & 1 & -1 & 0 \\ 0 & 1 & 1 & -1 \\ 0 & 1 & 0 & 1 \end{bmatrix}$$

In Sections 21.1 and 21.2 we shall discuss some further characterizations of total unimodularity. For more results and surveys on total unimodularity, see also Brown [1976, 1977], Cederbaum [1958], Chandrasekaran [1969], Commoner [1973], de Werra [1981], Heller [1963], Padberg [1975a], Tamir [1976], Truemper [1980], and Yannakakis [1980, 1985]. For combinatorial applications, see Hoffman [1960, 1976, 1979b]. For direct derivations of Ghouila-Houri's characterization from Camion's characterization, see Padberg [1976] and Tamir [1976]. For a survey, see Gondran [1973]. For results related to Gomory's characterization, see Kress and Tamir [1980].

19.3. THE BASIC EXAMPLES: NETWORK MATRICES

We here discuss some examples of totally unimodular matrices, in particular network matrices. In Section 19.4 we shall see Seymour's famous characterization of total unimodularity, showing that these network matrices form the building bricks for all totally unimodular matrices.

In describing the examples, we use some concepts from graph theory. The reader unfamiliar with graphs is recommended to consult Section 1.4.

Example 1. *Bipartite graphs* (cf. Motzkin [1956], Hoffman and Kruskal [1956], Hoffman and Kuhn [1956], and Heller and Tompkins [1956]). Let $G = (V, E)$ be an undirected graph, and let M be the $V \times E$-incidence matrix of G (i.e. M is the $\{0, 1\}$-matrix with rows and columns indexed by the vertices and edges of G, respectively, where $M_{v,e} = 1$ if and only if $v \in e$). Then:

(13)   *M is totally unimodular if and only if G is bipartite.*

So M is totally unimodular if and only if the rows of M can be split into two classes so that each column contains a 1 in each of these classes. Assertion (13) easily follows from Ghouila-Houri's characterization ((iv) in Theorem 19.3).

Application 19.1. With Hoffman and Kruskal's theorem, the total unimodularity of the incidence matrix of a bipartite graph implies several theorems, like König's theorems for matchings and coverings in bipartite graphs, and the Birkhoff–von Neumann theorem on doubly stochastic matrices.

Indeed, let M be the $V \times E$-incidence matrix of the bipartite graph $G = (V, E)$. Then by (13) and Corollary 19.2b, we have:

(14) $\max \{y1 \,|\, y \geqslant 0; yM \leqslant 1; y \text{ integral}\} = \min \{1x \,|\, x \geqslant 0; Mx \geqslant 1; x \text{ integral}\}$.

[Again, 1 stands for an all-one (row or column) vector.] This is equivalent to König's 'covering' theorem [1933]: the maximum cardinality of a coclique in a bipartite graph is equal to the minimum number of edges needed to cover all vertices. (Here we assume that each vertex of the graph is contained in at least one edge.)

Similarly one has:

(15) $\max \{1x \,|\, x \geqslant 0; Mx \leqslant 1; x \text{ integral}\} = \min \{y1 \,|\, y \geqslant 0; yM \geqslant 1; y \text{ integral}\}$.

This is equivalent to the König–Egerváry theorem (König [1931], Egerváry [1931]): the maximum cardinality of a matching in a bipartite graph is equal to the minimum cardinality of a set of vertices intersecting each edge.

More generally, for each $w : E \to \mathbb{Z}_+$,

(16) $\max \{wx \,|\, x \geqslant 0; Mx \leqslant 1; x \text{ integral}\} = \min \{1y \,|\, y \geqslant 0; yM \geqslant w; y \text{ integral}\}$.

If we consider w as a 'profit' function, (16) gives a min–max formula for the 'optimal assignment' problem. Also a most general application of total unimodularity to bipartite graphs:

(17) $\max \{wx \,|\, c \leqslant x \leqslant d; a \leqslant Mx \leqslant b; x \text{ integral}\}$

$= \min \{z'd - z''c + y'b - y''a \,|\, z', z'', y', y'' \geqslant 0;$

$(y' - y'')M + z' - z'' = w; z', z'', y', y'' \text{ integral}\}$

may be seen to correspond to an assignment problem.

It also follows that the polytopes $\{x \,|\, x \geqslant 0; Mx \leqslant 1\}$ and $\{x \,|\, x \geqslant 0; Mx = 1\}$ are integral. Therefore, a function $x : E \to \mathbb{R}_+$ is a convex combination of incidence vectors of (perfect) matchings in G, if and only if $\sum_{e \ni v} x(e) \leqslant 1$ ($\sum_{e \ni v} x(e) = 1$, respectively), for each vertex v. In particular, if G is the complete bipartite graph $K_{n,n}$, then the last result is equivalent to a theorem of Birkhoff [1946] and von Neumann [1953]: a doubly stochastic matrix is a convex combination of permutation matrices (a *doubly stochastic* matrix is a nonnegative matrix with all row sums and all column sums equal to 1)—cf. Section 8.10.

Example 2. *Directed graphs* (cf. Dantzig and Fulkerson [1956], Hoffman and Kruskal [1956], and Heller and Tompkins [1956]). Let $D = (V, A)$ be a directed graph, and let M be the $V \times A$-incidence matrix of D (i.e. M is a $\{0, \pm 1\}$-matrix with rows and columns indexed by the vertices and arcs of D, respectively, where $M_{v,a} = +1$ ($= -1$) if and only if a enters (leaves) v). Then M is totally unimodular. So

(18)   *a $\{0, \pm 1\}$-matrix with in each column exactly one $+1$ and exactly one -1 is totally unimodular.*

This fact was stated by Poincaré [1900]. Veblen and Franklin [1921] gave the following short inductive proof: any square submatrix N of M either contains a column with at most one nonzero entry (in which case expansion of the determinant by this column has determinant 0 or ± 1, by induction), or each column of N contains both a $+1$ and a -1 (in which case the determinant is 0).

Alternatively, (18) follows directly from Ghouila-Houri's characterization ((iv) in Theorem 19.3).

Application 19.2. Again with Hoffman and Kruskal's theorem, the total unimodularity of incidence matrices of digraphs implies several graph-theoretical results, like Menger's theorem, the max-flow min-cut theorem, Hoffman's circulation theorem, and Dilworth's theorem.

Let $D = (V, A)$ be a directed graph, and let M be its $V \times A$-incidence matrix. Let $x: A \rightarrow \mathbb{R}_+$ satisfies $Mx = 0$. The vector x may be considered as a circulation, i.e. x satisfies

(19) $\displaystyle\sum_{a \text{ enters } v} x(a) = \sum_{a \text{ leaves } v} x(a)$, for all $v \in V$.

This means that in any vertex the loss of flow is equal to the gain of flow. By Hoffman and Kruskal's theorem, the polytope $\{x \mid d \leqslant x \leqslant c; Mx = 0\}$ has all vertices integral, for all $c, d: A \rightarrow \mathbb{Z}_+$. This implies that if there exists a circulation x with $d \leqslant x \leqslant c$, then there exists an integral such circulation.

Hoffman and Kruskal's theorem also implies that for all $c, d, f: A \rightarrow \mathbb{Z} \cup \{\pm \infty\}$,

(20) $\max \{ fx \mid d \leqslant x \leqslant c; Mx = 0; x \text{ integral} \} =$
          $= \min \{ y'c - y''d \mid y', y'' \in \mathbb{R}_+^A; z \in \mathbb{R}^V; zM + y' - y'' = f; y', y'', z \text{ integral} \}$

(e.g. if a component of c is ∞, it gives no constraint in the maximum, while the corresponding component of y' in the minimum will be 0).

If we interpret $f(a)$ as the profit we make on having one flow unit through arc a (or if $-f(a)$ is the amount of cost), then (20) can be considered as a formula for the maximum profit (or minimum cost) of a circulation with prescribed lower and upper bounds.

If we take $f(a_0) = 1$ for a certain fixed arc $a_0 = (s, r)$, and $f(a) = 0$ for all other arcs a, and if $d = 0$, $c \geqslant 0$ such that $c(a_0) = \infty$, then the maximum in (20) is the maximum amount of r–s-flow in the digraph $D' = (V, A \setminus a_0)$, subject to the capacity function c, i.e. it is:

(21) $\qquad \max \sum_{a \text{ leaves } r} x(a) - \sum_{a \text{ enters } r} x(a)$

subject to:

$$\sum_{a \text{ leaves } v} x(a) = \sum_{a \text{ enters } v} x(a), \qquad \text{for } v \in V \setminus \{r, s\},$$

$$0 \leqslant x(a) \leqslant c(a), \qquad\qquad \text{for } a \in A \setminus \{a_0\}.$$

In that case, the minimum in (20) is equal to the minimum value of $y'c$ where $y' \in \mathbb{Z}^A_+$ such that there exists a $z \in \mathbb{Z}^V$ satisfying $y'(a) - z(v) + z(u) \geqslant 0$ if $a = (u, v) \neq (s, r)$, and $y'(a_0) - z(r) + z(s) \geqslant 1$, by the definition of f. As $c(a_0) = \infty$, we know $y'(a_0) = 0$, and hence $z(s) \geqslant z(r) + 1$. Letting $U := \{v \in V \mid z(v) \geqslant z(s)\}$, it follows that $y'(a) \geqslant 1$ if $a \in \delta^-(U)$ ($:=$ set of arcs entering U), and hence

(22) $\qquad y'c \geqslant \sum_{a \in \delta^-(U)} c(a).$

Therefore, the maximum amount of r–s-flow subject to c is at least the capacity of the r–s-cut $\delta^-(U)$. Since clearly this maximum cannot be more than this minimum, we know that they are equal.

So the total unimodularity of M gives the max-flow min-cut theorem of Ford and Fulkerson [1956] and Elias, Feinstein, and Shannon [1956]:

(23)　　the maximum value of an r–s-flow, subject to the capacity c, is equal to the minimum capacity of any r–s-cut. If all capacities are integers, the optimum flow can be taken to be integral.

The second line here is due to Dantzig [1951b].

If each capacity is 1, Menger's theorem [1927] follows:

(24)　　given a directed graph $D = (V, A)$ and vertices r and s of D, the maximum number of r–s-paths which pairwise do not have an arc in common is equal to the minimum size of any r–s-cut.

By taking f arbitrary one derives similarly a min–max formula for the minimum cost of an r–s-flow of a given value.

If we impose not only an upper bound c, but also a lower bound d for the flow function, where $0 \leqslant d \leqslant c$, then (20) gives: the maximum value of an r–s-flow, subject to the upper bound c and the lower bound d, is equal to the minimum value of

(25) $\qquad \sum_{a \in \delta^+(U)} c(a) - \sum_{a \in \delta^-(U)} d(a)$

where the minimum ranges over all subsets U of V with $r \in U$, $s \notin U$. ($\delta^+(U) =$ set of arcs leaving U.)

If we impose only lower bounds and no upper bounds (i.e. $c = \infty$), one can derive Dilworth's theorem [1950]:

(26)　　in a partially ordered set $(X, \leqslant)$ the maximum size of an antichain ($=$ set of pairwise incomparable elements) is equal to the minimum number of chains ($=$ sets of pairwise comparable elements) needed to cover X.

Dilworth's theorem can be derived also by taking negative capacities. To see this, define a digraph with capacities as follows. Split each element x of X into two points x' and x'', add two additional points r and s, make arcs and capacities as follows:

(27)　　(r, x')　　if $x \in X$,　　　 with capacity $c(r, x') = 0$

　　　　(x'', y')　　if $x, y \in X, x < y$,　　with capacity $c(x'', y') = 0$

(x', x'') if $x \in X$,   with capacity $c(x', x'') = -1$

(x'', s) if $x \in X$,   with capacity $c(x'', s) = 0$

Taking $d = -\infty$ we get the following. The maximum value of an r–s-flow subject to c is equal to:

(28)   — (minimum number of chains needed to cover X).

The minimum of (25) is equal to:

(29)   — (maximum size of an antichain in X).

So we know equality of (28) and (29), which is Dilworth's theorem.

Example 3. *A combination of Examples 1 and 2.* Heller and Tompkins [1956] and D. Gale showed the following. Let M be a $\{0, \pm 1\}$-matrix with exactly two nonzeros in each column. Then

(30)   *M is totally unimodular if and only if the rows of M can be split into two classes such that for each column: if the two nonzeros in the column have the same sign then they are in different classes, and if they have opposite sign then they are both in one and the same class.*

This follows easily from Example 2 (multiply all rows in one of the classes by -1, and we obtain a matrix of type (18)) and from (iv) of Theorem 19.3. Conversely, it includes Examples 1 and 2.

Characterization (30) directly yields a polynomial-time test for total unimodularity for matrices with most two nonzero entries in every column. Truemper [1977] observed that Camion's characterization ((v) in Theorem 19.3) implies: let M be a matrix with at most three nonzero entries in each column; then M is totally unimodular if and only if each submatrix of M with at most two nonzero entries in each column is totally unimodular.

Example 4. *Network matrices* (Tutte [1965]). Let $D = (V, A)$ be a directed graph and let $T = (V, A_0)$ be a directed tree on V. Let M be the $A_0 \times A$-matrix defined by, for $a = (v, w) \in A$ and $a' \in A_0$:

(31) $M_{a', a} := +1$ if the unique v–w-path in T passes through a' forwardly;

                    -1 if the unique v–w-path in T passes through a' backwardly;

                    0 if the unique v–w-path in T does not pass through a'.

Matrices arising in this way are called *network matrices*. If M, T, and D are as above, we say that T and D *represent* M. Note that the class of network matrices is closed under taking submatrices: deleting a column corresponds to deleting an arc of D, while deleting a row corresponds to contracting an arc of T.

In order to show that network matrices are totally unimodular, we first show the following:

(32)   *Let $D = (V, A)$ be a weakly connected digraph, let N be its $V \times A$-incidence matrix, and let $\tilde{N}$ arise from N by deleting one row of N. Then $\tilde{N}$ has full row rank. Moreover, a set K of columns of $\tilde{N}$ forms a basis of $\tilde{N}$ if*

and only if the arcs corresponding to columns in K form a spanning directed tree on V.

Proof. I. First let $A' \subseteq A$ form a spanning directed tree on V. Let B be the matrix formed by the columns of $\tilde{N}$ corresponding to A'. Suppose these columns are linearly dependent. Then $Bx = 0$ for some vector $x \neq 0$. Let A'' be the set of arcs a in A' which correspond to a nonzero component of x. As A' is a tree, A'' is a forest, and hence there are at least two vertices of degree 1 in (V, A''). Let u be such a vertex, not corresponding to the deleted row of N. Then the vector Bx is nonzero in the coordinate corresponding to u, as it is incident with exactly one arc in A''. So $Bx \neq 0$, contradicting our assumption.

Since $\tilde{N}$ has $|V| - 1$ rows, and any spanning tree has $|V| - 1$ arcs, it follows that $\tilde{N}$ has full row rank, and that each spanning tree yields a basis.

II. Conversely, let B be a basis of $\tilde{N}$. Then B consists of $|V| - 1$ linearly independent columns of N. Let A' be the arcs corresponding to the columns in B. We show that A' forms a spanning tree. As $|A'| = |V| - 1$, it suffices to show that A' contains no (undirected) circuit. Suppose it contains one, say C. Consider some orientation of C. Let x be the vector in $\{0, \pm 1\}^{A'}$ with, for $a \in A'$,

(33) $x(a) := +1$ if a occurs in forward direction in C;

$\quad\quad\quad x(a) := -1$ if a occurs in backward direction in C;

$\quad\quad\quad x(a) := 0$ if a does not occur in C.

One easily checks that $Bx = 0$, while $x \neq 0$, contradicting B being a basis of $\tilde{N}$. $\square$

We now show:

(34)   *Network matrices are totally unimodular.*

Proof. Let the network matrix M be represented by the tree $T = (V, A_0)$ and the digraph $D = (V, A)$. Let N be the $V \times A$-incidence matrix of D, and let L be the $V \times A_0$-incidence matrix of T. Let $\tilde{N}$ and $\tilde{L}$ arise by deleting one row (corresponding to the same vertex) from N and L. By (18), the matrix $[I\ \tilde{L}\ \tilde{N}]$ is unimodular, and by (32), $\tilde{L}$ is a basis for $[I\ \tilde{L}\ \tilde{N}]$. Hence also the matrix

(35) $\tilde{L}^{-1}[I\ \tilde{L}\ \tilde{N}] = [\tilde{L}^{-1}\ I\ \tilde{L}^{-1}\tilde{N}]$

is unimodular. This implies that $\tilde{L}^{-1}\tilde{N}$ is totally unimodular.

Now one easily checks that $LM = N$. Hence $\tilde{L}M = \tilde{N}$, and therefore $M = \tilde{L}^{-1}\tilde{N}$. So M is totally unimodular. $\square$

One similarly shows:

(36)   *If M is a network matrix of full row rank, and B is a basis of M, then $B^{-1}M$ is a network matrix again. If M is represented by the tree $T = (V, A_0)$ and the digraph $D = (V, A)$, then the columns in B correspond to a spanning tree, say A_1, in D, and $B^{-1}M$ is represented by (V, A_1) and D.*

Proof. Let $\tilde{L}, \tilde{N}$ be as in the previous proof. So $M = \tilde{L}^{-1}\tilde{N}$, and B is a basis of $[\tilde{L}^{-1}\tilde{I}\ \tilde{L}^{-1}\tilde{N}]$. Hence $\tilde{L}B$ is a basis of $[I\ \tilde{L}\ \tilde{N}]$. By (32), the columns of $\tilde{L}B$ correspond to a spanning tree, say A_1, in D. Now $\tilde{L}B(B^{-1}M) = \tilde{N}$. Hence $B^{-1}M$ is the network matrix represented by (V, A_1) and D. $\square$

(36) implies:

(37)   *the class of network matrices is closed under pivoting, i.e. under the operation*

$$\begin{bmatrix} \varepsilon & c \\ b & D \end{bmatrix} \rightarrow \begin{bmatrix} -\varepsilon & \varepsilon c \\ \varepsilon b & D - \varepsilon bc \end{bmatrix}$$

where $\varepsilon \in \{\pm 1\}$, b is a column vector, c is a row vector, and D is a matrix.

Finally note that the Examples 1, 2, and 3 are direct special cases of network matrices.

Example 5. *Cross-free families.* A family $\mathscr{C}$ of subsets of a set V is called *cross-free* if:

(38)   if $T, U \in \mathscr{C}$, then $T \subseteq U$, or $U \subseteq T$, or $T \cap U = \varnothing$, or $T \cup U = V$.

So a family is cross-free if and only if its Venn diagram can be drawn on the surface of a ball without crossings.

Edmonds and Giles [1977] observed the following. Let $T = (W, A_0)$ be a directed tree, and let $\phi: V \rightarrow W$. Define for each arc a of T:

(39) $V_a := \{u \in V \mid \phi(u)$ is in the same component of the graph $(W, A_0 \setminus \{a\})$
          as the head of a is$\}$.

Let

(40) $\mathscr{C} := \{V_a \mid a \in A_0\}$.

Then it is not difficult to see that $\mathscr{C}$ is a cross-free family. Moreover, each cross-free family arises in this way. The pair T, ϕ is called a *tree-representation* for $\mathscr{C}$.

For given cross-free family $\mathscr{C} \subseteq \mathscr{P}(V)$ and given digraph $D = (V, A)$, define the matrix M, with rows and columns indexed by $\mathscr{C}$ and A, respectively, by

(41) $M_{U,a} = +1$ if a enters U,

          $M_{U,a} = -1$ if a leaves U,

          $M_{U,a} = 0$ otherwise

for $U \in \mathscr{C}$ and $a \in A$. It can be shown with the tree-representation of $\mathscr{C}$ that M is a network matrix, represented by the tree T (of the tree-representation for $\mathscr{C}$) and D. Conversely, each network matrix can be described in this way. We leave the details to the reader. This equivalent representation of network matrices will be useful in the future.

Example 6. $\{0, 1\}$-*network matrices and alternating digraphs* (Hoffman and Kruskal [1956], Heller and Hoffman [1962]). By the definition of network matrices (Example 4), any nonnegative (i.e. $\{0, 1\}$-) network matrix comes from a directed tree T by choosing some directed paths in T, and by taking as columns of the matrix the incidence vectors of these paths (as vectors in $\mathbb{R}^{A_0}$, where A_0 is the set of arcs of T). Another characterization was given by Hoffman and Kruskal [1956] and Heller and Hoffman [1962]. A directed graph is called *alternating* if in each circuit the arcs are oriented alternatingly forwards and backwards. So each alternating digraph is acyclic, and has no odd circuits. Let $D = (V, A)$ be an alternating digraph. Make the matrix M with rows indexed by V, and with columns the incidence vectors of some directed paths in D. Then M is a network matrix, and each $\{0, 1\}$-network matrix arises in this way.

[To see that M is a network matrix, make a directed tree as follows. First replace each vertex v of D by two new vertices v^+ and v^-, and each arc $a = (v, w)$ by the 'red' arc $a = (v^-, w^+)$. Moreover, add the 'blue' arcs (v^+, v^-) for all $v \in V$. Next contract all red arcs. Then we are left with a blue directed tree T, with $|V|$ arcs. Each directed path in D (as a vertex set) coincides with a directed path in T (as an arc set) (not-necessarily conversely). So M is a network matrix.

To see that each $\{0, 1\}$-network matrix arises in this way, let M be such a matrix, coming from the directed tree $T = (V, A_0)$. Make the directed graph $D = (A_0, A)$, where for arcs a, a' in A_0, (a, a') is in A if and only if the head of a is the same as the tail of a'. Then D is alternating, and directed paths in D (as vertex sets) coincide with directed paths in T (as arc sets). So each $\{0, 1\}$-network matrix comes from an alternating digraph.]

Hoffman and Kruskal [1956] showed that if $D = (V, A)$ is a directed graph, and M is the matrix with rows indexed by V, and with columns the incidence vectors of all directed paths in D, then M is totally unimodular if and only if D is alternating.

Example 7. *Interval matrices.* If M is a $\{0, 1\}$-matrix, and each column of M has its 1's consecutively (assuming some linear ordering of the rows of M), then M is totally unimodular. These matrices are called *interval matrices*. In fact, each interval matrix is a network matrix, as follows by taking a directed path for the directed tree T in Example 4.

So all examples of totally unimodular matrices given above are (special cases of) network matrices. Clearly, also the transposes of network matrices are totally unimodular. In Section 19.4 we shall see two totally unimodular 5×5-matrices which are neither network matrices nor their transposes. Seymour proved that from these two 5×5-matrices, together with the network matrices, all totally unimodular matrices can be composed.

Note. Edmonds and Karp [1972] showed that a certain modification of Ford and Fulkerson's min-cost flow algorithm [1957] solves linear programs over network matrices in polynomial time. Also the simplex method yields a polynomial-time algorithm—see Ikura and Nemhauser [1983], and Orlin [1985a] (see Note 2 in Section 11.4). Tardos [1985a] showed that there exists a strongly polynomial algorithm for network problems—see Section 15.2.

19.4. DECOMPOSITION OF TOTALLY UNIMODULAR MATRICES

Network matrices form the basis for all totally unimodular matrices. This was shown by a deep and beautiful theorem of Seymour [1980] (cf. [1985]).

Not all totally unimodular matrices are network matrices or their transposes, as is shown by the matrices (cf. Hoffman [1960], Bixby [1977]):

$$(42) \quad \begin{bmatrix} 1 & -1 & 0 & 0 & -1 \\ -1 & 1 & -1 & 0 & 0 \\ 0 & -1 & 1 & -1 & 0 \\ 0 & 0 & -1 & 1 & -1 \\ -1 & 0 & 0 & -1 & 1 \end{bmatrix}, \begin{bmatrix} 1 & 1 & 1 & 1 & 1 \\ 1 & 1 & 1 & 0 & 0 \\ 1 & 0 & 1 & 1 & 0 \\ 1 & 0 & 0 & 1 & 1 \\ 1 & 1 & 0 & 0 & 1 \end{bmatrix}.$$

Seymour showed that each totally unimodular matrix arises, in a certain way, from network matrices and the matrices (42).

To describe Seymour's characterization, first observe that total unimodularity is preserved under the following operations:

(43)   (i) permuting rows or columns;

       (ii) taking the transpose;

       (iii) multiplying a row or column by -1;

       (iv) pivoting, i.e. replacing $\begin{bmatrix} \varepsilon & c \\ b & D \end{bmatrix}$ by $\begin{bmatrix} -\varepsilon & \varepsilon c \\ \varepsilon b & D - \varepsilon bc \end{bmatrix}$;

       (v) adding an all-zero row or column, or adding a row or column with one nonzero, being ± 1;

       (vi) repeating a row or column.

[In (iv), $\varepsilon \in \{\pm 1\}$, b is a column vector, c is a row vector, and D is a matrix.] Moreover, total unimodularity is preserved under the following compositions:

(44)   (i) $A \oplus_1 B := \begin{bmatrix} A & 0 \\ 0 & B \end{bmatrix}$ (1-sum);

       (ii) $[A \quad a] \oplus_2 \begin{bmatrix} b \\ B \end{bmatrix} := \begin{bmatrix} A & ab \\ 0 & B \end{bmatrix}$ (2-sum);

       (iii) $\begin{bmatrix} A & a & a \\ c & 0 & 1 \end{bmatrix} \oplus_3 \begin{bmatrix} 1 & 0 & b \\ d & d & B \end{bmatrix} := \begin{bmatrix} A & ab \\ dc & B \end{bmatrix}$ (3-sum).

(here A and B are matrices, a and d are column vectors, and b and c are row vectors, of appropriate sizes). It is not difficult to see that total unimodularity is maintained under these compositions, e.g. with Ghouila-Houri's characterization ((iv) in Theorem 19.3).

Theorem 19.6 (Seymour's decomposition theorem for totally unimodular matrices). *A matrix A is totally unimodular if and only if A arises from network matrices and the matrices (42) by applying the operations (43) and (44). Here the operations (44) are applied only if both for A and for B we have: the number of rows plus the number of columns is at least 4.*

We refer to Seymour [1980] for the proof, which is in terms of matroids. As a basis a 'splitter' technique for matroids is developed (if a regular matroid contains a certain matroid as a proper minor, then the matroid can be 'split'). Seymour's theorem immediately implies a good characterization of total unimodularity.

Corollary 19.6a. *The problem: 'given a matrix, is it totally unimodular?' has a good characterization (i.e. is in $\mathcal{NP} \cap \text{co-}\mathcal{NP}$).*

Proof. In order to show that a given matrix M is not totally unimodular, it suffices to specify a submatrix with determinant not equal to 0 or ± 1. To show that M is totally unimodular, we can specify a construction of M as described in Theorem 19.6. Such a construction can be described in polynomially bounded space (note that a series of pivoting operations can be described in a short way: if $A = \begin{bmatrix} E & C \\ B & D \end{bmatrix}$, where E is a square matrix of order k, and if we pivot successively on the elements in positions $(1,1), (2,2), \ldots, (k,k)$, then E is nonsingular, and we end up with the matrix

$$(45) \qquad \begin{bmatrix} -E^{-1} & E^{-1}C \\ BE^{-1} & D - BE^{-1}C \end{bmatrix}).$$ □

In Chapter 20 we shall see that in fact a polynomial algorithm for testing total unimodularity follows.

Here we describe another consequence.

Corollary 19.6b. *Let M be a totally unimodular matrix. Then we are in at least one of the following cases:*

(i) *M or its transpose is a network matrix,*

(ii) *M is one of the matrices (42), possibly after permuting rows or columns, or multiplying some rows and columns by -1;*

(iii) *M has a row or column with at most one nonzero, or M has two linearly dependent rows or columns;*

(iv) *the rows and columns of M can be permuted so that $M = \begin{bmatrix} A & B \\ C & D \end{bmatrix}$, with rank$(B)$ + rank$(C) \leqslant 2$, where both for A and for D we have: the number of rows plus the number of columns is at least 4.*

Proof. We may assume that each row and each column of M has at least two nonzeros, and that each two rows (columns) of M are linearly independent (otherwise we have conclusion (iii)). So if we construct M as in Theorem 19.6 we did not apply (43)(v) or (vi); either we did apply at least once one of the operations (44), in which case conclusion (iv) follows, or we did not apply (44), in which case M satisfies (i) or (ii). (Note that each of the conclusions (i), (ii), and (iv) is preserved under the operations (43)(i)–(iv).)   □

The fact that Seymour's theorem yields a polynomial-time test for total unimodularity is based on the observation that a matrix M can be tested to satisfy (i), (ii) or (iii) of Corollary 19.6b in polynomial time, and moreover that a decomposition of M as in (iv) can be found in polynomial time, if it exists. This last reduces the problem of testing M for total unimodularity to testing smaller matrices. Inductively, this gives a polynomial-time algorithm. We describe the algorithm in the next chapter.

20

Recognizing total unimodularity

Seymour's decomposition theorem, described in Section 19.4, yields a polynomial-time algorithm for testing whether a given matrix is totally unimodular. As ingredients one needs a polynomial-time algorithm for testing whether a given matrix is a network matrix—see Section 20.1—and a polynomial-time algorithm for finding a decomposition in the sense of Seymour—see Section 20.2. Combination of these tests yields a total unimodularity test—see Section 20.3.

20.1. RECOGNIZING NETWORK MATRICES

The first subroutine in recognizing total unimodularity is a polynomial-time algorithm to test if a given matrix is a network matrix. Such an algorithm was designed by Auslander and Trent [1959, 1961], Gould [1958], Tutte [1960, 1965, 1967], and Bixby and Cunningham [1980] (the latter paper gives more references—see also Bixby and Wagner [1988]). The algorithm we describe here is based on Tutte's method, and includes an idea of Bixby and Cunningham.

Testing a matrix for being a network matrix can be seen to contain as special case testing a graph for being planar: the *transpose* of the network matrix represented by the tree $T = (V, A_0)$ and digraph $D = (V, A)$, is a network matrix itself if and only if the digraph $(V, A_0 \cup A)$ is planar. This can be derived from Kuratowski's [1930] characterization of planar graphs—see Tutte [1958] (it can be derived also from the algorithm below).

It is convenient to keep in mind the usual algorithm for testing planarity of a (now undirected) graph $G = (V, E)$ (cf. Auslander and Parter [1961], Tutte [1963], Hopcroft and Tarjan [1974] (gives more references), Rubin [1975b]):

(1)   Choose a circuit C in G whose removal will split E into at least two 'pieces' (removal in the 'topological' sense: two different edges e and f are in the same piece iff there exists a path whose vertex set does not intersect C, but does intersect both e and f). If no such circuit exists, planarity testing is trivial. If C exists, then in any planar embedding of G, each piece will be either completely inside C or completely outside C. If two pieces P_1 and P_2 have 'intertwining' contact points with C, then these pieces will 'obstruct' each other if they are both inside C or both outside C. Now

make a graph H on the set of pieces, by letting P_1 and P_2 be adjacent iff the contact points of P_1 and P_2 with C intertwine. Then it is easy to see that G is planar if and only if H is bipartite and $C \cup P$ is planar for each piece P.

So planarity testing is reduced to bipartiteness testing together with planarity testing of smaller graphs. This recursively defines a (polynomial-time) algorithm. Hopcroft and Tarjan [1974] gave a linear-time implementation.

The algorithm for recognizing network matrices runs parallel. Recall that a matrix M is a network matrix if and only if there is a digraph $D = (V, A)$ and a cross-free family $\mathscr{C} \subseteq \mathscr{P}(V)$ such that M satisfies (41) of Section 19.3. Also recall that a collection of subsets of V is cross-free if and only if it has a 'planar' Venn diagram. So planarity crops up again, albeit in a different context, but it gives a link between testing matrices for being 'network' and graphs for being planar.

Algorithm for recognizing network matrices

Let M be an $m \times n - \{0, \pm 1\}$-matrix. If each column of M contains at most two nonzero entries, M is easily tested for being a network matrix.

To this end, we aim at splitting the row collection R of M into classes R_1 and R_2 so that if we multiply all rows in R_2 by -1, the resulting matrix has in each column at most one 1 and at most one -1 (cf. Example 3 in Section 19.3). If such a splitting is possible, M is a network matrix (easily). If it is not possible, M is not a network matrix, as it is not totally unimodular (by Ghouila-Houri's characterization (iv) of Theorem 19.3)).

To split R as above, define the undirected graph G as follows. Let $v_1, \ldots, v_m$ be vertices. Join two distinct vertices v_i and v_j by an edge if there is a column in M which has nonzero entries of the same sign in positions i and j. Join two distinct vertices v_i and v_j by a path of length two (with a new intermediate vertex) if there is a column in M which has nonzeros of opposite sign in positions i and j. Now R can be split as above if and only if G is bipartite. This last can be checked easily in linear time (cf. Section 1.4).

So we may assume that M has a column with at least three nonzeros. Define for any row index $i = 1, \ldots, m$ the undirected graph G_i as follows. The vertex set of G_i is $\{1, \ldots, m\} \backslash \{i\}$. Two vertices j and k are adjacent in G_i if and only if M has a column which has nonzeros in positions j and k and a zero in position i. Then:

(2)   *If M is a network matrix, there exists an i for which G_i is disconnected.*

To see this, let M be a network matrix, and let the directed tree T and the digraph D represent M, where T has arcs $a_1, \ldots, a_m$, corresponding to the row indexes $1, \ldots, m$. Then T contains a path of length three, as otherwise each column of M would have at most two nonzeros. Let a_i be the middle arc of such a path. Then if arc a_j and a_k of T are at different sides of a_i in T (i.e. in different components of $T \backslash a_i$), then j and k are not adjacent in G_i: any path in T containing a_j and a_k also contains a_i; so any column of M with nonzeros in positions j and k has a nonzero also in position i. This implies that G_i is disconnected.

So if each G_i is connected, we may conclude that M is not a network matrix. Otherwise, let G_1 be disconnected, without loss of generality. Let $C_1, \ldots, C_p$ be the connected components of G_1. Define

(3) $W :=$ the support of the first row of M;
        $W_i := W \cap$ the support of the ith row of M;
        $U_k := \bigcup \{ W_i | i \in C_k \}$;

for $i = 2, \ldots, m; \; k = 1, \ldots, p$ (the *support* of a vector is the set of coordinates in which the vector is nonzero).

Make the undirected graph H, with vertices $C_1, \ldots, C_p$, where distinct C_k and C_l are adjacent if and only if:

(4) $\exists i \in C_k: \quad U_l \nsubseteq W_i$ and $U_l \cap W_i \neq \varnothing, \quad$ and
        $\exists j \in C_l: \quad U_k \nsubseteq W_j$ and $U_k \cap W_j \neq \varnothing.$

For each component C_k, let M_k be the submatrix of M consisting of the first row of M and the rows of M with index in C_k.

As a final preliminary we show:

(5)   *M is a network matrix if and only if: H is bipartite and M_k is a network matrix for $k = 1, \ldots, p$.*

Inductively this implies that being a network matrix can be tested in polynomial time—see Theorem 20.1 below.

Proof of (5). **Necessity.** If M is a network matrix, clearly each M_k is a network matrix. To see that H is bipartite, let the directed tree T and the digraph D represent M, where T has arcs $a_1, \ldots a_m$ corresponding to the row indices $1, \ldots, m$. Let, without loss of generality, $\{a_2, \ldots, a_t\}$ and $\{a_{t+1}, \ldots, a_m\}$ be the arc sets of the two components of $T \backslash a_1$. So each C_k is either contained in $\{2, \ldots, t\}$ or in $\{t + 1, \ldots, m\}$. To show bipartiteness of H, we show that, for all k, l, if C_k and C_l are subsets of the same set (i.e., $C_k, C_l \subseteq \{2, \ldots, t\}$ or $C_k, C_l \subseteq \{t + 1, \ldots, m\}$), then they are not adjacent in H.

To see this, let $A_k := \{a_i | i \in C_k\}$ and $A_l := \{a_i | i \in C_l\}$. Then A_k and A_l form subtrees of T, both at the same side of a_1. Let v_k (v_l) be the (unique) vertex of T covered by A_k (A_l) nearest to a_1. Then we are in one or both of the following two cases.

(i) The path in T from v_k to a_1 contains no arc in A_l. Then for each arc a_i in A_k, either it is on the path from v_k to v_l in which case $U_l \subseteq W_i$, or it is not on this path, in which case $U_l \cap W_i = \varnothing$. Hence (4) is not satisfied, and C_k and C_l are not adjacent in H.

(ii) The path in T from v_l to a_1 contains no arc in A_k. This case is parallel to (i).

Sufficiency. Suppose H is bipartite and each M_k is a network matrix. Let, say, $C_1, \ldots, C_q$ form one colour class of H, and $C_{q+1}, \ldots, C_p$ the other.

Claim 1. Define a relation $\prec$ on $\{1, \ldots, q\}$ by: $k \prec l$ if and only if $k \neq l$ and

(6) $\exists j \in C_l: U_k \nsubseteq W_j$ and $U_k \cap W_j \neq \varnothing.$

Then $\prec$ is a partial order.

Proof of Claim 1. By (4), for no $k, l \in \{1, \dots, q\}$, both $k \prec l$ and $l \prec k$. So it suffices to show transitivity. Let $h \prec k$ and $k \prec l$. Then (6) is satisfied, and similarly:

(7) $\exists t \in C_k: U_h \nsubseteq W_t$ and $U_h \cap W_t \neq \varnothing$.

As $U_h \cap W_t \neq \varnothing$, we know that $W_s \cap U_k \neq \varnothing$ for some $s \in C_k$. Since h and k are not adjacent in H, it follows from (4) that $U_k \subseteq W_s$ (as (7) holds). Therefore $U_k \subseteq W_s \subseteq U_h$, and hence by (6),

(8) $U_h \nsubseteq W_j$ and $U_h \cap W_j \neq \varnothing$

which implies $h \prec l$. Since for no $k, l = 1, \dots, q$ we have both $k \prec l$ and $l \prec k$ (as C_k and C_l are non-adjacent in H (cf. (4))), we know that $\prec$ is a partial order.

End of Proof of Claim 1.

So we may assume that $k \prec l$ implies $k < l$.

Suppose now we have constructed a directed tree $T = (V, A_0)$ and a directed graph $D = (V, A)$ which represent the submatrix M_0 of M consisting of the rows with index in $\{1\} \cup C_1 \cup \dots \cup C_{k-1}$, in such a way that the tail, say v_0, of the arc of T corresponding to the first row is an end point of the tree (clearly, this is possible for $k = 1$). Let $a_1, \dots, a_n$ be the arcs of D, such that arc a_j corresponds to column j (possibly loops). Then index j belongs to W if and only if a_j connects v_0 with another vertex of T.

Let $T' = (V', A_0')$ and $D' = (V', A')$ represent the submatrix M_k of M. As C_k is a connected subgraph of G_1, we know that the arc of T' corresponding to row 1 is an end arc of T'. Since also after reversing the orientation of every arc of T' and D' they represent M_k, we may assume that the tail, say v_0', of this arc is an end point of T'. Again, let $a_1', \dots, a_n'$ be the arcs of D', such that a_j' corresponds to column j. Index j belongs to W if and only if a_j' connects v_0' with another vertex of T'.

If neither a_j nor a_j' is a loop, then column j of M has a nonzero in some of the positions $\{1\} \cup C_1 \cup \dots \cup C_{k-1}$ and in some of the positions $\{1\} \cup C_k$. Hence, by definition of W and $C_1, \dots, C_p$, we know that j is in W. We show:

Claim 2. There is a vertex w of T such that for all $j \in W$, if $a_j' \neq (v_0', v_1')$, then a_j connects v_0 and w.

Proof of Claim 2. The claim is equivalent to:

(9)   if the columns with index $j, j' \in W$ have at least one nonzero in C_k, then columns j and j' have the same nonzero positions in $C_1 \cup \dots \cup C_{k-1}$.

In order to prove (9), suppose j and j' contradict (9). From the premise it follows that $j, j' \in U_k$. The conclusion being not true gives that there is an $l < k$ and an $i \in C_l$ such that $j \in W_i$ and $j' \notin W_i$, or $j \notin W_i$ and $j' \in W_i$. This contradicts that (6) implies $k < l$.

End of Proof of Claim 2.

Now choose w as in Claim 2. Delete arc a_1' and vertex v_0' from T', and identify vertex v_1' of T' with vertex w of T. Thus we obtain a new tree T'', with vertex set $V'' = V \cup V' \setminus \{v_0', v_1'\}$.

Next, if $a_j' \neq (v_0', v_1')$ and $j \in W$, then either $a_j = (v_0, w)$ and $a_j' = (v_0', v_2')$ for some v_2', in which case we define $a_j'' := (v_0, v_2')$, or $a_j = (w, v_0)$ and $a_j' = (v_2', v_0')$ for some v_2', in which case we define $a_j'' := (v_2', v_0)$. For all other a_j we let $a_j'' := a_j$, or, if a_j is a loop, $a_j'' := a_j'$.

It is easy to see that T'' and $D'':=(V'',\{a_1'',\ldots,a_n''\})$ represent the submatrix of M consisting of the rows with index in $\{1\}\cup C_1\cup\ldots\cup C_k$, in such a way that the tail of the arc corresponding to row 1 is an end vertex.

So inductively we can represent the submatrix M^+ of M consisting of the rows with index in $\{1\}\cup C_1\cup\ldots\cup C_q$ by a tree T^+ and a digraph D^+ so that the tail, say v_0^+, of the arc, say a_1^+, of T^+ corresponding to row 1 is an end vertex.

Similarly, we can represent the submatrix M^- of M consisting of the rows with index in $\{1\}\cup C_{q+1}\cup\ldots\cup C_p$ by a tree T^- and a digraph D^- so that the *head*, say v_0^-, of the arc, say a_1^-, of T^+ corresponding to row 1 is an end vertex.

Now identifying arc a_1^+ of T^+ and arc a_1^- of T^- gives a new tree T. Moreover, if arcs (v_0^+,w) of D^+ and (w',v_0^-) of D^- correspond to the same column of M (which therefore has index in W), then we replace them by one new arc (w',w). Similarly, if arcs (w,v_0^+) of D^+ and (v_0^-,w') of D^- correspond to the same column of M (with index in W), then we replace them by one new arc (w,w'). Let these new arcs, together with the not replaced arcs of D^+ and D^-, form the digraph D. Then T and D represent M. □

This implies:

Theorem 20.1. *A given matrix can be tested for being a network matrix in polynomial time.*

Proof. It is not difficult to see that there exists a constant C and an exponent $t \geqslant 2$ such that, for each $m \times n$-matrix M, the following routines take time at most $C(mn)^t$:

(10)   (i) testing if each entry of M is 0, +1, or −1;
        (ii) testing if each column of M contains at most two nonzeros;
        (iii) if each column of M has at most two nonzeros, testing if M is a network matrix (see above);
        (iv) if M has a column with at least three nonzeros, constructing the graphs G_i $(i=1,\ldots,m)$, testing if one of the G_i is disconnected, constructing the graph H, testing if H is bipartite, and determining the matrices M_k (see above).

By (5), if M stands each of these tests, M is a network matrix if and only if each matrix M_k is a network matrix. This is a recursive definition of a test for being a network matrix. We show that its running time, for an $m \times n$-matrix, is bounded by $Cm^{t+1}n^t$, by induction on m.

If the matrices M_k are not made, then apparently the algorithm stops during the routines (10), and hence the time is bounded by $Cm^t n^t$.

If the matrices M_k are made, we need to test each M_k for being a network matrix. This requires time at most $Cm_k^{t+1}n^t$, by the induction hypothesis, where m_k is the number of rows of M_k. Moreover, the routines (10) require time at most $Cm^t n^t$. Altogether, the time is bounded by:

(11) $Cm^t n^t + Cm_1^{t+1}n^t + \cdots + Cm_p^{t+1}n^t \leqslant Cm^{t+1}n^t.$ □

Note that if M is a network matrix, the algorithm gives an explicit tree and a digraph representing M.

Note. The planarity test (1) can be extended to yield as a by-product Kuratowski's [1930] famous characterization of planar graphs (a graph is planar if and only if it does not contain a subgraph homeomorphic to K_5 or to $K_{3,3}$): if in (1) no appropriate circuit C exists or H is not bipartite, K_5 or $K_{3,3}$ can be derived.

In the same way, a further analysis of the above algorithm will give a characterization of Tutte [1958] of network matrices. To describe this characterization, consider the following operations on totally unimodular matrices:

(12)   (i) deleting rows or columns;

(ii) pivoting, i.e. replacing the matrix $\left[\begin{array}{c|c} \varepsilon & c \\ \hline b & D \end{array}\right]$ (where $\varepsilon \in \{\pm 1\}$, b is a column vector, c is a row vector, and D is a matrix), by

$$\left[\begin{array}{c|c} -\varepsilon & \varepsilon c \\ \hline \varepsilon b & D - \varepsilon b c \end{array}\right]$$

In Section 19.3 we saw that the class of network matrices is closed under these operations. Now Tutte showed that a totally unimodular matrix M is a network matrix, if and only if it cannot be transformed by the operations (12) to the *transpose* of one of the network matrices represented by a directed tree $T = (V, A_0)$ and digraph $D = (V, A)$, with $(V, A \cup A_0)$ being (an oriented version of) K_5 or $K_{3,3}$.

Most of these results can be formulated more smoothly, and extended, in the framework of matroids—see Welsh [1976], Seymour [1981], and Cunningham [1982].

20.2. DECOMPOSITION TEST

As the second ingredient for testing total unimodularity, we prove the following theorem due to Cunningham and Edmonds [1980].

Theorem 20.2. *There is a polynomial algorithm to check if the rows and columns of a given matrix M can be permuted in such a way that M can be decomposed as*

$$(13) \qquad M = \begin{bmatrix} A & B \\ C & D \end{bmatrix}$$

such that rank (B) + rank $(C) \leqslant 2$, *and such that both for A and for D we have: the number of rows plus the number of columns is at least* 4.

[Cunningham and Edmonds observed that such a decomposition can be found with the *matroid intersection algorithm*—we describe the method in terms of just matrices.]

Proof. Let X be the set of columns of the matrix $[I \ M]$. Let X_1 and X_2 denote the sets of columns of I and M, respectively.

Finding a decomposition as above is easily seen to be equivalent to finding a subset Y of X such that

$$(14) \qquad \rho(Y) + \rho(X \setminus Y) \leqslant \rho(X) + 2$$

(where, for any set Y of vectors,

(15) $\rho(Y) :=$ the rank of the vectors in Y $(:= \dim (\text{lin.hull } Y)))$

and such that $|Y| \geqslant 4$, $|X \backslash Y| \geqslant 4$, and $Y \cap X_1 \neq \varnothing \neq Y \cap X_2$, $X_1 \backslash Y \neq \varnothing \neq X_2 \backslash Y$.

Now such a Y can be found in polynomial time if we have a polynomial algorithm for the following problem:

(16)   given a set X of vectors, and $S, T \subseteq X$ with $S \cap T = \varnothing$, find a set Y such that $S \subseteq Y \subseteq X \backslash T$ and such that $\rho(Y) + \rho(X \backslash Y)$ is as small as possible.

To see that this suffices, in order to find Y as in (14), we can check, for all $S, T \subseteq X$ with $|S| = 4, |T| = 4, S \cap X_1 \neq \varnothing \neq S \cap X_2$ and $T \cap X_1 \neq \varnothing \neq T \cap X_2$, whether there exists a set Y with $S \subseteq Y \subseteq X \backslash T, \rho(Y) + \rho(X \backslash Y) \leqslant \rho(X) + 2$. This can be done by $O(|X|^8)$ repetitions of (16).

A polynomial method for (16) works as follows. Let $S, T \subseteq X, S \cap T = \varnothing$, $V := X \backslash (S \cup T)$. Suppose we have found a subset $Z \subseteq V$ satisfying

(17) $\rho(S \cup Z) = \rho(S) + |Z|$,

         $\rho(T \cup Z) = \rho(T) + |Z|$.

(At start, $Z = \varnothing$.) We try to make Z as large as possible. To this end, make a digraph $D = (V, E)$ as follows: for $u \in Z$, $v \in V \backslash Z$:

(18) $(u, v) \in E$ iff $\rho(S \cup (Z \backslash \{u\}) \cup \{v\}) = \rho(S) + |Z|$;

         $(v, u) \in E$ iff $\rho(T \cup (Z \backslash \{u\}) \cup \{v\}) = \rho(T) + |Z|$.

There are no other arcs in D. So D is a directed bipartite graph, with colour classes Z and $V \backslash Z$. Define:

(19) $U := \{v \in V \backslash Z \mid \rho(S \cup Z \cup v) = \rho(S) + |Z| + 1\}$,

         $W := \{v \in V \backslash Z \mid \rho(T \cup Z \cup v) = \rho(T) + |Z| + 1\}$.

Then we are in one of the following two cases.

Case 1. There is a directed path in D starting in U and ending in W. Let Π be a shortest such path. Define

(20) $Z' := Z \Delta \{v \in V \mid v \text{ occurs in } \Pi\}$

(Δ denoting symmetric difference). Then Z' satisfies

(21) $\rho(S \cup Z') = \rho(S) + |Z'|$,

         $\rho(T \cup Z') = \rho(T) + |Z'|$.

To see that $\rho(S \cup Z') = \rho(S) + |Z'|$, let $v_0, z_1, v_1, z_2, v_2, \ldots, z_t, v_t$ be the vertices in Π (in this order), with $v_0 \in U$, $v_t \in W$. So $Z' = (Z \backslash \{z_1, \ldots, z_t\}) \cup \{v_0, \ldots, v_t\}$. Since Π is a shortest path, we know $(v_i, z_j) \notin E$ if $j \geqslant i + 2$, and $(z_i, v_j) \notin E$ if $j \geqslant i + 1$.

Since v_0 is in U, $\rho(S \cup Z \cup v_0) = \rho(S) + |Z| + 1$. Hence there exists a subset S' of S such that $S' \cup Z \cup \{v_0\}$ forms a collection of $\rho(S) + |Z| + 1$ linearly independent columns. By applying row operations we may assume that the columns in $S' \cup Z \cup \{v_0\}$ are unit basis vectors. Then by permuting rows and columns and by multiplying columns by nonzero scalars, we may assume that the columns in $S' \cup Z \cup \{v_0, \ldots, v_t\}$ look as follows:

(22)

| | S' | | $Z\setminus\{z_1,\ldots,z_t\}$ | | $z_1 \cdots\cdots\cdots z_t$ | | v_0 | $v_1 \cdots\cdots\cdots v_t$ |

$$
\begin{array}{c}
1\ 0.....0 \quad 0......0 \quad 0......0 \quad 0 \\
0\ \cdot\ \cdot\ \ \ \vdots \quad\quad \vdots \quad\quad \vdots \quad\quad \vdots \\
\vdots\ \cdot\ \cdot\ \cdot\ \vdots \quad\quad \vdots \quad\quad \vdots \quad\quad \vdots \\
0....:0\ 1 \quad 0......0 \quad 0......0 \quad 0 \\
0......0 \quad 1\ 0.....0 \quad 0......0 \quad 0 \\
\vdots \quad\quad 0\ \cdot\ \cdot\ \ \vdots \quad \vdots \quad \vdots \\
\vdots \quad\quad \cdot\ \cdot\ \cdot\ 0 \quad \vdots \quad \vdots \\
0......0 \quad 0....:0\ 1 \quad 0......0 \quad 0 \\
0......0 \quad 0......0 \quad 1\ 0.....0 \quad 0 \quad 1\ 0.....0 \\
\vdots \quad\quad \vdots \quad\quad 0\ \cdot\ \cdot\ \ \vdots \quad \cdot\ \cdot\ \vdots \\
\vdots \quad\quad \vdots \quad\quad \cdot\ \cdot\ \cdot\ 0 \quad \vdots \quad \cdot\ \cdot\ 0 \\
0......0 \quad 0......0 \quad 0....:0\ 1 \quad 0 \quad \cdot\ 1 \\
0......0 \quad 0......0 \quad 0......0 \quad 1\ 0......0 \\
0......0 \quad 0......0 \quad 0......0 \quad 0 \\
\vdots \quad\quad \vdots \quad\quad \vdots \quad\quad \vdots \\
0......0 \quad 0......0 \quad 0......0 \quad 0
\end{array}
$$

From (22) it will be clear that the columns in $S' \cup Z'$ are linearly independent, which gives the first line of (21). Similarly the second line follows.

Now $|Z'| = |Z| + 1$, and we can repeat the iteration with Z replaced by Z'.

Case 2. There is no directed path in D from U to W. Define

(23) $Y := S \cup \{v \in V \mid$ there is a directed path in D from v to $W\}$.

Then Y minimizes $\rho(Y) + \rho(X \setminus Y)$ over all Y with $S \subseteq Y \subseteq X \setminus T$ (thus solving (16)). This follows from the fact that (17) implies that for all Y' with $S \subseteq Y' \subseteq X \setminus T$:

(24) $\rho(Y') \geqslant \rho((Y' \cap Z) \cup S) = \rho(S) + |Y' \cap Z|$,

$\rho(X \setminus Y') \geqslant \rho((Z \setminus Y') \cup T) = \rho(T) + |Z \setminus Y'|$.

This yields

(25) $\rho(Y') + \rho(X \setminus Y') \geqslant |Z| + \rho(S) + \rho(T)$.

Moreover, taking $Y' = Y$ gives equality in (25).

To show this, suppose that strict inequality holds in (25) for $Y' = Y$. Then we have also strict inequality in at least one of the inequalities (24).

First suppose $\rho(Y) > \rho((Y \cap Z) \cup S))$. Then there exists a v in $Y \backslash (Z \cup S)$ such that

$$(26) \qquad \rho((Y \cap Z) \cup S \cup v) = \rho((Y \cap Z) \cup S) + 1 = \rho(S) + |Y \cap Z| + 1.$$

Moreover,

$$(27) \qquad \rho(S \cup Z \cup v) = \rho(S) + |Z|$$

(since if $\rho(S \cup Z \cup v) = \rho(S) + |Z| + 1$, then v is in U, and hence there would be a directed path from U to W (by definition (23) of Y)).

(26) and (27) imply that $\rho(S \cup (Z \backslash u) \cup v) = \rho(S) + |Z|$ for some u in $Z \backslash Y$. This gives that (u, v) is an arc of D, with $u \notin Y$, $v \in Y$, contradicting definition (23) of Y.

One similarly shows that $\rho(X \backslash Y) > \rho((Z \backslash Y) \cup T)$ will lead to a contradiction, thus showing equality in (25) for $Y' = Y$. $\square$

20.3. TOTAL UNIMODULARITY TEST

The methods described in Sections 20.1 and 20.2 are subroutines in the total unimodularity test described in the following theorem.

Theorem 20.3. *The total unimodularity of a given matrix can be tested in polynomial time.*

Proof. We first check if all entries of M belong to $\{0, +1, -1\}$. Next we delete from M rows and columns which have at most one nonzero, and if two rows (columns) are linearly dependent (i.e. they are equal or each other's opposites), we delete one of them. (Total unimodularity is invariant under these operations.) Repeating this we end up with a matrix M in which all rows and columns have at least two nonzeros, and each pair of rows (columns) is linearly independent.

Next we test if M or M^T is a network matrix, or if M is one of the 5×5-matrices (42) in Section 19.4, possibly with rows or columns permuted or multiplied by -1. This can be done in polynomial time (Theorem 20.1). If one of these tests has a positive result, M is totally unimodular. Otherwise, we test with the algorithm of Theorem 20.2 whether M can be decomposed as in (13). If such a decomposition does not exist, M is not totally unimodular, by Seymour's theorem (Corollary 19.6b). If M can be decomposed as in (13), we are in one of the following six cases.

Case 1. rank $(B) =$ rank $(C) = 0$. Then $B = 0$ and $C = 0$, and hence trivially: M is totally unimodular if and only if A and D are totally unimodular.

Case 2. rank $(B) = 1$ and rank $(C) = 0$. Say $B = fg$, where f is a column $\{0, \pm 1\}$-vector and g is a row $\{0, +1\}$-vector. Then M is totally unimodular if and only if the matrices $[A \quad f]$ and $\begin{bmatrix} g \\ D \end{bmatrix}$ are totally unimodular. This is easy, e.g. by using Ghouila-Houri's characterization ((iv) in Theorem 19.3).

Case 3. rank $(B) = 0$ and rank $(C) = 1$. Similar to Case 2.

Case 4. rank $(B) =$ rank $(C) = 1$. We may assume that a decomposition (13) with rank (B) + rank $(C) \leq 1$ is not possible. By permuting rows or columns, and by multiplying rows or columns by -1, we may assume that the decomposition of M is:

$$(28) \qquad M = {}^{m_1}\Biggl\{ \begin{bmatrix} A_1 & A_2 & 0 & 0 \\ A_3 & A_4 & 1 & 0 \\ 0 & 1 & D_1 & D_2 \\ 0 & 0 & D_3 & D_4 \end{bmatrix} \Biggr\} m_2 \qquad \text{where}$$

$$A = \begin{bmatrix} A_1 & A_2 \\ A_3 & A_4 \end{bmatrix}, \quad D = \begin{bmatrix} D_1 & D_2 \\ D_3 & D_4 \end{bmatrix}$$

and where 0 and 1 stand for (appropriately sized) all-zero and all-one matrices.

Now find $\varepsilon_1, \varepsilon_2 \in \{+1, -1\}$ as follows. Make the bipartite graph G with as one colour class the rows of the matrix A and as the other colour class the columns of A, where a row r and a column c are adjacent if and only if A has a nonzero entry in the intersection of r and c. Let R and K be the sets of rows and of columns intersecting A_4. Then G has a path from R to K. Otherwise, the vertices of G can be split into two classes V_1 and V_2 such that $R \subseteq V_1$, $K \subseteq V_2$, and no edge connects V_1 and V_2. This yields

$$(29) \qquad M = \left[\begin{array}{cc|cc} 0 & & 0 & 0 \\ \hline 0 & 0 & 0 & 0 \\ 0 & A_4 & 1 & 0 \\ \hline 0 & 0 & 1 & D_1 & D_2 \\ 0 & 0 & 0 & D_3 & D_4 \end{array} \right]$$

with $A_4 = 0$. One can check that this gives a decomposition of M as in Case 2 or 3 above, which we have excluded by assumption.

Let Π be a shortest path from R to K, and let δ be the sum of the entries of A corresponding to edges in Π. As Π has odd length, δ is odd. Now define

$$(30) \qquad \begin{aligned} \varepsilon_1 &:= +1 \qquad \text{if } \delta \equiv 1 \ (\text{mod } 4) \\ \varepsilon_1 &:= -1 \qquad \text{if } \delta \equiv -1 \ (\text{mod } 4). \end{aligned}$$

(Π is an arbitrary shortest path, so that if A_4 has a nonzero entry, we can take ε_1 equal to this entry.) One similarly defines ε_2 with respect to the matrix D.

Now we claim: M is totally unimodular if and only if the matrices

$$(31) \qquad \begin{bmatrix} A_1 & A_2 & 0 & 0 \\ A_3 & A_4 & 1 & 1 \\ 0 & 1 & 0 & \varepsilon_2 \end{bmatrix} \quad \text{and} \quad \begin{bmatrix} \varepsilon_1 & 0 & 1 & 0 \\ 1 & 1 & D_1 & D_2 \\ 0 & 0 & D_3 & D_4 \end{bmatrix}$$

are totally unimodular.

Proof. Suppose the matrices (31) are totally unimodular. Then $\varepsilon_1 = \varepsilon_2$, as the submatrix of the first matrix in (31), formed by the rows in the shortest path Π together with the bottom row, and by the columns in Π together with the last column, has determinant $\pm(\varepsilon_1\varepsilon_2 - 1)$. As this matrix is totally unimodular, we know $\varepsilon_1 = \varepsilon_2$. The total unimodularity of M now follows with Ghouila-Houri's characterization ((iv) in Theorem 19.3).

Conversely, suppose M is totally unimodular. Then the matrices

$$
(32) \quad
\begin{bmatrix}
A_1 & A_2 & 0 \\
A_3 & A_4 & 1 \\
0 & 1 & \varepsilon_2
\end{bmatrix}
\quad \text{and} \quad
\begin{bmatrix}
\varepsilon_1 & 1 & 0 \\
1 & D_1 & D_2 \\
0 & D_3 & D_4
\end{bmatrix}
$$

are totally unimodular. For the second matrix in (32), this follows with Ghouila-Houri's characterization from the total unimodularity of the submatrix of M formed by the rows in Π and the rows intersecting D, and by the columns in Π and the columns intersecting D. Similarly, the first matrix in (32) is totally unimodular.

Also the matrices

$$
(33) \quad
\begin{bmatrix}
A_1 & A_2 & 0 \\
A_3 & A_4 & 1 \\
0 & 1 & 0
\end{bmatrix}
\quad \text{and} \quad
\begin{bmatrix}
0 & 1 & 0 \\
1 & D_1 & D_2 \\
0 & D_3 & D_4
\end{bmatrix}
$$

are totally unimodular. Indeed, if A_4 contains a zero entry, total unimodularity of the second matrix in (33) follows directly from the total unimodularity of M. If A_4 contains no zero elements, all its entries are equal to ε_2, as follows from the total unimodularity of the first matrix in (32). Then the bipartite graph G defined above will contain a path Π' from R to K not containing an edge directly connecting R and K. Otherwise, again, we could split the vertex set of G into two classes V_1 and V_2 such that $R \subseteq V_1$, $K \subseteq V_2$, and the only edges connecting V_1 and V_2 are the edges connecting R and K. This gives a decomposition as in (29), now with A_4 being an all-ε_2 matrix. This yields a decomposition of M as in Case 2 or 3 above, which we have excluded by assumption.

Let Π' be a shortest path in G from R to K not containing an edge directly connecting R and K. Let M' be the submatrix of M induced by the rows in Π' and the rows intersecting D, and by the columns in Π' and the columns intersecting D. The total unimodularity of M' implies the total unimodularity of the second matrix in (33). Similarly, the total unimodularity of the first matrix follows.

Clearly, the total unimodularity of the matrices in (32) and (33) implies the total unimodularity of the matrices in (31).

Case 5. rank $(B) = 2$ and rank $(C) = 0$. By pivoting among one of the nonzero entries of B we obtain a matrix M' as decomposed as in Case 4. As M is totally unimodular if and only if M' is totally unimodular, this reduces Case 5 to Case 4.

Case 6. rank $(B) = 0$ and rank $(C) = 2$. Similar to Case 5.

In each of the Cases 1 to 6 testing M for total unimodularity is reduced to testing smaller matrices for total unimodularity. Recursively this defines the algorithm. One easily checks that its running time is polynomial.   □

According to Bixby [1982], the implementation of Cunningham and Edmonds [1980] of the decomposition test yields a total unimodularity test with running time $O(m + n)^4 m$, where m and n are the number of rows and columns of the matrix.

R. G. Bland and J. Edmonds showed that Seymour's decomposition theorem also implies a polynomial algorithm to solve linear programs with totally unimodular constraint matrix. Decomposition reduces these linear programs to linear programs over (transposes of) network matrices, i.e. to network problems. These are known to be polynomially solvable (see the references in Section 19.3). Maurras, Truemper, and Akgül [1981] showed that linear programs over totally unimodular matrices are polynomially solvable, by a modification of the relaxation method (cf. Section 12.3). It clearly also follows from Khachiyan's method (Chapter 13).

Notes on Chapter 20. For more on testing total unimodularity, see Truemper [1982].

Bartholdi [1981–2] showed that it is $\mathcal{NP}$-hard to find the largest totally unimodular submatrix of a given matrix.

21

Further theory related to total unimodularity

In this chapter we collect some further results on total unimodularity and related topics. In Sections 21.1 and 21.2 we discuss the concept of *regular matroid* and the related concept of *chain group*, considering total unimodularity from a matroidal point of view. Fundamental results of Tutte and Seymour are formulated in terms of these concepts.

In Section 21.3 we prove an upper bound of Heller, stating that a totally unimodular matrix with m rows has at most $m^2 + m + 1$ different columns.

In Sections 21.4 and 21.5 we discuss two generalizations of totally unimodular matrices: *unimodular* matrices and *balanced* matrices, which each lead again to results in integer linear programming.

21.1. REGULAR MATROIDS AND SIGNING OF $\{0, 1\}$-MATRICES

Total unimodularity has been studied also in terms of regular matroids and chain groups, especially by Tutte [1958]. Here we review some of the results (cf. Welsh [1976: pp. 173–176]).

A pair $(S, \mathscr{I})$, consisting of a finite set S and a nonempty collection $\mathscr{I}$ of subsets of S is called a *matroid* if it satisfies:

(1)   (i) if $T \in \mathscr{I}$ and $S \subseteq T$ then $S \in \mathscr{I}$;
         (ii) if $T, U \in \mathscr{I}$ and $|T| < |U|$, then $T \cup \{u\} \in \mathscr{I}$ for some u in $U \setminus T$.

The sets in $\mathscr{I}$ are called the *independent* sets of the matroid.

We will restrict ourselves to linear matroids. A pair $(S, \mathscr{I})$ is called a *linear matroid* (*represented over* the field $\mathbb{F}$) if S is, or can be represented as, the set of columns of a matrix A over $\mathbb{F}$, where $T \subseteq S$ belongs to $\mathscr{I}$ if and only if the columns in T are linearly independent. It is not difficult to see that this always leads to a matroid. The matrix A is called a *representation* of $(S, \mathscr{I})$.

Clearly, a linear matroid always has a representation $[I \ B]$, where I is the identity matrix. Note that if B' arises from B by a series of pivotings, then $[I \ B]$ and $[I \ B']$ represent isomorphic linear matroids, as $[I \ B']$ arises from $[I \ B]$ by scaling rows or columns, by adding rows to other rows, and by permuting columns.

A linear matroid is called a *regular matroid* if it has a totally unimodular representation A (over $\mathbb{F} = \mathbb{Q}$). A linear matroid is called a *binary matroid* if it can be represented over $GF(2)$ (= the field {0, 1} with two elements).

It is not difficult to see that each regular matroid is a binary matroid: if the regular matroid is represented by the totally unimodular matrix A, let A' arise from A by replacing each -1 by $+1$. Then a submatrix of A is nonsingular over $\mathbb{Q}$, if and only if the corresponding submatrix of A' is nonsingular over $GF(2)$. So A' is a representation of the regular matroid over $GF(2)$.

Conversely, define *signing* of a {0, 1}-matrix as replacing some of the 1's by -1's. It is not difficult to see that for any {0, 1}-matrix B one has:

(2)   the binary matroid represented by $[I \ B]$ is regular if and only if B can be signed to become totally unimodular.

Not every binary matroid is a regular matroid, as is shown by the binary matroids represented by the $GF(2)$-matrices:

(3)
$$\begin{bmatrix} 1 & 0 & 0 & 1 & 1 & 0 & 1 \\ 0 & 1 & 0 & 1 & 0 & 1 & 1 \\ 0 & 0 & 1 & 0 & 1 & 1 & 1 \end{bmatrix} \quad \text{and} \quad \begin{bmatrix} 1 & 0 & 0 & 0 & 1 & 1 & 0 \\ 0 & 1 & 0 & 0 & 1 & 0 & 1 \\ 0 & 0 & 1 & 0 & 0 & 1 & 1 \\ 0 & 0 & 0 & 1 & 1 & 1 & 1 \end{bmatrix}.$$

These two binary matroids are called the *Fano matroid* and the *dual of the Fano matroid*, respectively.

Tutte [1958] gave a deep characterization of regular matroids. To describe this result, we need the concept of minor. Let $M = (S, \mathscr{I})$ be a linear matroid, represented by some matrix, and let S_1 and S_2 be disjoint subsets of S. Let

(4) $\mathscr{I}' := \{S' \subseteq S \backslash (S_1 \cup S_2) |$ the columns of S' are linearly independent modulo lin.hull $(S_1)\}$.

[A set V of vectors is *linearly independent modulo* a linear space L, if $V \cup B$ is linearly independent for some basis B of L (and hence for each basis B of L).] Then $(S \backslash (S_1 \cup S_2), \mathscr{I}')$ is a linear matroid, as one easily checks. A linear matroid arising in this way is called a *minor* of $(S, \mathscr{I})$, obtained by *contracting* S_1 and *deleting* S_2. Any minor of a regular (or binary) matroid is a regular (binary) matroid again. It follows that regular matroids do not have the Fano matroid or its dual as a minor (cf. (3)). Now Tutte's result is that this characterizes regular matroids among the binary matroids.

Theorem 21.1 (Tutte's regular matroid characterization). *A binary matroid is regular if and only if it has no minor isomorphic to the Fano matroid or its dual.*

For proofs, see also Seymour [1980], Truemper [1978b]. For an extension, see Bixby [1976].

The following can be seen to be equivalent to Tutte's theorem (cf. Tutte [1958, 1965, 1971], Truemper [1978b]).

Corollary 21.1a. *Let B be a* $\{0, 1\}$-*matrix. Then B can be signed to become a totally unimodular matrix, if and only if B cannot be transformed to one of the matrices*

(5) $\qquad \begin{bmatrix} 1 & 1 & 0 & 1 \\ 1 & 0 & 1 & 1 \\ 0 & 1 & 1 & 1 \end{bmatrix}$ *and* $\begin{bmatrix} 1 & 1 & 0 \\ 1 & 0 & 1 \\ 0 & 1 & 1 \\ 1 & 1 & 1 \end{bmatrix}$

by a series of the following operations: permuting or deleting rows or columns, and pivoting over GF(2) (i.e. replacing $\begin{bmatrix} 1 & c \\ b & D \end{bmatrix}$ *by* $\begin{bmatrix} 1 & c \\ b & D - bc \end{bmatrix}$ *and deleting minus-signs).*

Proof that Corollary 21.1a is equivalent to Theorem 21.1. By (3), it suffices to show that, for any $\{0, 1\}$-matrix B:

(6)   the binary matroid M represented by $[I \; B]$ has no minor isomorphic to the Fano matroid or its dual, if and only if B cannot be transformed to one of the matrices (5) by a series of operations as described in Corollary 21.1a.

First suppose that B can be transformed to one of the matrices (5). Deleting a column of B corresponds to deleting the corresponding element in M. Deleting the ith row of B corresponds to contracting the ith column of M as represented by $[I \; B]$. Since the matroid does not change by pivoting B (except for the order of its elements in the matrix representation), it follows that M has a minor isomorphic to the Fano matroid or its dual.

Conversely, suppose $M = (S, \mathscr{I})$ has a minor isomorphic to the Fano matroid or its dual. Let the minor arise by contracting S_1 and deleting S_2. Now one easily checks from (4) that we may assume that the columns in S_1 are linearly independent, and that lin.hull$(S) = $ lin.hull$(S \setminus S_2)$. Then by pivoting we may assume that S_1 corresponds to some columns in the I-part of $[I \; B]$, and that S_2 corresponds to some columns in the B-part of $[I \; B]$. This implies that contracting S_1 and deleting S_2 gives the binary matroid represented by $[I \; B']$, where B' arises from B by deleting the columns corresponding to S_2, and deleting the rows of B corresponding to S_1 (through the natural relation between columns of I and rows of B).

So the binary matroid $[I \; B']$ is isomorphic to the Fano matroid or to its dual. Then I must have order 3 or 4, and B' is a 3×4-matrix or a 4×3-matrix. It is finite checking that B' should be one of the matrices (5) (up to permutations of rows and columns). $\qquad \Box$

Camion [1963b] showed:

(7)   *if a* $\{0, 1\}$-*matrix can be signed to become totally unimodular, there is essentially only one such signing, up to multiplications of rows and columns by* -1.

Proof. To see this, let B be a $\{0, 1\}$-matrix, and let G be the bipartite graph with vertices the rows and columns of B, where a row r and a column c are adjacent if and only if B has a 1 in the intersection of r and c. Let T be a maximal forest in G. Let B' be a totally unimodular matrix arising by signing B. By multiplying rows and columns by -1, we may assume that B' has l's in positions corresponding to edges of T: if row r (column c) is an end vertex of T, with end edge e, by induction we can multiply rows and columns of B' by -1 so that B' has l's in positions corresponding to edges in $T \backslash e$. After that we can multiply row r (column c) by -1, if necessary, to obtain also a 1 in position e.

Camion's result now follows from the fact that for any circuit C in G without chords, the sum of the entries of B' corresponding to edges in C is divisible by 4 (as follows from the total unimodularity of B'). Let the edges of G not in T be ordered $e_1, \ldots, e_k$ so that if the circuit in $T \cup e_j$ has larger length than the circuit in $T \cup e_i$, then $j > i$. So for all j, all chords of the circuit in $T \cup e_j$ are edges e_i with $i < j$. Hence e_j belongs to a circuit C without chords, with edges from $T \cup \{e_1, \ldots, e_j\}$. Therefore, if $e_1, \ldots, e_{j-1}$ correspond to positions in B' which are uniquely determined in sign, then this is the case also for e_j. Inductively, the result follows. $\square$

This proof implicitly describes a polynomial algorithm to sign a $\{0, 1\}$-matrix B to a $\{0, \pm 1\}$-matrix B' so that: B can be signed to become totally unimodular if and only if B' is totally unimodular. Using the algorithm of Chapter 20 this implies a polynomial algorithm to test if a $\{0, 1\}$-matrix can be signed to become totally unimodular. Moreover, it implies a polynomial algorithm to test if a binary matroid, given by a $GF(2)$-matrix representing it, is regular.

Note. Fonlupt and Raco [1984] showed that each totally unimodular matrix can be transformed to a $\{0, 1\}$-matrix by pivoting and multiplying rows or columns by -1, by a polynomial-time algorithm.

21.2. CHAIN GROUPS

Related to regular matroids is Tutte's theory of regular chain groups (Tutte [1947, 1956, 1965], cf. Camion [1968]).

A *chain group* is a subgroup of the additive group $\mathbb{Z}^n$, i.e. it is a sublattice of $\mathbb{Z}^n$. A chain group C is *regular* if for each x in C there are $\{0, \pm 1\}$-vectors $x_1, \ldots, x_m$ in C such that $x = x_1 + \cdots + x_m$, and such that the supports of the x_i are contained in the support of x.

Now Tutte showed the following.

Theorem 21.2. *Let A be an integral matrix, let $M := [I\ A]$, and let C be the chain group $\{x \mid Mx = 0, x \text{ integral}\}$. Then A is totally unimodular if and only if C is regular.*

Proof. First suppose C is regular. We must show that each basis of M has determinant ± 1. Let B be a basis of M, and suppose $Bx' = e$, where e is some unit basis column vector. It suffices to show that x' is integral (for each choice of e). If e occurs among the columns of B, we are done. Otherwise, $Mx = 0$ for some vector x derived from x' by adding zero-components for all positions not occurring in B, except for a -1 in the position corresponding to column e of M. Let $x'' = \mu x$ for some $\mu \neq 0$ such that x'' is integral. Then $x'' \in C$, and $x'' = x_1 + \cdots + x_m$ for some nonzero $\{0, \pm 1\}$-vectors x_i in C, each having zeros where x has zeros. Since B is nonsingular, each x_i is a multiple of x'', and hence $x_1 = x$. This implies that x is integral.

Conversely, suppose A is totally unimodular. For any vector $x = (\xi_1, \ldots, \xi_n)^{\mathsf{T}}$, denote, as usual, $\|x\|_\infty := \max\{|\xi_1|, \ldots, |\xi_n|\}$. We show that each x in C is the required sum of x_i in C, by induction on $\|x\|_\infty$. The case $\|x\|_\infty = 0$ being trivial, assume $\|x\|_\infty > 0$. Now let $a := \|x\|_\infty^{-1} \cdot x$, and let $P := \{y \mid My = 0; \lfloor a \rfloor \leqslant y \leqslant \lceil a \rceil\}$. As $a \in P$, and as M is totally unimodular, P contains an integral vector, say x_1. As $\|a\|_\infty = 1$, we know that x_1 is a $\{0, \pm 1\}$-vector. Moreover, x_1 is 0 in coordinates where x is 0, and x_1 is ± 1 in coordinates where x is $\pm \|x\|_\infty$. Hence $\|x - x_1\|_\infty < \|x\|_\infty$. As $x - x_1$ belongs to C, by induction we know that $x - x_1 = x_2 + \cdots + x_m$ for some $\{0, \pm 1\}$-vectors $x_2, \ldots, x_m$ in C, each with support included in that of $x - x_1$. The result follows. $\square$

Note that this proof actually gives that if C is a regular chain group and x is in C, then $x = x_1 + \cdots + x_m$ for $\{0, \pm 1\}$-vectors x_i in C such that: if a component of any x_i is positive (negative), the corresponding component of x also is positive (negative). This last directly implies the following. Let $D := [I \ -I \ A \ -A]$, and let K be the cone

(8) $K := \{x \mid x \geqslant 0; Dx = 0\}$.

Then A is totally unimodular if and only if the integral vectors in K are nonnegative integral combinations of $\{0, 1\}$-vectors in K. This last is easily seen to be equivalent to K being generated by the $\{0, 1\}$-vectors contained in K. So A is totally unimodular if and only if the polar cone $K^* = \{w^{\mathsf{T}} \mid wx \leqslant 0 \text{ for all } x \in K\}$ satisfies:

(9) $K^* = \{w^{\mathsf{T}} \mid wx \leqslant 0 \text{ for each } \{0, 1\}\text{-vector } x \text{ with } Dx = 0\}$.

Since by Farkas' lemma (Corollary 7.1d) we know:

(10) $K^* = \{w^{\mathsf{T}} \mid wx \leqslant 0 \text{ for each } x \geqslant 0 \text{ with } Dx = 0\} = \{w^{\mathsf{T}} \mid \exists y: yD \geqslant w\}$

we have derived the following result of Hoffman [1976].

Theorem 21.3. *Let A be an integral matrix and let $D := [I \ -I \ A \ -A]$. Then A is totally unimodular if and only if for each vector w the following are equivalent:*

   (i) *there exists a vector y such that $yD \geqslant w$;*
   (ii) *$wx \leqslant 0$ for each $\{0, 1\}$-vector x with $Dx = 0$.*

Proof. See above. $\square$

(Note that Farkas' lemma asserts the equivalence of (i) and (ii) for each matrix, if we quantify in (ii) over all nonnegative vectors x with $Dx = 0$.)

Hoffman's result can be stated equivalently as:

Corollary 21.3a. *An integral matrix A is totally unimodular, if and only if the following are equivalent for all vectors a, b, c, d:*

(i) *the polytope* $\{x|c \leqslant x \leqslant d; a \leqslant Ax \leqslant b\}$ *is nonempty;*

(ii) $a \leqslant b$ *and* $c \leqslant d$, *and for all* $\{0, \pm 1\}$-*vectors* u, v *with* $uA = v$ *one has:*

$$(11) \qquad \sum_{\substack{i \\ v_i = -1}} \gamma_i + \sum_{\substack{j \\ u_j = 1}} \alpha_j \leqslant \sum_{\substack{i \\ v_i = 1}} \delta_i + \sum_{\substack{j \\ u_j = -1}} \beta_j.$$

[Denoting the components of a, b, c, and d by $\alpha_j, \beta_j, \gamma_i$, and δ_i, respectively.]

Proof. Apply Theorem 21.3 to the transpose of A. $\qquad\qquad\qquad\square$

Application 21.1. This result has the following application to directed graphs. Let $D = (V, A)$ be a directed graph, and let M be its $V \times A$-incidence matrix. Then M is totally unimodular (Section 19.3), and hence by Corollary 21.3a: the polyhedron $\{x|d \leqslant x \leqslant c; Mx = 0\}$ is nonempty, if and only if $d \leqslant c$ and for each vector $v \in \{0, \pm 1\}^V$ and each vector $w \in \{0, \pm 1\}^A$ such that $vM = w$ one has:

$$(12) \qquad \sum_{\substack{a \in A \\ w(a) = -1}} d(a) \leqslant \sum_{\substack{a \in A \\ w(a) = +1}} c(a).$$

This is equivalent to Hoffman's circulation theorem [1960] (*circulation* is defined in Section 1.4): there exists a circulation in D subject to given lower and upper bounds d and c, if and only if $d \leqslant c$ and for each $U \subseteq V$:

$$(13) \qquad \sum_{a \in \delta^-(U)} d(a) \leqslant \sum_{a \in \delta^+(U)} c(a).$$

As usual, $\delta^+(U)$ and $\delta^-(U)$ denote the sets of arcs leaving U and entering U, respectively.

A characterization related to Theorem 21.2 is due to Cederbaum [1958]: *an* $m \times n$-*matrix* A *(with* $m \leqslant n$) *is totally unimodular, if and only if, when fixing* $n - 1$ *variables from* $\eta_1, \ldots, \eta_m, \xi_1, \ldots, \xi_n$ *to be zero, the equation* $y = Ax$ *has a solution* $y = (\eta_1, \ldots, \eta_m)^T$, $x = (\xi_1, \ldots, \xi_n)^T$ *with* $x \neq 0$ *and both* x *and* y *being* $\{0, \pm 1\}$-*vectors.* (Here sufficiency follows as in the proof of Theorem 21.2, while necessity follows from Theorem 21.2 itself.)

21.3. AN UPPER BOUND OF HELLER

If M is the matrix with m rows, having as columns all possible column $\{0, \pm 1\}$-vectors containing at most one $+1$ and at most one -1, then M is totally unimodular, with $m^2 + m + 1$ distinct columns. Heller [1957] showed that this is an extreme case.

Bixby and Cunningham [1984] suggested the following proof. One first shows the following combinatorial lemma, due to Sauer [1972]. We follow the proof of Lovász [1979b: Problem 13.10 (c)].

Lemma. *Let* m *and* k *be natural numbers, and let* X *be a set with* m *elements. If* $\mathscr{E} \subseteq \mathscr{P}(X)$ *with*

$$(14) \qquad |\mathscr{E}| > \binom{m}{0} + \binom{m}{1} + \cdots + \binom{m}{k-1} + \binom{m}{k}$$

then X *has a subset* Y *with* $|Y| = k + 1$, *such that* $\{E \cap Y | E \in \mathscr{E}\} = \mathscr{P}(Y)$.

Proof. By induction on m, the case $m = 0$ being trivial. Let $m \geq 1$, and let $\mathscr{E} \subseteq \mathscr{P}(X)$ satisfy (14). Choose $x_0 \in X$ arbitrarily. Define

(15) $\mathscr{E}' := \{E \subseteq X \setminus \{x_0\} | E \in \mathscr{E} \text{ or } E \cup \{x_0\} \in \mathscr{E}\}$,

         $\mathscr{E}'' := \{E \subseteq X \setminus \{x_0\} | E \in \mathscr{E} \text{ and } E \cup \{x_0\} \in \mathscr{E}\}$.

Clearly, $|\mathscr{E}| = |\mathscr{E}'| + |\mathscr{E}''|$, so by (14):

(16)   (i) $|\mathscr{E}'| > \binom{m-1}{0} + \binom{m-1}{1} + \cdots + \binom{m-1}{k-1} + \binom{m-1}{k}$

         or (ii) $|\mathscr{E}''| > \binom{m-1}{0} + \binom{m-1}{1} + \cdots + \binom{m-1}{k-1}$.

If (i) holds, by induction we know that $X \setminus \{x_0\}$ has a subset Y with $|Y| = k + 1$ and $\{E \cap Y | E \in \mathscr{E}'\} = \mathscr{P}(Y)$. Hence, by (15), $\{E \cap Y | E \in \mathscr{E}\} = \mathscr{P}(Y)$.

If (ii) holds, by induction we know that $X \setminus \{x_0\}$ has a subset Y with $|Y| = k$ and $\{E \cap Y | E \in \mathscr{E}''\} = \mathscr{P}(Y)$. Hence, if we take $Y' := Y \cup \{x_0\}$, then $|Y'| = k + 1$ and, by (15), $\{E \cap Y' | E \in \mathscr{E}''\} = \mathscr{P}(Y')$. $\square$

Theorem 21.4. *Let M be a totally unimodular $m \times n$-matrix with no two equal columns. Then $n \leq m^2 + m + 1$.*

Proof. Suppose $n > m^2 + m + 1$. We may assume that one of the columns of M is all-zero (otherwise, we could add this column). Let $X := \{1, \ldots, m\}$ and define

(17) $\mathscr{E} :=$ the collection of supports of the columns of M.

As all columns are distinct, it follows that each $E \in \mathscr{E}$ is the support of at most two columns of M. Of course, the all-zero column is the only column with support $\varnothing$. Hence

(18) $|\mathscr{E}| \geq \frac{1}{2}(n - 1) + 1 > \frac{1}{2}(m^2 + m) + 1 = \binom{m}{0} + \binom{m}{1} + \binom{m}{2}$.

So by the Lemma, X has a subset Y with $|Y| = 3$ and $\{E \cap Y | E \in \mathscr{E}\} = \mathscr{P}(Y)$. It means that M has a submatrix (possibly after permuting rows or columns) of the form:

(19) $\begin{bmatrix} 0 & \pm 1 & 0 & 0 & \pm 1 & \pm 1 & 0 & \pm 1 \\ 0 & 0 & \pm 1 & 0 & \pm 1 & 0 & \pm 1 & \pm 1 \\ 0 & 0 & 0 & \pm 1 & 0 & \pm 1 & \pm 1 & \pm 1 \end{bmatrix}$.

However, such a matrix cannot be totally unimodular, as one easily checks. $\square$

By investigating the extreme cases in the Lemma and in the Theorem, one can show that if $n = m^2 + m + 1$, then M is a network matrix.

21.4. UNIMODULAR MATRICES MORE GENERALLY

In Section 4.3 a nonsingular matrix was defined to be unimodular if it is integral and has determinant ± 1. In Section 19.1 this definition was extended: a matrix of full row rank with, say, m rows is unimodular if it is integral and each basis (i.e. nonsingular submatrix of order m) has determinant ± 1. Now we extend the definition further: a matrix A of, say, rank r, is called *unimodular* if A is integral and if for each submatrix B consisting of r linearly independent columns of A, the g.c.d. of the subdeterminants of B of order r is 1. This extension does not conflict with the earlier definitions. It is easy to see that any column submatrix of a unimodular matrix is unimodular again.

Unimodular matrices were studied by Hoffman and Kruskal [1956], who showed the following (observe the parallel with Corollary 4.1c).

Theorem 21.5. *Let A be an integral matrix. Then the following are equivalent:*

   (i) *A is unimodular;*
  (ii) *for each integral vector b, the polyhedron $\{x \,|\, x \geqslant 0; Ax = b\}$ is integral;*
 (iii) *for each integral vector c, the polyhedron $\{y \,|\, yA \geqslant c\}$ is integral.*

Proof. (i)$\Rightarrow$(ii). Let b be an integral vector, and let x_0 be a vertex of the polyhedron $\{x \,|\, x \geqslant 0; Ax = b\}$. Let B be the column submatrix of A corresponding to the nonzero components of x_0. As x_0 is a vertex, B has full column rank. Let $\tilde{x}_0$ be the nonzero part of x_0. Then $B\tilde{x}_0 = b$. Corollary 4.1c (applied to B^{T}) implies that $\tilde{x}_0$ is integral, and hence x_0 is integral.

(i)$\Rightarrow$(iii). Let c be an integral vector, and let F be a minimal face of the polyhedron $\{y \,|\, yA \geqslant c\}$. Then, by Theorem 8.4, there is a submatrix B of A, consisting of linearly independent columns of A, such that

$$(20) \qquad F = \{y \,|\, yB = \tilde{c}\}$$

where $\tilde{c}$ is the part of c corresponding to B. Then by Corollary 4.1c (applied to B^{T}), F contains an integral vector.

(ii)$\Rightarrow$(i) and (iii)$\Rightarrow$(i). Suppose A is not unimodular. Let B be a submatrix of A consisting of r ($:=$ rank (A)) linearly independent columns of A, such that the subdeterminants of B of order r have a common denominator $k \in \mathbb{N}$, $k > 1$. By Corollary 4.1c there exists a non-integral vector $\tilde{x}_0$ and an integral vector $\tilde{c}$ such that

$$(21) \qquad b := B\tilde{x}_0 \text{ is integral, and}$$
$$yB = \tilde{c} \text{ has no integral solution } y.$$

We may assume that $\tilde{x}_0$ is nonnegative, as adding integers to the components of $\tilde{x}_0$ keeps $\tilde{x}_0$ non-integral and $B\tilde{x}_0$ integral. We can extend $\tilde{x}_0$ with zero-components to obtain the vector x_0, being a non-integral vertex of the polyhedron $\{x \,|\, x \geqslant 0; Ax = b\}$. This contradicts (ii).

Let y_0 satisfy $y_0 B = \tilde{c}$ (such y_0 exists as B has full column rank). Let P be the polyhedron

(22) $P := \{y \mid yA \geqslant \lfloor y_0 A \rfloor\}$

(where $\lfloor \quad \rfloor$ denotes component-wise lower integer parts). Then

(23) $F := \{y \in P \mid yB = \tilde{c}\}$

is a face of P (nonempty, as $y_0 \in F$), not containing integral vectors. This contradicts (iii). $\square$

So for any integral matrix A, the polyhedron $\{x \mid Ax \leqslant b\}$ is integral for each integral vector b, if and only if A^T is unimodular. Similarly we have:

Corollary 21.5a. *Let A be an integral matrix. Then A^T is unimodular, if and only if both sides of the linear programming duality equation*

(24) $\max \{cx \mid Ax \leqslant b\} = \min \{yb \mid y \geqslant 0; yA = c\}$

are attained by integral vectors x and y, for all integral vectors b, c.

Proof. Apply Theorem 21.5 to A^T. $\square$

Truemper [1978a] gave the following characterization of unimodular matrices. A *basis* of a matrix A of rank r is a submatrix of A consisting of r linearly independent columns of A.

Theorem 21.6. *Let A be an integral matrix. Then the following are equivalent:*

  (i)   *A is unimodular;*
  (ii) *there is a basis B of A satisfying: B is unimodular and the (unique) matrix C with $BC = A$ is totally unimodular;*
  (iii) *each basis B of A satisfies: B is unimodular and the (unique) matrix C with $BC = A$ is totally unimodular.*

Proof. Each of (i), (ii), and (iii) implies that A has at least one unimodular basis B. Since (i), (ii), and (iii) are invariant under elementary row operations (permuting rows, multiplying rows by -1, subtracting an integral multiple of one row from another), we may assume that B^T is in Hermite normal form (cf. Theorem 4.1). Then $B^\mathsf{T} = [I \ 0]$, as the full subdeterminants of B have g.c.d. ± 1. By permuting columns, we may assume

(25) $A = \begin{bmatrix} I & D \\ 0 & 0 \end{bmatrix}$.

Now the equivalence of (i), (ii), and (iii) follows easily from Theorem 19.5. $\square$

In a similar way as in this proof, other problems on unimodularity can also be reduced by elementary row operations to problems on total unimodularity. For example, it follows that unimodularity of an integral matrix A can be tested in polynomial time: first, find a basis B of A (with the Gaussian elimination method—see Section 3.3); second, test if B is unimodular (by determining the Hermite normal form of B^T, which should be $[I \ 0]$—see Section 5.3); moreover, find a nonsingular unimodular matrix U with $[I \ 0] = B^T U$ (Section 5.3); finally, A is unimodular if and only if $U^T A$ is totally unimodular (which can be tested in polynomial time—Theorem 20.3).

Truemper [1978a] derived similarly various other characterizations of unimodularity from those for total unimodularity.

Local unimodularity. Hoffman and Oppenheim [1978] and Truemper and Chandrasekaran [1978] introduced the following concepts (cf. also Hoffman [1979b]). Let A be an integral matrix, and let b be an integral vector. The pair (A, b) is called *locally unimodular* if for each column submatrix B of A for which B has full column rank and $Bx = b$ has a positive solution x, B is unimodular. The pair (A, b) is called *locally strongly unimodular* if such a matrix B contains a nonsingular submatrix of the same rank as B, with determinant ± 1. It is *locally totally unimodular* if such a submatrix B is totally unimodular. So: locally totally unimodular $\Rightarrow$ locally strongly unimodular $\Rightarrow$ locally unimodular. It is direct that an integral matrix A is unimodular if and only if the pair (A, b) is locally unimodular for each integral vector b.

Similarly as in the proof of Theorem 21.5 one shows that if the pair (A, b) is locally unimodular, then

(26) $\qquad \max \{cx \,|\, x \geq 0; Ax \leq b\}$

has an integral optimum solution, for each vector c. The dual problem $\min \{yb \,|\, y \geq 0; yA \geq c\}$ does not necessarily have an integral optimum solution, as is shown by taking:

(27) $\qquad A := \begin{bmatrix} 1 & 0 & 1 & 0 \\ 0 & 1 & 0 & 1 \\ 0 & 0 & 2 & -2 \end{bmatrix}, \quad b := \begin{bmatrix} 1 \\ 1 \\ 0 \end{bmatrix}, \quad c := (1, 1, 2, 0).$

21.5. BALANCED MATRICES

Another class of matrices related to total unimodularity is formed by balanced matrices, introduced by Berge [1969, 1970].

A $\{0, 1\}$-matrix M is *balanced* if M has no square submatrix of odd order, with in each row and in each column exactly two 1's. That is, if it has no submatrix of form (possibly after permutation of rows or columns):

(28) $\qquad \begin{bmatrix} 1 & 1 & 0 & \cdots\cdots & 0 \\ 0 & & & & \vdots \\ \vdots & & & & \\ \vdots & & & & 0 \\ 0 & & & & 1 \\ 1 & 0 & \cdots\cdots & 0 & 1 \end{bmatrix}$

of odd order. It follows that balanced matrices are closed under taking submatrices and transposes and under permuting rows or columns.

Since matrix (28) has determinant 2, totally unimodular $\{0,1\}$-matrices are
balanced. The converse is not true, as is shown by the balanced matrix

$$(29) \qquad \begin{bmatrix} 1 & 1 & 1 & 1 \\ 1 & 1 & 0 & 0 \\ 1 & 0 & 1 & 0 \\ 1 & 0 & 0 & 1 \end{bmatrix}$$

which has determinant -2.

Berge and Las Vergnas [1970], Berge [1972], and Fulkerson, Hoffman,
and Oppenheim [1974] characterized balanced matrices in terms of optimization. It implies that if M is balanced, then both optima in the LP-duality
equation

$$(30) \qquad \max\{cx\,|\,x \geqslant 0;\, Mx \leqslant b\} = \min\{yb\,|\,y \geqslant 0;\, yM \geqslant c\}$$

have integral optimum solutions, if b and c are integral, and if at least one of
b and c is an all-one vector. This property more or less characterizes balancedness—see Theorem 21.8. We first give a result of Fulkerson, Hoffman, and
Oppenheim [1974].

Theorem 21.7. *Let M be a balanced matrix. Then the polyhedron*

$$(31) \qquad P := \{x\,|\,x \geqslant 0;\, Mx = 1\}$$

is integral.

Proof. Let M be a balanced $m \times n$-matrix. The theorem is proved by induction
on m. We may assume that P consists of just one vector x, which is positive in all
coordinates (as each face of P arises by setting some coordinates equal to 0,
making another polyhedron of type (31)).

To see that P contains a $\{0,1\}$-vector (necessarily equal to x), we must find
columns $a_1,\ldots,a_t$ of M such that $a_1 + \cdots + a_t = 1$ (an all-one column vector).
Suppose we have found columns $a_1,\ldots,a_t$ of M such that:

$$(32) \qquad a_1 + \cdots + a_t \leqslant 1, \text{ and}$$
$$k := 1^{\mathsf{T}}(a_1 + \cdots + a_t) \text{ is as large as possible.}$$

If $k = m$ we are done. So assume $k < m$. We may assume that $a_1,\ldots,a_t$ are the
first t columns of M, and that

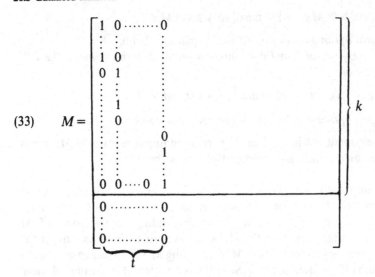

$$(33) \qquad M =$$

As $Mx = 1$, there exist indices i^* and j^* with $i^* > k$, $j^* > t$ and $M_{i^*j^*} = 1$. Let M_0 be the submatrix of M consisting of the first k rows of M. Then $M_0 x = 1$, and hence, by the induction hypothesis, x is a convex combination of $\{0, 1\}$-vectors z with $M_0 z = 1$. As x is positive, this implies $M_0 y = 1$ for some $\{0, 1\}$-vector $y = (\eta_1, \ldots, \eta_n)^\mathsf{T}$ with $\eta_{j^*} = 1$.

Let $V := \{1, \ldots, t\} \cup \{j \mid \eta_j = 1\}$. Let E be the collection of all doubletons $\{j, j'\}$ from V such that $j \neq j'$ and such that some row of M_0 has a 1 both in position j and in j'. Then the graph $G = (V, E)$ is bipartite, with colour classes $\{1, \ldots, t\}$ and $\{j > t \mid \eta_j = 1\}$. Let C be the vertex set of the component of G containing j^*. Now

(34)   for $i > k$, the ith row of M has at most one 1 in positions in C.

For suppose row $i_0 > k$ has 1's in more than one position in C. Let Π be the shortest path connecting two distinct j, j' in C with $M_{i_0 j} = M_{i_0 j'} = 1$. Then the columns of M indexed by column indices in Π, and the rows of M indexed by row indices representing an edge in Π, together with row i_0, induce a submatrix of M of type (28). This contradicts the balancedness of M.

Now (34) directly implies that if $b_1, \ldots, b_s$ are the columns of M with index in the set $\{1, \ldots, t\} \Delta C$ (Δ denoting symmetric difference), then $b_1 + \cdots + b_s \leqslant 1$. Moreover, $1^\mathsf{T}(b_1 + \cdots + b_s) > 1^\mathsf{T}(a_1 + \cdots + a_t)$, as j^* is in $C \setminus \{1, \ldots, t\}$ and $M_{i^*j^*} = 1$. This contradicts our choice of $a_1, \ldots, a_t$ (cf. (32)). □

Theorem 21.8. *Let M be a $\{0, 1\}$-matrix. Then the following are equivalent:*

(i) *M is balanced;*
(ii) *for each $\{0, 1\}$-vector b and each integral vector c, both optima in the LP-duality equation*

(35) $\min\{cx\,|\,x\geqslant 0;\,Mx\geqslant b\}=\max\{yb\,|\,y\geqslant 0;\,yM\leqslant c\}$

have integral optimum solutions (if the optima are finite);
(iii) for each $\{1,\infty\}$-vector b and each integral vector c, both optima in the LP-duality equation

(36) $\max\{cx\,|\,x\geqslant 0;\,Mx\leqslant b\}=\min\{yb\,|\,y\geqslant 0;\,yM\geqslant c\}$

have integral optimum solutions (if the optima are finite).

[In (36), if a component of b is ∞, then the corresponding constraint in $Mx\leqslant b$ is void, while the corresponding component of y must be 0.]

Proof. I. The implications (ii)$\Rightarrow$(i) and (iii)$\Rightarrow$(i) are not difficult. Suppose M contains a submatrix M' of type (28), of order, say, $2k+1$. Let b be the $\{0,1\}$-vector with 1's exactly in the positions corresponding to the rows of M intersecting M'. Similarly, let c be the $\{1,2k+2\}$-vector with 1's exactly in the positions corresponding to columns of M intersecting M'. Then one easily checks that the optimum (35) is equal to $k+\frac{1}{2}$. So (35) has no integral optimum solution.

Similarly, one makes a counterexample for (iii).

II. Let M be balanced and let b be a $\{0,1\}$-vector. Then any vertex x_0 of the polyhedron

(37) $P:=\{x\,|\,x\geqslant 0;\,Mx\geqslant b\}$

is determined by $M'x=\mathbf{1}$ for some row submatrix M' of M, and by requiring certain components of x to be 0. Hence, by Theorem 21.7, x_0 is an integral vector. As this is true for any vertex of P, the minimum in (35) has an integral optimum solution.

Similarly, the maximum in (36) has an integral optimum solution.

III. Let M be a balanced $m\times n$-matrix. In order to show that for $\{0,1\}$-vectors b the maximum in (35) has an integral optimum solution, it suffices to show that this is the case for $b=\mathbf{1}$. This is done by induction on m.

Let y attain the maximum in (35). If y is integral we are finished. Let $y=(\eta_1,\ldots,\eta_m)$, and assume, without loss of generality, $\eta_1\notin\mathbb{Z}$. Let $M=\begin{bmatrix}a\\N\end{bmatrix}$, where a is the top row of M. Now we have:

(38) $\max\{z\mathbf{1}\,|\,z\geqslant 0;\,zN\leqslant c-\lfloor\eta_1\rfloor a\}\geqslant y\mathbf{1}-\eta_1$

as $z:=(\eta_2,\ldots,\eta_m)$ is a feasible solution for this maximum. By induction, the maximum in (38) has an integral optimum solution, say, $(\zeta_2,\ldots,\zeta_m)$. Moreover, $y\mathbf{1}$ is an integer, as it is equal to the minimum in (35), which has, by II above, an integral optimum solution. Hence the maximum (38) is at least $\lceil y\mathbf{1}-\eta_1\rceil=y\mathbf{1}-\lfloor\eta_1\rfloor$. Therefore the vector $\tilde{y}:=(\lfloor\eta_1\rfloor,\zeta_2,\ldots,\zeta_m)$ satisfies: $\tilde{y}\mathbf{1}\geqslant y\mathbf{1}$ and $\tilde{y}M\leqslant c$. So, $\tilde{y}$ is an integral optimum solution for the maximum in (35).

IV. Let M be a balanced $m\times n$-matrix. In order to show that for $\{1,\infty\}$-vectors b

the minimum in (36) has an integral optimum solution, it again suffices to show that this is the case for $b = 1$. Let τ be the maximum value of the components of c. Then

$$(39) \quad \min\{y1 \mid y \geq 0; yM \geq c\} = \min\{y1 \mid 0 \leq y \leq \tau \cdot 1; yM \geq c\}$$
$$= \tau m - \max\{z1 \mid 0 \leq z \leq \tau \cdot 1; zM \leq \tau \cdot 1M - c\}$$
$$= \tau m - \max\{z1 \mid z \geq 0; z[I \quad M] \leq (\tau \cdot 1, \tau \cdot 1M - c)\}.$$

The middle equality follows by replacing y by $\tau \cdot 1 - z$.

Since $[I \quad M]$ is balanced again, the last maximum in (39) has an integral optimum solution z^* (by III above). Hence also the minimum has an integral optimum solution, viz. $y^* := \tau \cdot 1 - z^*$. $\qquad\square$

Berge and Las Vergnas [1970] proved the theorem for the case where c in (ii) is restricted to $\{1, \infty\}$-vectors and in (iii) to $\{0, 1\}$-vectors. Berge [1972] observed that Lovász's perfect graph theorem [1972] implies that (iii) holds for all integral c. Fulkerson, Hoffman, and Oppenheim [1974] showed that (ii) holds for all integral c. The proof method used in part III of the proof was communicated to me by A. J. Hoffman.

Several other characterizations can be derived similarly. It follows from part I of the proof above that the following are equivalent:

(40)   (i) M is balanced;

        (ii) the common value in (35) is an integer for each $\{0, 1\}$-vector b and each $\{1, \infty\}$-vector c;

       (iii) the common value in (36) is an integer for each $\{1, \infty\}$-vector b and each $\{0, 1\}$-vector c.

(Here (iii) arises from (ii) by replacing M by its transpose.) In particular, requiring only one of the optima in (35) (and similarly in (36)) to have integral optimum solutions suffices for characterizing balancedness.

Taking the transpose in (ii) and (iii) of Theorem 21.8 yields other corollaries. Moreover, similarly as in part IV of the proof above, one can derive that if M is balanced, then the optima

$$(41) \quad \min\{y1 \mid 0 \leq y \leq d; yM \geq c\}, \text{ and}$$
$$\max\{y1 \mid 0 \leq y \leq d; yM \leq c\}$$

have integral optimum solutions, for all integral vectors c and d (cf. Fulkerson, Hoffman, and Oppenheim [1974]). See also Berge [1980].

Clearly, the problem 'given a matrix, is it balanced?' belongs to the complexity class co-$\mathcal{NP}$, as it suffices to specify a submatrix of type (28). It is not known whether this problem is in $\mathcal{NP}$. In particular, no polynomial algorithm testing balancedness is known.

Remark 21.1. Each totally unimodular $\{0, 1\}$-matrix is balanced, but not conversely (cf. (29)). Truemper and Chandrasekaran [1978] linked total unimodularity and balancedness further through the following characterization: *for any pair (A, b) of a $\{0, 1\}$-matrix A and a nonnegative integral vector b, the following are equivalent:*

(42)   (i) *The polyhedron $\{x \mid x \geqslant 0; \, Ax \leqslant b'\}$ is integral for each integral vector b' with*
             $0 \leqslant b' \leqslant b$;
        (ii) *A does not have a square submatrix A_0 such that: $\det A_0 = \pm 2$, A_0^{-1} has*
             entries $\pm \frac{1}{2}$ only, and $A_0 1 \leqslant 2b_0$, where b_0 is the part of b corresponding to the
             rows intersecting A_0 and 1 stands for an all-one column vector.

For $b = 1$ this result is contained in Theorem 21.7, as then (ii) of (42) excludes just the
submatrices of type (28). If we take all components of b infinite (or sufficiently large), it gives
Hoffman and Kruskal's theorem (Corollary 19.2a).

For a proof of the equivalence of (i) and (ii), and for characterizations in terms of
local unimodularity, see the paper of Truemper and Chandrasekaran.

Remark 21.2. Another extension of the concept of balancedness was described by
Truemper [1978b]. A $\{0, \pm 1\}$-matrix A is *balanced* if for each submatrix B of A with
exactly two nonzeros in each row and in each column, the sum of the components in
B is divisible by 4. This extends the definition of balanced $\{0, 1\}$-matrices. Still each totally
unimodular matrix is balanced (even if the matrix contains 0's, 1's, and -1's). Truemper
characterized the $\{0, 1\}$-matrices having a signing (i.e. a resetting of some 1's by -1's)
which yields a balanced matrix.

Further notes for Chapter 21. For another generalization of total unimodularity, see
Rebman [1974], Bevis and Hall [1982], Batigne, Hall, and Katz [1978], and Hall and
Katz [1979, 1980, 1981] (integral generalized inverses of integral matrices). See also
Glover [1968b].

Sakarovitch [1975, 1976] studied 'quasi-balanced' matrices.

22

Integral polyhedra and total dual integrality

In Chapters 19–21 we investigated totally unimodular matrices, which are exactly those integral matrices A with the property that the polyhedron $P := \{x \geq 0 \mid Ax \leq b\}$ is integral for each integral vector b. (Recall that a rational polyhedron is *integral* if each face contains integral vectors, i.e. if the polyhedron is the convex hull of the integral vectors contained in it.)

In this chapter we fix A *and* b, and we study integral polyhedra and the related notion of total dual integrality. Fundamental is the following result of Edmonds and Giles [1977], extending earlier results of Fulkerson [1971] and Hoffman [1974].

Consider the LP-duality equation

(1) $\max \{cx \mid Ax \leq b\} = \min \{yb \mid y \geq 0; yA = c\}$

where A, b, c are rational. If the maximum is an integer or is ∞ for each integral vector c, then the maximum is attained by an integral vector x for each c for which the maximum is finite.

In particular, if b is integral and the minimum in (1) has an integral optimum solution y for each integral vector c for which the minimum is finite, then the maximum also has an integral optimum solution, for each such c.

This motivated Edmonds and Giles to define a rational system $Ax \leq b$ to be *totally dual integral*, abbreviated *TDI*, if the minimum in (1) has an integral optimum solution y for each integral vector c for which the optimum is finite. Thus if $Ax \leq b$ is TDI and b is integral, then the polyhedron $\{x \mid Ax \leq b\}$ is integral.

In Section 22.1 we describe these results, which we illustrate in Section 22.2 by two combinatorial applications. In Section 22.3 we discuss the relation of total dual integrality with 'Hilbert-bases', and we consider minimal TDI-systems. In Section 22.4 we discuss the stronger notion of box-total dual integrality. In Sections 22.5–7 we describe some further theory and characterizations of total dual integrality. In Section 22.8 we show that linear programs over TDI-systems with integral data can be solved in integers in polynomial time. In Section 22.9 we discuss the problem of recognizing total dual integrality; it can be done in polynomial time, if we fix the number of variables. Finally, in Section 22.10 we deal with the *integer decomposition property* and the *integer rounding property*, which can be studied in terms of total dual integrality.

22.1. INTEGRAL POLYHEDRA AND TOTAL DUAL INTEGRALITY

The basis for this section is the following result of Edmonds and Giles [1977]. (The simpler special case of P pointed occurs, explicitly or implicitly, in the papers of Chvátal [1973], Fulkerson [1971], Gomory [1963a], Hoffman [1974].)

Theorem 22.1. *A rational polyhedron P is integral, if and only if each rational supporting hyperplane of P contains integral vectors.*

Proof. Since the intersection of a supporting hyperplane with P is a face of P, necessity of the condition is trivial.

To prove sufficiency, suppose that each supporting hyperplane of P contains integral vectors. Let $P = \{x \mid Ax \leqslant b\}$ for matrix A and vector b. As P is rational, we may assume A and b to be integral. Let $F = \{x \mid A'x = b'\}$ be a minimal face of P, where $A'x \leqslant b'$ is a subsystem of $Ax \leqslant b$ (cf. Theorem 8.4). If F does not contain any integral vector, there exists, by Corollary 4.1a, a vector y such that $c := yA'$ is integral and $\delta := yb'$ is not an integer. We may suppose that all entries in y are positive (we may add integers to the components of y without violating the required properties of y). Now $H := \{x \mid cx = \delta\}$ is a supporting hyperplane of P, since $F = P \cap H$, as may be checked easily. As δ is not an integer, whereas c is integral, H does not contain an integral vector. This contradicts our assumption. $\square$

The following is an equivalent formulation.

Corollary 22.1a. *Let $Ax \leqslant b$ be a system of rational linear inequalities. Then* $\max \{cx \mid Ax \leqslant b\}$ *is achieved by an integral vector x for each vector c for which the maximum is finite, if and only if* $\max \{cx \mid Ax \leqslant b\}$ *is an integer for each integral vector c for which the maximum is finite.*

Proof. If c is integral, and $\max \{cx \mid Ax \leqslant b\}$ has an integral optimum solution x, then the maximum is an integer. So necessity of the condition is trivial.

To see sufficiency, suppose that for each integral vector c, $\max \{cx \mid Ax \leqslant b\}$ is an integer, if it is finite. Let $H = \{x \mid cx = \delta\}$ be a rational supporting hyperplane of the polyhedron $P := \{x \mid Ax \leqslant b\}$. We may assume that the components of c are relatively prime integers. Since $\delta = \max \{cx \mid Ax \leqslant b\}$ is an integer, the equation $cx = \delta$ has an integral solution x. So H contains an integral vector.

So each supporting hyperplane of P contains integral vectors, and hence by Theorem 22.1, P is integral. So if $\max \{cx \mid Ax \leqslant b\}$ is finite, it is attained by an integral solution. $\square$

As in the introduction of this chapter, call a rational system $Ax \leqslant b$ of linear inequalities *totally dual integral*, abbreviated *TDI*, if the minimum in the LP-duality equation

(2) $\max \{cx \mid Ax \leqslant b\} = \min \{yb \mid y \geqslant 0; yA = c\}$

has an integral optimum solution y for each integral vector c for which the minimum is finite. So by Hoffman and Kruskal's theorem (Corollary 19.2b), if A is totally unimodular, then $Ax \leqslant b$ is TDI for each rational vector b.

Edmonds and Giles [1977], generalizing results of Fulkerson [1971], Hoffman [1974], and Lovász [1976], showed that total dual integrality of $Ax \leqslant b$ implies that also the maximum has an integral optimum solution, if b is integral.

Corollary 22.1b. *Let $Ax \leqslant b$ be a TDI-system. If b is integral, and the maximum in* (2) *is finite, it has an integral optimum solution.*

Proof. Directly from Corollary 22.1a, since if b is integral, $Ax \leqslant b$ is TDI, and the optima in (2) are finite, then the optimum value is an integer.   □

In other words:

Corollary 22.1c. *If $Ax \leqslant b$ is a TDI-system and b is integral, the polyhedron $\{x \mid Ax \leqslant b\}$ is integral.*

Proof. Directly from Corollary 22.1b.   □

Note however that total dual integrality is *not* a property of just polyhedra. The systems

$$(3) \qquad \begin{bmatrix} 1 & 1 \\ 1 & 0 \\ 1 & -1 \end{bmatrix} \begin{pmatrix} \xi_1 \\ \xi_2 \end{pmatrix} \leqslant \begin{pmatrix} 0 \\ 0 \\ 0 \end{pmatrix} \quad \text{and} \quad \begin{bmatrix} 1 & 1 \\ 1 & -1 \end{bmatrix} \begin{pmatrix} \xi_1 \\ \xi_2 \end{pmatrix} \leqslant \begin{pmatrix} 0 \\ 0 \end{pmatrix}$$

define the same polyhedron, but the first system is TDI and the latter not. Generally a TDI-system contains more constraints than necessary for just defining the polyhedron.

One easily checks that if A is integral and b is rational, then the system $x \geqslant 0$; $Ax \leqslant b$ is TDI if and only if the minimum in

$$(4) \qquad \max \{cx \mid x \geqslant 0; Ax \leqslant b\} = \min \{yb \mid y \geqslant 0; yA \geqslant c\}$$

has an integral optimum solution y for each integral vector c with finite minimum. (Since, by definition $x \geqslant 0$; $Ax \leqslant b$ is TDI if and only if min $\{yb \mid y \geqslant 0$; $z \geqslant 0$; $yA - z = c\}$ has an integral optimum solution, for each integral vector c with finite minimum.)

Similarly, for integral A, the system $x \geqslant 0$; $Ax \geqslant b$ is TDI if and only if the maximum in

$$(5) \qquad \min \{cx \mid x \geqslant 0; Ax \geqslant b\} = \max \{yb \mid y \geqslant 0; yA \leqslant c\}$$

has an integral optimum solution y for each integral vector c with finite maximum.

Moreover, if A is integral, the system $x \geqslant 0$; $Ax = b$ is TDI (i.e. $x \geqslant 0$; $Ax \leqslant b$; $-Ax \leqslant -b$ is TDI), if and only if the minimum in

$$(6) \qquad \max \{cx \mid x \geqslant 0; Ax = b\} = \min \{yb \mid yA \geqslant c\}$$

has an integral optimum solution y for each integral c for which the minimum is finite.

In a sense, 'each face of a TDI-system is TDI again'. That is:

Theorem 22.2. *Let* $Ax \leqslant b$; $ax \leqslant \beta$ *be a TDI-system. Then the system* $Ax \leqslant b$; $ax = \beta$ *is also TDI.*

Proof. Let c be an integral vector, with

(7) $\qquad \max \{cx \,|\, Ax \leqslant b; ax = \beta\}$
$\qquad \qquad = \min \{yb + (\lambda - \mu)\beta \,|\, y \geqslant 0; \lambda, \mu \geqslant 0; yA + (\lambda - \mu)a = c\}$

finite. Let $x^*, y^*, \lambda^*, \mu^*$ attain these optima (possibly being fractional). Let $c' := c + Na$ where N is an integer satisfying $N \geqslant N^*$ and Na integer. Then the optima

(8) $\qquad \max \{c'x \,|\, Ax \leqslant b; ax \leqslant \beta\} = \min \{yb + \lambda\beta \,|\, y \geqslant 0; \lambda \geqslant 0; yA + \lambda a = c'\}$

are finite, since $x := x^*$ is a feasible solution for the maximum, and $y := y^*$, $\lambda := \lambda^* + N - \mu^*$ is a feasible solution for the minimum.

Since $Ax \leqslant b$; $ax \leqslant \beta$ is TDI, the minimum (8) has an integer optimum solution, say, $\tilde{y}, \tilde{\lambda}$. Then $y := \tilde{y}$, $\lambda := \tilde{\lambda}$, $\mu := N$ is an integer optimum solution for the minimum in (7): it is feasible in (7), and it is optimum as:

(9) $\qquad \tilde{y}b + (\tilde{\lambda} - N)\beta = \tilde{y}b + \tilde{\lambda}\beta - N\beta \leqslant y^*b + (\lambda^* + N - \mu^*)\beta - N\beta$
$\qquad \qquad = y^*b + (\lambda^* - \mu^*)\beta.$

(Here $\leqslant$ follows from the fact that y^*, $\lambda^* + N - \mu^*$ is a feasible solution for the minimum in (8), and $\tilde{y}, \tilde{\lambda}$ is an optimum solution for this minimum.) So the minimum in (7) has an integral optimum solution $\square$

22.2. TWO COMBINATORIAL APPLICATIONS

To illustrate the notion of total dual integrality, we describe here two combinatorial applications.

Application 22.1. *Arborescences.* Let $D = (V, A)$ be a directed graph, and let r be a fixed vertex of D. An *r-arborescence* is a set A' of $|V| - 1$ arcs forming a spanning tree such that each vertex $v \neq r$ is entered by exactly one arc in A'. So for each vertex v there is a unique directed path in A' from r to v. An *r-cut* is an arc set of the form $\delta^-(U)$, for some nonempty subset U of $V \setminus \{r\}$. [As usual, $\delta^-(U)$ denotes the set of arcs entering U.]

It is not difficult to see that the r-arborescences are the inclusion-wise minimal sets of arcs intersecting all r-cuts. Conversely, the inclusion-wise minimal r-cuts are the inclusion-wise minimal sets of arcs intersecting all r-arborescences.

Fulkerson [1974] showed:

Theorem 22.3 (Optimum arborescence theorem). *For any 'length' function* $l : A \to \mathbb{Z}_+$, *the minimum length of an r-arborescence is equal to the maximum number* t *of r-cuts* $C_1, \dots, C_t$ *(repetition allowed) so that no arc* a *is in more than* $l(a)$ *of these cuts.*

This result can be formulated in terms of integer linear programming as follows. Let C be the matrix with rows the incidence vectors of all r-cuts. So the columns of C are indexed by A, and the rows of C by the sets U for $\emptyset \neq U \subseteq V\backslash\{r\}$. Then the Optimum arborescence theorem is equivalent to both optima in the LP-duality equation

(10) $\min\{lx|x \geqslant 0; Cx \geqslant 1\} = \max\{y1|y \geqslant 0; yC \leqslant l\}$

having integral optimum solutions, for each $l \in \mathbb{Z}_+^A$. So in order to show the theorem, by Corollary 22.1b it suffices to show that the maximum in (10) has an integral optimum solution, for each function $l:A \to \mathbb{Z}$, i.e. that

(11)   *the system $x \geqslant 0$, $Cx \geqslant 1$ is TDI.*

Following Edmonds and Giles [1977] one can show this as follows.

Proof. If some component of l is negative, the maximum in (10) is infeasible. If all components of l are nonnegative, let vector y attain the maximum in (10), such that

(12) $\sum_U y_U |U|^2$

is as large as possible (where U ranges over all sets U with $\emptyset \neq U \subseteq V\backslash\{r\}$). Such a vector y exists by compactness arguments.

Then the collection

(13) $\mathscr{F} := \{U|y_U > 0\}$

is laminar, i.e. if T, $U \in \mathscr{F}$ then $T \subseteq U$ or $U \subseteq T$ or $T \cap U = \emptyset$. To see this, suppose that T, $U \in \mathscr{F}$ with $T \not\subseteq U \not\subseteq T$ and $T \cap U \neq \emptyset$. Let $\varepsilon := \min\{y_T, y_U\} > 0$. Next reset:

(14) $y_T := y_T - \varepsilon, \qquad y_{T \cap U} := y_{T \cap U} + \varepsilon,$
          $y_U := y_U - \varepsilon, \qquad y_{T \cup U} := y_{T \cup U} + \varepsilon$

while y does not change in the other coordinates. By this resetting, yC does not increase in any coordinate (since $\varepsilon\chi_{\delta^-(T)} + \varepsilon\chi_{\delta^-(U)} \geqslant \varepsilon\chi_{\delta^-(T \cap U)} + \varepsilon\chi_{\delta^-(T \cup U)}$, where χ denotes incidence vector), while $y1$ does not change. However, the sum (12) did increase, contradicting the maximality of (12). This shows that $\mathscr{F}$ is laminar.

Let C' be the submatrix of C consisting of the rows corresponding to r-cuts $\delta^-(U)$ with U in $\mathscr{F}$. Then

(15) $\max\{y'1|y' \geqslant 0; y'C' \leqslant l\} = \max\{y1|y \geqslant 0; yC \leqslant l\}$.

Here the inequality $\leqslant$ is trivial, as C' is a submatrix of C. The inequality $\geqslant$ follows from the fact that the vector y above attains the second maximum in (15), with 0's in positions corresponding to rows of C outside C'.

Now the matrix C' is totally unimodular. This can be derived as follows with Ghouila-Houri's criterion ((iv) of Theorem 19.3). Choose a set of rows of C', i.e. choose a sub-collection $\mathscr{G}$ of $\mathscr{F}$. Define for each U in $\mathscr{G}$ the 'height' $h(U)$ of U as the number of sets T in $\mathscr{G}$ with $T \supseteq U$. Now split $\mathscr{G}$ into $\mathscr{G}_{odd}$ and $\mathscr{G}_{even}$, according to $h(U)$ odd or even. One easily derives from the laminarity of $\mathscr{G}$ that, for any arc a of D, the number of sets in $\mathscr{G}_{odd}$ entered by a, and the number of sets in $\mathscr{G}_{even}$ entered by a, differ by at most 1. Therefore, we can split the rows corresponding to $\mathscr{G}$ into two classes fulfilling Ghouila-Houri's criterion. So C' is totally unimodular.

By Hoffman and Kruskal's theorem, the first maximum in (15) has an integral optimum solution y'. Extending this y' with components 0 gives an integral optimum solution for the second maximum in (15). So the maximum in (10) has an integral optimum solution. $\qquad\square$

Since the minimum in (10) has an integral optimum solution x for each function $l:A \to \mathbb{Q}_+$, we know that the polyhedron $\{x \geqslant 0|Cx \geqslant 1\}$ is integral. Its vertices are the incidence vectors, say $d_1, \ldots, d_t$, of r-arborescences. So

(16) $\{x \geqslant 0|Cx \geqslant 1\} = \{d_1, \ldots, d_t\}^\uparrow$

where, as in Section 9.2 on blocking polyhedra, $\{z_1,\ldots,z_k\}^\uparrow := \text{conv.hull}\,\{z_1,\ldots,z_k\} + \mathbb{R}^n_+$ (for $z_1,\ldots,z_k \in \mathbb{R}^n$).

Let $c_1,\ldots,c_m$ be the columns of C^T (i.e. they are the incidence vectors of r-cuts), and let matrix D be defined by having $d_1^\mathsf{T},\ldots,d_t^\mathsf{T}$ as its rows. It follows from the theory of blocking polyhedra that

(17) $\qquad \{x \geqslant 0 \,|\, Dx \geqslant 1\} = \{c_1,\ldots,c_m\}^\uparrow.$

That is, for each function $w: A \to \mathbb{Z}_+$, the minimum in

(18) $\qquad \min\,\{wx \,|\, x \geqslant 0;\, Dx \geqslant 1\} = \max\,\{y1 \,|\, y \geqslant 0;\, yD \leqslant w\}$

is attained by an integral vector x, i.e. by the incidence vector of an r-cut. So (18) gives a min–max relation for the minimum 'capacity' of an r-cut. Also the maximum in (18) is attained by an integral optimum solution y, i.e. the system $x \geqslant 0,\, Dx \geqslant 1$ is TDI (see Edmonds [1973]).

Application 22.2. *Orientations of undirected graphs.* As an application of polyhedral theory to non-optimizational combinatorics we derive a theorem of Nash–Williams [1969] with a method using total dual integrality due to Frank [1980] and Frank and Tardos [1984].

Theorem 22.4 (Nash–Williams orientation theorem). *A 2k-connected undirected graph can be oriented so as to become a k-connected directed graph.*

[An undirected graph is *2k-connected* if each nonempty proper subset of the vertex set is entered by at least $2k$ edges. A directed graph is *k-connected* if each nonempty proper subset of the vertices is entered by at least k arcs.]

Proof. Let $G = (V, E)$ be a 2k-connected undirected graph. Orient the edges of G arbitrarily, yielding the directed graph $D = (V, A)$. Consider the polyhedron in $\mathbb{R}^A$ determined by:

(19) $\qquad 0 \leqslant x_a \leqslant 1 \qquad\qquad\qquad (a \in A)$

$$\sum_{a \in \delta^-(U)} x_a - \sum_{a \in \delta^+(U)} x_a \leqslant |\delta^-(U)| - k \qquad (\varnothing \neq U \subset V).$$

[Here $\delta^-(U)\,(\delta^+(U))$ denotes the set of arcs of D entering U (leaving U).] Now if (19) has an integral solution x then G can be oriented so as to become a k-connected digraph: reversing the orientation on the arcs a of D with $x_a = 1$ gives a k-connected digraph $\tilde{D}$, since, if $\varnothing \neq U \neq V$, then:

(20) $\qquad$ the number of arcs of $\tilde{D}$ entering $U = \displaystyle\sum_{a \in \delta^+(U)} x_a + \sum_{a \in \delta^-(U)} (1 - x_a)$

$$= |\delta^-(U)| + \sum_{a \in \delta^+(U)} x_a - \sum_{a \in \delta^-(U)} x_a \geqslant k.$$

So it suffices to show that (19) has an integral solution. First observe that (19) has a solution, viz. $x_a = \frac{1}{2}$ $(a \in A)$, by the 2k-connectedness of G. Therefore, we are done if we have shown that (19) defines an integral polyhedron. This is done by showing that (19) is TDI.

To this end, let $c: A \to \mathbb{Z}$, and consider the linear program dual to maximizing $\sum_{a \in A} c_a x_a$ over (19). Let z_U $(\varnothing \neq U \subset V)$ be the dual variables corresponding to the second set of constraints in (19), and consider any dual optimum solution for which

(21) $\qquad \displaystyle\sum_{\varnothing \neq U \subset V} z_U |U| \cdot |V \setminus U|$

is as small as possible. Then the collection $\mathcal{F} := \{U \,|\, z_U > 0\}$ is cross-free (i.e. if U, $W \in \mathcal{F}$ then $U \subseteq W$ or $W \subseteq U$ or $U \cap W = \varnothing$ or $U \cup W = V$). For suppose $U, W \in \mathcal{F}$ with $U \nsubseteq W \nsubseteq U$, $U \cap W \neq \varnothing$ and $U \cup W \neq V$. Let $\varepsilon := \min\{z_U, z_W\} > 0$. Then reset

$$(22) \qquad z_U := z_U - \varepsilon, \qquad z_{U \cap W} := z_{U \cap W} + \varepsilon,$$
$$z_W := z_W - \varepsilon, \qquad z_{U \cup W} := z_{U \cup W} + \varepsilon$$

not changing the other components of z. Since the sum of the LHS's of the inequalities in (19) corresponding to U and W is equal to the sum of the LHS's of the inequalities corresponding to $U \cap W$ and $U \cup W$, the new z is again a feasible dual solution. Moreover, for the RHS's we have

$$(23) \qquad (|\delta^-(U)| - k) + (|\delta^-(W)| - k) \geqslant (|\delta^-(U \cap W)| - k) + (|\delta^-(U \cup W)| - k)$$

and so the new z is again an optimum solution. Moreover, the sum (21) has decreased, contradicting our assumption.

So $\mathcal{F}$ is cross-free. Therefore, the inequalities in (19) corresponding to positive dual variables form a totally unimodular matrix (cf. Example 5 in Section 19.3), and hence there is an integral optimum solution. So (19) is TDI, and it has an integral solution. $\qquad\square$

22.3. HILBERT BASES AND MINIMAL TDI-SYSTEMS

To understand total dual integrality better, Giles and Pulleyblank [1979] introduced the notion of a *Hilbert basis*. As we defined in Section 16.4, this is a set of vectors $a_1, \ldots, a_t$ with the property that each integral vector b in cone $\{a_1, \ldots, a_t\}$ is a nonnegative integral combination of $a_1, \ldots, a_t$. An *integral* Hilbert basis is one consisting of integral vectors only. Also in Section 16.4, we proved a result due to Gordan [1873] and van der Corput [1931b,c]: *each rational polyhedral cone C is generated by an integral Hilbert basis. If C is pointed, there exists a unique minimal integral Hilbert basis generating C.*

There is a close relation between Hilbert bases and total dual integrality.

Theorem 22.5. *The rational system $Ax \leqslant b$ is TDI if and only if for each face F of the polyhedron $P := \{x \,|\, Ax \leqslant b\}$, the rows of A which are active in F form a Hilbert basis.*

[A row of A is *active* in F if the corresponding inequality in $Ax \leqslant b$ is satisfied with equality by all vectors x in F.]

Proof. First suppose that $Ax \leqslant b$ is TDI. Let F be a face of P, and let $a_1, \ldots, a_t$ be the rows of A active in F. To prove that $a_1, \ldots, a_t$ form a Hilbert basis, choose an integral vector c in cone $\{a_1, \ldots, a_t\}$. Then the maximum in the LP-duality equation

$$(24) \qquad \max\{cx \,|\, Ax \leqslant b\} = \min\{yb \,|\, y \geqslant 0; yA = c\}$$

is attained by each vector x in F. The minimum has an integral optimum solution, say, y. This vector y has 0's in positions corresponding to rows not active in F (by complementary slackness). Hence c is an integral nonnegative combination of $a_1, \ldots, a_t$.

To prove the converse, let c be an integral vector for which the optima (24) are finite. Let F be a minimal face of P so that each vector in F attains the maximum in (24). Let $a_1, \ldots, a_t$ be the rows active in F. Then c belongs to cone $\{a_1, \ldots, a_t\}$ (since the minimum in (24) has a, possibly fractional, optimum solution y, with no positive component outside $a_1, \ldots, a_t$). Assuming the condition mentioned in the theorem, $a_1, \ldots, a_t$ form a Hilbert basis, and hence $c = \lambda_1 a_1 + \cdots + \lambda_t a_t$ for certain nonnegative integers $\lambda_1, \ldots, \lambda_t$. Extending the vector $(\lambda_1, \ldots, \lambda_t)$ with zero-components, we obtain an integral vector $y \geqslant 0$ such that $yA = c$ and $yb = yAx = cx$ for all x in F. So y attains the minimum in (24). As this is true for any integral c, $Ax \leqslant b$ is TDI. $\qquad\square$

Note that we in fact showed that we may restrict F in Theorem 22.5 to *minimal* faces. This gives a characterization of Hilbert bases in terms of total dual integrality as a direct consequence.

Corollary 22.5a. *The rows of the rational matrix A form a Hilbert basis, if and only if the system $Ax \leqslant 0$ is TDI.*

Proof. Directly by taking $b = 0$ in Theorem 22.5. $\qquad\square$

A TDI-system $Ax \leqslant b$ is called a *minimal TDI-system*, or *minimally TDI*, if any proper subsystem of $Ax \leqslant b$ which defines the same polyhedron as $Ax \leqslant b$ is not TDI. So the TDI-system $Ax \leqslant b$ is minimally TDI if and only if each constraint in $Ax \leqslant b$ determines a supporting hyperplane of $\{x \,|\, Ax \leqslant b\}$ and is not a nonnegative integral combination of other constraints in $Ax \leqslant b$.

Combining Theorems 16.4 and 22.5 gives the following result of Giles and Pulleyblank [1979] and Schrijver [1981].

Theorem 22.6. (i) *For each rational polyhedron P there exists a TDI-system $Ax \leqslant b$ with A integral and $P = \{x \,|\, Ax \leqslant b\}$. Here b can be chosen to be integral if and only if P is integral.*

(ii) *Moreover, if P is full-dimensional, there exists a unique minimal TDI-system $Ax \leqslant b$ with A integral and $P = \{x \,|\, Ax \leqslant b\}$. In that case, b is integral if and only if P is integral.*

Proof. (i) For each minimal face F of P, let C_F be the convex cone of all row vectors c for which $\max\{cx \,|\, x \in P\}$ is attained by all x in F. Given a minimal face F, let $a_1, \ldots a_t$ be an integral Hilbert basis generating C_F (cf. Theorem 16.4). Choose $x_0 \in F$, and let $\beta_i := a_i x_0$, for $i = 1, \ldots, t$. So $\beta_i = \max\{a_i x \,|\, x \in P\}$. Then the system Σ_F of the inequalities

(25) $a_1 x \leqslant \beta_1, \ldots, a_t x \leqslant \beta_t$

consists of valid inequalities for P. Moreover, let $Ax \leqslant b$ be the union of the systems Σ_F over all minimal faces F. Then $Ax \leqslant b$ determines P, and by Theorem 22.5, $Ax \leqslant b$ is TDI.

Clearly, if P is integral, the right-hand sides in (25) are integers, whence b is integral. If conversely b can be chosen to be integral, then P is integral by Corollary 22.1c.

(ii) If P is full-dimensional, each cone C_F is pointed. Hence there is a unique minimal integral Hilbert basis $a_1, \ldots, a_t$ generating C_F (Theorem 16.4). Let Σ_F be as above. By Theorem 22.5 each of the inequalities in Σ_F must be in each TDI-system $Ax \leqslant b$ with A integral and $P = \{x \mid Ax \leqslant b\}$.

Similarly as above it follows that P is integral if and only if b is integral. □

Corollary 22.6a. *A rational polyhedron P is integral if and only if there is a TDI-system $Ax \leqslant b$ with $P = \{x \mid Ax \leqslant b\}$ and b integral.*

Proof. Directly from Theorem 22.6. □

Part (ii) of Theorem 22.6 is parallel to the fact that each full-dimensional polyhedron is defined by a unique minimal system of inequalities, up to multiplication of inequalities by positive scalars. However, the minimal TDI-system can be larger, as is shown by the systems (3).

If $Ax \leqslant b$ is a minimal TDI-system, determining the full-dimensional polyhedron P, with A integral, then for each row of A, the components are relatively prime (cf. the remark after Theorem 16.4). So for each full-dimensional rational polyhedron P there exist unique minimal systems $Ax \leqslant b$ and $A'x \leqslant b'$ such that: (i) $P = \{x \mid Ax \leqslant b\} = \{x \mid A'x \leqslant b'\}$; (ii) each row of A and each row of A' contains relatively prime integers; (iii) $A'x \leqslant b'$ is TDI. Note that $Ax \leqslant b$ must be a subsystem of $A'x \leqslant b'$. Moreover, P is integral if and only if b' is integral. It would be interesting to characterize the polyhedra with $A = A'$.

For a full-dimensional rational polyhedron, one generally cannot find the minimal TDI-system, with integral left-hand sides, in polynomial time, plainly because this system itself can be exponentially large. For example, the minimal TDI-system for the polyhedron $\{(\xi, \eta) \mid \xi \geqslant 0; \xi + k\eta \geqslant 0\}$ (k integer) consists of the inequalities $\xi + j\eta \geqslant 0$, for $j = 0, \ldots, k$. So the TDI-system has size $O(k \cdot \log k)$, while the original system has size $O(\log k)$ only.

22.4. BOX-TOTAL DUAL INTEGRALITY

A rational system $Ax \leqslant b$ is *box-totally dual integral* or *box-TDI* if the system

(26)　　　$Ax \leqslant b; \quad l \leqslant x \leqslant u$

is TDI for each pair of rational vectors l and u; that is, if it is TDI for each intersection with a 'box'. Box-total dual integrality implies total dual integrality:

Theorem 22.7. *A box-TDI system is TDI.*

Proof. Let $Ax \leqslant b$ be box-TDI. Let c be integral so that max $\{cx | Ax \leqslant b\}$ is finite. Let it be attained by x^*. Choose l and u such that $l < x^* < u$. Then max $\{cx | Ax \leqslant b; l \leqslant x \leqslant u\}$ has an integral optimum dual solution, which must be zero in components corresponding to the constraints $l \leqslant x \leqslant u$ (by complementary slackness, as x^* is an optimum solution). Hence max $\{cx | Ax \leqslant b\}$ also has an integral optimum dual solution. $\qquad\qquad\qquad\qquad\qquad\qquad\qquad\qquad\qquad\square$

One similarly shows that if $Ax \leqslant b$ is box-TDI, then (26) is TDI also if some of the components of l or u are $-\infty$ or $+\infty$.

By Hoffman and Kruskal's theorem (Corollary 19.2b), A is totally unimodular if and only if $Ax \leqslant b$ is box-TDI for each vector b.

Call a polyhedron a *box-totally dual integral polyhedron* or a *box-TDI polyhedron* if it can be defined by a box-TDI system. The following theorem of Cook [1986] shows that box-total dual integrality essentially is a property of polyhedra.

Theorem 22.8. *Let P be a box-TDI polyhedron and let $Ax \leqslant b$ be any TDI-system such that $P = \{x | Ax \leqslant b\}$. Then $Ax \leqslant b$ is box-TDI.*

Proof. Let $P \subseteq \mathbb{R}^n$, and choose l, $u \in \mathbb{Q}^n$, $c \in \mathbb{Z}^n$. It suffices to show that the minimum in

$$(27) \qquad \max \{cx | Ax \leqslant b; l \leqslant x \leqslant u\}$$
$$= \min \{yb + vu - wl \,|\, y, v, w \geqslant 0; yA + v - w = c\}$$

has an integral optimum solution y, v, w, assuming the minimum is finite. Let $\tilde{A}x \leqslant \tilde{b}$ be a box-TDI system defining P. Then

$$(28) \qquad \min \{y\tilde{b} + vu - wl \,|\, y, v, w \geqslant 0; y\tilde{A} + v - w = c\}$$

has an integral optimum solution y^*, v^*, w^*. Now define $\tilde{c} := c - v^* + w^* = y^* \tilde{A}$, and consider

$$(29) \qquad \max \{\tilde{c}x | Ax \leqslant b\} = \min \{yb | y \geqslant 0; yA = \tilde{c}\}.$$

Since the maximum is feasible (as the optima (27) are finite) and since $\tilde{c}x = y^* \tilde{A} x \leqslant y^* \tilde{b}$ whenever $x \in P$, the optima (29) are finite. As $Ax \leqslant b$ is TDI, the minimum in (29) has an integral optimum solution, say $\bar{y}$. Then $\bar{y}, v^*, w^*$ is an integral optimum solution for the minimum in (27), since it is feasible, and since

$$(30) \qquad \bar{y}b + v^*u - w^*l = \min \{yb | y \geqslant 0; yA = \tilde{c}\} + v^*u - w^*l$$
$$= \min \{y\tilde{b} | y \geqslant 0; y\tilde{A} = \tilde{c}\} + v^*u - w^*l$$
$$= \min \{y\tilde{b} + v^*u - w^*l | y \geqslant 0; y\tilde{A} = \tilde{c}\}$$
$$= \min \{y\tilde{b} + vu - wl | y, v, w \geqslant 0; y\tilde{A} + v - w = c\}$$
$$= \min \{yb + vu - wl | y, v, w \geqslant 0; yA + v - w = c\}.$$

The first equality follows as $\bar{y}$ is an optimum solution for the minimum in (29).

The second equality follows as $Ax \leqslant b$ and $\tilde{A}x \leqslant \tilde{b}$ define the same polyhedron. The third equality is trivial. The fourth equality follows as y^*, v^*, w^* is an optimum solution for (28). The last equality follows as $Ax \leqslant b$ and $\tilde{A}x \leqslant \tilde{b}$ determine the same polyhedron. $\qquad\square$

Not every polyhedron is a box-TDI polyhedron. For example, $P := \left\{ \begin{pmatrix} \xi \\ \eta \end{pmatrix} \middle| 2\xi + \eta \leqslant 0 \right\}$ is not box-TDI (since $2\xi + \eta \leqslant 0$ is TDI, but $2\xi + \eta \leqslant 0, 0 \leqslant \xi \leqslant 1$, $-1 \leqslant \eta \leqslant 0$ does not give an integral polyhedron).

The following theorem of Cook [1986] characterizes box-TDI polyhedra.

Theorem 22.9. *A rational polyhedron P in $\mathbb{R}^n$ is box-TDI if and only if for each rational vector $c = (\gamma_1, \ldots, \gamma_n)$ there exists an integral vector $\tilde{c} = (\tilde{\gamma}_1, \ldots, \tilde{\gamma}_n)$ such that $\lfloor c \rfloor \leqslant \tilde{c} \leqslant \lceil c \rceil$ and such that each optimum solution of $\max \{cx | x \in P\}$ is also an optimum solution of $\max \{\tilde{c}x | x \in P\}$.*

Proof. Necessity. Let $Ax \leqslant b$ be a box-TDI system defining P. We show that for each $t = 0, \ldots, n$ the following holds:

(31)   for each rational vector $c = (\gamma_1, \ldots, \gamma_n)$ there exists an integral vector
     $\tilde{c} = (\tilde{\gamma}_1, \ldots, \tilde{\gamma}_n)$ such that each optimum solution of $\max \{cx | x \in P\}$ is also
     an optimum solution of $\max \{\tilde{c}x | x \in P\}$, and such that

$$\tilde{\gamma}_i \leqslant \lceil \gamma_i \rceil \qquad \text{for all } i$$
$$\tilde{\gamma}_i = \gamma_i \qquad \text{if } \gamma_i \text{ is an integer}$$
$$\tilde{\gamma}_i \geqslant \lfloor \gamma_i \rfloor \qquad \text{for } i = 1, \ldots, t.$$

Clearly, the strongest case $t = n$ is what we have to prove. We show (31) by induction on t. Let $x^* = (\xi_1^*, \ldots, \xi_n^*)^{\mathsf{T}}$ be a rational vector in the relative interior of the face F of optimum solutions of $\max\{cx | x \in P\}$ (i.e. $B(x^*, \varepsilon) \cap \text{aff.hull}(F) \subseteq F$ for some $\varepsilon > 0$).

First let $t = 0$. Consider

(32) $\max \{\lceil c \rceil x | Ax \leqslant b; \xi_i \leqslant \xi_i^* \text{ if } \gamma_i \notin \mathbb{Z}\}$

(denoting $x = (\xi_1, \ldots, \xi_n)^{\mathsf{T}}$). This maximum is attained by x^* (as for each feasible solution x of (32) one has $\lceil c \rceil (x^* - x) \geqslant c(x^* - x) \geqslant 0$). As $Ax \leqslant b$ is box-TDI, there exists integral row vectors $y \geqslant 0, z = (\zeta_1, \ldots, \zeta_n) \geqslant 0$ such that $yA + z = \lceil c \rceil$, $yb + zx^* = \lceil c \rceil x^*, \zeta_i = 0$ if $\gamma_i \in \mathbb{Z}$. So for $\tilde{c} := \lceil c \rceil - z$ we have that $\max \{\tilde{c}x | Ax \leqslant b\}$ is attained at x^* (as $\tilde{c}x^* = yb, yA = \tilde{c}, y \geqslant 0$). This shows (31) for $t = 0$.

Suppose we have proved (31) for a certain $t < n$. We prove (31) for $t + 1$. The induction hypothesis tells that there exists a $\tilde{c}$ with the properties as described for the case t, i.e. with $\tilde{\gamma}_i \geqslant \lfloor \gamma_i \rfloor$ for $i = 1, \ldots, t$. If also $\tilde{\gamma}_{t+1} \geqslant \lfloor \gamma_{t+1} \rfloor$ we are done; so suppose $\tilde{\gamma}_{t+1} < \lfloor \gamma_{t+1} \rfloor$. Make that convex combination $c' = \lambda c + (1 - \lambda)\tilde{c}$ of c and $\tilde{c}$ ($0 \leqslant \lambda \leqslant 1$) which is equal to $\lfloor \gamma_{t+1} \rfloor$ in coordinate $t + 1$. Then x^* is again an optimum solution of $\max \{c'x | Ax \leqslant b\}$. Hence, by our induction hypothesis again, there exists an integral vector $\tilde{c}' = (\tilde{\gamma}'_1, \ldots, \tilde{\gamma}'_n)$ such that x^* is an optimum solution of $\max\{\tilde{c}'x \mid Ax \leqslant b\}$ and such that $\tilde{\gamma}'_i \leqslant \lceil \gamma'_i \rceil \leqslant \lceil \gamma_i \rceil$ $(i = 1, \ldots, n)$, $\tilde{\gamma}'_i \geqslant \lfloor \gamma'_i \rfloor \geqslant \lfloor \gamma_i \rfloor$ $(i = 1, \ldots, t)$, $\tilde{\gamma}'_{t+1} = \gamma'_{t+1} = \lfloor \gamma_{t+1} \rfloor$, and $\tilde{\gamma}'_i = \tilde{\gamma}_i = \gamma_i$ if γ_i is an integer. Thus we have proved (31) for the case $t + 1$.

Sufficiency. Suppose P satisfies the condition mentioned in the theorem. Let $Ax \leqslant b$ be any TDI-system defining P. We prove that $Ax \leqslant b$ is box-TDI.

Choose $l = (\lambda_1, \ldots, \lambda_n)^\mathsf{T}, u = (v_1, \ldots, v_n)^\mathsf{T} \in \mathbb{Q}^n$ and $w = (\omega_1, \ldots, \omega_n) \in \mathbb{Z}^n$. Let x^* be an optimum solution of

(33) $\max\{wx \mid Ax \leqslant b; l \leqslant x \leqslant u\}$.

By LP-duality and complementary slackness, there exist row vectors $y, z' = (\zeta'_1, \ldots, \zeta'_n), z'' = (\zeta''_1, \ldots, \zeta''_n) \geqslant 0$ such that

(34) $yA + z' - z'' = w; yb + z'u - z''l = wx^*; \zeta'_i = 0$ or $\zeta''_i = 0$ $(i = 1, \ldots, n)$.

Now define $c = (\gamma_1, \ldots, \gamma_n) := yA = w - z' + z''$. By assumption, there exists an integral vector $\tilde{c} = (\tilde{\gamma}_1, \ldots, \tilde{\gamma}_n)$ such that $\lfloor c \rfloor \leqslant \tilde{c} \leqslant \lceil c \rceil$ and such that each optimum solution of $\max\{cx \mid Ax \leqslant b\}$ is also an optimum solution of $\max\{\tilde{c}x \mid Ax \leqslant b\}$. As x^* is an optimum solution of $\max\{cx \mid Ax \leqslant b\}$ (by complementary slackness: $cx^* = (w - z' + z'')x^* \geqslant wx^* - z'u + z''l = yb$), x^* is also an optimum solution of $\max\{\tilde{c}x \mid Ax \leqslant b\}$. As $Ax \leqslant b$ is TDI, there exists an integral vector $\tilde{y} \geqslant 0$ such that $\tilde{y}A = \tilde{c}$ and $\tilde{y}b = \tilde{c}x^*$. Define $\tilde{z}' := (w - \tilde{c})_+, \tilde{z}'' := (\tilde{c} - w)_+$ (where $(\cdot)_+$ arises from $(\cdot)$ by setting negative components to 0). Now if $\zeta^*_i < v_i$ then $\zeta'_i = 0$, hence $\gamma_i \geqslant \omega_i$, and hence $\tilde{\gamma}_i \geqslant \omega_i$. So $\tilde{z}'u = \tilde{z}'x^*$. Similarly, $\tilde{z}''l = \tilde{z}''x^*$. Therefore,

(35) $\tilde{y}A + \tilde{z}' - \tilde{z}'' = \tilde{y}A + w - \tilde{c} = w,$

            $\tilde{y}b + \tilde{z}'u - \tilde{z}''l = \tilde{c}x^* + \tilde{z}'u - \tilde{z}''l = \tilde{c}x^* + \tilde{z}'x^* - \tilde{z}''x^* = wx^*.$

Conclusion: (33) has an integral optimum dual solution.   □

Corollary 22.9a. *A rational system $Ax \leqslant b$ is box-TDI if and only if it is TDI and for each rational vector $c = (\gamma_1, \ldots, \gamma_n)$ there exists an integral vector $\tilde{c} = (\tilde{\gamma}_1, \ldots, \tilde{\gamma}_n)$ such that $\lfloor c \rfloor \leqslant \tilde{c} \leqslant \lceil c \rceil$ and such that each optimum solution of $\max\{cx \mid Ax \leqslant b\}$ is also an optimum solution of $\max\{\tilde{c}x \mid Ax \leqslant b\}$.*

Proof. Directly from Theorem 22.9.   □

It follows also from Theorem 22.9 that if P is a full-dimensional box-TDI polyhedron, then $P = \{x \mid Ax \leqslant b\}$ for some $\{0, \pm 1\}$-matrix A.

Proof. Let $ax \leqslant \beta$ define a facet F of P, let $k := \|a\|_\infty$ and $c := (1/k)a$. As $\max\{cx \mid x \in P\}$ is finite, by Theorem 22.9 there exists an integral vector $\tilde{c}$ such that $\lfloor c \rfloor \leqslant \tilde{c} \leqslant \lceil c \rceil$, and such that $\max\{\tilde{c}x \mid x \in P\}$ attains its maximum in each element of F. Hence $\tilde{c} = c$, and therefore c is a $\{0, \pm 1\}$-vector.   □

It was proved by Edmonds and Giles [1977, 1984] that this is also the case for non-full-dimensional polyhedra, as we shall see as Corollary 22.10a.

22.5. BEHAVIOUR OF TOTAL DUAL INTEGRALITY UNDER OPERATIONS

In this section we investigate how total dual integrality behaves under certain operations on the system of inequalities.

Scalar multiplications. Clearly, if $Ax \leqslant b$ is TDI, not-necessarily $(\alpha A)x \leqslant \alpha b$ is TDI for each $\alpha > 0$. This can be seen by taking α so that all components of αA are even integers.

However, if $Ax \leqslant b$ is TDI, then total dual integrality is maintained if we divide any inequality by a natural number. Moreover, it is maintained if we multiply all right-hand sides by a positive rational. In particular, if $Ax \leqslant b$ is TDI, also $(k^{-1}A)x \leqslant \alpha b$ is TDI, for each $k \in \mathbb{N}$ and $\alpha \geqslant 0$.

The same statements hold for box-total dual integrality.

Giles and Pulleyblank [1979] showed:

(36)   For each rational system $Ax \leqslant b$ there exists a natural number k such that $(k^{-1}A)x \leqslant k^{-1}b$ is totally dual integral.

Proof. (36) is equivalent to the fact that there exists a number k such that $\min\{yb|y \geqslant 0; yA = w\}$ is attained by a vector y with components in $k^{-1}\mathbb{Z}$, for each integral vector w for which the minimum is finite (since then $\min\{y(k^{-1}b)|y \geqslant 0; y(k^{-1}A) = w\}$ has an integral optimum solution). The existence of such a k can be shown as follows.

By Theorem 22.6 there exists a TDI-system $Cx \leqslant d$ with C integral, and with the same solution set as $Ax \leqslant b$. Let k be a natural number such that for each row c of C, $\min\{yb|y \geqslant 0; yA = c\}$ has an optimum solution y with components in $k^{-1}\mathbb{Z}$. (Such a k exists, as for each fixed row c in C the minimum is attained by a rational vector, say, y_c. For k we can take the l.c.m. of the denominators occurring in the vectors y_c.)

Now let w be an integral vector, such that

(37) $\max\{wx|Ax \leqslant b\} = \min\{yb|y \geqslant 0; yA = w\}$

is finite. Let the optimum value be δ. As $Cx \leqslant d$ is TDI, the inequality $wx \leqslant \delta$ is a nonnegative integral combination of inequalities in $Cx \leqslant d$. Each inequality in $Cx \leqslant d$ is a nonnegative integral combination of inequalities in $(k^{-1}A)x \leqslant k^{-1}b$. Therefore, $wx \leqslant \delta$ is a nonnegative integral combination of inequalities in $(k^{-1}A)x \leqslant k^{-1}b$. This means that $yb = \delta$ for some $y \geqslant 0$ with $yA = w$ and all components of y being in $(1/k)\mathbb{Z}$. So the minimum in (37) has a $(1/k)$-integral optimum solution y. $\square$

Cook [1983] showed:

(38)   Let $Ax \leqslant b$, $ax \leqslant \beta$ be a minimal TDI-system, with A, a integral, and defining a full-dimensional polyhedron. Then for $\alpha > 0$, the system $Ax \leqslant b$; $(\alpha a)x \leqslant \alpha \beta$ is TDI if and only if $\alpha = 1/k$ for some natural number k.

As in the proof of Theorem 22.6, this follows from the following: *let* $a_1, a_2, \ldots, a_t$ *form the minimal integral Hilbert basis generating the pointed cone* C; *then for each rational* $\alpha > 0$ *we have:* $\alpha a_1, a_2, \ldots, a_t$ *is again a Hilbert basis if and only if* $\alpha = 1/k$ *for some natural number* k.

Proof. Sufficiency in this last statement is immediate. To see necessity, let $\alpha a_1, a_2, \ldots, a_t$ form a Hilbert basis. As $\alpha > 0$, they generate C. Since a_1 is in C, there exist nonnegative integers $\lambda_1, \ldots, \lambda_t$ such that

(39) $a_1 = \lambda_1 \alpha a_1 + \lambda_2 a_2 + \cdots + \lambda_t a_t.$

Then $\lambda_1 \neq 0$, since otherwise $a_1, \ldots, a_t$ would not form a minimal Hilbert basis. Now $(1 - \lambda_1 \alpha)a_1 = \lambda_2 a_2 + \cdots + \lambda_t a_t$ is in C. Since C is pointed, $-a_1 \notin C$. Therefore, $1 - \lambda_1 \alpha \geqslant 0$, and hence $\lambda_1 \alpha \leqslant 1$. As $(\lambda_1 \alpha)a_1 = a_1 - \lambda_2 a_2 - \cdots - \lambda_t a_t$ is an integral vector, and the components of a_1 are relatively prime integers, we know that $\lambda_1 \alpha = 1$. So $\alpha = 1/\lambda_1$. $\square$

Given a polyhedron P and a natural number k, one may extend Theorem 22.1 straightforwardly to: each face of P contains a vector with all components in $(1/k)\mathbb{Z}$, if and only if each supporting hyperplane of P contains such a vector. This follows, for example, by replacing P by kP. The corollaries can be extended similarly.

Thus call a system $Ax \leqslant b$ *totally dual* $1/k$-*integral* if the minimum in

(40) $\max \{cx \,|\, Ax \leqslant b\} = \min \{yb \,|\, y \geqslant 0; yA = c\}$

has a $1/k$-integral optimum solution y, for each integral vector c for which the minimum is finite. Clearly, $Ax \leqslant b$ is totally dual $1/k$-integral, if and only if $(k^{-1}A)x \leqslant k^{-1}b$ is totally dual integral. This implies that if $Ax \leqslant b$ is totally dual $1/k$-integral and b is integral, the maximum in (40) has a $1/k$-integral optimum solution x.

Further operations (not) preserving total dual integrality. In Theorem 22.2 we saw that 'each face of a TDI-system is TDI again'. That is, total dual integrality is preserved if we set some inequalities to equality. This directly implies that the same holds for box-total dual integrality.

The following is also direct:

(41)   *If $A_1 x \leqslant b_1$ and $A_2 x \leqslant b_2$ define the same polyhedron, and each inequality in $A_1 x \leqslant b_1$ is a nonnegative integral combination of inequalities in $A_2 x \leqslant b_2$, then $A_1 x \leqslant b_1$ (box-)TDI implies $A_2 x \leqslant b_2$ (box-)TDI.*

Proof. The condition implies that for each integral $y \geqslant 0$ there exists an integral $y' \geqslant 0$ with $yA_1 = y'A_2$ and $yb_1 = y'b_2$. $\square$

(Box-)total dual integrality is preserved under adding and removing 'slack' variables (Cook [1983]):

(42)   *Let A be a rational matrix, b be a rational vector, a be an integral vector, and β be a rational. Then the system $Ax \leqslant b$; $ax \leqslant \beta$ is (box-)TDI, if and only if the system $Ax \leqslant b$; $ax + \eta = \beta$; $\eta \geqslant 0$ is (box-)TDI (where η is a new variable).*

Proof. For any integral vector c and integer γ, consider the LP-duality equations:

(43) $\max \{cx \,|\, Ax \leqslant b; ax \leqslant \beta\} = \min \{yb + \lambda\beta \,|\, y \geqslant 0; \lambda \geqslant 0; yA + \lambda a = c\}$, and
            $\max \{cx + \gamma\eta \,|\, Ax \leqslant b; ax + \eta = \beta; \eta \geqslant 0\}$
            $= \min \{yb + \mu\beta \,|\, y \geqslant 0; \mu \geqslant \gamma; yA + \mu a = c\}$.

Now y, λ form an optimum solution for the first minimum, if and only if $y, \mu = \lambda + \gamma$ form an optimum solution for the second minimum, with c replaced by $c + \gamma a$. Hence the first minimum has an integral optimum solution for each integral vector c for which the minimum is finite, if and only if the second minimum has an integral optimum solution for each integral vector (c, γ) for which the minimum is finite.   ☐

Total dual integrality is maintained also under the following elimination operation. Let A be an integral matrix, d, a_1, a_2 be integral vectors, ε be an integer, b be a rational vector, β_1, β_2 be rationals, x be a vector variable, and η be a single variable. Then:

(44)   the system $Ax + d\eta \leqslant b, a_1 x + \varepsilon \eta \leqslant \beta_1, a_2 x + \eta = \beta_2$ is TDI, if and only if
         the system $Ax + d\eta \leqslant b, (a_1 - \varepsilon a_2)x \leqslant \beta_1 - \varepsilon \beta_2, a_2 x + \eta = \beta_2$ is TDI.
         Similarly for box-TDI.

This follows directly from (41).

Combining (42) and (44) yields that if a variable has a coefficient $+1$ or -1 in an equality constraint in a (box-)TDI-system (i.e. in two opposite inequalities), this variable can be eliminated without spoiling (box-)total dual integrality.

Cook [1983] showed that (box-)total dual integrality is also maintained under 'Fourier–Motzkin elimination' of a variable η, if it occurs in each constraint with coefficient 0 or ± 1 (cf. Section 12.2). That is, *if the system*

(45)   (i) $a_i x + \eta \leqslant \beta_i$ $(i = 1, \ldots, m')$
         (ii) $a_i x - \eta \leqslant \beta_i$ $(i = m' + 1, \ldots, m'')$
         (iii) $a_i x \leqslant \beta_i$ $(i = m'' + 1, \ldots, m)$

is TDI, then also the system

(46) $(a_i + a_j)x \leqslant \beta_i + \beta_j$ $(i = 1, \ldots, m'; j = m' + 1, \ldots, m'')$
         $a_i x \leqslant \beta_i$ $(i = m'' + 1, \ldots, m)$

is TDI.

Proof. Observe that for each vector c,

(47) $\max \{cx \mid x \text{ satisfies } (46)\} = \max \left\{(c, 0) \binom{x}{\eta} \,\middle|\, \binom{x}{\eta} \text{ satisfies } (45) \right\}.$

Let y be an integral optimum solution to the dual of the second maximum in (47). As the last component of $(c, 0)$ is 0, the sum of the components of y corresponding to the inequalities (45)(i) is equal to the sum of the components of y corresponding to the inequalities (45)(ii). By appropriately pairing constraints we obtain an integral optimum solution for the dual problem of the first maximum in (47).   ☐

It is direct that if $a\xi_0 + Ax \leqslant b$ is box-TDI (where a is a column vector), then the system $Ax \leqslant b$ is also box-TDI (as it essentially means adding $0 \leqslant \xi_0 \leqslant 0$). So the intersection of a box-TDI polyhedron with a coordinate hyperplane is box-TDI again.

From Theorem 22.9 it also follows easily that the projection of a box-TDI polyhedron on a coordinate hyperplane is a box-TDI polyhedron again. That is, if P is a box-TDI polyhedron in $\mathbb{R}^{n+1}$, then also the polyhedron

$$(48) \qquad \left\{ x \,\middle|\, \exists \xi_0 \colon \begin{pmatrix} \xi_0 \\ x \end{pmatrix} \in P \right\}$$

is a box-TDI polyhedron.

Now we consider repeating variables. Total dual integrality is trivially preserved under replacing the system $Ax \leqslant b$ by $a\xi_0 + Ax \leqslant b$, where a is the first column of A, and ξ_0 is a new variable. This is also the case for box-total dual integrality (Edmonds and Giles [1984], cf. Cook [1986], whose proof we follow).

Theorem 22.10. *If $Ax \leqslant b$ is box-TDI, then $a\xi_0 + Ax \leqslant b$ is box-TDI again, where a is the first column of A and ξ_0 is a new variable.*

Proof. By Theorem 22.8 it suffices to show that if $P \subseteq \mathbb{R}^n$ is a box-TDI polyhedron, then the polyhedron

$$(49) \qquad Q := \{ (\xi_0, \xi_1, \xi_2, \ldots, \xi_n)^\mathsf{T} \,|\, (\xi_0 + \xi_1, \xi_2, \ldots, \xi_n)^\mathsf{T} \in P \}$$

is also a box-TDI polyhedron.

To show this, let $\max\{\gamma_0 \xi_0 + cx \,|\, (\xi_0, x^\mathsf{T})^\mathsf{T} = (\xi_0, \ldots, \xi_n)^\mathsf{T} \in Q\}$ be finite. Denote $c = (\gamma_1, \ldots, \gamma_n)$. It implies $\gamma_0 = \gamma_1$ (as otherwise we could replace ξ_0, ξ_1 by $\xi_0 \pm 1$, $\xi_1 \mp 1$, and increase the maximum). As $Ax \leqslant b$ is box-TDI, there exists by Theorem 22.9 an integral vector $\tilde{c} = (\tilde{\gamma}_1, \ldots, \tilde{\gamma}_n)$ such that $\lfloor c \rfloor \leqslant \tilde{c} \leqslant \lceil c \rceil$ and such that each optimum solution of $\max\{cx \,|\, Ax \leqslant b\}$ is also an optimum solution of $\max\{\tilde{c}x \,|\, Ax \leqslant b\}$. This directly implies that each optimum solution of $\max\{\gamma_0 \xi_0 + cx \,|\, a\xi_0 + Ax \leqslant b\}$ is also an optimum solution of $\max\{\tilde{\gamma}_1 \xi_0 + \tilde{c}x \,|\, a\xi_0 + Ax \leqslant b\}$. Hence, by Theorem 22.9, $a\xi_0 + Ax \leqslant b$ is box-TDI. $\qquad \square$

The following corollary is due to Edmonds and Giles [1984].

Corollary 22.10a. *Let P be a box-TDI polyhedron. Then $P = \{x \,|\, Ax \leqslant b\}$ for some $\{0, \pm 1\}$-matrix A and some vector b.*

Proof. Let $Cx \leqslant d$ be a box-TDI system defining P. Then any vector x^* belongs to P if and only if

$$(50) \qquad \max\{-\mathbf{1}x' + \mathbf{1}x'' \,|\, Cx + Cx' + Cx'' \leqslant d; x = x^*; x' \geqslant 0; x'' \leqslant 0\} \geqslant 0$$

(where $\mathbf{1}$ is an all-one row vector). As the system $Cx + Cx' + Cx'' \leqslant d$ is box-TDI (by Theorem 22.10), it follows that (50) is equivalent to

$$(51) \qquad \min\{yd - yCx^* \,|\, y \geqslant 0; yC \geqslant -\mathbf{1}; yC \leqslant \mathbf{1}; y \text{ integral}; yC \text{ integral}\} \geqslant 0.$$

So

$$(52) \qquad P = \{x \,|\, (yC)x \leqslant yd \text{ for each } y \geqslant 0 \text{ with } y \text{ and } yC \text{ integral and } -\mathbf{1} \leqslant yC \leqslant \mathbf{1}\}.$$

Hence P can be defined by linear inequalities with coefficients $0, \pm 1$ only. $\square$

Remark 22.1. Define

$$(53) \qquad A := \begin{bmatrix} 1 & 1 & 1 & 1 \\ 1 & 1 & 0 & 0 \\ 1 & 0 & 1 & 0 \\ 1 & 0 & 0 & 1 \\ 2 & 1 & 1 & 1 \end{bmatrix}.$$

Then the system $Ax \leqslant 0$ is minimally box-TDI. This shows that an integral minimally box-TDI system need not have $\{0, \pm 1\}$ left-hand side coefficients, not even if the feasible region is full-dimensional. Note moreover that $\{x \mid Ax \leqslant 0\}$ cannot be defined with a totally unimodular left-hand side matrix. On the other hand, it can be shown that if P is a box-TDI polyhedron, then aff.hull $(P) = \{x \mid Cx = d\}$ for some totally unimodular matrix C.

Remark 22.2. Let $Ax \leqslant b$ be box-TDI, and let $P := \{x \mid Ax \leqslant b\}$. If P is pointed, then each edge and each extremal ray of P is in the direction of a $\{0, \pm 1\}$-vector. To see this, we may assume that b is integral, and that each pair of vertices of P has distance at least $n + 1$, where n is the dimension of the space (otherwise multiply b by an appropriately large number). Let $z = (\zeta_1, \ldots, \zeta_n)^{\mathsf{T}}$ be a vertex of P. Then the polyhedron

$$(54) \qquad \{x \mid Ax \leqslant b; \zeta_i - 1 \leqslant \xi_i \leqslant \zeta_i + 1 \quad \text{for } i = 1, \ldots, n\}$$

is integral (by Corollary 22.1c), where $x = (\xi_1, \ldots, \xi_n)^{\mathsf{T}}$. This implies that each edge or extremal ray of P starting in z leaves the box $\{x \mid \zeta_i - 1 \leqslant \xi_i \leqslant \zeta_i + 1 \text{ for } i = 1, \ldots, n\}$ at an integral vector, say y. Then $y - z$ is a $\{0, \pm 1\}$-vector, determining the direction of the edge or extremal ray.

The following transformation, known from linear programming, generally does not preserve total dual integrality: replacing $Ax \leqslant b$ by $x' \geqslant 0$; $x'' \geqslant 0$; $Ax' - Ax'' \leqslant b$. This is shown by taking

$$(55) \qquad A := \begin{bmatrix} 1 & 5 \\ 1 & 6 \end{bmatrix}, \quad b := \begin{pmatrix} 1 \\ 1 \end{pmatrix}$$

(cf. Cook [1983]).

Another negative result: if $Ax \leqslant b_1$ and $Ax \leqslant b_2$ are TDI, then $Ax \leqslant b_1 + b_2$ is not always TDI (so the set $\{b \mid Ax \leqslant b$ TDI$\}$ is generally not a convex cone, although it is closed under nonnegative scalar multiplications). To see this take:

$$(56) \qquad A := \begin{bmatrix} 1 & 0 \\ 1 & 2 \\ 0 & 1 \\ 0 & -1 \end{bmatrix}, \quad b_1 := \begin{bmatrix} 0 \\ 2 \\ 1 \\ 0 \end{bmatrix}, \quad b_2 := \begin{bmatrix} 0 \\ 0 \\ 1 \\ 0 \end{bmatrix}.$$

Finally we show that the dominant of a box-TDI polyhedron is a box-TDI polyhedron again. Here the *dominant* of a polyhedron $P \subseteq \mathbb{R}^n$ is given by

$$(57) \qquad \text{dom}(P) := \{z \in \mathbb{R}^n \mid \exists x \in P : x \leqslant z\}.$$

Theorem 22.11. *The dominant of a box-TDI polyhedron is again a box-TDI polyhedron.*

Proof. We use Theorem 22.9. Let P be a box-TDI polyhedron in $\mathbb{R}^n$ and let $c = (\gamma_1, \ldots, \gamma_n)$ be a rational vector such that

(58) $\max\{cx \mid x \in \mathrm{dom}(P)\}$

is finite. It implies $c \leqslant 0$ (as the maximum can be increased in any coordinate direction in which c is positive). As P is a box-TDI polyhedron, there exists an integral vector $\tilde{c}$ such that $\lfloor c \rfloor \leqslant \tilde{c} \leqslant \lceil c \rceil$ and such that each optimum solution of $\max\{cx \mid x \in P\}$ is also an optimum solution of $\max\{\tilde{c}x \mid x \in P\}$. We show that each optimum solution x^* of $\max\{cx \mid x \in \mathrm{dom}(P)\}$ is also an optimum solution of of $\max\{\tilde{c}x \mid x \in \mathrm{dom}(P)\}$.

Indeed, we know that $x^* \geqslant z$ for some z in P. As x^* maximizes cx over $\mathrm{dom}(P)$, we know that x^* and z are the same in coordinates where c is negative. Hence $cz = cx^*$, and $\tilde{c}z = \tilde{c}x^*$ (as $\tilde{c} \geqslant \lfloor c \rfloor$). So z is an optimum solution of $\max\{cx \mid x \in P\}$. hence also of $\max\{\tilde{c}x \mid x \in P\}$. Therefore, z is an optimum solution of $\max\{\tilde{c}x \mid x \in \mathrm{dom}(P)\}$, and hence also x^* is an optimum solution of $\max\{\tilde{c}x \mid x \in \mathrm{dom}(P)\}$. $\square$

This implies a result of Edmonds and Giles [1977] that if P is a box-TDI polyhedron, then $\mathrm{dom}(P)$ can be defined by linear inequalities $Ax \geqslant b$ with coefficients 0 and 1 only.

22.6. AN INTEGER ANALOGUE OF CARATHÉODORY'S THEOREM

It follows from Carathéodory's theorem (cf. Corollary 7.1i) that if the optima in

(59) $\max\{cx \mid Ax \leqslant b\} = \min\{yb \mid y \geqslant 0; yA = c\}$

are finite, there is an optimum dual solution y with at most r positive components (r being the rank of A). Is this also true if $Ax \leqslant b$ is TDI, c is integral, and we require y to be integral? One easily finds TDI-systems $Ax \leqslant b$ with non-full-dimensional feasible region, for which the answer is negative $\left(\text{for example, } A = \begin{bmatrix} 2 \\ -3 \end{bmatrix}, b = \begin{pmatrix} 0 \\ 0 \end{pmatrix}, c = (1)\right).$
If we restrict ourselves to full-dimensional regions, the answer is not known.

In Cook, Fonlupt, and Schrijver [1986] the following bound is shown.

Theorem 22.12. *Let $Ax \leqslant b$ be a TDI-system, defining a full-dimensional polyhedron with A integral. Let c be an integral vector for which the optima (59) are finite. Then the minimum in (59) has an integral optimum solution y with at most $2r - 1$ positive components ($r := \mathrm{rank}(A)$).*

Proof. As before, this follows from the following result. *Let the vectors $a_1, \ldots, a_t$ from $\mathbb{Z}^n$ form a Hilbert basis, and let cone $\{a_1, \ldots, a_t\}$ be pointed. Then any integral vector c in cone $\{a_1, \ldots, a_t\}$ is a nonnegative integral combination of at most $2n - 1$ of the a_i.*

To prove this last statement, let $\lambda_1, \ldots, \lambda_t$ attain

(60) $\max\{\lambda_1 + \cdots + \lambda_t \mid \lambda_1, \ldots, \lambda_t \geqslant 0; c = \lambda_1 a_1 + \cdots + \lambda_t a_t\}.$

Since cone $\{a_1, \ldots, a_t\}$ is pointed, this maximum is finite. As (60) is a linear programming problem, we may assume that at most n of the λ_i are nonzero (by the 'fractional' Carathéodory's theorem). Now

(61) $c' := c - \lfloor \lambda_1 \rfloor a_1 - \cdots - \lfloor \lambda_t \rfloor a_t = (\lambda_1 - \lfloor \lambda_1 \rfloor)a_1 + \cdots + (\lambda_t - \lfloor \lambda_t \rfloor)a_t$

is an integral vector in cone $\{a_1, \ldots, a_t\}$. Hence

(62) $\qquad c' = \mu_1 a_1 + \cdots + \mu_t a_t$

for certain nonnegative integers $\mu_1, \ldots, \mu_t$. Now

(63) $\qquad \mu_1 + \cdots + \mu_t + \lfloor \lambda_1 \rfloor + \cdots + \lfloor \lambda_t \rfloor \leqslant \lambda_1 + \cdots + \lambda_t$

(as $\lambda_1, \ldots, j_t$ attain the maximum (60), and hence

(64) $\qquad \mu_1 + \cdots + \mu_t \leqslant \lambda_1 + \cdots + \lambda_t - \lfloor \lambda_1 \rfloor - \cdots - \lfloor \lambda_t \rfloor < n$

(as at most n of the λ_i are nonzero). Therefore, at most $n - 1$ of the μ_i are nonzero. This implies that the decomposition

(65) $\qquad c = (\lfloor \lambda_1 \rfloor + \mu_1) a_1 + \cdots + (\lfloor \lambda_t \rfloor + \mu_t) a_t$

has at most $2n - 1$ nonzero coefficients. $\qquad\qquad\qquad\qquad\qquad\qquad\qquad\square$

Theorem 22.12 implies that an integral optimum dual solution for a linear program over a TDI-system $Ax \leqslant b$, with A integral, can be described in space polynomial in the number of primal variables, even if the number of dual variables is exponentially large.

22.7. ANOTHER CHARACTERIZATION OF TOTAL DUAL INTEGRALITY

Inspired by results of Lovász [1976, 1977], Schrijver and Seymour [1977] (cf. Schrijver [1981]) described a stronger version of Edmonds and Giles' theorem (Corollary 22.1b), which can be useful in combinatorial applications.

Let A be a rational $m \times n$-matrix, and let b be an integral m-vector. Then for any rational vector c there are the following inequalities:

(66) $\qquad \max\{cx \,|\, Ax \leqslant b; x \text{ integral}\} \leqslant \max\{cx \,|\, Ax \leqslant b\}$

$\qquad = \min\{yb \,|\, y \geqslant 0; yA = c\} \leqslant \min\{yb \,|\, y \geqslant 0; yA = c; y \text{ half-integral}\}$

$\qquad \leqslant \min\{yb \,|\, y \geqslant 0; yA = c; y \text{ integral}\}.$

Corollary 22.1b asserts that if the last three optima are equal for each integral vector c, then all five optima are equal for each integral vector c. The following theorem asserts that it suffices to require that the last two optima in (66) are equal for each integral c.

Theorem 22.13. *The rational system $Ax \leqslant b$ is TDI, if and only if*

(67) $\qquad \min\{yb \,|\, y \geqslant 0; yA = c; y \text{ half-integral}\}$

is finite and is attained by an integral vector y, for each integral vector c for which $\min\{yb \,|\, y \geqslant 0; yA = c\}$ *is finite.*

Proof. By (66) it suffices to show the 'if' part. Suppose $Ax \leqslant b$ satisfies the condition stated in the theorem. Let C be the collection of all integral vectors c for which

(68) $\qquad \min\{yb \,|\, y \geqslant 0; yA = c\}$

is finite. We first show that for each c in C and each integer $k \geqslant 0$,

(69) $\min\{yb \,|\, y \geqslant 0; yA = c; 2^k y \text{ integral}\} = \min\{yb \,|\, y \geqslant 0; yA = c; y \text{ integral}\}.$

This is shown by induction on k, the case $k = 0$ being trivial. For $k \geqslant 0$ we have:

(70) $\min\{yb | y \geqslant 0; yA = c; 2^{k+1}y \text{ integral}\}$

$\qquad = 2^{-k}\min\{yb | y \geqslant 0; yA = 2^k c; 2y \text{ integral}\}$

$\qquad = 2^{-k}\min\{yb | y \geqslant 0; yA = 2^k c; y \text{ integral}\}$

$\qquad = \min\{yb | y \geqslant 0; yA = c; 2^k y \text{ integral}\}$

$\qquad = \min\{yb | y \geqslant 0; yA = c; y \text{ integral}\}.$

Here the first and third equality are straightforward, the second equality follows from the condition given in the theorem, while the last one follows from the induction hypothesis. This shows (69).

(69) implies that

(71) $\inf\{yb | y \geqslant 0; yA = c; y \text{ dyadic}\} = \min\{yb | y \geqslant 0; yA = c; y \text{ integral}\}$

for each c in C (a vector is *dyadic* if all its components are dyadic rationals, i.e. rationals with denominator a power of 2). We are finished if we have shown that the infimum in (71) is equal to the minimum (68), for all c in C.

To this end, let $c \in C$, and let $Q := \{y | y \geqslant 0; yA = c\}$. Then the dyadic vectors form a dense subset of Q. Indeed, as (67) is finite, Q contains at least one dyadic vector, say y_0. Let $H := \text{aff.hull}(Q)$. The dyadic vectors in the linear space $H - y_0$ form a dense subset of $H - y_0$ (as we are working in rational spaces). Therefore, as y_0 is dyadic, the dyadic vectors of H form a dense subset of H. Since Q is a polyhedron in H of the same dimension as H, it follows that the dyadic vectors in Q are dense in Q. Therefore, (71) equals (68). $\square$

Corollary 22.13a. *Let $Ax \leqslant b$ be a rational system of linear inequalities with at least one solution. Then $Ax \leqslant b$ is TDI if and only if:*

(i) *for each vector $y \geqslant 0$ with yA integral, there exists a dyadic vector $y' \geqslant 0$ with $y'A = yA$;*

(ii) *for each half-integral vector $y \geqslant 0$ with yA integral, there exists an integral vector $y' \geqslant 0$ with $y'A = yA$ and $y'b \leqslant yb$.*

Proof. If $Ax \leqslant b$ has a solution, (i) and (ii) are equivalent to the condition given in Theorem 22.13. $\square$

If A is an integral matrix, condition (ii) can be weakened.

Corollary 22.13b. *Let $Ax \leqslant b$ be a rational system of linear inequalities with at least one solution, and with A integral. Then $Ax \leqslant b$ is TDI if and only if:*

(i) *for each vector $y \geqslant 0$ with yA integral, there exists a dyadic vector $y' \geqslant 0$ with $y'A = yA$;*

(ii) *for each $\{0, \frac{1}{2}\}$-vector y with yA integral, there exists an integral vector $y' \geqslant 0$ with $y'A = yA$ and $y'b \leqslant yb$.*

Proof. It suffices to show that condition (ii) of Corollary 22.13b implies condition

(ii) of Corollary 22.13a. Let $y \geqslant 0$ be a half-integral vector with yA integral. Then the vector $y - \lfloor y \rfloor$ is a $\{0, \frac{1}{2}\}$-vector ($\lfloor \ \rfloor$ denoting componentwise lower integer parts). As yA is integral and A is integral, also $(y - \lfloor y \rfloor)A$ is integral. By (ii) of Corollary 22.13b, there exists an integral vector $y'' \geqslant 0$ with $y''A = (y - \lfloor y \rfloor)A$ and $y''b \leqslant (y - \lfloor y \rfloor)b$. Taking $y' := y'' + \lfloor y \rfloor$ gives y' as required in (ii) of Corollary 22.13a. $\square$

As corollaries there are analogous results for other types of linear programs, like the following.

Corollary 22.13c. *Let A be a nonnegative rational matrix and let b be an integral vector. Both optima in the LP-duality equation*

$$(72) \qquad \max \{cx \,|\, x \geqslant 0; Ax \leqslant b\} = \min \{yb \,|\, y \geqslant 0; yA \geqslant c\}$$

have integral optimum solutions x and y, for each nonnegative integral vector c for which these optima are finite, if and only if

$$(73) \qquad \min \{yb \,|\, y \geqslant 0; yA \geqslant c; y \text{ half-integral}\}$$

is finite and attained by an integral vector y, for each such vector c.

Proof. This corollary can be proved directly, in a way similar to the proof of Theorem 22.13. $\square$

Corollary 22.13d. *Let A be a nonnegative integral matrix, and let b be a rational vector. Then the system $x \geqslant 0$; $Ax \leqslant b$ is TDI, if and only if for each $\{0, \frac{1}{2}\}$-vector y there exists an integral vector $y' \geqslant 0$ with $y'A \geqslant \lfloor yA \rfloor$ and $y'b \leqslant yb$.*

Proof. From Corollary 22.13b we know that $x \geqslant 0$; $Ax \leqslant b$ is TDI if and only if:

(74)   (i) for all $y, z \geqslant 0$ with $yA - z$ integral, there exist dyadic vectors
        $y', z' \geqslant 0$ with $y'A - z' = yA - z$;
    (ii) for all $\{0, \frac{1}{2}\}$-vectors y, z with $yA - z$ integral, there exist integral
        vectors $y', z' \geqslant 0$ with $y'A - z' = yA - z$ and $y'b \leqslant yb$.

As A is nonnegative and integral, (i) is trivially satisfied, by taking $y' := \lceil y \rceil$ and $z' := (\lceil y \rceil - y)A + z$. Moreover, condition (ii) is equivalent to the condition stated in Corollary 22.13d (note that $yA - z = \lfloor yA \rfloor$ if y and z are $\{0, \frac{1}{2}\}$-vectors and $yA - z$ is integral). $\square$

We leave it to the reader to formulate similar results for the LP-duality equations

$$(75) \qquad \min \{cx \,|\, Ax \geqslant b\} = \max \{yb \,|\, y \geqslant 0; yA = c\}, \text{ and}$$
$$\min \{cx \,|\, x \geqslant 0; Ax \geqslant b\} = \max \{yb \,|\, y \geqslant 0; yA \leqslant c\}.$$

Application 22.3. *The König-Egerváry theorem.* As an illustration, we apply Corollary 22.13d to bipartite graphs, to obtain a theorem of König [1931] and Egerváry [1931].

Let $G = (V, E)$ be a bipartite graph, and let M be its $V \times E$-incidence matrix. Then from Corollary 22.13d it follows that the system $x \geqslant 0$, $Mx \leqslant 1$ is TDI. Indeed, let y be a $\{0, \frac{1}{2}\}$-vector in $\mathbb{R}^V$. Let V_1 and V_2 be the two colour classes of G. Without loss of generality, $\sum_{v \in V_1} y(v) \leqslant \frac{1}{2} \cdot y1$. Then let

(76) $y'(v) = 1$ if $v \in V_1$ and $y(v) = \frac{1}{2}$

              $y'(v) = 0$ otherwise.

One easily checks $y'M \geqslant \lfloor yM \rfloor$ and $y'1 \leqslant y1$.

So $x \geqslant 0$; $Mx \leqslant 1$ is TDI, and therefore both optima in

(77) $\max \{wx \,|\, x \geqslant 0; Mx \leqslant 1\} = \min \{y1 \,|\, y \geqslant 0; yM \geqslant w\}$

have integral optimum solutions. In the special case $w = 1$, the König–Egerváry theorem follows: the maximum cardinality of a matching in a bipartite graph is equal to the minimum number of vertices meeting each edge.

Remark 22.3. The following generalization of Theorem 22.13 is straightforward. Let A be a rational $m \times n$-matrix and let b be an integral vector. Moreover, let k, l be natural numbers, with $l \geqslant 2$. Then both optima in

(78) $\max \{cx \,|\, Ax \leqslant b\} = \min \{yb \,|\, y \geqslant 0; yA = c\}$

are attained by $1/k$-integral vectors x and y, for each integral vector c for which the optima are finite, if and only if

(79) $\min \{yb \,|\, y \geqslant 0; yA = c; y(1/kl)\text{-integral}\}$

is finite and is attained by a $1/k$-integral vector y for such vector c.

22.8. OPTIMIZATION OVER INTEGRAL POLYHEDRA AND TDI-SYSTEMS ALGORITHMICALLY

In Section 16.3 we saw that any integer linear programming problem over an integral polyhedron can be solved in polynomial time. That is, repeating Theorem 16.2,

Theorem 22.14. *There is a polynomial algorithm which, given a rational system $Ax \leqslant b$ defining an integral polyhedron, and given a rational vector c, finds an optimum solution for the ILP-problem* $\max \{cx \,|\, Ax \leqslant b; x \text{ integral}\}$ *(if it is finite).*

Proof. See Theorem 16.2.   □

Now let $Ax \leqslant b$ be a TDI-system. Consider the LP-duality equation

(80) $\max \{cx \,|\, Ax \leqslant b\} = \min \{yb \,|\, y \geqslant 0; yA = c\}$.

If b is integral, $Ax \leqslant b$ determines an integral polyhedron, and hence by Theorem 22.14 we can find an integral optimum solution for the maximum in (80) in polynomial time. With Khachiyan's method, we can find a (fractional) optimum solution for the minimum in (80) in polynomial time. Chandrasekaran [1981] (cf. Orlin [1982]) showed that, by combining a greedy technique with the algorithms of Khachiyan and of Frumkin and von zur Gathen and Sieveking, one can find an integral optimum solution y in polynomial time, if A is an integral matrix.

Theorem 22.15. *Let $Ax \leq b$ be a TDI-system with A integral, and let c be an integral vector. Then an integral optimum solution for the minimum in (80) can be found in polynomial time.*

Proof. With Khachiyan's method one can find an optimum solution x^* for the maximum in (80). Let $a_1, \ldots, a_t$ be the rows of A corresponding to constraints in $Ax \leq b$ which are satisfied with equality by x^*. As $Ax \leq b$ is TDI, $a_1, \ldots, a_t$ form an integral Hilbert basis. Let $C := \text{cone } \{a_1, \ldots, a_t\}$, and let F be the minimal face of C. Without loss of generality, let $a_{k+1}, \ldots, a_t$ be the vectors from $a_1, \ldots, a_t$ which belong to F. These vectors can be determined again with Khachiyan's method, as a_i is in F if and only if $-a_i$ is in C.

First, each integral vector d in F can be expressed as a nonnegative integral combination of $a_{k+1}, \ldots, a_t$, in polynomial time. Indeed, with Khachiyan's method, the all-zero vector 0 can be written as

(81) $\qquad 0 = \mu_{k+1} a_{k+1} + \cdots + \mu_t a_t$

with $\mu_{k+1}, \ldots, \mu_t$ positive. To see this, as $F = \text{cone } \{a_{k+1}, \ldots, a_t\}$, we know that for each $j > k$, we can find $v_{k+1}, \ldots, v_{j-1}, v_{j+1}, \ldots, v_t \geq 0$ such that

(82) $\qquad -a_j = v_{k+1} a_{k+1} + \cdots + v_{j-1} a_{j-1} + v_{j+1} a_{j+1} + \cdots + v_t a_t.$

So

(83) $\qquad 0 = v_{k+1} a_{k+1} + \cdots + v_{j-1} a_{j-1} + a_j + v_{j+1} a_{j+1} + \cdots + v_t a_t.$

Adding up (83) over all $j > k$, we obtain a decomposition of 0 as in (81).

Moreover, with the algorithm of Frumkin and von zur Gathen and Sieveking (Corollary 5.3b) d can be expressed as

(84) $\qquad d = \tau_{k+1} a_{k+1} + \cdots + \tau_t a_t$

with $\tau_{k+1}, \ldots, \tau_t$ integers. Hence, for suitable M,

(85) $\qquad d = (\tau_{k+1} + M\mu_{k+1})a_{k+1} + \cdots + (\tau_t + M\mu_t)a_t$

expresses d as a nonnegative integer combination of $a_{k+1}, \ldots, a_t$. This all can be done in time polynomially bounded by the sizes of A and d.

Now let σ_1 be the largest rational such that $c - \sigma_1 a_1$ belongs to C (σ_1 can be found with Khachiyan's method), and set $c^1 := c - \lfloor \sigma_1 \rfloor a_1$. Next let σ_2 be the largest rational such that $c^1 - \sigma_2 a_2$ belongs to C, and set $c^2 := c^1 - \lfloor \sigma_2 \rfloor a_2$. And so on, until we have found c^k.

Then c^k belongs to F. To see this, observe that $c^k = \lambda_1 a_1 + \cdots + \lambda_t a_t$ for certain nonnegative integers $\lambda_1, \ldots, \lambda_t$ (as c^k is an integral vector in C). By our choice of the σ_i, $c^k - a_i$ does not belong to C, for $i = 1, \ldots, k$. Hence $\lambda_1 = \cdots = \lambda_k = 0$, and c^k belongs to F.

As c^k is in F, by the method described above, we can find nonnegative integers $\sigma_{k+1}, \ldots, \sigma_t$ such that $c^k = \sigma_{k+1} a_{k+1} + \cdots + \sigma_t a_t$. Then

(86) $\qquad c = \lfloor \sigma_1 \rfloor a_1 + \cdots + \lfloor \sigma_k \rfloor a_k + \sigma_{k+1} a_{k+1} + \cdots + \sigma_t a_t.$

This yields an integral optimum solution y for the minimum in (80). $\qquad \square$

In fact, the algorithm described in this proof solves the following problem in polynomial time: given a system $Ax \leqslant b$, with A integral, and an integral vector c, either find an optimum solution for the ILP-problem min $\{yb | y \geqslant 0; yA = c;$ y integral$\}$ (if it is finite) or conclude that $Ax \leqslant b$ is not TDI.

22.9. RECOGNIZING INTEGRAL POLYHEDRA AND TOTAL DUAL INTEGRALITY

It is easy to see that the problem

(87)   Given a rational system $Ax \leqslant b$ of linear inequalities, does it determine an integral polyhedron?

belongs to the complexity class co-$\mathcal{NP}$. If $Ax \leqslant b$ does not determine an integral polyhedron, some minimal face does not contain an integral vector. This minimal face can be described by $A'x = b'$, where $A'x \leqslant b'$ is a subsystem of $Ax \leqslant b$, where A' has the same rank as A, and where $Ax_0 \leqslant b$ for some x_0 with $A'x_0 = b'$. The fact that $A'x = b'$ has no integral solution can be checked in polynomial time, with Corollary 5.3b.

We do not know if the problem (87) belongs to $\mathcal{NP}$. If we fix the rank of A, it is easy to test in polynomial time if $Ax \leqslant b$ determines an integral polyhedron.

Theorem 22.16. *For any fixed r, there exists a polynomial algorithm to test if a given rational system $Ax \leqslant b$, with rank $(A) = r$, determines an integral polyhedron.*

Proof. For each choice of r constraints $A'x \leqslant b'$ from $Ax \leqslant b$, test if the rows in A' are linearly independent; if so, find a solution x_0 of $A'x = b'$; test if $Ax_0 \leqslant b$; if so, $F := \{x | A'x = b'\}$ is a minimal face of $\{x | Ax \leqslant b\}$; test if F contains an integral vector (Corollary 5.3b).

If A has m rows, there are $\binom{m}{r}$ choices of $A'x \leqslant b'$. Each minimal face is tested

at least once for containing integral vectors. As $\binom{m}{r}$ is polynomially bounded by

m (for fixed r), this describes an algorithm as required. $\square$

A similar situation holds for total dual integrality. The problem:

(88)   Given a rational system $Ax \leqslant b$, is it TDI?

belongs to co-$\mathcal{NP}$, provided we restrict ourselves to integral matrices A. In that case, if $Ax \leqslant b$ is not TDI, there is an integral vector c for which the algorithm of Theorem 22.15 does not work. This c can be chosen of size polynomially bounded by the size of A: if y is a (possibly fractional) optimum solution for min $\{yb | y \geqslant 0;$ $yA = c\}$, then also $c' := (y - \lfloor y \rfloor) A$ is a counterexample for total dual integrality.

Problem (88) being in co-$\mathcal{NP}$ implies also that the problem:

(89)   Given a rational system $Ax \leqslant b$, is it box-TDI?

belongs to co-$\mathcal{NP}$ for integral A: If $Ax \leqslant b$ is not box-TDI, then $Ax \leqslant b, l \leqslant x \leqslant u$ is not TDI for some small column vectors l, u, which has a proof of polynomial length. Moreover, the problem:

(90)   Given a rational system $Ax \leqslant b$, does it determine a box-TDI polyhedron?

also belongs to co-$\mathcal{NP}$ (see Cook [1986]).

We do not know if problem (88) is in $\mathcal{NP}$. Edmonds and Giles [1984] conjecture that it is co-$\mathcal{NP}$-complete. In Corollary 22.17a below we shall see that if A is integral and if we fix the rank of A, then total dual integrality of $Ax \leqslant b$ can be tested in polynomial time.

The same holds for recognizing Hilbert bases. The problem: 'Given integral vectors $a_1, \ldots, a_t$, do they form a Hilbert basis?' belongs to co-$\mathcal{NP}$, and it is unknown whether or not it belongs to $\mathcal{NP}$. If we fix the rank of $a_1, \ldots, a_t$, there is a polynomial algorithm, based on a reduction to Lenstra's method for integer linear programming (Section 18.4). This reduction was described in Cook, Lovász, and Schrijver [1984] (extending work by Chandrasekaran and Shirali [1984] and Giles and Orlin [1981]).

Theorem 22.17. *For any fixed r, there exists a polynomial algorithm to test if the rows of a given integral matrix A of rank r form a Hilbert basis.*

Proof. I. Let $a_1, \ldots, a_m$ be the rows of the integral $m \times n$-matrix A. We first argue that we may assume that $C := \mathrm{cone}\{a_1, \ldots, a_m\}$ is pointed and full-dimensional. Let F be the unique minimal face of C, and let L be the linear hull of C. Without loss of generality, let $a_1, \ldots, a_t$ be the rows of A belonging to F (they can be determined in polynomial time as a row a_i of A belongs to F if and only if $-a_i$ belongs to C). Let $d := \dim F$ (and note that $r = \dim L$). Now there exists a unimodular matrix U such that $FU = \mathbb{R}^d \times 0^{n-d}$, and $LU = \mathbb{R}^r \times 0^{n-r}$. Such a matrix U can be found as follows. Choose d linearly independent vectors $v_1, \ldots, v_d$ from $a_1, \ldots a_t$, and after that choose $r - d$ vectors $v_{d+1}, \ldots, v_r$ from $a_{t+1}, \ldots, a_m$ such that $v_1, \ldots, v_r$ are linearly independent. Let V be the matrix with rows $v_1, \ldots, v_r$. We can find in polynomial time (cf. Corollary 5.3a) a unimodular matrix U such that VU is in (lower triangular) Hermite normal form. One easily checks that U has the required properties.

Now the rows of A form a Hilbert basis if and only if the rows of AU form a Hilbert basis. So we may assume that $A = AU$, and that $F = \mathbb{R}^d \times 0^{n-d}$ and $L = \mathbb{R}^r \times 0^{n-r}$. Since now the last $n - r$ rows of A are zero, we may assume that $n = r$. It is easy to see that the rows of A form a Hilbert basis if and only if:

(91)   (i) $a_1, \ldots, a_t$ form a Hilbert basis (generating F);
       (ii) the rows of A' form a Hilbert basis, where A' denotes the matrix consisting of the last $n - d$ columns of A.

Now

(92) $a_1, \ldots, a_t$ form a Hilbert basis if and only if $F \cap \mathbb{Z}^n$ is the lattice generated by $a_1, \ldots, a_t$.

Indeed, necessity of this condition is immediate. To see sufficiency, let z be an integral vector in F. Then, by assumption, $z = v_1 a_1 + \cdots + v_t a_t$ for certain integers $v_1, \ldots, v_t$. Moreover, for each $i = 1, \ldots, t$, the vector $-a_i$ belongs to F, and can hence be written as a nonnegative combination of $a_1, \ldots, a_t$. Therefore, $0 = \mu_1 a_1 + \cdots + \mu_t a_t$ for positive rationals $\mu_1, \ldots, \mu_t$. By choosing M appropriately, $z = (v_1 + M\mu_1)a_1 + \cdots + (v_t + M\mu_t)a_t$ is a decomposition of z as a nonnegative integral combination of $a_1, \ldots, a_t$.

The condition given in (92) can be checked in polynomial time by making the Hermite normal form of $A_0^\top$, where A_0 is the submatrix of A induced by the first t rows and the first d columns; then $F \cap \mathbb{Z}^n$ is the lattice generated by $a_1, \ldots, a_t$ if and only if $\mathbb{Z}^d$ is the lattice generated by the rows of A_0, which is the case if and only if $A_0^\top$ has Hermite normal form $[I \ 0]$.

So (91)(i) can be tested in polynomial time. It remains to test (91)(ii), i.e. if the rows of A' form a Hilbert basis. The cone generated by the rows of A' is pointed and full-dimensional. So without loss of generality we may assume that the cone C generated by $a_1, \ldots, a_m$ is pointed and full-dimensional.

II. Here we describe an algorithm to test whether $a_1, \ldots, a_m$ is a Hilbert basis for the cone C generated by $a_1, \ldots, a_m$, where C is pointed and full-dimensional.

We first observe that $a_1, \ldots, a_m$ form a Hilbert basis for C if and only if the only integral vector in the set

(93) $C_0 = \{z \in C \,|\, z - a_i \notin C \text{ for } i = 1, \ldots, m\}$

is the origin.

To prove necessity, let z be an integral vector in C_0. Then $z = \lambda_1 a_1 + \cdots + \lambda_m a_m$ for nonnegative integers $\lambda_1, \ldots, \lambda_m$. As $z - a_i \notin C$ we know that $\lambda_1 = \ldots = \lambda_m = 0$, i.e. that $z = 0$.

To prove sufficiency, let z be an integral vector in C. Let λ_1 be the highest rational number such that $z - \lambda_1 a_1$ belongs to C. Next, let λ_2 be the highest rational number such that $z - \lfloor \lambda_1 \rfloor a_1 - \lambda_2 a_2$ belongs to C. And so on, until we have found $\lambda_1, \ldots, \lambda_m$. Then $z - \lfloor \lambda_1 \rfloor a_1 - \cdots - \lfloor \lambda_m \rfloor a_m$ is an integral vector belonging to C_0, and hence the origin. This expresses z as a nonnegative integral combination of $a_1, \ldots, a_m$.

So it suffices to check whether the only integral vector in C_0 is the origin. To this end, let $b_1, \ldots, b_t$ be vectors such that

(94) $C = \{z \,|\, b_j z \geq 0 \text{ for } j = 1, \ldots, t\}$.

Since the rank $r = n$ of $a_1, \ldots, a_m$ is fixed, such $b_1, \ldots, b_t$ can be found in polynomial time (as each facet of C is determined by r linearly independent vectors from $a_1, \ldots, a_m$).

Now it follows trivially from (93) and (94) that

(95) $C_0 = \{z \in C \,|\, \text{for all } i = 1, \ldots, m \text{ there exists } j = 1, \ldots, t \text{ with } b_j z < b_j a_i\}$.

So if Φ denotes the collection of all functions $\phi : \{1, \ldots, m\} \to \{1, \ldots t\}$, then

(96) $C_0 = \bigcup_{\phi \in \Phi} \{z \,|\, b_j z \geq 0 \text{ for } j = 1, \ldots, t, \text{ and } b_{\phi(i)} z < b_{\phi(i)} a_i \text{ for } i = 1, \ldots, t\}$.

This expresses C_0 as a union of convex sets, and we have to test whether each of these convex sets contains no other integral vectors than the origin. We shall see that this can be done in polynomial time (for fixed dimension) with Lenstra's algorithm for integer linear programming (Corollary 18.7a). (Note that Φ generally has exponential size, even for fixed rank r of $a_1, \ldots, a_m$.)

Let Z be the collection of vectors z determined by n linearly independent equations from:

(97) $b_j z = 0 \qquad (j = 1, \ldots, t)$,

              $b_j z = b_j a_i \qquad (j = 1, \ldots, t; \, i = 1, \ldots, m)$.

Since n is fixed, we can enumerate and store Z in polynomial time. Next let Σ be the collection of all subsets $\{z_1, \ldots, z_n\}$ of Z such that:

(98)   (i) $z_1, \ldots, z_n$ are linearly independent,

         (ii) $z_1, \ldots, z_n$ belong to C,

         (iii) $\forall i = 1, \ldots, m \, \exists j = 1, \ldots, t \, \forall k = 1, \ldots, n: b_j z_k \leqslant b_j a_i$.

Again, Σ can be enumerated and stored in polynomial time. Define

(99) $\sigma(z_1, \ldots, z_n) := \text{conv.hull}\{0, z_1, \ldots, z_n\} \backslash \text{conv.hull}\{z_1, \ldots, z_n\}$

for $\{z_1, \ldots, z_n\}$ in Σ. We finally show:

(100) $C_0 = \bigcup\limits_{\{z_1, \ldots, z_n\} \in \Sigma} \sigma(z_1, \ldots, z_n)$.

We are finished as soon as we have proved (100): with Lenstra's algorithm we can test, for each $\{z_1, \ldots, z_n\}$ in Σ, whether $\sigma(z_1, \ldots, z_n) \backslash \{0\}$ contains integral vectors. Hence we can test whether C_0 contains integral vectors other than the origin.

To prove (100), first observe that C_0 is bounded. Indeed, $C_0 \subseteq \{\lambda_1 a_1 + \cdots + \lambda_m a_m | 0 \leqslant \lambda_i < 1 \text{ for } i = 1, \ldots, m\}$. Now let $w \in C_0$. Then by (96) there exists a function ϕ in Φ such that w belongs to the convex set

(101) $P = \{z | b_j z \geqslant 0 \text{ for } j = 1, \ldots, t, \text{ and } b_{\phi(i)} z < b_{\phi(i)} a_i \text{ for } i = 1, \ldots, m\}$.

Since P is bounded and nonempty, it keeps bounded if we replace in (101) the sign $<$ by $\leqslant$, thus obtaining the closure $\bar{P}$ of P. Since $0 \in \bar{P}$, there exists an $\varepsilon > 0$ such that $(1 + \varepsilon)w$ is a convex combination of some linearly independent vertices $z_1, \ldots, z_n$ of $\bar{P}$. Then $\{z_1, \ldots, z_n\} \in \Sigma$ and $w \in \sigma(z_1, \ldots, z_n)$.

To prove the reverse inclusion for (100), let $\{z_1, \ldots, z_n\} \in \Sigma$ and $w \in \sigma(z_1, \ldots, z_n)$. There exists $\varepsilon > 0$ such that $(1 + \varepsilon)w \in \sigma(z_1, \ldots, z_n)$. By (98)(ii) w belongs to C. Moreover, by (98)(iii), there exists a function ϕ in Φ such that $b_{\phi(i)} z_k \leqslant b_{\phi(i)} a_i$ for $i = 1, \ldots, m$ and $k = 1, \ldots, n$. Since $b_j z_k \geqslant 0$ for all $j = 1, \ldots, t$ and $k = 1, \ldots, n$, and since $z_1, \ldots, z_n$ are linearly independent, we know that $b_{\phi(i)} a_i > 0$ for all $i = 1, \ldots, m$. Therefore, $(1 + \varepsilon) b_{\phi(i)} w \leqslant b_{\phi(i)} a_i$ for $i = 1, \ldots, m$, implies that $b_{\phi(i)} w < b_{\phi(i)} a_i$ for $i = 1, \ldots, m$, and hence that w belongs to C_0.

$\square$

Corollary 22.17a. *For each fixed r, there exists a polynomial algorithm which tests any system $Ax \leqslant b$ for being TDI, where A is an integral matrix of rank r.*

Proof. Let $Ax \leqslant b$ be given, where A is an integral matrix of rank r. Let A have rows $a_1, \ldots, a_m$. By Theorem 22.5, we must test if, in each minimal face F of $P := \{x | Ax \leqslant b\}$, the active rows of A form a Hilbert basis. As in the proof of Theorem 22.16, we can enumerate all minimal faces of P. Let $F = \{x | A'x = b'\}$ be a minimal face of P, where $A'x \leqslant b'$ is a subsystem of $Ax \leqslant b$. Let, say, $a_1, \ldots, a_t$ be the rows of A which are active in F (they can be found by taking a solution x_0 of $A'x = b'$, and choosing all rows of A corresponding to constraints in $Ax \leqslant b$ satisfied with equality by x_0). Testing if $a_1, \ldots, a_t$ form a Hilbert basis can be done with the method of Theorem 22.17. $\square$

The question whether the problem 'Given a system $Ax \leqslant b$, is it TDI?' is in $\mathcal{NP}$ or co-$\mathcal{NP}$ (for not-necessarily integral A), seems unanswered as yet.

Cook [1986] derived from Theorem 22.17 that if A is integral and of fixed rank, there is a polynomial-time algorithm testing the box-total dual integrality of a given system $Ax \leqslant b$.

22.10. INTEGER ROUNDING AND DECOMPOSITION

Finally we study two properties related to integral polyhedra and total dual integrality: the integer rounding property for systems of linear inequalities, and the integer decomposition property for polyhedra.

A rational system $Ax \leqslant b$ of linear inequalities is said to have the *integer rounding property* if

$$(102) \quad \min\{yb \,|\, y \geqslant 0; \, yA = c; \, y \text{ integral}\} = \lceil \min\{yb \,|\, y \geqslant 0; \, yA = c\} \rceil$$

for each integral vector c for which $\min\{yb \,|\, y \geqslant 0; \, yA = c\}$ is finite.

Note that if b is integral, the inequality $\geqslant$ holds trivially in (102). Moreover, if b is integral, then $Ax \leqslant b$ is TDI if and only if $Ax \leqslant b$ has the integer rounding property and $\{x \,|\, Ax \leqslant b\}$ is an integral polyhedron. In fact, the integer rounding property can be characterized in terms of total dual integrality, more particularly in terms of Hilbert bases (Giles and Orlin [1981]).

Theorem 22.18. *Let $Ax \leqslant b$ be a feasible system of rational linear inequalities, with b integral. Then $Ax \leqslant b$ has the integer rounding property, if and only if the system $Ax - b\eta \leqslant 0, \eta \geqslant 0$ is TDI (η being a new variable), i.e. if and only if the rows of the matrix*

$$(103) \quad \begin{bmatrix} A & b \\ 0 & 1 \end{bmatrix}.$$

form a Hilbert basis.

Proof. First assume that $Ax \leqslant b$ has the integer rounding property. To see that the rows of (103) form a Hilbert basis, let (c, γ) be an integral vector in the cone generated by the rows of (103). So there exist $y_0 \geqslant 0$, $\zeta_0 \geqslant 0$ with $c = y_0 A, \gamma = y_0 b + \zeta_0$. Then the right-hand side in (102) is at most $\lceil y_0 b \rceil \leqslant \gamma$. Therefore, there exists an integral vector $y_1 \geqslant 0$ with $y_1 A = c$ and $y_1 b \leqslant \lceil y_0 b \rceil \leqslant \gamma$. Taking $\zeta_1 := \gamma - y_1 b$, gives $c = y_1 A, \gamma = y_1 b + \zeta_1$. So (c, γ) is a nonnegative integral combination of the rows of (103).

To see sufficiency, let c be an integral vector such that $\delta := \min\{yb \,|\, y \geqslant 0; \, yA = c\}$ is finite. Then $(c, \lceil \delta \rceil)$ is an integral vector in the cone generated by the rows of (103). Assuming they form a Hilbert basis, there exist an integral vector $y_1 \geqslant 0$ and an integer $\zeta_1 \geqslant 0$ such that $y_1 A = c, y_1 b + \zeta_1 = \lceil \delta \rceil$. This implies the inequality $\leqslant$ in (102). The reverse inequality is trivial. $\square$

It follows from this theorem, together with Theorem 22.17, that if A and b are integral, and if the rank of A is fixed, the integer rounding property can be tested in polynomial time (cf. Baum and Trotter [1982]). Moreover, with Theorem 22.15, if $Ax \leqslant b$ has the integer rounding property, and A, b, and c are integral, we can solve the ILP-problem

$$(104) \quad \min\{yb \,|\, y \geqslant 0; \, yA = c; \, y \text{ integral}\}$$

in polynomial time (Chandrasekaran [1981]).

It is easy to derive the meaning of the integer rounding property for other types of inequality systems. Thus, if A is integral, the system $x \geqslant 0; Ax \geqslant b$ has the integer rounding property if and only if

(105) $\max \{yb | y \geqslant 0; yA \leqslant c; y \text{ integral}\} = \lfloor \max \{yb | y \geqslant 0; yA \leqslant c\} \rfloor$

for each integral vector c for which the right-hand side in (105) is finite.

Baum and Trotter [1981] showed that the integer rounding property is related through polarity to the integer decomposition property. A rational polyhedron is said to have the *integer decomposition property* if for each natural number k and for each vector y in $kP(:= \{kx | x \in P\})$, y is the sum of k integral vectors in P. Equivalently, if for each natural number k and for each vector x in P, whose components all are multiples of $1/k$, there exist integral vectors $x_1, \ldots, x_m$ in P and $\lambda_1, \ldots, \lambda_m$ such that

(106) $x = \lambda_1 x_1 + \cdots + \lambda_m x_m, \lambda_1 + \cdots + \lambda_m = 1, \lambda_1, \ldots, \lambda_m \geqslant 0$,

each λ_i is a multiple of $1/k$.

So if P has the integer decomposition property, P is integral.

Several polyhedra occurring in combinatorics turn out to have the integer decomposition property. Baum and Trotter [1977] showed that a matrix A is totally unimodular if and only if the polyhedron $\{x | x \geqslant 0; Ax \leqslant b\}$ has the integer decomposition property for each integral vector b (see Theorem 19.4).

A relation between the integer rounding and decomposition properties was given by Baum and Trotter [1981], through the polarity relation of 'blocking' and 'anti-blocking' polyhedra (cf. Sections 9.2 and 9.3).

Theorem 22.19. *Let A be a nonnegative integral matrix.*

(i) *The system $x \geqslant 0; Ax \geqslant 1$ has the integer rounding property, if and only if: the blocking polyhedron $B(P)$ of $P := \{x \geqslant 0 | Ax \geqslant 1\}$ has the integer decomposition property and all minimal integral vectors of $B(P)$ are rows of A (minimal with respect to $\leqslant$).*

(ii) *The system $x \geqslant 0; Ax \leqslant 1$ has the integer rounding property, if and only if: the anti-blocking polyhedron $A(P)$ of $P := \{x \geqslant 0 | Ax \leqslant 1\}$ has the integer decomposition property and all maximal integral vectors in $A(P)$ are rows of A (maximal with respect to $\leqslant$).*

Proof. To see necessity in (i), let k be a natural number. Let c be an integral vector in $kB(P)$. So $cx \geqslant k$ for all x in P. This implies:

(107) $k \leqslant \min \{cx | x \geqslant 0; Ax \geqslant 1\} = \max \{y1 | y \geqslant 0; yA \leqslant c\}$.

If $x \geqslant 0; Ax \geqslant 1$ has the integer rounding property, there is an integral vector $y_1 \geqslant 0$ such that $y_1 A \leqslant c$ and $y_1 1 \geqslant k$ (cf. (105)). Since the rows of A are integral vectors in $B(P)$, it follows that c is the sum of k integral vectors in $B(P)$.

If c is a minimal integral vector in $B(P)$, let

(108) $k := \lfloor \min \{cx | x \in P\} \rfloor$.

Then $k = 1$: as c is in $kB(P)$, by the above we know that $c = c_1 + \cdots + c_k$ for certain integral vectors $c_1, \ldots, c_k$ in $B(P)$; as c is minimal, $k = 1$.

Since $x \geqslant 0$; $Ax \geqslant 1$ has the integer rounding property,

(109) $1 = k = \lfloor \min \{cx \,|\, x \geqslant 0; Ax \geqslant 1\} \rfloor = \lfloor \max \{y1 \,|\, y \geqslant 0; yA \leqslant c\} \rfloor$

$\qquad = \max \{y1 \,|\, y \geqslant 0; yA \leqslant c; y \text{ integral}\}.$

Therefore $y1 = 1$ and $yA \leqslant c$ for some integral vector y. This means that $c \geqslant a$ for some row a of A. As c is a minimal integral vector in $B(P)$, c itself is a row of A.

To see sufficiency in (i), let c be an integral vector such that

(110) $\max \{y1 \,|\, y \geqslant 0; yA \leqslant c\}$

is finite. Let k be the lower integer part of (110). Then c belongs to $kB(P)$. Assuming that $B(P)$ has the integer decomposition property, there are integral vectors in $B(P)$ such that $c = c_1 + \cdots + c_k$. Hence there are rows $a_1, \ldots, a_k$ of A such that $c \geqslant a_1 + \cdots + a_k$. Therefore, $k \leqslant \max \{y1 \,|\, y \geqslant 0; yA \leqslant c; y \text{ integral}\}$.

Part (ii) is shown similarly.   □

Baum and Trotter [1981] said a matrix A has the *integer round-down property* if the system $x \geqslant 0$; $Ax \geqslant 1$ has the integer rounding property. Similarly, A has the *integer round-up property* if the system $x \geqslant 0$, $Ax \leqslant 1$ has the integer rounding property. So Theorem 22.19 concerns the integer round-down and -up properties.

McDiarmid [1983] defined a polyhedron P to have the *middle integer decomposition property* if, for each natural number k and for each integral vector c, the polyhedron $P \cap (c - kP)$ is integral. This property implies that P has the integer decomposition property: if $c \in kP$, then $k^{-1}c \in P \cap (c - (k-1)P)$, and hence there is an integral vector, say, c_k in $P \cap (c - (k-1)P)$. In particular, $c - c_k \in (k-1)P$, and inductively we find integral vectors $c_k, \ldots, c_1$ in P with $c = c_1 + \cdots + c_k$.

Further references include Orlin [1982] and Chandrasekaran and Shirali [1984].

23

Cutting planes

This chapter considers the integralization of polyhedra by *cutting planes*. The cutting plane method was developed at the end of the 1950's by Gomory to solve integer linear programs with the simplex method. The method turns out to be also theoretically interesting, yielding a characterization of the integer hull of a polyhedron.

In Section 23.1 we describe the geometric basis of cutting plane theory. It implies that certain implications in integer linear programming have so-called *cutting plane proofs*, which we explain in Section 23.2. The number of cutting planes, and the length of cutting plane proofs, can be seen as a measure for the complexity of an ILP-problem. We shall give bounds for this in Section 23.3. In Section 23.4 we discuss the *Chvátal rank* of a matrix, which is a bound on the number of rounds of cutting planes to be added to any integer linear program over this matrix. In Section 23.5 we illustrate cutting planes with two examples from combinatorics. In Section 23.6 we describe a relation of cutting planes with $\mathcal{NP}$-theory, and in Section 23.7 we relate cutting planes with duality. Finally, in Section 23.8 we describe Gomory's cutting plane method, an algorithm solving integer linear programming problems, and we discuss some related methods.

23.1. FINDING THE INTEGER HULL WITH CUTTING PLANES

We describe here the theoretical background of the cutting plane method, as given by Chvátal [1973] and Schrijver [1980].

As before, let P_I denote the *integer hull* of P, i.e., the convex hull of the integral vectors in P. Obviously, if H is a rational affine half-space $\{x \mid cx \leqslant \delta\}$, where c is a nonzero vector whose components are relatively prime integers (each rational affine half-space can be represented in this way), then

(1) $H_I = \{x \mid cx \leqslant \lfloor \delta \rfloor\}$.

Geometrically, H_I arises from H by shifting the bounding hyperplane of H until it contains integral vectors. Now define for any polyhedron P:

(2) $P' := \bigcap_{H \supseteq P} H_I$

where the intersection ranges over all rational affine half-spaces H with $H \supseteq P$. (Clearly, we may restrict the intersection to half-spaces whose bounding hyperplane is a supporting hyperplane of P.) As $P \subseteq H$ implies $P_I \subseteq H_I$, it follows that $P_1 \subseteq P'$. So $P \supseteq P' \supseteq P'' \supseteq P''' \supseteq \cdots \supseteq P_1$. Below we shall show that if P is rational, then P' is a polyhedron again, and that $P^{(t)} = P_1$ for some natural number t. (Here: $P^{(0)} := P$, $P^{(t+1)} := P^{(t)'}$.)

This is the theory behind Gomory's famous *cutting plane method* [1958, 1960, 1963a] for solving integer linear programs. The successive half-spaces H_1 (or more strictly, their bounding hyperplanes) are called *cutting planes*.

The cutting plane method solves linear programs over P, P', P'', P''', ... until a program has an integral optimum solution. In fact, the respective polyhedra are not determined completely, as in the ILP-problem only one linear functional is maximized.

Below we shall see that the range of the intersection (2) may be restricted to a finite number of half-spaces H, namely to those corresponding to any TDI-system defining P (cf. Theorem 22.6). They can be found in finite, but generally exponential, time from the linear inequalities defining P.

Theorem 23.1. *Let $P = \{x \mid Ax \leqslant b\}$ be a polyhedron, with $Ax \leqslant b$ TDI and A integral. Then $P' = \{x \mid Ax \leqslant \lfloor b \rfloor\}$. In particular, for any rational polyhedron P, the set P' is a polyhedron again.*

[$\lfloor \quad \rfloor$ denotes component-wise lower integer parts.]

Proof. The theorem being trivial if $P = \varnothing$, we may assume $P \neq \varnothing$. First observe, $P' \subseteq \{x \mid Ax \leqslant \lfloor b \rfloor\}$, as each inequality in $Ax \leqslant b$ gives an affine half-space H, while the corresponding inequality in $Ax \leqslant \lfloor b \rfloor$ contains H_1.

To show the reverse inclusion, let $H = \{x \mid cx \leqslant \delta\}$ be a rational affine half-space with $H \supseteq P$. Without loss of generality, the components of c are relatively prime integers. Then $H_1 = \{x \mid cx \leqslant \lfloor \delta \rfloor\}$. Now

(3) $\delta \geqslant \max \{cx \mid Ax \leqslant b\} = \min \{yb \mid y \geqslant 0; yA = c\}$.

As $Ax \leqslant b$ is TDI, we know that the minimum in (3) is attained by an integral vector, say, y_0. Now $Ax \leqslant \lfloor b \rfloor$ implies

(4) $cx = y_0 Ax \leqslant y_0 \lfloor b \rfloor \leqslant \lfloor y_0 b \rfloor \leqslant \lfloor \delta \rfloor$.

So $\{x \mid Ax \leqslant \lfloor b \rfloor\} \subseteq H_1$. As this is true for each rational affine half-space $H \supseteq P$, we have shown $\{x \mid Ax \leqslant \lfloor b \rfloor\} \subseteq P'$.   □

We next show a lemma.

Lemma. *Let F be a face of the rational polyhedron P. Then $F' = P' \cap F$.*

Proof. Let $P = \{x \mid Ax \leqslant b\}$, with A integral and $Ax \leqslant b$ TDI. Let $F = \{x \mid Ax \leqslant b; ax = \beta\}$ be a face of P, with $ax \leqslant \beta$ a valid inequality for P, a integral and β an integer. As $Ax \leqslant b$; $ax \leqslant \beta$ is TDI, by Theorem 22.2, also the system $Ax \leqslant b$; $ax = \beta$ is TDI. Then, as β is an integer,

(5) $P' \cap F = \{x \mid Ax \leqslant \lfloor b \rfloor; ax = \beta\} = \{x \mid Ax \leqslant \lfloor b \rfloor, ax \leqslant \lfloor \beta \rfloor; ax \geqslant \lceil \beta \rceil\} = F'$.

                                                                                                □

So if F' is nonempty, then F' is a face of P' (since $F = P \cap H$ for some supporting hyperplane H, and hence $F' = P' \cap F = P' \cap H$).

The lemma immediately implies that for any face F of P and any t:

(6) $\qquad F^{(t)} = P^{(t)} \cap F$.

Theorem 23.2. *For each rational polyhedron P there exists a number t such that $P^{(t)} = P_1$.*

Proof. Let $P \subseteq \mathbb{R}^n$ be a rational polyhedron. We prove the theorem by induction on the dimension d of P, the cases $d = -1$ (i.e. $P = \varnothing$) and $d = 0$ (i.e. P is a singleton) being trivial.

If aff.hull$P$ does not contain integral vectors, by Corollary 4.1a, there exists an integral vector c and a non-integer δ such that aff.hull$P \subseteq \{x \mid cx = \delta\}$. Hence

(7) $\qquad P' \subseteq \{x \mid cx \leqslant \lfloor \delta \rfloor; cx \geqslant \lceil \delta \rceil\} = \varnothing$

implying $P' = P_1$.

If aff.hull$P$ contains integral vectors, we may assume aff.hull$P = \{0\}^{n-d} \times \mathbb{R}^d$. Indeed, we may assume that the origin belongs to aff.hull$P$ (as the theorem is invariant under translation over an integral vector). Hence there exists a rational matrix C of full row rank such that aff.hull$P = \{x \mid Cx = 0\}$. So, by Corollary 4.3b, there exists a unimodular matrix U bringing C to Hermite normal form: $CU = [B\ 0]$, where B is nonsingular. As the theorem is invariant under the operation $x \mapsto U^{-1}x$ (as this transformation brings $\mathbb{Z}^n$ onto $\mathbb{Z}^n$), we may assume $C = [B\ 0]$. Hence, aff.hull$P = \{0\}^{n-d} \times \mathbb{R}^d$. As for each rational affine half-space H of $\mathbb{R}^d$ we have $(\mathbb{R}^{n-d} \times H)_1 \cap (\{0\}^{n-d} \times \mathbb{R}^d) = \{0\}^{n-d} \times H_1$, we may assume that $n - d = 0$, i.e. that P is full-dimensional.

By Theorem 17.4 there exists an integral matrix A and rational column vectors b, b' such that $P = \{x \mid Ax \leqslant b\}$, $P_1 = \{x \mid Ax \leqslant b'\}$. Let $ax \leqslant \beta'$ be an inequality from $Ax \leqslant b'$, and let $H := \{x \mid ax \leqslant \beta'\}$. We show that $P^{(s)} \subseteq H$ for some s. As this holds for each inequality from $Ax \leqslant b'$, this will prove the theorem.

Let $ax \leqslant \beta$ be the corresponding inequality from $Ax \leqslant b$. Suppose that $P^{(s)} \not\subseteq H$ for all s. Since $P' \subseteq \{x \mid ax \leqslant \lfloor \beta \rfloor\}$, there exists an integer β'' and an integer r such that:

(8) $\qquad \beta' < \beta'' \leqslant \lfloor \beta \rfloor, P^{(s)} \subseteq \{x \mid ax \leqslant \beta''\}$ for all $s \geqslant r$,

$\qquad\quad P^{(s)} \not\subseteq \{x \mid ax \leqslant \beta'' - 1\}$ for all $s \geqslant r$.

Let $F := P^{(r)} \cap \{x \mid ax = \beta''\}$. Then $\dim(F) < \dim(P)$. Moreover, F does not contain integral vectors. Hence, by our induction hypothesis, $F^{(u)} = \varnothing$ for some u. Therefore, by (6),

(9) $\qquad \varnothing = F^{(u)} = P^{(r+u)} \cap F = P^{(r+u)} \cap \{x \mid ax = \beta''\}$.

So $P^{(r+u)} \subseteq \{x \mid ax < \beta''\}$ and hence $P^{(r+u+1)} \subseteq \{x \mid ax \leqslant \beta'' - 1\}$, contradicting (8). $\qquad \square$

A direct consequence of this theorem is Theorem 22.1: if each rational supporting hyperplane of a rational polyhedron P contains integral vectors, then $P = P'$, and hence $P = P_1$.

The theorem also implies a result of Chvátal [1973] for not-necessarily rational polytopes:

Corollary 23.2a. *For each polytope P there exists a number t such that $P^{(t)} = P_1$.*

Proof. As P is bounded, there exists a rational polytope $Q \supseteq P$ such that $Q_1 = P_1$ (as $P \subseteq \{x \mid \|x\|_\infty \leqslant M\}$ for some integer M, and as for each integral vector z in $\{x \mid \|x\|_\infty \leqslant M\}\backslash P$ there exists a rational affine half-space containing P but not containing z; then we can take for Q the intersection of $\{x \mid \|x\|_\infty \leqslant M\}$ with all these half-spaces). Hence, by Theorem 23.2, $Q^{(t)} = Q_1$ for some t. Therefore, $P_1 \subseteq P^{(t)} \subseteq Q^{(t)} = Q_1 = P_1$, implying $P^{(t)} = P_1$. $\square$

Note. One easily derives from definition (2), that:

(10)   if $P = \{x \mid Ax \leqslant b\}$, then $P' = \{x \mid uAx \leqslant \lfloor ub \rfloor$ for all $u \geqslant 0$ with uA integral$\}$;

if $P = \{x \mid x \geqslant 0; Ax \leqslant b\}$, then $P' = \{x \mid x \geqslant 0; \lfloor uA \rfloor x \leqslant \lfloor ub \rfloor$ for all $u \geqslant 0\}$;

if $P = \{x \mid x \geqslant 0; Ax = b\}$, then $P' = \{x \mid x \geqslant 0; \lfloor uA \rfloor x \leqslant \lfloor ub \rfloor$ for all $u\}$.

Remark 23.1. *Cutting planes for mixed integer linear programming.* A cutting plane theory for mixed integer linear programming was developed by Blair and Jeroslow [1977, 1979, 1984, 1985]—it seems to be considerably more complicated than for the pure integer case. Here we restrict ourselves to the following note.

The cutting planes for pure integer linear programming described above suggest the following cutting plane approach for the mixed case, which however will be shown not to work. Let P be a rational polyhedron in $\mathbb{R}^n \times \mathbb{R}^k$. We wish to characterize

(11) $\text{conv.hull}\left\{\begin{pmatrix} x \\ y \end{pmatrix} \in P \;\middle|\; x \text{ integral}\right\}.$

To this end, determine a sequence of convex sets $P_0, P_1, P_2, \ldots$ such that $P_0 := P$, and P_{t+1} arises from P_t as follows: P_{t+1} is the set of vectors $\begin{pmatrix} x \\ y \end{pmatrix}$ in P_t satisfying the 'cut inequalities'

(12) $cx + dy \leqslant \varepsilon$

for each integral vector c and each rational vector d, where

(13) $\varepsilon := \max\left\{ cx + dy \;\middle|\; \begin{pmatrix} x \\ y \end{pmatrix} \in P_t; cx \in \mathbb{Z} \right\}.$

If $k = 0$ (the pure integer case), then $P_t = P^{(t)}$, as one easily checks, and hence P_t is equal to (11) for some t by Theorem 23.2. This is generally not true for the mixed integer case: take $P \subseteq \mathbb{R}^2 \times \mathbb{R}$ defined by

(14) $\xi_1 + \eta \leqslant \frac{1}{2}, \quad \xi_2 + \eta \leqslant 1$

($n = 2$, $k = 1$, so ξ_1, ξ_2 are supposed to be integer). Now $P = P_0 = P_1$. Indeed, let γ_1, γ_2 be integral, δ be rational, and

(15) $\varepsilon = \max\{\gamma_1\xi_1 + \gamma_2\xi_2 + \delta\eta \mid (\xi_1, \xi_2, \eta) \in P; \gamma_1\xi_1 + \gamma_2\xi_2 \in \mathbb{Z}\}.$

As this maximum is finite,

(16) $\max\{\gamma_1\xi_1 + \gamma_2\xi_2 + \delta\eta\,|\,(\xi_1,\xi_2,\eta)^\mathsf{T}\in P\}$

is also finite (as otherwise maximum (16) could be increased along a ray of char.cone P, and hence also the maximum (15) could be increased). Maximum (16) is attained by all vectors on the (unique) minimal face

(17) $F = \{(\xi_1,\xi_2,\eta)^\mathsf{T}\,|\,\xi_1 + \eta = \frac{1}{2}, \xi_2 + \eta = 1\}$

of P. As γ_1, $\gamma_2 \geqslant 0$ (since (16) is finite), there exists an η^* such that $\frac{1}{2}\gamma_1 + \gamma_2 - (\gamma_1 + \gamma_2)\eta^* = 0$. Now $(\xi_1^*,\xi_2^*,\eta^*):=(\frac{1}{2} - \eta^*, 1 - \eta^*, \eta^*)$ is a point of F with $\gamma_1\xi_1 + \gamma_2\xi_2\in\mathbb{Z}$. So ε is equal to (16). As this holds for each such choice of γ_1, γ_2, δ, it follows that $P_1 = P_0$.

23.2. CUTTING PLANE PROOFS

Let $Ax \leqslant b$ be a system of linear inequalities, and let $cx \leqslant \delta$ be an inequality. We say that a sequence of linear inequalities $c_1x \leqslant \delta_1, c_2x \leqslant \delta_2,\ldots,c_mx \leqslant \delta_m$ is a *cutting plane proof* of $cx \leqslant \delta$ (*from $Ax \leqslant b$*), if each of the vectors $c_1,\ldots,c_m$ is integral, if $c_m = c$, $\delta_m = \delta$, and if for each $i = 1,\ldots,m$:

(18) $c_ix \leqslant \delta_i'$ is a nonnegative linear combination of the inequalities $Ax \leqslant b, c_1x \leqslant \delta_1,\ldots,c_{i-1}x \leqslant \delta_{i-1}$ for some δ_i' with $\lfloor \delta_i' \rfloor \leqslant \delta_i$.

The number m will be called the *length* of the cutting plane proof.

Clearly, if $cx \leqslant \delta$ has a cutting plane proof from $Ax \leqslant b$, then $cx \leqslant \delta$ is valid for each integral solution of $Ax \leqslant b$. This implication also holds the other way around, if we assume that $Ax \leqslant b$ has at least one integral solution. This is a consequence of Theorem 23.2. Some similar results also hold (cf. Chvátal [1984, 1985]).

Corollary 23.2b. *Let $P = \{x\,|\,Ax \leqslant b\}$ be a nonempty polyhedron which is rational or bounded.*

(i) *If $P_1 \neq \varnothing$ and $cx \leqslant \delta$ holds for P_1 (c being integral), then there is a cutting plane proof of $cx \leqslant \delta$ from $Ax \leqslant b$.*

(ii) *If $P_1 = \varnothing$, then there is a cutting plane proof of $0x \leqslant -1$ from $Ax \leqslant b$.*

Proof. Let t be such that $P^{(t)} = P_1$ (t exists by Theorem 23.2 and Corollary 23.2a). For each $i = 1,\ldots,t$, by definition of $P^{(i)}$, there exists a system $A_ix \leqslant b_i$ defining $P^{(i)}$ so that $A_0 = A, b_0 = b$, and so that for each inequality $ax \leqslant \beta$ in $A_ix \leqslant b_i$, a is integral and there exists a vector $y \geqslant 0$ with $yA_{i-1} = a$, $\lfloor yb_{i-1} \rfloor = \beta$.

Hence, if $P_1 \neq \varnothing$, and $cx \leqslant \delta$ is valid for P_1 (with c integral), then $yA_t = c$, $\delta \geqslant \lfloor yb_t \rfloor$ for some $y \geqslant 0$ (by the affine form of Farkas' lemma). Hence

(19) $A_1x \leqslant b_1, A_2x \leqslant b_2,\ldots, A_tx \leqslant b_t, cx \leqslant \delta$

gives a cutting plane proof of $cx \leqslant \delta$ from $Ax \leqslant b$.

If $P_1 = \varnothing$, then $yA_t = 0$, $yb_t = -1$ for some $y \geqslant 0$ (by Farkas' lemma). Hence

(20) $A_1x \leqslant b_1, A_2x \leqslant b_2,\ldots, A_tx \leqslant b_t, 0x \leqslant -1$

forms a cutting plane proof of $0x \leqslant -1$ from $Ax \leqslant b$. $\square$

This corollary can be seen as an analogue of Farkas' lemma for integer linear programming. However, it generally does not lead to good characterizations, as the length of the cutting plane proof can be rather high, as we shall see in the next section.

23.3. THE NUMBER OF CUTTING PLANES AND THE LENGTH OF CUTTING PLANE PROOFS

Theorems 23.1 and 23.2 give a procedure for determining P_I (assuming P_I to be full-dimensional). Starting with P, find the minimal TDI-system $Ax \leqslant b$, with A integral, defining P. Let $P' = \{x \mid Ax \leqslant \lfloor b \rfloor\}$. Next find the minimal TDI-system $\tilde{A}x \leqslant \tilde{b}$, with $\tilde{A}$ integral, defining P'. Let $P'' = \{x \mid \tilde{A}x \leqslant \lfloor \tilde{b} \rfloor\}$. And so on. Finally, we obtain a polyhedron, say, $P^{(t)}$, for which the minimal TDI-system (with integral LHS's) has integral RHS's. Then rounding down will not change the system, and hence $P^{(t)} = P_I$.

The smallest t for which $P^{(t)} = P_I$ gives a certain measure on the complexity of the corresponding ILP-problems. There is no upper bound on this t in terms of the dimension of the polyhedron. For example, let

(21) $P := \text{conv.hull} \{(0,0),(0,1),(k,\tfrac{1}{2})\}$

in two-dimensional Euclidean space (cf. Figure 7). Then $P_I = \text{conv.hull}\{(0,0),(0,1)\}$. Now P' contains the vector $(k-1,\tfrac{1}{2})$. Inductively it follows that $(k-t,\tfrac{1}{2})$ is in $P^{(t)}$ for $t < k$, and hence $P^{(t)} \neq P_I$ for $t < k$. This also shows that t is not polynomially bounded by the size of the system defining the polyhedron. So the number of cutting planes, and the length of cutting plane proofs, can be high.

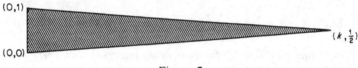

Figure 7

On the other hand, if $P_I = \varnothing$, there is a bound on the number of cutting planes in terms of the dimension, as was shown by Cook, Coullard, and Turán [1985].

Theorem 23.3. *For each natural number d there exists a number $t(d)$ such that if P is a rational polyhedron of dimension d, containing no integral vectors, then $P^{(t(d))} = \varnothing$.*

Proof. The theorem is proved by induction on d, the cases $d = -1$ (i.e. $P = \varnothing$) and $d = 0$ (i.e. P is a singleton) being trivial. As in the proof of Theorem 23.2 we may assume that P is full-dimensional.

Theorem 18.7 implies that if P is a rational polyhedron in $\mathbb{R}^d$, containing no integral vectors, then there exists a row vector c such that

(22) $\max \{cx \mid x \in P\} - \min \{cx \mid x \in P\} \leqslant l(d)$

for a certain constant $l(d)$ only depending on d, such that the components of c are relatively prime integers. Let $\delta := \lfloor \max \{cx \mid x \in P\} \rfloor$. Now for each $k = 0$, $1, \dots, l(d) + 1$ we have:

(23) $\qquad P^{(k+1+k \cdot t(d-1))} \subseteq \{x \mid cx \leqslant \delta - k\}$.

For $k = 0$ this follows from the definition of P'. Suppose we know (23) for a certain k. Then it follows from our induction hypothesis that

(24) $\qquad P^{(k+1+(k+1) \cdot t(d-1))} \subseteq \{x \mid cx < \delta - k\}$.

Indeed, consider the face

(25) $\qquad F := P^{(k+1+k \cdot t(d-1))} \cap \{x \mid cx = \delta - k\}$.

By our induction hypothesis, as $\dim(F) < d$, $F^{(t(d-1))} = \varnothing$ (assuming without loss of generality $t(d-1) \geqslant t(d')$ for all $d' \leqslant d - 1$). Hence, by (6),

(26) $\qquad (P^{(k+1+k \cdot t(d-1))})^{(t(d-1))} \cap \{x \mid cx = \delta - k\} = F^{(t(d-1))} = \varnothing$.

Therefore we have (24).

(24) implies

(27) $\qquad (P^{(k+1+(k+1) \cdot t(d-1))})' \subseteq \{x \mid cx \leqslant \delta - k - 1\}$,

which is (23) for the case $k + 1$.

Taking $k = l(d) + 1$ in (23) we have

(28) $\qquad P^{(l(d)+2+(l(d)+1) \cdot t(d-1))} \subseteq \{x \mid cx \leqslant \delta - l(d) - 1\}$.

Since

(29) $\qquad P \subseteq \{x \mid cx > \delta - l(d) - 1\}$

it follows that for $t(d) := l(d) + 2 + (l(d) + 1) \cdot t(d-1)$ we have $P^{(t(d))} = \varnothing$. $\qquad \square$

As in the proof of Corollary 23.2a, one can derive that Theorem 23.3 also holds for, not-necessarily rational, bounded polyhedra.

Theorem 23.3 is used for proving a result of Cook, Gerards, Schrijver, and Tardos [1986] stating that the number of cutting planes necessary for an integer linear program has an upper bound depending only on the constraint matrix. It can be derived also from a theorem of Blair and Jeroslow [1982] (cf. Corollary 23.4b).

Theorem 23.4. *For each rational matrix A there exists a number t such that for each column vector b one has:* $\{x \mid Ax \leqslant b\}^{(t)} = \{x \mid Ax \leqslant b\}_I$.

Proof. Without loss of generality, A is integral. Let A have n columns, and let Δ be an upper bound on the absolute values of the subdeterminants of A. We show that we can take in the theorem:

(30) $\qquad t := \max \{t(n), n^{2n+2}\Delta^{n+1} + 1 + n^{2n+2}\Delta^{n+1}t(n-1)\}$

(where $t(n)$ is as in Theorem 23.3).

Let b be a volumn vector, and let $P := \{x \mid Ax \leqslant b\}$. If $P_I = \varnothing$, then $P^{(t)} = P_I$

by Theorem 23.3. So assume $P_1 \neq \emptyset$. By Theorem 17.4, $P_1 = \{x \mid Mx \leqslant d\}$, where M is an integer matrix with all entries at most $n^{2n}\Delta^n$ in absolute value. Let $mx \leqslant \delta$ be an inequality from $Mx \leqslant d$. Without loss of generality, $\delta = \max\{mx \mid x \in P_1\}$. Let $\delta' := \lfloor \max\{mx \mid x \in P\}\rfloor$. Then by Theorem 17.2, $\delta' - \delta \leqslant \|m\|_1 n\Delta \leqslant n^{2n+2}\Delta^{n+1}$. Now for each $k = 0, 1, \ldots, \delta' - \delta$,

(31) $P^{(k+1+k \cdot t(n-1))} \subseteq \{x \mid mx \leqslant \delta' - k\}$

which may be proved as in Theorem 23.3. Hence, by taking $k = \delta' - \delta$, and using (30), we see

(32) $P^{(t)} \subseteq \{x \mid mx \leqslant \delta\}$.

As this is true for each inequality in $Mx \leqslant d$, it follows that $P^{(t)} = P_1$. $\square$

Corollary 23.4a. *For each rational matrix A there exists a number $t(A)$ such that: if $Ax \leqslant b$ is a rational system of linear inequalities with at least one integral solution, and if $cx \leqslant \delta$ holds for each integral solution of $Ax \leqslant b$ (c being integral), then $cx \leqslant \delta$ has a cutting plane proof from $Ax \leqslant b$, of length at most $t(A)$.*

Proof. Directly from Theorem 23.4. $\square$

Another consequence is a theorem of Blair and Jeroslow [1982], which in fact can be seen to be equivalent to Theorem 23.4 (see Cook, Gerards, Schrijver, and Tardos [1986]). A function $f: \mathbb{Q}^m \to \mathbb{Q}$ is a *Gomory function* if there exist rational matrices $M_1, \ldots, M_t$ so that $M_2, \ldots, M_t$ are nonnegative, and so that for each $b \in \mathbb{Q}^m$ one has

(33) $f(b) = \max_j (M_t\lceil M_{t-1}\lceil \cdots \lceil M_2\lceil M_1 b\rceil\rceil \cdots \rceil\rceil)_j$

(Here $\lceil\ \rceil$ denotes component-wise upper integer parts; M_1 has m columns, and M_i has the same number of rows as M_{i+1} has columns ($i = 1, \ldots, t-1$); the maximum ranges over all coordinates j of the column vector $M_t\lceil M_{t-1}\lceil \cdots \lceil M_2\lceil M_1 b\rceil\rceil \cdots \rceil\rceil$.)

Blair and Jeroslow [1982: Thms. 5.1 and 5.2] showed:

Corollary 23.4b. *For each rational $m \times n$-matrix A and row vector $c \in \mathbb{Q}^n$ with $\min\{cx \mid x \geqslant 0; Ax = 0\}$ finite, there exist Gomory functions $f, g: \mathbb{Q}^m \to \mathbb{Q}$ so that for each $b \in \mathbb{Q}^m$ one has:*

(34)   (i) $f(b) \leqslant 0$ if and only if $\{x \mid x \geqslant 0; Ax = b; x \text{ integral}\}$ is nonempty;
          (ii) $g(b) = \min\{cx \mid x \geqslant 0; Ax = b; x \text{ integral}\}$ if $f(b) \leqslant 0$.

Proof. Choose A and c. Theorem 23.4 implies that there exist matrices $P_1, \ldots, P_t$ so that $P_2, \ldots, P_t$ are nonnegative, and so that for each $b \in \mathbb{Q}^m$:

(35) $\{x \mid x \geqslant 0; Ax = b\}_I$
          $= \{x \mid x \geqslant 0; \lceil P_t\lceil \cdots \lceil P_2\lceil P_1 A\rceil\rceil \cdots \rceil\rceil x \geqslant \lceil P_t\lceil \cdots \lceil P_2\lceil P_1 b\rceil\rceil \cdots \rceil\rceil\}$.

Hence, with LP-duality,

(36) $\min \{cx | x \geqslant 0; Ax = b; x \text{ integral}\} =$

$\min \{cx | x \geqslant 0; \lceil P_t \lceil \cdots \lceil P_2 \lceil P_1 A \rceil \rceil \cdots \rceil \rceil x \geqslant \lceil P_t \lceil \cdots \lceil P_2 \lceil P_1 b \rceil \rceil \cdots \rceil \rceil \} =$

$\max \{y \lceil P_t \lceil \cdots \lceil P_2 \lceil P_1 b \rceil \rceil \cdots \rceil \rceil | y \geqslant 0; y \lceil P_t \lceil \cdots \lceil P_2 \lceil P_1 A \rceil \rceil \cdots \rceil \rceil \leqslant c \}.$

Let M be the matrix whose rows are the vertices of the polyhedron $\{y \geqslant 0 | y \lceil P_t \lceil \cdots \lceil P_2 \lceil P_1 A \rceil \rceil \cdots \rceil \rceil \leqslant c \}$, and let N be the matrix whose rows are the extremal rays of this polyhedron. Then the maximum in (36) is equal to:

(37) $\max \{uN \lceil P_t \lceil \cdots \lceil P_2 \lceil P_1 b \rceil \rceil \cdots \rceil \rceil$
$+ vM \lceil P_t \lceil \cdots \lceil P_2 \lceil P_1 b \rceil \rceil \cdots \rceil \rceil | u \geqslant 0; v \geqslant 0; v\mathbf{1} = 1 \}.$

Hence defining, for $b \in \mathbb{Q}^m$,

(38) $f(b) := \max_j (N \lceil P_t \lceil \cdots \lceil P_2 \lceil P_1 b \rceil \rceil \cdots \rceil \rceil)_j,$

$g(b) := \max_j (M \lceil P_t \lceil \cdots \lceil P_2 \lceil P_1 b \rceil \rceil \cdots \rceil \rceil)_j,$

gives Gomory functions satisfying (34).   □

23.4. THE CHVÁTAL RANK

Theorem 23.4 motivates the following definitions. Let A be a rational matrix. The *Chvátal rank* of A is the smallest number t such that $\{x | Ax \leqslant b\}^{(t)} = \{x | Ax \leqslant b\}_I$ for each integral vector b. The *strong Chvátal rank* of A is the Chvátal rank of the matrix

(39) $\begin{bmatrix} I \\ -I \\ A \\ -A \end{bmatrix}.$

So Theorem 23.4 states that the (strong) Chvátal rank is a well-defined integer.

It follows from Hoffman and Kruskal's theorem (Corollary 19.2a) that an integral matrix A has strong Chvátal rank 0 if and only if A is totally unimodular. Moreover, Theorem 21.5 tells that an integral matrix A has Chvátal rank 0 if and only if A^T is unimodular (in the sense of Section 21.4, i.e. if and only if for each submatrix B consisting of r ($:= \text{rank}(A)$) linearly independent rows of A, the g.c.d. of the subdeterminants of B of order r is 1).

Similar characterizations for higher Chvátal ranks are not known. Classes of matrices known to have Chvátal rank at most 1 are as follows.

(i) If A is totally unimodular, then A and $2A$ have strong Chvátal rank at most 1 (this is not difficult to derive from Hoffman and Kruskal's theorem).

(ii) If $A = (\alpha_{ij})$ is an integral matrix such that

(40) $\sum_i |\alpha_{ij}| \leqslant 2$

for each column index j, then A has strong Chvátal rank at most 1 (Edmonds and Johnson [1973]—see Example 23.1 below).

(iii) Let $A = (\alpha_{ij})$ be an integral matrix such that

(41) $\sum_j |\alpha_{ij}| \leqslant 2$

for each row index i. Then A has strong Chvátal rank at most 1 if and only if A cannot be transformed to the matrix

(42)
$$\begin{bmatrix} 1 & 1 & 0 & 0 \\ 1 & 0 & 1 & 0 \\ 1 & 0 & 0 & 1 \\ 0 & 1 & 1 & 0 \\ 0 & 1 & 0 & 1 \\ 0 & 0 & 1 & 1 \end{bmatrix}$$

by a series of the following operations:

(43)   (a) deleting or permuting rows or columns, or multiplying them by -1;

(b) replacing $\begin{bmatrix} 1 & c \\ b & D \end{bmatrix}$ by $D - bc$, where b is a column vector, c is a row vector and D is a matrix

(Gerards and Schrijver [1986]—see Example 23.2 below).

23.5. TWO COMBINATORIAL ILLUSTRATIONS

We illustrate cutting plane theory by two examples from combinatorics.

Example 23.1. *The matching polytope.* Let $G = (V, E)$ be an undirected graph, and let $P_{\text{mat}}(G)$ be the *matching polytope* of G, i.e. $P_{\text{mat}}(G)$ is the convex hull of the incidence vectors (in $\mathbb{R}^E$) of matchings in G. Let P be the polytope in $\mathbb{R}^E$ defined by:

(44) $x(e) \geqslant 0$ $(e \in E)$

$\sum_{e \ni v} x(e) \leqslant 1$ $(v \in V)$.

Since the integral vectors in P are exactly the incidence vectors of matchings in G, we know

(45) $P_I = P_{\text{mat}}(G)$.

Now the system

(46) $x(e) \geqslant 0$ $(e \in E)$

$\sum_{e \ni v} x(e) \leqslant 1$ $(v \in V)$

$\sum_{e \subseteq U} x(e) \leqslant \frac{1}{2}|U|$ $(U \subseteq V)$

also defines P, as the third row in (46) consists of redundant inequalities. Moreover, the system (46) can be proved to be TDI. Therefore, the polytope P' is determined by:

(47) $\quad x(e) \geqslant 0 \qquad\qquad (e \in E)$

$$\sum_{e \ni v} x(e) \leqslant 1 \qquad\qquad (v \in V)$$

$$\sum_{e \subseteq U} x(e) \leqslant \lfloor \tfrac{1}{2}|U| \rfloor \qquad (U \subseteq V).$$

Edmonds [1965c] showed that in fact $P' = P_1 = P_{mat}(G)$—see Corollary 8.7a. That is, (47) determines the matching polytope. So the matching polytope arises from (44) by adding one 'round' of cutting planes.

Edmonds and Johnson [1973] derived from this that more generally if $A = (\alpha_{ij})$ is an integral matrix such that for each column index j

(48) $\quad \sum_i |\alpha_{ij}| \leqslant 2$

then for all integral column vectors b_1, b_2, d_1, d_2, the integer hull of

(49) $\quad d_1 \leqslant x \leqslant d_2, b_1 \leqslant Ax \leqslant b_2$

is obtained by adding to (49) one 'round' of cutting planes; i.e. adding all inequalities

(50) $\quad cx \leqslant \lfloor \delta \rfloor$

where c is integral and $cx \leqslant \delta$ is valid for each feasible solution of (49). Equivalently: A has strong Chvátal rank at most 1.

Example 23.2. *The coclique polytope* (cf. Chvátal [1973, 1975a, b, 1984, 1985], Padberg [1974], Nemhauser and Trotter [1974, 1975]).

I. Let $G = (V, E)$ be an undirected graph. Let $P_{cocl}(G)$ be the *coclique polytope* of G, being the convex hull of the incidence vectors (in $\mathbb{R}^V$) of cocliques in G. It seems to be a difficult problem to determine a set of linear inequalities defining $P_{cocl}(G)$ for graphs G in general. If $\mathcal{NP} \neq$ co-$\mathcal{NP}$, then $P_{cocl}(G)$ will have 'difficult' facets, in the sense of Theorem 18.3.

To approach the coclique polytope, let $P(G)$ be the polytope in $\mathbb{R}^V$ defined by

(51) $\quad x(v) \geqslant 0 \qquad\qquad (v \in V)$

$$\sum_{v \in C} x(v) \leqslant 1 \qquad (C \subseteq V, C \text{ clique}).$$

(So $P(G)$ is the anti-blocking polyhedron of the clique polytope—see Example 9.2.) Clearly, $P_{cocl}(G) \subseteq P(G)$, as each coclique intersects each clique in at most one vertex. In fact, as the integral solutions of (51) are exactly the incidence vectors of cocliques,

(52) $\quad P_{cocl}(G) = P(G)_I$.

Now we can ask: given a graph G, for which t do we have $P(G)^{(t)} = P(G)_I$?

II. There is no natural number t such that $P(G)^{(t)} = P(G)_I$ for all graphs G (if there were, $\mathcal{NP} =$ co-$\mathcal{NP}$ would follow—see Corollary 23.5a below). This was shown by Chvátal [1973]. Chvátal [1984] gave a different proof based on Erdős [1961], which implies that even if we restrict ourselves to graphs G with $\alpha(G) = 2$, then there is no such t ($\alpha(G)$ is the *coclique number* of G, i.e. the maximum size of a coclique in G). The proof is as follows.

Erdős showed that there exists a constant C and arbitrarily large graphs G with $\alpha(G) = 2$ and without cliques of size larger than $k := \lfloor C \cdot n^{1/2} \ln n \rfloor$ (where n is the number of vertices of G). For such graphs, $P(G)$ contains the vector $a := (1/k, \ldots, 1/k)^T$. It is easy to show, by induction on t, that $P(G)^{(t)}$ contains the vector $2^{-t}a$. (*Proof.* Let $P(G)^{(t)}$ contain $2^{-t}a$. Let $cx \leqslant \delta$ be a valid inequality for $P(G)^{(t)}$, with c integral. As 0 is in $P(G)^{(t)}$, we know $\delta \geqslant 0$. If $\delta \geqslant 1$ then $c(2^{-t-1}a) = \tfrac{1}{2}c(2^{-t}a) \leqslant \tfrac{1}{2}\delta \leqslant \lfloor \delta \rfloor$. If $0 \leqslant \delta < 1$, then c has no positive components (as all unit basis vectors belong to $P(G)^{(t)}$). Then $c(2^{-t-1}a) \leqslant 0 = \lfloor \delta \rfloor$. This shows

$2^{-t-1}a \in P(G)^{(t+1)}$.) Now $P(G)_1$ does not contain $(\beta, \ldots, \beta)^\top$ if $\beta > 2/n$, since the inequality $1x \le 2$ is valid for $P_{\text{cocl}}(G) = P(G)_1$, as $\alpha(G) \le 2$. Hence $P(G)^{(t)} \ne P(G)_1$ if $2^{-t}/k > 2/n$, i.e. if

$$(53) \qquad t < \log_2\left(\frac{n}{2k}\right) \approx \log_2\left(\frac{n^{1/2}}{2C.\ln n}\right).$$

III. The class of graphs G with $P(G)_1 = P(G)$ turns out to be exactly the class of 'perfect' graphs, as was shown by Lovász [1972], Fulkerson [1970b, 1973], Chvátal [1975a] (cf. Trotter [1973]). In Example 23.1 above we mentioned the result of Edmonds [1965c], that if G is the line graph of some graph H, then $P(G)' = P(G)_1$, and $P(G)_1$ is the matching polytope of H.

IV. The smallest t for which $P(G)^{(t)} = P(G)_1$ is an indication of the computational complexity of the coclique number $\alpha(G)$ ($:=$ the maximum cardinality of a coclique in G). Chvátal [1973] raised the problem whether there exists, for each fixed t, a polynomial-time algorithm determining $\alpha(G)$ for graphs G with $P(G)^{(t)} = P(G)_1$. This is true for $t = 0$ (i.e. for the class of perfect graphs)—see Grötschel, Lovász, and Schrijver [1981]. Edmonds [1965a, c] described a polynomial-time algorithm for $\alpha(G)$ if G is the line graph of some graph. In fact, he gave a polynomial-time algorithm for finding a maximum weighted coclique in a line graph (i.e. a maximum weighted matching in a graph).

V. Minty [1980] and Sbihi [1978, 1980] extended Edmonds' result, by describing polynomial-time algorithms for finding a maximum weighted coclique in a $K_{1,3}$-free graph (i.e. graphs without $K_{1,3}$ as an induced subgraph). Since line graphs are $K_{1,3}$-free, this generalizes Edmonds' result. So in the terminology of Section 14.2, the optimization problem for the class of polyhedra

$$(54) \qquad (P_{\text{cocl}}(G) \mid G\ K_{1,3}\text{-free undirected graph})$$

is polynomially solvable. Hence, by Corollary 14.1c, also the separation problem for the class (54) is polynomially solvable. In particular, any facet of the coclique polytope of a $K_{1,3}$-free graph can be proved to be a facet in polynomial time. Yet no explicit description of a linear inequality system defining $P_{\text{cocl}}(G)$ for $K_{1,3}$-free graphs has been found. This would extend Edmonds' description (47) of the matching polytope. It follows from Chvátal's result described in II above that there is no t such that $P(G)^{(t)} = P(G)_1$ for all $K_{1,3}$-free graphs. (See Giles and Trotter [1981].)

VI. It was shown by Gerards and Schrijver [1986] that if $G = (V, E)$ is an undirected graph with the following property:

$$(55) \qquad \begin{array}{l} G \text{ has no subgraph } H \text{ which arises from } K_4 \text{ by replacing edges by paths such that} \\ \text{each triangle in } K_4 \text{ becomes an odd circuit in } H, \end{array}$$

then $P_{\text{cocl}}(G) = Q(G)'$, where $Q(G)$ is the polyhedron in $\mathbb{R}^V$ defined by

$$(56) \qquad \begin{array}{ll} x(v) \ge 0 & (v \in V) \\ x(v) + x(w) \le 1 & (\{v, w\} \in E). \end{array}$$

So $P_{\text{cocl}}(G)$ arises in this case by adding to (56) one round of cutting planes. Elaborated it means that $P_{\text{cocl}}(G)$ is determined by (56) together with the inequalities

$$(57) \qquad \sum_{v \in C} x(v) \le \frac{|C| - 1}{2} \qquad (C \text{ is the vertex set of an odd circuit}).$$

Graphs with the property that (56) and (57) are enough to determine $P_{\text{cocl}}(G)$ are introduced by Chvátal [1975a] and called *t-perfect*. Gerards and Schrijver showed more generally the result mentioned in (iii) of Section 23.4.

VII. Chvátal [1973] showed that if G is the complete graph K_n, and $Q(G)$ is determined by (56), then the smallest t for which $Q(G)^{(t)} = P_{\text{cocl}}(G)$ is about $\log n$.

Chvátal [1984] showed that for any graph $G = (V, E)$, there is a cutting plane proof of

$$(58) \qquad \sum_{v \in V} x_v \leqslant \alpha(G)$$

from (56) of length at most $\binom{|V|}{\alpha(G)}$. Moreover, Chvátal showed that if $Q(G)$ is the polytope defined by (56) and if $\alpha(G) \leqslant \frac{1}{2}(|V| - 1)$, then $Q(G)^{(t)} = P_{\text{cocl}}(G)$ for

$$(59) \qquad t = \alpha(G) + \left\lceil (2\alpha(G) + 1) \ln \left(\frac{|V(G)|}{2\alpha(G) + 1} \right) \right\rceil \leqslant \alpha(G) + \left\lceil \frac{|V(G)|}{e} \right\rceil.$$

23.6. CUTTING PLANES AND $\mathcal{NP}$-THEORY

We do not know whether there is a polynomial algorithm for the following problem:

(60)   given rational matrix A and rational vectors b and x, does x belong to $\{x \mid Ax \leqslant b\}$?

More generally, we do not know whether, in the terminology of Section 14.2, for any polynomially solvable class $(P_i \mid i \in \mathbb{N})$ of rational polyhedra, the class $(P_i' \mid i \in \mathbb{N})$ is polynomially solvable again. If this is the case, the polynomial solvability of the matching problem follows from the fact that the class of matching polyhedra comes in this way from a polynomially solvable class of polyhedra (cf. Example 23.1).

What can be proved is that $\mathcal{NP}$-membership is preserved under the operation $P \to P'$, in the sense of Theorem 23.5 below. Again we consider classes $(P_i \mid i \in \mathbb{N})$ of rational polyhedra satisfying the condition:

(61)   for each given $i \in \mathbb{N}$, we can compute natural numbers n_i and φ_i such that $P_i \subseteq \mathbb{R}^{n_i}$ and such that P_i has facet-complexity at most φ_i, in time polynomially bounded by $\log i$.

Theorem 23.5. *Let $(P_i \mid i \in \mathbb{N})$ be a class of polyhedra satisfying (61). If the problem*

(62)   *given $i \in \mathbb{N}$, $c \in \mathbb{Q}^{n_i}$, $\delta \in \mathbb{Q}$; is it true that $\max \{cx \mid x \in P_i\} \leqslant \delta$?*

is in $\mathcal{NP}$, then also the following problem belongs to $\mathcal{NP}$:

(63)   *given $i \in \mathbb{N}$, $c \in \mathbb{Q}^{n_i}$, $\delta \in \mathbb{Q}$; is it true that $\max \{cx \mid x \in P_i'\} \leqslant \delta$?*

Proof. Let i, c, δ as in (63) be given. Consider all inequalities

$$(64) \qquad (\lambda_1 c_1 + \cdots + \lambda_{n_i} c_{n_i}) x \leqslant (\lambda_1 \delta_1 + \cdots + \lambda_{n_i} \delta_{n_i})$$

where

(65) $c_j x \leqslant d_j$ is valid for P_i with c_j integral and $c_j x \leqslant d_j$ having size at most $n_i \varphi_i$ $(j = 1, \ldots, n_i)$;

$$0 \leqslant \lambda_j \leqslant 1 \qquad (j = 1, \ldots, n_i);$$

$\lambda_1 c_1 + \cdots + \lambda_{n_i} c_{n_i}$ is integral.

It is easy to see that all inequalities (64) form a TDI-system defining P. Hence, by Theorem 23.1, the inequalities

(66) $(\lambda_1 c_1 + \cdots + \lambda_{n_i} c_{n_i})x \leqslant \lfloor \lambda_1 \delta_1 + \cdots + \lambda_{n_i} \delta_{n_i}) \rfloor$

(under the conditions (65)) define P_i'. Now a positive answer to (63) implies that there is a $\delta' \leqslant \delta$ such that $cx \leqslant \delta'$ can be written as a nonnegative combination of inequalities (66). The scalars here have size polynomially bounded by $\log i$ and size(c). Thus a positive answer can be proved in polynomial time.   □

A consequence is a result of Boyd and Pulleyblank [1984].

Corollary 23.5a. *Let* $(P_i | i \in \mathbb{N})$ *be a polynomially solvable class of rational polyhedra, such that the problem*

(67)   *given* $i \in \mathbb{N}$, $c \in \mathbb{Q}^{n_i}$, $\delta \in \mathbb{Q}$; *is* max $\{cx | x \in P_i; x \text{ integral}\} > \delta$?

is $\mathcal{NP}$-*complete. Then, assuming* $\mathcal{NP} \neq$ co-$\mathcal{NP}$, *there is no fixed* t *such that* $P_i^{(t)} = (P_i)_{\mathrm{I}}$ *for all* $i \in \mathbb{N}$.

Proof. Suppose such a t exists. As $(P_i | i \in \mathbb{N})$ is a polynomially solvable class, problem (62) belongs to $\mathcal{NP}$. Hence, by Theorem 23.5, also after replacing P by $P^{(t)}$, problem (62) belongs to $\mathcal{NP}$. As $P^{(t)} = P_{\mathrm{I}}$ it means that problem (67) belongs to co-$\mathcal{NP}$. This contradicts our assumptions that (67) is $\mathcal{NP}$-complete and $\mathcal{NP} \neq$ co-$\mathcal{NP}$.   □

Example 23.3. *Travelling salesman problem.* The *(asymmetric) traveling salesman problem* is the following:

(68)   given a directed graph $D = (V, A)$, a length function $l: A \to \mathbb{Q}$, and a rational δ, does there exist a Hamilton cycle of length less than δ?

For any directed graph $D = (V, A)$, let the polyhedron $P_D \subseteq \mathbb{R}^A$ be defined by the linear inequalities:

(69)   (i) $\xi_a \geqslant 0$ $(a \in A)$,

         (ii) $\sum_{a \in \delta^-(v)} \xi_a = 1$ $(v \in V)$,

         (iii) $\sum_{a \in \delta^+(v)} \xi_a = 1$ $(v \in V)$,

         (iv) $\sum_{a \in \delta^+(U)} \xi_a \geqslant 1$ $(U \subseteq V, \varnothing \neq U \neq V)$.

[Here $\delta^-(U)$ and $\delta^+(U)$ denote the sets of arcs entering U and leaving U, respectively. $\delta^-(v) := \delta^-(\{v\}), \delta^+(v) := \delta^+(\{v\})$.] The integral solutions of (69) are exactly the incidence vectors of Hamilton cycles in D. So the polyhedron $(P_D)_{\mathrm{I}}$ is exactly the *traveling salesman polytope*, being the convex hull of the incidence vectors of Hamilton cycles in D. Hence, the traveling salesman problem (68) is equivalent to:

(70)   given a directed graph $D = (V, A)$, a vector $l \in \mathbb{Q}^A$ and $\delta \in \mathbb{Q}$, is max $\{lx | x \in (P_D)_{\mathrm{I}}\} > \delta$?

So problem (70) is $\mathcal{NP}$-complete. On the other hand, the class $(P_D | D$ digraph) is polynomially solvable, as the separation problem for this class is polynomially solvable: we can test a given vector $x \in \mathbb{Q}^A$ to satisfy (69) in polynomial time as follows (although there are exponentially many constraints):

(71)   Conditions (i), (ii), and (iii) of (69) are easily checked in polynomial time, namely one by one. If x does not satisfy (i), (ii), or (iii), we obtain a violated inequality. If x satisfies (i), (ii), and (iii), we test (iv) as follows. Consider x as a capacity function on A. For each pair $r, s \in V$ $(r \neq s)$, find an r–s-cut of minimum capacity, say $\delta^+(U_{r,s})$, with $U_{r,s} \subseteq V, r \notin U_{r,s}, s \in U_{r,s}$ (this can be done in polynomial time, with Ford–Fulkerson's max-flow min-cut algorithm—see Application 12.1).

If for one of the pairs r, s the capacity of $\delta^+(U_{r,s})$ is less than 1, we know that (iv) is not satisfied, and $U_{r,s}$ gives us a violated inequality. If for each pair r, s the capacity of $\delta^+(U_{r,s})$ is at least 1, we know that also all constraints (iv) are satisfied.

So $(P_D | D$ digraph) is a polynomially solvable class of polyhedra, and Corollary 23.5a applies: assuming $\mathcal{NP} \neq \text{co-}\mathcal{NP}$, there is no fixed number t such that for each digraph the polyhedron $(P_D)^{(t)}$ is the traveling salesman polytope of D.

See also Section 24.4.

23.7. CHVÁTAL FUNCTIONS AND DUALITY

The second statement in (10) in fact says that if $P = \{x \mid x \geqslant 0; Ax \leqslant b\}$ with A, b rational, then there exists a nonnegative rational matrix M such that

(72) $$P' = \{x \mid x \geqslant 0; \lfloor MA \rfloor x \leqslant \lfloor Mb \rfloor\}$$

($\lfloor \ \rfloor$ denoting component-wise lower integer parts). Therefore, Theorem 23.2 implies that if $P = \{x \mid x \geqslant 0; Ax \leqslant b\}$, then there exist nonnegative rational matrices $M_1, \ldots, M_t$ such that

(73) $$P_1 = \{x \mid x \geqslant 0; \lfloor M_t \cdots \lfloor M_2 \lfloor M_1 A \rfloor \rfloor \cdots \rfloor x \leqslant \lfloor M_t \cdots \lfloor M_2 \lfloor M_1 b \rfloor \rfloor \cdots \rfloor\}.$$

Blair and Jeroslow [1982] derived the following duality result for integer linear programming. Let $m \geqslant 0$. A function $\phi \colon \mathbb{R}^m \to \mathbb{R}$ is called a *Chvátal function* if there exist nonnegative rational matrices $M_1, \ldots, M_t$ and a nonnegative rational vector u such that

(74) $$\phi(z) = u \lfloor M_t \cdots \lfloor M_2 \lfloor M_1 z \rfloor \rfloor \cdots \rfloor$$

for each (column) vector z in $\mathbb{R}^m$. Then (73) implies:

(75) $\max \{cx \mid x \geqslant 0; Ax \leqslant b; x \text{ integral}\}$

   $= \min \{\phi(b) \mid \phi \text{ Chvátal function}; \phi(A) \geqslant c\}$.

Here $\phi(A)$ is the vector $[\phi(a_1), \ldots, \phi(a_n)]$, where $a_1, \ldots, a_n$ are the columns of A.

Observe the similarity to the linear programming duality equation, which is obtained from (75) by deleting 'x integral' and by replacing 'ϕ Chvátal function' by 'ϕ monotonically nondecreasing linear functional'.

To see $\leqslant$ in (75), if $x \geqslant 0$; $Ax \leqslant b$; x integral and ϕ is a Chvátal function with $\phi(A) \geqslant c$, then

(76) $\quad cx \leqslant \phi(A)x = u\lfloor M_t \cdots \lfloor M_2 \lfloor M_1 A \rfloor \rfloor \cdots \rfloor x \leqslant u\lfloor M_t \cdots \lfloor M_2 \lfloor M_1 Ax \rfloor \rfloor \cdots \rfloor$

$\qquad\qquad \leqslant u\lfloor M_t \cdots \lfloor M_2 \lfloor M_1 b \rfloor \rfloor \cdots \rfloor = \phi(b)$

where $u, M_1, \ldots, M_t$ are as in (74).

To see $\geqslant$ in (75), let $M_1, \ldots, M_t$ be nonnegative matrices satisfying (73). Then

(77) $\quad \min \{\phi(b) | \phi \text{ Chvátal function; } \phi(A) \geqslant c\}$

$\qquad \leqslant \min \{u\lfloor M_t \cdots \lfloor M_2 \lfloor M_1 b \rfloor \rfloor \cdots \rfloor | u \geqslant 0; u\lfloor M_t \cdots \lfloor M_2 \lfloor M_1 A \rfloor \rfloor \cdots \rfloor \geqslant c\}$

$\qquad = \max \{cx | x \geqslant 0; \lfloor M_t \cdots \lfloor M_2 \lfloor M_1 A \rfloor \rfloor \cdots \rfloor x \leqslant \lfloor M_t \cdots \lfloor M_2 \lfloor M_1 b \rfloor \rfloor \cdots \rfloor \}$

$\qquad = \max \{cx | x \in P_1\} = \max \{cx | x \geqslant 0; Ax \leqslant b; x \text{ integral}\}$

(the first equality is an LP-duality equation).

The min–max relation (75) does not give a good characterization for the ILP-problem: the number t and the sizes of the matrices M_j can be exponentially large.

Notes for Section 23.7 For some generalizations, see Jeroslow [1975, 1977, 1978a, 1979a, b,c], Tind and Wolsey [1981], and Wolsey [1981a,b]. See also Balas [1970b], Blair [1978], Blair and Jeroslow [1977, 1979, 1982], Bell and Shapiro [1977], Fisher and Shapiro [1974], Alcaly and Klevorick [1966], Balas [1970a], Gorry, Shapiro, and Wolsey [1971–2], Nemhauser and Ullmann [1968], Shapiro [1971, 1977, 1979], Gomory and Baumol [1960], Wolsey [1977, 1979 (survey), 1981a], Geoffrion and Nauss [1976–7], Klein and Holm [1979], Holm and Klein [1984], Bachem, Johnson, and Schrader [1981–2]. Bachem and Schrader [1979, 1980], Johnson [1981a,b], Fisher, Northup, and Shapiro [1975], Geoffrion [1974], Nauss [1979], Bowman [1972b], Karwan and Rardin [1979], and Schrage and Wolsey [1985].

23.8. GOMORY'S CUTTING PLANE METHOD

We now describe Gomory's *cutting plane method* [1958, 1960, 1963a] for solving integer linear programming problems. Above we gave the theoretical background of the method, but our description below does not use the foregoing theory.

Suppose we wish to solve the ILP-problem

(78) $\quad \max \{cx | x \geqslant 0; Ax \leqslant b; x \text{ integral}\}$

where A is an $m \times n$-matrix, and A, b, c are integral. Then (78) is equivalent to

(79) $\quad \max \{cx | x, \tilde{x} \geqslant 0; Ax + \tilde{x} = b; x, \tilde{x} \text{ integral}\}$.

Now first solve the *LP-relaxation*

(80) $\quad \max \{cx | x \geqslant 0; Ax \leqslant b\} = \max \{cx | x, \tilde{x} \geqslant 0; Ax + \tilde{x} = b\}$

with the (primal) simplex method. Let

(81)

	d_0	φ_0	ξ_0
			$\xi_{\sigma(1)}$
	D	f	$\vdots$
			$\xi_{\sigma(m)}$

be the optimal tableau. So $d_0 \geqslant 0$ and $f \geqslant 0$. Denote $f =: (\varphi_1, \ldots, \varphi_m)^\mathsf{T}$. If φ_0 and f are integral, the corresponding basic feasible solution is integral, which is hence an optimum solution for the ILP-problem (79). If φ_0 or f is not integral, choose an index i_0 in $\{0, \ldots, m\}$ such that φ_{i_0} is not integral. Let

$$(82) \qquad \delta_{i_0,1}, \ldots, \delta_{i_0,n+m} \quad \varphi_{i_0} \quad \xi_{\sigma(i_0)}$$

be the i_0th row in (81).

If $i_0 = 0$ then

$$(83) \qquad (\delta_{0,1} + \gamma_1)\xi_1 + \cdots + (\delta_{0,n} + \gamma_n)\xi_n + \delta_{0,n+1}\xi_{n+1} + \cdots + \delta_{0,n+m}\xi_{n+m} = \varphi_0$$

is a valid equation for all $(\xi_1, \ldots, \xi_{n+m})^\mathsf{T} = \begin{pmatrix} x \\ \tilde{x} \end{pmatrix}$ in the feasible region of (80) (writing $c = (\gamma_1, \ldots, \gamma_n)$).

If $i_0 > 0$, then

$$(84) \qquad \delta_{i_0,1}\xi_1 + \cdots + \delta_{i_0,n+m}\xi_{n+m} = \varphi_{i_0}$$

is a valid equation for all $(\xi_1, \ldots, \xi_{n+m})^\mathsf{T}$ in the feasible region of (80).

Let $\{\alpha\}$ denote the fractional part of a number α (i.e. $\{\alpha\} = \alpha - \lfloor \alpha \rfloor$). Then (83) and (84) imply

$$(85) \qquad \{\delta_{i_0,1}\}\xi_1 + \cdots + \{\delta_{i_0,n+m}\}\xi_{n+m} \equiv \{\varphi_{i_0}\} \,(\mathrm{mod}\ 1), \text{ and}$$
$$\{\delta_{i_0,1}\}\xi_1 + \cdots + \{\delta_{i_0,n+m}\}\xi_{n+m} \geqslant 0$$

for all $(\xi_1, \ldots, \xi_{n+m})^\mathsf{T}$ in the feasible region of the ILP-problem (79). (Here $\alpha \equiv \beta \,(\mathrm{mod}\ 1)$ means that $\alpha - \beta$ is an integer.)

Now the two assertions in (85) imply:

$$(86) \qquad \{\delta_{i_0,1}\}\xi_1 + \cdots + \{\delta_{i_0,n+m}\}\xi_{n+m} \geqslant \{\varphi_{i_0}\}$$

which is called a *cut (constraint)* or *cutting plane*, obtained from *source row* (82). So adding this constraint to (80) will not cut off any integral feasible solution, but cuts off the optimum basic feasible solution found in (81). (Since, if $(\xi_1^*, \ldots, \xi_{n+m}^*)^\mathsf{T}$ is the optimum basic feasible solution, then for each j, $\xi_j^* = 0$ or $\delta_{i_0,j} \in \{0, 1\}$, so that the left-hand side in (86) is 0, while the right-hand side is, by assumption, not 0.)

Therefore, we can add to (79) the extra constraint

$$(87) \qquad -\{\delta_{i_0,0}\}\xi_0 - \cdots - \{\delta_{i_0,n+m}\}\xi_{n+m} + \xi_{n+m+1} = -\{\varphi_{i_0}\}$$

where ξ_{n+m+1} is a new nonnegative variable (which can be required to be integer, by (85)). So we add the row corresponding to (87) to the optimal tableau (81), and apply the dual simplex method (cf. Section 11.7). When we have found the optimum tableau for this new LP-problem, we check again if the last f-column is integral. If so, we obtain an integral solution to the LP-relaxation (80), which is hence an optimum solution for the original ILP-problem (78). If not, we repeat the above procedure.

This adding of cuts is repeated until we find an optimum tableau for which the last column is integral. In that case the corresponding (optimum) basis feasible solution is integral. This is an optimum solution for the original ILP-problem (79).

We decide that the ILP-problem (79) is infeasible if, at the dual simplex steps, there is a row i with $\varphi_i < 0$ and all other components of that row being nonnegative. (Then the dual problem is unbounded (Section 11.7), and hence the primal problem is infeasible.)

Termination of the method. The above method can be shown to terminate if we take care that throughout the procedure of adding cut constraints and applying the dual simplex method the following conditions are satisfied:

(88)   (i)   each column in each tableau is lexicographically positive;
        (ii) the right-most column of the subsequent tableaux decreases lexicographically, at each dual pivot iteration;
        (iii) we always take the highest possible row of the tableau as source row for the cut.

[Here a column vector $y = (\eta_0, \ldots, \eta_m)^\mathsf{T}$ is *lexicographically less* than the column vector $z = (\zeta_0, \ldots, \zeta_{m'})^\mathsf{T}$, if $\eta_0 = \zeta_0, \ldots, \eta_{i-1} = \zeta_{i-1}$, $\eta_i < \zeta_i$ for some $i = 0, \ldots, m$. y is *lexicographically positive* if y is lexicographically larger than the all-zero vector, i.e. $y \neq 0$, and the topmost nonzero component of y is positive.]

Let T_0 be the optimal tableau obtained by applying, initially, the primal simplex method to solve the LP-relaxation (80). Then condition (88) (i) can be achieved, e.g. by adding the extra constraint

$$(89) \qquad \sum_{i=0}^{n+m} \xi_i \leqslant M$$

where M is chosen so that (89) is valid for some optimum solution for the ILP-problem (79). Such a number M can be found in practice often easily, and in general with the techniques described in Section 17.1. The extra constraint (89), and extra slack variable, added as first row in T_0 (just under the topmost row), will make all columns lexicographically positive.

Alternatively, one could try to make all columns lexicographically positive by reordering rows, or by applying some additional pivot steps in the columns with a 0 in the top row of the tableau.

In order to fulfil conditions (88) (i) and (ii) throughout the further iterations, we choose the pivot element in the dual simplex method as follows. The variable ξ_{i_0} leaving the basis is chosen arbitrarily from those $\xi_{\sigma(i)}$ with $\varphi_i < 0$. For the variable ξ_{j_0} entering the basis is chosen that j for which

$$(90) \qquad \frac{1}{-\delta_{i_0 j}} \begin{pmatrix} \delta_{0j} \\ \delta_{1j} \\ \vdots \\ \delta_{m'j} \end{pmatrix}$$

is lexicographically minimal, over all j with $\delta_{i_0 j} < 0$. Then one easily checks that by pivoting the last column of the tableau is lexicographically decreased, and that after pivoting all columns are lexicographically positive again.

Theorem 23.6. *Under the conditions* (88), *the cutting plane method terminates.*

Proof. Define for any tableau T:

(91) $v(T) :=$ number of columns of T;

        $\mu(T) := v(T) -$ number of rows of T.

Note that $\mu(T)$ does not change if we add cuts. Let T_0 be the optimal tableau of the initial primal simplex method, solving the LP-relaxation (80), where T_0 satisfies (88) (i). We show that the method terminates by induction on $\mu(T_0)$.

If $\mu(T_0) = 0$, the method clearly terminates: either all φ_i are integers, and then the optimum LP-solution is integral; or not all φ_i are integers, and then the ILP-problem (79) is infeasible, which will emerge immediately, when we choose a row (82), with φ_i non-integer, as source row for the cut, and we try to apply the dual simplex method.

So assume $\mu(T_0) > 0$. We now do, as an 'inner' induction, induction on $v(T_0)$. First assume the top row of T_0 is all-zero, i.e. $c = 0$. Let $\sigma(1) = j_0$. So the j_0th column of T_0 is

(92)
$$\begin{array}{c} 0 \\ \overline{1} \\ 0. \\ \vdots \\ 0 \end{array}$$

Then we can forget the top row of T_0 together with the j_0th column of T_0, yielding tableau T_0': applying the cutting plane method starting with T_0' gives the same sequence as when starting with T_0, if we delete in each tableau of the latter sequence the top row and the j_0th column. As $v(T_0') < v(T_0)$, termination follows.

Next assume that the top row of T_0 is not all-zero, i.e. $c \neq 0$. Then, in the notation (81), the part d_0 of T_0 and of all further tableaux is nonzero. The component φ_0 in this tableau-sequence will be monotonically nonincreasing, by condition (88) (ii). Let α be the infimum of the φ_0 during the further procedure. Then α is an integer, and $\varphi_0 = \alpha$ after a finite number of steps.

[*Proof.* After a finite number of steps, $\varphi_0 < \lfloor \alpha \rfloor + 1$. If $\varphi_0 > \lfloor \alpha \rfloor$, then the next cut added has source row 0 (by (88) (iii)). Then the dual simplex method will make $\varphi_0 = \lfloor \alpha \rfloor$ (or will imply infeasibility of the ILP-problem). Hence ultimately $\varphi_0 = \lfloor \alpha \rfloor = \alpha$.]

Let $\varphi_0 = \alpha$ in, say, tableau T_k. As α is the infimum of the φ_0 over all tableaux, it follows that after tableau T_k the value of φ_0 will not change any more. So at the further pivotings, we will not choose our pivot element in a column j with $\delta_{0,j} \neq 0$. Hence row 0 will not change any more at all.

Now let $j_0 = \sigma_k(1)$, where σ_k is the function σ corresponding to T_k. So the j_0th column of T_k is like (92). Now delete from tableau T_k, and all further tableaux, row 0 and all columns j with $\delta_{0,j} \neq 0$, and also column j_0. Let T_0' be the tableau coming from T_k by this deletion. Then all further pivot steps and cut addings in the original tableau sequence are also allowed in the sequence of smaller tableaux. But starting from T_0', our induction hypothesis gives that the sequence of smaller tableaux terminates, as

(93) $\mu(T_0') < \mu(T_0)$.

This follows directly from the fact that $\delta_{0,j} \neq 0$ for at least one index $j \neq j_0$ (as $\delta_{0,j_0} = 0$, and $d_0 \neq 0$ in T_k).

Hence also the original sequence starting from T_0 terminates.   □

The termination of Gomory's cutting plane method yields another proof of Theorem 23.2 for polyhedra P in the nonnegative orthant. Let P be given by

(94) $P = \{x \mid x \geq 0; Ax \leq b\}$

with A and b integral. Let $cx \leq \delta$ define a supporting half-space for P_I, such that max $\{cx \mid x \in P\}$ is finite. We show that:

(95) $P^{(s)} \subseteq \{x \mid cx \leq \delta\}$

where s is the number of cuts sufficient in the cutting plane method when solving the ILP-problem

(96)   max $\{cx \mid x \geq 0; Ax \leq b; x \text{ integral}\}$.

By Theorem 16.1 this shows that $P^{(t)} = P_I$ for some t.

The inclusion (95) is shown by induction on s. We start with solving the LP-relaxation

(97)   max $\{cx \mid x, \tilde{x} \geq 0; Ax + \tilde{x} = b\}$.

If $s = 0$, that is, if we need not add any cut, we know max $\{cx \mid x \in P\} = \delta$. Hence (95) holds.

If $s > 0$, consider the first cut added. As the source row is a linear combination of rows in the original tableau, we know that the source row corresponds to an equality

(98) $(yA)x + y\tilde{x} = yb$

for a certain vector y. Hence the cut constraint added is:

(99) $(yA - \lfloor yA \rfloor)x + (y - \lfloor y \rfloor)\tilde{x} \geq (yb - \lfloor yb \rfloor)$.

With the new slack variable:

(100) $(\lfloor yA \rfloor - yA)x + (\lfloor y \rfloor - y)\tilde{x} + \xi_{n+m+1} = (\lfloor yb \rfloor - yb)$.

Now we can transform this constraint back to the initial tableau, i.e. we replace $\tilde{x}$ by $b - Ax$:

(101) $(\lfloor yA \rfloor - \lfloor y \rfloor A)x + \xi_{n+m+1} = \lfloor yb \rfloor - \lfloor y \rfloor b$.

So if we had added the constraint

(102) $(\lfloor yA \rfloor - \lfloor y \rfloor A)x \leq (\lfloor yb \rfloor - \lfloor y \rfloor b)$

right at the beginning of the whole procedure, we could solve the ILP-problem with one cut less to add. Let Q be the polyhedron

(103) $Q := P \cap \{x \mid (\lfloor yA \rfloor - \lfloor y \rfloor A)x \leq (\lfloor yb \rfloor - \lfloor y \rfloor b)\}$.

Then max $\{cx \mid x \in Q; x \text{ integral}\}$ can be solved with the cutting plane method using $s - 1$ added cuts. By induction it follows that

(104) $Q^{(s-1)} \subseteq \{x \mid cx \leq \delta\}$.

Now $P' \subseteq Q$, since taking for H the affine half-space

(105) $H := \{x | (\lfloor yA \rfloor - \lfloor y \rfloor A)x \leqslant yb - \lfloor y \rfloor b\}$

we have $P \subseteq H$. This follows from the fact that $x \geqslant 0$; $Ax \leqslant b$ implies:

(106) $(\lfloor yA \rfloor - \lfloor y \rfloor A)x \leqslant (yA - \lfloor y \rfloor A)x = (y - \lfloor y \rfloor)Ax \leqslant (y - \lfloor y \rfloor) b$.

Now

(107) $H_1 \subseteq \{x | (\lfloor yA \rfloor - \lfloor y \rfloor A)x \leqslant (\lfloor yb \rfloor - \lfloor y \rfloor b\}$.

Hence $P' \subseteq P \cap H_1 = Q$. Therefore, $P^{(s)} \subseteq Q^{(s-1)}$, which combined with (104) gives (95).

In practice, the cutting plane method as described above turns out to be time and memory consuming. The calculations should be made very precise, as we must be able to distinguish integers from non-integers. However, we can at any stage stop the method (e.g. if not much is gained any more in the objective value by adding cutting planes), and use the current objective value as upper bound, e.g. in a branch-and-bound procedure (cf. Section 24.1).

Intermediate stopping does not yield a feasible integral solution. Therefore, alternative methods have been designed (Ben-Israel and Charnes [1962], Young [1965, 1968a], Glover [1968a]), which keep all solutions in the simplex-tableaux primal feasible, and have all entries in these tableaux integral. Then generally there are no rounding problems, and any intermediate stopping gives an integral solution, hopefully near-optimal (cf. Gomory [1963b], Garfinkel and Nemhauser [1972a: Ch. 5]).

A combination of cutting planes with a branch-and-bound procedure, designed by Crowder, Johnson, and Padberg [1983], turns out to be able to solve large-scale ILP-problems with sparse constraint matrix—see Section 24.6.

Notes for Chapter 23. For *Dantzig cuts*, see Dantzig [1959], Ben-Israel and Charnes [1962], Bowman and Nemhauser [1970], Charnes and Cooper [1961: Vol. II, pp. 700–701], Gomory and Hoffman [1963], and Rubin and Graves [1972].

For other cutting plane methods, see Balas [1971, 1972, 1975b, 1980], Young [1968, 1971b], Glover [1966–7, 1972, 1973], Charnes, Granot, and Granot [1977], Kianfar [1971], Srinivasan [1965], Blair [1980], Blair and Jeroslow [1982] (mixed integer linear programming), Chervak [1971], Bowman and Nemhauser [1971], Wolsey [1979 (survey)], and Balas and Jeroslow [1980].

The relation between cutting plane theory and algorithm is also discussed by Rosenberg [1975].

Computational results with cutting plane methods are given by Balas and Ho [1980], Steinmann and Schwinn [1969], and Trauth and Woolsey [1968–9]. The number of cuts in Gomory's cutting plane method is estimated by Finkel'shtein [1970] and Nourie and Venta [1981–2], while Jeroslow and Kortanek [1971] study the worst-case behavior in two dimensions.

For ALGOL 60 implementations of Gomory's cutting plane method, see Bauer [1963], Langmaack [1965], and Proll [1970].

Salkin [1971] comments on Gomory's cutting plane method, and Kondor [1965] gives a termination proof. Bowman [1972a] studies the method in relation to 'Hermitian basic solutions'. Martin [1963] gives a variant of Gomory's method. Piehler [1975] studies the 'cut rank' of Gomory cuts.

24

Further methods in integer linear programming

Algorithms for solving ILP-problems were described in Sections 18.4 (Lenstra's method), 18.5 and 18.6 (dynamic programming), and 23.8 (Gomory's cutting plane method). In this last chapter of the book we discuss some further methods in integer linear programming, centered around branching and bounding integer linear programs.

The idea of *branch-and-bound* is described in Section 24.1. It was developed by Land and Doig and by Dakin. Having good upper bounds for the maximum value of an ILP-problem is essential in branch-and-bound methods: good in the sense of not difficult to compute and not far from the optimum value. As we saw in Section 23.8, Gomory's cutting plane method is one method of obtaining an upper bound.

In Section 24.2 we describe another method: Gomory's method of *corner polyhedra*, which is related to the *group problem*.

In Section 24.3 we discuss the *Lagrangean relaxation method*, again a method for obtaining bounds, which we illustrate in Section 24.4 with the traveling salesman problem. Related to Lagrangean relaxation is *Benders' decomposition method* for solving mixed ILP-problems, discussed in Section 24.5. Both methods are based on decomposing the problem, and applying LP-duality to one of the parts.

Finally in Section 24.6 we discuss briefly the recent work of Crowder, Johnson, and Padberg on large-scale $\{0, 1\}$-linear programming, with sparse constraint matrices. It is based on cleverly combining branch-and-bound with ideas from cutting plane theory for the knapsack problem and with heuristics.

24.1. BRANCH-AND-BOUND METHODS FOR INTEGER LINEAR PROGRAMMING

The following *branch-and-bound* method for integer linear programming was designed by Dakin [1965], and is a modification of an earlier method of Land and Doig [1960] (see below).

Let A be a rational $m \times n$-matrix, and let $b \in \mathbb{Q}^m$, $c \in \mathbb{Q}^n$. We wish to solve the following ILP-problem:

(1) $\max \{cx \mid Ax \leqslant b; x \text{ integral}\}$.

The method consists of a number of iterations. At stage 1 we have $\Pi_1 := \{P\}$, where $P := \{x \mid Ax \le b\}$. Suppose at stage k we have a collection $\Pi_k = \{P_1, \ldots, P_k\}$ such that

(2)   (i) $P_1, \ldots, P_k$ are pairwise disjoint polyhedra in $\mathbb{R}^n$ (each given by a system of linear inequalities);
        (ii) all integral vectors in P are contained in $P_1 \cup \cdots \cup P_k$.

Then determine

(3) $\mu_j := \max \{cx \mid x \in P_j\}$

for $j = 1, \ldots, k$. These μ_j can be found with any LP-method. Choose j^* such that

(4) $\mu_{j^*} = \max \{\mu_j \mid j = 1, \ldots, k\}$

and let $x^* = (\xi_1^*, \ldots, \xi_n^*)^\mathsf{T}$ attain maximum (3) for $j = j^*$.

Case 1. x^* is not integral. Choose a non-integral component, say ξ_i^*, of x^* and define

(5) $Q_1 := \{x = (\xi_1, \ldots, \xi_n)^\mathsf{T} \in P_{j^*} \mid \xi_i \le \lfloor \xi_i^* \rfloor\}$
        $Q_2 := \{x = (\xi_1, \ldots, \xi_n)^\mathsf{T} \in P_{j^*} \mid \xi_i \ge \lceil \xi_i^* \rceil\}.$

Let $\Pi_{k+1} := \{P_1, \ldots, P_{j^*-1}, Q_1, Q_2, P_{j^*+1}, \ldots, P_k\}$. This collection satisfies (2), and we can start stage $k + 1$.

Note that at the next iteration one needs to determine (3) only for the two new polyhedra Q_1 and Q_2—for the other polyhedra one knows the μ_j from the previous iteration. If we determine (3) with the simplex method, we can use the last tableau used for finding μ_{j^*}, add the constraint $\xi_i \le \lfloor \xi_i^* \rfloor$, and apply the dual simplex method in order to find max $\{cx \mid x \in Q_1\}$. Similarly for Q_2.

Case 2. x^* is integral. Then x^* attains the maximum (1).

If each P_j is empty, the ILP-problem (1) is infeasible.

Theorem 24.1 *If P is bounded, the method above terminates.*

Proof. If P is bounded, there exists a number T such that

(6) $P \subseteq \{x = (\xi_1, \ldots, \xi_n)^\mathsf{T} \in \mathbb{R}^n \mid -T \le \xi_i \le T \quad \text{for } i = 1, \ldots, n\}$

(by Theorem 10.2 we could take $T := 2^{4n^2\varphi}$, where φ is the maximum row size of the matrix $[A \ b]$). Then at each iteration, each polyhedron P_j in $\Pi_k = \{P_1, \ldots, P_k\}$ is defined by the system $Ax \le b$, added with some inequalities taken from:

(7) $\xi_i \le \tau \quad (i = 1, \ldots, n; \tau = -T, \ldots, T-1)$
        $\xi_i \ge \tau \quad (i = 1, \ldots, n; \tau = -T+1, \ldots, T)$

(where $x = (\xi_1, \ldots, \xi_n)^\mathsf{T}$). Now if $j' \ne j''$, the polyhedra $P_{j'}$ and $P_{j''}$ are defined by different subsets of the $4nT$ inequalities in (7), as may be seen by induction on k. Therefore, $k \le 2^{4nT}$, and so after at most 2^{4nT} iterations the algorithm terminates.

$\square$

If we do not know that $P = \{x \mid Ax \leqslant b\}$ is bounded, we can proceed as follows. Let φ be the maximum size of the inequalities in $Ax \leqslant b$. By Corollary 17.1b, if P contains an integral vector, it contains an integral vector of size at most $6n^3\varphi$. Moreover, by Corollary 17.1c, if the maximum (1) is finite, it has an optimum solution of size at most $6n^3\varphi$. Define

(8) $P_0 := P \cap \{x = (\xi_1, \ldots, \xi_n)^\mathsf{T} \in \mathbb{R}^n \mid -2^{6n^3\varphi} \leqslant \xi_i \leqslant 2^{6n^3\varphi} \text{ for } i = 1, \ldots, n\}$

and apply the algorithm above to find

(9) $\max \{cx \mid x \in P_0; x \text{ integral}\}$.

If (9) is infeasible, (1) is also infeasible. If (9) has an optimum solution x^*, then check if the 'LP-relaxation' of (1), i.e. $\max \{cx \mid Ax \leqslant b\}$, is finite or unbounded (e.g. with the simplex method). If it is finite, (1) is also finite and x^* is an optimal solution. If it is unbounded, (1) is also unbounded.

The running time of the branch-and-bound method is not polynomially bounded by the input size. Indeed, consider the following ILP-problems ILP_t, for $t = 1, 2, \ldots$:

(10) $ILP_t : \max \{\eta \mid 2^t\xi = (2^t + 1)\eta; 0 \leqslant \xi \leqslant 2^t; \xi, \eta \text{ integers}\}$.

The only feasible solution for ILP_t is the origin $(\xi, \eta) = (0, 0)$. The branch-and-bound method needs at least 2^t iterations.

Proof. The μ_j, attaining the maximum in (4) at the kth iteration is at least $2^t - k$. This follows from the fact that, by indication on k, in the kth stage one of the polyhedra P_j in (2) contains

(11) $\left\{ \begin{pmatrix} \xi \\ \eta \end{pmatrix} \middle| 2^t\xi = (2^t + 1)\eta; 0 \leqslant \eta \leqslant 2^t - k \right\}$

as a subset: one easily shows that if this polyhedron is split as in (5), one of the two new polyhedra will contain

(12) $\left\{ \begin{pmatrix} \xi \\ \eta \end{pmatrix} \middle| 2^t\xi = (2^t + 1)\eta; 0 \leqslant \eta \leqslant 2^t - k - 1 \right\}$

as a subset.   □

As the size of ILP_t is linear in t, there is no polynomial upper bound on the running time.

Chvátal [1980] showed that a sharpened version of the branch-and-bound procedure as described above requires $2^{n/10}$ iterations for almost all knapsack problems of type

(13) $\max \{cx \mid cx \leqslant \frac{1}{2}c\mathbf{1}; 0 \leqslant x \leqslant \mathbf{1}; x \text{ integral}\}$

where c is a positive integral n-vector with all components at most $10^{n/2}$ (and $\mathbf{1}$ is the all-one column n-vector).

Jeroslow [1974] showed that, for n odd, a branch-and-bound procedure for

(14) $\max \{\xi_1 \mid 2\xi_1 + \cdots + 2\xi_n = n; 0 \leqslant \xi_i \leqslant 1 \text{ and integer } (i = 1, \ldots, n)\}$

requires at least $2^{(n+1)/2}$ iterations before infeasibility of (14) is concluded.

For more negative results, see Fulkerson, Nemhauser, and Trotter [1974] and Avis [1980a].

Land and Doig's method. Dakin's method described above is a modification of a method proposed earlier by Land and Doig [1960]. Suppose we know that P is bounded, and that we know (6). In Dakin's iteration the polyhedron P_{j*} is replaced by the polyhedra Q_1 and Q_2 (cf. (5)). In Land and Doig's version, P_{j*} is replaced by the $2T + 1$ polyhedra

$$(15) \qquad R_\lambda := P_{j*} \cap \{x = (\xi_1, \ldots, \xi_n)^\mathsf{T} \in \mathbb{R}^n \mid \xi_i = \lambda\}$$

for $\lambda = -T, \ldots, +T$. So Π_k generally contains many more than exactly k polyhedra.

This might suggest a huge bookkeeping, but at the next iteration we need only to evaluate

$$(16) \qquad v_\lambda := \max\{cx \mid x \in R_\lambda\}$$

for $\lambda = \lfloor \xi_i^* \rfloor$ and $\lambda = \lceil \xi_i^* \rceil$. This follows from the fact that we are interested only in finding the maximum of the μ_j, and that the maximum of the v_λ is attained for $\lambda = \lfloor \xi_i^* \rfloor$ or $\lambda = \lceil \xi_i^* \rceil$ (as v_λ is a concave function of λ, with maximum attained at $\lambda = \xi_i^*$). We only evaluate v_λ for other values of λ when necessary during the further branching process.

Note. In the description above of the branch-and-bound method we have used the LP-relaxation for the upper bound—cf. (3). However, we could take any other upper bound for the ILP-optimum, i.e. any μ_j satisfying

$$(17) \qquad \mu_j \geqslant \max\{cx \mid x \in P_j; \ x \text{ integral}\}.$$

As the maximum here again forms an ILP-optimum, this raises the quest for good upper bounds for ILP-optima—good in the sense of quickly computable and not too far from the optimum.

Upper bounds are discussed in Section 23.8 (Gomory's cutting plane method), and in Sections 24.2, 24.3, and 24.5.

Notes for Section 24.1. Another enumerative method for $\{0, 1\}$-problems, the *additive algorithm*, was designed by Balas [1963, 1964, 1965] and Glover [1965] (cf. Balas [1967], Brauer [1967], Ellwein [1974], Freeman [1966], Fleischmann [1967], Geoffrion [1967, 1969], Glover and Zionts [1965], Greenberg and Hegerich [1969–70], Hrouda [1971], Lemke and Spielberg [1967], Loehman, Nghiem, and Whinston [1970]).

See also Balas and Ho [1980], Breu and Burdet [1974], Garfinkel [1979], Gilmore and Gomory [1961, 1963, 1965], Kolesar [1966–7], Krolak [1969], Lawler and Bell [1966], Steinmann and Schwinn [1969], Tomlin [1971], Trotter and Shetty [1974], and Wolsey [1980]. For surveys, see Lawler and Wood [1966] and Spielberg [1979].

24.2. THE GROUP PROBLEM AND CORNER POLYHEDRA

The following is due to Gomory [1965, 1967, 1969]. Suppose we are to solve the following ILP-problem:

$$(18) \qquad \max\{cx \mid Ax = b; \ x \geqslant 0; \ x \text{ integral}\}$$

where A, b are integral, and we have found for the LP-relaxation of (18)

(19) $\max \{cx \mid Ax = b; \, x \geqslant 0\}$

the optimum basic solution

(20) $x^* = \begin{pmatrix} B^{-1}b \\ 0 \end{pmatrix}$

where $A = [B \; N]$. Split

(21) $x = \begin{pmatrix} x_B \\ x_N \end{pmatrix}, \quad c = (c_B, c_N)$

correspondingly. So

(22) $cx^* = \max \{cx \mid [B \; N] \begin{pmatrix} x_B \\ x_N \end{pmatrix} = b; \, x_N \geqslant 0\}$,

since for the optimum dual solution u^* to (19) we have $u^*B = c_B$, $u^*N \geqslant c_N$, $u^*b = cx^*$, and hence for any feasible solution x of the maximum in (22):

(23) $cx^* = u^*b = u^*[B \; N] \begin{pmatrix} x_B \\ x_N \end{pmatrix} \geqslant c_B x_B + c_N x_N = cx.$

Now consider the problem

(24) $\max \{cx \mid [B \; N] \begin{pmatrix} x_B \\ x_N \end{pmatrix} = b; \, x_N \geqslant 0; \, x_B, x_N \text{ integral}\}.$

So compared with (18), we have dropped the nonnegativity of x_B. Hence (24) is an upper bound for (18), generally better than (19) (by (22)), and it can be used in any branch-and-bound procedure.

To study (24), observe that (24) is equal to

(25) $\max \{c_B B^{-1}b + (c_N - c_B B^{-1}N)x_N \mid$
          $B^{-1}Nx_N \equiv B^{-1}b \, (\text{mod } 1); \, x_N \geqslant 0; \, x_N \text{ integral}\}.$

Here $e \equiv f \, (\text{mod } 1)$ means $e - f$ being an integral vector. So when solving $B^{-1}Nx_N \equiv B^{-1}b \, (\text{mod } 1)$, we only have to consider the fractional parts of the components of $B^{-1}b$. As A and b are integral, each of these fractional parts is one of

(26) $0, \dfrac{1}{|\det B|}, \dfrac{2}{|\det B|}, \ldots, \dfrac{|\det B| - 1}{|\det B|}.$

Calculating modulo 1 means that we are actually working in a product of cyclic groups. Therefore, (25) is called the *group problem*.

(25) is a problem in fewer variables than the original ILP-problem, and we could try to solve it, for example, by adapting the dynamic programming method as described in Sections 18.5 and 18.6.

By renaming (forgetting the original A, b, c, x), problem (25) can be described as

(27) $\max \{cx \mid Ax \equiv b \,(\text{mod } 1); x \geqslant 0; x \text{ integral}\}$

where $b \not\equiv 0 \,(\text{mod } 1)$. Gomory defined the corresponding *corner polyhedron* as:

(28) $P(A, b) := \text{conv.hull} \{x \mid Ax \equiv b \,(\text{mod } 1); x \geqslant 0; x \text{ integral}\}.$

So (27) is equivalent to maximizing cx over $P(A, b)$. Gomory studied the facets of $P(A, b)$. To describe his results, let A have order $m \times n$, and let Δ_i be the l.c.m. of the denominators occurring in the ith row of the matrix $[A \; b]$ $(i = 1, \ldots, m)$. Let $\tilde{A}$ be the matrix whose columns are all possible nonzero column m-vectors

(29) $\begin{pmatrix} \gamma_1/\Delta_1 \\ \vdots \\ \gamma_m/\Delta_m \end{pmatrix}$

where $\gamma_1, \ldots, \gamma_m$ are integers (not all 0) with $0 \leqslant \gamma_i < \Delta_i$ $(i = 1, \ldots, m)$. So $\tilde{A}$ has $N := \Delta_1 \cdots \Delta_m - 1$ columns. Now $P(\tilde{A}, b)$ is called the *master corner polyhedron*. The facets of $P(\tilde{A}, b)$ yield linear inequalities defining $P(A, b)$, as A arises from $\tilde{A}$ by

(30)   (i) deleting a column (which means deleting the corresponding term in the inequality);

        (ii) duplicating a column (which means duplicating the corresponding term in the inequality);

        (iii) adding integers to entries of the matrix (which does not change the corner polyhedron);

        (iv) permuting columns (which means permuting the corresponding terms in the inequality).

Note that $P(\tilde{A}, b)$ is a polyhedron of 'blocking type' (in the sense of Section 9.2), meaning that all facets have one of the following forms:

(31) $\xi_i \geqslant 0$ $(i = 1, \ldots, N)$,

        $dx \geqslant 1$ (d some nonnegative vector).

Theorem 24.2 (Gomory's corner polyhedron theorem). *The inequality* $y^\mathsf{T} x \geqslant 1$ *determines a facet of* $P(\tilde{A}, b)$ *if and only if* y *is a vertex of the polytope*

(32) $Q := \{y \mid y = (\eta_1, \ldots, \eta_N)^\mathsf{T} \geqslant 0; \quad \eta_i + \eta_j \geqslant \eta_k \quad \text{if } \tilde{a}_i + \tilde{a}_j \equiv \tilde{a}_k \,(\text{mod } 1);$
                    $\eta_i + \eta_j = 1 \quad \text{if } \tilde{a}_i + \tilde{a}_j \equiv b \,(\text{mod } 1);$
                    $\eta_k = 1 \quad \text{if } \tilde{a}_k \equiv b \,(\text{mod } 1)\},$

where $\tilde{a}_i$ *denotes the i-th column of* $\tilde{A}$.

Proof. I. We first show

(33) $Q = \{y \geqslant 0 \mid$ (i) $y^\mathsf{T} x \geqslant 1,$ if $x \in P(\tilde{A}, b), x$ integral;

                    (ii) $\eta_i + \eta_j = 1,$ if $\tilde{a}_i + \tilde{a}_j \equiv b \,(\text{mod } 1);$

                    (iii) $\eta_i = 1,$ if $\tilde{a}_i \equiv b \,(\text{mod } 1)\}.$

To show $\subseteq$, it suffices to show that if $y \in Q$ then (i) in (33) holds. Suppose $y^T x < 1$ for some integral x in $P(\tilde{A}, b)$, and choose x such that $1x$ (= sum of components of x) is as small as possible. Suppose $x = (\xi_1, \ldots, \xi_N)^T$ has more than one nonzero component, say $\xi_i \geqslant 1$, $\xi_j \geqslant 1$. If $\tilde{a}_i + \tilde{a}_j \equiv \tilde{a}_k$ (mod 1). let x' arise from x by decreasing ξ_i and ξ_j by 1, and increasing ξ_k by 1. Then $x' \in P(\tilde{A}, b)$, and hence $y^T x' = y^T x - \eta_i - \eta_j + \eta_k \leqslant y^T x < 1$, and $1x' < 1x$, contradicting the minimality of $1x$. If $\tilde{a}_i + \tilde{a}_j \equiv 0$ (mod 1), let x' arise from x by decreasing ξ_i and ξ_j by 1. Then again $x' \in P(\tilde{A}, b)$ and hence $y^T x' = y^T x - \eta_i - \eta_j \leqslant y^T x < 1$, and $1x' < 1x$, contradicting the minimality of $1x$. We conclude that x has only one nonzero component, which must be the component ξ_i corresponding to column b of $\tilde{A}$. Hence $y^T x = \eta_i \xi_i \geqslant 1$.

To show $\supseteq$, it suffices to show that if y is in the RHS of (33), then $\eta_i + \eta_j - \eta_k \geqslant 0$ whenever $\tilde{a}_i + \tilde{a}_j \equiv \tilde{a}_k$ (mod 1). If $\tilde{a}_k \equiv b$ (mod 1), this follows from (ii) and (iii) in (33). If $\tilde{a}_k \not\equiv b$ (mod 1), let $\tilde{a}_l \equiv b - \tilde{a}_k$ (mod 1). So $\eta_k + \eta_l = 1$ by (ii) and $\eta_i + \eta_j + \eta_l \geqslant 1$ by (i). Then $\eta_i + \eta_j - \eta_k = (\eta_i + \eta_j + \eta_l) - (\eta_k + \eta_l) \geqslant 1 - 1 = 0$.

II. Next we show:

$$(34) \qquad Q + \mathbb{R}_+^N = \{ y \geqslant 0 \mid y^T x \geqslant 1 \text{ if } x \in P(\tilde{A}, b), x \text{ integral} \}.$$

Clearly, $\subseteq$ holds, as if $y \in Q$, $z \in \mathbb{R}_+^N$ then $(y + z)^T x = y^T x + z^T x \geqslant 1$ (since $z \geqslant 0$, $x \geqslant 0$).

To see $\supseteq$, let y be a minimal element in the RHS of (34), i.e. if $y' \leqslant y$, $y' \neq y$ then y' does not belong to the RHS of (34). It suffices to show that y belongs to Q, i.e. that (ii) and (iii) in (33) hold.

Ad (ii). Let $\tilde{a}_i + \tilde{a}_j \equiv b$ (mod 1). Then from (34) we know that $\eta_i + \eta_j \geqslant 1$. We know also that by lowering either η_i or η_j we leave the RHS of (34). If $\eta_i > 0$ and $\eta_j > 0$, this implies that there exist integral vectors x', x'' in $P(\tilde{A}, b)$ such that $\xi_i' \geqslant 1$, $\xi_j'' \geqslant 1$, and $y^T x' = y^T x'' = 1$. Let x arise from $x' + x''$ by decreasing its ith and jth component by 1. Then $x \in P(\tilde{A}, b)$, and hence $\eta_i + \eta_j = y^T x' + y^T x'' - y^T x \leqslant 1$. If one of η_i and η_j is 0, say $\eta_i = 0$, then there exists an integral vector x in $P(\tilde{A}, b)$ such that $\xi_j \geqslant 1$ and $y^T x = 1$. Hence $\eta_i + \eta_j = \eta_i \geqslant 1$.

Ad (iii). Let $\tilde{a}_i \equiv b$ (mod 1). From (34) we know $\eta_i \geqslant 1$. As y is a minimal element in the RHS of (34), there exists an integral vector x in $P(\tilde{A}, b)$ such that $\xi_i \geqslant 1$ and $y^T x = 1$. Hence $\eta_i \leqslant 1$.

III. Equation (34) means that $Q + \mathbb{R}_+^N$ is the blocking polyhedron of $P(\tilde{A}, b)$. Hence the facets of $P(\tilde{A}, b)$ correspond to the vertices of $Q + \mathbb{R}_+^N$, which are exactly the vertices of Q. To show this last, let y be a vertex of $Q + \mathbb{R}_+^N$. Then by the Decomposition theorem for polyhedra (Corollary 7.1b), y is a vertex of Q. Conversely, let y be a vertex of Q. Suppose $y = \frac{1}{2}(z' + z'')$ with $z', z'' \in Q + \mathbb{R}_+^N$, $z', z'' \neq y$. Let $y', y'' \in Q$ be such that $y' \leqslant z'$, $y'' \leqslant z''$. Then $\frac{1}{2}(y' + y'') \leqslant y$. As $y \in Q$ and $\frac{1}{2}(y' + y'') \in Q$ we know by (33) (ii) that $y = \frac{1}{2}(y' + y'')$. As y is a vertex, $y = y' = y''$. But then $y = z' = z''$, contradicting our assumption.

$\square$

Notes for Section 24.2. Further studies of the group problem and corner polyhedra (including extensions to mixed ILP) include Aráoz [1973], Aráoz and Johnson [1981], Bachem [1976a, b], Bachem, Johnson, and Schrader [1981–2], Bachem and Schrader [1979], Balas [1973], Bell [1979], Bell and Fisher [1975], Burdet and Johnson [1974, 1977], Chen and Zionts [1972, 1976], Crowder and Johnson [1973], Devine and Glover [1973], Fisher and Shapiro [1974], Frieze [1976], Glover [1969, 1970, 1975b], Gomory [1970], Gomory and Johnson [1972a, b, 1973], Gorry, Northup, and Shapiro [1973], Gorry and Shapiro [1970–1], Gorry, Shapiro, and Wolsey [1971–2], Greenberg [1969], Halfin [1972], Hu [1969, 1970a, b], Jeroslow [1978a], Johnson [1973, 1974b, 1978, 1979, 1980a, b, 1981a, b], Shapiro [1968a, b, 1971], and Wolsey [1971, 1971–2, 1973, 1974a, b, c, 1979] (the last paper gives a survey).

24.3. LAGRANGEAN RELAXATION

In Sections 24.3 and 24.5 we give a short review of certain decomposition techniques in integer linear programming, based on applying LP-duality to a subset of the variables or the constraints. They are related to a decomposition technique for large-scale linear programming, due to Dantzig and Wolfe [1960, 1961], which we do not discuss in this book.

In this section we describe the *Lagrangean relaxation method*, a method for obtaining upper bounds for integer linear programming problems, especially for well-structured, e.g. combinatorial, problems. Good upper bounds (i.e. which are not difficult to calculate and for which the gap is not too large) are essential in branch-and-bound methods. The Lagrangean relaxation method was developed by Held and Karp [1970, 1971].

Suppose we wish to solve the integer linear programming problem

$$(35) \qquad \max \{cx \,|\, A_1 x \leqslant b_1; A_2 x \leqslant b_2; x \text{ integral}\}.$$

Let A_1 and A_2 have order $m_1 \times n$ and $m_2 \times n$, respectively. An upper bound for the maximum value of (35) is given as follows:

$$(36) \qquad \max_{\substack{A_1 x \leqslant b_1 \\ A_2 x \leqslant b_2 \\ x \text{ integral}}} cx \leqslant \min_{y \geqslant 0} \left[yb_1 + \max_{\substack{A_2 x \leqslant b_2 \\ x \text{ integral}}} (c - yA_1)x \right].$$

The minimum here is called the *Lagrangean relaxation (value)*, and the components of y the *Lagrange multipliers*.

The inequality can be shown easily: if x^* attains the first maximum, and y^* attains the minimum, then $cx^* = y^*A_1x^* + (c - y^*A_1)x^* \leqslant y^*b_1 + (c - y^*A_1)x^* \leqslant y^*b_1 + \max \{(c - y^*A_1)x \,|\, A_2x \leqslant b_2; x \text{ integral}\}$.

Now define $F : \mathbb{R}^{m_1} \to \mathbb{R}$ by

$$(37) \qquad F(y) := yb_1 + \max_{\substack{A_2 x \leqslant b_2 \\ x \text{ integral}}} (c - yA_1)x.$$

Then F is a convex function, whose minimum over $\mathbb{R}^{m_1}_+$ is equal to the minimum in (36). If it is the case that we can solve the ILP-problem

$$(38) \qquad \max \{\tilde{c}x \,|\, A_2 x \leqslant b_2; x \text{ integral}\}$$

quickly, for each objective function $\tilde{c}$, then we can calculate $F(y)$ quickly, for each given y. Quite often in applications one can make the splitting $A_1 x \leqslant b_1$, $A_2 x \leqslant b_2$ so that (38) can be solved in polynomial, or pseudo-polynomial, time. (Fisher [1981] wrote that this is the case for all applications he was aware of.)

Suppose for a given $y^0 \in \mathbb{R}_+^{m_1}$ we have calculated $F(y^0)$, and we have found that the maximum in (37) is attained by x^0. Then $u^0 := b_1 - A_1 x^0$ is a *subgradient* for F in y^0, i.e. it satisfies

(39) $F(y^0 + z) - F(y^0) \geqslant z u^0$ for each $z \in \mathbb{R}^{m_1}$.

This follows straightforwardly from: $F(y^0 + z) - F(y^0) \geqslant (y^0 + z) b_1 + (c - (y^0 + z) A_1) x^0 - (y^0 b_1 + (c - y^0 A_1) x^0) = z u^0$.

Hence, if y^* attains the minimum value of F over $\mathbb{R}_+^{m_1}$, then $(y^* - y^0) u^0 \leqslant 0$. This suggests a procedure for approximating the minimum value. Start with an arbitrary y^0, and create a sequence $y^0, y^1, y^2, \ldots$, recursively as follows. If y^k has been found, determine x^k attaining $\max \{(c - y^k A_1) x \,|\, A_2 x \leqslant b_2; x \text{ integral}\}$. Then $y^{k+1} := y^k - \lambda_k (b_1 - A_1 x^k)$, where λ_k is the 'step length', still to be specified— it could be motivated by heuristic or empirical reasons.

This method is the essence of the *subgradient method*, proposed by Shor [1962, 1964] as a generalization of the gradient method for differentiable functions. It follows from the work of Polyak [1967] that if F as above is bounded from below, and if $\lim_{k \to \infty} \lambda_k = 0$ while $\sum_{k=0}^{\infty} \lambda_k = \infty$, the sequence $F(y^0), F(y^1), \ldots$ converges to the minimum. The subgradient method was developed by Held and Karp [1970, 1971] for approximating the Lagrangean relaxation. It has led to solving larger instances of the traveling salesman problem than one could manage before.

Note that at any stage, $F(y^k)$ is an upper bound on the minimum value of F, and hence on the Lagrangean relaxation, therefore on the maximum value of the original ILP-problem. So if, while determining $y^0, y^1, \ldots$, we do not have significant decrease in $F(y^0), F(y^1), F(y^2), \ldots$, we could break off the iterations and take the current $F(y^k)$ as upper bound for the ILP.

The system $A_2 x \leqslant b_2$ in (35) can be given implicitly, by a separation algorithm, in the sense of Chapter 14, as long as we are able to solve (38) efficiently. In fact we have:

Theorem 24.3. *If for a class of ILP-problems of type* (35) *we can solve the maximization problem* (38) *in polynomial time (for each objective function $\tilde{c}$), then we can find the value of the Lagrangean relaxation in polynomial time.*

Proof. Consider the polyhedron

(40) $P := \{(y, \lambda) \,|\, \lambda \geqslant y b_1 + (c - y A_1) x \text{ for all } x \text{ with } A_2 x \leqslant b_2, x \text{ integral}\}$.

Then the value of the Lagrangean relaxation is equal to $\min \{\lambda \,|\, (y, \lambda) \in P\}$. It can be determined in polynomial time, as the separation problem for P is polynomially solvable: to check if (y^0, λ^0) belongs to P, find x^* maximizing $(c - y^0 A_1) x$

over $A_2x \leqslant b_2$, x integral, and check if $\lambda^0 \geqslant y^0b_1 + (c - y^0A_1)x^*$ holds. If so, (y^0, λ^0) belongs to P. If not, $\lambda \geqslant yb_1 + (c - yA_1)x^*$ is an inequality separating (y^0, λ^0) from P.   □

We leave the details of this theorem to the reader.

Often we can delete the integrality condition in (38) without changing the maximum value. That is, the polyhedron determined by $A_2x \leqslant b_2$ is integral. (A theoretical note: by the equivalence of 'optimization' and 'separation' discussed in Chapter 14, as soon as we can solve (38) in polynomial time, we may assume that $\{x \,|\, A_2x \leqslant b_2\}$ is integral, as the separation problem for $\{x \,|\, A_2x \leqslant b_2\}_1$ is polynomially solvable.)

If $A_2x \leqslant b_2$ determines an integral polyhedron, the Lagrangean relaxation satisfies (using LP-duality):

$$(41) \quad \min_{y \geqslant 0}\left[yb_1 + \max_{\substack{A_2x \leqslant b_2 \\ x\,\text{integral}}}(c - yA_1)x\right] = \min_{y \geqslant 0}\left[yb_1 + \max_{A_2x \leqslant b_2}(c - yA_1)x\right]$$

$$= \min_{y \geqslant 0}\left[yb_1 + \min_{\substack{u \geqslant 0 \\ uA_2 = c - yA_1}} ub_2\right] = \min_{\substack{y, u \geqslant 0 \\ yA_1 + uA_2 = c}}(yb_1 + ub_2) = \max_{\substack{A_1x \leqslant b_1 \\ A_2x \leqslant b_2}} cx.$$

So now the Lagrangean relaxation has the same value as the LP-relaxation of the original ILP-problem. So we could apply now direct LP-techniques to find the value of the Lagrangean relaxation. This was an essential idea in the recent successful solution of large-scale traveling salesman problems—see below.

A similar approach can be adopted for other types of ILP-problems. For example, we have

$$(42) \quad \min_{\substack{A_1x \geqslant b_1 \\ A_2x \geqslant b_2 \\ x \geqslant 0 \\ x\,\text{integral}}} cx \geqslant \max_y\left[yb_1 + \min_{\substack{A_2x \geqslant b_2 \\ x \geqslant 0 \\ x\,\text{integral}}}(c - yA_1)x\right]$$

So the range of the maximum now is not restricted to nonnegative y.

Note. The minimum in (36) is equal to

$$(43) \quad \min yb_1 + \lambda$$

subject to (i) $y(A_1x) + \lambda \geqslant cx$ (for all integral x with $A_2x \leqslant b_2$)
                    (ii) $y \geqslant 0$

which is equal to

$$(44) \quad \min \mu$$

subject to (i) $y(A_1x - b_1) + \mu \geqslant cx$ (for all integral x with $A_2x \leqslant b_2$)
                    (ii) $y \geqslant 0$.

So the problem can be interpreted as minimizing the errors in the inequality system: $y(A_1x - b_1) \geqslant cx$ (for all integral x with $A_2x \leqslant b_2$).

In the subgradient method, if we have found y^k, then $y^{k+1} = y^k - \lambda_k(b_1 - A_1 x^k)$ for some $x = x^k$ maximizing $cx - y^k(A_1 x - b_1)$. Therefore, the subgradient method can be viewed as an application of the relaxation method (Section 12.3). This was observed by A. J. Hoffman.

Formulation (43) also connects to Benders' decomposition to be described in Section 24.5. Held and Karp [1970, 1971] proposed a column generation technique for determining the minimum (44), which has much in common with Benders' method.

Notes for Section 24.3. For surveys on the Lagrangean relaxation method, see Fisher [1981], Geoffrion [1974], Shapiro [1979], and Burkard [1980]. See also Balas and Christofides [1981], Balas and Ho [1980], Brooks and Geoffrion [1966], Everett [1963], Gavish [1978], Gorry, Shapiro, and Wolsey [1971–2], Held, Wolfe, and Crowder [1974], Karwan and Rardin [1979], Nemhauser and Ullmann [1968], Polyak [1969], Shapiro [1971], and Shor [1979].

24.4. APPLICATION: THE TRAVELING SALESMAN PROBLEM

The traveling salesman problem (cf. Example 23.3) can be described as an integer linear programming problem as follows:

$$(45) \qquad \min \sum_{i,j=1}^{n} \gamma_{ij}\xi_{ij}$$

$$\text{such that} \quad \text{(i)} \quad \sum_{j=1}^{n} \xi_{ij} = 1 \qquad (i = 1,\ldots,n)$$

$$\text{(ii)} \quad \sum_{i=1}^{n} \xi_{ij} = 1 \qquad (j = 1,\ldots,n)$$

$$\text{(iii)} \quad \sum_{i \notin U} \sum_{j \in U} \xi_{ij} \geqslant 1 \quad (\varnothing \neq U \subseteq \{2,\ldots,n\})$$

$$\text{(iv)} \quad \xi_{ij} \geqslant 0 \qquad (i,j = 1,\ldots,n)$$

$$\text{(v)} \quad \xi_{ij} \text{ integral} \qquad (i,j = 1,\ldots,n).$$

Here the γ_{ij} $(i,j = 1,\ldots,n)$ are given rationals. Now a Lagrangean relaxation of (45) is (cf. (42)):

$$(46) \qquad \max_{y=(\eta_1,\ldots\eta_n)} \left[\eta_1 + \cdots + \eta_n + \min_{\substack{\xi_{ij} \text{ satisfying} \\ \text{(ii), (iii), (iv), (v)}}} \sum_{i,j=1}^{n} (\gamma_{ij} - \eta_i)\xi_{ij} \right].$$

This means maximizing a concave function over $\mathbb{R}^n$. The minimization problem in (46) means (denoting $\bar{\gamma}_{ij} := \gamma_{ij} - \eta_i$):

$$(47) \qquad \min \sum_{i,j=1}^{n} \bar{\gamma}_{ij}\xi_{ij}$$

such that (ii), (iii), (iv), (v) hold.

A brief analysis of this ILP-problem shows that it amounts to finding a so-called 'shortest 1-arborescence'. Here a 1-*arborescence* is a collection of arcs forming a rooted spanning tree, with root 1. Note that the variables ξ_{i1} do not occur in (iii), and only once in (ii), so that they can be treated isolatedly (they do not correspond to arcs in the 1-arborescence).

Now it is a result of Edmonds [1967a] that one can find a shortest 1-arborescence in polynomial time—this was also shown in Application 14.1. So (47) can be solved in polynomial time, and hence, by Theorem 24.3, the Lagrangean relaxation value can be found in polynomial time. Moreover, we can incorporate Edmonds' algorithm in a subgradient approximation of (46), as was proposed by Held and Karp [1970, 1971].

In fact, the minimum (47) does not change if we delete the integrality condition (v)—this is also a result of Edmonds [1967a] (proved in Section 22.2 above). So, by (41), the Lagrangean relaxation (46) has the same optimum value as the LP-relaxation. Hence we could equally well solve this LP-relaxation. This was the approach initiated by Dantzig, Fulkerson, and Johnson [1954, 1959], and developed and sharpened by Miliotis [1978], Grötschel and Padberg [1979a, b], Grötschel [1980], Crowder and Padberg [1980], and Padberg and Hong [1980] (see Grötschel and Padberg [1985] and Padberg and Grötschel [1985] for a survey).

These authors proposed the following method to solve the LP-relaxation of (45). First minimize, with the simplex method, $\sum_{i,j=1}^{n} \gamma_{ij}\xi_{ij}$ over (i), (ii), (iv). Next test if the optimum solution, say x^*, satisfies (iii). To this end, consider x^* as a 'capacity' function defined on the arcs (i, j) $(i, j = 1, \ldots, n)$. Find, with Ford–Fulkerson's minimum cut algorithm (cf. Application 12.1), a minimum capacitated 1-j-cut C_{1j}, for each $j \neq 1$. If each of the C_{1j} has capacity at least 1, then x^* also satisfies (iii), and hence x^* is an optimum solution for the LP-relaxation. If one of the C_{1j} has capacity less than 1, it gives an inequality in (iii) violated by x^*. This inequality can be added to the simplex scheme, after which we determine a new optimum solution, with the dual simplex method. Again we test if the new optimum solution satisfies (iii). Again we find that it is an optimum solution for the LP-relaxation, or we find a new violated inequality, which we add to the scheme. Repeating this, we finally will have the optimum value of the LP-relaxation.

This process terminates, as there are 'only' finitely many constraints (iii), but it may take exponential time. We could build in some stopping criterion, however, e.g. if the increase in the optimum values of the subsequent LP's is getting small: the current optimum value can still serve as a lower bound for the ILP-optimum value.

This approach is called a *cutting plane* approach. In fact, the references given above mainly study the *symmetric* traveling salesman problem, where $\gamma_{ij} = \gamma_{ji}$ for all i, j. This allows some simplifications of the above. Crowder and Padberg [1980] solved with this method a 318-city traveling salesman problem in about 2300 CPU seconds on an IBM 370/168 MVS.

24.5. BENDERS' DECOMPOSITION

Another decomposition method was proposed by Benders [1960, 1962] to solve mixed ILP-problems. It has analogies with Lagrangean relaxation—actually it is in some respects the opposite of the Lagrangean approach.

Suppose we are given the mixed ILP-problem

$$(48) \qquad \max\{cx + dy \,|\, Ax + By \leqslant b; \, x \text{ integral}\}$$

where A and B have order $m \times n_1$ and $m \times n_2$, respectively. Then, using LP-duality,

$$(49) \qquad \max_{\substack{Ax + By \leqslant b \\ x \text{ integral}}} cx + dy = \max_{x \text{ integral}} \left[cx + \max_{By \leqslant b - Ax} dy \right]$$

$$= \max_{x \text{ integral}} \left[cx + \min_{\substack{uB = d \\ u \geqslant 0}} u(b - Ax) \right].$$

This last form means maximizing a concave function over integral variables. Its optimum value is equal to that of

(50) $\max cx + \lambda$

such that (i) $uAx + \lambda \leqslant ub$ (for all $u \geqslant 0$ with $uB = d$)
        (ii) x integral

(assuming $u \geqslant 0$, $uB = d$ is feasible). So (48) is equivalent to a mixed ILP-problem with only one non-integer variable λ.

Benders proposed the following method to solve maximum (50). First determine a number M such that if (50) has an optimum solution, it has one with all components at most M in absolute value. Next solve $\max\{cx + \lambda | x$ integral; $\|x\|_\infty \leqslant M; |\lambda| \leqslant M\}$. Let x^0, λ^0 be an optimum solution. Test if it satisfies (50) (i), by testing, with LP-techniques, if

(51) $\min\{u(b - Ax^0)|u \geqslant 0; uB = d\} \geqslant \lambda^0$

holds. If (51) holds, then (x^0, λ^0) is an optimum solution of (50). If (51) does not hold, let $u^0 \geqslant 0$ be such that $u^0 B = d$ and $u^0(b - Ax^0) < \lambda^0$. Then u^0 gives a violated inequality in (50) (i) (a *Benders cut*). Next solve $\max\{cx + \lambda | u^0 Ax + \lambda \leqslant u^0 b$; x integral; $\|x\|_\infty \leqslant M$; $|\lambda| \leqslant M\}$. Let x^1, λ^1 be an optimum solution. If it satisfies (50) (i) it forms an optimum solution of (50). If it does not satisfy (50) (i), let u^1 give a violated inequality. Then solve $\max\{cx + \lambda | u^0 Ax + \lambda \leqslant u^0 b$; $u^1 Ax + \lambda \leqslant u^1 b$; x integral $\|x\|_\infty \leqslant M; |\lambda| \leqslant M\}, \ldots$. The procedure is clear: at each iteration we either find on optimum solution of (50) or we find a new violated inequality.

If we take care that each time we solve a linear program of type (51) we get a vertex of $\{u \geqslant 0 | uB = d\}$ or an extremal ray of $\{u \geqslant 0 | uB = 0\}$ for our violated inequality, then we can find only finitely many violated inequalities, and hence the procedure will terminate.

So Benders' method alternately solves a mixed integer linear programming problem with only one non-integral variable (λ), and a linear programming problem of type (51).

The method can be used also as an upper bound method in a branch-and-bound procedure: for each x^k, λ^k, the current value $cx^k + \lambda^k$ is an upper bound for (50), and hence for (48).

Further references: Balas [1970a], Fleischmann [1973], Geoffrion [1972], Hrouda [1975, 1976], Johnson and Suhl [1980].

24.6. SOME NOTES ON INTEGER LINEAR PROGRAMMING IN PRACTICE

In fact, although being $\mathcal{NP}$-complete, the knapsack problem turns out to be solvable quite efficiently in practice, by cleverly combining branch-and-bound, cutting plane, and dynamic programming methods with heuristics. Balas and Zemel [1980] report solving randomly generated problems of up to 10 000 variables and with coefficients up to 10 000, in time less than a second on the average, and about three seconds in the worst case. See also Balas [1967], Balas and Zemel [1978, 1980], Fayard and Plateau [1975, 1977–8], Geoffrion [1967, 1969], Gilmore [1979], Gilmore and Gomory [1961, 1963, 1966], Greenberg and Hegerich [1969–70], Guignard and Spielberg [1972], Ingargiola and Korsh [1977], Laurière [1978], Lifschitz [1980], Martello and Toth [1977, 1978, 1979], Nauss [1976–7], Salkin and de Kluyver [1975] (survey), Saunders and Schinzinger [1970], Toth [1980], and Zoltners [1978].

Next consider more general problems. Crowder, Johnson, and Padberg [1983] designed a method which is able to solve large-scale problems of type

(52) $\max \{cx \mid Ax \leqslant b; x \{0, 1\}\text{-vector}\},$

in which A is a sparse matrix, say at most approximately 5% of the entries are nonzero. Then, with high probability, any two rows of A have almost disjoint supports. (If each two rows of A have completely disjoint supports, (52) is just a union of independent knapsack problems.) Now when applying a branch-and-bound method to (52), at each iteration we have to find a good upper bound for a problem of type (52) (as fixing some of the variables to 0 or 1 gives again a problem of type (52)).

In order to solve (52), Crowder, Johnson, and Padberg solve first (after some preprocessing) the LP-relaxation of (52):

(53) $\max \{cx \mid Ax \leqslant b; 0 \leqslant x \leqslant 1\}.$

If x^* is an optimum solution of (53), we try to find, for each inequality $ax \leqslant \beta$ in $Ax \leqslant b$, whether there is an inequality satisfied by the *knapsack polytope*

(54) $\text{conv.hull} \{x \mid ax \leqslant \beta; x \{0, 1\}\text{-vector}\}$

but not by x^*. That is, we wish to find a 'cutting plane' cutting off x^* from (54). If we find such an inequality, we add it to $Ax \leqslant b$, find the new LP-optimum x^{**}, and repeat the iteration for x^{**}.

This is done for each inequality $ax \leqslant \beta$ in $Ax \leqslant b$. Each time we obtain a better upper bound for (52). We stop improving this upper bound if the ratio cost: improvement becomes too bad.

As the knapsack problem is $\mathcal{NP}$-complete, we may not hope to be able to detect all cutting planes for the knapsack polytope (54). Therefore, Crowder, Johnson, and Padberg content themselves with subclasses of all inequalities necessary for (54), and they give some heuristics.

One subclass of inequalities consists of the following, where it is assumed that $a \geqslant 0$ (otherwise we reset component ξ_j of x by $1 - \xi_j$, whenever the corresponding component of a is negative):

(55) $\sum_{j \in S} \xi_j \leqslant |S| - 1$

where S is a *minimal cover* for $ax \leqslant \beta$, i.e. $S \subseteq K := \{j \mid \alpha_j > 0\}$ (denoting $a = (\alpha_1, \ldots, \alpha_n)$), with $\sum_{j \in S} \alpha_j > \beta$, and $\sum_{j \in S} \alpha_j - \alpha_i \geqslant \beta$ for all $i \in S$.

An inequality (55) violated by a given $x^* = (\xi_1^*, \ldots, \xi_n^*)$ can be found by solving the knapsack problem

$$(56) \qquad \min\left\{\sum_{j \in K} (1 - \xi_j^*)\eta_j \,\middle|\, \sum_{j \in K} \alpha_j \eta_j > \beta; \eta_j \in \{0, 1\}(j \in K)\right\}.$$

Crowder, Johnson, and Padberg conjecture that this problem can be solved in polynomial time.

Other cutting planes for (54) are obtained by a further analysis of the knapsack polytope, based on earlier work of Balas [1975a], Hammer, Johnson, and Peled [1975], and Wolsey [1975].

Crowder, Johnson, and Padberg report on solving, with the above method, sparse $\{0, 1\}$-linear programs in up to 2756 variables in time less than one CPU hour, mostly less than ten CPU minutes (on an IBM 370/168).

See also the notes on the related 'cutting plane method' described in Section 24.4 (the traveling salesman problem). (Johnson and Suhl [1980] is a predecessor to Crowder, Johnson, and Padberg's paper.)

Historical and further notes on integer linear programming

HISTORICAL NOTES

The history of solving linear equations in positive integers goes back to ancient times. We refer to the historical notes on linear diophantine equations (Part II). The lack of good characterizations and efficient algorithms is reflected by a lack of structural results. Mainly special examples of linear equations were studied—see the references given in Dickson [1920: Ch. II, III].

We here discuss some topics briefly.

Enumeration of solutions

It was observed essentially by Euler [1748: Ch. 16; 1770a: 2. Abschnitt, Cap. 2; 1770b: Sec. 27] that the number of nonnegative integral solutions of a system $Ax = b$ of linear equations is equal to the coefficient of $\zeta_1^{\beta_1} \dots \zeta_m^{\beta_m}$ in the expansion of

(1) $\qquad (1 - \zeta_1^{\alpha_{11}} \dots \zeta_m^{\alpha_{m1}})^{-1} \dots (1 - \zeta_1^{\alpha_{1n}} \dots \zeta_m^{\alpha_{mn}})^{-1}$

where $A = (\alpha_{ij})$ and $b = (\beta_1, \dots, \beta_m)^{\mathsf{T}}$ are nonnegative and integral. This is easy to see, using $(1 - y)^{-1} = 1 + y + y^2 + y^3 + \cdots$.

Counting formulas were derived by, *inter alia*, Cayley [1855b, 1860], Sylvester [1857a, b, 1858a, b, 1882], and Laguerre [1876–7] (cf. Tegnér [1932–3]). Cayley [1860] and Sylvester [1858b] derived a counting formula reducing r equations to one equation (different from aggregation). Another consequence is the following asymptotic formula. Let $\alpha_1, \dots, \alpha_n$ be relatively prime positive integers, and for $\beta \in \mathbb{N}$, let $N(\beta)$ denote the number of nonnegative integral solutions of $\alpha_1 \xi_1 + \cdots + \alpha_n \xi_n = \beta$. Then

(2) $\qquad \lim_{\beta \to \infty} \dfrac{N(\beta)}{\beta^{n-1}} = \dfrac{1}{(n-1)! \, \alpha_1 \cdots \alpha_n}.$

For a discussion, see Netto [1901: Ch. 6] and Bachmann [1910: Ch. 3, 4]. For a more extensive historical treatment, see Dickson [1920: Ch. II, III].

Frobenius' problem

It is known that if $\alpha_1, \ldots, \alpha_n$ are positive, relatively prime integers, then there exists a smallest integer β_0 such that if β is an integer with $\beta \geq \beta_0$, then the equation

$$(3) \qquad \alpha_1 \xi_1 + \cdots + \alpha_n \xi_n = \beta$$

has a nonnegative integral solution $\xi_1, \ldots, \xi_n$. According to Brauer [1942], G. Frobenius mentioned occasionally in his lectures the problem of determining β_0. It is known that for $n = 2$, $\beta_0 = \alpha_1 \alpha_2 - \alpha_1 - \alpha_2 + 1$. (Sylvester [1884] and Curran Sharp [1884] showed that there are exactly $\frac{1}{2}(\alpha_1 - 1)(\alpha_2 - 1)$ natural numbers $\beta < \beta_0$ for which (3) has no solution.) Brauer [1942] also mentions that I. Schur, in his last lecture in Berlin in 1935, proved that if $\alpha_1 < \alpha_2 < \cdots < \alpha_n$, then $\beta_0 \leq \alpha_1 \alpha_n - \alpha_1 - \alpha_n + 1$.

Finite basis theorems

Gordan [1873] defined a positive integral solution of a system $Ax = 0$ of homogeneous linear equations to be *irreducible* if it is not the sum of other positive integral solutions. He argued that any given rational system has only finitely many irreducible solutions. Similar statements were given by Grace and Young [1903: pp. 102–107] and van der Corput [1931b, c], which amount to each rational polyhedral cone having a finite integral Hilbert basis. Van der Corput observed moreover that a pointed rational polyhedral cone has a unique minimal integral Hilbert basis. Elliott [1903] studied several special cases.

The relation of Hilbert bases for cones to Hilbert's famous finite basis theorem for algebraic forms [1890] is studied by Grace and Young [1903], MacMahon [1916: Sec. VIII], and Weitzenböck [1930].

Aggregation

Mathews [1897] showed that for any system $Ax = b$ of linear equations with A nonnegative, there exists a vector u such that

$$(4) \qquad \{x | Ax = b; x \geq 0; x \text{ integral}\} = \{x | (uA)x = ub; x \geq 0; x \text{ integral}\}.$$

This is a first result on the aggregation of linear equations. For further references, see Leveque [1974: Vol. 2, pp. 7–16], Kastning [1976: pp. 252–253], Hausmann [1978: p. 174] and von Randow [1982: p. 182; 1985: p. 206].

Geometry of numbers

The theme of the geometry of numbers is about the same as that of integer linear programming: both search for integral vectors in given convex sets. The following famous result of Minkowski [1893] was a starting point for the geometry of numbers:

(5)   *Let $C \subseteq \mathbb{R}^n$ be a full-dimensional compact convex set, symmetric with respect to the origin, and with volume at least 2^n. Then C contains a nonzero integral vector.*

Another fundamental result was obtained by Blichfeldt [1914]:

(6)   *Let $C \subseteq \mathbb{R}^n$ be a full-dimensional compact convex set, with volume at least 1. Then there exist x, y in C such that $x - y$ is a nonzero integral vector.*

For surveys on the geometry of numbers, see Cassels [1959] and Lekkerkerker [1969]. See also the historical notes at the end of Part II.

Combinatorial optimization and integer linear programming (until 1960)

Most integer linear programming problems studied before 1950 had the form of a combinatorial optimization problem. One of the earliest studies in combinatorial optimization was made by Monge [1784] (cf. Taton [1951]), who investigated the problem of transporting earth, which he considered as the discontinuous, combinatorial problem of transporting molecules:

Étant données sur un même plan deux aires égales *ABCD*, & *abcd*, terminées par des contours quelconques, continus ou discontinus, trouver la route que doit suivre chaque molécule *M* de la première, & le point *m* où elle doit arriver dans la seconde, pour que tous les points étant semblablement transportés, ils remplissent exactement la seconde aire, & que la somme des produits de chaque molécule multipliée par l'espace parcouru soit un *minimum*.

Si par un point *M* quelconque de la première aire, on mène une droite *Bd*, telle que le segment *BAD* soit égal au segment *bad*, je dis que pour satisfaire à la question, il faut que toutes les molécules du segment *BAD*, soient portées sur le segment *bad*, & que par conséquent les molécules du segment *BCD* soient portées sur le segment égal *bcd*; car si un point *K* quelconque du segment *BAD*, étoit porté sur un point *k* de *bcd*, il faudroit nécessairement qu'un point égal *L*, pris quelque part dans *BCD*, fût transporté dans un certain point *l* de *bad*, ce qui ne pourroit pas se faire sans que les routes *Kk*, *Ll*, ne se coupassent entre leurs extrémités & la somme des produits des molécules par les espaces parcourus ne seroit pas un *minimum*. Pareillement, si par un point *M'* infiniment proche du point *M*, on mène la droite *B' d'*, telle qu'on ait encore le segment *B' A' D'*, égal au segment *b' a' d'*, il faut pour que la question soit satisfaite, que les molécules du segment *B' A' D'* soient transportées sur *b' a' d'*. Donc toutes les molécules de l'élément *B B' D' D* doivent être transportées sur l'élément égal *bb' d' d*. Ainsi en divisant le déblai & le remblai en une infinité d'élémens par des droites qui coupent dans l'un & dans l'autre des segmens égaux entr'eux, chaque élément du déblai doit être porté sur l'élément correspondant du remblai.

Les droites *Bd* & *B'd'* etant infiniment proches, il est indifferent dans quel ordre les molécules de l'élément *B B' D'D* se distribuent sur l'élément *bb'd'd*; de quelque manière en effet que se fasse cette distribution, la somme des produits des molécules par les espaces parcourus, est toujours la même; mais si l'on remarque que dans la pratique il convient de débleyer premièrement les parties qui se trouvent sur le passage des autres, & de n'occuper que les dernières les parties du remblai qui sont dans le même cas; la molécule *M M'* ne devra se transporter que lorsque toute la partie *M M' D' D* qui la précêde, aura été transportée en *mm'd'd*; donc dans cette hypothèse, si l'on fait *mm'd'd = MM' D'D*, le point *m* sera celui sur lequel le point *M* sera transporté.

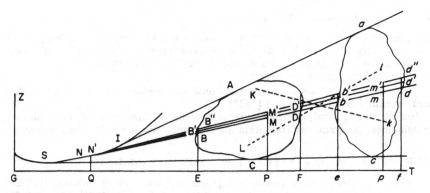

Monge also discusses the problem when all the earth must be transported via given points, and he describes an extension to three dimensions.

Interest in the transportation problem, and the related assignment problem, returned in the 1930's and 1940's, when it was studied by König [1931, 1933], Egerváry [1931], Kantorovich [1939] (including an *integer* transportation problem), [1942], Tolstoĭ [1939], Hitchcock [1941], Easterfield [1946], and Koopmans [1948]. See also the historical notes on linear programming at the end of Part III.

Disjoint optimal packings of paths in networks were studied by Menger [1927], leading to 'Menger's theorem', a combinatorial form of the max-flow min-cut theorem. In his Kolloquium of February 5, 1930, Menger [1932] posed the traveling salesman problem (*Das Botenproblem*).

In the 1950's the proper study of integer linear programming started. First, Dantzig [1951b] showed that the transportation problems form LP-problems which have automatically integral optimum solutions, developing a special form of the simplex method for transportation problems.

Next, several other combinatorial problems were studied as integer linear programs, like the fixed charge problem (Hirsch and Dantzig [1954]), the traveling salesman problem (Dantzig, Fulkerson, and Johnson [1954, 1959], Heller [1953a, b, 1955a, b, c], Kuhn [1955b], Norman [1955], who searched for facets of the traveling salesman polytope), the trim problem (cutting newsprint paper—Paull and Walter [1955]), the caterer problem (Gaddum, Hoffman, and Sokolowsky [1954], Jacobs [1954], Prager [1956–7]), flow and transportation problems (Dantzig and Fulkerson [1956], Elias, Feinstein, and Shannon [1956], Ford and Fulkerson [1956, 1956-7, 1957]), and assignment problems (Koopmans and Beckmann [1957], Kuhn [1955a, 1956b], Motzkin [1956], Motzkin and Straus [1956], Munkres [1957], Egerváry [1958]). See Dantzig [1960] for a survey.

A cutting plane approach for the traveling salesman problem was described by Dantzig, Fulkerson, and Johnson [1954, 1959]. Markowitz and Manne [1957] proposed cutting planes to solve general ILP-problems. Gomory [1958] gave a systematic procedure of adding cutting planes which he could prove to terminate with a solution of the ILP-problem. Beale [1958] extended the cutting plane method to mixed integer linear programming.

Dynamic programming approaches to integer linear programming were given by Bellman [1956–7, 1957a, b, 1957–8], Dantzig [1957], and Dreyfus [1957], while Land and Doig [1960] designed an enumerative (branch-and-bound) method.

Total unimodularity

Poincaré [1900] proved the following theorem: if $A = (\alpha_{ij})$ is a $\{0, \pm 1\}$-matrix, such that for each $k \geq 2$ and for each 'cycle'

(7) $\alpha_{i_1 j_1}, \alpha_{i_2 j_1}, \alpha_{i_2 j_2}, \ldots, \alpha_{i_k j_k - 1}, \alpha_{i_k j_k}, \alpha_{i_1 j_k}$

the product of these terms is $(-1)^k$ or 0, then A is totally unimodular (see p. 366 of Poincaré's Oeuvres, Tome VI). He derived as a consequence that $\{0, \pm 1\}$-matrices with in each column exactly one $+1$ and exactly one -1 are totally unimodular (p. 369—'tableau' T_1 is the matrix in question).

Poincaré's work on network matrices and total unimodularity was extended by Veblen and Franklin [1921] and Chuard [1922].

Hoffman and Kruskal [1956] showed that total unimodularity is exactly the reason why certain linear programs like transportation problems have integral optimum solutions.

FURTHER NOTES ON INTEGER LINEAR PROGRAMMING

Books on integer linear programming are: Garfinkel and Nemhauser [1972a], Greenberg [1971], Hu [1969], Kaufmann and Henry-Labordère [1977], Murty [1976], Nemhauser and Wolsey [1988], Plane and McMillan [1971], Salkin [1975], Taha

[1975], Zionts [1974], Brucker [1975], and Burkard [1972]. See also Papadimitriou and Steiglitz [1982]. For a more general approach, see Saaty [1970]. Relations of polyhedral theory with combinatorial optimization are described by Ford and Fulkerson [1962] (a standard work on network flows), Girlich and Kowaljow [1981] and Yemelichev, Kovalev, and Kravstov [1981]. See also the monographs by Grötschel [1977] and Johnson [1980a].

Survey papers are: Balinski [1965–6, 1967, 1970], Beale [1965, 1977], Dantzig [1957], Balinski and Spielberg [1969], Garfinkel and Nemhauser [1973], Geoffrion and Marsten [1971–2], Geoffrion [1976], Jeroslow [1977], Korte [1976], and Salkin [1973].

Two books on combinatorial optimization are: Lawler [1976] and Papadimitriou and Steiglitz [1982].

For an extensive classified bibliography on integer programming and related areas, see Kastning [1976] and its supplements Hausmann [1978] and von Randow [1982, 1985].

Lovász [1977, 1979a] gives a survey of polyhedral methods in combinatorial optimization. For another survey, see Frumkin [1982]. See also Hoffman [1979a].

Ibaraki [1976] considers the general problem of formulating a combinatorial optimization problem as an integer linear program.

Meyer [1974] studies the existence of optimum solutions in (mixed) ILP, thereby showing Theorem 16.1. Meyer and Wage [1978] showed that also for irrational A, b, the convex hull of $\{x\,|\,x \geqslant 0;\ Ax = b;\ x \text{ integral}\}$ is a polyhedron. Balas and Padberg [1979] study the adjacency of vertices of polytopes with $\{0, 1\}$-vertices.

Stanley [1974] made an interesting study of solving linear equations in nonnegative or in positive integers, relating the nonnegative integral solutions of $Ax = b$ to the positive integral solutions of $Ax = -b$. For relations between solving linear equations in nonnegative integers, magic labelings of graphs, and Hilbert's syzygy theorem, see Stanley [1973] (cf. Ehrhart [1967a, b, 1977], Stanley [1980]). Stanley [1982] considers solving linear equations in nonnegative integers with commutative algebra.

Achou [1974], Beged-Dov [1972], Buzytsky [1982], Faaland [1972], Lambe [1974], Padberg [1971] and Raghavachari and Sabharwal [1978] study the number of feasible solutions of a knapsack problem. The number of nonnegative solutions to a system of linear diophantine equations was studied by Frumkin [1980, 1981]. The number of integral vectors in an integral polytope was studied by Bernshtein [1976]. Also Ehrhart [1967a, b, 1973, 1977] makes a study of the enumeration of lattice points in polyhedra, by means of polynomials in the way of Euler (see the historical notes above). See also McMullen [1975], Israilov [1981], and Buzytsky and Freiman [1983].

Frobenius' problem (see the historical notes above) was studied by Bateman [1958], Brauer [1942], Brauer and Seelbinder [1954], Brauer and Shockley [1962], Byrnes [1974, 1975], Clarke [1977], Dulmage and Mendelsohn [1964], Ehrhart [1979], Erdös and Graham [1972], Goldberg [1976], Greenberg [1980], Heap and Lynn [1964, 1965], Hofmeister [1966], Huang [1981], Israilov [1981], Johnson [1960], Lewin [1972, 1973, 1975], Lex [1977], Mendelsohn [1970], Milanov [1984], Nijenhuis and Wilf [1972], Roberts [1956, 1957], Rödseth [1978, 1979], Selmer [1977], Selmer and Beyer [1978], Siering [1974], and Vitek [1975, 1976].

Cabot [1970], Dantzig and Eaves [1973], Bradley and Wahi [1973], and Williams [1976, 1983] studied extensions of the Fourier–Motzkin elimination method to ILP.

Algorithms for finding all integer solutions to linear inequalities were given by Fiorot [1969, 1970a, b, 1972]. See also Perl [1975]. Huet [1978] studies the problem of finding a Hilbert basis for $\{x \geqslant 0\,|\,ax = 0\}$, where a is a row vector.

Rubin [1971–2] studies the problem of identifying redundant constraints in integer programming. Bradley, Hammer, and Wolsey [1974] study for which pairs of linear inequalities $ax \leqslant \beta$ and $\tilde{a}x \leqslant \tilde{\beta}$ one has $\{x\,|\,ax \leqslant \beta;\ x\ \{0, 1\}\text{-vector}\} = \{x\,|\,\tilde{a}x \leqslant \tilde{\beta};\ x\ \{0, 1\}\text{-vector}\}$.

Kelley [1960] is an early reference where the cutting plane approach is applied in convex programming.

The structure of integer linear programs under the Smith and Hermite normal forms is studied by Bradley [1970b, 1970–1] and Bowman [1974] (cf. Faaland [1973–4]).

{0, 1}-problems (including set covering, set packing, vertex packing, set partitioning, facets of the corresponding polytopes) were studied by Balas [1975a, 1977, 1980, 1984], Balas and Ho [1980], Balas and Padberg [1972, 1975, 1976], Balas and Zemel [1977, 1984], Christofides and Korman [1974–5], Chvátal [1975a, 1979], Fulkerson, Nemhauser, and Trotter [1974], Garfinkel and Nemhauser [1969, 1972b (survey)], Geoffrion [1969], Gondran and Laurière [1974, 1975], Guha [1973], Hammer, Johnson, and Peled [1975], Hochbaum [1982], Houck and Vemuganti [1977], Hwang, Sun and Yao [1985], Johnson [1974a, 1980b], Lawler [1966], Lemke, Salkin, and Spielberg [1971], Lifschitz and Pittel [1983], Lovász [1975], Marsten [1973–4], Michaud [1972], Nemhauser and Trotter [1974, 1975], Nemhauser, Trotter, and Nauss [1973–4], Padberg [1973, 1974, 1975a, b, 1977, 1979], Picard and Queyranne [1977], Roth [1969], Salkin and Koncal [1973], Thiriez [1971], Trotter [1975], Wolsey [1975, 1976a, b, 1977], and Zemel [1978].

Hammer [1977] collected unsolved problems concerning lattice points.

References

Numbers at the end of each reference are pages on which the reference occurs.

Abadie, J. (1964), The dual of Fourier's method for solving linear inequalities, in: *Abstracts, Papers presented at 1964 International Symposium on Mathematical Programming,* London, 1964 (2 pages). [157]

Achou, O. (1974), The number of feasible solutions to a knapsack problem, *SIAM Journal on Applied Mathematics* **27** (1974) 606–610. [379]

Adams, W. W. (1967), Simultaneous asymptotic diophantine approximations, *Mathematika* **14** (1967) 173–180. [74]

Adams, W. W. (1969a), Simultaneous diophantine approximations and cubic irrationals, *Pacific Journal of Mathematics* **30** (1969) 1–14. [74]

Adams, W. W. (1969b), Simultaneous asymptotic diophantine approximations to a basis of a real cubic number field, *Journal of Number Theory* **1** (1969) 179–194. [74]

Adams, W. W. (1971), Simultaneous asymptotic diophantine approximations to a basis of a real number field, *Nagoya Mathematical Journal* **42** (1971) 79–87. [74]

Adleman, L. M. (1983a), On breaking generalized knapsack public key cryptosystems (abstract), in: *Proceedings of the Fifteenth Annual ACM Symposium on Theory of Computing* (Boston, Mass., 1983), The Association for Computing Machinery, New York, 1983, pp. 402–412. [74]

Adleman, L. (1983b), On breaking the iterated Merkle-Hellman public key cryptosystem, in: *Advances in Cryptology: Proceedings of Crypto '82* (D. Chaum, R. L. Rivest and A. T. Sherman, eds.), Plenum Press, New York, 1983, pp. 303–308. [74]

Adler, I. (1974), Lower bounds for maximum diameters of polytopes, *Mathematical Programming Study* **1** (1974) 11–19. [142]

Adler, I. (1983), *The expected number of pivots needed to solve parametric linear programs and the efficiency of the self-dual simplex method,* manuscript, Department of Industrial Engineering and Operations Research, University of California, Berkeley, Cal., 1983. [147]

Adler, I. and Berenguer, S. E. (1981), *Random linear programs,* ORC Report 81–4, Universiy of California, Berkeley, Cal., 1981. [147]

Adler, I. and Berenguer, S. E. (1983), *Generating random linear programs,* manuscript, Department of Industrial Engineering and Operations Research, University of California, Berkeley, 1983 (submitted to *Mathematical Programming*). [147]

Adler, I. and Dantzig, G. B. (1974), Maximum diameter of abstract polytopes, *Mathematical Programming Study* **1** (1974) 20–40. [142]

Adler, I., Dantzig, G., and Murty, K. (1974), Existence of A–avoiding paths in abstract polytopes, *Mathematical Programming Study* **1** (1974) 41–42. [142]

Adler, I., Karp, R., and Shamir, R. (1986), A family of simplex variants solving an $m \times d$ linear program in expected number of pivot steps depending on d only, *Mathematics of Operations Research* **11** (1986) 570–590. [147]

Adler, I., Karp, R. M., and Shamir, R. (1987), A simplex variant solving an $m \times d$ linear program in $O(\min(m^2, d^2))$ expected number of pivot steps, *Journal of Complexity* **3** (1987) 372–387. [147]

Adler, I. and Megiddo, N. (1983), *A simplex-type algorithm solves linear programs of order $m \times n$ in only $O((\min(m, n))^2)$ steps on the average*, preliminary report, 1983. [147]

Adler, I. and Megiddo, N. (1984), A simplex algorithm whose average number of steps is bounded between two quadratic functions of the smaller dimension, in: *Proceedings of the Sixteenth Annual ACM Symposium on Theory of Computing* (Washington, 1984), The Association for Computing Machinery, New York, 1984, pp. 312–323 [final version in: *Journal of the Association for Computing Machinery* **32** (1985) 871–895]. [147]

Adler, I. and Saigal, R. (1976), Long monotone paths in abstract polytopes, *Mathematics of Operations Research* **1** (1976) 89–95. [142]

Agmon, S. (1954), The relaxation method for linear inequalities, *Canadian Journal of Mathematics* **6** (1954) 382–392. [157, 158]

Aho, A. V., Hopcroft, J. E., and Ullman, J. D. (1974), *The Design and Analysis of Computer Algorithms*, Addison-Wesley, Reading, Mass., 1974. [14, 16, 41, 199, 246]

Akl, S. G. (1979), Two remarks on a convex hull algorithm, *Information Processing Letters* **8** (1979) 108–109. [224].

Akl, S. G. and Toussaint, G. T. (1978), A fast convex hull algorithm, *Information Processing Letters* **7** (1978) 219–222. [224].

Alcaly, R. E. and Klevorick, A. K. (1966), A note on the dual prices for integer programs, *Econometrica* **34** (1966) 206–214. [354]

[Aleksandrov, A. D. (1958)=] Alexandrow, A. D., *Konvexe Polyeder* (translated from the Russian), Akademie-Verlag, Berlin (DDR), 1958. [223]

Ali, A. I., Helgason, R. V., Kennington, J. L., and Lall, H. S. (1978), Primal simplex network codes: state-of-the-art implementation technology, *Networks* **8** (1978) 315–339. [142]

Altherr, W. (1975), An algorithm for enumerating all vertices of a convex polyhedron, *Computing* **15** (1975) 181–193. [224]

Altherr, W. (1978), Eckpunktbestimmung konvexer Polyeder, in: *Kombinatorische Entscheidungsprobleme: Methoden und Anwendungen* (T. M. Liebling and M. Rössler, eds.), Lecture Notes in Economics and Mathematical Systems 153, Springer, Berlin, 1978, pp. 184–195. [224]

Altshuler, A. (1985), The Mani-Walkup spherical counterexamples to the W_v-path conjecture are not polytopal, *Mathematics of Operations Research* **10** (1985) 158–159. [142]

Andrew, A. M. (1979), Another efficient algorithm for convex hulls in two dimensions, *Information Processing Letters* **9** (1979) 216–219. [224]

Aráoz, J. (1973), *Polyhedral neopolarities*, Doctoral Thesis, University of Waterloo, Waterloo, Ontario, 1973. [119, 367]

Araoz, J. (1979), Blocking and anti-blocking extensions, *Operations Research Verfahren/Methods of Operations Research* **32** (III Symposium on Operations Research, Mannheim, 1978; W. Oettli and F. Steffens, eds.), Athenäum/Hain/Scriptor/Hanstein, Königstein/Ts., 1979, pp. 5–18. [119]

Aráoz D., J., Edmonds, J., and Griffin, V. J. (1983), Polarities given by systems of bilinear inequalities, *Mathematics of Operations Research* **8** (1983) 34–41. [119]

Aráoz, J. and Johnson, E. L. (1981), Some results on polyhedra of semigroup problems, *SIAM Journal on Algebraic and Discrete Methods* **2** (1981) 244–258. [367]

Aspvall, B. (1980), *Efficient algorithms for certain satisfiability and linear programming problems*, Ph.D. dissertation, Dept. of Computer Science, Stanford University, Stanford, Cal., 1980. [195]

Aspvall, B. and Shiloach, Y. (1980a), A polynomial time algorithm for solving systems of linear inequalities with two variables per inequality, *SIAM Journal on Computing* **9** (1980) 827–845. [195]

Aspvall, B. and Shiloach, Y. (1980b), A fast algorithm for solving systems of linear equations with two variables per equation, *Linear Algebra and Its Applications* **34** (1980) 117–124. [36]

Atkinson, K. E. (1978), *An Introduction to Numerical Analysis*, Wiley, New York, 1978. [41]

Auslander, L. and Parter, S. V. (1961), On imbedding graphs in the sphere, *Journal of Mathematics and Mechanics* **10** (1961) 517–523. [282]

Auslander, L. and Trent, H. M. (1959), Incidence matrices and linear graphs, *Journal of Mathematics and Mechanics* **8** (1959) 827–835. [282]

Auslander, L. and Trent, H. M. (1961), On the realization of a linear graph given its algebraic specification, *Journal of the Acoustical Society of America* **33** (1961) 1183–1192. [282]

Avis, D. (1980a), A note on some computationally difficult set covering problems, *Mathematical Programming* **18** (1980) 138–145. [362]

Avis, D. (1980b), Comments on a lower bound for convex hull determination, *Information Processing Letters* **11** (1980) 126. [224]

Avis, D. (1982), On the complexity of finding the convex hull of a set of points, *Discrete Applied Mathematics* **4** (1982) 81–86. [224]

Avis, D. and Chvátal, V. (1978), Notes on Bland's pivoting rule, *Mathematical Programming Study* **8** (1978) 24–34. [139, 141]

Babaev, Dzh. A. and Mamedov, K. Sh. (1978), Aggregation of a class of systems of integer-valued equations (in Russian), *Zhurnal Vychislitel'noi Matematiki i Matematicheskoi Fiziki* **18** (3) (1978) 614–619 [English translation: *U.S.S.R. Computational Mathematics and Mathematical Physics* **18** (3) (1978) 86–91]. [236]

Babai, L. (1984), *On Lovász' lattice reduction and the nearest lattice point problem*, preprint, 1984 (to appear in *Combinatorica*) [shortened version in: *STACS 85 2nd Annual Symposium on Theoretical Aspects of Computer Science* (Saarbrücken, 1985; K. Mehlhorn, ed.), Lecture Notes in Computer Science 182, Springer, Berlin, 1985, pp. 13–20]. [261]

Babayev, D. A. and Glover, F. (1984), Aggregation of nonnegative integer-valued equations, *Discrete Applied Mathematics* **8** (1984) 125–130. [236]

Bachem, A. (1976a), *Beiträge zur Theorie der Corner Polyeder*, Verlag Anton Hain, Meisenheim am Glan (F.R.G.), 1976. [367]

Bachem, A. (1976b), Approximation ganzzahliger Polyeder durch Corner Polyeder, *Zeitschrift für angewandte Mathematik und Mechanik* **56** (1976) T332–T333. [367]

Bachem, A. (1978), The theorem of Minkowski for polyhedral monoids and aggregated linear diophantine systems, in: *Optimization and Operations Research* (Proceedings of a Workshop held at the University of Bonn, 1977; R. Henn, B. Korte and W. Oettli, eds.), Lecture Notes in Economics and Mathematical Systems 157, Springer, Berlin, 1978, pp. 1–13. [234]

Bachem, A. and Grötschel, M. (1981), Characterizations of adjacency of faces of polyhedra, *Mathematical Programming Study* **14** (1981) 1–22. [224]

Bachem, A., Johnson, E. L., and Schrader, R. (1981–2), A characterization of minimal valid inequalities for mixed integer programs, *Operations Research Letters* **1** (1981–2) 63–66. [354, 367]

Bachem, A. and Kannan, R. (1979), Applications of polynomial Smith normal form calculations, in: *Numerische Methoden bei graphentheoretischen und kombinatorischen Problemen–Band 2* (L. Collatz, G. Meinardus and W. Wetterling, eds.), Birkhäuser, Basel, 1979, pp. 9–21. [59]

Bachem, A. and Kannan, R. (1984), *Lattices and the basis reduction algorithm*, Report CMU-CS-84-112, Department of Computer Science, Carnegie-Mellon University, Pittsburg, Pennsylvania, 1984. [75]

Bachem, A. and von Randow, R. (1979), Integer theorems of Farkas lemma type, *Operations Research Verfahren/Methods of Operations Research* **32** (III Symposium on

Operations Research, Mannheim, 1978; W. Oettli and F. Steffens, eds.), Athenäum/Hain/Scriptor/Hanstein, Königstein/Ts., 1979, pp. 19–28. [51,224]

Bachem, A. and Schrader, R. (1979), A duality theorem and minimal inequalities in mixed integer programming, *Zeitschrift für angewandte Mathematik und Mechanik* **59** (1979) T88-T89. [354,367]

Bachem, A. and Schrader, R. (1980), Minimal inequalities and subadditive duality, *SIAM Journal on Control and Optimization* **18** (1980) 437–443. [354]

Bachet, Sr de Méziriac, C.-G. (1612), *Problèms plaisans et délectables, qui se font par les nombres, partie recueillis de divers autheurs, et inventez de nouveau, avec leur démonstration,* P. Rigaud, Lyon, 1612 [reprinted: A. Blanchard, Paris, 1959]. [77]

Bachmann, P. (1898), *Die Arithmetik der quadratischen Formen, Erste Abteilung,* Teubner, Leipzig, 1898. [82]

Bachmann, P. (1902), *Niedere Zahlentheorie, Vol. I,* Teubner, Leipzig, 1902 [reprinted with Volume II: Chelsea, New York, 1968]. [82]

Bachmann, P. (1910), *Niedere Zahlentheorie, Vol. II,* Teubner, Leipzig, 1910 [reprinted with Volume I: Chelsea, New York, 1968]. [82,375]

Bachmann, P. (1923), *Die Arithmetik der quadratischen Formen, Zweite Abteilung,* Teubner, Leipzig, 1923. [82]

Balas, E. (1963), Programare liniară cu variabile bivalente (Rumanian) [Linear programming with zero-one variables], *Proceedings Third Scientific Session on Statistics,* Bucharest, December 5–7, 1963. [363]

Balas, E. (1964), Un algorithme additif pour la résolution des programmes linéaires en variables bivalentes, *Comptes Rendus Hebdomadaires des Séances de l'Académie des Sciences (Paris)* **258** (1964) 3817–3820. [363]

Balas, E. (1965), An additive algorithm for solving linear programs with zero-one variables, *Operations Research* **13** (1965) 517–546. [363]

Balas, E. (1967), Discrete programming by the filter method, *Operations Research* **15** (1967) 915–957. [363,373]

Balas, E. (1970a), Minimax and duality for linear and nonlinear mixed-integer programming, in: *Integer and Nonlinear Programming* (J. Abadie, ed.), North-Holland, Amsterdam, 1970, pp. 353–365. [354, 372]

Balas, E. (1970b), Duality in discrete programming, in: *Proceedings of the Princeton Symposium on Mathematical Programming* (Princeton, 1967; H. W. Kuhn, ed.), Princeton University Press, Princeton, N.J., 1970, pp. 179–197. [354]

Balas, E. (1971), Intersection cuts—a new type of cutting planes for integer programming, *Operations Research* **19** (1971) 19–39. [359]

Balas, E. (1972), Integer programming and convex analysis: intersection cuts from outer polars, *Mathematical Programming* **2** (1972) 330–382. [359]

Balas, E. (1973), A note on the group theoretic approach to integer programming and the 0–1 case, *Operations Research* **21** (1973) 321–322. [367]

Balas, E. (1975a), Facets of the knapsack polytope, *Mathematical Programming* **8** (1975) 146–164. [253,374,380]

Balas, E. (1975b), Disjunctive programming: cutting planes from logical conditions, in: *Nonlinear Programming* 2 (Proc. Symp. Madison, Wisconsin, 1974; O. L. Mangasarian, R. R. Meyer and S. M. Robinson, eds.), Academic Press, New York, 1975, pp. 279–312. [359]

Balas, E. (1977), Some valid inequalities for the set partitioning problem, [in: *Studies in Integer Programming* (P. L. Hammer, et. al., eds.),] *Annals of Discrete Mathematics* **1** (1977) 13–47. [380]

Balas, E. (1980), Cutting planes from conditional bounds: a new approach to set covering, *Mathematical Programming Study* **12** (1980) 19–36. [359,380]

Balas, E. (1984), A sharp bound on the ratio between optimal integer and fractional covers, *Mathematics of Operations Research* **9** (1984) 1–5. [380]

Balas, E. and Christofides, N. (1981), A restricted Lagrangean approach to the traveling salesman problem, *Mathematical Programming* **21** (1981) 19–46. [370]

Balas, E. and Ho, A. (1980), Set covering algorithms using cutting planes, heuristics, and subgradient optimization: a computational study, *Mathematical Programming Study* **12** (1980) 37–60. [359, 363, 370, 380]

Balas, E. and Jeroslow, R. G. (1980), Strengthening cuts for mixed integer programs, *European Journal of Operational Research* **4** (1980) 224–234. [359]

Balas, E. and Padberg, M. W. (1972), On the set-covering problem, *Operations Research* **20** (1972) 1152–1161. [380]

Balas, E. and Padberg, M. (1975), On the set-covering problem: II. An algorithm for set partitioning, *Operations Research* **23** (1975) 74–90. [380]

Balas, E. and Padberg, M. W. (1976), Set partitioning: a survey, *SIAM Review* **18** (1976) 710–760. [380]

Balas, E. and Padberg, M. W. (1979), Adjacent vertices of the all 0–1 programming polytope, *Revue Française d'Automatique, d'Informatique et de Recherche Opérationelle* [*R.A.I.R.O.*] (*Recherche Opérationelle*) **13** (1979) 3–12. [379]

Balas, E. and Zemel, E. (1977), Critical cutsets of graphs and canonical facets of set-packing polytopes, *Mathematics of Operations Research* **2** (1977) 15–19. [380]

Balas, E. and Zemel, E. (1978), Facets of the knapsack polytope from minimal covers, *SIAM Journal on Applied Mathematics* **34** (1978) 119–148. [373]

Balas, E. and Zemel, E. (1980), An algorithm for large zero-one knapsack problems, *Operations Research* **28** (1980) 1130–1154. [373]

Balas, E. and Zemel, E. (1984), Lifting and complementing yields all the facets of positive zero-one programming polytopes, in: *Mathematical Programming* (R. W. Cottle, M. L. Kelmanson and B. Korte, eds.), North-Holland, Amsterdam, 1984, pp. 13–24. [380]

Balinski, M. L. (1961), An algorithm for finding all vertices of convex polyhedral sets, *Journal of the Society for Industrial and Applied Mathematics* **9** (1961) 72–88. [224]

Balinski, M. L. (1965–6), Integer programming: methods, uses, computation, *Management Science* **12** (A) (1965-6) 253–313. [379]

Balinski, M. L. (1967), Some general methods in integer programming, in : *Nonlinear Programming* (J. Abadie, ed.), North-Holland, Amsterdam, 1967, pp. 221–247. [379]

Balinski, M. L. (1970), On recent developments in integer programming, in: *Proceedings of the Princeton Symposium on Mathematical Programming* (Princeton, 1967; H. W. Kuhn, ed.), Princeton University Press, Princeton, N.J., 1970, pp. 267–302. [379]

Balinski, M. (1983), Signatures des points extrêmes du polyèdre dual du problème de transport, *Comptes Rendus des Séances de l'Académie des Sciences [Paris] Série I* **296** (1983) 457–459. [142]

Balinski, M. (1984), The Hirsch conjecture for dual transportation polyhedra, *Mathematics of Operations Research* **9** (1984) 629–633. [142]

Balinski, M. L. (1985), Signature methods for the assignment problem, *Operations Research* **33** (1985) 527–536. [142]

Balinski, M. L. and Spielberg, K. (1969), Methods for integer programming: algebraic, combinatorial, and enumerative, in: *Progress in Operations Research, Relationship between Operations Research and the Computer, Volume III* (J. S. Aranofsky, ed.), Wiley, New York, 1969, pp. 195–292. [379]

Bareiss, E. H. (1968), Sylvester's identity and multistep integer-preserving Gaussian elimination, *Mathematics of Computation* **22** (1968) 565–578. [35]

Barker, G. P. (1973), The lattice of faces of a finite dimensional cone, *Linear Algebra and Its Applications* **7** (1973) 71–82. [224]

Barker, G. P. (1978), Faces and duality in convex cones, *Linear and Multilinear Algebra* **6** (1978) 161-169. [224]

Barnes, E. S. (1959), The construction of perfect and extreme forms II, *Acta Arithmetica* **5** (1959) 205–222. [81]

Barnette, D. (1974), An upper bound for the diameter of a polytope, *Discrete Mathematics* **10** (1974) 9–13. [142]

Barnette, D. (1979), Path problems and extremal problems for convex polytopes, in: *Relations between Combinatorics and Other Parts of Mathematics* (Proceedings of Symposia in Pure Mathematics Volume XXXIV [Columbus, Ohio, 1978]; D. K. Ray-Chaudhuri, ed.), American Mathematical Society, Providence, R.I., 1979, pp. 25–34. [223]

Barr, R. S., Glover, F., and Klingman, D. (1977), The alternating basis algorithm for assignment problems, *Mathematical Programming* 13 (1977) 1–13. [142]

Bartels, H. G. (1973), *A priori Informationen zur linearen Programmierung, Über Ecken und Hyperflächen auf Polyedern*, Verlag Anton Hain, Meisenheim am Glan, 1973. [223]

Bartholdi, III, J. J. (1981–2), A good submatrix is hard to find, *Operations Research Letters* 1 (1981–2) 190–193. [293]

Bateman, P. T. (1958), Remark on a recent note on linear forms, *The American Mathematical Monthly* 65 (1958) 517–518. [379]

Batigne, D. R., Hall, F. J. and Katz, I. J. (1978), Further results on integral generalized inverses of integral matrices, *Linear and Multilinear Algebra* 6 (1978) 233–241. [308]

Bauer, F. L. (1963), Algorithm 153 Gomory, *Communications of the ACM* 6 (1963) 68. [359]

Baum, S. and Trotter, Jr, L. E. (1977), Integer rounding and polyhedral decomposition of totally unimodular systems, in: *Optimization and Operations Research* (Proc. Bonn 1977; R. Henn, B. Korte and W. Oettli, eds.), Lecture Notes in Economics and Math. Systems 157, Springer, Berlin, 1977, pp. 15–23. [271, 337]

Baum, S. and Trotter, Jr, L. E. (1981), Integer rounding for polymatroid and branching optimization problems, *SIAM Journal on Algebraic and Discrete Methods* 2 (1981) 416–425. [337, 338]

Baum, S. and Trotter, Jr, L. E. (1982), Finite checkability for integer rounding properties in combinatorial programming problems, *Mathematical Programming* 22 (1982) 141–147. [336]

Bazaraa, M. S. and Jarvis, J. J. (1977), *Linear Programming and Network Flows*, Wiley, New York, 1977. [224]

Beale, E. M. L. (1954), An alternative method for linear programming, *Proceedings of the Cambridge Philosophical Society* 50 (1954) 513–523. [148, 152]

Beale, E. M. L. (1955), Cycling in the dual simplex algorithm, *Naval Research Logistics Quarterly* 2 (1955) 269–275. [138]

Beale, E. M. L. (1958), *A method of solving linear programming problems when some but not all of the variables must take integral values*, Technical Report No. 19, Statistical Techniques Research Group, Princeton University, Princeton, N. J., 1958. [378]

Beale, E. M. L. (1965), Survey of integer programming, *Operational Research Quarterly* 16 (1965) 219–228. [379]

Beale, E. M. L. (1968), *Mathematical Programming in Practice*, Pitman, London, 1968. [224]

Beale, E. M. L. (1977), Integer programming, in: *The State of the Art in Numerical Analysis* (Proceedings Conference York, 1976; D. Jacobs, ed.), Academic Press, London, 1977, pp. 409–448. [379]

Beckenbach, E. F. and Bellman, R. (1983), *Inequalities, Fourth Printing*, Springer, Berlin, 1983. [47]

Becker, O. (1933), Eudoxos-Studien I. Eine voreudoxische Proportionenlehre und ihre Spuren bei Aristoteles und Euklid, *Quellen und Studien zur Geschichte der Mathematik Astronomie und Physik, Abteilung B: Studien* 2 (1933) 311–333. [76]

Beckmann, M. J. (1970), Dynamic programming of some integer and nonlinear programming problems, in: *Integer and Nonlinear Programming* (J. Abadie, ed.), North-Holland, Amsterdam, 1970, pp. 463–472. [265]

Beged-Dov, A. G. (1972), Lower and upper bounds for the number of lattice points in a simplex, *SIAM Journal on Applied Mathematics* 22 (1972) 106–108. [379]

Bell, D. E. (1977), A theorem concerning the integer lattice, *Studies in Applied Mathematics* **56** (1977) 187–188. [234]

Bell, D. E. (1979), Efficient group cuts for integer programs, *Mathematical Programming* **17** (1979) 176–183. [367]

Bell, D. E. and Fisher, M. L. (1975), Improved integer programming bounds using intersections of corner polyhedra, *Mathematical Programming* **8** (1975) 345–368. [367]

Bell, D. E. and Shapiro, J. F. (1977), A convergent duality theory for integer programming, *Operations Research* **25** (1977) 419–434. [354]

Bellman, R. (1956), Notes on the theory of dynamic programming, IV—maximization over discrete sets, *Naval Research Logistics Quarterly* **3** (1956) 67–70. [261]

Bellman, R. (1956–7), On a dynamic programming approach to the caterer problem—I, *Management Science* **3** (1956–7) 270–278. [378]

Bellman, R. (1957a), Comment on Dantzig's paper on discrete variable extremum problems, *Operations Research* **5** (1957) 723–724. [261, 378]

Bellman, R. (1957b), *Dynamic Programming*, Princeton University Press, Princeton, N. J., 1957. [261, 378]

Bellman, R. (1957–8), Notes on the theory of dynamic programming—transportation models, *Management Science* **4** (1957–8) 191–195. [378]

Bellman, R. (1960), *Introduction to Matrix Analysis*, McGraw-Hill, New York, 1960. [40]

Ben-Israel, A. (1969), Linear equations and inequalities on finite dimensional, real or complex, vector spaces: a unified theory, *Journal of Mathematical Analysis and Applications* **27** (1969) 367–389. [98]

Ben-Israel A. and Charnes, A. (1962), On some problems of diophantine programming, *Cahiers du Centre d'Études de Recherche Opérationelle* **4** (1962) 215–280. [358]

Benders, J. F. (1960), *Partitioning in mathematical programming*, Thesis, University of Utrecht, Utrecht, 1960. [371]

Benders, J. F. (1962), Partitioning procedures for solving mixed-variables programming problems, *Numerische Mathematik* **4** (1962) 238–252. [371]

Benichou, M., Gauthier, J. M., Hentges, G., and Ribière, G. (1977), The efficient solution of large-scale linear programming problems—some algorithmic techniques and computational results, *Mathematical Programming* **13** (1977) 280–322. [139]

Berenguer, S. E. (1978), *Randomly generated linear programs*, Ph.D. Thesis, University of California, Berkeley, Cal., 1978. [147]

Berenguer, S. E. and Smith, R. L. (1986), The expected number of extreme points of a random linear program, *Mathematical Programming* **35** (1986) 129–134. [147]

Berge, C. (1969), The rank of a family of sets and some applications to graph theory, in: *Recent Progress in Combinatorics* (Proceedings of the Third Waterloo Conference on Combinatorics, Waterloo, Ontario, 1968; W. T. Tutte, ed.), Academic Press, New York, 1969, pp. 49–57. [303]

Berge, C. (1970), Sur certains hypergraphes généralisant les graphes bipartites, in: *Combinatorial Theory and Its Applications I* (Proceedings Colloquium on Combinatorial Theory and its Applications, Balatonfüred, Hungary, 1969; P. Erdös, A. Rényi and V. T. Sós, eds.), North-Holland, Amsterdam, 1970, pp. 119–133. [303]

Berge, C. (1972), Balanced matrices, *Mathematical Programming* **2** (1972) 19–31. [304, 307]

Berge, C. (1980), Balanced matrices and property (G), *Mathematical Programming Study* **12** (1980) 163–175. [307]

Berge, C. and Las Vergnas, M. (1970), Sur un théorème du type König pour hypergraphes, [in: *International Conference on Combinatorial Mathematics* (New York, 1970; A. Gewirtz and L. V. Quintas, eds.),] *Annals of the New York Academy of Sciences* **175** [Article 1] (1970) 32–40. [304, 307]

Bergström, V. (1935), *Beiträge zur Theorie der endlichdimensionalen Moduln und der diophantischen Approximationen*, Meddelanden från Lunds Universitets Matematiska Seminarium, Band 2, Håkan Ohlssons Buchdruckerei, Lund, 1935. [82]

Berman, A. (1973), *Cones, Matrices and Mathematical Programming*, Lecture Notes in Economics and Mathematical Systems 79, Springer, Berlin, 1973. [98]

Berman, A. and Ben-Israel, A. (1971a), Linear inequalities, mathematical programming and matrix theory, *Mathematical Programming* 1 (1971) 291–300. [98]

Berman, A. and Ben-Israel, A. (1971b), More on linear inequalities with applications to matrix theory, *Journal of Mathematical Analysis and Applications* 33 (1971) 482–496. [98]

Berman, A. and Ben-Israel, A. (1973), Linear equations over cones with interior: a solvability theorem with applications to matrix theory, *Linear Algebra and Its Applications* 7 (1973) 139–149. [98]

Bernshtein, D. N. (1976), The number of integral points in integral polyhedra (in Russian), *Funktsional'nyi Analiz i Ego Prilozheniya* 10 (3) (1976) 72–73 [English translation: *Functional Analysis and Its Applications* 10 (1976) 223–224]. [379]

Bernstein, L. (1971), *The Jacobi-Perron Algorithm—Its Theory and Application*, Lecture Notes in Mathematics 207, Springer, Berlin, 1971. [74]

Bertsekas, D. P. (1981), A new algorithm for the assignment problem, *Mathematical Programming* 21 (1981) 152–171. [142]

Bevis, J. H. and Hall, F. J. (1982), Some classes of integral matrices, *Linear Algebra and Its Applications* 48 (1982) 473–483. [308]

Bézout, E. (1767), Recherches sur le degré des équations résultantes de l'évanouissement des inconnus, et sur les moyens qu'il convient d'employer pour trouver ces équations, *Histoire de l'Académie Royale des Sciences avec les Mémoires de Mathématique et de Physique (Paris)* [année 1764] (1767) 288–338. [39]

Bézout, E. (1779), *Théorie générale des équations algébriques*, Impr. de P.-D. Pierres, Paris, 1779. [39]

Binet, J. P. M. (1813), Mémoire sur un système de formules analytiques, et leur application à des considérations géométriques [lu le 30 Novembre 1812], *Journal de l'École Polytechnique* tome 9, cahier 16 (1813) 280–354. [40]

Birkhoff, G. (1946), Tres observaciones sobre el algebra lineal, *Revista Facultad de Ciencias Exactas, Puras y Aplicadas Universidad Nacional de Tucuman, Serie A (Matematicas y Fisica Teorica)* 5 (1946) 147–151. [108, 274]

Birkhoff, G. and Mac Lane, S. (1977), *A Survey of Modern Algebra (fourth edition)*, Macmillan, New York, 1977. [4, 41]

Bixby, R. E. (1976), A strengthened form of Tutte's characterization of regular matroids, *Journal of Combinatorial Theory (B)* 20 (1976) 216–221. [295]

Bixby, R. E. (1977), Kuratowski's and Wagner's theorems for matroids, *Journal of Combinatorial Theory (B)* 22 (1977) 31–53. [279]

Bixby, R. E. (1982), Matroids and operations research, in: *Advanced Techniques in the Practice of Operations Research* (6 tutorials presented at the Semi-Annual joint ORSA/TIMS meeting, Colorado Springs, 1980; H. J. Greenberg, F. H. Murphy and S. H. Shaw, eds.), North-Holland, New York, 1982, pp. 333–458. [292]

Bixby, R. E. and Cunningham, W. H. (1980), Converting linear programs to network problems, *Mathematics of Operations Research* 5 (1980) 321–357. [299].

Bixby, R. E. and Cunningham, W. H. (1984), *Short cocircuits in binary matroids*, Report No. 84344-OR, Institut für Ökonometrie und Operations Research, Universität Bonn, 1984 (to appear in *European Journal of Combinatorics*). [282]

Bixby, R. E. and Wagner, D. K. (1988), An almost linear-time algorithm for graph realization, *Mathematics of Operations Research* 13 (1988) 99–123. [282]

Blair, C. E. (1978), Minimal inequalities for mixed integer programs, *Discrete Mathematics* 24 (1978) 147–151. [354].

Blair, C. E. (1980), Facial disjunctive programs and sequences of cutting-planes, *Discrete Applied Mathematics* 2 (1980) 173–179. [359]

Blair, C. (1986), Random linear programs with many variables and few constraints, *Mathematical Programming* **34** (1986) 62–71. [147]

Blair, C. E. and Jeroslow, R. G. (1977), The value function of a mixed integer program: I, *Discrete Mathematics* **19** (1977) 121–138. [241,342,354]

Blair, C. E. and Jeroslow, R. G. (1979), The value function of a mixed integer program: II, *Discrete Mathematics* **25** (1979) 7–19. [239,342,354]

Blair, C. E. and Jeroslow, R. G. (1982), The value function of an integer program, *Mathematical Programming* **23** (1982) 237–273. [242, 243, 345, 346, 353, 354, 359].

Blair, C. E. and Jeroslow, R. G. (1984), Constructive characterizations of the value-function of a mixed-integer program I, *Discrete Applied Mathematics* **9** (1984) 217–233. [342]

Blair, C. E. and Jeroslow, R. G. (1985), Constructive characterizations of the value function of a mixed-integer program II, *Discrete Applied Mathematics* **10** (1985) 227–240. [342]

Bland, R. G. (1977a), New finite pivoting rules for the simplex method, *Mathematics of Operations Research* **2** (1977) 103–107. [129]

Bland, R. G. (1977b), A combinatorial abstraction of linear programming, *Journal of Combinatorial Theory (B)* **23** (1977) 33–57.[150]

Bland, R. G. (1978), Elementary vectors and two polyhedral relaxations, *Mathematical Programming Study* **8** (1978) 159–166. [119,150].

Bland, R. G., Goldfarb, D., and Todd, M. J. (1981), The ellipsoid method: a survey, *Operations Research* **29** (1981) 1039–1091. [163,171]

Blankinship, W. A. (1963), A new version of the Euclidean algorithm, *The American Mathematical Monthly* **70** (1963) 742–745. [54]

Blankinship, W. A. (1966a), Algorithm 287, Matrix triangulation with integer arithmetic [F1], *Communications of the ACM* **9** (1966) 513. [56]

Blankinship, W. A. (1966b), Algorithm 288, Solution of simultaneous linear diophantine equations [F4], *Communications of the ACM* **9** (1966) 514. [56]

Blichfeldt, H. F. (1914), A new principle in the geometry of numbers, with some applications, *Transactions of the American Mathematical Society* **15** (1914) 227–235.[81,376]

Blichfeldt, H. F. (1935), The minimum values of positive quadratic forms in six, seven and eight variables, *Mathematische Zeitschrift* **39** (1935) 1–15. [81]

Blum, M., Floyd, R. W., Pratt, V., Rivest, R. L., and Tarjan, R. E. (1973), Time bounds for selection, *Journal of Computer and System Sciences* **7** (1973) 448–461. [199]

Bodewig, E. (1956), *Matrix Calculus*, North-Holland, Amsterdam, 1956. [56,59]

Bond, J. (1967), Calculating the general solution of a linear diophantine equation, *The American Mathematical Monthly* **74** (1967) 955–957. [54]

Bondy, J. A. and Murty, U. S. R. (1976), *Graph Theory with Applications*, Macmillan, London, 1976. [8]

Bonnesen, T. and Fenchel, W. (1934), *Theorie der konvexen Körper*, Springer, Berlin, 1934 [reprinted: Chelsea, New York, 1948]. [217,224]

Borel, É. (1921), La théorie du jeu et les équations intégrales à noyau symétrique, *Comptes Rendus Hebdomadaires des Séances de l'Académie des Sciences* **173** (1921) 1304–1308 [reprinted in: *Oeuvres de Emile Borel, Tome II*, Centre National de la Recherche Scientifique, Paris, 1972, pp. 901–904] [English translation by L. J. Savage in: *Econometrica* **21** (1953) 97–100]. [218]

Borel, É. (1924), *Éléments de la Théorie des Probabilités (3e édition revue et augmentée)*, J. Hermann, Paris, 1924 [English translation of Note IV ('Sur les jeux où interviennent le hasard et l'habileté des joueurs') by L. J. Savage: On games that involve chance and the skill of the players, *Econometrica* **21** (1953) 101–115. [218]

Borel, É. (1927), Sur les systèmes de formes linéaires à déterminant symétrique gauche et la théorie générale du jeu, *Comptes Rendus Hebdomadaires des Séances de l'Académie des*

Sciences **184** (1927) 52–54 [reprinted in: *Oeuvres de Emile Borel, Tome II,* Centre National de la Recherche Scientifique, Paris, 1972, pp. 1123–1124] [English translation by L. J. Savage in: *Econometrica* **21** (1953) 116–117]. [218]

Borgwardt, K.-H. (1977a), *Untersuchungen zur Asymptotik der mittleren Schrittzahl von Simplexverfahren in der linearen Optimierung,* Diss. Universität Kaiserlautern, 1977. [147]

Borgwardt, K.-H. (1977b), Untersuchungen zur Asymptotik der mittleren Schrittzahl von Simplexverfahren in der linearen Optimierung, *Operations Research Verfahren* **28** (1977) 332–345. [139, 147]

Borgwardt, K.-H. (1978), Zum Rechenaufwand von Simplexverfahren, *Operations Research Verfahren* **31** (1978) 83–97. [147]

Borgwardt, K.-H. (1979), Die asymptotische Ordnung der mittleren Schrittzahl von Simplexverfahren, *Methods of Operations Research* **37** (IV. Symposium on Operations Research, Universität des Saarlandes, 1979; H. König and V. Steinmetz, eds.), Athenäum/Hain/Scriptor/Hanstein, Königstein/Ts., 1979, pp. 81–95. [147]

Borgwardt, K.-H. (1982a), Some distribution-independent results about the asymptotic order of the average number of pivot steps of the simplex method, *Mathematics of Operations Research* **7** (1982) 441–462. [141, 142, 144]

Borgwardt, K.-H. (1982b), The average number of pivot steps required by the simplex-method is polynomial, *Zeitschrift für Operations Research* **26** (1982) 157–177. [139, 141, 142]

Borodin, A. and Munro, I. (1975), *The Computational Complexity of Algebraic and Numeric Problems,* American Elsevier, New York, 1975. [41]

Borosh, I. and Treybig, L. B. (1976), Bounds on positive integral solutions of linear diophantine equations, *Proceedings of the American Mathematical Society* **55** (1976) 299–304. [239]

Borosh, I. and Treybig, L. B. (1979), Bounds on positive integral solutions of linear diophantine equations II, *Canadian Mathematical Bulletin* **22** (1979) 357–361. [239]

Bowman, V. J. (1972a), A structural comparison of Gomory's fractional cutting planes and Hermitian basic solutions, *SIAM Journal on Applied Mathematics* **23** (1972) 460–462. [359]

Bowman, Jr, V. J. (1972b), Sensitivity analysis in linear integer programming, *AIIE Transactions* **4** (1972) 284–289. [354]

Bowman, V. J. (1974), The structure of integer programs under the Hermitian normal form, *Operations Research* **22** (1974) 1067–1080. [380]

Bowman, Jr., V. J. and Nemhauser, G. L. (1970), A finiteness proof for modified Dantzig cuts in integer programming, *Naval Research Logistics Quarterly* **17** (1970) 309–313. [359]

Bowman, V. J. and Nemhauser, G. L. (1971), Deep cuts in integer programming, *Opsearch* **8** (1971) 89–111. [359]

Boyd, S. C. and Pulleyblank, W. R. (1984), *Facet generating techniques,* to appear. [352]

Boyer, C. B. (1968), *A History of Mathematics,* Wiley, New York, 1968. [41]

Bradley, G. H. (1970a), Algorithm and bound for the greatest common divisor of n integers, *Communications of the ACM* **13** (1970) 433–436. [54]

Bradley, G. H. (1970b), Equivalent integer programs, in: *OR 69* (Proceedings of the Fifth International Conference on Operational Research, Venice, 1969; J. Lawrence, ed.), Tavistock Publications, London, 1970, pp. 455–463. [380]

Bradley, G. H. (1970–1), Equivalent integer programs and canonical problems, *Management Science* **17** (1970–1) 354–366. [380]

Bradley, G. H. (1971a), Algorithms for Hermite and Smith normal matrices and linear diophantine equations, *Mathematics of Computation* **25** (1971) 897–907. [56, 59]

Bradley, G. H. (1971b), Transformation of integer programs to knapsack problems, *Discrete Mathematics* **1** (1971) 29–45. [236]

Bradley, G. H., Hammer, P. L., and Wolsey, L. (1974), Coefficient reduction for inequalities in 0—1 variables, *Mathematical Programming* **7** (1974) 263–282. [379]

Bradley, G. H. and Wahi, P. N. (1973), An algorithm for integer linear programming: a combined algebraic and enumeration approach, *Operations Research* **21** (1973) 45–60. [379]

Brauer, A. (1942), On a problem of partitions, *American Journal of Mathematics* **64** (1942) 299–312. [376,379]

Brauer, A. and Seelbinder, B. M. (1954), On a problem of partitions II, *American Journal of Mathematics* **76** (1954) 343–346. [379]

Brauer, A. and Shockley, J. E. (1962), On a problem of Frobenius, *Journal für die reine und angewandte Mathematik* **211** (1962) 215–220. [379]

Brauer, K. M. (1967), A note on the efficiency of Balas' algorithm, *Operations Research* **15** (1967) 1169–1171. [363]

Brentjes, A. J. (1981), A two-dimensional continued fraction algorithm for best approximations with an application in cubic number fields, *Journal für die reine und angewandte Mathematik* **326** (1981) 18–44. [74]

Breu, R. and Burdet, C.-A. (1974), Branch and bound experiments in zero-one programming, *Mathematical Programming Study* **2** (1974) 1–50. [363]

Brezinski, C. (1981), The long history of continued fractions and Padé approximants, in: *Padé Approximation and its Applications Amsterdam 1980* (Proceedings of a Conference Held in Amsterdam, 1980; M. G. de Bruin and H. van Rossum, eds.), Lecture Notes in Mathematics 888, Springer, Berlin, 1981, pp. 1–27. [82]

Brøndsted, A. (1983), *An Introduction to Convex Polytopes*, Springer, New York, 1983. [142,223]

Brooks, R. and Geoffrion, A. (1966), Finding Everett's Lagrange multipliers by linear programming, *Operations Research* **14** (1966) 1149–1153. [370]

Brown, D. P. (1976), Circuits and unimodular matrices, *SIAM Journal on Applied Mathematics* **31** (1976) 468–473. [272]

Brown, D. P. (1977), Compound and unimodular matrices, *Discrete Mathematics* **19** (1977) 1–5. [272]

Broyden, C. G. (1975), *Basic Matrices–An Introduction to Matrix Theory and Practice*, Macmillan, London, 1975. [41]

Brucker, P. (1975), *Ganzzahlige lineare Programmierung mit ökonomischen Anwendungen*, Mathematical Systems in Economics 16, Anton Hain, Meisenheim am Glan, 1975. [379]

Brückner, M. (1900), *Vielecke und Vielflache, Theorie und Geschichte*, Teubner, Leipzig, 1900. [223]

Brun, V. (1959), Mehrdimensionale Algorithmen, welche die Eulersche Kettenbruchentwicklung der Zahl e verallgemeinern, in: *Sammelband der zu Ehren des 250. Geburtstages Leonhard Eulers der Deutschen Akademie der Wissenschaften zu Berlin vorgelegten Abhandlungen* (K. Schröder, ed.), Akademie-Verlag, Berlin, 1959, pp. 87–100. [74]

Brun, V. (1961), Musikk og Euklidske algoritmer, *Nordisk Matematisk Tidskrift* **9** (1961) 29–36. [74]

Bundschuh, P. (1975), On a theorem of L. Kronecker, *Tamkang Journal of Mathematics* **6** (1975) 173–176. [74]

Bunt, L. N. H. (1934), *Bijdrage tot de theorie der convexe puntverzamelingen*, [Ph.D. Thesis, University Groningen,] Noord-Hollandsche Uitgeversmaatschappij, Amsterdam, 1934. [217]

Burdet, C.-A. (1974), Generating all the faces of a polyhedron, *SIAM Journal on Applied Mathematics* **26** (1974) 479–489. [224]

Burdet, C.-A. and Johnson, E. L. (1974), A subadditive approach to the group problem of integer programming, *Mathematical Programming Study* **2** (1974) 51–71. [367]

Burdet, C.-A. and Johnson, E. L. (1977), A subadditive approach to solve linear integer programs, [in: *Studies in Integer Programming* (P. L. Hammer, et al., eds),] *Annals of Discrete Mathematics* **1** (1977) 117–143. [367]

Burger, E. (1956), Über homogene lineare Ungleichungssysteme, *Zetschrift für angewandte Mathematik und Mechanik* **36** (1956) 135–139 [224]

Burkard, R. E. (1972), *Methoden der ganzzahligen Optimierung*, Springer, Vienna, 1972. [379]

Burkard, R. E. (1980), Subgradient methods in combinatorial optimization. in: *Discrete Structures and Algorithms* (Proceedings Workshop WG 79 5th Conference on Graphtheoretic Concepts in Computer Science, Berlin (West), 1979; U. Pape, ed.), Carl Hanser Verlag, München, 1980, pp. 141–151. [370]

Buzytsky, P. L. (1982), An effective formula for the number of solutions of linear Boolean equations, *SIAM Journal on Algebraic and Discrete Methods* **3** (1982) 182–186. [379]

Buzytsky, P. L. and Freiman, G. A. (1983), An effective formula for the number of solutions of a system of two 0, 1-equations, *Discrete Applied Mathematics* **6** (1983) 127–133. [379]

Bykat, A. (1978), Convex hull of a finite set of points in two dimensions, *Information Processing Letters* **7** (1978) 296–298. [224]

Byrnes, J. S. (1974), On a partition problem of Frobenius, *Journal of Combinatorial Theory (A)* **17** (1974) 162–166. [379]

Byrnes, J. S. (1975), A partition problem of Frobenius, II, *Acta Arithmetica* **28** (1975) 81–87. [379]

Cabot, A. V. (1970), An enumeration algorithm for knapsack problems, *Operations Research* **18** (1970) 306–311. [379]

Cajori, F. (1895), *A History of Mathematics*, Macmillan, New York, 1895 [third edition: Chelsea, New York, 1980]. [41]

Camion, P. (1965), Characterization of totally unimodular matrices, *Proceedings of the American Mathematical Society* **16** (1965) 1068–1073. [269]

Camion, P. (1963b), *Matrices totalement unimodulaires et problèmes combinatoires*, Ph.D. Thesis, Université Libre de Bruxelles, Brussels, 1963. [269,296]

Camion, P. (1965), Characterizations of totally unimodular matrices, *Proceedings of the American Mathematical Society* **16** (1965) 1068–1073. [269]

Camion, P. (1968), Modules unimodulaires, *Journal of Combinatorial Theory* **4** (1968) 301–362. [98,297]

Carathéodory, C. (1911), Über den Variabilitätsbereich der Fourierschen Konstanten von positiven harmonischen Funktionen, *Rendiconto del Circolo Matematico di Palermo* **32** (1911) 193–217 [reprinted in: *Constantin Carathéodory, Gesammelte Mathematische Schriften, Band III* (H. Tietze, ed.), C. H. Beck'sche Verlagsbuchhandlung, München, 1955, pp. 78–110]. [85,94,215]

Carnal, H. (1970), Die konvexe Hülle von *n* rotationssymmetrisch verteilten Punkten, *Zeitschrift für Wahrscheinlichkeitstheorie und verwandte Gebiete* **15** (1970) 168–176. [147]

Carver, W. B. (1921-2), Systems of linear inequalities, *Annals of Mathematics* **23** (1921-2) 212–220. [95,217]

Cassel, G. (1918), *Theoretische Sozialökonomie*, C. F. Winter, Leipzig, 1918 [English translation: *The Theory of Social Economy*, Harcourt, Brace & Co., New York, 1932]. [218]

Cassels, J. W. S. (1955), Simultaneous diophantine approximation, *The Journal of The London Mathematical Society* **30** (1955) 119–121. [74]

Cassels, J. W. S. (1957), *An Introduction to Diophantine Approximation*, [Cambridge Tracts in Mathematics 45,] Cambridge University Press, Cambridge, 1957. [82]

Cassels, J. W. S. (1959), *An Introduction to the Geometry of Numbers*, Springer, Berlin, 1959. [82,376]

Cauchy, A. L. (1815), Mémoire sur les fonctions qui ne peuvent obtenir que deux valeurs égales et de signes contraires par suite des transpositions opérées entre les variables qu'elles renferment, *Journal de l'École Polytechnique* tome **10**, cahier 17 [lu le 30 Novembre 1812] (1815) 29–112 [reprinted in: *Oeuvres Complètes d'Augustin Cauchy Ser. II, Vol. I*, Gauthier-Villars, Paris, 1905, pp. 91–169]. [40]

Cauchy, A. (1847), Mémoire sur les lieux analytiques, *Compte[s] Rendu[s Hebdo-madaires] des Séances de l'Académie des Sciences* **24** [24 Mai 1847] (1847) 885–887 [reprinted in: *Oeuvres Complètes d'Augustin Cauchy Ser. I, Vol. X*, Gauthier-Villars, Paris, 1897, pp. 292–295] [English translation (partially) in: D. E. Smith, *A Source Book in Mathematics*, McGraw-Hill, New York, 1929, pp. 530–531]. [40]

Cayley, A. (1843), Chapters in the analytic geometry of (n) dimensions, *Cambridge Mathematical Journal* **4** (1843) 119–127 [reprinted in: *The Collected Mathematical Papers of Arthur Cayley, Vol. I*, The University Press, Cambridge, 1889, [reprinted: Johnson Reprint Corp., New York, 1963] pp. 55–62]. [40]

Cayley, A. (1846), Sur quelques théorèmes de la géométrie de position, *Journal für die reine und angewandte Mathematik* **31** (1846) 213–226 [reprinted in: *The Collected Mathematical Papers of Arthur Cayley, Vol. I*, The University Press, Cambridge, 1889 [reprinted: Johnson Reprint Corp., New York, 1963,] pp. 317–328] [English translation (partially) in: D. E. Smith, *A Source Book in Mathematics*, McGraw-Hill, New York, pp. 527–529]. [40]

Cayley, A. (1855a), [Sept différents mémoires d'analyse:] No. 3 Remarques sur la notation des fonctions algébriques, *Journal für die reine und angewandte Mathematik* **50** (1855) 282–285 [reprented in: *The Collected Mathematical Papers of Arthur Cayley, Vol. II*, The University Press, Cambridge, 1889 [reprinted: Johnson Reprint Corp., New York, 1963], pp. 185–188]. [40]

Cayley, A. (1855b), Researches on the partition of numbers, *Philosophical Transactions of the Royal Society of London* **145** (1855) 127–140 [reprinted in: *The Collected Mathematical Papers of Arthur Cayley, Vol. II*, The University Press, Cambridge, 1889 [reprinted: Johnson Reprint Corp., New York, 1963], pp. 235–249]. [375]

Cayley, A. (1858), A memoir on the theory of matrices, *Philosophical Transactions of the Royal Society of London* **148** (1858) 17–37 [reprinted in: *The Collected Mathematical Papers of Arthur Cayley, Vol. II*, The University Press, Cambridge, 1889 [reprinted: Johnson Reprint Corp., New York, 1963], pp. 475–496]. [40]

Cayley, A. (1860), On a problem of double partitions, *Philosophical Magazine* **20** (1860) 337–341 [reprinted in: *The Collected Mathematical Papers of Arthur Cayley, Vol. IV*, The University Press, Cambridge, 1891, [reprinted: Johnson Reprint Corp., New York, 1963], pp. 166–170]. [375]

Cederbaum, I. (1958), Matrices all of whose elements and subdeterminants are 1, − 1 or 0, *Journal of Mathematical Physics* **36** (1958) 351–361. [272,299]

Censor, Y. and Elfving, T. (1982), New methods for linear inequalities, *Linear Algebra and Its Applications* **42** (1982) 199–211. [162]

Chandrasekaran, R. (1969), Total unimodularity of matrices, *SIAM Journal on Applied Mathematics* **17** (1969) 1032–1034. [272]

Chandrasekaran, R. (1981), Polynomial algorithms for totally dual integral systems and extensions, [in: *Studies on Graphs and Discrete Programming* (P. Hansen, ed.),] *Annals of Discrete Mathematics* **11** (1981) 39–51. [330,336]

Chandrasekaran, R., Kabadi, S. N., and Murty, K. G. (1981–2), Some NP-complete problems in linear programming, *Operations Research Letters* **1** (1981–2) 101–104. [224]

Chandrasekaran, R. and Shirali, S. (1984), Total weak unimodularity: testing and applications, *Discrete Mathematics* **51** (1984) 137–145. [333,338]

Charnes, A. (1952), Optimality and degeneracy in linear programming, *Econometrica* **20** (1952) 160–170. [138]

Charnes, A. and Cooper, W. W. (1961), *Management Models and Industrial Applications of Linear Programming*, Vol. I, II, Wiley, New York, 1961. [224,359]

Charnes, A. and Cooper, W. W. (1961–2), On some works of Kantorovich, Koopmans and others, *Management Science* **8** (1961–2) 246–263. [223]

Charnes, A., Cooper, W. W., and Henderson, A. (1953), *An Introduction to Linear Programming*, Wiley, New York, 1953. [224]

Charnes, A., Granot, D., and Granot, F. (1977), On intersection cuts in interval integer linear programming, *Operations Research* **25** (1977) 352–355. [358]

Châtelet, A. (1913), *Leçons sur la Théorie des Nombres (modules, entiers algébriques, réduction continuelle)*, Gauthier-Villars, Paris, 1913. [80, 81, 82]

Châtelet, A. (1914), Sur une communication de M. Georges Giraud, *Comptes Rendus des Séances, Société Mathématique de France* 1914 (1914) 46–48. [80]

Chaundy, T. W. (1946), The arithmetic minima of positive quadratic forms (I), *Quarterly Journal of Mathematics (Oxford)* **17** (1946) 166–192. [81]

Chebyshev, P. L. (1866), On an arithmetical question (in Russian), *Zapiski Imperatorskoj Akademii Nauk* **10** (1866) Suppl. No. 4 [French translation: Sur une question arithmétique, in: *Oeuvres de P. L. Tchebychef, Tome I* (A. A. Markoff and N. Ya. Sonin, eds.), Commissionaires de l'Académie Impériale des Sciences, St. Petersburg, 1899 [reprinted: Chelsea, New York, 1962] pp. 637–684]. [80]

Cheema, M. S. (1966), Integral solution of a system of linear equations, *The American Mathematical Monthly* **73** (1966) 487–490. [56]

Chen, D.-S. and Zionts, S. (1972), An exposition of the group theoretic approach to integer linear programming, *Opsearch* **9** (1972) 75–102. [367]

Chen, D.-S. and Zionts, S. (1976), Comparison of some algorithms for solving the group theoretic integer programming problem, *Operations Research* **24** (1976) 1120–1128. [367]

Cheng, H. C. and Pollington, A. D. (1981), A result from diophantine approximation which has applications in economics, *Manuscripta Mathematica* **35** (1981) 271–276. [67]

Chernikov, S. N. (1961), The solution of linear programming problems by elimination of unknowns (in Russian), *Doklady Akademii Nauk SSSR* **139** (1961) 1314–1317 [English translation: *Soviet Mathematics Doklady* **2** (1961) 1099–1103]. [157]

Chernikov, S. N. (1965), Contraction of finite systems of linear inequalities (in Russian), *Zhurnal Vychislitel'noi Matematiki i Matematicheskoi Fiziki* **5** (1965) 3–20. [English translation: *U.S.S.R. Computational Mathematics and Mathematical Physics* **5** (1) (1965) 1–24]. [157]

Chernikova, N. V. (1964), Algorithm for finding a general formula for the non-negative solutions of a system of linear equations (in Russian), *Zhurnal Vychislitel'noi Matematiki i Matematicheskoi Fiziki* **4** (1964) 733–738 [English translation: *U.S.S.R. Computational Mathematics and Mathematical Physics* **4** (4) (1964) 151–158]. [224]

Chernikova, N. V. (1965), Algorithm for finding a general formula for the non-negative solutions of a system of linear inequalities (in Russian), *Zhurnal Vychislitel'noi Matematiki i Matematicheskoi Fiziki* **5** (1965) 334–337 [English translation: *U.S.S.R. Computational Mathematics and Mathematical Physics* **5** (2) (1965) 228–233]. [224]

Chernikova, N. V. (1968), Algorithm for discovering the set of all the solutions of a linear programming problem (in Russian), *Zhurnal Vychislitel'noi Matematiki i Matematicheskoi Fiziki* **8** (1968) 1387–1395 [English translation: *U.S.S.R. Computational Mathematics and Mathematical Physics* **8** (6) (1968) 282–293]. [224]

Chervak, Yu. Yu. (1971), A cutting-plane method for discrete problems (in Russian), *Ukrainskii Matematicheskii Zhurnal* **23** (6) (1971) 839–843 [English translation: *Ukrainian Mathematical Journal* **23** (1971) 691–694]. [359]

Chou, T.-W. J. and Collins, G. E. (1982), Algorithms for the solution of systems of linear diophantine equations, *SIAM Journal on Computing* **11** (1982) 687–708. [59]

Christofides N. and Korman, S. (1974–5), A computational survey of methods for the set covering problem, *Management Science* **21** (1974–5) 591–599. [370]

Chuard, J. (1922), Questions d' analysis situs, *Rendiconti del Circolo Matematico di Palermo* **46** (1922) 185–224. [378]

Chvátal, V. (1973), Edmonds polytopes and a hierarchy of combinatorial problems, *Discrete Mathematics* **4** (1973) 305–337. [310, 339, 342, 349, 350, 351]

Chvátal, V. (1975a), On certain polytopes associated with graphs, *Journal of Combinatorial Theory (B)* **18** (1975) 138–154. [118, 255, 349, 350, 380]

Chvátal, V. (1975b), Some linear programming aspects of combinatorics, in: *Proceedings of the Conference on Algebraic Aspects of Combinatorics* (Toronto, 1975; D. Corneil and E. Mendelsohn, eds.), Congressus Numerantium XIII, Utilitas, Winnipeg, 1975, pp. 2–30. [349]

Chvátal, V. (1979), A greedy heuristic for the set-covering problem, *Mathematics of Operations Research* **4** (1979) 233–235. [370]

Chvátal, V. (1980), Hard knapsack problems, *Operations Research* **28** (1980) 1402–1411. [262, 362]

Chvátal, V. (1983), *Linear Programming*, Freeman, New York, 1983. [41, 224]

Chvátal, V. (1984), *Cutting-plane proofs and the stability number of a graph*, Report No. 84326, Institut für Ökonometrie und Operations Research, Rheinische Friedrich-Wilhelms-Universität, Bonn, 1984. [343, 349, 351]

Chvátal, V. (1985), Cutting planes in combinatorics, *European Journal of Combinatorics* **6** (1985) 217–226. [343, 349]

Chvátal V. and Hammer, P. L. (1977), Aggregation of inequalities in integer programming, [in: *Studies in Integer Programming* (P. L. Hammer, et al., eds.),] *Annals of Discrete Mathematics* **1** (1977) 145–162. [236]

Clarke, J. H. (1977), Conditions for the solution of a linear diophantine equation, *The New Zealand Mathematics Magazine* **14** (1977) 45–47. [379]

Clasen, B. J. (1888), Sur une nouvelle méthode de résolution des équations linéaires et sur l'application de cette méthode au calcul des déterminants, *Annales de la Sociéte Scientifique de Bruxelles* (2) **12** (1888) 251–281 [also: *Mathesis* (*Recueil Mathématique à l'Usage des Écoles Spéciales*) **9** (Suppl. 2) (1889) 1–31]. [40]

Cobham, A. (1965), The intrinsic computational difficulty of functions, in: *Logic, Methodology and Philosophy of Science* (Proc. Intern. Congress 1964; Y. Bar-Hillel, ed.), North-Holland, Amsterdam, 1965, pp. 24–30. [14, 22]

Cohn, H. (1962), *Advanced Number Theory*, Dover, New York, 1962. [51, 82]

Cohn, J. H. E. (1973), Hurwitz' theorem, *Proceedings of the American Mathematical Society* **38** (1973) 436. [67]

Collins, G. E. (1974), The computing time of the Euclidean algorithm, *SIAM Journal on Computing* **3** (1974) 1–10. [54]

Commoner, F. G. (1973), A sufficient condition for a matrix to be totally unimodular, *Networks* **3** (1973) 351–365. [272]

Conte, S. D. and de Boor, C. (1965), *Elementary Numerical Analysis, an Algorithmic Approach*, McGraw-Hill, New York, 1965. [41]

Cook, S. A. (1971), The complexity of theorem-proving procedures, in: *Proceedings of Third Annual ACM Symposium on Theory of Computing* (Shaker Heights, Ohio, 1971), ACM, New York, 1971, pp. 151–158. [14, 18, 21, 245]

Cook, S. A. (1976), *A short proof that the linear diophantine problem is in NP*, unpublished manuscript, 1976. [239]

Cook, W. (1983), Operations that preserve total dual integrality, *Operations Research Letters* **2** (1983) 31–35. [321, 322, 323, 325]

Cook, W. (1986), On box totally dual integral polyhedra, *Mathematical Programming* **34** (1986) 48–61. [318, 319, 324, 333, 335]

Cook, W., Coullard, C., and Turán, Gy. (1985), *On the complexity of cutting-plane proofs*, to appear. [344]

Cook, W., Fonlupt, J. and Schrijver, A. (1986), An integer analogue of Carathéodory's theorem, *Journal of Combinatorial Theory* (*B*) **40** (1986) 63–70. [326]

Cook, W., Gerards, A. M. H., Schrijver, A., and Tardos, É. (1986), Sensitivity theorems in integer linear programming, *Mathematical Programming* **34** (1986) 251–264. [126, 239, 241, 242, 243, 345, 346]

Cook, W., Lovász, L., and Schrijver, A. (1984), A polynomial-time test for total dual integrality in fixed dimension, *Mathematical Programming Study* **22** (1984) 64–69. [333]

van der Corput, J. G. (1931a), Ueber Systeme von linear-homogenen Gleichungen und

Ungleichungen, *Proceedings Koninklijke Akademie van Wetenschappen te Amsterdam* **34** (1931) 368–371. [217]

van der Corput, J. G. (1931b), Ueber Diophantische Systeme von linear-homogenen Gleichungen und Ungleichungen, *Proceedings Koninklijke Akademie van Wetenschappen te Amsterdam* **34** (1931) 372–382. [233, 376]

van der Corput, J. G. (1931c), Konstruktion der Minimalbasis für spezielle Diophantische Systeme von linear-homogenen Gleichungen und Ungleichungen, *Proceedings Koninklijke Akademie van Wetenschappen te Amsterdam* **34** (1931) 515–523. [233, 376]

Cottle, R. W. and Dantzig, G. B. (1968), Complementary pivot theory of mathematical programming, *Linear Algebra and Its Applications* **1** (1968) 103–125. [150]

[Cournot, A. A. (1826) =] A. C., Sur le calcul des conditions d'inégalité, annoncé par M. Fourier, *Bulletin des Sciences Mathématiques, Astronomiques, Physiques et Chimiques* **6** (1826) 1–8. [212]

[Cournot, A. A. (1827) =] A.C., Extension du principe des vitesses virtuelles au cas où les conditions de liaison du système sont exprimées par des inégalités, *Bulletin des Sciences Mathématiques, Astronomiques, Physiques et Chimiques* **8** (1827) 165–170. [210]

Cournot, A. A. (1830a), Mémoire sur le mouvement d'un corps rigide, soutenu par un plan fixe, *Journal für die reine und angewandte Mathematik* **5** (1830) 133–162. [211]

Cournot, A. A. (1830b), Du mouvement d'un corps sur un plan fixe, quand on a égard à la résistance du frottement, et qu'on ne suppose qu'un seul point de contact, *Journal für die reine und angewandte Mathematik* **5** (1830) 223–249. [211]

Cournot, A. A. (1832), Du mouvement d'un corps sur un plan fixe, quand on a égard à la résistance du frottement, *Journal für die reine und angewandte Mathematik* **8** (1832) 1–12. [211]

Cover, T. M. and Efron, B. (1967), Geometrical probability and random points on a hypersphere, *The Annals of Mathematical Statistics* **38** (1967) 213–220. [147]

Coveyou, R. R. and Macpherson, R. D. (1967), Fourier analysis of uniform random number generators, *Journal of the Association for Computing Machinery* **14** (1967) 100–119. [72]

Coxeter, H. S. M. (1947), [Review of] Chaundy, T. W. The arithmetic minima of positive quadratic forms. I., *Mathematical Reviews* **8** (1947) 137–138. [81]

Coxeter, H. S. M. (1948), *Regular Polytopes*, Methuen & Co., London, 1948 [second edition: Collier-Macmillan, London. 1963 [reprinted: Dover, New York, 1973]]. [223]

Cramer, G. (1750), *Introduction à l'analyse des lignes courbes algébriques*, Les frères Cramer et C. Philibert, Genève, 1750. [translated partially in: H. Midonick, *The Treasury of Mathematics: 2*, A Pelican Book, Penguin Books, Harmondsworth, 1968, pp. 311–319]. [29, 39]

Crowder, H. and Hattingh, J. M. (1975), Partially normalized pivot selection in linear programming, *Mathematical Programming Study* **4** (1975) 12–25. [139, 150]

Crowder, H. P. and Johnson, E. L. (1973), Use of cyclic group methods in branch and bound, in: *Mathematical Programming* (T. C. Hu and S. M. Robinson, eds.), Academic Press, New York, 1973, pp. 213–226. [367]

Crowder, H., Johnson, E. L., and Padberg, M. W. (1983), Solving large-scale zero-one linear programming problems, *Operations Research* **31** (1983) 803–834. [359, 373]

Crowder, H. and Padberg, M. W. (1980), Solving large-scale symmetric travelling salesman problems to optimality, *Management Science* **26** (1980) 495–509. [371]

Cunningham, W. H. (1976), A network simplex method, *Mathematical Programming* **11** (1976) 105–116. [142]

Cunningham, W. H. (1979), Theoretical properties of the network simplex method, *Mathematics of Operations Research* **4** (1979) 196–208. [142]

Cunningham, W. H. (1982), Separating cocircuits in binary matroids, *Linear Algebra and Its Applications* **43** (1982) 69–86. [287]

Cunningham, W. H. and Edmonds, J. (1980), A combinatorial decomposition theory, *Canadian Journal of Mathematics* **32** (1980) 734–765. [287, 292]

Cunningham, W. H. and Klincewicz, J. G. (1983), On cycling in the network simplex method, *Mathematical Programming* **26** (1983) 182–189. [142]

Curran Sharp, W. J. (1884), [Solution to Problem 7382,] *Mathematics from the Educational Times with Solutions* **41** (1884) 21. [376]

Cusick, T. W. (1971), Diophantine approximation of ternary linear forms, *Mathematics of Computation* **25** (1971) 163–180. [74]

Cusick, T. W. (1972a), Formulas for some diophantine approximation constants, *Mathematische Annalen* **197** (1972) 182–188. [74]

Cusick, T. W. (1972b), Simultaneous diophantine approximation of rational numbers, *Acta Arithmetica* **22** (1972) 1–9. [74]

Cusick, T. W. (1974), Formulas for some diophantine approximation constants, II, *Acta Arithmetica* **26** (1974) 117–128. [74]

Cusick, T. W. (1977a), The Szekeres multidimensional continued fraction, *Mathematics of Computation* **31** (1977) 280–317. [74]

Cusick, T. W. (1977b), Dirichlet's diophantine approximation theorem, *Bulletin of the Australian Mathematical Society* **16** (1977) 219–224. [74]

Cusick, T. W. (1980), Best diophantine approximations for ternary linear forms, *Journal für die reine und angewandte Mathematik* **315** (1980) 40–52. [74]

Dakin, R. J. (1965), A tree-search algorithm for mixed integer programming problems, *The Computer Journal* **8** (1965) 250–255. [360]

Dantzig, G. B. (1951a), Maximization of a linear function of variables subject to linear inequalities, in: *Activity Analysis of Production and Allocation* (Tj. C. Koopmans, ed.), Wiley, New York, 1951, pp. 339–347. [129, 137, 138, 223]

Dantzig, G. B. (1951b), Application of the simplex method to a transportation problem, in: *Activity Analysis of Production and Allocation* (Tj. C. Koopmans, ed.), Wiley, New York, 1951, pp. 359–373. [98, 142, 275, 378]

Dantzig, G. B. (1951c), A proof of the equivalence of the programming problem and the game problem, in: *Activity Analysis of Production and Allocation* (Tj. C. Koopmans, ed.), Wiley, New York, 1951, pp. 330–335. [218]

Dantzig, G. B. (1953), *Computational algorithm of the revised simplex method*, Report RM 1266, The Rand Corporation, Santa Monica, Cal., 1953. [148]

Dantzig, G. B. (1957), Discrete-variable extremum problems, *Operations Research* **5** (1957) 266–277. [261, 378, 379]

Dantzig, G. B. (1959), Note on solving linear programs in integers, *Naval Research Logistics Quarterly* **6** (1959) 75–76. [359]

Dantzig, G. B. (1960), On the significance of solving linear programming problems with some integer variables, *Econometrica* **28** (1960) 30–44. [378]

Dantzig, G. B. (1963), *Linear Programming and Extensions*, Princeton University Press, Princeton, N.J., 1963. [139, 142, 223, 224]

Dantzig, G. B. (1980), *Expected number of steps of the simplex method for a linear program with a convexity constraint*, Tech. Report SOL 80–3, Systems Optimization Laboratory, Dept. of Operations Research, Stanford University, Stanford, Cal., 1980. [147]

Dantzig, G. B. (1982), Reminiscences about the origins of linear programming, *Operations Research Letters* **1** (1982) 43–48 [also in: *Mathematical Programming — The State of the Art, Bonn 1982* (A. Bachem, M. Grötschel and B. Korte, eds.), Springer, Berlin, 1983, pp. 78–86]. [223]

Dantzig, G. B., Dempster, M. A. H., and Kallio, M. (eds.) (1981), *Large-scale Linear Programming*, International Institute for Applied Systems Analysis, Laxenburg (Austria), 1981. [224]

Dantzig, G. B. and Eaves, B. C. (1973), Fourier-Motzkin elimination and its dual, *Journal of Combinatorial Theory (A)* **14** (1973) 288–297. [157, 379]

Dantzig, G. B., Ford, L. R., and Fulkerson, D. R. (1956), A primal-dual algorithm for linear programs, in: *Linear Inequalities and Related Systems* (H. W. Kuhn and A. W. Tucker, eds.), Princeton University Press, Princeton, N.J., 1956, pp. 171–181. [151]

Dantzig, G. B. and Fulkerson, D. R. (1956), On the max-flow min-cut theorem of networks, in: *Linear Inequalities and Related Systems* (H. W. Kuhn and A. W. Tucker, eds.), Princeton University Press, Princeton, N.J., 1956, pp. 215–221. [274, 378]

Dantzig, G., Fulkerson, R., and Johnson, S. (1954), Solution of a large-scale traveling-salesman problem, *Operations Research* **2** (1954) 393–410. [371, 378].

Dantzig, G. B., Fulkerson, D. R., and Johnson, S. M. (1959), On a linear-programming, combinatorial approach to the traveling-salesman problem, *Operations Research* **7** (1959) 58–66. [370, 378]

Dantzig, G. B. and Orchard-Hays, W. (1954), The product form for the inverse in the simplex method, *Mathematical Tables and Other Aids to Computation* **8** (1954) 64–67. [148]

Dantzig, G. B., Orden, A., and Wolfe, P. (1955), The generalized simplex method for minimizing a linear form under linear inequality restraints, *Pacific Journal of Mathematics* **5** (1955) 183–195. [138]

Dantzig, G. B. and Wolfe, Ph. (1960), Decomposition principle for linear programs, *Operations Research* **8** (1960) 101–111. [367]

Dantzig, G. B. and Wolfe, Ph. (1961), The decomposition algorithm for linear programs, *Econometrica* **29** (1961) 767–778. [367]

Danzer, L., Grünbaum, B., and Klee, V. (1963), Helly's theorem and its relatives, in: *Convexity* (V. L. Klee, ed.), American Mathematical Society, Providence, R.I., 1963, pp. 101–180. [205]

Danzer, L., Laugwitz, D., and Lenz, H. (1957), Über das Löwnersche Ellipsoid und sein Analogon unter den einem Eikörper einbeschriebenen Elliposoiden, *Archiv der Mathematik (Basel)* **8** (1957) 214–219. [205]

Davenport, H. (1952), *The Higher Arithmetic, An Introduction to the Theory of Numbers*, Hutchinson, New York, 1952. [82]

Davenport, H. (1955), On a theorem of Furtwängler, *The Journal of The London Mathematical Society* **30** (1955) 186–195 [reprinted in: *The Collected Works of Harold Davenport, Vol. II* (B. J. Birch, H. Halberstam and C. A. Rogers, eds.), Academic Press, London, 1977, pp. 659–668]. [74]

Davenport, H. and Mahler, K. (1946), Simultaneous diophantine approximation, *Duke Mathematical Journal* **13** (1946) 105–111 [reprinted in: *The Collected Works of Harold Davenport, Vol. I* (B. J. Birch, H. Halberstam and C. A. Rogers, eds.), Academic Press, London, 1977, pp. 143–149]. [74]

Davenport, H. and Schmidt, W. M. (1967), Approximation to real numbers by quadratic irrationals, *Acta Arithmetica* **13** (1967) 169–176 [reprinted in: *The Collected Works of Harold Davenport, Vol. II* (B. J. Birch, H. Halberstam and C. A. Rogers, eds.), Academic Press, London, 1977, pp. 770–777]. [74]

Davenport, H. and Schmidt, W. M. (1969–70), Dirichlet's theorem on diophantine approximation. II, *Acta Arithmetica* **16** (1969–70) 413–424 [reprinted in: *The Collected Works of Harold Davenport, Vol. II* (B. J. Birch, H. Halberstam and C. A. Rogers, eds.), Academic Press, London, 1977, pp. 849–860]. [74]

Davenport, H. and Schmidt, W. M. (1970), Dirichlet's theorem on diophantine approxim-ation, in: *Teoria dei Numeri* (9–12 dicembre 1968), Symposia Mathematica Vol. IV, Academic Press, London, 1970, pp. 113–132 [reprinted in: *The Collected Works of Harold Davenport, Vol. II* (B. J. Birch, H. Halberstam and C. A. Rogers, eds.), Academic Press, London, 1977, pp. 829–848]. [74]

Debreu, G. (1964), Nonnegative solutions of linear inequalities, *International Economic Review* **5** (1964) 178–184. [86]

Dévai, F. and Szendrényi, T. (1979), Comments on convex hull of a finite set of points in two dimensions, *Information Processing Letters* **9** (1979) 141–142. [224]

Devine, M. and Glover, F. (1973), Computational design and numerical results for generating the nested faces of the Gomory polyhedron, *Opsearch* **10** (1973) 143–160. [367]

Devroye, L. (1980), A note on finding convex hulls via maximal vectors, *Information*

Processing Letters **11** (1980) 53–56. [224]

Dickson, J. C. and Frederick, F. P. (1960), A decision rule for improved efficiency in solving linear programming problems with the simplex algorithm, *Communications of the ACM* **3** (1960) 509–512. [138]

Dickson, L. E. (1920), *History of the Theory of Numbers, Vol. II, Diophantine Analysis,* Carnegie Institution of Washington, Washington, 1920 [reprinted: Chelsea, New York, 1966]. [77,82,375]

Dickson, L. E. (1923), *History of the Theory of Numbers, Vol. III, Quadratic and Higher Forms,* Carnegie Institution of Washington, Washington, 1923 [reprinted: Chelsea, New York, 1966]. [82]

Dickson, L. E. (1929), *Introduction to the Theory of Numbers,* The University of Chicago Press, Chicago, Ill., 1929. [82]

Dickson, L. E. (1930), *Studies in the Theory of Numbers,* The University of Chicago Press, Chicago, Ill., 1930 [reprinted: Chelsea, New York, 1957]. [82]

Dickson, L. E. (1939), *Modern Elementary Theory of Numbers,* The University of Chicago Press, Chicago, Ill., 1939. [82]

Dieter, U. (1975), How to calculate the shortest vectors in a lattice, *Mathematics of Computation* **29** (1975) 827–833. [72]

Dijkstra, E. W. (1959), A note on two problems in connexion with graphs, *Numerische Mathematik* **1** (1959) 269–271. [12]

Dilworth, R. P. (1950), A decomposition theorem for partially ordered sets, *Annals of Mathematics* (2) **51** (1950) 161–166. [275]

Dines, L. L. (1917), Concerning preferential voting, *The American Mathematical Monthly* **24** (1917) 321–325. [157,217]

Dines, L. L. (1918–9), Systems of linear inequalities, *Annals of Mathematics* (2) **20** (1918–9) 191–199. [155,157,217]

Dines, L. L. (1925), Definite linear dependence, *Annals of Mathematics* (2) **27** (1925) 57–64. [157,217]

Dines, L. L. (1926–7), Note on certain associated systems of linear equalities and inequalities, *Annals of Mathematics* (2) **28** (1926–7) 41–42. [217]

Dines, L. L. (1927a), On positive solutions of a system of linear equations, *Annals of Mathematics* (2) **28** (1927) 386–392. [157,217]

Dines, L. L. (1927b), On sets of functions of a general variable, *Transactions of the American Mathematical Society* **29** (1927) 463–470. [217]

Dines, L. L. (1927c), On completely signed sets of functions, *Annals of Mathematics* (2) **28** (1927) 393–395. [217]

Dines, L. L. (1930), Linear inequalities and some related properties of functions, *Bulletin of the American Mathematical Society* **36** (1930) 393–405. [217]

Dines, L. L. (1936), Convex extension and linear inequalities, *Bulletin of the American Mathematical Society* **42** (1936) 353–365. [217]

Dines, L. L. and McCoy, N. H. (1933), On linear inequalities, *Proceedings and Transactions of the Royal Society of Canada* (3) **27** (1933) 37–70. [157, 217, 223]

Dinits, E. A. (1970), Algorithm for solution of a problem of maximum flow in a network with power estimation (in Russian), *Doklady Akademii Nauk SSSR* **194** (1970) 754–757 [English translation: *Soviet Mathematics Doklady* **11** (1970) 1277–1280]. [154]

Dirichlet, G. Lejeune (1842), Verallgemeinerung eines Satzes aus der Lehre von den Kettenbrüchen nebst einigen Anwendungen auf die Theorie der Zahlen, *Bericht über die zur Bekanntmachung geeigneten Verhandlungen der Königlich Preussischen Akademie der Wissenschaften zu Berlin* (1842) 93–95 [reprinted in: *G. Lejeune Dirichlet's Werke, Vol. I* (L. Kronecker, ed.), G. Reimer, Berlin, 1889, [reprinted: Chelsea, New York, 1969,] pp. 635–638]. [60,72,78]

Dirichlet, G. Lejeune (1850), Über die Reduction der positiven quadratischen Formen mit drei unbestimmten ganzen Zahlen, *Journal für die reine und angewandte Mathematik* **40** (1850) 209–227 [reprinted in: *G. Lejeune Dirichlet's Werke, Vol. II* (L. Kronecker, ed.),

G. Reimer, Berlin, 1889, [reprinted: Chelsea, New York, 1969,] pp. 29–48]. [78]

Dirichlet, P. G. Lejeune (1879), *Vorlesungen über Zahlentheorie, herausgegeben und mit Zusätzen versehen von R. Dedekind*, Dritte Auflage, Vieweg, Braunschweig, 1879 [reprinted: Chelsea, New York, 1968]. [82]

Dixon, J. D. (1970), The number of steps in the Euclidean algorithm, *Journal of Number Theory* 2 (1970) 414–422. [54,67]

Dixon, J. D. (1971), A simple estimate for the number of steps in the Euclidean algorithm, *The American Mathematical Monthly* 78 (1971) 374–376. [54]

Dixon, J. D. (1981), On duality theorems, *Linear Algebra and Its Applications* 39 (1981) 223–228. [224]

Dobkin, D., Lipton, R. J., and Reiss, S. (1979), Linear programming is log-space hard for P, *Information Processing Letters* 8 (1979) 96–97. [225]

Doignon, J.-P. (1973), Convexity in cristallographical lattices, *Journal of Geometry* 3 (1973) 71–85. [234]

Domich, P. D. (1983), *Three new polynomially-time bounded Hermite normal form algorithms*, M.Sc. Thesis, School of Operations Research and Industrial Engineering, Cornell University, Ithaca, N.Y., 1983. [56]

Domich, P. D., Kannan, R., and Trotter, Jr, L. E. (1985), *Hermite normal form computation using modulo determinant arithmetic*, CORE Discussion Paper 8507, Center for Operations Research and Econometrics, Louvain-la-Neuve (Belgium), 1985. [56]

Donaldson, J. L. (1979), Minkowski reduction of integral matrices, *Mathematics of Computation* 33 (1979) 201–216. [56]

Dorfman, R. (1984), The discovery of linear programming, *Annals of the History of Computing* 6 (1984) 283–295. [223]

Dorfman, R., Samuelson P. A., and Solow, R. M. (1958), *Linear Programming and Economic Analysis*, McGraw-Hill, New York, 1958. [219]

Dreyfus, S. E. (1957), *Dynamic programming solution of allocation problems*, RAND Paper P-1083, The RAND Corporation, Santa Monica, Cal., 1957. [378]

Drezner, Z. (1983), The nested ball principle for the relaxation method, *Operations Research* 31 (1983) 587–590. [162]

Driebeek, N. J. (1969), *Applied Linear Programming*, Addison-Wesley, Reading, Mass., 1969. [224]

Dubois, E. and Rhin, G. (1980), Approximations simultanées de deux nombres réels, in: *Séminaire Delange-Pisot-Poitou, 20e année: 1978/79* —: — *Théorie des nombres* —: — *Fascicule 1: Exposés 1 à 21*, Université Pierre et Marie Curie, Institut Henri Poincaré, Secrétariat mathématique, Paris, 1980, [Exposé 9] pp. 9–01—9–13. [74]

Dubois, E. and Rhin, G. (1982), Meilleures approximations d'une forme linéaire cubique, *Acta Arithmetica* 40 (1982) 197–208. [74]

Duffin, R. J. (1967), An orthogonality theorem of Dines related to moment problems and linear programming, *Journal of Combinatorial Theory* 2 (1967) 1–26. [157]

Duffin, R. J. (1974), On Fourier's analysis of linear inequality systems, *Mathematical Programming Study* 1 (1974) 71–95. [157]

Dulmage, A. L. and Mendelsohn, N. S. (1964), Gaps in the exponent set of primitive matrices, *Illinois Journal of Mathematics* 8 (1964) 642–656. [379]

Dunham, J. R., Kelly, D. G., and Tolle, J. W. (1977), *Some experimental results concerning the expected number of pivots for solving randomly generated linear programs*, Technical Report 77–16, Operations Research and System Analysis Department, University of North Carolina at Chapel Hill, 1977. [139]

Dyer, M. E. (1983), The complexity of vertex enumeration methods, *Mathematics of Operations Research* 8 (1983) 381–402. [224]

Dyer, M. E. (1984a), Linear time algorithms for two- and three-variable linear programs, *SIAM Journal on Computing* 13 (1984) 31–45. [199]

Dyer, M. E. (1984b), An $O(n)$ algorithm for the multiple-choice knapsack linear program, *Mathematical Programming* 29 (1984) 57–63. [199]

Dyer, M. E. and Proll, L. G. (1977), An algorithm for determining all extreme points of a convex polytope, *Mathematical Programming* 12 (1977) 81–96. [224]

Easterfield, T. E. (1946), A combinatorial algorithm, *The Journal of the London Mathematical Society* **21** (1946) 219–226. [378]

Eberhard, V. (1891), *Zur Morphologie der Polyeder*, Teubner, Leipzig, 1891. [223]

Eddy, W. F. (1977a), A new convex hull algorithm for planar sets, *ACM Transactions on Mathematical Software* **3** (1977) 398–403. [224]

Eddy, W. F. (1977b), Algorithm 523 CONVEX, A new convex hull algorithm for planar sets, *ACM Transactions on Mathematical Software* **3** (1977) 411–412. [224]

Edmonds, J. (1965a), Paths, trees, and flowers, *Canadian Journal of Mathematics* **17** (1965) 449–467. [14, 18, 22, 350]

Edmonds, J. (1965b), Minimum partition of a matroid into independent subsets, *Journal of Research of the National Bureau of Standards (B)* **69** (1965) 67–72. [18, 22]

Edmonds, J. (1965c), Maximum matching and a polyhedron with 0, 1-vertices, *Journal of Research of the National Bureau of Standards (B)* **69** (1965) 125–130. [109, 118, 251, 349, 350]

Edmonds, J. (1967a), Optimum branchings, *Journal of Research of the National Bureau of Standards (B)* **71** (1967) 233–240. [182, 371]

Edmonds, J. (1967b), Systems of distinct representatives and linear algebra, *Journal of Research of the National Bureau of Standards (B)* **71** (1967) 241–245. [32, 34]

Edmonds, J. (1973), Edge-disjoint branchings, in: *Combinatorial Algorithms* (Courant Computer Science Symposium, Monterey, Cal., 1972; R. Rustin, ed.), Academic Press, New York, 1973, pp. 91–96. [314]

Edmonds, J. and Giles, R. (1977), A min-max relation for submodular functions on graphs, [in: *Studies in Integer Programming* (P. L. Hammer, et al., eds.),] *Annals of Discrete Mathematics* **1** (1977) 185–204. [278, 309, 310, 311, 313, 320, 326]

Edmonds, J. and Giles, R. (1984), Total dual integrality of linear inequality systems, in: *Progress in Combinatorial Optimization* (Jubilee Conference, University of Waterloo, Waterloo, Ontario, 1982; W. R. Pulleyblank, ed.), Academic Press, Toronto, 1984, pp. 117–129. [320, 324, 333]

Edmonds, J. and Johnson, E. L. (1973), Matching, Euler tours and the Chinese postman, *Mathematical Programming* **5** (1973) 88–124. [348, 349]

Edmonds, J. and Karp, R. M. (1972), Theoretical improvements in algorithmic efficiency for network flow problems, *Journal of the Association for Computing Machinery* **19** (1972) 248–264. [142, 154, 195, 279]

Edmonds, J., Lovász, L., and Pulleyblank, W. R. (1982), Brick decompositions and the matching rank of graphs, *Combinatorica* **2** (1982) 247–274. [183, 186]

Egerváry, E. (1931), Matrixok kombinatorius tulajdonságairól (Hungarian) [On combinatorial properties of matrices], *Matematikai és Fizikai Lapok* **38** (1931) 16–28. [273, 329, 378]

Egerváry E. (1958), Bemerkungen zum Transportproblem, *MTW Mitteilungen* **5** (1958) 278–284. [378]

Eggleston, H. G. (1958), *Convexity*, Cambridge University Press, Cambridge, 1958. [224]

Ehrhart, E. (1967a), Sur un problème de géométrie diophantienne linéaire. I, Polyèdres et réseaux, *Journal für die reine und angewandte Mathematik* **226** (1967) 1–29. [379]

Ehrhart, E. (1967b), Sur un problème de géométrie diophantienne linéaire. II, Systèmes diophantiens linéaires, *Journal für die reine und angewandte Mathematik* **227** (1967) 25–49. [379]

Ehrhart, E. (1973), Sur les systèmes d'inéquations diophantiennes linéaires, *Journal für die reine und angewandte Mathematik* **262/263** (1973) 45–57. [379]

Ehrhart, E. (1977), *Polynômes Arithmétiques et Méthode des Polyèdres en Combinatoire*, International Series of Numerical Mathematics 35, Birkhäuser, Basel, 1977. [379]

Ehrhart, E. (1979), Sur les équations diophantiennes linéaires, *Comptes Rendus des Séances de l'Académie des Sciences (Paris) Séries A-B* **288** (1979) A785-A787. [379]

Eisemann, K. (1955), Linear programming, *Quarterly of Applied Mathematics* **13** (1955) 209–232. [148]

Eisenstein, G. (1847), Neue Theoreme der höheren Arithmetik, *Journal für die reine und angewandte Mathematik* **35** (1847) 117–136, [reprinted in: *Mathematische Werke—Gotthold Eisenstein, Vol. I*, Chelsea, New York, 1975, pp. 483–502]. [79]

Elias, P., Feinstein, A., and Shannon, C. E. (1956), A note on the maximum flow through a network, *IRE Transactions on Information Theory* IT **2** (1956) 117–119. [97, 115, 275, 378]

Elliott, E. B. (1903), On linear homogeneous diophantine equations, *Quarterly Journal of Pure and Applied Mathematics* **34** (1903) 348–377. [376]

Ellwein, L. B. (1974), A flexible enumeration scheme for zero-one programming, *Operations Research* **22** (1974) 144–150. [363]

van Emde Boas, P. (1980), On the $\Omega(n\log n)$ lower bound for convex hull and maximal vector determination, *Information Processing Letters* **10** (1980) 132–136. [224]

van Emde Boas, P. (1981), *Another NP-complete partition problem and the complexity of computing short vectors in a lattice*, Report 81–04, Mathematical Institute, University of Amsterdam, Amsterdam, 1981. [72]

Erdös, P. (1961), Graph theory and probability. II, *Canadian Journal of Mathematics* **13** (1961) 346–352 [reprinted in: *Paul Erdös, The Art of Counting, Selected Writings* (J. Spencer, ed.), The MIT Press, Cambridge, Mass., 1973, pp. 411–417]. [349]

Erdös, P. and Graham, R. L. (1972), On a linear diophantine problem of Frobenius, *Acta Arithmetica* **21** (1972) 399–408. [379]

Euler, L. (1740), Solutio problematis arithmetici de inveniendo numero qui per datos numeros divisus, relinquat data residua, *Commentarii Academiae Scientiarum Imperialis Petropolitanae* **7** [1734–5] (1740) 46–66 [reprinted in: *Leonhardi Euleri Opera Omnia, Ser. I Vol. II* [*Commentationes Arithmeticae, Vol. I*] (F. Rudio, ed.), Teubner, Leipzig, 1915, pp. 18–32]. [77]

Euler, L. (1744), De fractionibus continuis, dissertatio, *Commentarii Academiae Scientiarum Imperialis Petropolitanae* **9** [1737] (1744) 98–137 [reprinted in: *Leonhardi Euleri Opera Omnia, Ser. I Vol. XIV* [*Commentationes Analyticae, ad theoriam serierum infinitarum pertinentes, Vol. I*] (C. Boehm and G. Faber, eds.), Teubner, Leipzig, 1924, pp. 187–215]. [77]

Euler, L. (1748), *Introductio in Analysin Infinitorum, Vol. I*, M.-M. Bousquet, Lausanne, 1748 [reprinted as: *Leonhardi Euleri Opera Omnia, Ser. I, Vol. VIII* (A. Krazer and F. Rudio, eds.), Teubner, Leipzig, 1922] [German translation by H. Maser: *Einleitung in die Analysis des Unendlichen, Erster Teil*, Springer, Berlin, 1885 [reprinted: 1983]]. [375]

Euler, L. (1770a), *Vollstaendige Anleitung zur Algebra, 2. Theil, Von Auflösung algebraischer Gleichungen und der unbestimmten Analytic*, Kays. Acad. der Wissenschaften, St. Petersburg, 1770 [reprinted in: *Leonhardi Euleri Opera Omnia, Ser. I, Vol. I* [*Vollständige Anleitung zur Algebra, mit den Zusätzen von Joseph Louis Lagrange*] (H. Weber, ed.), Teubner, Leipzig, 1911, pp. 209–498] [English translation (with 1. Theil): *Elements of Algebra*, J. Johnson, London, 1797 [fifth edition (1840) reprinted: Springer, New York, 1984]]. [77, 375]

Euler, L. (1770b), De partitione numerorum in partes tam numero quam specie dates, *Novi Commentarii Academiae Scientiarum Imperialis Petropolitanae* **14** [1769]: I (1770) 168–187 [reprinted in: *Leonhardi Euleri Opera Omnia, Ser. I, Vol. III* [*Commentationes Arithmeticae, Vol. II*] (F. Rudio, ed.), Teubner, Leipzig, 1917, pp. 132–147]. [375]

Everett III, H. (1963), Generalized Lagrange multiplier method for solving problems of optimum allocation of resources, *Operations Research* **11** (1963) 399–417. [370]

Eves, H. (1964), *An Introduction to the History of Mathematics*, Holt, Rinehart and Winston, New York, 1964. [41]

Eves, H. (1966), *Elementary Matrix Theory*, Allyn and Bacon, Boston, Mass., 1966. [40]

Faaland, B. (1972), On the number of solutions to a diophantine equation, *Journal of Combinatorial Theory (A)* **13** (1972) 170–175. [379]

Faaland, B. (1973–4), Generalized equivalent integer programs and canonical problems,

Management Science **20** (1973-4) 1554-1560. [380]

Faddeev, D. K. and Faddeeva, V. N. (1963), *Computational Methods of Linear Algebra*, Freeman, San Francisco, 1963. [41]

Faddeeva, V. N. (1959), *Computational Methods of Linear Algebra*, Dover, New York, 1959. [41]

Farkas, Gy. (1894), A Fourier-féle mechanikai elv alkalmazásai (Hungarian) [On the applications of the mechanical principle of Fourier], *Mathematikai és Természettudományi Értesitö* **12** [1893-4] (1894) 457-472 [German translation, with slight alterations: Farkas [1895]]. [85, 89, 214]

Farkas, J. (1895), Über die Anwendungen des mechanischen Princips von Fourier, *Mathematische und naturwissenschaftliche Berichte aus Ungarn* **12** (1895) 263-281. [214]

Farkas, Gy. (1896), A Fourier-féle mechanikai elv alkalmazásainak algebrai alapjáról (Hungarian) [On the algebraic foundation of the applications of the mechanical principle of Fourier], *Mathematikai és Physikai Lapok* **5** (1896) 49-54 [German translation, with some additions: Section I of Farkas [1897-9]]. [156, 214]

Farkas, J. (1897-9), Die algebraischen Grundlagen der Anwendungen des Fourier'schen Princips in der Mechanik, *Mathematische und naturwissenschaftliche Berichte aus Ungarn* **15** (1897-9) 25-40. [214]

Farkas, Gy. (1898a), Paraméteres módszer Fourier mechanikai elvéhez (Hungarian) [A parametric method for the mechanical principle of Fourier], *Mathematikai és Physikai Lapok* **7** (1898) 63-71 [German translation: Section II of Farkas [1897-9]]. [85, 87, 89, 157, 214]

Farkas, Gy. (1898b), A Fourier-féle mechanikai elv alkalmazásának algebrai alapja (Hungarian) [The algebraic foundation of the applications of the mechanical principle of Fourier], *Mathematikai és Természettudományi Értesitö* **16** (1898) 361-364 [German translation: Farkas [1898c]]. [214]

Farkas, J. (1898c), Die algebraische Grundlage der Anwendungen des mechanischen Princips von Fourier, *Mathematische und naturwissenschaftliche Berichte aus Ungarn* **16** (1898) 154-157. [214]

Farkas, J. (1900a), Allgemeine Principien für die Mechanik des Aethers, *Archives Néerlandaises des Sciences Exactes et Naturelles* (2) **5** (1900) 56-75. [214]

Farkas, Gy. (1900b), *Vectortan és az egyszerü inaequatiok tana* (Hungarian) [Theory of vectors and simple inequalities], Kolozsvár, 1900. [214]

Farkas, J. (1902), Theorie der einfachen Ungleichungen, *Journal für die reine und angewandte Mathematik* **124** (1902) 1-27. [87, 214]

Farkas, J. (1906), Beiträge zu den Grundlagen der analytischen Mechanik, *Journal für die reine und angewandte Mathematik* **131** (1906) 165-201. [214]

Farkas, Gy. (1914), *A Mechanika Alaptanai* (Hungarian) [The Foundations of Mechanics], Kolozsvári egyetemi elöadások könyomata, Kolozsvár, 1914. [214]

Farkas, Gy. (1917a), Nemvonalas egyenlötlenségek vonalassá tétele (Hungarian) [Conversion of nonlinear inequalities into linear ones], *Mathematikai és Természettudományi Értesitö* **35** (1917) 41-50. [214]

Farkas, Gy. (1917b), Multiplicatoros módszer négyzetes alakokhoz (Hungarian) [Method of multipliers for quadratic forms], *Mathematikai és Természettudományi Értesitö* **35** (1917) 51-53. [214]

Farkas, Gy. (1918a), Egyenlötlenségek alkalmazásának új módjai (Hungarian) [New ways of applying inequalities], *Mathematikai és Természettudományi Értesitö* **36** (1918) 297-308. [214]

Farkas, Gy. (1918b), A linearis egyenlötlenségek következményei (Hungarian) [Consequences of linear inequalities], *Mathematikai és Természettudományi Értesitö* **36** (1918) 397-408. [214]

Farkas, J. (1923), Stabiles Gleichgewicht ohne Potential, *Mathematische und naturwissenschaftliche Berichte aus Ungarn* **32** (1923) 43-50. [214]

Farkas, Gy. (1926), Alapvetés az egyszerü egyenlötlenségek vektorelméletéhez (Hungarian) [Foundations of the vectorial theory of simple inequalities], *Mathematikai és Természettudományi Értesitö* **43** (1926) 1–3. [214]

Fayard, D. and Plateau, G. (1973), Résolution d'un problème d'affectation, *Électricité de France, Bulletin de la Direction des Études et Recherches, Série C (Mathématiques* * *Informatique*) 1973 No. 1 (1973) 83–108. [236]

Fayard, D. and Plateau, G. (1975), Resolution of the 0–1 knapsack problem: comparison of methods, *Mathematical Programming* **8** (1975) 272–307. [373]

Fayard, D. and Plateau, G. (1977–8), On "An efficient algorithm for the 0–1 knapsack problem, by Robert M. Nauss", *Management Science* **24** (1977–8) 918–919. [373]

Feit, S. D. (1984), A fast algorithm for the two-variable integer programming problem, *Journal of the Association for Computing Machinery* **31** (1984) 99–113. [261]

Fenchel, W. (1953), *Convex Cones, Sets, and Functions* [from notes by D. W. Blackett of lectures at Princeton University, 1951], report, Princeton University, Princeton, N.J., 1953. [224]

Fenchel, W. (1983), Convexity through the ages, in: *Convexity and Its Applications* (P. M. Gruber and J. M. Wills, eds.), Birkhäuser, Basel, 1983, pp. 120–130. [224]

Ferguson, H. R. P. and Forcade, R. W. (1979), Generalization of the Euclidean algorithm for real numbers to all dimensions higher than two, *Bulletin (New Series) of the American Mathematical Society* **1** (1979) 912–914. [74]

Ferguson, R. O. and Sargent, L. F. (1958), *Linear Programming: Fundamentals and Applications*, McGraw-Hill, New York, 1958. [224]

Finkel'shtein, Yu. Yu. (1970), Estimation of the number of iterations for Gomory's all-integer algorithm (in Russian), *Doklady Akademii Nauk SSSR* **193** (1970) 543–546 [English translation: *Soviet Mathematics Doklady* **11** (1970) 988–992]. [359]

Fiorot, J.-Ch. (1969), Génération des points entiers d'un cône polyédrique, *Comptes Rendus Hebdomadaires des Séances de l'Académie des Sciences Sér. A* **269** (1969) 215–217. [379]

Fiorot, J.-Ch. (1970a), Algorithme de génération des points entiers d'un cône polyédrique, *Électricité de France, Bulletin de la Direction des Études et Recherches, Série C (Mathématiques* * *Informatique*) 1970 No. 1 (1970) 5–28. [379]

Fiorot, J.-Ch. (1970b), Génération des points entiers d'un parallélotope de R^n, *Comptes Rendus Hebdomadaires des Séances de l'Académie des Sciences Sér. A* **270** (1970) 395–398. [379]

Fiorot, J. C. (1972), Generation of all integer points for given sets of linear inequalities, *Mathematical Programming* **3** (1972) 276–295. [379]

Fiorot, J.-Ch. and Gondran, M. (1969), Résolution des systèmes linéaires en nombres entiers, *Électricité de France, Bulletin de la Direction des Études et Recherches, Série C (Mathématiques* * *Informatique*) 1969 No. 2 (1969) 65–116. [59]

Fischer, R. and Schweiger, F. (1975), The number of steps in a finite Jacobi algorithm, *Manuscripta Mathematica* **17** (1975) 291–308. [74]

Fisher, I. (1938), Cournot forty years ago, *Econometrica* **6** (1938) 198–202. [211]

Fisher, M. L. (1981), The Lagrangian relaxation method for solving integer programming problems, *Management Science* **27** (1981) 1–18. [368, 370]

Fisher, M. L., Northup, W. D., and Shapiro, J. F. (1975), Using duality to solve discrete optimization problems: theory and computational experience, *Mathematical Programming Study* **3** (1975) 56–94. [354]

Fisher, M. L. and Shapiro, J. F. (1974), Constructive duality in integer programming, *SIAM Journal on Applied Mathematics* **27** (1974) 31–52. [354, 367]

Fleischmann, B. (1967), Computational experience with the algorithm of Balas, *Operations Research* **15** (1967) 153–155. [363]

Fleischmann, B. (1973), Eine primale Version des Benders'schen Dekompositionsverfahrens und seine Anwendung in der gemischt-ganzzahligen Optimierung, in: *Numerische*

Methoden bei Optimierungsaufgaben (L. Collatz and W. Wetterling, eds.), Birkhäuser, Basel, 1973, pp. 37–49. [372]

Fonlupt, J. and Raco, M. (1984), Orientation of matrices, *Mathematical Programming Study* **22** (1984) 86–98. [297]

Ford, Jr, L. R. and Fulkerson, D. R. (1956), Maximal flow through a network, *Canadian Journal of Mathematics* **8** (1956) 399–404. [97, 115, 275, 378]

Ford, Jr, L. R. and Fulkerson, D. R. (1956–7), Solving the transportation problem, *Management Science* **3** (1956–7) 24–32. [378]

Ford, Jr, L. R. and Fulkerson, D. R. (1957), A simple algorithm for finding maximal network flows and an application to the Hitchcock problem, *Canadian Journal of Mathematics* **9** (1957) 210–218. [154, 279, 378]

Ford, Jr, L. R. and Fulkerson, D. R. (1958–9), A suggested computation for maximal multi-commodity network flows, *Management Science* **5** (1958–9) 97–101. [148]

Ford, Jr, L. R. and Fulkerson, D. R. (1962), *Flows in Networks*, Princeton University Press, Princeton, N.J., 1962. [379]

Forsythe, G. E. and Moler, C. B. (1967), *Computer Solution of Linear Algebraic Systems*, Prentice-Hall, Englewood Cliffs, N.J., 1967. [41]

Fourier, J. B. J. (1798), Mémoire sur la statique, contenant la démonstration du principe des vîtesses virtuelles, et la théorie des momens, *Journal de l'École Polytechnique, tome 2* cahier 5 ([an vi =] 1798) 20–60 [reprinted in: *Oeuvres de Fourier, Tome II* (G. Darboux, ed.), Gauthier-Villars, Paris, 1890, [reprinted: G. Olms, Hildesheim, 1970] pp. 477–521]. [210]

Fourier, J. B. J. (1826a), Solution d'une question particulière du calcul des inégalités, *Nouveau Bulletin des Sciences par la Société philomathique de Paris* (1826) 99–100 [reprinted in: *Oeuvres de Fourier, Tome II* (G. Darboux, ed.), Gauthier-Villars, Paris, 1890, [reprinted: G. Olms, Hildesheim, 1970] pp. 317–319]. [94]

Fourier, J. B. J. (1826b), [reported in:] Analyse des travaux de l'Académie Royale des Sciences, pendant l'année 1823, Partie mathématique, *Histoire de l'Académie Royale des Sciences de l'Institut de France* **6** [1823] (1826) xxix–xli [partially reprinted as: Premier extrait, in: *Oeuvres de Fourier, Tome II* (G. Darboux, ed.), Gauthier-Villars, Paris, 1890, [reprinted: G. Olms, Hildesheim, 1970] pp. 321–324.]. [90, 129, 211, 212]

Fourier, J. B. J. (1827), [reported in:] Analyse des travaux de l'Académie Royale des Sciences, pendant l'année 1824, Partie mathématique, *Histoire de l'Académie Royale des Sciences de l'Institut de France* **7** [1824] (1827) xlvii–lv [reprinted as: Second extrait, in: *Oeuvres de Fourier, Tome II* (G. Darboux, ed.), Gauthier-Villars, Paris, 1890,[reprinted: G.Olms, Hildesheim, 1970,] pp. 325–328] [English translation (partially) in: D. A. Kohler, Translation of a report by Fourier on his work on linear inequalities, *Opsearch* **10** (1973) 38–42]. [90, 155, 157, 211, 212]

Fournier, A. (1979), Comments on convex hull of a finite set of points in two dimensions, *Information Processing Letters* **8** (1979) 173. [224]

Fox, L. (1964), *An Introduction to Numerical Linear Algebra*, Clarendon Press, Oxford, 1964. [41]

Frank, A. (1980), On the orientation of graphs, *Journal of Combinatorial Theory (B)* **28** (1980) 251–261. [314]

Frank, A. and Tardos, É. (1984), Matroids from crossing families, in: *Finite and Infinite Sets (Vol. I)* (Proceedings Sixth Hungarian Combinatorial Colloquium (Eger, 1981; A. Hajnal, L. Lovász and V. T. Sós, eds.), North-Holland, Amsterdam, 1984, pp. 295–304. [314]

Frank, A. and Tardos, É. (1985), An application of simultaneous approximation in combinatorial optimization, in: *26th Annual Symposium on Foundations of Computer Science*, IEEE, New York, 1985, pp. 459–463. [198]

Franklin, J. N. (1968), *Matrix Theory*, Prentice-Hall, Englewood Cliffs, N. J., 1968. [40]

Fredholm, I. (1903), Sur une classe d'équations fonctionelles, *Acta Mathematica* **27** (1903)

365-390 [reprinted in: *Oeuvres Complètes de Ivar Fredholm*, Litos Reprotryck, Malmö, 1955, pp. 81-106]. [28, 39]

Freeman, R. J. (1966), Computational experience with a 'Balasian' integer programming algorithm, *Operations Research* **14** (1966) 935-941. [363]

Friedberg, S. H., Insel, A. J., and Spence, L. E. (1979), *Linear Algebra*, Prentice-Hall, Englewood Cliffs, N.J., 1979. [40]

Frieze, A. M. (1976), Shortest path algorithms for knapsack type problems, *Mathematical Programming* **11** (1976) 150-157. [367]

Frobenius, F. G. (1879), Theorie der linearen Formen mit ganzen Coefficienten, *Journal für die reine und angewandte Mathematik* **86** (1879) 146-208 [reprinted in: *Ferdinand Georg Frobenius, Gesammelte Abhandlungen, Band I* (J.-P. Serre, ed.), Springer, Berlin, 1968, pp. 482-544]. [40, 49, 51, 80]

Frobenius, F. G. (1880), Theorie der linearen Formen mit ganzen Coefficienten (Fortsetzung), *Journal für die reine und angewandte Mathematik* **88** (1880) 96-116 [reprinted in: *Ferdinand Georg Frobenius, Gesammelte Abhandlungen, Band I* (J.-P. Serre, ed.), Springer, Berlin, 1968, pp. 591-611]. [40, 49, 51, 80]

Frobenius, F. G. (1912), Über Matrizen aus nicht negativen Elementen, *Sitzungsberichte der Königlich Preussischen Akademie der Wissenschaften zu Berlin* (1912) 456-477 [reprinted in: *Ferdinand Georg Frobenius, Gesammelte Abhandlungen, Band III* (J.-P. Serre, ed.), Springer, Berlin, 1968, pp. 546-567]. [108]

Frobenius, F. G. (1917), Über zerlegbare Determinanten, *Sitzungsberichte der Königlich Preussischen Akademie der Wissenschaften zu Berlin* (1917) 274-277 [reprinted in: *Ferdinand Georg Frobenius, Gesammelte Abhandlungen, Band III* (J.-P. Serre, ed.), Springer, Berlin, 1968, pp. 701-704]. [108]

Frumkin, M. A. (1975), Power algorithms in the theory of systems of linear diophantine equations (in Russian), *Uspekhi Matematicheskikh Nauk* **30**(4) (184) (1975) 263-264. [56].

Frumkin, M. A. (1976a), An application of modular arithmetic to the construction of algorithms for solving systems of linear equations (in Russian), *Doklady Akademii Nauk SSSR* **229** (1976) 1067-1070 [English translation: *Soviet Mathematics Doklady* **17** (1976) 1165-1168]. [56, 58]

Frumkin, M. A. (1976b), Algorithms for the solution in integers of systems of linear equations (in Russian), in: *Issledovaniya po diskretnoi optimizatsii* [Studies in Discrete Optimization] (A. A. Fridman, ed.), Izdat. "Nauka", Moscow, 1976, pp. 97-127. [56, 58]

Frumkin, M. A. (1976c), An algorithm for the reduction of a matrix of integers to triangular form with power complexity of the computations (in Russian), *Èkonomika i Matematicheskie Metody* **12** (1976) 173-178. [56]

Frumkin, M. A. (1977), Polynomial time algorithms in the theory of linear diophantine equations, in: *Fundamentals of Computation Theory* (M. Karpinski, ed.), Lecture Notes in Computer Science 56, Springer, New York, 1977, pp. 386-392. [56]

Frumkin, M. A. (1980), Number of natural solutions of a system of linear diophantine equations (in Russian), *Matematicheskie Zametki* **28** (3) (1980) 321-334 [English translation: *Mathematical Notes of the Academy of Sciences of the USSR* **28** (1980) 627-634]. [379]

Frumkin, M. A. (1981), On the number of nonnegative integer solutions of a system of linear diophantine equations, [in: *Studies on Graphs and Discrete Programming* (P. Hansen, ed.),] *Annals of Discrete Mathematics* **11** (1981) 95-108. [379]

Frumkin, M. A. (1982), Complexity questions in number theory (in Russian), [in: *The Theory of the Complexity of Computations I* (in Russian) (D. Yu. Grigoryeva and A. O. Slisenko, eds.),] *Zapiski Nauchnykh Seminarov LOMI* **118** (1982) 188-210 [English translation: *Journal of Soviet Mathematics* **29** (1985) 1502-1527]. [82, 379]

Fujiwara, M. (1928), On the system of linear inequalities and linear integral inequalities, *Proceedings of the Imperial Academy of Japan* **4** (1928) 330-333. [217]

Fujiwara, M. (1930), On the system of linear inequalities, *Proceedings of the Imperial Academy of Japan* **6** (1930) 297–298. [217]

Fulkerson, D. R. (1968), Networks, frames, blocking systems, in: *Mathematics of the Decision Sciences, Part 1* (G. B. Dantzig and A. F. Veinott, Jr, eds.), Lectures in Applied Mathematics Vol. 11, American Mathematical Society, Providence, R.I., 1968, pp. 303–334. [98]

Fulkerson, D. R. (1970a), Blocking polyhedra, in: *Graph Theory and Its Applications* (B. Harris, ed.), Academic Press, New York, 1970, pp. 93–112. [113, 114]

Fulkerson, D. R. (1970b), The perfect graph conjecture and pluperfect graph theorem, in: *Proceedings of the Second Chapel Hill Conference on Combinatorial Mathematics and Its Applications* (1970; R. C. Bose, et al., eds.), University of North Carolina, Chapel Hill, N.C., 1970, pp. 171–175. [118, 350]

Fulkerson, D. R. (1971), Blocking and anti-blocking pairs of polyhedra, *Mathematical Programming* **1** (1971) 168–194. [113, 114, 117, 309, 310, 311]

Fulkerson, D. R. (1972), Anti-blocking polyhedra, *Journal of Combinatorial Theory (B)* **12** (1972) 50–71. [113, 117]

Fulkerson, D. R. (1973), On the perfect graph theorem, in: *Mathematical Programming* (T. C. Hu and S. M. Robinson, eds.), Academic Press, New York, 1973, pp. 69–76. [118, 350]

Fulkerson, D. R. (1974), Packing rooted directed cuts in a weighted directed graph, *Mathematical Programming* **6** (1974) 1–13. [312]

Fulkerson, D. R., Hoffman, A. J., and Oppenheim, R. (1974), On balanced matrices, *Mathematical Programming Study* **1** (1974) 120–132. [304, 307]

Fulkerson, D. R., Nemhauser, G. L., and Trotter, Jr, L. E. (1974), Two computationally difficult set covering problems that arise in computing the 1-width of incidence matrices of Steiner triple systems, *Mathematical Programming Study* **2** (1974) 72–81. [362, 380]

Furtwängler, Ph. (1927), Über die simultane Approximation von Irrationalzahlen, *Mathematische Annalen* **96** (1927) 169–175. [74]

Furtwängler, Ph. (1928), Über die simultane Approximation von Irrationalzahlen. Zweite Mitteilung, *Mathematische Annalen* **99** (1928) 71–83. [74]

Gács, P. and Lovász, L. (1981), Khachiyan's algorithm for linear programming, *Mathematical Programming Study* **14** (1981) 61–68. [163, 171]

Gaddum, J. W., Hoffman, A. J., and Sokolowsky, D. (1954), On the solution of the caterer problem, *Naval Research Logistics Quarterly* **1** (1954) 223–229. [378]

Gale, D. (1951), Convex polyhedral cones and linear inequalities, in: *Activity Analysis of Production and Allocation* (Tj. C. Koopmans, ed.), Wiley, New York, and Chapman & Hall, London, 1951, pp. 287–297. [224]

Gale, D. (1960) *The Theory of Linear Economic Models*, McGraw-Hill, New York, 1960. [224]

Gale, D. (1963), Neighborly and cyclic polytopes, in: *Convexity* (Proceedings of Symposia in Pure Mathematics VII; V. L. Klee, ed.), American Mathematical Society, Providence, R. I., 1963, pp. 225–232. [142]

Gale, D. (1964), On the number of faces of a convex polytope, *Canadian Journal of Mathematics* **16** (1964) 12–17. [142]

Gale, D., Kuhn, H. W., and Tucker, A. W. (1951), Linear programming and the theory of games, in: *Activity Analysis of Production and Allocation* (Tj. C. Koopmans, ed.), Wiley, New York, 1951, pp. 317–329. [90, 218, 223]

Galperin, A. M. (1976), The general solution of a finite system of linear inequalities, *Mathematics of Operations Research* **1** (1976) 185–196. [224]

Gantmacher, F. R. (1959), *The Theory of Matrices, Volumes I, II* (translated from the Russian), Chelsea, New York, 1959. [4, 40]

Garcia, C. B., Gould, F. J., and Guler, O. (1984), *Experiments with a variant of Dantzig's self-dual method*, report, Graduate School of Business, University of Chicago, Chicago, 1984. [139]

Garey, M. R. and Johnson, D. S. (1979), *Computers and Intractability: a Guide to the Theory of NP-completeness*, Freeman, San Francisco, 1979. [14, 16, 18, 21, 263]

Garfinkel, R. S. (1979), Branch and bound methods for integer programming, in: *Combinatorial Optimization* (N. Christofides, A. Mingozzi, P. Toth and C. Sandi, eds.), Wiley, Chichester, 1979, pp. 1–20. [363]

Garfinkel, R. S. and Nemhauser, G. L. (1969), The set-partitioning problem: set covering with equality constraints, *Operations Research* 17 (1969) 848–856. [380]

Garfinkel, R. S. and Nemhauser, G. L. (1972a), *Integer Programming*, Wiley, New York, 1972. [59, 359, 378]

Garfinkel, R. and Nemhauser, G. L. (1972b), Optimal set covering: a survey, in: *Perspectives on Optimization; A Collection of Expository Articles* (A. M. Geoffrion, ed.), Addison-Wesley, Reading, Mass., 1972, pp. 164–193. [380]

Garfinkel, R. S. and Nemhauser, G. L. (1973), A survey of integer programming emphasizing computation and relations among models, in: *Mathematical Programming* (Proc. Advanced Seminar, Madison, Wis., 1972; T. C. Hu and S. M. Robinson, eds.), Academic Press, New York, 1973, pp. 77–155. [379]

Gass, S. I. (1975), *Linear Programming Methods and Applications (fourth edition)*, McGraw-Hill, New York, 1975. [224]

Gassner, B. J. (1964), Cycling in the transportation problem, *Naval Research Logistics Quarterly* 11 (1964) 43–58. [142]

von zur Gathen, J. and Sieveking, M. (1976), Weitere zum Erfüllungsproblem polynomial äquivalente kombinatorische Aufgaben, in: *Komplexität von Entscheidungsproblemen* (E. Specker and V. Strassen, eds.), Lecture Notes in Computer Science 43, Springer, Heidelberg, 1976, pp. 49–71. [55, 56, 58]

von zur Gathen, J. and Sieveking, M. (1978), A bound on solutions of linear integer equalities and inequalities, *Proceedings of the American Mathematical Society* 72 (1978) 155–158. [55, 239]

Gauss, C. F. (1801), *Disquisitiones Arithmeticae*, Gerh. Fleischer, Leipzig, 1801 [English translation: Yale University Press, New Haven, 1966] [reprinted as: *Carl Friedrich Gauss Werke, Vol. I*, Königlichen Gesellschaft der Wissenschaften zu Göttingen, Göttingen, 1863 [reprinted: G. Olms, Hildesheim, 1973]]. [78]

Gauss, C. F. (1809), *Theoria Motus Corporum Coelestium in Sectionibus Conicis Solem Ambientium*, F. Perthes & J. H. Besser, Hamburg, 1809 [reprinted in: *Carl Friedrich Gauss Werke, Vol. VII*, Königlichen Gesellschaft der Wissenschaften zu Göttingen, Göttingen, 1871 [reprinted: G. Olms, Hildesheim, 1973], pp. 3–280] [English translation: *Theory of the Motion of the Heavenly Bodies Moving about the Sun in Conic Sections*, Little, Brown & Co., Boston, 1857 [reprinted: Dover, New York, 1963]] [German translation (partially) in: C. F. Gauss, *Abhandlungen zur Methode der kleinsten Quadrate* (A. Börsch and P. Simon, eds), Stankiewicz, Berlin, 1887 [reprinted: Physica-Verlag, Würzburg, 1964], pp. 92–117]. [28, 36, 39]

Gauss, C. F. (1811), Disquisitio de elementis ellipticis Palladis ex oppositionibus annorum 1803, 1804, 1805, 1807, 1808, 1809, *Commentationes Societatis Regiae Scientiarum Gottingensis (Classis Mathematicae)* 1 [read 25 November 1810] (1811) 3–26 [reprinted in: *Carl Friedrich Gauss Werke, Vol. VI*, Königlichen Gesellschaft der Wissenschaften zu Göttingen, Göttingen, 1874 [reprinted: G. Olms, Hildesheim, 1973,] pp. 3–24] [German translation in: C. F. Gauss, *Abhandlungen zur Methode der kleinsten Quadrate*, (A. Börsch and P. Simon, eds.), Stankiewicz, Berlin, 1887 [reprinted: Physica-Verlag, Würzburg, 1964,] pp. 118–128. [39]

Gauss, C. F. (1823), *Brief an Gerling, Göttingen, December 26, 1823* [reproduced in: *Carl Friedrich Gauss Werke, Vol. IX*, Königlichen Gesellschaft der Wissenschaften zu Göttingen, Göttingen, (in commission bei B. G. Teubner, Leipzig,) 1903 [reprinted: G. Olms, Hildesheim, 1973], pp. 278–281] [English translation in: Forsythe [1951]]. [36, 39]

Gauss, C. F. (1828), Supplementum Theoriae Combinationis Observationum Erroribus

Minimis Obnoxiae, *Commentationes Societatis Regiae Scientiarum Gottingensis* [*Recentiores*] 6 [read 16 September 1826] (1828) 57–98 [reprinted in: *Carl Friedrich Gauss Werke, Vol. IV*, Königlichen Gesellschaft der Wissenschaften zu Göttingen, Göttingen, 1873 [reprinted: G. Olms, Hildesheim, 1973], pp. 57–93] [German translation in: *C. F. Gauss, Abhandlungen zur Methode der kleinsten Quadrate*, (A. Börsch and P. Simon, eds.), Stankiewicz, Berlin, 1887 [reprinted: Physica-Verlag, Würzburg, 1964], pp. 54–91]. [39]

Gauss, C. F. (1829), Über ein neues allgemeines Grundgesetz der Mechanik, *Journal für die reine und angewandte Mathematik* 4 (1829) 232–235 [reprinted in: *Carl Friedrich Gauss Werke, Vol. V*, Königlichen Gesellschaft der Wissenschaften zu Göttingen, Göttingen, 1877 [reprinted: G. Olms, Hildesheim, 1973], pp. 25–28] [English translation: On a new general principle of mechanics, *Philosophical Magazine* 8 (1830) 137–140]. [211]

[Gauss, C. F. (1831) =] Anonymous, "Untersuchungen über die Eigenschaften der positiven ternären quadratischen Formen von Ludwig August Seeber, Dr. d. Philos. ordentl. Prof. der Physik an der Univers. in Freiburg. 1831. 248 S. in 4", *Göttingische gelehrte Anzeigen* [*unter der Aufsicht der Königl. Gesellschaft der Wissenschaften*] (1831) [2. Band, 108. Stück, Den 9. Julius 1831] 1065–1077 [reprinted in: *Journal für die reine und angewandte Mathematik* 20 (1840) 312–320] [reprinted also in: *Carl Friedrich Gauss Werke, Vol. II*, Königlichen Gesellschaft der Wissenschaften zu Göttingen, Göttingen, 1863, [reprinted: G. Olms, Hildesheim, 1973,] pp. 188–196]. [78]

Gauss, C. F. (1832), Principia Generalia Theoriae Figurae Fluidorum in Statu Aequilibrii, *Commentationes Societatis Regiae Scientiarum Gottingensis Recentiores* 7 [read 28 September 1829] (1832) 39–88 [reprinted in: *Carl Friedrich Gauss Werke, Vol. V*, Königlichen Gesellschaft der Wissenschaften zu Göttingen, Göttingen, 1867, [reprinted: G. Olms, Hildesheim, 1973,] pp. 30–77] [German translation:*Allgemeine Grundlagen einer Theorie der Gestalt von Flüssigkeiten im Zustand des Gleichgewichts* [translated by R. H. Weber, edited by H. Weber], Wilhelm Engelmann, Leipzig, 1903]. [211]

Gavish, B. (1978), On obtaining the 'best' multipliers for a Lagrangean relaxation for integer programming, *Computers and Operations Research* 5 (1978) 55–71. [370]

Geoffrion, A. M. (1967), Integer programming by implicit enumeration and Balas' method, *SIAM Review* 9 (1967) 178–190. [363, 373]

Geoffrion, A. M. (1969), An improved implicit enumeration approach for integer programming, *Operations Research* 17 (1969) 437–454. [362, 373, 380]

Geoffrion, A. M. (1972), Generalized Benders decomposition, *Journal of Optimization Theory and Applications* 10 (1972) 237–260. [372]

Geoffrion, A. M. (1974), Lagrangean relaxation for integer programming, *Mathematical Programming Study* 2 (1974) 82–114. [354, 370]

Geoffrion, A. M. (1976), A guided tour of recent practical advances in integer linear programming, *Omega* 4 (1976) 49–57. [379]

Geoffrion, A. M. and Marsten, R. E. (1971–2), Integer programming algorithms: a framework and state-of-the-art survey, *Management Science* 18 (1971–2) 465–491. [379]

Geoffrion, A. M. and Nauss, R. (1976–7), Parametric and postoptimality analysis in integer linear programming, *Management Science* 23 (1976–7) 453–466. [354]

Gerards, A. M. H. and Schrijver, A. (1986), Matrices with Edmonds–Johnson property, *Combinatorica* 6 (1986) 365–379. [348, 350]

Gergonne, J. D. (1820–1), Démonstration des deux théorèmes de géométrie énoncés à la page 289 du IX.e volume de ce recueil, *Annales de mathématiques pures et appliquées* 11 (1820–1) 326–336. [212]

Gergonne, J. D. (1824–5), Recherche de quelques-unes des lois générales qui régissent les polyèdres, *Annales de mathématiques pures et appliquées* 15 (1824–5) 157–164. [212]

Gergonne, J. D. (1825–6), Géométrie de situation. Philosophie mathématique. Considérations philosophiques sur les élémens de la science de l'étendue, *Annales de mathématiques pures et appliquées* **16** (1825–6) 209–231. [40, 212]

Gerhardt, K. I. (1891), Leibniz über die Determinanten, *Sitzungsberichte der Königlich Preussischen Akademie der Wissenschaften zu Berlin* (1891) 407–423. [39]

Gerling, C. L. (1843), *Die Ausgleichungs-Rechnungen der practischen Geometrie oder die Methode der kleinsten Quadrate mit ihren Anwendungen für geodätische Aufgaben*, F. & A. Perthes, Hamburg, 1843. [36, 40]

Gerstenhaber, M. (1951), Theory of convex polyhedral cones, in: *Activity Analysis of Production and Allocation* (Tj. C. Koopmans, ed.), Wiley, New York, and Chapman & Hall, London, 1951, pp. 298–316. [224]

Gewirtz, A., Sitomer, H., and Tucker, A. W. (1974), *Constructive Linear Algebra*, Prentice-Hall, Englewood Cliffs, N.J., 1974. [40]

Ghouila-Houri, A. (1962), Caractérisation des matrices totalement unimodulaires, *Comptes Rendus Hebdomadaires des Séances de l'Académie des Sciences (Paris)* **254** (1962) 1192–1194. [269]

Giles, R. and Orlin, J. B. (1981), *Verifying total dual integrality*, manuscript, 1981. [333, 336]

Giles, F. R. and Pulleyblank, W. R. (1979), Total dual integrality and integer polyhedra, *Linear Algebra and Its Applications* **25** (1979) 191–196. [232, 315, 316, 321]

Giles, R. and Trotter, Jr, L. E. (1981), On stable set polyhedra for $K_{1,3}$-free graphs, *Journal of Combinatorial Theory (B)* **31** (1981) 313–326. [350]

Gill, P. E., Murray, W., and Wright, M. H. (1981), *Practical Optimization*, Academic Press, London, 1981. [224]

Gilmore, P. C. (1979), Cutting stock, linear programming, knapsacking, dynamic programming and integer programming, some interconnections, [in: *Discrete Optimization I* (P. L. Hammer, E. L. Johnson and B. H. Korte, eds.),] *Annals of Discrete Mathematics* **4** (1979) 217–235. [373]

Gilmore, P. C. and Gomory, R. E. (1961), A linear programming approach to the cutting-stock problem, *Operations Research* **9** (1961) 849–859. [148, 363, 373]

Gilmore, P. C. and Gomory, R. E. (1963), A linear programming approach to the cutting-stock problem–Part II, *Operations Research* **11** (1963) 863–888. [363, 373]

Gilmore, P. C. and Gomory, R. E. (1965), Multistage cutting stock problems of two and more dimensions, *Operations Research* **13** (1965) 94–120. [363]

Gilmore, P. C. and Gomory, R. E. (1966), The theory and computation of knapsack functions, *Operations Research* **14** (1966) 1045–1074 [erratum: **15** (1967) 366]. [263, 373]

Giraud, G. (1914), Sur la résolution approchée en nombres entiers d'un système d'équations linéaires non homogènes, *Comptes Rendus des Séances, Société Mathématique de France* (1914) 29–32. [80]

Girlich, E. and Kowaljow, M. M. (1981), *Nichtlineare diskrete Optimierung*, Akademie-Verlag, Berlin (DDR), 1981. [379]

Glover, F. (1965), A multiphase-dual algorithm for the zero-one integer programming problem, *Operations Research* **13** (1965) 879–919. [363]

Glover, F. (1966–7), Generalized cuts in diophantine programming, *Management Science* **13** (1966–7) 254–268. [359]

Glover, F. (1968a), A new foundation for a simplified primal integer programming algorithm, *Operations Research* **16** (1968) 727–740. [359]

Glover, F. (1968b), A note on linear programming and integer feasibility, *Operations Research* **16** (1968) 1212–1216. [308]

Glover, F. (1969), Integer programming over a finite additive group, *SIAM Journal on Control* **7** (1969) 213–231. [367]

Glover, F. (1970), Faces of the Gomory polyhedron, in: *Integer and Nonlinear Programming* (J. Abadie, ed.), North-Holland, Amsterdam, 1970, pp. 367–379. [367]

Glover, F. (1972), Cut search methods in integer programming, *Mathematical Programming* **3** (1972) 86–100. [359]

Glover, F. (1973), Convexity cuts and cut search, *Operations Research* **21** (1973) 123–134. [359]

Glover, F. (1975a), New results on equivalent integer programming formulations, *Mathematical Programming* **8** (1975) 84–90. [236]

Glover, F. (1975b), Polyhedral annexation in mixed integer and combinatorial programming, *Mathematical Programming* **9** (1975) 161–188. [367]

Glover, F. and Klingman, D. (1976), A practitioner's guide to the state of large scale network and network-related problems, in: *AFIPS Conference Proceedings [Volume 45] 1976 National Computer Conference* (S. Winkler, ed.), AFIPS Press, Montvale, N. J., 1976, pp. 945–950. [142]

Glover, F. and Woolsey, R. E. (1972), Aggregating diophantine constraints, *Zeitschrift für Operations Research* **16** (1972) 1–10. [236]

Glover, F. and Zionts, S. (1965), A note on the additive algorithm of Balas, *Operations Research* **13** (1965) 546–549. [363]

Goedseels, P. J. E. (1902), *Théorie des erreurs d'observation*, Charles Peeters, Leuven, and Gauthier-Villars, Paris, 1902. [216]

Goffin, J. L. (1980), The relaxation method for solving systems of linear inequalities, *Mathematics of Operations Research* **5** (1980) 388–414. [157, 158]

Goffin, J. L. (1982), On the non-polynomiality of the relaxation method for systems of linear inequalities, *Mathematical Programming* **22** (1982) 93–103. [161]

Goffin, J.-L. (1984), Variable metric relaxation methods, Part II: The ellipsoid method, *Mathematical Programming* **30** (1984) 147–162. [161, 205]

Goldberg, E. L. (1976), On a linear diophantine equation, *Acta Arithmetica* **31** (1976) 239–246. [379]

Goldfarb, D. (1983), *Worst-case complexity of the shadow vertex simplex algorithm*, report, Department of Industrial Engineering and Operations Research, Columbia University, 1983. [141]

Goldfarb, D. (1985), Efficient dual simplex algorithms for the assignment problem, *Mathematical Programming* **33** (1985) 187–203. [142]

Goldfarb, D. and Reid, J. K. (1977), A practicable steepest-edge simplex algorithm, *Mathematical Programming* **12** (1977) 361–371. [139]

Goldfarb, D. and Sit, W. Y. (1979), Worst case behavior of the steepest edge simplex method, *Discrete Applied Mathematics* **1** (1979) 277–285. [141]

Goldfarb, D. and Todd, M. J. (1982), Modifications and implementation of the ellipsoid algorithm for linear programming, *Mathematical Programming* **23** (1982) 1–19. [171]

Goldman, A. J. (1956), Resolution and separation theorems for polyhedral convex sets, in: *Linear Inequalities and Related Systems* (H. W. Kuhn and A. W. Tucker, eds.), Princeton Univ. Press, Princeton, N. J., 1956, pp. 41–51. [99]

Goldman, A. J. and Kleinman, D. (1964), Examples relating to the simplex method, *Operations Research* **12** (1964) 159–161. [142]

Goldman, A. J. and Tucker, A. W. (1956), Polyhedral convex cones, in: *Linear Inequalities and Related Systems* (H. W. Kuhn and A. W. Tucker, eds.), Princeton University Press, Princeton, N. J., 1956, pp. 19–40. [99]

Goldstine, H. H. (1977), *A History of Numerical Analysis*, Springer, New York, 1977. [39, 41]

Golub, G. H. and van Loan, C. F. (1983), *Matrix Computations*, North Oxford Academic, Oxford, 1983. [41]

Gomory, R. E. (1958), Outline of an algorithm for integer solutions to linear programs, *Bulletin of the American Mathematical Society* **64** (1958) 275–278. [340, 354, 378]

Gomory, R. E. (1960), Solving linear programming problems in integers, in: *Combinatorial Analysis* (R. Bellman and M. Hall, Jr, eds.), Proceedings of Symposia in Applied Mathematics X, American Mathematical Society, Providence, R. I., 1960, pp. 211–215. [340, 354]

Gomory, R. E. (1963a), An algorithm for integer solutions to linear programs, in: *Recent*

Advances in Mathematical Programming (R. L. Graves and P. Wolfe, eds.), McGraw-Hill, New York, 1963, pp. 269–302. [310, 340, 354]

Gomory, R. E. (1963b), An all-integer integer programming algorithm, in: *Industrial Scheduling* (J. F. Muth and G. L. Thompson, eds.), Prentice-Hall, Englewood Cliffs, N.J., 1963, pp. 193–206. [359]

Gomory, R. E. (1965), On the relation between integer and noninteger solutions to linear programs, *Proceedings of the National Academy of Sciences of the United States of America* **53** (1965) 260–265. [241, 363]

Gomory, R. E. (1967), Faces of an integer polyhedron, *Proceedings of the National Academy of Sciences of the United States of America* **57** (1967) 16–18. [363]

Gomory, R. E. (1969), Some polyhedra related to combinatorial problems, *Linear Algebra and Its Applications* **2** (1969) 451–558. [363]

Gomory, R. E. (1970), Properties of a class of integer polyhedra, in: *Integer and Nonlinear Programming* (J. Abadie, ed.), North-Holland, Amsterdam, 1970, pp. 353–365. [367]

Gomory, R. E. and Baumol, W. J. (1960), Integer programming and pricing, *Econometrica* **28** (1960) 521–550. [354]

Gomory, R. E. and Hoffman, A. J. (1963), On the convergence of an integer-programming process, *Naval Research Logistics Quarterly* **10** (1963) 121–123. [359]

Gomory, R. E. and Johnson, E. L. (1972a), Some continuous functions related to corner polyhedra, *Mathematical Programming* **3** (1972) 23–85. [367]

Gomory, R. E. and Johnson, E. L. (1972b), Some continuous functions related to corner polyhedra, II, *Mathematical Programming* **3** (1972) 359–389. [367]

Gomory, R. E. and Johnson, E. L. (1973), The group problems and subadditive functions, in: *Mathematical Programming* (T. C. Hu and S. M. Robinson, eds.), Academic Press, New York, 1973, pp. 157–184. [367]

Gondran, M. (1973), Matrices totalement unimodulaires, *Électricité de France, Bulletin de la Direction des Études et Recherches, Série C (Mathématiques * Informatique)* 1973 No. 1 (1973) 55–73. [272]

Gondran, M. and Laurière, J. L. (1974), Un algorithme pour le problème de partitionnement, *Revue Française d'Automatique Informatique Recherche Opérationelle [R. A. I. R. O.]* **8** (*Recherche Opérationelle* V–1) (1974) 27–40. [380]

Gondran, M. and Laurière, J. L. (1975), Un algorithme pour les problèmes de recouvrement, *Revue Française d'Automatique Informatique Recherche Opérationelle [R. A. I. R. O.]* **9** (*Recherche Opérationelle* V–2) (1975) 33–51. [380]

Good, R. A. (1959), Systems of linear relations, *SIAM Review* **1** (1959) 1–31. [223]

Gordan, P. (1873), Ueber die Auflösung linearer Gleichungen mit reellen Coefficienten, *Mathematische Annalen* **6** (1873) 23–28. [95, 213, 233, 315, 376]

Gorry, G. A., Northup, W. D., and Shapiro, J. F. (1973), Computational experience with a group theoretic integer programming algorithm, *Mathematical Programming* **4** (1973) 171–192. [367]

Gorry, G. A. and Shapiro, J. F. (1970–1), An adaptive group theoretic algorithm for integer programming problems, *Management Science* **17** (1970–1) 285–306. [367]

Gorry, G. A., Shapiro, J. F., and Wolsey, L. A. (1971–2), Relaxation methods for pure and mixed integer programming problems, *Management Science* **18** (1971–2) 229–239. [354, 367, 370]

Gould, R. (1958), Graphs and vector spaces, *Journal of Mathematics and Physics* **37** (1958) 193–214. [282]

Goult, R. J., Hoskins, R. F., Milner, J. A., and Pratt, M. J. (1974), *Computational Methods in Linear Algebra*, Stanley Thornes, London, 1974. [41]

Grace, J. H. and Young, A. (1903), *The Algebra of Invariants*, Cambridge University Press, Cambridge, 1903 [reprinted: Chelsea, New York, 1965]. [376]

Graham, R. (1972), An efficient algorith[m] for determining the convex hull of a finite planar set, *Information Processing Letters* **1** (1972) 132–133. [224]

Grassmann, H. (1844), [*Die Wissenschaft der extensiven Grösse, oder die Ausdehnungslehre, eine neue mathematische Disciplin dargestellt und durch Anwendungen erläutert, Erster Theil, die lineale Ausdehnungslehre enthaltend:] Die lineale Ausdehnungslehre, ein neuer Zweig der Mathematik, O.* Wigand, Leipzig, 1844 [abstract: Grassmann [1845]] [second edition: *Die Ausdehnungslehre von 1844 oder Die lineale Ausdehnungslehre, ein neuer Zweig der Mathematik, O.* Wigand, Leipzig, 1878 [reprinted in: *Hermann Grassmanns Gesammelte mathematische und physikalische Werke, Vol. I, Part I* (F. Engel, ed.), Teubner, Leipzig, 1894, pp. 5–319] [reprinted: Chelsea, New York, 1969]] [English translation (partially) in: D. E. Smith, *A Source Book in Mathematics*, McGraw-Hill, New York, 1929, pp. 684–696]. [40]

Grassmann, H. (1845), Ueber die Wissenschaft der extensiven Grösse oder die Ausdehnungslehre, *Archiv der Mathematik und Physik* **6** (1845) 337–350 [reprinted in: *Hermann Grassmanns Gesammelte mathematische und physikalische Werke, Vol. I, Part I* (F. Engel, ed.), Teubner, Leipzig, 1894, [reprinted: Chelsea, New York, 1969,] pp. 297–312]. [40]

Grattan-Guinness, I. (1970), Joseph Fourier's anticipation of linear programming, *Operational Research Quarterly* **21** (1970) 361–364. [211]

Graver, J. E. (1975), On the foundations of linear and integer linear programming I, *Mathematical Programming* **9** (1975) 207–226. [234, 242]

Green, P. J. and Silverman, B. W. (1979), Constructing the convex hull of a set of points in the plane, *The Computer Journal* **22** (1979) 262–266. [224]

Greenberg, H. (1969), A dynamic programming solution to integer linear programs, *Journal of Mathematical Analysis and Applications* **26** (1969) 454–459. [367]

Greenberg, H. (1971), *Integer Programming*, Academic Press, New York, 1971. [378]

Greenberg, H. (1975), An algorithm for determining redundant inequalities and all solutions to convex polyhedra, *Numerische Mathematik* **24** (1975) 19–26. [224]

Greenberg, H. (1980), An algorithm for a linear diophantine equation and a problem of Frobenius, *Numerische Mathematik* **34** (1980) 349–352. [379]

Greenberg, H. and Hegerich, R. L. (1969–70), A branch search algorithm for the knapsack problem, *Management Science* **16** (1969–70) 327–332. [363, 373]

Greub, W. (1975), *Linear Algebra (fourth edition)*, Springer, New York, 1975. [40]

Griffin, V. (1977), *Polyhedral polarity*, Doctoral Thesis, University of Waterloo, Waterloo, Ontario, 1977. [119]

Griffin, V., Aráoz, J., and Edmonds, J. (1982), Polyhedral polarity defined by a general bilinear inequality, *Mathematical Programming* **23** (1982) 117–137. [119]

Grötschel, M. (1977), *Polyedrische Charakterisierungen kombinatorischer Optimierungsprobleme*, Verlag Anton Hain, Meisenheim am Glan, 1977. [379]

Grötschel, M. (1980), On the symmetric travelling salesman problem: solution of a 120–city problem, *Mathematical Programming Study* **12** (1980) 61–77. [371]

Grötschel, M., Lovász, L., and Schrijver, A. (1981), The ellipsoid method and its consequences in combinatorial optimization, *Combinatorica* **1** (1981) 169–197 [corrigendum: **4** (1984) 291–295]. [163, 172, 350]

Grötschel, M., Lovász, L., and Schrijver, A. (1984), Geometric methods in combinatorial optimization, in: *Progress in Combinatorial Optimization* (Jubilee Conference, University of Waterloo, Waterloo, Ontario, 1982; W. R. Pulleyblank, ed.), Academic Press, Toronto, 1984, pp. 167–183. [173, 256]

Grötschel, M., Lovász, L., and Schrijver, A. (1988), *Geometric Algorithms and Combinatorial Optimization*, Springer, Berlin, 1988. [173, 189, 205, 208, 224, 256, 260]

Grötschel, M. and Padberg, M. W. (1979a), On the symmetric travelling salesman problem I: Inequalities, *Mathematical Programming* **16** (1979) 265–280. [371]

Grötschel, M. and Padberg, M. W. (1979b), On the symmetric travelling salesman problem II: Lifting theorems and facets, *Mathematical Programming* **16** (1979) 281–302. [371]

Grötschel, M. and Padberg, M. W. (1985), Polyhedral theory, in: *The Traveling Salesman*

Problem, A Guided Tour of Combinatorial Optimization (E. L. Lawler, et al., eds.), Wiley, Chichester, 1985, pp. 251–305. [371]

Gruber, P. M. and Wills, J. M. (eds.) (1983), *Convexity and Its Applications*, Birkhäuser, Basel, 1983. [224]

Grünbaum, B. (1967), *Convex Polytopes*, Interscience-Wiley, London, 1967. [99, 223]

Grünbaum, B. and Motzkin, T. S. (1962), Longest simple paths in polyhedral graphs, *Journal of the London Mathematical Society* **37** (1962) 152–160 [reprinted in: *Theodore S. Motzkin: Selected Papers* (D. Cantor, B. Gordon and B. Rothschild, eds.), Birkhäuser, Boston, 1983, pp. 292–300]. [142]

Grünbaum, B. and Shephard, G. C. (1969), Convex polytopes, *The Bulletin of the London Mathematical Society* **1** (1969) 257–300. [223]

Guha, D. K. (1973), The set-covering problem with equality constraints, *Operations Research* **21** (1973) 348–351. [380]

Guignard, M. M. and Spielberg, K. (1972), Mixed-integer algorithms for the (0,1) knapsack problem, *IBM Journal of Research and Development* **16** (1972) 424–430. [373]

Gummer, C. F. (1926), Sets of linear equations in positive unknowns, *Abstract lecture Mathematical Association of America*, Columbus, Ohio, September, 1926 [see: *The American Mathematical Monthly* **33** (1926) 487–488]. [217]

Güting, R. (1975), Zur Verallgemeinerung des Kettenbruchalgorithmus. I, *Journal für die reine und angewandte Mathematik* **278/279** (1975) 165–173. [74]

Güting, R. (1976a), Zur Verallgemeinerung des Kettenbruchalgorithmus. II, *Journal für die reine und angewandte Mathematik* **281** (1976) 184–198. [74]

Güting, R. (1976b), Zur Verallgemeinerung des Kettenbruchalgorithmus. III, *Journal für die reine und angewandte Mathematik* **283/284** (1976) 384–387. [74]

Haar, A. (1918), A linearis egyenlötlenségekröl (Hungarian) [On the linear inequalities], *Mathematikai és Természettudományi Értesitö* **36** (1918) 279–296 [reprinted in: *Alfred Haar Gesammelte Arbeiten* (B. Szökefalvi-Nagy, ed.), Akadémiai Kiadó, Budapest, 1959, pp. 421–438] [German translation: Haar [1924–6]]. [93, 216, 217]

Haar, A. (1924–6), Über lineare Ungleichungen, *Acta Litterarum ac Scientiarum Regiae Universitatis Hungaricae Francisco-Josephinae [Szeged] Sectio Scientiarum Mathematicarum* **2** (1924–6) 1–14 [reprinted in: *Alfred Haar, Gesammelte Arbeiten* (B. Szökefalvi-Nagy, ed.), Akadémiai Kiadó, Budapest, 1959, pp. 439–452]. [93, 217]

Hadley, G. (1962), *Linear Programming*, Addison-Wesley, Reading, Mass., 1962. [224]

Hadwiger, H. (1955), *Altes und Neues über konvexe Körper*, Birkhäuser, Basel, 1955. [224]

Haimovich, M. (1983), *The simplex method is very good!—On the expected number of pivot steps and related properties of random linear programs*, preprint, 1983. [139, 143, 146]

Haimovich, M. (1984a), *On the expected behavior of variable dimension simplex algorithms*, preprint, 1984. [147]

Haimovich, M. (1984b), *A short proof of results on the expected number of steps in Dantzig's self dual algorithm*, preprint, 1984. [147]

Halfin, S. (1972), Arbitrarily complex corner polyhedra are dense in R^n, *SIAM Journal on Applied Mathematics* **23** (1972) 157–163. [255, 367]

Halfin, S. (1983), The sphere method and the robustness of the ellipsoid algorithm, *Mathematical Programming* **26** (1983) 109–116. [170]

Hall, F. J. and Katz, I. J. (1979), On ranks of integral generalized inverses of integral matrices, *Linear and Multilinear Algebra* **7** (1979) 73–85. [308]

Hall, F. J. and Katz, I. J. (1980), More on integral generalized inverses of integral matrices, *Linear and Multilinear Algebra* **9** (1980) 201–209. [308]

Hall, F. J. and Katz, I. J. (1981), Nonnegative integral generalized inverses, *Linear Algebra and Its Applications* **39** (1981) 23–39. [308]

Halmos, P. R. (1974), *Finite-dimensional Vector Spaces*, Springer, New York, 1974. [41]

Hammer, J. (1977), *Unsolved Problems Concerning Lattice Points*, Pitman, London, 1977. [82, 380]

Hammer, P. L., Johnson, E. L., and Peled, U. N. (1975), Facets of regular 0–1 polytopes,

Mathematical Programming **8** (1975) 179–206. [374, 380]

Hancock, H. (1917), *Theory of Maxima and Minima*, Ginn and Co., New York, 1917 [reprinted: Dover, New York, 1960]. [221]

Hancock, H. (1939), *Development of the Minkowski Geometry of Numbers*, The Macmillan Company, New York, 1939 [reprinted (in two volumes) by: Dover, New York, 1964]. [82]

Hardy, G. H. and Wright, E. M. (1979), *An Introduction to the Theory of Numbers (fifth edition)*, Oxford University Press, Oxford, 1979. [82]

Harris, V. C. (1970a), An algorithm for finding the greatest common divisor, *The Fibonacci Quarterly* **8** (1970) 102–103. [54]

Harris, V. C. (1970b), Note on the number of divisions required in finding the greatest common divisor, *The Fibonacci Quarterly* **8** (1970) 104. [54]

Hausmann, D. (ed.) (1978), *Integer Programming and Related Areas, A Classified Bibliography 1976-1978*, Lecture Notes in Economics and Math. Systems 160, Springer, Berlin, 1978. [236, 376, 379]

Hayes, A. C. and Larman, D. G. (1983), The vertices of the knapsack polytope, *Discrete Applied Mathematics* **6** (1983) 135–138. [256]

Heap, B. R. and Lynn, M. S. (1964), A graph-theoretic algorithm for the solution of a linear diophantine problem of Frobenius, *Numerische Mathematik* **6** (1964) 346–354. [379]

Heap, B. R. and Lynn, M. S. (1965), On a linear diophantine problem of Frobenius: an improved algorithm, *Numerische Mathematik* **7** (1965) 226–231. [379]

Heger, J. (1856), Über die Auflösung eines Systemes von mehreren unbestimmten Gleichungen des ersten Grades in ganzen Zahlen, welche eine grössere Anzahl von Unbekannten in sich schliessen, als sie zu bestimmen vermögen, *Sitzungsberichte Oesterreichische Akademie der Wissenschaften (mathematisch-naturwissenschaftliche Klasse)* **21** (1856) 550–560. [80]

Heger, I. (1858), Über die Auflösung eines Systemes von mehreren unbestimmten Gleichungen des ersten Grades in ganzen Zahlen, *Denkschriften der Königlichen Akademie der Wissenschaften (Wien), Mathematisch-naturwissenschaftliche Klasse* **14** (2. Abth.) (1858) 1–122 [also published by: Gerold's Sohn, Wien, 1858]. [51, 80]

Heilbronn, H. (1968), On the average length of a class of finite continued fractions, in: *Abhandlungen aus Zahlentheorie und Analysis, zur Erinnerung an Edmund Landau* (P. Turán, ed.), VEB Deutscher Verlag der Wissenschaften, Berlin (DDR) and Plenum Press, New York, 1968, pp. 87–96. [54, 67]

Held, M. and Karp, R. M. (1970), The traveling-salesman problem and minimum spanning trees, *Operations Research* **18** (1970) 1138–1162. [367, 368, 370, 371]

Held, M. and Karp, R. M. (1971), The traveling-salesman problem and minimum spanning trees: part II, *Mathematical Programming* **1** (1971) 6–25. [367, 368, 370, 371]

Held, M., Wolfe, P., and Crowder, H. P. (1974), Validation of subgradient optimization, *Mathematical Programming* **6** (1974) 62–88. [370]

Heller, I. (1953a), On the problem of shortest path between points. I, *Bulletin of the American Mathematical Society* **59** (1953) 551. [378]

Heller, I. (1953b), On the problem of shortest path between points. II, *Bulletin of the American Mathematical Society* **59** (1953) 551–552. [378]

Heller, I. (1955a), On the travelling salesman's problem, in: *Proceedings of the Second Symposium in Linear Programming, Vol. II* (H. A. Antosiewitz, ed.), National Bureau of Standards, Washington, 1955, pp. 643–665. [378]

Heller, I. (1955b), Geometric characterization of cyclic permutations (Abstract), *Bulletin of the American Mathematical Society* **61** (1955) 227. [378]

Heller, I. (1955c), Neighbor relations on the convex of cyclic permutations (Abstract), *Bulletin of the American Mathematical Society* **61** (1955) 440 [full article: *Pacific Journal of Mathematics* **6** (1956) 467–477]. [378]

Heller, I. (1957), On linear systems with integral valued solutions, *Pacific Journal of Mathematics* **7** (1957) 1351–1364. [299]

Heller, I. (1963), On unimodular sets of vectors, in: *Recent Advances in Mathematical Programming* (R. L. Graves and P. Wolfe, eds.), McGraw-Hill, New York, 1963, pp. 39–53. [272]

Heller, I. and Hoffman, A. J. (1962), On unimodular matrices, *Pacific Journal of Mathematics* **12** (1962) 1321–1327. [278]

Heller, I. and Tompkins, C. B. (1956), An extension of a theorem of Dantzig's, in: *Linear Inequalities and Related Systems* (H. W. Kuhn and A. W. Tucker, eds.), Princeton University Press, Princeton, N.J., 1956, pp. 247–254. [272, 274, 276]

Henrici, P. (1977), *Applied and Computational Complex Analysis, Vol. 2, Special Functions—Integral Transforms—Asymptotics—Continued Fractions*, Wiley, New York, 1977. [82]

Hermite, Ch. (1849), Sur une question relative à la théorie des nombres, *Journal de Mathématiques Pures et Appliquées* **14** (1849) 21–30 [reprinted in: *Oeuvres de Charles Hermite, Tome I* (É. Picard, ed.), Gauthier-Villars, Paris, 1905, pp. 265–273]. [78]

Hermite, Ch. (1850a), Extraits de lettres de M. Ch. Hermite à M. Jacobi sur différents objets de la théorie des nombres, *Journal für die reine und angewandte Mathematik* **40** (1850) 261–278, 279–290, 291–307, 308–315 [reprinted in: *Oeuvres de Charles Hermite, Tome I* (É. Picard, ed.), Gauthier-Villars, Paris, 1905, pp. 100–121, 122–135, 136–155, 155–163]. [67, 78]

Hermite, Ch. (1850b), Sur la théorie des formes quadratiques ternaires, *Journal für die reine und angewandte Mathematik* **40** (1850) 173–177 [reprinted in: *Oeuvres de Charles Hermite, Tome I* (É. Picard, ed.), Gauthier-Villars, Paris, 1905, pp. 94–99]. [78]

Hermite, Ch. (1851), Sur l'introduction des variables continues dans la théorie des nombres, *Journal für die reine und angewandte Mathematik* **41** (1851) 191–216 [reprinted in: *Oeuvres de Charles Hermite, Tome III* (K. Hensel, ed.), Gauthier-Villars, Paris, 1905, pp. 164–192]. [80]

Hermite, Ch. (1880), Sur une extension donnée à la théorie des fractions continues par M. Tchebychef (Extrait d'une lettre de M. Ch. Hermite à M. Borchardt [22 mars 1879]), *Journal für die reine und angewandte Mathematik* **88** (1880) 10–15 [reprinted in: *Oeuvres de Charles Hermite, Tome III* (É. Picard, ed.), Gauthier-Villars, Paris, 1912, pp. 513–519]. [80]

Hilbert, D. (1890), Ueber die Theorie der algebraischen Formen, *Mathematische Annalen* **36** (1890) 473–534 [reprinted in: *David Hilbert Gesammelte Abhandlungen, Vol. II*, Springer, Berlin, 1933 [reprinted: Chelsea, New York, 1965], pp. 199–257]. [233, 376]

Hirsch, W. M. and Dantzig, G. B. (1954), *The fixed charge problem*, Report P–648, The RAND Corporation, Santa Monica, Cal., 1954. [378]

Hirschberg, D. S. and Wong, C. K. (1976), A polynomial-time algorithm for the knapsack problem with two variables, *Journal of the Association for Computing Machinery* **23** (1976) 147–154. [256]

Hitchcock, F. L. (1941), The distribution of a product from several sources to numerous localities, *Journal of Mathematics and Physics* **20** (1941) 224–230. [222, 378]

Ho, J. K. and Loute, E. (1980), A comparative study of two methods for staircase linear programs, *ACM Transactions on Mathematical Software* **6** (1980) 17–30. [139]

Ho, J. K. and Loute, E. (1983), Computational experience with advanced implementation of decomposition algorithms for linear programming, *Mathematical Programming* **27** (1983) 283–290. [139]

Hochbaum, D. S. (1982), Approximation algorithms for the set covering and vertex cover problems, *SIAM Journal on Computing* **11** (1982) 555–556. [380]

Hoffman, A. J. (1952), On approximate solutions of systems of linear inequalities, *Journal of Research of the National Bureau of Standards* **49** (1952) 263–265. [126, 158]

Hoffman, A. J. (1953), *Cycling in the simplex algorithm*, Report No. 2974, Nat. Bur. Standards, Washington, D.C., 1953. [138]

Hoffman, A. J. (1960), Some recent applications of the theory of linear inequalities to extremal combinatorial analysis, in: *Combinatorial Analysis* (Proceedings of Symposia in Applied Mathematics, Volume X (R. Bellman and M. Hall, Jr, eds.)), American Mathematical Society, Providence, R.I., 1960, pp. 113–127. [272, 279, 299]

Hoffman, A. J. (1974), A generalization of max flow-min cut, *Mathematical Programming* **6** (1974) 352–359. [309, 310, 311]

Hoffman, A. J. (1976), Total unimodularity and combinatorial theorems, *Linear Algebra and Its Applications* **13** (1976) 103–108. [272, 298]

Hoffman, A. J. (1979a), Linear programming and combinatorics, in: *Relations Between Combinatorics and Other Parts of Mathematics* (Proceedings of Symposia in Pure Mathematics Vol. XXXIV; D. K. Ray-Chaudhuri, ed.), American Mathematical Society, Providence, R.I., 1979, pp. 245–253. [379]

Hoffman, A. J. (1979b), The role of unimodularity in applying linear inequalities to combinatorial theorems, [in: *Discrete Optimization I* (P. L. Hammer, E. L. Johnson and B. H. Korte, eds.),] *Annals of Discrete Mathematics* **4** (1979) 73–84. [272, 303]

Hoffman, A. J. and Kruskal, J. B. (1956), Integral boundary points of convex polyhedra, in: *Linear Inequalities and Related Systems* (H. W. Kuhn and A. W. Tucker, eds.), Princeton Univ. Press, Princeton, N.J., 1956, pp. 223–246. [99, 104, 267, 268, 269, 273, 274, 278, 279, 301, 378]

Hoffman, A. J. and Kuhn, H. W. (1956), Systems of distinct representatives and linear programming, *The American Mathematical Monthly* **63** (1956) 455–460. [273]

Hoffman, A., Mannos, M., Sokolowsky, D., and Wiegmann, N. (1953), Computational experience in solving linear programs, *Journal of the Society for Industrial and Applied Mathematics* **1** (1953) 17–33. [138, 139]

Hoffman, A. J. and Oppenheim, R. (1978), Local unimodularity in the matching polytope, [in: *Algorithmic Aspects of Combinatorics* (B. Alspach, P. Hell and D. J. Miller, eds.),] *Annals of Discrete Mathematics* **2** (1978) 201–209. [303]

Hofmeister, G. R. (1966), Zu einem Problem von Frobenius, *Det Kongelige Norske Videnskabers Selskabs Skrifter* 1966 Nr. 5 (1966) 1–37. [379]

Holm, S. and Klein, D. (1984), Three methods for postoptimal analysis in integer linear programming, *Mathematical Programming Study* **21** (1984) 97–109. [354]

Hopcroft, J. and Tarjan, R. (1974), Efficient planarity testing, *Journal of the Association for Computing Machinery* **21** (1974) 549–568. [282, 283]

Houck, D. J. and Vemuganti, R. R. (1977), An algorithm for the vertex packing problem, *Operations Research* **25** (1977) 773–787. [380]

Householder, A. S. (1964), *The Theory of Matrices in Numerical Analysis*, Blaisdell, New York, 1964 [reprinted: Dover, New York, 1975]. [41]

Hrouda, J. (1971), A contribution to Balas' algorithm, *Aplikace Matematiky* **16** (1971) 336–353. [363]

Hrouda, J. (1975), Modifikovaný Bendersův algoritmus, *Ekonomicko-Matematický Obzor* **11** (1975) 423–443. [372]

Hrouda, J. (1976), The Benders method and parametrization of the right-hand sides in the mixed integer linear programming problem, *Aplikace Matematiky* **21** (1976) 327–364. [372]

Hu, T. C. (1969), *Integer Programming and Network Flows*, Addison-Wesley, Reading, Mass., 1969. [56, 59, 367, 378]

Hu, T. C. (1970a), On the asymptotic integer algorithm, in: *Integer and Nonlinear Programming* (J. Abadie, ed.), North-Holland, Amsterdam, 1970, pp. 381–383. [367]

Hu, T. C. (1970b), On the asymptotic integer algorithm, *Linear Algebra and Its Applications* **3** (1970) 279–294. [367]

Hua, L. K. (1982), *Introduction to Number Theory*, Springer, Berlin, 1982. [82]

Huang, H.-C. and Trotter, Jr, L. E. (1980), A technique for determining blocking and anti-blocking polyhedral descriptions, *Mathematical Programming Study* **12** (1980) 197–205. [119]

Huang, H.-S. (1981), An algorithm for the solution of a linear diophantine problem of Frobenius, *Chinese Journal of Mathematics* **9** (1981) 67–74. [379]

Huber, A. (1930), Eine Erweiterung der Dirichletschen Methode des Diskontinuitätsfaktors und ihre Anwendung auf eine Aufgabe der Wahrscheinlichkeitsrechnung, *Monatshefte für Mathematik und Physik* **37** (1930) 55–72. [217]

Huet, G. (1978), An algorithm to generate the basis of solutions to homogeneous linear diophantine equations, *Information Processing Letters* **7** (1978) 144–147. [234, 379]

Hung, M. S. (1983), A polynomial simplex method for the assignment problem, *Operations Research* **31** (1983) 595–600. [142]

Hung, M. S. and Rom, W. O. (1980), Solving the assignment problem by relaxation, *Operations Research* **28** (1980) 969–982. [142]

Hurwitz, A. (1891), Ueber die angenäherte Darstellung der Irrationalzahlen durch rationale Brüche, *Mathematische Annalen* **39** (1891) 279–284 [reprinted in: *Mathematische Werke von Adolf Hurwitz, Zweiter Band*, Birkhäuser, Basel, 1963, pp. 122–128]. [67, 81]

Hwang, F. K., Sun, J. and Yao, E. Y. (1985), Optimal set partitioning, *SIAM Journal on Algebraic and Discrete Methods* **6** (1985) 163–170. [380]

Ibaraki, T. (1976), Integer programming formulation of combinatorial optimization problems, *Discrete Mathematics* **16** (1976) 39–52. [379]

Ibarra, O. H. and Kim, C. E. (1975), Fast approximation algorithms for the knapsack and sum of subset problems, *Journal of the Association for Computing Machinery* **22** (1975) 463–468. [262]

Ikura, Y. and Nemhauser, G. L. (1983), *A polynomial-time dual simplex algorithm for the transportation problem*, Technical Report 602, School of Operations Research and Industrial Engineering, Cornell University, Ithaca, N.Y., 1983. [142, 279]

Ingargiola, G. P. and Korsh, J. F. (1977), A general algorithm for one-dimensional knapsack problems, *Operations Research* **25** (1977) 752–759. [373]

Israilov, M. I. (1981), Numbers of solutions of linear diophantine equations and their applications in the theory of invariant cubature formulas (in Russian), *Sibirskii Matematicheskii Zhurnal* **22** (2) (1981) 121–136 [English translation: *Siberian Mathematical Journal* **22** (1981) 260–273]. [379]

Jacobi, C. G. J. (1845), Über eine neue Auflösungsart der bei der Methode der kleinsten Quadrate vorkommenden linearen Gleichungen, *Astronomische Nachrichten* **22** (1845) 297–303 [reprinted in: *C. G. J. Jacobi's Gesammelte Werke, 3. Band* (K. Weierstrass, ed.), G. Reimer, Berlin, 1884 [reprinted: Chelsea New York, 1969], pp. 469–478]. [36, 40]

Jacobi, C. G. J. (1868), Allgemeine Theorie der kettenbruchähnlichen Algorithmen, in welchen jede Zahl aus drei vorhergehenden gebildet wird (Aus den hinterlassenen Papieren von C. G. J. Jacobi mitgetheilt durch Herrn E. Heine), *Journal für die reine und angewandte Mathematik* **69** (1868) 29–64 [reprinted in: *C. G. J. Jacobi's Gesammelte Werke, 6. Band* (K. Weierstrass, ed.), G. Reimer, Berlin, 1891 [reprinted: Chelsea, New York, 1969], pp. 385–426]. [76]

Jacobs, W. (1954), The caterer problem, *Naval Research Logistics Quarterly* **1** (1954) 154–165. [378]

Jaromczyk, J. W. (1981), Linear decision trees are too weak for convex hull problem, *Information Processing Letters* **12** (1981) 138–141. [224]

Jarvis, R. A. (1973), On the identification of the convex hull of a finite set of points in the plane, *Information Processing Letters* **2** (1973) 18–21. [224]

Jeroslow, R. G. (1969), *On the unlimited number of faces in integer hulls of linear programs with two constraints*, Tech. Report No. 67, Dept. of Operations Research, College of Engineering, Cornell University, Ithaca, N.Y., 1969. [255]

Jeroslow, R. G. (1971), Comments on integer hulls of two linear constraints, *Operations Research* **19** (1971) 1061–1069. [256]

Jeroslow, R. G. (1973), The simplex algorithm with the pivot rule of maximizing criterion improvement, *Discrete Mathematics* **4** (1973) 367–377. [141]

Jeroslow, R. G. (1974), Trivial integer programs unsolvable by branch-and-bound,

Mathematical Programming **6** (1974) 105–109. [362]

Jeroslow, R. G. (1975), A generalization of a theorem of Chvátal and Gomory, in: *Nonlinear Programming*, 2 (O. L. Mangasarian, R. R. Meyer and S. M. Robinson, eds.), Academic Press, New York, 1975, pp. 313–331. [354]

Jeroslow, R. G. (1977), Cutting-plane theory: disjunctive methods, [in: *Studies in Integer Programming* (P. L. Hammer, et al., eds.),] *Annals of Discrete Mathematics* **1** (1977) 293–330. [354, 379]

Jeroslow, R. G. (1978a), Cutting-plane theory: algebraic methods, *Discrete Mathematics* **23** (1978) 121–150. [354, 367]

Jeroslow, R. G. (1978b), Some basis theorems for integral monoids, *Mathematics of Operations Research* **3** (1978) 145–154. [234]

Jeroslow, R. (1979a), The theory of cutting-planes, in: *Combinatorial Optimization* (N. Christofides, A. Mingozzi, P. Toth and C. Sandi, eds.), Wiley, Chichester, 1979, pp. 21–72. [354]

Jeroslow, R. G. (1979b), Minimal inequalities, *Mathematical Programming* **17** (1979) 1–15. [354]

Jeroslow, R. (1979c), An introduction to the theory of cutting-planes, [in: *Discrete Optimization II* (P. L. Hammer, E. L. Johnson and B. H. Korte, eds.),] *Annals of Discrete Mathematics* **5** (1979) 71–95. [354]

Jeroslow, R. G. (1979d), Some relaxation methods for linear inequalities, *Cahiers du Centre d'Études de Recherche Opérationnelle* **21** (1979) 43–53. [158]

Jeroslow, R. G. and Kortanek, K. O. (1971), On an algorithm of Gomory, *SIAM Journal on Applied Mathematics* **21** (1971) 55–60. [255, 359]

John, F. (1948), Extremum problems with inequalities as subsidiary conditions, in: *Studies and Essays, presented to R. Courant on his 60th birthday January 8, 1948* [preface by K. O. Friedrichs, O. E. Neugebauer and J. J. Stoker,] Interscience, New York, 1948, pp. 187–204 [reprinted in: *Fritz John, Collected Papers Volume 2* (J. Moser, ed.), Birkhäuser, Boston, 1985, pp. 543–560]. [161, 205, 221]

John, F. (1966), *Lectures on Advanced Numerical Analysis*, Gordon and Breach, New York, 1966. [41]

Johnson, D. S. (1974a), Approximation algorithms for combinatorial problems, *Journal of Computer and System Sciences* **9** (1974) 256–278. [263, 380]

Johnson, E. L. (1973), Cyclic groups, cutting planes, shortest paths, in: *Mathematical Programming* (T. C. Hu and S. M. Robinson, eds.), Academic Press, New York, 1973, pp. 185–211. [367]

Johnson, E. L. (1974b), On the group problem for mixed integer programming, *Mathematical Programming Study* **2** (1974) 137–179. [367]

Johnson, E. L. (1978), Support functions, blocking pairs, and anti-blocking pairs, *Mathematical Programming Study* **8** (1978) 167–196. [119, 367]

Johnson, E. L. (1979), On the group problem and a subadditive approach to integer programming, [in: *Discrete Optimization II* (P. L. Hammer, E. L. Johnson and B. H. Korte, eds.),] *Annals of Discrete Mathematics* **5** (1979) 97–112. [367]

Johnson, E. L. (1980a), *Integer Programming–Facets, Subadditivity, and Duality for Group and Semi-group Problems*, [CBMS-NSF Regional Conference Series in Applied Mathematics 32,] Society for Industrial and Applied Mathematics, Philadelphia, Penn., 1980. [367, 379]

Johnson, E. L. (1980b), Subadditive lifting methods for partitioning and knapsack problems, *Journal of Algorithms* **1** (1980) 75–96. [367, 380]

Johnson, E. L. (1981a), On the generality of the subadditive characterization of facets, *Mathematics of Operations Research* **6** (1981) 101–112. [354, 367]

Johnson, E. L. (1981b), Characterization of facets for multiple right-hand choice linear programs, *Mathematical Programming Study* **14** (1981) 112–142. [354, 367]

Johnson, E. L. and Suhl, U. H. (1980), Experiments in integer programming, *Discrete Applied Mathematics* **2** (1980) 39–55. [372, 374]

Johnson, S. M. (1960), A linear diophantine problem, *Canadian Journal of Mathematics* **12** (1960) 390–398. [379]

Jones, W. B. and Thron, W. J. (1980), *Continued Fractions–Analytic Theory and Applications*, Addison-Wesley, Reading, Mass., 1980. [77, 82]

Jordan, W. (1904), *Handbuch der Vermessungskunde, Erster Band: Ausgleichungs-Rechnung nach der Methode der kleinsten Quadrate (5th edition)*, J. B. Metzlersche Verlagsbuchhandlung, Stuttgart, 1904. [40]

Jurkat, W. B., Kratz, W., and Peyerimhoff, A. (1977), Explicit representations of Dirichlet approximations, *Mathematische Annalen* **228** (1977) 11–25. [74]

Jurkat, W., Kratz, W., and Peyerimhoff, A. (1979), On best two-dimensional Dirichlet-approximations and their algorithmic calculation, *Mathematische Annalen* **244** (1979) 1–32. [74]

Jurkat, W. and Peyerimhoff, A. (1976), A constructive approach to Kronecker approximations and its application to the Mertens conjecture, *Journal für die reine und angewandte Mathematik* **286/287** (1976) 322–340. [74–75]

Kabatyanskiĭ, G. A. and Levenshteĭn, V. I. (1978), Bounds for packings on a sphere and in space (in Russian), *Problemy Peredachi Informatsii* **14** (1) (1978) 3–25 [English translation: *Problems of Information Transmission* **14** (1978) 1–17]. [81]

Kakeya, S. (1913–4), On a system of linear forms with integral variables, *The Tôhoku Mathematical Journal* **4** (1913–4) 120–131. [80, 217]

Kannan, R. (1976), *A proof that integer programming is in NP*, unpublished, 1976. [239]

Kannan, R. (1980), A polynomial algorithm for the two-variable integer programming problem, *Journal of the Association for Computing Machinery* **27** (1980) 118–122. [256]

Kannan, R. (1983a), Polynomial-time aggregation of integer programming problems, *Journal of the Association for Computing Machinery* **30** (1983) 133–145. [236, 263]

Kannan, R. (1983b), Improved algorithms for integer programming and related lattice problems, in: *Proceedings of the Fifteenth Annual ACM Symposium on Theory of Computing*, The Association for Computing Machinery, New York, 1983, pp. 193–206. [71, 72, 261]

Kannan, R. and Bachem, A. (1979), Polynomial algorithms for computing the Smith and Hermite normal forms of an integer matrix, *SIAM Journal on Computing* **8** (1979) 499–507. [56, 59]

Kannan, R. and Monma, C. L. (1978), On the computational complexity of integer programming problems, in: *Optimization and Operations Research* (Proceedings Bonn, 1977; R. Henn, B. Korte and W. Oettli, eds.), Lecture Notes in Economics and Mathematical Systems 157, Springer, Berlin, 1978, pp. 161–172. [239]

Kantorovich, L. V. (1939), *Mathematical Methods of Organizing and Planning Production* (in Russian), Publication House of the Leningrad State University, Leningrad, 1939 [English translation: *Management Science* **6** (1959–60) 366–422]. [221, 378]

[Kantorovich L. V. (1942) =] Kantorovitch, L. V., On the translocation of masses, *Comptes Rendus de l'Académie des Sciences de l'U.R.S.S.* [= *Doklady Akademii Nauk SSSR*] **37** (1942) 199–201 [reprinted: *Management Science* **5** (1958–9) 1–4]. [378]

Karmarkar, N. (1984), A new polynomial-time algorithm for linear programming, in: *Proceedings of the Sixteenth Annual ACM Symposium on Theory of Computing* (Washington, 1984), The Association for Computing Machinery, New York, 1984, pp. 302–311 [also: *Combinatorica* **4** (1984) 373–395]. [190]

Karp, R. M. (1972), Reducibility among combinatorial problems, in: *Complexity of Computer Computations* (R. E. Miller and J. W. Thatcher, eds.), Plenum Press, New York, 1972, pp. 85–103. [14, 18, 21, 249]

Karp, R. M. (1975), On the computational complexity of combinatorial problems, *Networks* **5** (1975) 45–68. [14]

Karp, R. M. and Papadimitriou, C. H. (1980), On linear characterizations of combinatorial optimization problems, in: *21st Annual Symposium on Foundations of Computer Science*, IEEE, New York, 1980, pp. 1–9 [also in: *SIAM Journal on Computing* **11** (1982) 620–632]. [163, 172, 173, 237, 252]

Karush, W. (1939), *Minima of Functions of Several Variables with Inequalities as Side Constraints*, M.Sc. Dissertation, Department of Mathematics, University of Chicago, Chicago, Ill., 1939. [220]

Karwan, M. H. and Rardin, R. L. (1979), Some relationships between Lagrangian and surrogate duality in integer programming, *Mathematical Programming* 17 (1979) 320–324. [354, 370]

Kastning, C. (ed.) (1976), *Integer Programming and Related Areas, A Classified Bibliography*, Lecture Notes in Economics and Mathematical Systems 128, Springer, Berlin, 1976. [236, 376, 379]

Kaufmann, A. and Henry-Labordère, A. (1977), *Integer and Mixed Programming: Theory and Applications*, Academic Press, New York, 1977. [378]

Kelley, Jr, J. E. (1960), The cutting-plane method for solving convex programs, *Journal of the Society for Industrial and Applied Mathematics* 8 (1960) 703–712. [379]

Kellison, S. G. (1975), *Fundamentals of Numerical Analysis*, Irwin, Homewood, Ill., 1975. [41]

Kelly, D. G. (1981), *Some results on random linear programs*, Department of Statistics, Univ. of North Carolina, Chapel Hill, N.C., 1981. [147]

Kelly, D. G. and Tolle, J. W. (1979), Expected simplex algorithm behaviour for random linear programs, *Operations Research Verfahren/Methods of Operations Research* 31 (III Symposium on Operations Research, Mannheim, 1978; W. Oettli and F. Steffens, eds.), Athenäum/Hain/Scriptor/Hanstein, Königstein/Ts., 1979, pp. 361–367. [147]

Kelly, D. G. and Tolle, J. W. (1981), Expected number of vertices of a random convex polyhedron, *SIAM Journal on Algebraic and Discrete Methods* 2 (1981) 441–451. [147]

Kendall, K. E. and Zionts, S. (1977), Solving integer programming problems by aggregating constraints, *Operations Research* 25 (1977) 346–351. [236]

Kersey, J. (1673), *The Elements of That Mathematical Art Commonly Called Algebra, Vol. I*, T. Passinger, London, 1673. [77]

Kertzner, S. (1981), The linear diophantine equation, *The American Mathematical Monthly* 88 (1981) 200–203. [54]

Khachiyan, L. G. (1979), A polynomial algorithm in linear programming (in Russian), *Doklady Akademii Nauk SSSR* 244 (1979) 1093–1096 [English translation: *Soviet Mathematics Doklady* 20 (1979) 191–194]. [126, 163, 171]

Khachiyan, L. G. (1980), Polynomial algorithms in linear programming (in Russian), *Zhurnal Vychislitel'noi Matematiki i Matematicheskoi Fiziki* 20 (1980) 51–68 [English translation: *U.S.S.R. Computational Mathematics and Mathematical Physics* 20 (1980) 53–72]. [163, 171]

Khintchine, A. (1956), Kettenbrüche [translated from the Russian, published 1935], Teubner, Leipzig, 1956 [English translation: *Continued Fractions*, Noordhoff, Groningen (The Netherlands), 1963]. [82]

Kianfar, F. (1971), Stronger inequalities for 0,1 integer programming using knapsack functions, *Operations Research* 19 (1971) 1374–1392. [359]

Kim, C. (1971), *Introduction to Linear Programming*, Holt, Rinehart, and Winston, New York, 1971. [224]

Klee, V. (1959), Some characterizations of convex polyhedra, *Acta Mathematica* (Uppsala) 102 (1959) 79–107. [223]

Klee, V. L. (ed.) (1963), *Convexity*, Proceedings of Symposia in Pure Mathematics Vol. VII, American Mathematical Society, Providence, R.I., 1963. [224]

Klee, V. (1964a), On the number of vertices of a convex polytope, *Canadian Journal of Mathematics* 16 (1964) 701–720. [142, 223]

Klee, V. (1964b), Diameters of polyhedral graphs, *Canadian Journal of Mathematics* 16 (1964) 602–614. [142]

Klee, V. (1965a), Heights of convex polytopes, *Journal of Mathematical Analysis and Applications* 11 (1965) 176–190. [142]

Klee, V. (1965b), Paths on polyhedra. I, *Journal of Society for Industrial and Applied Mathematics* **13** (1965) 946–956. [142]

Klee, V. (1966a), Convex polytopes and linear programming, *Proceedings of the IBM Scientific Computing Symposium on Combinatorial Problems* (Yorktown Heights, N.Y., 1964), I.B.M. Data Processing Division, White Plains, N.Y., 1966, pp. 123–158. [142]

Klee, V. (1966b), Paths on polyhedra. II, *Pacific Journal of Mathematics* **17** (1966) 249–262. [142]

Klee, V. (1969), Separation and support properties of convex sets – a survey, in: *Control Theory and the Calculus of Variations* (A.V. Balakrishnan, ed.), Academic Press, New York, 1969, pp. 235–303. [223]

Klee, V. (1973), *Vertices of convex polytopes*, Lecture Notes, Department of Mathematics, University of Washington, Seattle, Wash., 1973. [157]

Klee, V. (1974), Polytope pairs and their relationship to linear programming, *Acta Mathematica (Uppsala)* **133** (1974) 1–25. [142]

Klee, V. (1978), Adjoints of projective transformations and face-figures of convex polytopes, *Mathematical Programming Study* **8** (1978) 208–216. [224]

Klee, V. and Minty, G. J. (1972), How good is the simplex algorithm?, in: *Inequalities, III* (O. Shisha, ed.), Academic Press, New York, 1972, pp. 159–175. [139]

Klee, V. and Walkup, D. W. (1967), The d-step conjecture for polyhedra of dimension $d < 6$, *Acta Mathematica (Uppsala)* **117** (1967) 53–78. [142]

Klein, D. and Holm, S. (1979), Integer programming post-optimal analysis with cutting planes, *Management Science* **25** (1979) 64–72. [354]

Kline, M. (1972), *Mathematical Thought from Ancient to Modern Times*, Oxford University Press, New York, 1972. [41, 77]

Kloyda, M. Th. à K. (1937), Linear and quadratic equations 1550–1660, *Osiris* **3** (1937) 165–192. [41, 82]

Knobloch, E. (1980), *Der Beginn der Determinantentheorie, Leibnizens nachgelassene Studien zum Determinantenkalkül*, Gerstenberg Verlag, Hildesheim, 1980. [39]

Knuth, D. E. (1968), *The Art of Computer Programming, Vol. 1: Fundamental Algorithms*, Addison-Wesley, Reading, Mass., 1968. [16]

Knuth, D. E. (1969), *The Art of Computer Programming, Vol. 2: Seminumerical Algorithms*, Addison-Wesley, Reading, Mass., 1969. [54, 72, 82]

Kohler, D. A. (1967), *Projections of convex polyhedral sets*, Oper. Research Center Report 67–29, University of California, Berkeley, Cal., 1967. [157]

Kohler, D. A. (1973), Translation of a report by Fourier on his work on linear inequalities, *Opsearch* **10** (1973) 38–42. [157, 212]

Koksma, J. F. (1936), *Diophantische Approximationen*, Springer, Berlin, 1936 [reprinted: Chelsea, New York, 1950; Springer, Berlin, 1974]. [46, 82]

Koksma, J. F. and Meulenbeld, B. (1941), Ueber die Approximation einer homogenen Linearform an die Null, *Proceedings Koninklijke Nederlandse Akademie van Wetenschappen [Section of Sciences]* **44** (1941) 62–74. [75]

Kolesar, P. J. (1966–7), A branch and bound algorithm for the knapsack problem, *Management Science* **13** (1966–7) 723–735. [363]

Kondor, G. (1965), Comments on the solution of the integer linear programming problem, in: *Colloquium on Applications of Mathematics to Economics, Budapest, 1963* (A. Prékopa, ed.), Akadémiai Kiadó, Budapest, 1965, pp. 193–201. [359]

König, D. (1916), Gráfok és alkalmazásuk a determinánsok és halmazok elméletében (Hungarian), *Mathematikai és Természettudományi Értesitö* **34** (1916) 104–119 [German translation: Über Graphen und ihre Anwendung auf Determinantentheorie und Mengenlehre, *Mathematische Annalen* **77** (1916) 453–465]. [108]

König, D. (1931), Graphok és matrixok (Hungarian) [Graphs and matrices], *Matematikai és Fizikai Lapok* **38** (1931) 116–119. [273, 329, 378]

König, D. (1933), Über trennende Knotenpunkte in Graphen (nebst Anwendungen auf Determinanten und Matrizen), *Acta Litterarum ac Scientiarum Regiae Universitatis*

Hungaricae Francisco-Josephinae (Szeged), Sectio Scientiarum Mathematicarum **6** (1933) 155–179. [273, 378]

König, H. and Pallaschke, D. (1981), On Khachian's algorithm and minimal ellipsoids, *Numerische Mathematik* **36** (1981) 211–223. [208]

Koopmans, Tj. C. (1948), Optimum utilization of the transportation system, in: *The Econometric Society Meeting* (Washington, D.C., 1947; D. H. Leavens, ed.), [Proceedings of the International Statistical Conferences, Volume V,] 1948, pp. 136–146 [reprinted in: *Econometrica* **17** (Supplement) (1949) 136–146] [reprinted in: *Scientific Papers of Tjalling C. Koopmans*, Springer, Berlin, 1970, pp. 184–193]. [223, 378]

Koopmans, Tj. C. (ed.) (1951), *Activity Analysis of Production and Allocation*, Wiley, New York, 1951. [224]

Koopmans, Tj. C. (1959–60), A note about Kantorovich's paper, 'Mathematical methods of organizing and planning production', *Management Science* **6** (1959–60) 363–365. [222, 223]

Koopmans, Tj. C. (1961–2), On the evaluation of Kantorovich's work of 1939, *Management Science* **8** (1961–2) 264–265. [223]

Koopmans, Tj. C. and Beckmann, M. (1957), Assignment problems and the location of economic activities, *Econometrica* **25** (1957) 53–76. [378]

Koplowitz, J. and Jouppi, D. (1978), A more efficient convex hull algorithm, *Information Processing Letters* **7** (1978) 56–57. [224]

Korkine A. and Zolotareff, G. (1872), Sur les formes quadratiques positives quaternaires, *Mathematische Annalen* **5** (1872) 581–583. [81]

Korkine, A. and Zolotareff, G. (1873), Sur les formes quadratiques, *Mathematische Annalen* **6** (1873) 366–389. [81]

Korkine, A. and Zolotareff, G. (1877), Sur les formes quadratiques positives, *Mathematische Annalen* **11** (1877) 242–292. [81]

Körner, F. (1980), *Über die optimale Grösse von Teiltableaus beim Simplexverfahren*, preprint, Technische Universität Dresden, 1980. [147]

Kortanek, K. O. and Rom, W. O. (1971), Classification schemes for the strong duality of linear programming over cones, *Operations Research* **19** (1971) 1571–1585. [98]

Kortanek, K. O. and Soyster, A. L. (1972), On refinements of some duality theorems in linear programming over cones, *Operations Research* **20** (1972) 137–142. [98]

Korte, B. (1976), *Integer programming*, Report No 7649-OR, Institut für Ökonometrie und Operations Research, Bonn, 1976. [379]

Kotiah, T. C. T. and Steinberg, D. I. (1978), On the possibility of cycling with the simplex method, *Operations Research* **26** (1978) 374–376. [138]

Kress, M. and Tamir, A. (1980), The use of Jacobi's lemma in unimodularity theory, *Mathematical Programming* **18** (1980) 344–348. [272]

Krol, Y. and Mirman, B. (1980), *Some practical modifications of ellipsoidal methods for LP problems*, Arcon Inc., Boston, Mass., 1980. [208]

Krolak, P. D. (1969), Computational results of an integer programming algorithm, Operations Research, Bonn, 1976. [379]

Kronecker, L. (1877), Über Abelsche Gleichungen (Auszug aus der am 16. April 1877 gelesenen Abhandlung), *Monatsberichte der Königlich Preussischen Akademie der Wissenschaften zu Berlin* (1877) 845–851 [reprinted in: *Leopold Kronecker's Werke, Band IV* (K. Hensel, ed.), Teubner, Leipzig, 1929 [reprinted: Chelsea, New York, 1968] pp. 63–71]. [51]

Kronecker, L. (1883), Sur les unités complexes, *Comptes Rendus Hebdomadaires des Séances de l'Académie des Sciences* **96** (1883) 93–98, 148–152, 216–221 [reprinted in: *Leopold Kronecker's Werke, Band III-1* (K. Hensel, ed.), Teubner, Leipzig, 1899 [reprinted: Chelsea, New York, 1968], pp. 3–19]. [80]

Kronecker, L. (1884a), Die Periodensysteme von Functionen reeller Variabeln, *Monatsberichte der Königlich Preussischen Akademie der Wissenschaften zu Berlin* (1884) 1071–1080 [reprinted in: *Leopold Kronecker's Werke, Band III-1* (K. Hensel, ed.), Teubner,

Leipzig, 1899 [reprinted: Chelsea, New York, 1968], pp. 33–46]. [80]

Kronecker, L. (1884b), Näherungsweise ganzzahlige Auflösung linearer Gleichungen, *Monatsberichte der Königlich Preussischen Akademie der Wissenschaften zu Berlin* (1884) 1179–1193, 1271–1299 [reprinted in: *Leopold Kronecker's Werke, Band III-1* (K. Hensel, ed.), Teubner, Leipzig, 1899 [reprinted: Chelsea, New York, 1968], pp. 49–109]. [46, 80]

Kuhn, H. W. (1955a), The Hungarian method for the assignment problem, *Naval Research Logistics Quarterly* **2** (1955) 83–97. [378]

Kuhn, H. W. (1955b), On certain convex polyhedra (Abstract), *Bulletin of the American Mathematical Society* **61** (1955) 557–558. [378]

Kuhn, H. W. (1956a), Solvability and consistency for linear equations and inequalities, *The American Mathematical Monthly* **63** (1956) 217–232. [89, 94, 156, 157, 223]

Kuhn, H. W. (1956b), Variants of the Hungarian method for assignment problems, *Naval Research Logistics Quarterly* **3** (1956) 253–258. [378]

Kuhn, H. W. (1956c), On a theorem of Wald, in: *Linear Inequalities and Related Systems* (H. W. Kuhn and A. W. Tucker, eds.), Annals of Mathematics Studies 38, Princeton University Press, Princeton, N.J., 1956, pp. 265–274 [reprinted in: *Readings in Mathematical Economics, Vol. I, Value Theory* (P. Newman, ed.), The Johns Hopkins Press, Baltimore, 1968, pp. 106–115]. [219]

Kuhn, H. W. and Quandt, R. E. (1963), An experimental study of the simplex method, in: *Experimental Arithmetic, High-Speed Computing and Mathematics* (Proceedings of Symposia in Applied Mathematics XV; N. C. Metropolis, et al., eds.), American Mathematical Society, Providence, R. I., 1963, pp. 107–124. [138, 139]

Kuhn, H. W. and Tucker, A. W. (1951), Nonlinear programming, in: *Proceedings of the Second Berkeley Symposium on Mathematical Statistics and Probability* (Berkeley, 1950; J. Neyman, ed.), University of California Press, Berkeley, California, 1951, pp. 481–492 [reprinted in: *Readings in Mathematical Economics, Vol. I, Value Theory* (P. Newman, ed.), The Johns Hopkins Press, Baltimore, 1968, pp. 3–14]. [99, 221]

Kuhn, H. W. and Tucker, A.W. (eds.) (1956), *Linear Inequalities and Related Systems*, Annals of Mathematics Studies 38, Princeton University Press, Princeton, N.J., 1956. [223, 224]

Kuratowski, C. (1930), Sur le problème des courbes gauches en topologie, *Fundamenta Mathematicae* **15** (1930) 271–283. [9, 282, 287]

La Menza, F. (1930), Los sistemas de inecuaciones lineales y la división del hiperespacio, in: *Atti del Congresso Internazionale dei Matematici, Vol. II* (Bologna, 1928), N. Zanichelli, Bologna, [1930,] pp. 199–209. [217]

La Menza, F. (1932a), Sobre los sistemas de inecuaciones lineales, *Boletin Matematico* **5** (1932) 127–130. [217]

La Menza, F. (1932b), Sobre los sistemas de inecuaciones lineales (Conclusión), *Boletin Matematico* **5** (1932) 149–152. [217]

Lagarias, J. C. (1980a), Some new results in simultaneous diophantine approximation, in: *Proceedings of the Queen's Number Theory Conference, 1979* (P. Ribenboim, ed.), Queen's Papers in Pure and Applied Mathematics 54, Queen's University, Kingston, Ontario, 1980, pp. 453–474. [75]

Lagarias, J. C. (1980b), Worst-case complexity bounds for algorithms in the theory of integral quadratic forms, *Journal of Algorithms* **1** (1980) 142–186. [82]

Lagarias, J. C. (1982), The computational complexity of simultaneous diophantine approximation problems, in: *23rd Annual Symposium on Foundations of Computer Science*, IEEE, New York, 1982, pp. 32–39 [final version: *SIAM Journal on Computing* **14** (1985) 196–209]. [74, 251]

Lagarias, J. C. and Odlyzko, A. M. (1983), Solving low-density subset sum problems, in: *24th Annual Symposium on Foundations of Computer Science*, IEEE, New York, 1983, pp. 1–10 [final version: *Journal of the Association for Computing Machinery* **32** (1985) 229–246]. [74, 249]

de Lagny, T. F. (1697), *Nouveaux Elémens d'Arithmétique et d'Algèbre, ou Introduction aux Mathématiques*, J. Jombert, Paris, 1697. [77]

de Lagny, T. F. (1733), *Analyse générale, ou méthodes nouvelles pour résoudre les problêmes de tous les genres et de tous les degrez à l'infini*, La Compagnie des Libraires, Paris, 1733 [also: *Mémoires de l'Académie Royale des Sciences, depuis 1666 jusqu'à 1699*, **11** [année 1720] (1733) 1–612 (+ 12p.)]. [77]

[Lagrange, J. L. (1769a) =] de la Grange, Sur la solution des problemes indéterminés du second degré, *Histoire de l'Académie Royale des Sciences et Belles-Lettres (Berlin)* **23** [année 1767] (1769) 165–310 [reprinted in: *Oeuvres de Lagrange, Vol. II* (J.-A. Serret, ed.), Gauthier-Villars, Paris, 1868, [reprinted: G. Olms, Hildesheim, 1973,] pp. 377–535]. [77]

[Lagrange, J. L. (1769b) =] de la Grange, Sur la résolution des équations numériques, *Histoire de l'Académie Royale des Sciences et Belles-Lettres (Berlin)* **23** [année 1767] (1769) 311–352 [reprinted in: *Oeuvres de Lagrange, Vol. II* (J.-A. Serret, ed.), Gauthier-Villars, Paris, 1868, [reprinted: G. Olms, Hildesheim, 1973,] pp. 539–578]. [77]

[Lagrange, J. L. (1770) =] de la Grange, Nouvelle méthode pour résoudre les problemes indéterminés en nombres entiers, *Histoire de l'Académie Royale des Sciences et Belles-Lettres (Berlin)* **24** [année 1768] (1770) 181–250 [reprinted in: *Oeuvres de Lagrange, Vol. II* (J.-A. Serret, ed.), Gauthier-Villars, Paris, 1868, [reprinted: G. Olms, Hildesheim, 1973,] pp. 655–726]. [77]

de Lagrange, J. L. (1774), Additions aux éléments d'algèbre d'Euler—Analyse indéterminée, in: *L. Euler, Éléments d'Algèbre, traduits de l'allemand avec des Notes et Additions [par J. Bernoulli et Lagrange]*, J.-M. Bruyset, Lyon, 1774, pp. 517–523 [reprinted in: *Oeuvres de Lagrange, Vol. VII* (J.-A. Serret, ed.), Gauthier-Villars, Paris, 1877, [reprinted: G. Olms, Hildesheim, 1973,] pp. 5–180] [English translation in: *Elements of Algebra*, by Leonard Euler, J. Johnson, London, 1797 [fifth edition (1840) reprinted by: Springer, New York, 1984]]. [77]

[Lagrange, J. L. (1775a) =] de la Grange, Recherches d'arithmétique [Première partie], *Nouveaux Mémoires de l'Académie Royale des Sciences et Belles-Lettres (Berlin)* [année 1773] (1775) 265–312 [reprinted in: *Oeuvres de Lagrange, Vol. III* (J.-A. Serret, ed.) Gauthier-Villars, Paris, 1869, [reprinted: G. Olms, Hildesheim, 1973] pp. 695–758]. [77]

[Lagrange, J. L. (1775b) =] de la Grange, Nouvelle solution du probleme du mouvement de rotation d'un corps de figure quelconque qui n'est animé par aucune force accélératrice, *Nouveaux Mémoires de l'Académie Royale des Sciences et Belles-Lettres (Berlin)* [année 1773] (1775) 85–120 [reprinted in: *Oeuvres de Lagrange, Vol. III* (J.-A. Serret, ed.), Gauthier-Villars, Paris, 1869, [reprinted: G. Olms, Hildesheim, 1973,] pp. 579–616]. [39]

[Lagrange, J. L. (1775c) =] de la Grange, Solutions analytiques de quelques problemes sur les pyramides triangulaires, *Nouveaux Mémoires de l'Académie Royale des Sciences et Belles-Lettres (Berlin)* [année 1773] (1775) 149–176 [reprinted in: *Oeuvres de Lagrange, Vol. III* (J.-A. Serret, ed.), Gauthier-Villars, Paris, 1869, [reprinted: G. Olms, Hildesheim, 1973,] pp. 661–692]. [39]

Lagrange, J. L. (1778), *Sur les interpolations*, manuscript, [lu par l'Auteur à l'Académie des Sciences de Berlin le 3 septembre 1778,] Bibliothèque de l'Institut de France [German translation by Schulze: Über das Einschalten, nebst Tafeln und Beispielen, *Astronomisches Jahrbuch oder Ephemeriden für das Jahr 1783*, Dummler, Berlin, 1780, pp. 35–61] [reprinted in: *Oeuvres de Lagrange, Vol. VII* (J.-A. Serret, ed.), Gauthier-Villars, Paris, 1877, [reprinted: G. Olms, Hildesheim, 1973,] pp. 535–553]. [39]

[Lagrange, J. L. (1788) =] de la Grange, *Méchanique analitique*, Vve Desaint, Paris, 1788 [second, revised edition: *Mécanique analytique*, Vve Courcier, Paris, 1811–1815] [fourth edition as: *Oeuvres de Lagrange, Vols. XI (Première partie: la statique) and XII Seconde partie: la dynamique)* (J.-A. Serret and G. Darboux, eds.), Gauthier-Villars, Paris, 1888 and 1889 [reprinted in one volume: G. Olms, Hildesheim, 1973]] [reprinted in two volumes: A. Blanchard, Paris, 1965]. [210]

Lagrange, J. L. (1797), *Théorie des fonctions analytiques*, Impr. de la République, Paris, [an. v =] 1797 [second, revised edition: Ve Courcier, Paris, 1813 [reprinted as: *Oeuvres de Lagrange, Vols. IX* (J.-A. Serret, ed.), Gauthier-Villars, Paris, 1881 [reprinted by: G. Olms, Hildesheim, 1973]]]. [210]

Lagrange, J. L. (1798), Essai d'analyse numérique sur la transformation des fractions, *Journal de l'École Polytechnique*, tome 2, cahier 5 ([an.vi =] 1798) 93–114 [reprinted in: *Oeuvres de Lagrange, Vol. VII* (J.-A. Serret, ed.), Gauthier-Villars, Paris, 1877, [reprinted: G. Olms, Hildesheim, 1973,] pp. 291–313]. [77].

Laguerre, E. (1876–7), Sur la partition des nombres, *Bulletin de la Société mathématique de France* 5 (1876–7) 76–78 [reprinted in: *Oeuvres de Laguerre, Tome I* (Ch. Hermite, H. Poincaré and E. Rouché, eds.), Gauthier-Villars, Paris, 1898, [reprinted: Chelsea, New York, 1972,] pp. 218–220]. [375]

Lambe, T. A. (1974), Bounds on the number of feasible solutions to a knapsack problem, *SIAM Journal on Applied Mathematics* 26 (1974) 302–305. [379]

Lamé, G. (1844), Note sur la limite du nombre des divisions dans la recherche du plus grand commun diviseur entre deux nombres entiers, *Compte[s] Rendu[s] Hebdomadaires] des Séances de l'Académie des Sciences (Paris)* 19 (1844) 867–870. [21, 54]

Lancaster, P. and Tismenetsky, M. (1985), *The Theory of Matrices, Second Edition, with Applications*, Academic Press, Orlando, 1985. [4, 41]

Land, A. H. and Doig, A. G. (1960), An automatic method of solving discrete programming problems, *Econometrica* 28 (1960) 497–520. [360, 363, 378]

Lang, S. (1966a), *Linear Algebra*, Addison-Wesley, Reading, Mass., 1966. [4, 41]

Lang, S. (1966b), *Introduction to Diophantine Approximations*, Addison-Wesley, Reading, Mass., 1966. [82]

Langmaack, H. (1965), Algorithm 263 Gomory 1 [H], *Communications of the ACM* 8 (1965) 601–602. [359]

Langmayr, F. (1979a), Zur simultanen Diophantischen Approximation. I, *Monatshefte für Mathematik* 86 (1979) 285–300. [75]

Langmayr, F. (1979b), Zur simultanen Diophantischen Approximation. II, *Monatshefte für Mathematik* 87 (1979) 133–143. [75]

[Laplace, P. S. (1776) =] de la Place, Recherches sur le calcul intégral et sur le système du monde, *Histoire de l'Académie Royale des Sciences, avec les Mémoires de Mathématique et de Physique* [année 1772] (2e partie) (1776) 267–376 [reprinted in: *Oeuvres Complètes de Laplace, Tome VIII*, Gauthier-Villars, Paris, 1891, pp. 369–501]. [39]

Laplace, P. S. (1798), *Traité de mécanique céleste, Tome II*, J. B. M. Duprat, Paris, [an. vii =] 1798 [English translation: *Mécanique Céleste, Vol. II*, Hillard, Little and Wilkins, Boston, 1832 [reprinted: *Celestial Mechanics, Vol. II*, Chelsea, New York, 1966]]. [212]

Larman, D. (1970), Paths on polytopes, *Proceedings of the London Mathematical Society* (3) 20 (1970) 161–178. [142]

Lasdon, L. S. (1970), *Optimization Theory for Large Systems*, Macmillan, New York, 1970. [224].

Laurière, M. (1978), An algorithm for the 0/1 knapsack problem, *Mathematical Programming* 14 (1978) 1–10. [373]

Lawler, E. L. (1966), Covering problems: duality relations and a new method of solution, *SIAM Journal on Applied Mathematics* 14 (1966) 1115–1132. [380]

Lawler, E. L. (1976), *Combinatorial Optimization: Networks and Matroids*, Holt, Rinehart and Winston, New York, 1976. [379]

Lawler, E. L. (1977), Fast approximation algorithms for knapsack problems, in: *18th Annual Symposium on Foundation of Computer Science* (Providence, R.I., 1977), IEEE, New York, 1977, pp. 206–213. [263]

Lawler, E. L. (1979), Fast approximation algorithms for knapsack problems, *Mathematics of Operations Research* 4 (1979) 339–356. [263]

Lawler, E. L. and Bell, M. D. (1966), A method for solving discrete optimization problems, *Operations Research* 14 (1966) 1098–1112 [errata: 15 (1967) 578]. [363]

Lawler, E. L. and Wood, D. E. (1966), Branch-and-bound methods: a survey, *Operations Research* **14** (1966) 699-719. [363]

Lawrence, Jr, J. A. (1978), Abstract polytopes and the Hirsch conjecture, *Mathematical Programming* **15** (1978) 100-104. [142]

Lay, S. R. (1982), *Convex Sets and Their Applications*, Wiley, New York, 1982. [224]

Legendre, A. M. (1798), *Essai sur la théorie des nombres*, J. B. M. Duprat, Paris, 1798 [4th edition reprinted: *Théorie des nombres*, A. Blanchard, Paris, 1979]. [65, 77]

Legendre, A.-M. (1805), *Nouvelles méthodes pour la détermination des orbites des comètes*, F. Didot, Paris, 1805 [English translation of parts of pp. 72–75 in: D. E. Smith, *A Source Book in Mathematics*, McGraw-Hill, New York, 1929, pp. 576–579]. [36, 39, 212]

Lehman, A. (1965), *On the width-length inequality*, mimeographic notes, 1965 [published: *Mathematical Programming* **16** (1979) 245–259 (without proof corrections) and **17** (1979) 403–417 (with proof corrections)]. [116, 117]

Lehmer, D. N. (1919), The general solution of the indeterminate equation: $Ax + By + Cz + \cdots = r$, *Proceedings of the National Academy of Sciences* **5** (1919) 111–114. [54]

Leibniz, G. W. (1678?), *manuscript*, 1678? [cf. Gerhardt [1891]] [printed in: *Leibnizens Mathematische Schriftem, 2. Abth., Band III* (C. I. Gerhardt, ed.) H. W. Schmidt, Halle, 1856, [reprinted: G. Olms, Hildesheim, 1971], pp. 5–6] [English translation (partially) in: D. E. Smith, *A Source Book in Mathematics*, McGraw-Hill, New York, 1929, pp. 269–270]. [39]

Leibniz, G. W. (1693), *Letter to De l'Hospital, April 28, 1693* [reprinted in: *Leibnizens Mathematische Schriften, 1.Abth., Band II: Briefwechsel zwischen Leibniz, Hugens van Zulichem und dem Marquis de l'Hospital* (C. I. Gerhardt, ed.), A. Asher, Berlin, 1850 [reprinted, in one volume with Band I: G. Olms, Hildesheim, 1971,] pp. 236–241] [English translation (partially) in: D. E. Smith, *A Source Book in Mathematics*, McGraw-Hill, New York, 1929, pp. 267–269]. [38]

Leichtweiss, K. (1980), *Konvexe Mengen*, VEB Deutscher Verlag der Wissenschaften/Springer, Berlin, 1980. [224]

Lekkerkerker, C. G. (1969), *Geometry of Numbers*, Wolters–Noordhoff, Groningen and North-Holland, Amsterdam, 1969. [82, 376]

Lemke, C. E. (1954), The dual method of solving the linear programming problem, *Naval Research Logistics Quarterly* **1** (1954) 36–47. [148]

Lemke, C. E., Salkin, H. M., and Spielberg, K. (1971), Set covering by single-branch enumeration with linear-programming subproblems, *Operations Research* **19** (1971) 998–1022. [380]

Lemke, C. E. and Spielberg, K. (1967), Direct search algorithms for zero-one and mixed-integer programming, *Operations Research* **15** (1967) 892–914. [363]

Lenstra, A. K., Lenstra, Jr, H. W., and Lovász, L. (1982), Factoring polynomials with rational coefficients, *Mathematische Annalen* **261** (1982) 515–534. [68, 74]

Lenstra, Jr, H. W. (1983), Integer programming with a fixed number of variables, *Mathematics of Operations Research* **8** (1983) 538–548. [71, 205, 256, 260]

Leveque, W. J. (ed.) (1974), *Reviews in Number Theory* (6 vols), American Mathematical Society, Providence, R. I., 1974. [236, 375]

Levin, A. Yu. (1965), On an algorithm for the minimization of convex functions (in Russian), *Doklady Akademii Nauk SSSR* **160** (1965) 1244–1247 [English translation: *Soviet Mathematics Doklady* **6** (1965) 286–290]. [171]

Levit, R. J. (1956), A minimum solution of a diophantine equation, *The American Mathematical Monthly* **63** (1956) 646–651. [54]

Lewin, M. (1972), A bound for a solution of a linear diophantine problem, *Journal of the London Mathematical Society (2)* **6** (1972) 61–69. [379]

Lewin, M. (1973), On a linear diophantine problem, *The Bulletin of the London Mathematical Society* **5** (1973) 75–78. [379]

Lewin, M. (1975), An algorithm for a solution of a problem of Frobenius, *Journal für die reine und angewandte Mathematik* **276** (1975) 68–82. [379]

Lex, W. (1977), Über Lösungsanzahlen linearer diophantischer Gleichungen, *Mathematisch-physikalische Semesterberichte* **24** (1977) 254–279. [379]

Liebling, Th. M. (1973), On the number of iterations of the simplex method, in: *Operations Research-Verfahren/Methods of Operations Research* XVII (V. Oberwolfach-Tagung über Operations Research, Teil 2; 1972; R. Henn, H. P. Künzi and H. Schubert, eds.), Verlag Anton Hain, Meisenheim am Glan, 1973, pp. 248–264. [147]

Lifschitz, V. (1980), The efficiency of an algorithm of integer programming: a probabilistic analysis, *Proceedings of the American Mathematical Society* **79** (1980) 72–76. [373]

Lifschitz, V. and Pittel, B. (1983), The worst and the most probable performance of a class of set-covering algorithms, *SIAM Journal on Computing* **12** (1983) 329–346. [380]

Loehman, E., Nghiem, Ph. T., and Whinston, A. (1970), Two algorithms for integer optimization, *Revue Française d'Informatique et de Recherche Opérationelle* **4** (V-2) (1970) 43–63. [363]

Lovász, L. (1972), Normal hypergraphs and the perfect graph conjecture, *Discrete Mathematics* **2** (1972) 253–267. [118, 307, 350]

Lovász, L. (1975), On the ratio of optimal integral and fractional covers, *Discrete Mathematics* **13** (1975) 383–390. [380]

Lovász, L. (1976), On two minimax theorems in graph, *Journal of Combinatorial Theory (B)* **21** (1976) 96–103. [311, 327]

Lovász, L. (1977), Certain duality principles in integer programming, [in: *Studies in Integer Programming* (P. L. Hammer, et al., eds.),] *Annals of Discrete Mathematics* **1** (1977) 363–374. [327, 379]

Lovász, L. (1979a), Graph theory and integer programming, [in: *Discrete Optimization I* (P. L. Hammer, E. L. Johnson and B. H. Korte, eds.),] *Annals of Discrete Mathematics* **4** (1979) 141–158. [379]

Lovász, L. (1979b), *Combinatorial Problems and Exercises*, Akadémiai Kiadó, Budapest and North-Holland, Amsterdam, 1979. [299]

Lovitt, W. V. (1916), Preferential voting, *The American Mathematical Monthly* **23** (1916) 363–366. [217]

Lueker, G. S. (1975), *Two polynomial complete problems in nonnegative integer programming*, Report TR-178, Dept. of Computer Science, Princeton University, Princeton, N.J., 1975. [250].

MacDuffee, C. C. (1933), *The Theory of Matrices*, Springer, Berlin, 1933 [reprinted: Chelsea, New York, 1946]. [41]

Mack, J. M. (1977), Simultaneous diophantine approximation, *The Journal of the Australian Mathematical Society* **24** (Series A) (1977) 266–285. [75]

MacLaurin, C. (1748), *A Treatise of Algebra in three parts, to which is added an Appendix concerning the general properties of geometrical lines*, A. Millar and J. Nourse, London, 1748. [39]

MacMahon, P. A. (1916), *Combinatory Analysis, Vol. II*, The University Press, Cambridge, 1916 [reprinted, with Vol. I in one volume: Chelsea, New York, 1960]. [375]

Mahler, K. (1976), A theorem on diophantine approximations, *Bulletin of the Australian Mathematical Society* **14** (1976) 463–465. [75]

Maňas, M. and Nemoda, J. (1968), Finding all vertices of a convex polyhedron, *Numerische Mathematik* **12** (1968) 226–229. [224]

Mangasarian, O. L. (1979), Uniqueness of solution in linear programming, *Linear Algebra and Its Applications* **25** (1979) 151–162. [225]

Mangasarian, O. L. (1981a), A stable theorem of the alternative: an extension of the Gordan theorem, *Linear Algebra and Its Applications* **41** (1981) 209–223. [98, 225]

Mangasarian, O. L. (1981b), A condition number for linear inequalities and linear programs, in: *Methods of Operations Research* **43** (6. Symposium über Operations Research, Augsburg, 1981; G. Bamberg and O. Opitz, eds.), Athenäum/ Hain/Scriptor/Hanstein, Königstein/Ts., 1981, pp. 3–15. [126]

Mani, P. and Walkup, D. W. (1980), A 3-sphere counterexample to the W_v-path conjecture, *Mathematics of Operations Research* **5** (1980) 595–598. [142]

Manne, A. S. (1957–8), Programming of economic lot sizes, *Management Science* **4** (1957–8) 115–135. [148]

Manning, H. P. (1914), *Geometry of Four Dimensions*, Macmillan, New York, 1914 [reprinted: Dover, New York, 1956]. [40]

Marcus, M. and Minc, H. (1964), *A Survey of Matrix Theory and Matrix Inequalities*, Allyn and Bacon, Boston, Mass., 1964. [41]

Marcus, M. and Underwood, E. E. (1972), A note on the multiplicative property of the Smith normal form, *Journal of Research of the National Bureau of Standards* **76B** (1972) 205–206. [51]

Markowitz, H. M. and Manne, A. S. (1957), On the solution of discrete programming problems, *Econometrica* **25** (1957) 84–110. [378]

Marsten, R. E. (1973–4), An algorithm for large set partitioning problems, *Management Science* **20** (1973–4) 774–787. [380]

Martello, S. and Toth, P. (1977), An upper bound for the zero-one knapsack problem and a branch and bound algorithm, *European Journal of Operational Research* **1** (1977) 169–175. [373]

Martello, S. and Toth, P. (1978), Algorithm 37, Algorithm for the solution of the 0–1 single knapsack problem, *Computing* **21** (1978) 81–86. [373]

Martello, S. and Toth, P. (1979), The 0–1 knapsack problem, in: *Combinatorial Optimization* (N. Christofides, A. Mingozzi, P. Toth and C. Sandi, eds.), Wiley, Chichester, 1979, pp. 237–279. [373]

Martello, S. and Toth, P. (1984), Worst-case analysis of greedy algorithms for the subset-sum problem, *Mathematical Programming* **28** (1984) 198–205. [263]

Marti, J. T. (1977), *Konvexe Analysis*, Birkhäuser, Basel, 1977. [224]

Martin, G. T. (1963), An accelerated Euclidean algorithm for integer linear programming, in: *Recent Advances in Mathematical Programming* (R. L. Graves and P. Wolfe, eds.), McGraw-Hill, New York, 1963, pp. 311–317. [359]

Mathews, J. B. (1897), On the partition of numbers, *Proceedings of the London Mathematical Society* **28** (1897) 486–490. [235, 376]

Mattheiss, T. H. (1973), An algorithm for determining irrelevant constraints and all vertices in systems of linear inequalities, *Operations Research* **21** (1973) 247–260. [224]

Mat[t]heiss, T. H. and Rubin, D. S. (1980), A survey and comparison of methods for finding all vertices of convex polyhedral sets, *Mathematics of Operations Research* **5** (1980) 167–185. [224]

Mattheiss, T. H. and Schmidt, B. K. (1980), Computational results on an algorithm for finding all vertices of a polytope, *Mathematical Programming* **18** (1980) 308–329. [224]

Maurras, J. F., Truemper, K., and Akgül, M. (1981), Polynomial algorithms for a class of linear programs, *Mathematical Programming* **21** (1981) 121–136. [162, 293]

May, J. H. and Smith, R. L. (1982), Random polytopes: their definition, generation and aggregate properties, *Mathematical Programming* **24** (1982) 39–54. [147]

McCallum, D. and Avis, D. (1979), A linear algorithm for finding the convex hull of a simple polygon, *Information Processing Letters* **9** (1979) 201–206. [224]

McDiarmid, C. (1983), Integral decomposition of polyhedra, *Mathematical Programming* **25** (1983) 183–198. [338]

McMullen, P. (1970), The maximum numbers of faces of a convex polytope, *Mathematika* **17** (1970) 179–184. [142]

McMullen, P. (1975), Non-linear angle-sum relations for polyhedral cones and polytopes, *Mathematical Proceedings of the Cambridge Philosophical Society* **78** (1975) 247–261. [379]

McMullen, P. and Shephard, G. C. (1971), *Convex Polytopes and the Upper Bound Conjecture*, Cambridge University Press, Cambridge, 1971. [142, 223]

McRae, W. B. and Davidson, E. R. (1973), An algorithm for the extreme rays of a pointed convex polyhedral cone, *SIAM Journal on Computing* **2** (1973) 281–293. [224]

Megiddo, N. (1982), Linear-time algorithms for linear programming in $\mathbb{R}^3$ and related problems, in: *23rd Annual Symposium on Foundations of Computer Science*, IEEE, New York, 1982, pp. 329–338 [also: *SIAM Journal on Computing* **12** (1983) 759–776]. [199]

Megiddo, N. (1984), Linear programming in linear time when the dimension is fixed, *Journal of the Association for Computing Machinery* **31** (1984) 114–127. [199]

Megiddo, N. (1983), Towards a genuinely polynomial algorithm for linear programming, *SIAM Journal on Computing* **12** (1983) 347–353. [195]

Megiddo, N. (1986), Improved asymptotic analysis of the average number of steps performed by the self-dual simplex algorithm, *Mathematical Programming* **35** (1986) 140–172. [147]

Mendelsohn, N. S. (1970), A linear diophantine equation with applications to non-negative matrices, [in: [*Proceedings*] *International Conference on Combinatorial Mathematics* (New York, 1970; A. Gewirtz and L. V. Quintas, eds),] *Annals of the New York Academy of Sciences* **175** (article 1) (1970) 287–294. [379]

Menger, K. (1927), Zur allgemeinen Kurventheorie, *Fundamenta Mathematicae* **10** (1927) 96–115. [275, 378]

Menger, K. (1932), Bericht über das Kolloquium 1929/30, 9. Kolloquium (5. II. 1930), *Ergebnisse eines mathematischen Kolloquiums* [K. Menger, ed.] **2** (1932) 11–12. [378]

Merkes, E. P. and Meyers, D. (1973), On the length of the Euclidean algorithm, *The Fibonacci Quarterly* **11** (1973) 56–62. [54]

Meyer, R. R. (1974), On the existence of optimal solutions to integer and mixed-integer programming problems, *Mathematical Programming* **7** (1974) 223–235. [230, 379]

Meyer, R. R. and Wage, M. L. (1978), On the polyhedrality of the convex hull of the feasible set of an integer program, *SIAM Journal on Control and Optimization* **16** (1978) 682–687. [379]

Michaud, P. (1972), Exact implicit enumeration method for solving the set partitioning problem, *IBM Journal of Research and Development* **16** (1972) 573–578. [380]

Midonick, H. (1968), *The Treasury of Mathematics: 1*, A Pelican Book, Penguin Books, Harmondsworth, 1968. [38, 76, 77]

Mikami, Y. (1913), *The Development of Mathematics in China and Japan*, Teubner, Leipzig, 1913 [reprinted: Chelsea, New York, 1974]. [38]

Mikami, Y. (1914), On the Japanese theory of determinants, *Isis, International Review devoted to the History of Science and Civilization* **2** (1914) 9–36. [38]

Milanov, P. B. (1984), *Polynomial solvability of some linear diophantine equations in nonnegative integers*, preprint, Institute of Mathematics, Bulgarian Academy of Sciences, Sofia, 1984. [379]

Miliotis, P. (1978), Using cutting planes to solve the symmetric travelling salesman problem, *Mathematical Programming* **15** (1978) 177–188. [371]

Minkowski, H. (1891), Ueber die positiven quadratischen Formen und über ketten-bruchähnliche Algorithmen, *Journal für die reine und angewandte Mathematik* **107** (1891) 278–297 [reprinted in: *Gesammelte Abhandlungen von Hermann Minkowski, Band I* (D. Hilbert, ed.), Teubner, Leipzig, 1911 [reprinted: Chelsea, New York, 1967,] pp. 243–260]. [71, 81]

Minkowski, H. (1893), Extrait d'une lettre adressée à M. Hermite, *Bulletin des Sciences Mathématiques* (2) **17** (1893) 24–29 [reprinted in: *Gesammelte Abhandlungen von Hermann Minkowski, Band II* (D. Hilbert, ed.), Teubner, Leipzig, 1911, [reprinted: Chelsea, New York, 1967,] pp. 266–270]. [81, 376]

Minkowski, H. (1896), *Geometrie der Zahlen (Erste Lieferung)*, Teubner, Leipzig, 1896 [reprinted: Chelsea, New York, 1953]. [67, 81, 82, 85, 87, 88, 89, 214, 217]

Minkowski, H. (1897), Allgemeine Lehrsätze über die konvexen Polyeder, *Nachrichten von der königlichen Gesellschaft der Wissenschaften zu Göttingen, mathematisch-physikalische Klasse* (1897) 198–219 [reprinted in: *Gesammelte Abhandlungen von Hermann Minkow-*

ski, Band II (D. Hilbert, ed.), Teubner, Leipzig, 1911, [reprinted: Chelsea, New York, 1967,] pp. 103–121]. [214]

Minkowski, H. (1907), *Diophantische Approximationen, Eine Einführung in die Zahlentheorie*, Teubner, Leipzig, 1907 [reprinted: Chelsea, New York, 1957]. [82]

Minkowski, H. (1911a), *Gesammelte Abhandlungen, Band I, II*, Teubner, Leipzig, 1911 [reprinted: Chelsea, New York, 1967]. [214]

Minkowski, H. (1911b), Theorie der konvexen Körper, insbesondere Begründung ihres Oberflächenbegriffs, [first published in:] *Gesammelte Abhandlungen von Hermann Minkowski, Band II* (D. Hilbert, ed.), Teubner, Leipzig, 1911, [reprinted: Chelsea, New York, 1967,] pp. 131–229. [214,217]

Minty, G. J. (1974), A "from scratch" proof of a theorem of Rockafellar and Fulkerson, *Mathematical Programming* 7 (1974) 368–375. [98]

Minty, G. J. (1980), On maximal independent sets of vertices in claw-free graphs, *Journal of Combinatorial Theory* (B) **28** (1980) 284–304. [350]

Möbius, A. F. (1827), *Der barycentrische Calcul, ein neues Hülfsmittel zur analytischen Behandlung der Geometrie*, J. A. Barth, Leipzig, 1827 [reprinted: G. Olms, Hildesheim, 1976] [reprinted in: *August Ferdinand Möbius, Gesammelte Werke, Band I* (R. Baltzer, ed.), S. Hirzel, Leipzig, 1885 [reprinted: M. Sändig, Wiesbaden, 1967] pp. 1–388] [English translation (partially) in: D. E. Smith, *A Source Book in Mathematics*, McGraw-Hill, New York, 1929, pp. 525–526 and 670–677]. [40]

Monge, G. (1784), Mémoire sur la théorie des déblais et des remblais, *Histoire de l'Académie Royale des Sciences, avec les Mémoires de Mathematique et de Physique* [année 1781] (2e partie) (1784) [*Histoire*: 34–38, *Mémoire*:] 666–704. [221, 377]

Mordell, L. J. (1969), *Diophantine Equations*, Academic Press, London, 1969. [51,82]

Morito, S. and Salkin, H. M. (1979), Finding the general solution of a linear diophantine equation, *The Fibonacci Quarterly* **17** (1979) 361–368. [54]

Morito, S. and Salkin, H. M. (1980), Using the Blankinship algorithm to find the general solution of a linear diophantine equation, *Acta Informatica* **13** (1980) 379–382. [54]

Motzkin, T. S. (1936), *Beiträge zur Theorie der linearen Ungleichungen*, (Inaugural Dissertation Basel,) Azriel, Jerusalem, 1936 [English translation: *Contributions to the theory of linear inequalities*, RAND Corporation Translation 22, Santa Monica, Cal., 1952 [reprinted in: *Theodore S. Motzkin: Selected Papers* (D. Cantor, B. Gordon and B. Rothschild, eds.), Birkhäuser, Boston, 1983, pp. 1–80]].[88, 94, 155, 157, 217]

Motzkin, T. S. (1951), Two consequences of the transposition theorem on linear inequalities, *Econometrica* **19** (1951) 184–185. [95]

Motzkin, T. S. (1956), The assignment problem, in: *Numerical Analysis* (Proceedings of Symposia in Applied Mathematics Vol. VI; J. H. Curtiss, ed.), McGraw-Hill, New York, 1956, pp. 109–125 [reprinted in: *Theodore S. Motzkin: Selected Papers* (D. Cantor, B. Gordon and B. Rothschild, eds.), Birkhäuser, Boston, 1983, pp. 121–137]. [273,378]

Motzkin, T. S., Raiffa, H., Thompson, G. L., and Thrall, R. M. (1953), The double description method, in: *Contributions to the Theory of Games Vol. II* (H. W. Kuhn and A. W. Tucker, eds.), Princeton University Press, Princeton, N.J., 1953, pp. 51–73 [reprinted in: *Theodore S. Motzkin: Selected Papers* (D. Cantor, B. Gordon and B. Rothschild, eds.), Birkhäuser, Boston, 1983, pp. 81–103]. [224]

Motzkin, T. S. and Schoenberg, I. J. (1954), The relaxation method for linear inequalities, *Canadian Journal of Mathematics* 6 (1954) 393–404 [reprinted in: *Theodore S. Motzkin: Selected Papers* (D. Cantor, B. Gordon and B. Rothschild, eds.), Birkhäuser, Boston, 1983, pp. 104–115]. [157,158]

Motzkin, T. S. and Straus, E. G. (1956), Some combinatorial extremum problems, *Proceedings of the American Mathematical Society* 7 (1956) 1014–1021 [reprinted in: *Theodore S. Motzkin: Selected Papers* (D. Cantor, B. Gordon and B. Rothschild, eds.), Birkhäuser, Boston, 1983, pp. 279–286]. [378]

Mueller, R. K. and Cooper, L. (1965), A comparison of the primal-simplex and primal-dual algorithms for linear programming, *Communications of the ACM* **8** (1965) 682–686. [139]

Muir, Th. (1890), *The Theory of Determinants in the Historical Order of Development, Vol. I*, Macmillan, London, 1890 [reprinted (with Volume II in one volume): Dover, New York, 1960]. [41]

Muir, Th. (1898), List of writings on the theory of matrices, *American Journal of Mathematics* **20** (1898) 225–228. [41]

Muir, Th. (1911), *The Theory of Determinants in the Historical Order of Development, Vol. II, The period 1840 to 1860*, Macmillan, London, 1911 [reprinted (with Volume I in one volume): Dover, New York, 1960]. [41]

Muir, Th. (1920), *The Theory of Determinants in the Historical Order of Development, Vol. III, The period 1861 to 1880*, Macmillan, London, 1920 [reprinted (with Volume IV in one volume): Dover, New York, 1960]. [41]

Muir, Th. (1923), *The Theory of Determinants in the Historical Order of Development, Vol. IV, The period 1880 to 1900*, Macmillan, London, 1923 [reprinted (with Volume III in one volume): Dover, New York, 1960]. [41]

Muir, Th. (1930), *Contributions to the Theory of Determinants 1900–1920*, Blackie & Son, London, 1930. [41]

Mullender, P. (1950), Simultaneous approximation, *Annals of Mathematics* **52** (1950) 417–426. [75]

Munkres, J. (1957), Algorithms for the assignment and transportation problems, *Journal of the Society for Industrial and Applied Mathematics* **5** (1957) 32–38. [378]

Murtagh, B. A. (1981), *Advanced Linear Programming: Computation and Practice*, McGraw-Hill, New York, 1981. [224]

Murty, K. G. (1971), Adjacency on convex polyhedra, *SIAM Review* **13** (1971) 377–386. [224]

Murty, K. G. (1972), A fundamental problem in linear inequalities with applications to the travelling salesman problem, *Mathematical Programming* **2** (1972) 296–308. [224]

Murty, K. G. (1976), *Linear and Combinatorial Programming*, Wiley, New York, 1976. [224, 378]

Murty, K. G. (1983), *Linear Programming*, Wiley, New York, 1983. [224]

Nash-Williams, C. St.J. A. (1969), Well-balanced orientations of finite graphs and unobtrusive odd-vertex-pairings, in: *Recent Progress in Combinatorics* (W. T. Tutte, ed.), Academic Press, New York, 1969, pp. 133–149. [314]

Nathanson, M. B. (1974), Approximation by continued fractions, *Proceedings of the American Mathematical Society* **45** (1974) 323–324. [67]

Nauss, R. M. (1976–7), An efficient algorithm for the 0–1 knapsack problem, *Management Science* **23** (1976–7) 27–31. [373]

Nauss, R. M. (1979), *Parametric Integer Programming*, University of Missouri Press, Columbia, 1979. [354]

[Navier, C. L. M .H. (1825)=] Nav., Sur le calcul des conditions d'inégalité, *Bulletin des Sciences de la Société philomatique de Paris* (1825) 66–68. [212]

Neisser, H. (1932), Lohnhöhe und Beschäftigungsgrad in Marktgleichgewicht, *Weltwirtschaftliches Archiv* **36** (1932) 415–455. [219]

Nelson, C. G. (1982), *An $n^{\log n}$ algorithm for the two-variable-per-constraint linear programming satisfiability problem*, preprint. [157, 195]

Nemhauser, G. L. and Trotter, Jr, L. E. (1974), Properties of vertex packing and independence system polyhedra, *Mathematical Programming* **6** (1974) 48–61. [349, 380]

Nemhauser, G. L. and Trotter, Jr, L. E. (1975), Vertex packings: structural properties and algorithms, *Mathematical Programming* **8** (1975) 232–248. [349, 380]

Nemhauser, G. L., Trotter, Jr, L. E., and Nauss, R. M. (1973–4), Set partitioning and chain decomposition, *Management Science* **20** (1973–4) 1413–1423. [380]

Nemhauser, G. L. and Ullmann, Z. (1968), A note on the generalized Lagrange multiplier solution to an integer programming problem, *Operations Research* **16** (1968) 450-453. [354, 370]

Nemhauser, G. L. and Ullmann, Z. (1968-9), Discrete dynamic programming and capital allocation, *Management Science* **15** (1968-9) 494-505. [263]

Nemhauser, G. L. and Wolsey, L. A. (1988), *Integer and Combinatorial Optimization*, Wiley, New York, 1988. [378]

Nemirovsky, A. S. and Yudin, D. B. (1983), *Problem Complexity and Method Efficiency in Optimization*, Wiley, Chichester, 1983 [translation from the Russian *Slozhnost' zadach i effektivnost' metodov optimizatsii*, Izdat. 'Nauka', Moscow, 1979]. [224]

Nering, E. D. (1963), *Linear Algebra and Matrix Theory*, Wiley, New York, 1963. [4, 41]

Netto, E. (1901), *Lehrbuch der Combinatorik*, Teubner, Leipzig, 1901. [375].

von Neumann, J. (1928), Zur Theorie der Gesellschaftsspiele, *Mathematische Annalen* **100** (1928) 295-320 [reprinted in: *John von Neumann, Collected Works, Vol. VI* (A. H. Taub, ed.) Pergamon Press, Oxford, 1963, pp. 1-26] [English translation: On the theory of games of strategy, in: *Contributions to the Theory of Games, Vol. IV* (A. W. Tucker and R. D. Luce, eds.), Annals of Mathematics Studies 40, Princeton University Press, Princeton, N.J., 1959, pp. 13-42]. [218]

von Neumann, J. (1937), Über ein ökonomisches Gleichungssystem und eine Verallgemeinerung des Brouwerschen Fixpunktsatzes, *Ergebnisse eines mathematischen Kolloquiums* **8** [1935-6] (1937) 73-87 [English translation: A model of general economic equilibrium, *The Review of Economic Studies* **13** (1945-6) 1-9 [reprinted in: *John von Neumann, Collected Works, Vol. VI* (A. H. Taub, ed.), Pergamon Press, Oxford, 1963, pp. 29-37] [also reprinted in: *Readings in Mathematical Economics, Volume II: Capital and Growth* (P. Newman, ed.), The Johns Hopkins Press, Baltimore, 1968, pp. 221-229]]. [219]

von Neumann, J. (1945), Communication on the Borel notes, *Econometrica* **21** (1945) 124-125 [reprinted in: *John von Neumann, Collected Works, Vol. VI* (A. H. Taub, ed.), Pergamon Press, Oxford, 1963, pp. 27-28]. [218]

von Neumann, J. (1947), *Discussion of a maximum problem*, unpublished working paper, Institute for Advanced Study, Princeton, N.J., 1947 [reprinted in: *John von Neumann, Collected Works, Vol. VI* (A. H. Taub, ed.), Pergamon Press, Oxford, 1963, pp. 89-95]. [90, 223]

von Neumann, J. (1953), A certain zero-sum two-person game equivalent to the optimal assignment problem, in: *Contributions to the Theory of Games, II* (H. W. Kuhn and A. W. Tucker, eds.), Annals of Mathematics Studies 28, Princeton University Press, Princeton, N.J., 1953, pp. 5-12 [reprinted in: *John von Neumann, Collected Works, Vol. VI*, (A. H. Taub, ed.) Pergamon Press, Oxford, 1963, pp. 44-49]. [14, 21, 108, 274]

Newman, D. J. (1965). Location of the maximum on unimodal surfaces, *Journal of the Association for Computing Machinery* **12** (1965) 395-398. [171]

Newman, M. (1972), *Integral Matrices*, Academic Press, New York, 1972. [51, 82]

Nichol, A. J. (1938), Tragedies in the life of Cournot, *Econometrica* **6** (1938) 193-197. [211]

Niedringhaus, W. P. and Steiglitz, K. (1978), Some experiments with the pathological linear programs of N. Zadeh, *Mathematical Programming* **15** (1978) 352-354. [142]

Nijenhuis, A. and Wilf, H. S. (1972), Representations of integers by linear forms in nonnegative integers, *Journal of Number Theory* **4** (1972) 98-106. [379]

Niven, I. (1967), *Irrational Numbers*, Mathematical Association of America [distributed by: Wiley, New York], 1967. [82]

Nonweiler, T. R. F. (1984), *Computational Mathematics - An Introduction to Numerical Approximation*, Ellis Horwood, Chichester, 1984. [82]

Norman, R. Z. (1955), On the convex polyhedra of the symmetric traveling salesman problem (Abstract), *Bulletin of the American Mathematical Society* **61** (1955) 559. [378]

Nourie, F. J. and Venta, E. R. (1981-2), An upper bound on the number of cuts needed in Gomory's method of integer forms, *Operations Research Letters* **1** (1981-2) 129-133. [359]

Nowak, W. G. (1981), A note on simultaneous diophantine approximation, *Manuscripta Mathematica* **36** (1981) 33–46. [75]

Odlyzko, A. M. and te Riele, H. J. J. (1985), Disproof of the Mertens conjecture, *Journal für die reine und angewandte Mathematik* **357** (1985) 138–160. [74]

Olds, C. D. (1963), *Continued Fractions*, Random House, New York, 1963. [82]

Orchard-Hays, W. (1954), *Background, development and extensions of the revised simplex method*, Report RM 1433, The Rand Corporation, Santa Monica, Cal., 1954. [148]

Orchard-Hays, W. (1968), *Advanced Linear-Programming Computing Techniques*, McGraw-Hill, New York, 1968. [224]

Orden, A. (1955–6), The transhipment problem, *Management Science* **2** (1955–6) 276–285. [142]

Orden, A. (1971), On the solution of linear equation/inequality systems, *Mathematical Programming* **1** (1971) 137–152. [147]

Orden, A. (1976), Computational investigation and analysis of probabilistic parameters of convergence of a simplex algorithm, in: *Progress in Operations Research, Vol. II* (Proc. Conf. Eger (Hungary), 1974; A. Prékopa ed.), North-Holland, Amsterdam, 1976, pp. 705–715. [139, 147]

Orden, A. (1979), A study of pivot probabilities in LP tableaus, in: *Survey of Mathematical Programming, Vol. 2* (Proceedings of the 9th International Mathematical Programming Symposium, Budapest, 1976; A. Prékopa, ed.), Akadémia Kiadó, Budapest, 1979, pp. 141–154. [147].

Orden, A. (1980), A step toward probabilistic analysis of simplex method convergence, *Mathematical Programming* **19** (1980) 3–13. [147]

Orlin, J. B. (1982), A polynomial algorithm for integer programming covering problems satisfying the integer round-up property, *Mathematical Programming* **22** (1982) 231–235. [330, 338]

Orlin, J. B. (1985a), On the simplex algorithm for networks and generalized networks, *Mathematical Programming*, to appear. [142, 279]

Orlin, J. B. (1985b), Genuinely polynomial simplex and non-simplex algorithms for the minimum cost flow problem, *Operations Research*, to appear. [142]

Ostrogradsky, M. (1838a), Considérations générales sur les momens des forces [lu le 7 Novembre 1834], *Mémoires de l'Académie Impériale des Sciences de Saint-Pétersbourg, Série VI: Sciences mathématiques et physiques* **1** (1838) 129–150. [211]

Ostrogradsky, M. (1838b), Mémoire sur les déplacemens instantanés des systèmes assujettis à des conditions variables [lu le 20 Avril 1838], *Mémoires de l'Académie Impériale des Sciences de Saint-Pétersbourg, Série VI: Sciences mathématiques et physiques* **1** (1838) 565–600. [211]

Pace, I. S. and Barnett, S. (1974), Efficient algorithms for linear system calculations, Part I—Smith form and common divisor of polynomial matrices, *International Journal of Systems Science* **5** (1974) 403–411. [56]

Padberg, M. W. (1971), A remark on 'An inequality for the number of lattice points in a simplex', *SIAM Journal on Applied Mathematics* **20** (1971) 638–641. [379]

Padberg, M. W. (1972), Equivalent knapsack-type formulations of bounded integer linear programs: an alternative approach, *Naval Research Logistics Quarterly* **19** (1972) 699–708. [236]

Padberg, M. W. (1973), On the facial structure of set packing polyhedra, *Mathematical Programming* **5** (1973) 199–215. [380]

Padberg, M. W. (1974), Perfect zero-one matrices, *Mathematical Programming* **6** (1974) 180–196. [349, 380]

Padberg, M. W. (1975a), Characterisations of totally unimodular, balanced and perfect matrices, in: *Combinatorial Programming: Methods and Applications* (B. Roy, ed.), Reidel, Dordrecht (Holland), 1975, pp. 275–284. [272, 380]

Padberg, M. W. (1975b), A note on zero-one programming, *Operations Research* **23** (1975) 833–837. [380]

Padberg, M. W. (1976), A note on the total unimodularity of matrices, *Discrete Mathematics* 14 (1976) 273–278. [269, 272]

Padberg, M. W. (1977), On the complexity of set packing polyhedra, [in: *Studies in Integer Programming* (P. L. Hammer, et al., eds.),] *Annals of Discrete Mathematics* 1 (1977) 421–434. [380]

Padberg, M. W. (1979), Covering, packing and knapsack problems, [in: *Discrete Optimization I* (P. L. Hammer, E. L. Johnson and B. H. Korte, eds.),] *Annals of Discrete Mathematics* 4 (1979) 265–287. [380]

Padberg, M. W. and Grötschel, M. (1985), Polyhedral computations, in: *The Traveling Salesman Problem, A Guided Tour of Combinatorial Optimization* (E. L. Lawler, et al., eds.), Wiley, Chichester, 1985, pp. 307–360. [371]

Padberg, M. W. and Hong, S. (1980), On the symmetric travelling salesman problem: a computational study, *Mathematical Programming Study* 12 (1980) 78–107. [371]

Padberg, M. W. and Rao, M. R. (1980), *The Russian method and integer programming*, GBA Working paper, New York University, New York, 1980 (to appear in *Annals of Operations Research*). [163, 172]

Padberg, M. W. and Rao, M. R. (1982), Odd minimum cut-sets and *b*-matchings, *Mathematics of Operations Research* 7 (1982) 67–80. [252]

Paley, R. E. A. C. and Ursell, H. D. (1930), Continued fractions in several dimensions, *Proceedings of the Cambridge Philosophical Society* 26 (1930) 127–144. [75]

van de Panne, C. (1971), *Linear Programming and Related Techniques*, North-Holland, Amsterdam, 1971. [224].

Papadimitriou, C. H. (1978), The adjacency relation on the traveling salesman polytope is NP-complete, *Mathematical Programming* 14 (1978) 312–324. [252, 255]

Papadimitriou, C. H. (1979), Efficient search for rationals, *Information Processing Letters* 8 (1979) 1–4. [66, 124]

Papadimitriou, C. H. (1981), On the complexity of integer programming, *Journal of the Association for Computing Machinery* 28 (1981) 765–768. [239, 261, 264]

Papadimitriou, C. H. and Steiglitz, K. (1982), *Combinatorial Optimization: Algorithms and Complexity*, Prentice-Hall, Englewood Cliffs, N.J., 1982. [153, 379]

Papadimitriou, C. H. and Yannakakis, M. (1982), The complexity of facets (and some facets of complexity), in: *Proceedings of the Fourteenth Annual ACM Symposium on Theory of Computing* (San Francisco, Cal., 1982), The Association for Computing Machinery, New York, 1982, pp. 255–260 [also: Journal of Computer and System Science 28 (1984) 244–259]. [252, 253]

Paull, A. E. and Walter, J. R. (1955), The trim problem: an application of linear programming to the manufacture of newsprint paper (Abstract), *Econometrica* 23 (1955) 336. [378]

Peck, L. G. (1961), Simultaneous rational approximations to algebraic numbers, *Bulletin of the American Mathematical Society* 67 (1961) 197–201. [75]

Perl, J. (1975), On finding all solutions of the partitioning problem, in: *Mathematical Foundations of Computer Science 1975* (Proc. 4th Symposium, Mariánské Lázně, 1975; J. Bečvář, ed.), Lecture Notes in Computer Science 32, Springer, Berlin, 1975, pp. 337–343. [379]

Perron, O. (1907), Grundlagen für eine Theorie des Jacobischen Kettenbruch-algorithmus, *Mathematische Annalen* 64 (1907) 1–76. [75]

Perron, O. (1913), *Die Lehre von den Kettenbrüchen*, Teubner, Leipzig, 1913 [2nd revised edition: 1929 [reprinted: Chelsea, New York, 1950], 3rd revised edition: Teubner, Stuttgart, 1954 (Vol. I), 1957 (Vol. II)]. [82]

Perron, O. (1921), *Irrationalzahlen*, W. de Gruyter, Berlin, 1921 [2nd revised editon: 1939 [reprinted: Chelsea, New York, 1948]]. [80, 82]

Picard, J.-C. and Queyranne, M. (1977), On the integer-valued variables in the linear vertex packing problem, *Mathematical Programming* 12 (1977) 97–101. [380]

Piehler, J. (1975), Einige Bemerkungen zum Schnittrang in der rein-ganzzahligen linearen Optimierung, *Mathematische Operationsforschung und Statistik* 6 (1975) 523–533. [359]

Pipping, N. (1921), Ein Kriterium für die reellen algebraischen Zahlen auf eine direkte Verallgemeinerung des Euklidischen Algorithmus gegründet, *Acta Academiae Aboensis Mathematica et Physica [Ser. B]* 1 (1921) 1–16. [75]

Pipping, N. (1957), Approximation zweier reellen Zahlen durch rationale Zahlen mit gemeinsamem Nenner, *Acta Academiae Aboensis Mathematica et Physica [Ser. B]* 21 (1957) 1–17. [75]

Plane, D. R. and McMillan, C. (1971), *Discrete Optimization—Integer Programming and Network Analysis for Management Decisions*, Prentice-Hall, Englewood Cliffs, N.J., 1971. [378]

Plücker, J. (1829), Über ein neues Coordinatensystem, *Journal für die reine und angewandte Mathematik* 5 (1829) 1–36 [reprinted in: *Julius Plückers Gesammelte Wissenschaftliche Abhandlungen, Band I: Gesammelte Mathematische Abhandlungen* (A. Schoenflies, ed.), Teubner, Leipzig, 1895 [reprinted: Johnson Reprint Corp., New York, 1972,] pp. 124–158]. [40]

Plücker, J. (1830a), Über ein neues Princip der Geometrie und den Gebrauch allgemeiner Symbole und unbestimmter Coëfficienten, *Journal für die reine und angewandte Mathematik* 5 (1830) 268–286 [reprinted in: *Julius Plückers Gesammelte Wissenschaftliche Abhandlungen. Band I: Gesammelte Mathematische Abhandlungen* (A. Schoenflies, ed.), Teubner, Leipzig, 1895 [reprinted: Johnson Reprint Corp., New York, 1972], pp. 159–177]. [40]

Plücker, J. (1830b), Über eine neue Art, in der analytischen Geometrie Punkte und Curven durch Gleichungen darzustellen, *Journal für die reine und angewandte Mathematik* 6 (1830) 107–146 [reprinted in: *Julius Plückers Gesammelte Wissenschaftliche Abhandlungen, Band I: Gesammelte Mathematische Abhandlungen* (A. Schoenflies, ed.), Teubner, Leipzig, 1895 [reprinted: Johnson Reprint Corp., New York, 1972], pp. 178–219]. [40]

Plücker, J. (1832), Note sur une théorie générale et nouvelle des surfaces courbes, *Journal für die reine und angewandte Mathematik* 9 (1832) 124–134 [reprinted in: *Julius Plückers Gesammelte Wissenschaftliche Abhandlungen, Band I: Gesammelte Mathematische Abhandlungen* (A. Schoenflies, ed.), Teubner, Leipzig, 1895 [reprinted: Johnson Reprint. Corp., New York, 1972,] pp. 224–234]. [40]

Podsypanin, E. V. (1977), A generalization of the algorithm for continued fractions related to the algorithm of Viggo Brun (in Russian), [in: *Investigations in Number Theory, Part* 4 (A. V. Malyshev, ed.),] *Zapiski Nauchnykh Seminarov LOMI* 67 (1977) 184–194 [English translation: *Journal of Soviet Mathematics* 16 (1981) 885–893]. [75]

Poincaré, H. (1880), Sur un mode nouveau de représentation géométrique des formes quadratiques définies ou indéfinies, *Journal de l'École Polytechnique* 47 (1880) 177–245 [reprinted in: *Oeuvres de Henri Poincaré, Tome V*, Gauthier-Villars, Paris, 1950, pp. 117–180]. [81]

Poincaré, H. (1884), Sur une généralisation des fractions continues, *Comptes Rendus Hebdomadaires des Séances de l'Académie des Sciences* 99 (1884) 1014–1016 [reprinted in: *Oeuvres de Henri Poincaré, Tome V*, Gauthier-Villars, Paris, 1950, pp. 185–187]. [75]

Poincaré, H. (1900), Second complément à l'analysis situs, *Proceedings of the London Mathematical Society* 32 (1900) 277–308 [reprinted in: *Oeuvres de Henri Poincaré, Tome VI*, Gauthier-Villars, Paris, 1953, pp. 338–370]. [274, 378]

Polyak, B. T. (1967), A general method of solving extremum problems (in Russian), *Doklady Akademii Nauk SSSR* 174 (1) (1967) 33–36 [English translation: *Soviet Mathematics Doklady* 8 (1967) 593–597]. [368]

Polyak, B. T. (1969), Minimization of unsmooth functionals (in Russian), *Zhurnal Vychislitel'noĭ Matematiki i Matematicheskoĭ Fiziki* 9 (1969) 509–521 [English translation: *U.S.S.R. Computational Mathematics and Mathematical Physics* 9 (3) (1969) 14–29]. [370]

Poncelet, J. V. (1822), *Traité des propriétés projectives des figures*, Bachelier, Paris, 1822 [English translation (partially) in: D. E. Smith, *A Source Book in Mathematics*, McGraw-Hill, New York, 1929, pp. 315–323]. [40, 212]

Poncelet, J. V. (1829), Mémoire sur la théorie générale des polaires réciproques, *Journal für die reine und angewandte Mathematik* **4** (1829) 1–71. [40, 212]

Prager, W. (1956–7), On the caterer problem, *Management Science* **3** (1956–7) 15–23. [378]

Prékopa, A. (1972), On the number of vertices of random convex polyhedra, *Periodica Mathematica Hungarica* **2** (1972) 259–282. [147]

Prékopa, A. (1980), On the development of optimization theory, *The American Mathematical Monthly* **87** (1980) 527–542. [214, 223]

Preparata, F. P. and Hong, S. J. (1977), Convex hulls of finite sets of points in two and three dimensions, *Communications of the ACM* **20** (1977) 87–93. [224]

Preparata, F. P. and Shamos, M. I. (1985), *Computational Geometry – An Introduction*, Springer, New York, 1985. [224]

Proll, L. G. (1970), Certification of algorithm 263A [H] Gomory 1, *Communications of the ACM* **13** (1970) 326–327. [359]

Pulleyblank, W. and Edmonds, J. (1974), Facets of 1-matching polyhedra, in: *Hypergraph Seminar* (Ohio State University, Columbus, Ohio, 1972; C. Berge and D. Ray-Chaudhuri, eds.), Lecture Notes in Mathematics 411, Springer, Berlin, 1974, pp. 214–242. [111]

Quandt, R. E. and Kuhn, H. W. (1964), On upper bounds for the number of iterations in solving linear programs, *Operations Research* **12** (1964) 161–165. [142]

Raghavachari, M. and Sabharwal, Y. P. (1978), On the number of non-negative integral solutions to the knapsack problem, *SCIMA* **7** (1978) 81–87. [379]

Raisbeck, G. (1950), Simultaneous diophantine approximation, *Canadian Journal of Mathematics* **2** (1950) 283–288. [75]

von Randow, R. (ed.) (1982), *Integer Programming and Related Areas, A Classified Bibliography 1978–1981*, Lecture Notes in Economics and Mathematical Systems 197, Springer, Berlin, 1982. [236, 376, 379]

von Randow, R. (ed.) (1985), *Integer Programming and Related Areas, A Classified Bibliography 1981–1984*, Lecture Notes in Economics and Mathematical Systems 243, Springer, Berlin, 1985. [236, 376, 379]

Ravindran, A. (1973), A comparison of the primal-simplex and complementary pivot methods for linear programming, *Naval Research Logistics Quarterly* **20** (1973) 95–100. [139]

Rayward-Smith, V. J. (1979), On computing the Smith normal form of an integer matrix, *ACM Transactions on Mathematical Software* **5** (1979) 451–456. [59]

Rebman, K. R. (1974), Total unimodularity and the transportation problem: a generalization, *Linear Algebra and Its Applications* **8** (1974) 11–24. [308]

Reiss, S. P. (1979), Rational search, *Information Processing Letters* **8** (1979) 89–90. [66]

Rényi, A. and Sulanke, R. (1963), Über die konvexe Hülle von *n* zufällig gewählten Punkten, *Zeitschrift für Wahrscheinlichkeitstheorie und verwandte Gebiete* **2** (1963) 75–84 [reprinted in: *Selected Papers of Alfréd Rényi 3, 1962–1970* (P. Turán, ed.), Akadémiai Kiadó, Budapest, 1976, pp. 143–152]. [147]

Rényi, A. and Sulanke, R. (1968), Zufällige konvexe Polygone in einem Ringgebiet, *Zeitschrift für Wahrscheinlichkeitstheorie und verwandte Gebiete* **9** (1968) 146–157 [reprinted in: *Selected Papers of Alfréd Rényi 3, 1962–1970* (P. Turán, ed.), Akadémiai Kiadó, Budapest, 1976, pp. 502–513]. [147]

Richards, I. (1981), Continued fractions without tears, *Mathematics Magazine* **54** (1981) 163–171. [67]

Rieger, G. J. (1976), On the Harris modification of the Euclidean algorithm, *The Fibonacci Quarterly* **14** (1976) 196, 200. [54]

Riesz, F. (1904), Sur la résolution approchée de certaines congruences, *Comptes Rendus Hebdomadaires des Séances de l'Académie des Sciences* **139** (1904) 459–462 [reprinted in: *Riesz Frigyes összegyüjtött munkái – Friedrich Riesz Gesammelte Arbeiten, Vol. II* (A. Császár, ed.), Akadémiai Kiadó, Budapest, 1960, pp. 1423–1426]. [80]

Riley, V. and Gass, S. I. (1958), *Linear Programming and Associated Techniques, A comprehensive bibliography in linear, nonlinear, and dynamic programming*, The Johns Hopkins Press, Baltimore, 1958. [224]

Roberts, J. B. (1956), Note on linear forms, *Proceedings of the American Mathematical Society* 7 (1956) 465–469. [379]

Roberts, J. B. (1957), On a diophantine problem, *Canadian Journal of Mathematics* 9 (1957) 219–222. [379]

Robinson. S. M. (1973), Bounds for error in the solution set of a perturbed linear program, *Linear Algebra and Its Applications* 6 (1973) 69–81. [128]

Robinson. S. M. (1975), Stability theory for systems of inequalities. Part I: Linear systems, *SIAM Journal on Numerical Analysis* 12 (1975) 754–769. [225]

Robinson, S. M. (1977), A characterization of stability in linear programming, *Operations Research* 25 (1977) 435–447. [225]

Rockafellar, R. T. (1969), The elementary vectors of a subspace of R^N, in: *Combinatorial Mathematics and Its Applications* (Proceedings North Carolina Conference, Chapel Hill, 1967; R. C. Bose and T. A. Dowling, eds.), The University of North Carolina Press, Chapel Hill, North Carolina, 1969, pp. 104–127. [150]

Rockafellar, R. T. (1970), *Convex Analysis*, Princeton University Press, Princeton, N.J., 1970. [98, 99, 224]

Rödseth, Ö. J. (1978), On a linear diophantine problem of Frobenius, *Journal für die reine und angewandte Mathematik* 301 (1978) 171–178. [379]

Rödseth, Ö. J. (1979), On a linear diophantine problem of Frobenius. II, *Journal für die reine und angewandte Mathematik* 307/308 (1979) 431–440. [379]

Rohde, F. V. (1957), Bibliography on linear programming, *Operations Research* 5 (1957) 45–62. [224]

Rolle, M. (1690), *Traité d'Algèbre, ou Principes Généraux pour Résoudre les Questions de Mathématique*, E. Michallet, Paris, 1690. [77]

Roohy-Laleh, E. (1981), *Improvements to the theoretical efficiency of the network simplex method*, Ph.D. Dissertation, Carleton University, Ottawa, 1981. [142]

Rosenberg, I. G. (1974), Aggregation of equations in integer programming, *Discrete Mathematics* 10 (1974) 325–341. [236]

Rosenberg, I. (1975), On Chvátal's cutting planes in integer linear programming, *Mathematische Operationsforschung und Statistik* 6 (1975) 511–522. [359]

Ross, S. M. (1981), *A simple heuristic approach to simplex efficiency*, Report ORC 81–21, Department of Industrial Engineering and Operations Research, University of California, Berkeley, Cal., 1981. [147]

Rosser, [J.] B. (1942), A generalization of the Euclidean algorithm to several dimensions, *Duke Mathematical Journal* 9 (1942) 59–95. [59, 72]

Rosser, J. B. (1952), A method of computing exact inverses of matrices with integer coefficients, *Journal of Research of the National Bureau of Standards* 49 (1952) 349–358. [35]

Roth, B. (1969), Computer solutions to minimum-cover problems, *Operations Research* 17 (1969) 455–465. [380]

Roy, R. (1939), L'oeuvre économique d'Augustin Cournot, *Econometrica* 7 (1939) 134–144. [211]

Rubin, D. S. (1970), On the unlimited number of faces in integer hulls of linear programs with a single constraint, *Operations Research* 18 (1970) 940–946. [255]

Rubin, D. S. (1971–2), Redundant constraints and extraneous variables in integer programs, *Management Science* 18 (1971–2) 423–427. [379]

Rubin. D. S. (1975a), Vertex generation and cardinality constrained linear programs, *Operations Research* 23 (1975) 555–565. [224]

Rubin, D. S. (1984–5), Polynomial algorithms for $m \times (m + 1)$ integer programs and $m \times (m + k)$ diophantine systems, *Operations Research Letters* 3 (1984–5) 289–291. [260]

Rubin, D. S. and Graves, R. L. (1972), Strengthened Dantzig cuts for integer programming, *Operations Research* 20 (1972) 178–182. [359]

Rubin, F. (1975b), An improved algorithm for testing the planarity of a graph, *IEEE Transactions on Computers* C-24 (1975) 113–121. [282]

Saaty, T. L. (1955), The number of vertices of a polyhedron, *The American Mathematical Monthly* 62 (1955) 326–331. [142]

Saaty, T. L. (1963), A conjecture concerning the smallest bound on the iterations in linear programming, *Operations Research* 11 (1963) 151–153. [142]

Saaty, T. L. (1970), *Optimization in Integers and Related Extremal Problems*, McGraw-Hill, New York, 1970. [379]

Sahni, S. (1975), Approximate algorithms for the 0/1 knapsack problem, *Journal of the Association for Computing Machinery* 22 (1975) 115–124. [263]

Saigal, R. (1969), A proof of the Hirsch conjecture on the polyhedron of the shortest route problem, *SIAM Journal on Applied Mathematics* 17 (1969) 1232–1238. [142]

Saigal, R. (1983), *On some average results for linear complementary problems*, manuscript, Department of Industrial Engineering and Management Sciences, Northwestern University, Evanston, Ill., 1983. [147]

Sakarovitch, M. (1975), Quasi-balanced matrices, *Mathematical Programming* 8 (1975) 382–386. [308]

Sakarovitch, M. (1976), Quasi-balanced matrices—an addendum, *Mathematical Programming* 10 (1976) 405–407. [308]

Salkin, H. M. (1971), A note on Gomory fractional cuts, *Operations Research* 19 (1971) 1538–1541. [359]

Salkin, H. M. (1973), A brief survey of algorithms and recent results in integer programming, *Opsearch* 10 (1973) 81–123. [379]

Salkin, H. M. (1975), *Integer Programming*, Addison–Wesley, Reading, Mass., 1975. [378]

Salkin, H. M. and de Kluyver, C. A. (1975), The knapsack problem: a survey, *Naval Research Logistics Quarterly* 22 (1975) 127–144. [373]

Salkin, H. M. and Koncal, R. D. (1973), Set covering by an all integer algorithm: computational experience, *Journal of the Association for Computing Machinery* 20 (1973) 189–193. [380]

Sauer, N. (1972), On the density of families of sets, *Journal of Combinatorial Theory (A)* 13 (1972) 145–147. [299]

Saunders, R. M. and Schinzinger, R. (1970), A shrinking boundary algorithm for discrete system models, *IEEE Transactions on Systems Science and Cybernetics* SSC-6 (1970) 133–140. [373]

Saunderson, N. (1741), *The Elements of Algebra in ten books, Vol. I*, The University Press, Cambridge, 1741. [77]

Savage, J. E. (1976), *The Complexity of Computing*, Wiley, New York, 1976. [14, 41]

Sbihi, N. (1978), *Étude des stables dans les graphes sans étoile*, M.Sc. Thesis, Université Scientifique et Médicale de Grenoble (Mathématiques Appliquées), Grenoble, 1978. [350]

Sbihi, N. (1980), Algorithme de recherche d'un stable de cardinalité maximum dans un graphe sans étoile, *Discrete Mathematics* 29 (1980) 53–76. [350]

Scarf, H. E. (1977), An observation on the structure of production sets with indivisibilities, *Proceedings of the National Academy of Sciences of the United States of America* 74 (1977) 3637–3641. [234]

Scarf, H. E. (1981a), Production sets with indivisibilities—part I: generalities, *Econometrica* 49 (1981) 1–32. [256]

Scarf, H. E. (1981b), Production sets with indivisibilities—part II: the case of two activities, *Econometrica* 49 (1981) 395–423. [256]

Scharlau, W. and Opolka, H. (1980), *Von Fermat bis Minkowski, Eine Vorlesung über Zahlentheorie und ihre Entwicklung*, Springer, Berlin, 1980 [English translation: *From Fermat to Minkowski, Lectures on the Theory of Numbers and Its Historical Development*, Springer, New York, 1985]. [82]

Scherk, H. F. (1825), *Mathematische Abhandlungen*, Reimer, Berlin, 1825. [40]

Schläfli, L. (1901), *Theorie der vielfachen Kontinuität*, Druck von Zürcher und Furrer, Zürich, 1901 [also: *Neue Denkschriften der allgemeinen schweizerischen Gesellschaft für die gesamten Naturwissenschaften* **38** (1901) 1–237] [reprinted in: *Ludwig Schläfli, 1814–1895, Gesammelte Mathematische Abhandlungen, Vol. I*, Birkhäuser, Basel, 1950, pp. 167–387]. [223]

Schlauch, H. M. (1932), Mixed systems of linear equations and inequalities, *The American Mathematical Monthly* **39** (1932) 218–222. [217]

Schlesinger, K. (1935), Über die Produktionsgleichungen der ökonomischen Wertlehre, *Ergebnisse eines mathematischen Kolloquiums* **6** [1933–4] (1935) 10–12. [219]

Schmidt, E. (1913), Zum Hilbertschen Beweise des Waringschen Theorems (Aus einem an Herrn Hilbert gerichteten Briefe), *Mathematische Annalen* **74** (1913) 271–274. [215]

Schmidt, W. M. (1966), Simultaneous approximation to a basis of a real numberfield, *American Journal of Mathematics* **88** (1966) 517–527. [75]

Schmidt, W. M. (1968), Some results in probabilistic geometry, *Zeitschrift für Wahrscheinlichkeitstheorie und verwandte Gebiete* **9** (1968) 158–162. [147]

Schmidt, W. M. (1980), *Diophantine Approximation*, Lecture Notes in Mathematics 785, Springer, Berlin, 1980. [82]

Schnorr, C. P. (1985), A hierarchy of polynomial time basis reduction algorithms, in: *Theory of Algorithms* (Proceedings Conference Pécs (Hungary); L. Lovász and E. Szemerédi, eds.), North-Holland, Amsterdam, 1985, pp. 375–386. [68, 71]

Schoenberg, I. J. (1932), On finite-rowed systems of linear inequalities in infinitely many variables, *Transactions of the American Mathematical Society* **34** (1932) 594–619. [217]

Schönhage, A. (1971), Schnelle Berechnung von Kettenbruchentwicklungen, *Acta Informatica* **1** (1971) 139–144. [54]

Schönhage, A., Paterson, M., and Pippenger, N. (1976), Finding the median, *Journal of Computer and System Sciences* **13** (1976) 184–199. [199]

Schrader, R. (1982), Ellipsoid methods, in: *Modern Applied Mathematics—Optimization and Operations Research* (B. Korte, ed.), North-Holland, Amsterdam, 1982, pp. 265–311. [163, 171]

Schrader, R. (1983), The ellipsoid method and its implications, *OR Spektrum* **5** (1983) 1–13. [163, 171]

Schrage, L. and Wolsey, L. (1985), Sensitivity analysis for branch and bound integer programming, *Operations Research* **33** (1985) 1008–1023. [354]

Schrijver, A. (1980), On cutting planes, [in: *Combinatorics 79 Part II* (M. Deza and I. G. Rosenberg, eds.),] *Annals of Discrete Mathematics* **9** (1980) 291–296. [339]

Schrijver, A. (1981), On total dual integrality, *Linear Algebra and Its Applications* **38** (1981) 27–32. [316, 327]

Schrijver, A. and Seymour, P. D. (1977), *A proof of total dual integrality of matching polyhedra*, Mathematical Centre report ZN 79/77, Mathematical Centre, Amsterdam, 1977. [327]

Schwartz, J. T. (1961), *Introduction to Matrices and Vectors*, McGraw-Hill, New York, 1961. [41]

Schweiger, F. (1973), *The metrical theory of Jacobi-Perron algorithm*, Lecture Notes in Mathematics 334, Springer, Berlin, 1973. [75]

Schweiger, F. (1977), Über einen Algorithmus von R. Güting, *Journal für die reine und angewandte Mathematik* **293/294** (1977) 263–270. [75]

Seeber, L. A. (1831), *Untersuchungen über die Eigenschaften der positiven ternären quadratischen Formen*, Loeffler, Mannheim, 1831. [78]

Seidel, P. L. (1874), Über ein Verfahren, die Gleichungen, auf Welche die Methode der kleinsten Quadrate führt, sowie lineäre Gleichungen überhaupt, durch successive Annäherung aufzulösen, *Abhandlungen der Bayerischen Akademie der Wissenschaften (München), Mathematisch-naturwissenschaftliche Abteilung* **2** (1874) 81–108. [36, 40]

Selmer, E. S. (1961), Om flerdimensjonal kjedebrøk, *Nordisk Matematisk Tidskrift* **9** (1961) 37–43. [75]

Selmer, E. S. (1977), On the linear diophantine problem of Frobenius, *Journal für die reine und angewandte Mathematik* **293/294** (1977) 1–17. [379]

Selmer, E. S. and Beyer, Ö. (1978), On the linear diophantine problem of Frobenius in three variables, *Journal für die reine und angewandte Mathematik* **301** (1978) 161–170. [379]

Seymour, P. D. (1979), On multi-colourings of cubic graphs, and conjectures of Fulkerson and Tutte, *Proceedings of the London Mathematical Society* (3) **38** (1979) 423–460. [119]

Seymour, P. D. (1980), Decomposition of regular matroids, *Journal of Combinatorial Theory* (B) **28** (1980) 305–359. [279, 280, 295]

Seymour, P.D. (1981), Recognizing graphic matroids, *Combinatorica* **1** (1981) 75–78. [287]

Seymour, P. D. (1985), Applications of the regular matroid decomposition, in: *Matroid Theory* (Proceedings Matroid Theory Colloquium, Szeged, 1982; L. Lovász and A. Recski, eds.), North-Holland, Amsterdam, 1985, pp. 345–357. [279]

Shachtman, R. (1974), Generation of the admissible boundary of a convex polytope, *Operations Research* **22** (1974) 151–159. [224]

Shamir, A. (1982), A polynomial time algorithm for breaking the basic Merkle-Hellman cryptosystem, in: *23rd Annual Symposium on Foundations of Computer Science*, IEEE, New York, 1982, pp. 145–152 [revised version: *IEEE Transactions on Information Theory* IT-**30** (1984) 699–704]. [74]

Shamir, R. (1987), *The efficiency of the simplex method: a survey*, Management Science **33** (1987) 301–334. [139–147]

Shamos, M. I. (1978), *Computational geometry*, Ph.D. Thesis, Yale University, New Haven, 1978. [147]

Shamos, M. I. and Hoey, D. (1975), Closest-point problems, in: *16th Annual Symposium on Foundations of Computer Science*, IEEE, New York, 1975, pp. 151–162. [224]

Shapiro, J. F. (1968a), Dynamic programming algorithms for the integer programming problem—I: The integer programming problem viewed as a knapsack type problem, *Operations Research* **16** (1968) 103–121. [367]

Shapiro, J. F. (1968b), Group theoretic algorithms for the integer programming problem II: extension to a general algorithm, *Operations Research* **16** (1968) 928–947. [367]

Shapiro, J. F. (1971), Generalized Lagrange multipliers in integer programming, *Operations Research* **19** (1971) 68–76. [354, 370]

Shapiro, J. F. (1977), Sensitivity analysis in integer programming, [in: *Studies in Integer Programming* (P. L. Hammer, et al., eds.),] *Annals of Discrete Mathematics* **1** (1977) 467–477. [354]

Shapiro, J. F. (1979), A survey of Lagrangean techniques for discrete optimization, [in: *Discrete Optimization II* (P. L. Hammer, E. L. Johnson and B. H. Korte, eds.),] *Annals of Discrete Mathematics* **5** (1979) 113–138. [354, 370]

Shea, D. D. (1973), On the number of divisions needed in finding the greatest common divisor, *The Fibonacci Quarterly* **11** (1973) 508–510. [54]

Shevchenko, V. N. (1976), Discrete analog of the Farkas theorem and the problem of aggregation of a system of linear integer equations (in Russian), *Kibernetika (Kiev)* 1976 (2) (1976) 99–101 [English translation: *Cybernetics* **12** (1976) 276–279]. [235]

Shor, N. Z. (1962), *Application of the gradient method for the solution of network transportation problems* (in Russian), Notes Scientific Seminar on Theory and Applications of Cybernetics and Operations Research, Academy of Sciences of the Ukrainian SSR, Kiev, 1962. [368]

Shor, N. Z. (1964), *On the structure of algorithms for the numerical solution of optimal planning and design problems* (in Russian), Dissertation, Cybernetics Institute, Academy of Sciences of the Ukrainian SSR, Kiev, 1964. [171, 368]

Shor, N. Z. (1970a), Utilization of the operation of space dilatation in the minimization of convex functions (in Russian), *Kibernetika (Kiev)* 1970 (1) (1970) 6–12 [English translation: *Cybernetics* **6** (1970) 7–15]. [163, 171]

Shor, N. Z. (1970b), Convergence rate of the gradient descent method with dilatation of the space (in Russian), *Kibernetika (Kiev)* 1970 (2) (1970) 80–85 [English translation: *Cybernetics* **6** (1970) 102–108]. [163, 171]

Shor, N. Z. (1977), Cut-off method with space extension in convex programming problems (in Russian), *Kibernetika* (*Kiev*) 1977 (1) (1977) 94–95 [English translation: *Cybernetics* **13** (1977) 94–96]. [163, 171]

Shor, N. Z. (1979), *Minimization Methods for Non-differentiable Functions* (in Russian), Naukova Dumka, Kiev, 1979 [English translation: Springer, Berlin, 1985]. [171, 224, 370]

Shor, N. Z. and Gershovich, V. I. (1979), Family of algorithms for solving convex programming problems (in Russian), *Kibernetika* (*Kiev*) 1979 (4) (1979) 62–67 [English translation: *Cybernetics* **15** (1979) 502–508]. [171, 208]

Shostak, R. (1981), Deciding linear inequalities by computing loop residues, *Journal of the Association for Computing Machinery* **28** (1981) 769–779. [195]

Siering, E. (1974), Über lineare Formen und ein Problem von Frobenius. I, *Journal für die reine und angewandte Mathematik* **271** (1974) 177–202. [379]

Simonnard, M. (1966), *Linear Programming*, Prentice-Hall, Englewood Cliffs, N.J., 1966. [224]

Skolem, Th. (1938), *Diophantische Gleichungen*, Springer, Berlin, 1938 [reprinted: Chelsea, New York, 1950]. [51, 82]

Slater, M. L. (1951), A note on Motzkin's transposition theorem, *Econometrica* **19** (1951) 185–187. [95]

Smale, S. (1983a), The problem of the average speed of the simplex method, in: *Mathematical Programming, The State of the Art—Bonn 1982* (A. Bachem, M. Grötschel and B. Korte, eds.), Springer, Berlin, 1983, pp. 530–539. [143]

Smale, S. (1983b), On the average number of steps in the simplex method of linear programming, *Mathematical Programming* **27** (1983) 241–262. [143]

Smith, D. E. (1925), *History of Mathematics, Vol. II: Special topics of elementary mathematics*, Ginn and Company, Boston, 1925 [reprinted: Dover, New York, 1953]. [41, 77, 82]

Smith, D. E. (1929), *A Source Book in Mathematics*, McGraw-Hill, New York, 1929. [40]

Smith, D. M. and Orchard-Hays, W. (1963), Computational efficiency in product form LP codes, in: *Recent Advances in Mathematical Programming* (R. L. Graves and P. Wolfe, eds.), McGraw-Hill, New York, 1963, pp. 211–218. [148]

Smith, H. J. S. (1861), On systems of linear indeterminate equations and congruences, *Philosophical Transactions of the Royal Society of London* (*A*) **151** (1861) 293–326 [reprinted in: *The Collected Mathematical Papers of Henry John Stephen Smith, Vol. I* (J. W. L. Glaisher, ed.), The Clarendon Press, Oxford, 1894 [reprinted: Chelsea, New York, 1965], pp. 367–409]. [40, 49, 50, 51, 80]

Solodovnikov, A. S. (1977), *Systems of Linear Inequalities* (in Russian), Izdat. 'Nauka', Moscow, 1977 [English translation published by: Mir Publishers, Moscow, 1979; and: The University of Chicago Press, Chicago, 1980]. [223]

Spielberg, K. (1979), Enumerative methods in integer programming, [in: *Discrete Optimization II* (P. L. Hammer, E. L. Johnson and B. H. Korte, eds.),] *Annals of Discrete Mathematics* **5** (1979) 139–183. [363]

Spivey, W. A. and Thrall, R. M. (1970), *Linear Optimization*, Holt, Rinehart and Winston, New York, 1970. [224]

Spohn, W. G. (1968), Blichfeldt's theorem and simultaneous diophantine approximation, *American Journal of Mathematics* **90** (1968) 885–894. [75]

Srinivasan, A. V. (1965), An investigation of some computational aspects of integer programming, *Journal of the Association for Computing Machinery* **12** (1965) 525–535. [359]

von Stackelberg, H. (1933), Zwei kritische Bermerkungen zur Preistheorie Gustav Cassels, *Zeitschrift für Nationalökonomie* **4** (1933) 456–472. [219]

Stanley, R. P. (1973), Linear homogeneous diophantine equations and magic labelings of graphs, *Duke Mathematical Journal* **40** (1973) 607–632. [379]

Stanley, R. P. (1974), Combinatorial reciprocity theorems, *Advances in Mathematics* **14** (1974) 194–253. [379] ·

Stanley, R. P. (1980), Decompositions of rational convex polytopes, [in: *Combinatorial Mathematics, Optimal Designs and Their Applications* (J. Srivastava, ed.),] *Annals of Discrete Mathematics* **6** (1980) 333–342. [379]

Stanley, R. P. (1982), Linear diophantine equations and local cohomology, *Inventiones Mathematicae* **68** (1982) 175–193. [379]

Stark, H. M. (1970), *An Introduction to Number Theory*, Markham, Chicago, 1970. [82]

von Staudt, G. K. C. (1847), *Geometrie der Lage*, Bauer und Raspe, Nürnberg, 1847. [40, 212]

Steiner, J. (1832), *Systematische Entwickelung der Abhängigkeit geometrischer Gestalten von einander, mit Berücksichtigung der Arbeiten alter und neuer Geometer über Porismen, Projections-Methoden, Geometrie der Lage, Transversalen, Dualität und Reciprocität, etc.*, Fincke, Berlin, 1832 [reprinted in: *Jacob Steiner's Gesammelte Werke, Vol. I* (K. Weierstrass, ed.), G. Reimer, Berlin, 1881 [reprinted: Chelsea, New York, 1971,] pp. 229–460]. [40]

Steinitz, E. (1913), Bedingt konvergente Reihen und konvexe Systeme, *Journal für die reine und angewandte Mathematik* **143** (1913) 128–175. [215]

Steinitz, E. (1914), Bedingt konvergente Reihen und konvexe Systeme (Fortsetzung), *Journal für die reine und angewandte Mathematik* **144** (1914) 1–40. [215]

Steinitz, E. (1916), Bedingt konvergente Reihen und konvexe Systeme (Schluss), *Journal für die reine und angewandte Mathematik* **146** (1916) 1–52. [88, 89, 215]

Steinitz, E. (1934), *Vorlesungen über die Theorie der Polyeder* (herausgegeben unter Mitarbeitung von H. Rademacher), Springer, Berlin, 1934 [reprinted: 1976]. [217, 223]

Steinmann, H. and Schwinn, R. (1969), Computational experience with a zero-one programming problem, *Operations Research* **17** (1969) 917–920. [359, 363]

Stewart, G. W. (1973), *Introduction to Matrix Computations*, Academic Press, New York, 1973. [41]

Stiemke, E. (1915), Über positive Lösungen homogener linearer Gleichungen, *Mathematische Annalen* **76** (1915) 340–342. [95, 216]

Stoelinga, Th. G. D. (1932), *Convexe puntverzamelingen*, [Ph.D Thesis, University Groningen,] H. J. Paris, Amsterdam, 1932. [217]

Stoer, J. and Bulirsch, R. (1980), *Introduction to Numerical Analysis*, Springer, New York, 1980. [41]

Stoer, J. and Witzgall, C. (1970), *Convexity and Optimization in Finite Dimensions I*, Springer, Berlin, 1970. [94, 99, 224]

Stokes, R. W. (1931), A geometric theory of solution of linear inequalities, *Transactions of the American Mathematical Society* **33** (1931) 782–805. [217]

Stouff, M. (1902), Remarques sur quelques propositions dues à M. Hermite, *Annales Scientifiques de l'École Normale Supérieure* (3) **19** (1902) 89–118. [79]

Strang, G. (1980), *Linear Algebra and Its Applications* (*2nd edition*), Academic Press, New York, 1980. [4, 41]

Strassen, V. (1969), Gaussian elimination is not optimal, *Numerische Mathematik* **13** (1969) 354–356. [41]

Swart, G. (1985), Finding the convex hull facet by facet, *Journal of Algorithms* **6** (1985) 17–48. [224]

Sylvester, J. J. (1850), Additions to the articles in the September number of this Journal, "On a new class of theorems," and on Pascal's theorem, *The London, Edinburgh, and Dublin Philosophical Magazine and Journal of Science* (3) **37** (1850) 363–370 [reprinted in: *The Collected Mathematical Papers of James Joseph Sylvester, Vol. I*, The University Press, Cambridge, 1904, [reprinted: Chelsea, New York, 1973,] pp. 145–151]. [40]

Sylvester, J. J. (1857a), On a discovery in the partition of numbers, *Quarterly Journal of Mathematics* **1** (1857) 81–84 [reprinted in: *The Collected Mathematical Papers of James Joseph Sylvester, Vol. II*, The University Press, Cambridge, 1908, [reprinted: Chelsea, New York, 1973,] pp. 86–89]. [375]

Sylvester, J. J. (1857b), On the partition of numbers, *Quarterly Journal of Mathematics* **1**

(1857) 141–152 [reprinted in: *The Collected Mathematical Papers of James Joseph Sylvester, Vol. II*, The University Press, Cambridge, 1908, [reprinted: Chelsea, New York, 1973,] pp. 90–99]. [375]

Sylvester, J. J. (1858a), On the problem of the virgins, and the general theory of compound partition, *Philosophical Magazine* **16** (1858) 371–376 [reprinted in: *The Collected Mathematical Papers of James Joseph Sylvester, Vol. II*, The University Press, Cambridge, 1908 [reprinted: Chelsea, New York, 1973,] pp. 113–117]. [375]

Sylvester, J. J. (1858b), Note on the equation in numbers of the first degree between any number of variables with positive coefficients, *Philosophical Magazine* **16** (1858) 369–371 [reprinted in: *The Collected Mathematical Papers of James Joseph Sylvester, Vol. II*, The University Press, Cambridge, 1908, [reprinted: Chelsea, New York, 1973,] pp. 110–112]. [375]

Sylvester, J. J. (1882), On subinvariants, that is, semi-invariants to binary quantics of an unlimited order, *American Journal of Mathematics* **5** (1882) 79–136 [reprinted in: *The Collected Mathematical Papers of James Joseph Sylvester, Vol. II*, The University Press, Cambridge, 1908, [reprinted: Chelsea, New York, 1973,] pp. 568–622]. [375]

Sylvester, J. J. (1884), [Problem] 7382, *Mathematics from the Educational Times with Solutions* **41** (1884) 21. [376]

Szekeres, G. (1970), Multidimensional continued fractions, *Annales Universitatis Scientiarum Budapestinensis de Rolando Eötvös Nominatae, Sectio Mathematica* **13** (1970) 113–140. [75]

Taha, H. A. (1975), *Integer Programming. Theory, Applications, and Computations*, Academic Press, New York, 1975. [378]

Takahashi, S. (1932), On the system of linear forms, *Japanese Journal of Mathematics* **9** (1932) 19–26. [217]

Tamir, A. (1976), On totally unimodular matrices, *Networks* **6** (1976) 373–382. [272]

Tardos, É. (1985), A strongly polynomial minimum cost circulation algorithm, *Combinatorica* **5** (1985) 247–255. [195, 279].

Tardos, É. (1986), A strongly polynomial algorithm to solve combinatorial linear programs, *Operations Research* **34** (1986) 250–256. [195]

Taton, R. (1951), *L' Oeuvre Scientifique de Monge*, Presses universitaires de France, Paris, 1951, [221, 377]

Tegnér, H. (1932–3), Von dem Sylvesterschen Denumeranten. *Arkiv för Matematik, Astronomi och Fysik* **23A** (7) (1932–3) 1–58. [375]

Thiriez, H. (1971), The set covering problem: a group theoretic approach, *Revue Française d'Informatique et de Recherche Opérationelle* [*R.I.R.O.*] **5** (*Recherche Opérationelle* V-3) (1971) 83–103. [380]

Thompson, G. E. (1971), *Linear Programming—An Elementary Introduction*, Macmillan, New York, 1971. [224]

Thurnheer, P. (1981), Un raffinement du théorème de Dirichlet sur l'approximation diophantienne, *Comptes Rendus des Séances de l'Académie des Sciences (Paris) Série I Mathématique* **293** (1981) 623–624. [75]

Tichy, R. F. (1979), Zum Approximationssatz von Dirichlet, *Monatshefte für Mathematik* **88** (1979) 331–333. [75]

Tind, J. (1974), Blocking and antiblocking sets, *Mathematical Programming* **6** (1974) 157–166. [119]

Tind, J. (1977), On antiblocking sets and polyhedra, [in: *Studies in Integer Programming* (P. L. Hammer, et al., eds.),] *Annals of Discrete Mathematics* **1** (1977) 507–515. [119]

Tind, J. (1979), Blocking and antiblocking polyhedra, [in: *Discrete Optimization I* (P. L. Hammer, E. L. Johnson and B. H. Korte, eds.,),] *Annals of Discrete Mathematics* **4** (1979) 159–174. [119]

Tind, J. (1980), Certain kinds of polar sets and their relation to mathematical programming, *Mathematical Programming Study* **12** (1980) 206–213. [119]

Tind, J. and Wolsey, L. A. (1981), An elementary survey of general duality theory in mathematical programming, *Mathematical Programming* **21** (1981) 241–261. [354]

Tind, J. and Wolsey, L. A. (1982), On the use of penumbras in blocking and antiblocking theory, *Mathematical Programming* **22** (1982) 71–81. [119]

Todd, M. J. (1976), A combinatorial generalization of polytopes, *Journal of Combinatorial Theory (B)* **20** (1976) 229–242. [150]

Todd, M. J. (1977), *The number of necessary constraints in an integer program: A new proof of Scarf's theorem*, Technical Report 355, School of Operations Research and Industrial Engineering, Cornell University, Ithaca, N. Y.,. 1977. [234]

Todd, M. J. (1979), *Some remarks on the relaxation method for linear inequalities*, Tech. Rep. 419, School of Operations Research and Industrial Engineering, Cornell University, Ithaca, N.Y., 1979. [158, 161]

Todd, M. J. (1980), The monotonic bounded Hirsch conjecture is false for dimension at least 4, *Mathematics of Operations Research* **5** (1980) 599–601. [142]

Todd, M. J. (1982), On minimum volume ellipsoids containing part of a given ellipsoid, *Mathematics of Operations Research* **7** (1982) 253–261. [208]

Todd, M. J. (1986), Polynomial expected behavior of a pivoting algorithm for linear complementarity and linear programming problems, *Mathematical Programming* **35** (1986) 173–192. [147]

Tolstoi, A. N. (1939), Methods of removing irrational shipments in planning (in Russian), *Sotsialisticheskii Transport* **9** (1939) 28–51. [378]

Tomlin, J. A. (1971), An improved branch-and-bound method for integer programming, *Operations Research* **19** (1971) 1070–1075. [363]

Tonkov, T. (1974), On the average length of finite continued fractions, *Acta Arithmetica* **26** (1974) 47–57. [67]

Toth, P. (1980), Dynamic programming algorithms for the zero-one knapsack problem, *Computing* **25** (1980) 29–45. [373]

Traub, J. F. and Woźniakowski, H. (1981–2), Complexity of linear programming, *Operations Research Letters* **1** (1981–2) 59–62. [225]

Trauth, Jr., C. A. and Woolsey, R. E. (1968–9), Integer linear programming: a study in computational efficiency, *Management Science* **15** (1968–9) 481–493. [359]

van, Trigt, C. (1978), Worst-case analysis of algorithms, A. Some g.c.d. algorithms, *Philips Journal of Research* **33** (1978) 66–77. [54]

Trotter, Jr, L. E. (1973), *Solution characteristics and algorithms for the vertex–packing problem*, Thesis Cornell University, Ithaca, N.Y., 1973. [350]

Trotter, Jr, L. E. (1975), A class of facet producing graphs for vertex packing polyhedra, *Discrete Mathematics* **12** (1975) 373–388. [380]

Trotter, Jr, L. E. and Shetty, C. M. (1974), An algorithm for the bounded variable integer programming problem, *Journal of the Association for Computing Machinery* **21** (1974) 505–513. [363]

Truemper, K. (1977), Unimodular matrices of flow problems with additional constraints, *Networks* **7** (1977) 343–358. [276]

Truemper, K. (1978a), Algebraic characterizations of unimodular matrices, *SIAM Journal on Applied Mathematics* **35** (1978) 328–332. [302, 303]

Truemper, K. (1978b), *On balanced matrices and Tutte's characterization of regular matroids*, preprint, 1978. [295, 308]

Truemper, K. (1980), Complement total unimodularity, *Linear Algebra and Its Applications* **30** (1980) 77–92. [272]

Truemper, K. (1982), On the efficiency of representability tests for matroids, *European Journal of Combinatorics* **3** (1982) 275–291. [293]

Truemper, K. and Chandrasekaran, R. (1978), Local unimodularity of matrix-vector pairs,

Linear Algebra and Its Applications **22** (1978) 65–78. [303, 307]

Tschernikow, S. N. (1971), *Lineare Ungleichungen*, Deutscher Verlag der Wissenschaften, Berlin (DDR), 1971. [223]

Tucker, A. W. (1956), Dual systems of homogeneous linear relations, in: *Linear inequalities and Related Systems* (H. W. Kuhn and A. W. Tucker, eds.), Princeton University Press, Princeton, N.J., 1956, pp. 3–18. [223]

Turing, A. M. (1936–7), On computable numbers, with an application to the Entscheidungsproblem, *Proceedings of the London Mathematical Society* (2) **42** (1936–7) 230–265 [correction: **43** (1937) 544–546]. [16]

Tutte, W. T. (1947), A ring in graph theory, *Proceedings of the Cambridge Philosophical Society* **43** (1947) 26–40 [reprinted in: *Selected Papers of W. T. Tutte, Vol. I* (D. McCarthy and R. G. Stanton, eds.), Charles Babbage Research Centre, St. Pierre, Manitoba, 1979, pp. 51–69]. [297]

Tutte, W. T. (1956), A class of abelian groups, *Canadian Journal of Mathematics* **8** (1956) 13–28 [reprinted in: *Selected Papers of W. T. Tutte, Vol. I* (D. McCarthy and R. G. Stanton, eds.), Charles Babbage Research Centre, St. Pierre, Manitoba, 1979, pp. 176–198]. [297]

Tutte, W. T. (1958), A homotopy theorem for matroids I, II, *Transactions of the American Mathematical Society* **88** (1958) 144–160, 161–174. [282, 287, 294, 295]

Tutte, W. T. (1960), An algorithm for determining whether a given binary matroid is graphic, *Proceedings of the American Mathematical Society* **11** (1960) 905–917. [282]

Tutte, W. T. (1963), How to draw a graph, *Proceedings of the London Mathematical Society* (3) **13** (1963) 743–767 [reprinted in: *Selected Papers of W. T. Tutte, Vol. I* (D. McCarthy and R. G. Stanton, eds.), Charles Babbage Research Centre, St. Pierre, Manitoba, 1979, pp. 360–388]. [282]

Tutte, W. T. (1965), Lectures on matroids, *Journal of Research of the National Bureau of Standards* (B) **69** (1965) 1–47 [reprinted in: *Selected Papers of W. T. Tutte, Vol. II* (D. McCarthy and R. G. Stanton, eds.), Charles Babbage Research Centre, St. Pierre, Manitoba, 1979, pp. 439–496]. [276, 282, 295, 297]

Tutte, W. T. (1967), On even matroids, *Journal of Research of the National Bureau of Standards* (B) **71** (1967) 213–214. [282]

Tutte, W. T. (1971), *Introduction to the Theory of Matroids*, American Elsevier, New York, 1971. [295]

Ursic, S. (1982), The ellipsoid algorithm for linear inequalities in exact arithmetic, in: *23rd Annual Symposium on Foundations of Computer Science*, IEEE, New York, 1982, pp. 321–326. [171]

Ursic, S. and Patarra, C. (1983), Exact solution of systems of linear equations with iterative methods, *SIAM Journal on Algebraic and Discrete Methods* **4** (1983) 111–115. [37, 67]

Uspensky, J. V. and Heaslet, M. A. (1939), *Elementary Number Theory*, McGraw-Hill, New York, 1939. [82]

Vajda, S. (1962–3), Communication to the editor, *Management Science* **9** (1962–3) 154–156. [223]

Valentine, F. A. (1964), *Convex Sets*, McGraw-Hill, New York, 1964. [224]

de la Vallée Poussin, Ch. (1910–1), Sur la méthode de l'approximation minimum, *Annales de la Société Scientifique de Bruxelles* **35** (2) (1910–1) 1–16. [216]

Vandermonde, A.-T. (1776), Mémoire sur l'élimination, *Histoire de l'Académie Royale des Sciences, avec les Mémoires de Mathématique et de Physique (Paris)* [année 1772] (1776) 516–532. [39]

Varga, R. S. (1962), *Matrix Iterative Analysis*, Prentice-Hall, Englewood Cliffs, N.J., 1962. [41]

Vaughan, T. P. (1978), A generalization of the simple continued fraction algorithm, *Mathematics of Computation* **32** (1978) 537–558. [75]

Veblen, O. and Franklin, Ph. (1921–2), On matrices whose elements are integers, *Annals of Mathematics* **23** (1921-2) 1–15. [49, 51, 274, 378]

Veinott, Jr, A. F. and Dantzig, G. B. (1968), Integral extreme points, *SIAM Review* **10** (1968) 371–372. [267]

Vershik, A. M. and Sporyshev, P. V. (1983), An estimate of the average number of steps in the simplex method, and problems in asymptotic integral geometry (in Russian), *Doklady Akademii Nauk SSSR* **271** (1983) 1044–1048 [English translation: *Soviet Mathematics Doklady* **28** (1983) 195–199]. [147]

Veselov, S. I. and Shevchenko, V. N. (1978), Exponential growth of coefficients of aggregating equations (in Russian), *Kibernetika (Kiev)* 1978 (4) (1978) 78–79 [English translation: *Cybernetics* **14** (1978) 563–565]. [236]

Ville, J. (1938), Sur la théorie générale des jeux où intervient l'habileté des joueurs, in: *Traité du Calcul des Probabilités et de ses Applications*, by É. Borel, *Tome IV, Fascicule II, Applications aux jeux de hasard* (J. Ville, ed.), Gauthier-Villars, Paris, 1938, pp. 105–113. [218]

Vitek, Y. (1975), Bounds for a linear diophantine problem of Frobenius, *Journal of the London Mathematical Society* (2) **10** (1975) 79–85. [379]

Vitek, Y. (1976), Bounds for a linear diophantine problem of Frobenius, II, *Canadian Journal of Mathematics* **28** (1976) 1280–1288. [379]

Vizing, V. G. (1964), On an estimate of the chromatic class of a p-graph (in Russian), *Diskretnyĭ Analiz* **3** (1964) 25–30. [119]

Vogel, K. (translation) (1968), *CHIU CHANG SUAN SHU = Neun Bücher arithmetischer Technik, Ein chinesisches Rechenbuch für den praktischen Gebrauch aus der frühen Hanzeit (202 v. Chr. bis 9 n.Chr.), übersetzt und erläutert von Kurt Vogel*, F. Vieweg & Sohn, Braunschweig, 1968. [38]

Voronoï, G. (1908), Nouvelles applications des paramètres continus à la théorie des formes quadratiques—Premier Mémoire. Sur quelques propriétés des formes quadratiques positives parfaites, *Journal für die reine und angewandte Mathematik* **133** (1908) 97–178 [errata: 242]. [215]

Votyakov, A. A. and Frumkin, M. A. (1976), An algorithm for finding the general integer solution of a system of linear equations (in Russian), in: *Issledovaniya po diskretnoi optimizatsii [Studies in Discrete Optimization]* (A. A. Fridman, ed.), Izdat. 'Nauka', Moscow, 1976, pp. 128–140. [56, 58]

van der Waerden, B. L. (1956), Die Reduktionstheorie der positiven quadratischen Formen, *Acta Mathematica* **96** (1956) 265–309. [82]

van der Waerden, B. L. (1983), *Geometry and Algebra in Ancient Civilizations*, Springer, Berlin, 1983. [41, 76, 77]

Wagner, H. M. (1957a), A comparison of the original and revised simplex methods, *Operations Research* **5** (1957) 361–369. [148]

Wagner, H. M. (1957b), A supplementary bibliography on linear programming, *Operations Research* **5** (1957) 555–563. [224]

Wald, A. (1935), Über die eindeutige positive Lösbarkeit der neuen Produktionsgleichungen, *Ergebnisse eines mathematischen Kolloquiums* **6** [1933–4] (1935) 12–18. [219]

Wald, A. (1936a), Über die Produktionsgleichungen der ökonomischen Wertlehre, *Ergebnisse eines mathematischen Kolloquiums* **7** [1934–5] (1936) 1–6. [219]

Wald, A. (1936b), Über einige Gleichungssysteme der mathematischen Ökonomie, *Zeitschrift für Nationalökonomie* **7** (1936) 637–670 [English translation: On some systems of equations of mathematical economics, *Econometrica* **19** (1951) 368–403]. [219]

Walkup, D. W. (1978), The Hirsch conjecture fails for triangulated 27-spheres, *Mathematics of Operations Research* **3** (1978) 224–230. [142]

Wall, H. S. (1948), *Analytic Theory of Continued Fractions*, Van Nostrand, New York, 1948 [reprinted: Chelsea, New York, 1973]. [82]

Walras, L. (1874), *Éléments d'économie politique pure ou théorie de la richesse sociale*, L. Corbaz & Cie, Lausanne, 1874 [English translation: *Elements of Pure Economics, or: The Theory of Social Wealth*, Allen and Unwin, London, 1954]. [218]

Watson, G. L. (1966), On the minimum of a positive quadratic form in n ($\leqslant 8$) variables (verification of Blichfeldt's calculations), *Proceedings of the Cambridge Philosophical Society (Mathematical and Physical Sciences)* **62** (1966) 719. [81]

Weidner, H. G. (1976), A periodicity lemma in linear diophantine analysis, *Journal of Number Theory* **8** (1976) 99–108. [234].

Weinberg, F. (1977), A necessary and sufficient condition for the aggregation of linear diophantine equations, *Zeitschrift für angewandte Mathematik und Physik* (*ZAMP*) **28** (1977) 680–696. [236]

Weinberg, F. (1979a), A spectral algorithm for sequential aggregation of m linear diophantine constraints, *Zeitschrift für angewandte Mathematik und Physik* (*ZAMP*) **30** (1979) 677–698. [236]

Weinberg, F. (1979b), A necessary and sufficient condition for the aggregation of linear diophantine equations, in: *Survey of Mathematical Programming Vol. 2* (Proceedings of the 9th International Mathematical Programming Symposium, Budapest, 1976; A. Prékopa, ed.), Akadémiai Kiadó, Budapest, 1979, pp. 559–565. [236]

Weingartner, H. M. and Ness, D. N. (1967), Methods for the solution of the multi-dimensional 0/1 knapsack problem, *Operations Research* **15** (1967) 83–103. [263]

Weinstock, R. (1960), Greatest common divisor of several integers and an associated linear diophantine equation, *The American Mathematical Monthly* **67** (1960) 664–667. [54]

Weitzenböck, R. (1930), Die Endlichkeit der Invarianten von kontinuierlichen Gruppen linearer Tranformationen, *Proceedings Koninklijke Akademie van Wetenschappen te Amsterdam* **33** (1930) 232–242. [376]

Welsh, D. J. A. (1976), *Matroid Theory*, Academic Press, London, 1976. [287, 294]

de Werra, D. (1981), On some characterisations of totally unimodular matrices, *Mathematical Programming* **20** (1981) 14–21. [272]

Westlake, J. R. (1968), *A Handbook of Numerical Matrix Inversion and Solution of Linear Equations*, Wiley, New York, 1968. [41]

Wets, R. (1976), *Grundlagen konvexer Optimierung*, Lecture Notes in Economics and Mathematical Systems 137, Springer, Berlin, 1976. [224]

Wets, R. J. -B. and Witzgall, C. (1968), Towards an algebraic characterization of convex polyhedral cones, *Numerische Mathematik* **12** (1968) 134–138. [224]

Weyl, H. (1935), Elementare Theorie der konvexen Polyeder, *Commentarii Mathematici Helvetici* **7** (1935) 290–306 [English translation: The elementary theory of convex polyhedra, in: *Contributions to the Theory of Games I* (H. W. Kuhn and A. W. Tucker, eds.), Princeton University Press, Princeton, N. J., 1950, pp. 3–18] [reprinted in: *H. Weyl, Gesammelte Abhandlungen, Band III* (K. Chandrasekharan, ed.), Springer, Berlin, 1968, pp. 517–533]. [85, 87, 88, 89, 93, 214, 217]

Wieleitner, H. (1925), Zur Frühgeschichte der Räume von mehr als drei Dimensionen, *Isis* **7** (1925) 486–489. [40]

Wiener, C. (1864), *Über Vielecke and Vielflache*, Teubner, Leipzig, 1864. [223]

Wilkinson, J. H. and Reinsch, C. (1971), *Linear Algebra*, Springer, Berlin, 1971. [41]

Williams, H. P. (1976), Fourier-Motzkin elimination extension to integer programming problems, *Journal of Combinatorial Theory* (*A*) **21** (1976) 118–123. [379]

Williams, H. P. (1983), A characterisation of all feasible solutions to an integer program, *Discrete Applied Mathematics* **5** (1983) 147–155. [379]

Williams, H. P. (1984), A duality theorem for linear congruences, *Discrete Applied Mathematics* **7** (1984) 93–103. [51]

Wills, J. M. (1968a), Zur simultanen homogenen diophantischen Approximation I, *Monatshefte für Mathematik* **72** (1968) 254–263. [75]

Wills, J. M. (1968b), Zur simultanen homogenen diophantischen Approximation II, *Monatshefte für Mathematik* **72** (1968) 368–381. [75]

Wills, J. M. (1970), Zur simultanen homogenen diophantischen Approximation III, *Monatshefte für Mathematik* **74** (1970) 166–171. [75]

Wilson, R. J. (1972), *Introduction to Graph Theory*, Oliver and Boyd, Edinburgh, 1972 [3rd edition: Longman, Harlow (Essex, England), 1985]. [8]

Winograd, S. (1980), *Arithmetic Complexity of Computations*, Society for Industrial and Applied Mathematics, Philadelphia, Pennsylvania, 1980. [41]

Wolfe, P. (1963), A technique for resolving degeneracy in linear programming, *Journal of the Society for Industrial and Applied Mathematics* 11 (1963) 205–211. [139]

Wolfe, Ph. (1976), Finding the nearest point in a polytope, *Mathematical Programming* 11 (1976) 128–149. [224]

Wolfe, P. and Cutler, L. (1963), Experiments in linear programming, in: *Recent Advances in Mathematical Programming* (R. L. Graves and P. Wolfe, eds.), McGraw-Hill, New York, 1963, pp. 177–200. [139]

Wolsey, L. A. (1971), Group-theoretic results in mixed integer programming, *Operations Research* 19 (1971) 1691–1697. [367]

Wolsey, L. A. (1971-2), Extensions of the group theoretic approach in integer programming, *Management Science* 18 (1971-2) 74–83. [367]

Wolsey, L. A. (1973), Generalized dynamic programming methods in integer programming, *Mathematical Programming* 4 (1973) 222–232. [367]

Wolsey, L. A. (1974a), A view of shortest route methods in integer programming, *Cahiers du Centre d'Études de Recherche Opérationelle* 16 (1974) 317–335. [265, 367]

Wolsey, L. A. (1974b), A number theoretic reformulation and decomposition method for integer programming, *Discrete Mathematics* 7 (1974) 393–403. [367]

Wolsey, L. A. (1974c), Groups, bounds and cuts for integer programming problems, in: *Mathematical Programs for Activity Analysis* (P. van Moeseke, ed.), North-Holland, Amsterdam, 1974, pp. 73–78. [367]

Wolsey, L. A. (1975), Faces for a linear inequality in 0–1 variables, *Mathematical Programming* 8 (1975) 165–178. [253, 374, 380]

Wolsey, L. A. (1976a), Facets and strong valid inequalities for integer programs, *Operations Research* 24 (1976) 367–372. [380]

Wolsey, L. A. (1976b), Further facet generating procedures for vertex packing polytopes, *Mathematical Programming* 11 (1976) 158–163. [380]

Wolsey, L. A. (1977), Valid inequalities and superadditivity for 0–1 integer programs, *Mathematics of Operations Research* 2 (1977) 66–77. [354, 380]

Wolsey, L. A. (1979), Cutting plane methods, in: *Operations Research Support Methodology* (A. G. Holzman, ed.), Marcel Dekker, New York, 1979, pp. 441–466. [354, 359, 367]

Wolsey, L. A. (1980), Heuristic analysis, linear programming and branch and bound, *Mathematical Programming Study* 13 (1980) 121–134. [363]

Wolsey, L. A. (1981a), Integer programming duality: price functions and sensitivity analysis, *Mathematical Programming* 20 (1981) 173–195. [354]

Wolsey, L. A. (1981b), The *b*-hull of an integer program, *Discrete Applied Mathematics* 3 (1981) 193–201. [241, 243, 354]

Yamnitsky, B. (1982), *Notes on linear programming*, Master's thesis, Boston University, Boston, Mass., 1982. [171]

Yamnitsky, B. and Levin, L. A. (1982), An old linear programming algorithm runs in polynomial time, in: *23rd Annual Symposium on Foundations of Computer Science*, IEEE, New York, 1982, pp. 327–328. [171]

Yannakakis, M. (1980), On a class of totally unimodular matrices, in: *21st Annual Symposium on Foundations of Computer Science*, IEEE, New York, 1980, pp. 10–16. [272]

Yannakakis, M. (1985), On a class of totally unimodular matrices, *Mathematics of Operations Research* 10 (1985) 280–304. [272]

Yao, A. C.-C. (1981), A lower bound to finding convex hulls, *Journal of the Association for Computing Machinery* 28 (1981) 780–787. [224]

Yao, A. C. and Knuth, D. E. (1975), Analysis of the subtractive algorithm for greatest common divisors, *Proceedings of the National Academy of Sciences of the United States of America* **72** (1975) 4720–4722. [54]

Yemelichev, V. A., Kovalev, M. M., and Kravtsov, M. K. (1981), *Mnogogranniki grafy optimizatsiya*, Izdat. 'Nauka', Moscow, 1981 [English translation: *Polytopes, Graphs and Optimisation*, Cambridge University Press, Cambridge, 1984]. [223, 379]

Young, D. M. (1971a), *Iterative Solution of Large Linear Systems*, Academic Press, New York, 1971. [41]

Young, R. D. (1965), A primal (all-integer) integer programming algorithm, *Journal of Research of the National Bureau of Standards (B)* **69** (1965) 213–250. [359]

Young, R. D. (1968), A simplified primal (all-integer) integer programming algorithm, *Operations Research* **16** (1968) 750–782. [359]

Young, R. D. (1971b), Hypercylindrically deduced cuts in zero-one integer programs, *Operations Research* **19** (1971) 1393–1405. [359]

Yudin, D. B. and Gol'shtein, E. G. (1965), *Linear Programming*, Israel Program for Scientific Translations, Jerusalem, 1965 [English translation of: *Teoriya i metody lineinoe programmirovanie*, Izdat. 'Sovetskoe Radio', Moscow, 1963 [German translation: *Lineare Optimierung I, II*, Akademie-Verlag, Berlin (DDR), 1968, 1970]]. [224]

Yudin D. B. and Nemirovskiĭ, A. S. (1976a), Evaluation of the informational complexity of mathematical programming problems (in Russian), *Èkonomika i Matematicheskie Metody* **12** (1976) 128–142 [English translation: *Matekon* **13** (2) (1976–7) 3–25]. [163, 171]

Yudin D. B. and Nemirovskiĭ, A. S. (1976b), Informational complexity and efficient methods for the solution of convex extremal problems (in Russian), *Èkonomika i Matematicheskie Metody* **12** (1976) 357–369 [English translation: *Matekon* **13** (3) (1977) 25–45]. [163, 171, 189, 205]

Zadeh, N. (1973a), More pathological examples for network flow problems, *Mathematical Programming* **5** (1973) 217–224. [142]

Zadeh, N. (1973b), A bad network problem for the simplex method and other minimum cost flow algorithms, *Mathematical Programming* **5** (1973) 255–266. [142]

Zemel, E. (1978), Lifting the facets of zero-one polytopes, *Mathematical Programming* **15** (1978) 268–277. [380]

Zemel, E. (1981–2), On search over rationals, *Operations Research Letters* **1** (1981–2) 34–38. [66]

Zeuthen, F. (1933), Das Prinzip der Knappheit, technische Kombination und ökonomische Qualität, *Zeitschrift für Nationalökonomie* **4** (1933) 1–24. [219]

Zionts, S. (1974), *Linear and Integer Programming*. Prentice-Hall, Englewood Cliffs, N. J., 1974. [224, 379]

Zoltners, A. A. (1978), A direct descent binary knapsack algorithm, *Journal of the Association for Computing Machinery* **25** (1978) 304–311. [373]

Notation index

in the order of first occurrence in the text

452

Author index

Subject index

(bold numbers refer to pages on which items are introduced)